SPECTROPHOTOMETRIC DETERMINATION OF ELEMENTS

ELLIS HORWOOD SERIES IN ANALYTICAL CHEMISTRY
General Editor: Dr. R. A. Chalmers
University of Aberdeen

Founded as a library of fundamental books on important and growing subject areas in analytical chemistry, this series will serve chemists in industrial work and research, and in teaching or advanced study

Published or in active preparation

AUTOMATIC METHODS IN CHEMICAL ANALYSIS
J. K. FOREMAN } *Laboratory of the Government Chemist, London*
P. S. STOCKWELL }

THEORETICAL FOUNDATIONS OF CHEMICAL ELECTROANALYSIS
Z. GALUS, *Warsaw University*

ELECTROANALYTICAL CHEMISTRY
G. F. REYNOLDS, *University of Reading*

ANALYSIS OF ORGANIC SOLVENTS
V. ŠEDIVEC } *Institute of Hygiene and Epidemiology, Prague*
J. FLEK }

HANDBOOK OF PROCESS STREAM ANALYSIS
K. J. CLEVETT, *Crest Engineering (U.K.) Inc.*

METHODS OF CATALYTIC ANALYSIS
G. SVEHLA, *Queen's University of Belfast*
H. THOMPSON, *University of New York*

ORGANIC REAGENTS IN INORGANIC ANALYSIS
Z. HOLZBECHER, et al., *Institute of Chemical Technology, Prague*

ANALYSIS OF SYNTHETIC POLYMERS
J. URBANSKI, et al., *Warsaw Technical University*

SPECTROPHOTOMETRIC DETERMINATION OF ELEMENTS
Z. MARCZENKO, *Warsaw Technical University*

OPERATIONAL AMPLIFIERS IN CHEMICAL INSTRUMENTATION
R. KALVODA, *J. Heyrovský Institute of Physical Chemistry and Electrochemistry, Czechoslovak Academy of Sciences, Prague*

ANALYTICAL APPLICATIONS OF COMPLEX EQUILIBRIA
J. INCZÉDY, *University of Chemical Engineering, Veszprém*

ELECTROCHEMICAL STRIPPING ANALYSIS
F. VYDRA, *J. Heyrovský Institute of Physical Chemistry and Electrochemistry, Prague*
K. ŠTULÍK, *Charles University, Prague*
E. JULÁKOVÁ, *The State Institute for Control of Drugs, Prague*

GRADIENT LIQUID CHROMATOGRAPHY
C. LITEANU and S. GOCAN, *University of Cluj*

SPECTROPHOTOMETRIC DETERMINATION OF ELEMENTS

ZYGMUNT MARCZENKO
Professor of Analytical Chemistry
Warsaw Technical University

Translation Editor:
CAMERON G. RAMSAY
University of Aberdeen

ELLIS HORWOOD LIMITED
Chichester

Halsted Press, a Division of
JOHN WILEY & SONS Inc.
NEW YORK · LONDON · SYDNEY · TORONTO

English Edition first published in 1976 by
ELLIS HORWOOD Ltd., Coll House, Westergate, Chichester, Sussex, England
in co-edition with
WYDAWNICTWA NAUKOWO-TECHNICZNE, Warsaw, Poland

Distributed in:
Australia, New Zealand, South-east Asia by
JOHN WILEY & SONS AUSTRALASIA PTY LIMITED
110 Alexander Street, Crow's Nest, N.S.W., Australia

Canada by JOHN WILEY & SONS CANADA LIMITED
22 Worcester Road, Rexdale, Ontario, Canada

Europe, Africa by
JOHN WILEY & SONS LIMITED
Baffins Lane, Chichester, Sussex, England

North and South America and the rest of the world by
Halsted Press, a division of
JOHN WILEY & SONS INC.
605 Third Avenue, New York, N. Y. 10016, U.S.A.

Translated by Marian Jurecki from the Polish
Kolorymetryczne oznaczanie pierwiastków
Published by Wydawnictwa Naukowo-Techniczne, Warsaw

Library of Congress Cataloging in Publication Data

Marczenko, Zygmunt,
   Spectrophotometric Determination of Elements.
   Translation of Kolorymetryczne oznaczanie pierwiastków
   1. Colorimetry. 2. Chemical elements.
I. Title.
QD113.M3713 1976   545'.812   74-33186
ISBN 0-470-56865-8   (Halsted Press)
  " 85312-021-8   (Ellis Horwood Limited)

© 1976 Wydawnictwa Naukowo-Techniczne and Ellis Horwood, Publisher

All Rights Reserved. No part of this publication may be reproduced, stored in a retrieval system, or transmitted, in any form or by any means, electronic, mechanical, photocopying, recording or otherwise, without prior permission of Wydawnictwa Naukowo-Techniczne

PRINTED IN POLAND

CONTENTS

Foreword	vii
Preface	ix
Abbreviations Used	xi

Part I GENERAL

1 PRINCIPLES AND PRACTICE OF SPECTROPHOTOMETRIC ANALYSIS	3
2 SPECTROPHOTOMETRIC REAGENTS	40
3 PRECONCENTRATION AND SEPARATION OF ELEMENTS	67

Part II METHODS FOR SEPARATION AND DETERMINATION OF INDIVIDUAL ELEMENTS

SOME PRACTICAL NOTES	103	33	NICKEL	369
4 ALKALI METALS	105	34	NIOBIUM	380
5 ALUMINIUM	110	35	NITROGEN	391
6 ANTIMONY	121	36	OSMIUM	403
7 ARSENIC	131	37	OXYGEN	408
8 BERYLLIUM	142	38	PALLADIUM	412
9 BISMUTH	149	39	PHOSPHORUS	421
10 BORON	159	40	PLATINUM	431
11 BROMINE	171	41	RARE EARTH ELEMENTS	438
12 CADMIUM	176	42	RHENIUM	448
13 CALCIUM	182	43	RHODIUM	457
14 CARBON	191	44	RUTHENIUM	462
15 CERIUM	198	45	SCANDIUM	468
16 CHLORINE	204	46	SELENIUM	474
17 CHROMIUM	213	47	SILICON	481
18 COBALT	224	48	SILVER	490
19 COPPER	238	49	STRONTIUM AND BARIUM	499
20 FLUORINE	254	50	SULPHUR	504
21 GALLIUM	267	51	TANTALUM	516
22 GERMANIUM	274	52	TELLURIUM	521
23 GOLD	281	53	THALLIUM	529
24 INDIUM	288	54	THORIUM	537
25 IODINE	296	55	TIN	546
26 IRIDIUM	302	56	TITANIUM	555
27 IRON	305	57	TUNGSTEN	567
28 LEAD	322	58	URANIUM	574
29 MAGNESIUM	329	59	VANADIUM	588
30 MANGANESE	338	60	ZINC	601
31 MERCURY	350	61	ZIRCONIUM AND HAFNIUM	609
32 MOLYBDENUM	358		INDEX	623

FOREWORD

Chemical analysis by means of the intensity of colour produced in certain reactions has been known for well over 100 years, and it is not surprising that a large literature has developed, describing the determination of most elements in the periodic table, with a formidable array of reagents. Even for the experienced worker in the field it is often difficult to select the best or most appropriate method for a particular application, and the time factor invariably precludes the possibility of comparing the various procedures available. It is therefore particularly valuable to have a book which will not only map the major (and many minor) roads to the desired goal, but will also indicate those which experience has shown to be amongst the most reliable. Professor Marczenko has distilled into such a book his many years of experience and research in colorimetric analysis, and it is now presented in translation into English. Among its many valuable features is the extensive coverage of the Russian and Eastern European literature, which is often difficult to come by in the Western world. The emphasis throughout is on practicality, based on careful testing at the laboratory bench, and its author is to be congratulated on his care and assiduity. It is a pleasure and a privilege to be associated with its presentation in English.

R. A. CHALMERS
University of Aberdeen

PREFACE

Visual spectrophotometry is an instrumental method of analysis widely adopted because of the high precision, sensitivity, and availability of the instruments used. Spectrophotometric methods enable nearly all the elements to be determined over a wide range of concentrations in any material. Spectrophotometric methods (especially when combined with preconcentration) are of particular importance in trace analysis, testing of high-purity materials, environmental studies, biochemistry and agrochemistry.

The recent rapid developments in spectrophotometric and separation methods for the elements have become feasible thanks to the comprehensive studies of complexation chemistry, especially of new organic chelating reagents and their reactions with inorganic ions. Besides the older reagents such as dithizone, oxine, and dithiocarbamates, the very promising new analytical reagents such as Arsenazo III, PAN, and Xylenol Orange have become available. There are also the ternary ion-association complexes of anionic complexes with basic triphenylmethane and xanthene dyes, etc.

This monograph aims at presenting in reasonable compass spectrophotometric (colorimetric) methods for determination of metals and nonmetals, and also methods for their separation. Many of these are discussed at length, the rest are summarized, as systematically as possible. Keeping the book reasonably short precludes detailed description of the determination of elements in particular materials, but this is compensated by inclusion of appropriate references to the world literature. The book includes about 7000 references covering mainly the last 15 years up to early 1974.

The sections devoted to individual elements are preceded by a general account of spectrophotometric methods, covering the principles and practice of the technique, groups of reagents of major importance, and methods of preconcentration and separation of elements.

The monograph is a revised and updated version of the Polish (WNT, Warsaw, 1968) and Russian (Mir, Moscow, 1971) editions.

The detailed procedures of the separation and determination of individual elements have almost all been checked in practice by myself and my co-workers, Drs. K. Kasiura, M. Mojski, M. Krasiejko, Ł. Lenarczyk, A. Ramsza, and others, to all of whom my thanks are due.

I am indebted to Professor J. Minczewski for discussions which have been helpful in setting the concept of the book.

I am very grateful to Dr. R. A. Chalmers for his valuable advice and suggestions which permitted me to introduce final improvements and corrections in the English version of the monograph.

I should like to thank Mr. C. G. Ramsay for his painstaking work in editing the translation of the book and for assistance with the updating.

Cordial thanks are due to my wife Sophie for technical assistance with the preparative work.

Z. MARCZENKO

Warsaw

ABBREVIATIONS USED

BPHA	N-benzoyl-N-phenylhydroxylamine
DAM	diantipyrylmethane
DAPM	diantipyrylpropylmethane
DCTA	1,2-diaminocyclohexane-N,N,N'',N'-tetra-acetic acid, Complexone IV
DDTC	diethyldithiocarbamate
DMSO	dimethyl sulphoxide
EDTA	ethylenediaminetetra-acetic acid (or its disodium salt), Complexone II (III)
EGTA	ethylene glycol bis(β-aminoethyl ether)-$N,N,N'N'$-tetra-acetic acid
HDEHP	di-(2-ethylhexyl)phosphoric acid
MEDTA	1,2-diaminopropanetetra-acetic acid
MIBK	methyl isobutyl ketone, hexone
NTA	nitrilotriacetic acid, Complexone I
oxine	8-hydroxyquinoline, 8-quinolinol
PAN	1-(2-pyridylazo)-2-naphthol
PAR	4-(2-pyridylazo)resorcinol
phen	1,10-phenanthroline, o-phenanthroline
REE	rare earth elements
SNADNS	2-(4-sulphonaphthylazo)chromotropic acid
SPADNS	2-(4-sulphophenylazo)chromotropic acid
TAN	1-(2-thiazolylazo)-2-naphthol
TAR	4-(2-thiazolylazo)resorcinol
TBP	tri-n-butyl phosphate
TEA	triethanolamine
TEHPO	tris(2-ethylhexyl)phosphine oxide
TOPO	tri-n-octylphosphine oxide
TTA	thenoyltrifluoroacetone

Part I

GENERAL

Chapter 1

PRINCIPLES AND PRACTICE OF SPECTROPHOTOMETRIC ANALYSIS

1.1 Introduction

The purpose of this book is a systematic presentation of spectrophotometric methods for the determination of the elements, that is to say, of the methods based upon the absorption of visible radiation. Until recently, visible spectrophotometry was more often called colorimetry and even at present such definitions as colorimetric, photometric or absorptiometric methods are sometimes used in the literature, besides the term spectrophotometric method used in the identical sense.

The basis of spectrophotometric methods is the simple relationship between the absorption of radiation by a solution and the concentration of coloured species in the solution. In order to determine a species (the determinand) spectrophotometrically (colorimetrically) it is usually converted into a coloured complex. The colour of the determinand itself is utilized much less often. When the determinand is not coloured, or forms no coloured compounds, indirect spectrophotometric methods may be used for its determination.

Spectrophotometric methods are remarkable for their versatility, sensitivity, and precision. Almost all are direct and can be used for all the elements, except for the noble gases. A very extensive range of concentrations may be covered, from macroquantities (1–50%) (especially by the differential method) to traces (10^{-8}–10^{-6}%) (after preconcentration). Spectrophotometric methods are among the most precise instrumental methods of analysis.

Furthermore, the basic apparatus required, a spectrophotometer or photoelectric colorimeter, is now fairly cheap, and certainly much cheaper than most other instruments used in analysis.

Several of the books dealing with spectrophotometry and spectrophotometric methods for determining the elements are classed among treatises of major importance [1–11a]. For the practising analyst, one of the most useful, and certainly the most entertaining, is the book by Edisbury [11a].

1.2 Historical Outline

Attempts to utilize the colour of substances for their quantitative estimation are mentioned in reports dating back to ancient times and the Middle

Ages. Certain writers [12–14] believe the origin of colorimetry to date from 1852, when Beer formulated the fundamental law of colorimetry (spectrophotometry), taking advantage of earlier works by Bouguer and Lambert. The colorimetric methods for determining bromide in natural waters by oxidation and extraction of the resulting bromine with ether, of iron with thiocyanate or ferrocyanide, of ammonia by Nessler's method, of titanium by the peroxide method, and of molybdenum by the thiocyanate method, all came from about the middle of the nineteenth century. All are still used today.

Originally, colorimetric determinations were done in colorimetric test-tubes. Towards the end of the last century, visual colorimeters with filters began to be used. In the 1930's the first photoelectric colorimeters and spectrophotometers were introduced into laboratories. Ever since, the apparatus based on the photoelectric effect has been incessantly improved.

In the last 25 years the development of organic reagents and of complex chemistry has entailed a tremendous increase in the number of spectrophotometric methods covering almost all the periodic table [15–18].

At first, the term colorimetry denoted a method of analysis in which elements were determined by comparison, by estimation of the colours of sample and standard solutions in test-tubes or visual colorimeters (comparators). In photoelectric instruments the colour is neither measured nor compared; instead the fraction of the incident radiation that is absorbed by coloured solutions is measured. Therefore, "spectrophotometry", "absorptiometry", and "photometry" have become more proper definitions.

1.3 Colour and Molecular Structure

In Table 1.1 the colours of visible radiation are shown. Visible light represents a very small part of the electromagnetic spectrum and is generally considered to extend from 380 to 780 nm. A solution or object appears coloured when it transmits or absorbs only part of the radiation in the visible spectrum. The optical characteristic of the substance is its absorption

Table 1.1 Colours of visible radiation

Observed colour	Complementary hue, filter colour (absorbed radiation, nm)	
Green-yellow	Violet	(380–420)
Yellow	Violet-blue	(420–440)
Orange	Blue	(440–470)
Red	Blue-green	(470–500)
Purple	Green	(500–520)
Violet	Yellow-green	(520–550)
Violet-blue	Yellow	(550–580)
Blue	Orange	(580–620)
Blue-green	Red	(620–680)
Green	Purple	(680–780)

spectrum. Description of a colour by name (e.g. yellow, red) is a very rough means of characterization. Figure 1.1 presents the absorption spectrum of a yellow solution absorbing in the ultraviolet and visible ranges. The demarcation line between the two ranges is only approximate.

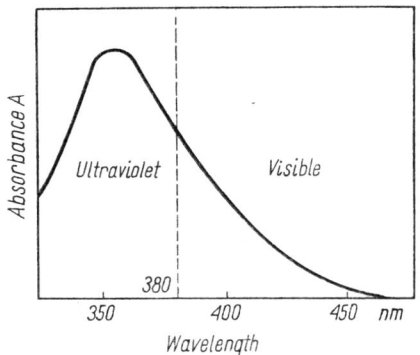

Fig. 1.1 Absorption spectrum of a yellow solution (maximum in near ultraviolet)

There is a close relation between the colour of a substance and it electronic structure [19–22]. A molecule (or ion) exhibits absorption in the visible or ultraviolet range, when radiation (photons) causes an electronic transition, raising the molecule (ion) from the ground state to an excited state. The production, or change, of a colour is connected with deformation of the normal electronic structure of the molecule. Irradiation causes variations in the electronic energy of molecules containing one or more *chromophoric groups*, i.e. atomic groupings with unsaturated linkages. Some examples of the commoner chromophoric (literally "colour-bearing") groups are:

$$\diagup\!\!\!\!C=O, \quad -N=N-, \quad -N=O, \quad =\!\!\bigcirc\!\!=, \quad -C\equiv N, \quad \diagup\!\!\!\!C=S$$

Two or more chromophoric groups in the molecule often enhance one another's effect, to deepen the colour by displacing the absorption maximum (λ_{max}) towards longer wavelengths (from the ultraviolet towards the red). This is called a bathochromic shift. The displacement of the absorption maximum from the red towards the ultraviolet is known as a hypsochromic shift.

The colour of a molecule may be intensified by substituents called auxochromic groups. These groups may also effect bathochromic shifts. Examples of auxochromic groups, when substituted into phenyl, naphthyl, furyl or piperidyl rings, are:

—OH, —NH$_2$, —SH, —CH$_3$, —Cl, and —Br.

The colour-determining factor in a number of molecules is the introduction of conjugation of double bonds by means of electron-donor and/or electron-acceptor groups.

As the pH is increased, spectrophotometric reagents ionize and their electronic structures become deformed, which often leads to a bathochromic shift of the absorption maximum. Ionization causes polarization of the chromophoric system. The formation of a chelate complex disturbs the electronic state of the organic molecule to produce, as a rule, a bathochromic shift, though some hypsochromic shifts are known.

Most transition metals with an incomplete d-electron subshell have chromophoric properties. These metals may occur in various oxidation states. They can give colour reactions with colourless reagents containing no chromophoric groups.

1.4 Absorption Laws

When a beam of radiant energy of intensity I_0 impinges upon a layer of coloured solution, some of this energy (I_a) is absorbed, some (I_t) is transmitted, and some (I_r) is reflected and scattered by the walls of the cell (cuvette):

$$I_0 = I_a + I_t + I_r$$

Since measurements are always made with respect to a reference solution in a similar cell, I_r is usually regarded as constant, and neglected.

Absorption of radiation depends on the thickness of the coloured layer and on the concentration of the solution [23–26]. In 1729, Bouguer established the relationship between the absorption and the thickness of the absorbing medium. This relationship was formulated by Lambert in 1760 in a more accurate, mathematical way.

Beer (1852) settled in turn the relationship between the absorption of radiation and the concentration of the light-absorbing component of the solution. The formula derived by him (the *Bouguer–Lambert–Beer law*) relates the absorption to both the concentration of the solution and the thickness of the layer.

When a parallel beam of monochromatic radiation of intensity I impinges upon a layer of solutions of thickness dl, some radiant energy is absorbed. The fraction absorbed increases exponentially with linear increase in the layer thickness:

$$\frac{dI}{I} = -k\,dl$$

where k is a constant and the minus sign denotes that the intensity of the radiation transmitted decreases as the thickness of the layer increases.

Integration of this expression gives:

$$\ln \frac{I_0}{I_t} = kl$$

where I_0 is the initial intensity of the beam (i.e. $l = 0$).

Conversion into Briggsian logarithms gives:

$$\log \frac{I_0}{I_t} = 0\cdot 434 \ln \frac{I_0}{I_t} = 0\cdot 434 kl = Kl = A$$

where K is a new constant and A is called the *absorbance*. This formula (*Bouguer–Lambert law*) represents the dependence of the absorption on the thickness of the layer.

If the concentration c of the absorbing species of the solution is doubled and the thickness of the layer is reduced by a factor of two, in other words, if the total number of the absorbing centres remains the same, the absorbance A will also remain the same. The absorbance is therefore a function of the number of absorbing centres in the light-beam, i.e. of the product $c\,.\,l$, and the equation for the absorbance can be given the form:

$$A = \log \frac{I_0}{I_t} = \varepsilon\,.\,c\,.\,l$$

where ε is a new constant called the *molar absorptivity* (or *absorption coefficient*); c is the concentration of coloured species (mole/l.) and l is the thickness of the absorbing layer (cm). When the concentration is expressed in g/l., the constant is called the *specific absorption coefficient*. This equation is a mathematical expression of the fundamental law of spectrophotometry (the Bouguer–Lambert–Beer law), which states that the absorption of radiation depends on the total number of the absorbing centres, i.e. on the product of the concentration and the optical path-length.

In some countries the absorbance A is also called the extinction E or optical density D.

The ratio of the transmitted radiant power to the incident radiant power (I_t/I_0) shows the fraction of radiation transmitted through the solution, and is known as the *transmittance T*. Hence

$$A = \log \frac{1}{T}$$

It is much to be regretted that so far the literature on spectrophotometry shows little or no uniformity whatever in terminology and symbols used [5, 27].

Usually the same optical path-lengths are used for both the sample and the reference solutions in spectrophotometric measurements, so that only the law which relates the absorption to the concentration of the solute is of practical significance.

If the solution contains more than one absorbing species, the total absorbance is equal to the sum of all the component absorbances. This law of additivity (provided the optical path-length is constant) is expressed by:

$$A = (\varepsilon_1 c_1 + \varepsilon_2 c_2 + \ldots + \varepsilon_n c_n) l$$

When the coloured species obeys Beer's law, a straight line passing through the origin is obtained in the absorbance vs. concentration plot. Curve 1 in Fig. 1.2 indicates a positive deviation from Beer's law, and curve 2 a negative one. From a practical point of view it is desirable that the solution should follow Beer's law for concentrations corresponding to absorbances up to at least 1·0.

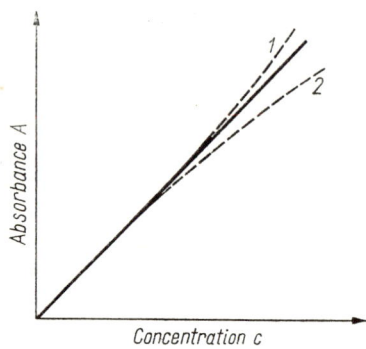

Fig. 1.2 Deviations from Beer's law: 1 positive, 2 negative

Deviations from Beer's law may be produced by either chemical or physical causes [28–30]. Chemical deviations are caused by reactions induced in the solution as the concentration of the component determined is increased. These may include condensation, polymerization, or hydrolysis. Orange dichromate ions, for example, are transformed into yellow chromate ions as the solution is diluted: $Cr_2O_7^{2-} + H_2O \rightarrow 2CrO_4^{2-} + 2H^+$. When the solution of a weak complex is diluted, the complex dissociates, which leads to positive deviations from Beer's law at higher concentrations. When extraction systems are used, there may be positive or negative deviations from Beer's law, caused by polymerization in the organic phase or the aqueous phase [30a].

Systems in which stepwise formation of complexes takes place do not, in general, obey Beer's law. The equilibria of the reactions and the concentration ratios of the resulting individual complexes depend on the reactant concentration ratio and on the pH of the solution.

The most important physical cause of deviations from Beer's law is the polychromaticity of radiation. Most coloured systems failing to obey Beer's law in measurements by means of a photocolorimeter equipped with filters, do adhere to it when spectrophotometric measurements are used. This phenomenon can be explained by reference to Fig. 1.3 which presents the absorption spectra of two solutions of the same coloured solute (B more dilute and C less dilute) and the band-width of the radiation transmitted through the filter used for absorbance measurements. The area 1–2–2′–1′ underneath the absorption curve B lies within the band-width of the radiation transmitted through the filter. The sub-areas 2–3–4 and

2′–3′–4′ of the area 1–3–3′–1′ underneath the absorption curve C extend outside the band-width transmitted through the filter. Consequently the absorbance increases more slowly than does the concentration of the coloured substance, and the non-conformity to Beer's law is obvious. This phenomenon does not occur when the beam of light transmitted through

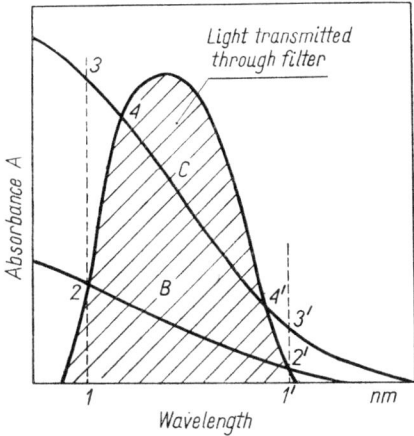

Fig. 1.3 Deviations from Beer's law owing to the polychromaticity of the radiation used

the filter encompasses the absorption maximum in the spectrum of the coloured substance. Deviations from Beer's law due to failure to use monochromatic light occur most often when filters are used in measuring the absorbance of yellow solutions.

Deviations from Beer's law may be produced by failure of the light-sensitive element to respond strictly linearly to the radiation incident upon it, which happens sometimes in poor-quality absorptiometers or at high intensities of incident light.

Stray light may also cause deviations from Beer's law at higher absorbances. What is measured is $A = \log(I_0 + I_{\text{stray}})/(I_t + I_{\text{stray}})$ and if I_{stray} becomes of the same order as I_t, there is a marked diminution in A. This is only likely to happen when I_t is small, i.e. A is large.

The optical medium must be homogeneous. Turbid solutions give deviations from Beer's law by scattering radiation. Such deviations may occur in two-phase systems insufficiently homogenized by protective colloids.

Apparent non-conformity to Beer's law occurs in two-colour systems where the reference solution absorbance is considerable at the wavelength chosen for measuring the absorbance of the complex solution. As the concentration of the complexed metal is increased, the concentration of the free reagent decreases in the solution, therefore its absorbance will be lower than that of the reference solution. If a large enough excess of the colour reagent is applied, this phenomenon does not occur.

Despite so many possibilities of deviation from Beer's law in the absorbance range of practical interest for analytical purposes, colour systems not conforming to Beer's law are of fairly rare occurrence (provided a spectrophotometer or a photoelectric colorimeter with suitably selected filters is used).

1.5 Methods of Visual Colorimetry

In visual methods the human eye compares the colour of the sample with that of the standard. Visual measurements are subjective and not very precise. The human eye quickly tires and the precision deteriorates if too many measurements are made one after the other.

Simple colorimetric comparisons are performed in Nessler cylinders. These are narrow, flat-bottomed glass cylinders, 50 or 100 ml in capacity for macro and 5 or 10 ml for smaller scale determinations. Hehner tubes, which are more precisely graduated and have a stopcock on the side at the base, are no longer used.

In the *standard-series method* the colour of the sample solution in a colorimetric tube is matched against a series of standard solutions in identical tubes. The standards are prepared by adding to separate tubes increasing amounts of standard solution of the element to be determined, and all the reagents in the same sequence and quantity as for the sample solution. The solutions are diluted to the same level and well mixed. The sample solution is matched in colour against the colour standards by viewing axially against a white background in a well and uniformly lit place. If the system obeys Beer's law it is not necessary to equalize the solution levels, since $c_1 . l_1 = c_2 . l_2$.

The standard-series method is simple and rapid, especially when the standards are stable for long periods of time, and can be stored and used more or less permanently. If the reaction products are not stable with time, synthetic standards may be used, made of stable coloured substances, such as iron(III) solution in hydrochloric acid (yellow), $CoCl_2$ solutions (pink), and $CuSO_4$ solutions (blue). The precision of the standard-series method depends on the number of standards (3–6) in the scale and on the concentration range covered.

In the method of *colorimetric titration* the colour reaction with the sample solution is carried out in one colorimetric tube while to a second tube all the reagents are added in the same proportions as to the first. Then a standard solution of the element to be determined is added from a burette until the colours in both tubes are judged to match. The comparison solution is mixed each time after the solution from the burette has been added. Colour matching is carried out against a white background in a well-lit place. If the colour system does not obey Beer's law, the levels of both solutions should be alike. The method of colorimetric titration may be employed only when the colour reaction proceeds rapidly.

Before the advent of photoelectric instruments, visual colorimeters [6] were prevalent, but now are scarcely ever used. In the Duboscq colorimeter (comparator) the colours of standard and sample solutions are matched by changing the depth of layer in one of them. In a more precise visual colorimeter, the Pulfrich photometer, instead of layer depth the intensities of the light-beams compared are changed by means of suitably adjusted diaphragms. Visual colorimeters are sometimes fitted with filters which permit partial monochromation of the light.

1.6 Photoelectric Apparatus*

The fundamental apparatus used for measuring the absorbance of coloured solutions comprises spectrophotometers and photoelectric colorimeters with filters. The basis of operation for both types is provided by the photoelectric effect. A block diagram is presented in Fig. 1.4.

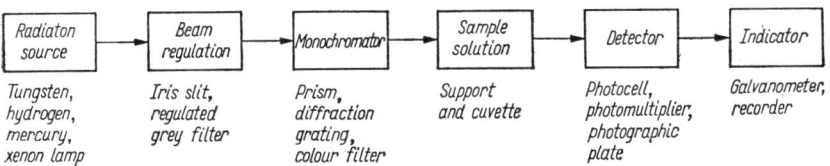

Fig. 1.4 Block diagram of spectrophotometer or photoelectric colorimeter

A *photovoltaic cell* (barrier-layer cell), shown schematically in Fig. 1.5, consists of a metal (e.g. iron) plate with a semi-conductor layer deposited on it, usually copper or selenium oxide. This layer is coated with a very thin translucent silver or gold film in order to obtain electric conduction. The radiation impinging upon the semi-conductor layer produces a potential difference between the metal plate and the semi-conductor layer, the latter becoming negatively charged. Thus a cell is formed which generates an electromotive force and is incorporated in a circuit with a galvanometer which responds in proportion to the intensity of radiation impinging upon the surface of the cell. The selenium photovoltaic cell is sen-

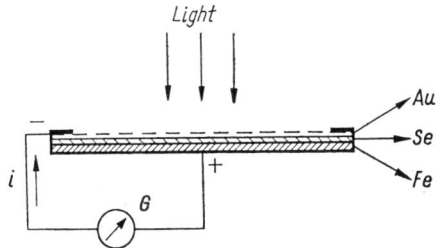

Fig. 1.5 Scheme of selenium photovoltaic cell

---

* A full-length discussion will be found in the books by Świętosławska [5] and by Kortüm [6].

sitive over the wavelength range from 300 to 800 nm, the most sensitive range extending from 500 to 600 nm. Such cells have the disadvantage of differing in sensitivity with change in the energy (wavelength) of the incident radiation. They show fatigue effects, and on aging become less sensitive. They are generally used only for the visible region of radiation.

The *photoemissive cell* (photocell), a schematic view of which is given in Fig. 1.6, consists of two metal electrodes sealed within an evacuated glass or quartz envelope, the anode being maintained at a positive potential by means of an auxiliary battery. Under the influence of incident radiation,

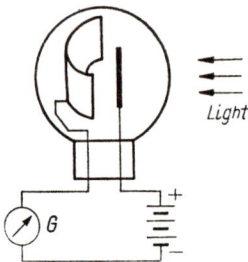

Fig. 1.6 Electric circuit with photoemissive cell

photoelectrons are emitted by the cathode and accordingly, the current flowing in the system is changed. The envelope is either filled with an inert gas at low pressure or is evacuated. A metal plate is used as the cathode, coated with a film of material capable of emitting electrons under the effect of radiation. The wire anode collects the photoelectrons. A galvanometer is connected into the circuit. The wavelength range within which the photocell is sensitive depends on the material coating the cathode. Often alkali metals (Cs, K, Na) serve the purpose. A very thin film is deposited by sublimation onto the base metal (Ag, Sb, or Bi). Photocells are suitable for the ultraviolet and the visible regions.

The *photomultiplier* (Fig. 1.7), which greatly amplifies the very weak primary photocurrent generated in the photocell, has been more and more frequently employed as a detector in instruments for measuring absorbance. The radiation impinging upon the photocathode liberates electrons which

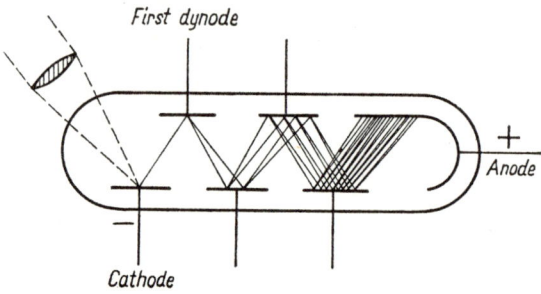

Fig. 1.7 Scheme of photomultiplier

are accelerated towards the first secondary-emitting electrode (dynode) which is at a positive potential relative to the cathode potential, and knock out of it several secondary electrons for each incident electron. These electrons impinge, in turn, upon the second dynode, thus liberating further electrons. The number of electrons emitted by the successive dynodes grows in avalanche. By means of a photomultiplier it is possible to detect incident radiation producing primary photocurrents (between the cathode and the first dynode) as low as 1 pA.

The radiation sources used should give a continuous spectrum over the widest possible wavelength range. For the visible region use is made of tungsten filament lamps. The spectrum given by them is continuous but the radiant energy emitted is unequally distributed within the operating range. For this reason and the changes in detector sensitivity with change in wavelength, a zero adjustment is necessary. The incandescent tungsten lamp may be used at wavelengths down to 350 nm.

A continuous ultraviolet radiation source is provided by an electric gas-discharge. At low gas pressures discharges produce a discontinuous line spectrum. When, however, the gas pressure is increased, the lines broaden and overlap; at high enough pressure the spectra of some gases become continuous. Usually hydrogen, xenon, and mercury vapour are used. Hydrogen discharge lamps need a lower pressure than mercury lamps do. Quartz envelopes are used to contain the gas.

The light intensity from the radiation source is regulated by means of appropriate diaphragms or slits in the optical path.

Partial monochromation of radiation in photoelectric colorimeters is obtained by inserting light filters in the beam of the white light. These filters are usually made of coloured glass, or occasionally of coloured plastic or a coloured solution in a sealed glass cell. A filter is characterized by its spectral transmittance curve (Fig. 1.8). In appraisal of a filter the band-pass corresponding to $\geqslant 50\%$ of maximum transmittance ($\lambda'_{1/2max} - \lambda_{1/2max}$) is of importance. This difference in wavelengths for the most selective filters is 20–30 nm.

The light filters are chosen so that the wavebands for the maximum absorbance of the sample solution and maximum filter transmittance coincide; in other words, the absorption maximum of the solution should correspond to the transmittance maximum of the filter. The absorbance measurement is the more precise and sensitive, the narrower is the waveband isolated by the filter and the greater the transmittance achieved at λ_{max}. Colours of filters and corresponding colours of solutions are listed in Table 1.1.

Interference filters have much narrower band-widths and higher peak transmittances than glass filters do. This type of filter consists of a layer of transparent magnesium fluoride sandwiched between semireflecting silver films. The thickness of magnesium fluoride corresponds to half the wavelength of the radiation transmitted through the filter; the silver films reflect only part of the incident radiation. The multiple reflections give reinforcement of the wavelengths desired and suppression of others. Harmonics of the desired wavelength will also be transmitted but can be eliminated by inclusion of a coloured glass filter.

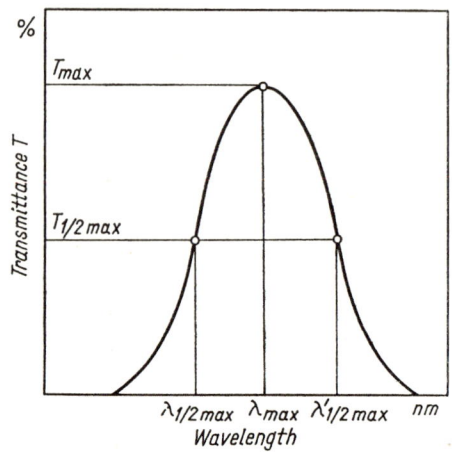

Fig. 1.8 Curve of the filter transmittance

Spectrophotometers are fitted with monochromators capable of isolating a radiation band that for all practical purposes is monochromatic (0·1–2 nm band-width). The monochromator consists of an element for dispersing light (a prism, diffraction grating, or an interference wedge) and two narrow slits, through which the white light enters and the monochromatic light leaves. In common spectrophotometers the width of the slits varies from 0·1 to 1 mm, in high-grade instruments from 0·01 to 0·1 mm.

Instruments for use over the visible region are equipped with glass optics. If, however, the region is to be extended to include the ultraviolet (the case with most spectrophotometers) the optical parts should be made of quartz.

Spectrophotometers and *photoelectric colorimeters* are divided into single-beam and double-beam instruments.

In the single-beam instrument presented schematically in Fig. 1.9

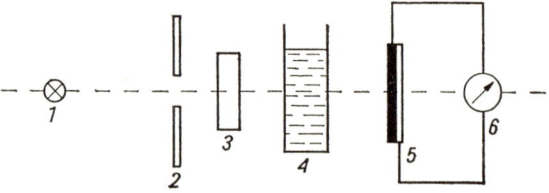

Fig. 1.9 Scheme of single-beam spectrophotometer or photoelectric colorimeter: 1 lamp, 2 iris, 3 monochromator or filter, 4 sample or reference solution, 5 photocell, 6 galvanometer

the reference solution and the sample solution are inserted successively into the radiation beam, and the radiant powers transmitted are measured.

In the double-beam instrument (Fig. 1.10) the beam of radiation is divided into two parallel beams, of which one passes through the reference solution and the other traverses the sample solution. Each beam impinges on a separate detector and a galvanometer indicates the difference in the currents generated in the two detectors.

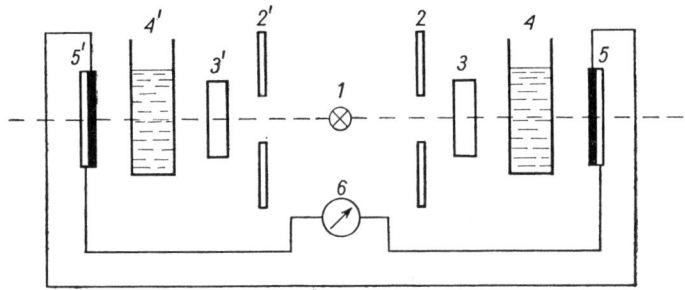

Fig. 1.10 Scheme of double-beam spectrophotometer or photoelectric colorimeter: 1 lamp, 2 and 2' irises, 3 and 3' monochromators or filters, 4 sample solution, 4' reference solution, 5 and 5' photocells, 6 galvanometer

Spectrophotometers and absorptiometers are usually calibrated in both transmittance units and absorbance units.

The response of a single-beam instrument is reproducible when the intensity of the light-source does not change, in other words when the voltage supplied to the lamp is constant. In double-beam instruments changes in supply voltage do not matter, since both beams are affected equally.

For accurate calibration of the spectrophotometer wavelength scale, narrow maxima (peaks) in the ultraviolet and the visible region are used, e.g. those of samarium and neodymium chlorides [31].

Spectrophotometers equipped with recorders and automatic spectrum scan are very useful for obtaining absorption spectra quickly, but are not so accurate as manually operated instruments for absorbance measurements.

For trace analysis microcuvettes are used, which are constructed so that small volumes of solution can be used without sacrifice in optical path-length [32–34]. Photoelectric colorimeters can be modified for determination of smaller trace amounts [35]. Ultramicro trace amounts of elements can be determined spectrophotometrically [36,37].

1.7 Spectrophotometric Methods

In the *standard curve* (*analytical curve*) *method* the relationship between the absorbance and the concentration of the substance to be determined is ascertained by using standard solutions and is expressed graphically.

The standard curve is a straight line if Beer's law is followed under experimental conditions. The absorbance of a sample solution can be translated into concentration by means of the standard curve. In indirect standard curve methods, the element determined reduces the absorbance of a coloured system [38].

The standard curve method is the one most generally used in spectrophotometric analysis. If Beer's law is obeyed by the system used, only one standard is really necessary. For details see p. 33.

Mahr [39, 40] suggests the addition method for spectrophotometric determination. The element is determined from the change in absorbance when a known amount of the determinand is added. The method compensates for matrix effects, but is more time-consuming.

Differential spectrophotometry [41,42] is based on measuring the absorbance of the test solution against a standard of accurately known concentration instead of a reagent blank or pure solvent. In this way the error in measurement is confined to the difference in the two concentrations and so can be minimized.

There are several variants of differential spectrophotometry [43–51]. The measurement error [52–56] is much smaller that in ordinary spectrophotometry and may be as low as 0·2%. This makes spectrophotometry almost as precise as gravimetric and titrimetric methods and permits its use for determining major components (5–50%) of test samples. For typical applications of the method see references [42,57–59].

For each system there is an optimum concentration of the reference solution with respect to precision. This concentration is found experimentally.

In differential spectrophotometry it is necessary to take heed of some sources of error which are often left out of account in the ordinary method, e.g. cuvettes must be intercompared [60]. Particular care is required in the preparation of the standard solutions. Temperature may also exert an influence, and in making more precise measurements it is indispensable to use a thermostat for securing constant temperature ($\pm 0.2°C$).

In differential spectrophotometry the absorbance is usually high (> 1) and a wider slit width is necessary to permit sufficient light to be transmitted to the detector [61]. Stray light then becomes an important source of error since A is large (see p. 9). Photoelectric colorimeters can be adapted for use with the differential method [62–65]. Consideration of the equations for the error in differential spectrophotometry [53] shows that there is an advantage in using the technique even for absorbances as low as 0·2 [66].

Spectrophotometric titration [67–76] involves monitoring the absorbance during titration of the sample solution. This method can be applied whether Beer's law is obeyed or not, provided that near the end-point there is an almost linear relation between the absorbance and the concentration of the light-absorbing compound in the solution. The species to be monitored can be a reactant, a product, or an indicator. The titration is represented graphically by two intersecting lines in a plot of absorbance *vs.* volume of titrant added. In order to find the end-point it is enough to determine two points before and after it. For the most accurate work it is necessary to apply a correction for the dilution of the system, unless a special apparatus could be used in which the light-beam traversed the cell vertically so that the dilution was automatically compensated by the increase in path-length if the cuvette had a constant cross-section (though there would be difficulty in using it).

The titration is carried out either at a definite wave-length or with a suitable filter. Figure 1.11 shows a schematic view of the assembly for spectrophotometric titration. Conventional instruments must be modified to accommodate a titration vessel of convenient size, burette jet, and stirrer [77-80]. The cell can be mounted at any convenient point in the light-beam e.g. outside the spectrophotometer and between the lamp and the optical system. The method can be automated [81]. Spectrophotometric titration allows the concentration of a substance to be determined in the presence of other substances partly absorbing at the wavelength chosen for titration. It is the change in absorbance during the titration that is important, not the absolute values.

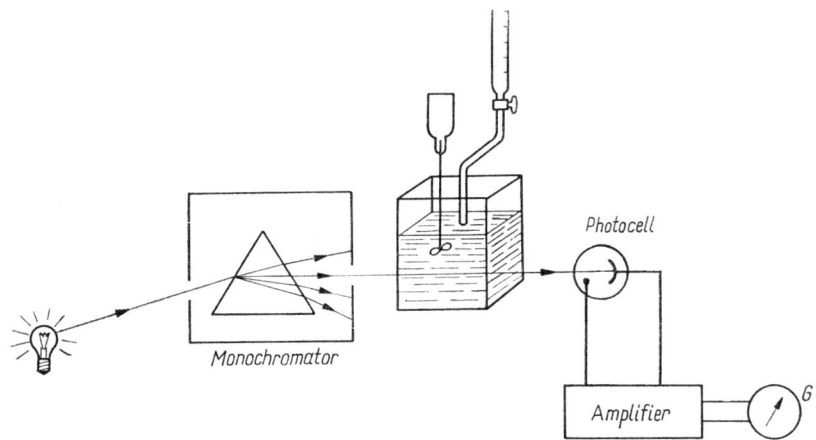

Fig. 1.11 Scheme of apparatus for spectrophotometric titration

To reduce the dilution effect, it is recommended to use concentrated titrant solutions and a micrometer syringe.

Spectrophotometric titration yields various types of titration curve, as shown in Fig. 1.12. The method is especially useful when the end-point is sluggish or difficult to discern visually. Complex-formation, redox, neutralization, and precipitation reactions are all used (with or without indicators). Extraction systems have also been used [82]. When precipitation reactions are used, a collimating slit should be added to the apparatus to avoid errors caused by back-scatter of radiation by the precipitate.

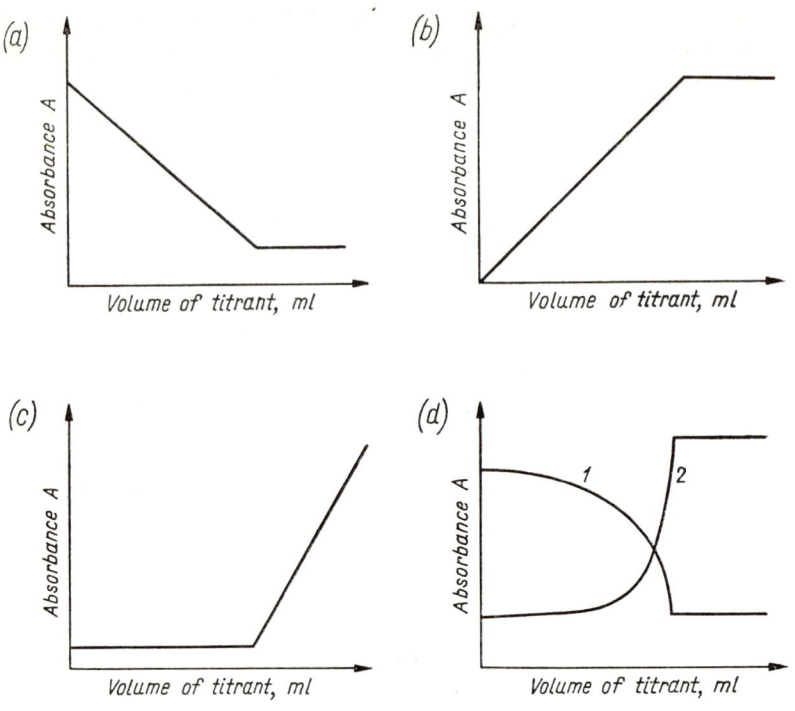

Fig. 1.12 Spectrophotometric titration. (a) Titrand absorbs and forms a non-absorbing product; (b) product formed absorbs; (c) titrant absorbs; (d) indicator present which is measured (1) in the form present at start of titration, (2) in form present at end of titration

Automated spectrophotometry [83–88] is of ever-increasing significance in chemical analysis. Automation is applied to spectrophotometric methods for routine analyses and for on-line control in industrial processes [87,88]. Typical examples are the determination of chloride, nitrate, and iron in water [83], silicon, phosphorus, and manganese in steel [89], trace metal contaminants in a wide range of chemicals [90], and major components in nickel, cobalt, iron, and manganese oxide mixtures [91]. Other examples are given in the chapters dealing with individual elements.

Turbidimetry involves measurement of the light-absorption (or scatter) by suspensions of sparingly soluble compounds [92–94]. It refers to truly turbid systems and not to colloidal pseudosolutions stabilized with protective colloids. Turbidimetric determinations are often performed visually, with use of a set of standards in colorimetric tubes, e.g. the determination of sulphate as $BaSO_4$ or chloride as AgCl. The light-absorption by the suspension depends on the particle-size distribution of the precipitate, which in turn depends on a number of factors such as the concentration of the ion to be precipitated, rate of precipitant addition, mixing, temperature, ionic strength of the solution, and presence of organic solvents miscible with water.

The turbid systems used do not obey Beer's law and the precision is low because of difficulty in keeping experimental conditions strictly reproducible. The sensitivity is higher if the suspension is coloured (e.g. metal sulphides).

Turbidimetric titrations are also possible [95].

1.8 Sensitivity of Spectrophotometric Methods

Strictly speaking, sensitivity refers to the slope of a calibration curve, but is frequently used to mean the least determinable concentration or amount of the species of interest.

The objective numerical expression of the sensitivity of spectrophotometric methods [96–98] is the *molar absorptivity* (ε) at the wavelength (λ_{max}) of maximum absorbance of the coloured species

$$\varepsilon = \frac{A}{c.l}$$

where A is the absorbance, c the concentration of the coloured species (mole/l.), and l the light path-length. If l is in cm, the molar absorptivity (ε) is expressed in l.mole^{-1}.cm^{-1}. If SI units are used, however, l is in mm and ε is in l.mole^{-1}.mm^{-1}, and has 1/10 the value it has if expressed in l.mole^{-1}.cm^{-1}.

Molar absorptivity was formerly called the molar absorption coefficient or molar extinction coefficient; ε is the slope of a calibration curve in terms of absorbance *vs.* molar concentration.

Molar absorptivity should be expressed so as to take into account the number of significant digits, i.e. the precision of the measurements. Therefore, it is more appropriate to write $\varepsilon = 2\cdot 4 \times 10^4$ or $\varepsilon = 2\cdot 40 \times 10^4$ rather than $\varepsilon = 24\,000$. Until quite recently most authors (and editors) paid no attention to this point.

For sensitive spectrophotometric methods ε is $> 1 \times 10^4$ l.mole.cm^{-1}, and values of ε below 1×10^3 correspond to less sensitive methods. The molar absorptivity cannot exceed $\sim 1\cdot 5 \times 10^5$, according to quantum theory. Higher molar absorptivities are only possible in certain indirect methods such as amplification methods (e.g. determination of bismuth, p. 153). The sensitivities of spectrophotometric methods for elements with similar atomic weights can be intercompared in terms of ε.

It is convenient to express and compare the sensitivities of spectrophotometric methods in terms of specific absorptivity (a) [99]. This is obtained by dividing ε by the atomic weight of the element and by 1000

$$a = \frac{\varepsilon}{\text{at.wt.} \times 1000}$$

The value a (in ml.g^{-1}.cm^{-1}) corresponds to the absorbance of a 1-µg/ml (= 1 ppm) solution of the determinand in a cuvette with an optical path-length of 1 cm. This is similar to the $E_{1cm}^{1\%}$ terminology formerly used by organic chemists (to express the absorbance of a 1-cm thick layer of a 1% solution of the determinand). The quantity a gives a comparison that is independent of atomic weight.

The sensitivity of spectrophotometric methods is often expressed in terms of the expression (*sensitivity index*) given by Sandell [1] which represents the number of micrograms of the determinand per ml of a solution having an absorbance of 0·001 for a path-length of 1 cm. The sensitivity (S) according to Sandell is expressed in μg . cm^{-2} and is therefore equal to $10^{-3}/a$.

The sensitivity of the dithizone method for determining copper (λ_{max} = 550 nm, at. wt. of copper = 63·54) may be expressed as follows:

molar absorptivity (ε)	4·52 × 10^4
specific absorptivity (a)	0·71
Sandell sensitivity (S)	0·0014

In the 1,10-phenanthroline method for determining iron(II) (λ_{max} = 512 nm) the respective figures are: $\varepsilon = 1 \cdot 11 \times 10^4$; $a = 0 \cdot 20$; $S = 0 \cdot 0050$; and in the thiocyanate method for determining tungsten (λ_{max} = 403 nm) they are: $\varepsilon = 1 \cdot 56 \times 10^4$; $a = 0 \cdot 09$; $S = 0 \cdot 012$.

The sensitivity of spectrophotometric measurements depends very much on the monochromaticity of the radiation. The molar absorptivity diminishes as the band-width increases. Particularly great differences occur between the values measured with a spectrophotometer and with a filter

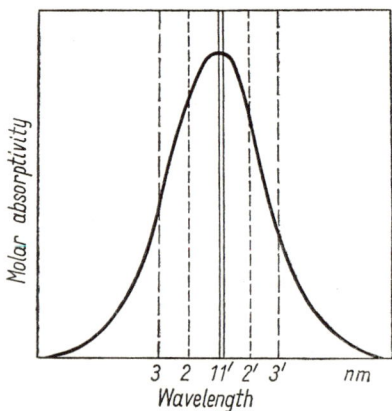

Fig. 1.13 Dependence of molar absorptivity on radiation band-width

photocolorimeter, if the absorption peak of the coloured compound is fairly narrow. If a wide-band filter is used (3–3' in Fig. 1.13) the absorbance is averaged over the wavelength interval concerned. With a narrower filter (2–2') the mean absorbance is higher. When the absorbance is measured by means of a spectrophotometer with monochromatic light of very narrow band-width (1–1') corresponding exactly to a wavelength of λ_{max}, the maximum value of the molar absorptivity is obtained.

In the determination of ε the absorbance measured should be within the range over which the coloured system conforms to Beer's law and within the range of values for which the measurement error is minimal. In addition, the slit-width used is of importance. A full account can be found in the literature [100].

It is easiest to determine ε when the spectrophotometric reagent has practically zero absorbance at λ_{max} for the complex and only one, stable, coloured complex is formed in the system. If the complex has low stability a large excess of reagent is used to shift the equilibrium. If stepwise complexation occurs (e.g. Fe–SCN system), special methods must be used to evaluate the several ε values. Determination of the molar absorptivity when the reagent also absorbs at λ_{max} for the complex is more difficult, but if only one stable coloured complex is formed, the absorbance can be measured against a reference solution in which the reagent concentration is the same as the concentration of uncombined reagent in the test solution (this concentration can be calculated from the stability constant and concentrations used).

There are computational and graphical methods for determining ε when the coloured reagent absorbs at λ_{max} of the complex. Both methods give consistent results, as demonstrated for the thorium complex with 4-(2-pyridylazo)resorcinol (PAR) [101].

If more than one coloured complex is formed that absorbs in the same region, it is necessary to add a very large excess of reagent in order to form the highest complex exclusively and completely and to determine its absorption spectrum and hence its ε values at the wavelengths of interest. If the stability constants of the successive complexes are known, then the composition of any mixture of the complexes can be calculated if the total concentrations of the reactants are known. If the reagent excess is adjusted so that essentially only the two highest complexes are present, then by measurement of the overall spectrum and calculation of the contribution to it made by the highest complex, the spectrum of the second complex can be calculated. By this process of "spectrum stripping" the ε_{max} values of all the complexes can be obtained. If there is overlap of spectral peaks they can be resolved by assuming they are Gaussian in shape, and starting from one end of the spectrum, "stripping" each peak in turn [102]. Naturally such a process is not very accurate, but it does provide information that is impossible to obtain directly. Another technique is to use a spectrum synthesizer which generates Gaussian or other functions and adds them, with display on a cathode-ray tube, so that the simulated spectrum can be adjusted until it matches the observed spectrum, and the individual contributory peaks can then be read off.

In some indirect methods the determinand decolorizes the reagent. The apparent molar absorptivity is then calculated on the basis of the change in absorbance produced by a given concentration of the determinand.

The determination of ε in extraction spectrophotometric methods is generally simple. The extraction usually involves a single complex whereas several may coexist in aqueous solution. Extraction often enhances the sensitivity of the method, presumably by virtue of the effect of solute-solvent interactions.

When ε is determined from a reaction system in which colour-formation is incomplete or there is incomplete extraction, allowance must be made for this in the calculations, and the true concentration of the complex calculated or determined by another method.

Difficulty may arise from the definition of the mole—for example 127 µg of iodine is 1 µmole of I or 0·5 µmole of I_2, and the species referred to must be clearly stated.

An advantage of the Irving reversion technique [103] is that it is always an equivalent amount of reagent that is measured, and so one value of ε and one calibration curve serve for all applications of that reagent.

In some spectrophotometric methods the sensitivity depends on the quality of the reagent used, particularly with reagents which are natural products. Various curcumin preparations, for example, give considerably different sensitivities in the method for determining boron. In the case of certain synthetic organic reagents differences in sensitivity have also been found, e.g. when using α-furildioxime for determining nickel and rhenium, diphenylcarbazide for determining chromium, or metallochromic indicators such as Xylenol Orange or Methylthymol Blue for various metals. Such differences in the sensitivity result from the presence of foreign substances in the preparation, which affect the reactions concerned or give competing reactions (e.g. Semixylenol Orange). In the case of α-furildioxime such a substance is γ-furildioxime, which gives with nickel a complex of a weaker colour than that of the complex with the α-isomer.

The sensitivity of spectrophotometric methods is sometimes discussed in the literature in terms of the least possible determinable concentration (*lower limit of determination*). In such cases additional information should be given, such as the solution volume, optical path-length of the cuvette, type of instrument, and minimum value of the absorbance to be measured.

The sensitivity can also be determined visually in colorimetric tubes and expressed in concentration units. It refers, strictly speaking, to the lowest concentration of the coloured compound which can be distinguished (at some acceptable confidence level, say 90 or 95%) from a blank test solution.

The *lowest concentration* that can be determined spectrophotometrically [104,105] can be calculated from the fundamental formula $A = \varepsilon c l$. If the minimum absorbance to be measured is 0·02, and l is 2 cm, and the molar absorptivity of an averagely sensitive spectrophotometric method is taken as 1×10^4, the corresponding concentration is $c = 0·02/2 \times 10^4 = 10^{-6} M$. For an element with atomic weight 100 (specific absorptivity $a = 10^4/10^2 \times 10^3 = 0·1$), this corresponds to $10^{-6} \times 10^2/10^3 = 10^{-7}$ g/ml $= 0·1$ µg/ml.

If the absorbance is to be measured in an ordinary 2-cm cuvette (i.e. the optical path-length is 2 cm), at least 6 ml of solution will be used in the cuvette, so the minimum size of standard flask that can be used is 10 ml, which corresponds to 1 μg of the element. If we take it that saturated solutions of easily soluble salts average 10% (w/v), then 10 ml of saturated solution corresponds to 1 g of solute. If the 1 μg of the determinand is present in this 1 g of test sample, its concentration in the sample is 10^{-4}%. Therefore the limit of determination by averagely sensitive spectrophotometric methods is 10^{-4}%; in other words, concentrations not less than 10^{-4}% can be determined directly without preconcentration of trace amounts of elements. For the most sensitive methods with ε about 10^5 ($a \sim 1\cdot 0$) the limit is lowered to 10^{-5}%.

Trace concentrations smaller than 10^{-4}% are below the sensitivities of most spectrophotometric methods where the colour is developed directly in the aqueous phase. In order to determine them spectrophotometrically, either an extraction procedure utilizing a phase-volume ratio that will give a concentration effect, or *preconcentration* (*enrichment*) *step* is necessary (p. 67). Depending on the sample type and the size of the weighed portion (e.g. 10 g, 100 g, or more) this operation may enhance the sensitivity by one, two, or more orders. In this way the "normal" limit of determination by spectrophotometric methods is enhanced to 10^{-6}–10^{-7}%. In principle it would be possible to achieve enhancement by a series of alternating extraction and stripping reactions, each step being a concentration process.

Further increase in the sensitivity is practicable by reducing the minimum absolute amount of determinand needed. This may be achieved by using microcuvettes in which the minimum amount of determinand may be about 0·1 μg thus increasing the sensitivity by one order. A further small increase can be achieved by using longer cuvettes (e.g. 5 cm).

Quantities below 1 μg may be estimated visually by comparison in tubes 7–9 mm in diameter and 4–5 ml in capacity, the volume of the coloured solution totalling 3–4 ml. Typical limits of determination by this method are:

lead (dithizone, CCl_4)	0·10 μg
nickel (α-furildioxime, $CHCl_3$)	0·08 μg
chromium (diphenylcarbazide, pH $\sim$ 1)	0·04 μg
iron (SCN^-, methyl isobutyl ketone)	0·04 μg
zinc (dithizone, CCl_4)	0·02 μg
boron (curcumin, ethanol)	0·002 μg

The limits attainable depend on the tube diameter, the molar absorptivity, the response of the eye to a given wavelength range, and the shape of the absorption spectrum of the coloured compound.

In trace analysis an important role is played by the blank test, especially in determination of the commoner elements such as Fe, Zn, Ca, Al, and Si. The amount of determinand that is present in the blank test sets a limit to the amount that can be determined in a given material. Therefore reducing the blank (e.g. by purification of reagents, use of quartz and polyethylene laboratory ware instead of glassware) can enhance the sensitivity.

1.9 Precision and Accuracy of Spectrophotometric Methods

In analytical chemistry the term precision is used to indicate reproducibility, scatter, consistency of results. The term accuracy, on the other hand, indicates how close the results are to the true value*.

The precision of spectrophotometric methods [53,106–113] depends on the concentration of the determinand and on the measuring technique adopted. Visual methods give results with a precision of 5–10%. The precision of the objective photoelectric method is, of course, higher and varies from 0.5 to 2% under suitable measuring conditions. By the differential method a precision of 0.2–0.5% is obtainable, which enables macroquantities of components to be determined.

In the photoelectric methods the measurement error is of importance for precision. The precision attainable is a function of the absorbance measured. When very low concentrations are determined the error is very large, because the error in setting the galvanometer needle to zero in null-balance methods is a substantial fraction of the absorbance being measured. When intensely coloured solutions are being measured only an insignificant part of the radiation is transmitted and on the logarithmic absorbance scale the graduations are so close that the reading error is very high. For example, a transmittance of 1% corresponds to an absorbance of 2.0, a transmittance of 0.1% to an absorbance of 3.0. Thus 0.9% of the total scale length corresponds to a 50% change in concentration. To find the theoretical value of the absorbance which can be measured with the maximum precision, the following relationships are used (where c is concentration, x is I_0/I_t, and a is a constant).

$$c = a\log x$$

$$dc = \frac{(a\log e)dx}{x}$$

$$\frac{dc}{c} = \frac{(a\log e)dx}{ax\log x} = \frac{0.434\,dx}{x\log x}$$

Differentiation of this equation and equating the second derivative to zero shows that the error is minimal when $\log x(= A) = 0.434$.

Experimental results for the dependence of relative measuring error on measured absorbance values deviate considerably from the theoretical curve (Fig. 1.14). The minimum error obtained experimentally lies above

---

* In common parlance accuracy has a broader sense which includes precision.

the value $A = 0.43$, and the range of absorbances suitable for reasonably precise measurements extends much higher than the value $A = 1.0$ when certain instruments are used. When a null-point method is used, an estimate of reading error can be obtained by setting the galvanometer needle just perceptibly off zero, on both sides, and reading the corresponding absorbances. These can be treated as the limits of a square distribution, so that the standard deviation can be taken as $\Delta A/\sqrt{12}$. Another method is

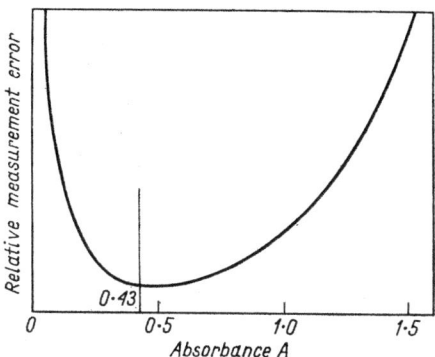

Fig. 1.14 Dependence of relative measurement error on absorbance

to take at least ten measurements on the same solution in the same cell, resetting all the controls (including wavelength) each time, and calculating the standard deviation. A typical result is a standard deviation of 0·0008 absorbance units at $A \sim 0.35$ [114].

Errors connected with the measurement of absorbance are usually smaller than those associated with the chemical operations in the determination. In some spectrophotometric methods the colour reaction is not reproducible. In others the colour produced is not stable with respect to time, and the absorbance must be measured at a fixed time from the start of the colour reaction. In some systems temperature variations as small as 3–5°C can cause differences in colour. Some colour reactions are very sensitive to pH changes. A change of as little as 0·1 in the pH may sometimes cause errors of 5% (of course, methods as pH-sensitive as this are best avoided). Other possibilities of error are associated with competitive reactions in the system and with too great a variation in ionic strength in the sample solution.

The overall error of the determination is the summation of the errors committed at each stage of the procedure, viz. in sampling, dissolution of the sample, preconcentration, separation of elements, and spectrophotometric determination. In the course of these operations, determinand may be introduced from outside the system or may be lost. An accurate blank determination (in trace analysis) is also of vital importance. The influence of the blank on the accuracy increases as the determinand concentration decreases.

In determination of contents of about 10^{-3}–10^{-4}%, an error of up to ± 10% is usually obtained, whereas in determination of traces at the 10^{-6}–10^{-7}% level, with preconcentration, the error may be as high as ± 30%.

The error in determining an element by a given method may be estimated by comparing the results with the values considered to be true. If no standard samples of accurately known composition are available, the accuracy of determination is estimated by analysing the sample with and without a "spike" of accurately known amount of the determinand added to a weighed sample. The amounts of spike added should be close to those contained in the sample itself. The spike is added in a form which behaves similarly (e.g. as regards volatility) to the form in the sample.

1.10 Selectivity of Spectrophotometric Methods

According to the definitions recommended by the Analytical Chemistry Division of the International Union of Pure and Applied Chemistry a reagent which reacts with a limited number of elements is considered to be *selective*; a reagent may be called *specific* if it gives a reaction with only one element in certain circumstances. Accordingly, we may speak of selective and specific spectrophotometric methods.

The selectivity of colour reactions and the corresponding spectrophotometric methods depends on the nature of the reagent used, the oxidation state of the element, the pH of the medium, and the nature of the complexing agents used to mask interfering ions. If, in spite of use of all these factors, some ions can still interfere in a determination, the species to be determined has to be separated from the interfering species or vice versa. Methods for separation of elements are discussed in Chapter 3.

There exist rather few specific reagents and reactions. As rare examples cuproine for copper(I) and bathophenanthroline for iron(II) may be mentioned.

A change in oxidation state of certain ions is enough to prevent their reaction with certain reagents. In the determination of niobium by the thiocyanate method, for example, iron does not interfere if reduced to Fe(II). Change in oxidation state is rather seldom used to enhance the selectivity of spectrophotometric reactions.

The selectivity of most methods is improved by choosing a suitable pH for the reaction medium. The optimum pH range for reaction with a given element is related to the nature of the element and the reagent. For reagents of the R.OH type there is a relation between the colour reactions and the hydrolysis of given elements. In strongly acid solutions the colour reactions are given by elements having cations which readily hydrolyse, i.e. Zr, Hf, Th, U(IV), and Ti. In moderately acid solutions,

Fe(III), Al, and U(VI) also react, in weakly acid and neutral solutions rare earth elements (REE), Fe(II), Cu, and Mn react, and in alkaline solutions Ca, Sr, and Mg. The nature of the reagent used has a specific influence in reactions with particular elements but the general tendency remains the same. The more readily hydrolysed elements react in more acidic media, and other elements in less acidic media.

In acid medium the metal ions of the hydrogen sulphide group that form the most stable sulphides react with R.SH type reagents, whereas in slightly acidic and neutral media the Analytical Group III metals do so.

The acidity of the medium has an important influence on the state of certain ions in the solution (e.g. Zr, Ti, Nb, and V) and on this state (e.g. for zirconium, Zr^{4+}, ZrO^{2+}, polymerized forms) may depend the course of reaction of the elements with spectrophotometric reagents [115].

In some spectrophotometric methods the most suitable pH range is rather narrow, owing to the nature of the colour reaction and/or the effect of other ions. For many spectrophotometric procedures universal indicator paper (even narrow range) is not good enough and it is necessary to control the adjustment of the acidity of the medium by means of a pH-meter.

Increased selectivity of spectrophotometric methods is mainly obtained by masking the interfering ions [116–120]. The masking consists in transforming the interfering ion into a stable complex with a complex-forming agent so that it can neither react with the spectrophotometric reagent, nor otherwise interfere in the colour reaction of interest.

Some masking agents of importance are listed in Table 1.2 for particular elements. Additional data on masking can be found in Table 3.2 (p. 72).

High selectivity is obtainable by a suitable combination of masking agents and pH. The effective stability of complexes is not constant but varies with the pH and other parameters, e.g. the concentration of the masking agent and the presence of other complex-forming substances [119].

The release of an ion from the complex formed as a result of a masking reaction is called demasking. Zinc and cadmium, for example, are demasked from their cyanide complexes by formaldehyde which converts cyanide ions into cyanohydrin. Some metals are released (demasked) from fluoride complexes by means of beryllium, aluminium, or boric acid.

Thus, notwithstanding the paucity of specific reagents, there exist a certain number of specific methods for the determination of elements by use of group reagents (reacting normally with a great many elements) which in the presence of suitable masking agents and at a given pH react specifically. The determination of zinc with dithizone at pH 4–5 in the presence of thiosulphate, or the determination of aluminium with 8-hydroxyquinoline at pH ~ 9, in the presence of EDTA, cyanide, and H_2O_2 as masking agents, may be quoted by way of example.

Table 1.2 Some more important masking agents

Element	masked by	Element	masked by
Ag	CN^-, $S_2O_3^{2-}$, Br^-, I^-, Cl^-, thiourea, NH_3	Nb	F^-, citr, tartr, oxal, H_2O_2
Al	F^-, oxal, acet, citr, tartr, EDTA, OH^-, TEA, sulphosal	Ni	CN^-, EDTA, NH_3
As	S^{2-}, OH^-	Os	CN^-, SCN^-, Cl^-, thiourea
Au	CN^-, Cl^-, Br^-, $S_2O_3^{2-}$	Pb	acet, $S_2O_3^{2-}$, I^-, citr, tartr, EDTA, SO_4^{2-}
B	F^-, hydroxy acids, glycols	Pd	CN^-, I^-, Cl^-, NH_3, NO_2^-, SCN^-, $S_2O_3^{2-}$, EDTA
Ba	EDTA, citr, tartr, SO_4^{2-}	Pt	I^-, CN^-, Cl^-, NO_2^-, SCN^-, NH_3, $S_2O_3^{2-}$, oxal
Be	F^-, citr, tartr, sulphosal	Rh	Cl^-, thiourea, citr, tartr
Bi	citr, tartr, EDTA, I^-, thiourea	Ru	CN^-, Cl^-, thiourea
Ca	EDTA, citr, tartr, $P_2O_7^{4-}$	Sb	citr, tartr, I^-, S^{2-}, OH^-, F^-
Cd	EDTA, CN^-, $S_2O_3^{2-}$, SCN^-, I^-, citr, tartr	Sc	citr, tartr, EDTA
Ce	F^-, EDTA, citr, tartr	Sn	citr, tartr, oxal, OH^-, S^{2-}, F^-
Co	NH_3, NO_2^-, SCN^-, CN^-, $S_2O_3^{2-}$, EDTA	Sr	SO_4^{2-}, EDTA, citr, tartr
Cr(III)	EDTA, citr, tartr, oxal, acet	Ta	F^-, citr, oxal, tartr
Cu	NH_3, I^-, CN^-, $S_2O_3^{2-}$, thiourea, EDTA, citr, tartr	Th	F^-, EDTA, citr, tartr, acet, CO_3^{2-}
F	H_3BO_3, Al, Be, Zr, Ti, Nb	Ti	SO_4^{2-}, H_2O_2, F^-, citr, tartr, sulphosal
Fe(III)	F^-, PO_4^{3-}, $P_2O_7^{4-}$, EDTA, citr, TEA, tartr, oxal, sulphosal	Tl	Cl^-, oxal, EDTA, citr, tartr
Fe(II)	CN^-, $S_2O_3^{2-}$, phen	U	F^-, CO_3^{2-}, oxal, H_2O_2, citr, tartr
Ga	EDTA, tartr, citr, oxal	V	H_2O_2, EDTA, F^-, oxal
Ge	oxal, F^-	W	F^-, Cl^-, H_2O_2, oxal, tartr, citr
Hg	I^-, CN^-, Cl^-, NO_2^-	Zn	CN^-, EDTA, SCN^-, OH^-, NH_3
In	EDTA, oxal, Cl^-, citr	Zr	F^-, SO_4^{2-}, oxal, citr, tartr, H_2O_2, $P_2O_7^{4-}$, PO_4^{3-}
Ir	Cl^-, SCN^-, thiourea, NH_3		
Mg	EDTA, oxal, citr, tartr		
Mn	oxal, $P_2O_7^{4-}$, EDTA, TEA, citr, tartr		
Mo	F^-, Cl^-, oxal, H_2O_2, citr, tartr, EDTA		

Abbreviations: tartr — tartrate, citr — citrate, oxal — oxalate, acet — acetate, sulphosal — sulphosalicylate, TEA — triethanolamine

Kuznetsov and Petrova [121] have increased selectivity by measuring the absorbance of the solution at two different temperatures. Complexes of Arsenazo I with thorium and REE behave differently on heating. The absorbance of the thorium complex does not change with solution temperature over the range 20–80°C. The absorbance of the REE complexes increases by a factor of 2–3 over this range of temperature. Changes in the colour of copper, cobalt, and nickel chloride solutions with temperature provide the basis for thermospectrophotometric determination of these metals simultaneously [122].

In discussion of the selectivity of spectrophotometric methods, mention should also be made of simultaneous determination of two coloured complexes by measuring the absorbances at two different wavelengths [123–127].

1.11 Colour Systems Used in Spectrophotometry

Almost all the colour systems used in spectrophotometric methods involve complex formation of some kind [128–131]. Particular reagents are dealt with in Chapter 2.

The largest group is made up of the methods using bifunctional organic reagents to form *chelate complexes* with metal ions. These complexes are usually electrically neutral and soluble in inert sovents. To this group belong the extraction-spectrophotometric methods using dithizone, 8-hydroxyquinoline, 1-nitroso-2-naphthol, dioximes, dithiocarbamates, and 1-(2-pyridylazo)-2-naphthol (PAN). In most cases only the coloured complex is obtained in the extract. Less often the extract also contains the excess of the coloured reagent, e.g. PAN (see determination of Mn, p. 344).

Another large group uses organic reagents forming water-soluble *anionic* or *cationic chelate complexes* with metal ions. These reagents have hydrophilic groups in their structure, usually sulphonic acid groups. Anionic complexes are formed by Arsenazo III (with Th, Zr, U), Arsenazo I (with U, REE), nitroso-R-salt (with Co), Chrome Azurol S (with Al, Be), Xylenol Orange (with Bi, Zr, Sc), chromotropic acid (with Ti), and sulphosalicylic acid (with Fe). Cationic chelates, soluble in water, are formed by 1,10-phenanthroline [with Fe(II)], cuproine [with Cu(I)], and thiourea (with Bi).

Some organic reagents form with metal ions compounds insoluble in water as well as in organic solvents. These are either *polynuclear complexes*, e.g. phenylfluorone with Sn(IV) and Ge, or *adsorption compounds* (usually called lakes) e.g. Titan Yellow with Mg. In these cases the absorbance of the coloured suspension, sol, or pseudosolution stabilized with a protective colloid [e.g. gum arabic, gelatine, or poly(vinyl alcohol)] is measured.

Sometimes, instead of extraction of a sparingly soluble chelate, its aqueous pseudosolution is stabilized with a protective colloid and measured, e.g. determination of tin with dithiol or of copper with sodium diethyldithiocarbamate. This course may be chosen either to avoid an extraction or because the colour of the pseudosolution is more intense than that of the complex dissolved in the organic solvent.

Basic dyes of the xanthene (e.g. Rhodamine B), triphenylmethane (e.g. Methyl Violet, Brilliant Green), thionine (e.g. Methylene Blue), and other groups form *ion-association complexes* (ion-pairs, ternary complexes) with anionic halide complexes of some metals (e.g. $SbCl_6^-$, CdI_4^{2-}, and TaF_6^-), which are extractable into inert solvents, and this provides numerous sensitive spectrophotometric methods for determining metals [132–136]. Other anionic metal complexes are also used. The free basic dye remains, as a rule, in the acid aqueous phase during extraction. Conversely, coloured cationic complexes [e.g. the iron(II) phenanthroline complex] can be used to determine anions.

Complexes of metals with inorganic reagents are also frequently used in spectrophotometric methods, e.g. thiocyanate (Fe, Co, Nb, Mo, Re, W, U, Ti), iodide (Pd, Bi, Sb, Pt), bromide (Au), and peroxide (Ti, V, U) complexes. Determinations based on these complexes are carried out in aqueous medium or after extraction with organic solvents having oxygen-containing groups.

A group of elements (P, As, Si, Ge, V, W, Mo etc) forms yellow *heteropoly acids* which can be reduced to an intensely coloured blue form. Heteropoly acids and their reduced forms are put to good use in spectrophotometric methods for determining these elements. They can be extracted with oxygen-containing solvents.

In several instances the coloured system is formed as a result of a *redox reaction*, e.g. determination of manganese as MnO_4^-, chromium as CrO_4^{2-}, oxidation of dimethylnaphthidine with vanadium(V) or chromium(VI), and of *o*-tolidine with cerium(IV) or chlorine. Oxidation reactions can also be exemplified by iodide methods, in which iodide is oxidized with bromine to iodate which, in turn is reacted with added potassium iodide to produce elemental iodine (p. 297). Examples of reduction reactions include determination of selenium and tellurium as the coloured sols formed when Se(IV) and Te(IV) are reduced to their elemental forms.

There are also spectrophotometric methods in which the colour system is formed as a result of *synthesis*. The following reactions can be mentioned by way of example: formation of indophenol in the method for ammonia, synthesis of azo dyes in determining nitrite, formation of Methylene Blue in determining sulphide, the pararosaniline method for sulphite, and the benzidine–pyridine method for cyanide.

Other organic reactions may be turned to account. The determination of nitrate, for example, consists in nitration of phenoldisulphonic acid to yield a coloured product, and the determination of bromine is based on bromination of Phenol Red.

Indirect spectrophotometric methods [38, 137] involve a change, brought about by the determinand, in the colour of another system. For example, by forming stable colourless complexes with numerous metals, fluoride ions decompose certain coloured complexes of these metals (p. 256). Another example of the indirect method is the determination of phosphate with lanthanum chloranilate. The phosphate forms the less soluble lanthanum phosphate, releasing the coloured chloranilate ion into the solution.

Indirect spectrophotometric methods include *amplification methods* [138–140]. The amplification or multiplication reactions underlying these methods can be defined as those in which the normal equivalence is altered in some way so that a more favourable measurement can be made. Thanks to their exceptionally high sensitivity, amplification methods are widely used for determining traces of some species, e.g. halides and some metals (Mo, Cr, Bi, and others). Belcher [138] has given an exhaustive review of these methods.

Spectrophotometric methods can be divided into two groups according to whether measurements are made on aqueous solutions or on extracts in organic solvents. In general, the *extraction methods* [141, 142] are considered more useful and are preferred. The extraction step may not only enhance the selectivity of the method but also provide enrichment and so increase the sensitivity.

Babko *et al.* [143–145] have worked out a technique for intercomparing various spectrophotometric methods for the determination of a given element. The following parameters must be quoted for systems involving a coloured reagent and a coloured complex (both water-soluble): λ_{max} of reagent and λ_{max} of complex, $\Delta\lambda$ (difference between the wavelengths of the absorption maxima of the complex and the reagent), molar absorptivities of the reagent and the complex at λ_{max} of the complex, and pH range most suitable for the colour reaction.

The development of a spectrophotometric method is often preceded by the investigation of the nature of the coloured species (composition and stability of the complex) involved. Physico-chemical examination of the coloured complexes in the solution enables the optimum parameters of the method to be rationally and precisely established. The composition of the complex (the ligand:metal ratio) in the solution is most often determined via the continuous variations [146, 147], mole-ratio [148], slope-ratio [149], and isosbestic point [150] procedures and their variations, and more rarely by other methods such as pH-absorbance plots [151]. There is an extensive literature on study of complexes but only some, e.g. [152–161], is worth looking at.

Development and publication of new spectrophotometric methods for analysis have been discussed in the literature [162, 163].

1.12 Spectrophotometric Determination

1.12.1 STANDARD SOLUTIONS

Spectrophotometric methods all use comparison with a reference material. Their accuracy depends to a considerable extent on the proper preparation of the standards, whether these are used for preparation of calibration curves, standard series, or as solutions for colorimetric titration.

Standard solutions are divided into stock and working solutions. Sufficiently concentrated stock solutions may usually be stored for unlimited periods of time. The working solutions obtained by suitable dilution of stock solutions cannot, in general, be stored for a long time.

Stock solutions, usually containing 1 mg of element (or ion) per ml (less often 10 mg/ml), are obtained by dissolving the element or sufficiently pure salts of definite composition. Water is often used as solvent, or dilute acids if metals are used or the metal ions would hydrolyse in non-acidified solution or precipitate as carbonates with atmospheric carbon dioxide.

The concentrations of the working solutions are related to the sensitivity of the spectrophotometric method: 0·1 mg, 10 μg, or 1 μg of element per ml. Solutions of 10 μg/ml, and especially 1 μg/ml, concentration are freshly prepared on the day of use. Stock solutions are diluted with water in the case of non-hydrolysing species or with adequately diluted acids in the case of hydrolysable metals. The instability with time of more dilute working solutions is caused principally by adsorption of ions on the surface of the vessel [164–168].

When preparing standard working solutions it is best to apply the rule of tenfold dilution by means of adequately accurate pipettes and volumetric flasks (e.g. a 10-ml pipette and a 100-ml flask). For really accurate work the glassware must be calibrated [169].

Unfortunately, very few chemical compounds are available in sufficient purity to be taken from a reagent bottle and used without further ado. Indeed, some compounds may not even be what the label on the bottle purports them to be [170]. Hydrates, oxides, and hydroxides are particularly suspect. Standards have recently been reviewed [169].

As a result of this state of affairs, it is preferable to use the pure element (there is no sacrifice in accuracy if sufficient stock solution is prepared), or failing this, it is necessary to prepare a stock solution rather more concentrated than that desired, determine its titre by an accurate gravimetric or titrimetric method, and then dilute appropriately. It is more convenient in practice to prepare a solution of approximately the desired concentration, standardize, and use the solution as it is.

Sometimes it is necessary for the standard solution to contain no complexing anions, and therefore perchlorates or nitrates are used. A perchlorate or nitrate solution can be obtained by starting with any other salt available (e.g. chloride, sulphate). The salt is dissolved in water or diluted acid, the metal is precipitated as hydroxide with ammonia or alkali, the precipitate is filtered off, dissolved in perchloric or nitric acid and diluted with water to a definite volume.

1.12.2 ABSORPTION SPECTRUM, CALIBRATION CURVE

In order to select the most suitable wavelength for the spectrophotometric measurements, it is necessary to know the absorption spectra of the coloured complex and the reagent.

The absorption spectrum is a plot of absorbance *vs.* the wavelength (Fig. 1.1, p. 5), and is either obtained with a recording spectrophotometer or plotted manually from measurements made on a non-recording instrument at 10–20 nm intervals, and at 2–5 nm intervals near the absorption maximum. The solvent is used as reference.

The absorbance measurements are usually made at λ_{max} for the coloured compound, where a small error in wavelength setting will cause least error in the absorbance. However, if the reagent also absorbs in the same region as the coloured species, it is customary to choose a wavelength at which the difference in the molar absorptivities of product and reagent is greatest, but in such cases it is imperative always to add exactly the same volume of reagent to standards and samples.

The calibration curve relating absorbance to concentration of the determinand is prepared by applying the colour reaction procedure to a set of standard solutions evenly spaced over a concentration range that will give a maximum absorbance of 0·8–1·0. The conditions laid down in the procedure must be strictly adhered to. The curve is plotted with the scale arranged so that the curve is at $\sim 45°$ to the abscissa, as shown in Fig. 1.15.

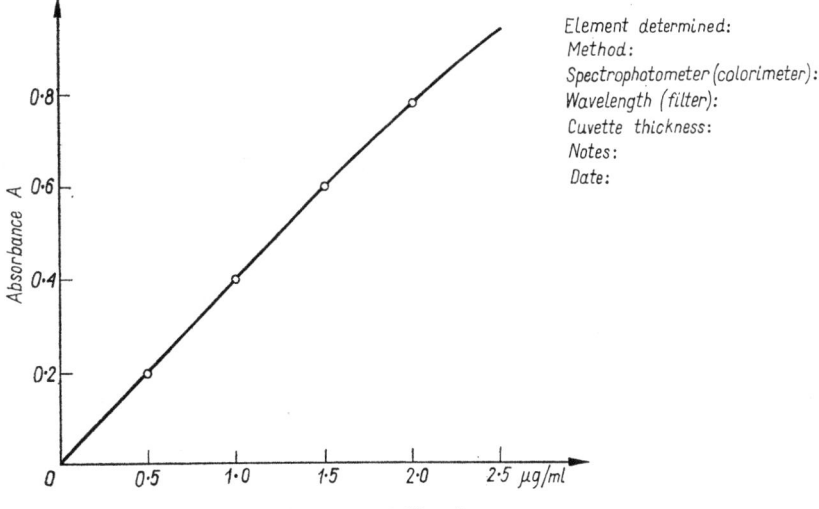

Fig. 1.15 Calibration curve

The absorbances of standards and samples are measured with the solutions in cuvettes of the same path-length, usually 1 or 2 cm, although 0·1, 0·5, 4, and 5 cm cells are sometimes used. The choice of cuvette depends on the sensitivity of the method, the quantity of determinand and the behaviour of the solution in relation to Beer's law. If at low values of absorbance deviations from Beer's law are observed, then the use of a longer path-length cuvette is recommended so that higher absorbance values are obtained for the more dilute solutions.

The cuvettes must always be used the same way round and should be marked for ease of orientation. They must also be intercompared to allow for any differences in reflectance characteristics (and appropriate corrections applied) and their path-lengths must also be checked by measurement of the same solution in each of them (this is especially important when the shorter cells are used, but is all too often forgotten) [60].

The conditions of the determination, such as the weight of sample, size of volumetric flask, are selected (if possible) so that the absorbance measured is 0·2–0·9 (or higher for differential spectrophotometry). For trace amounts, however, lower absorbances may be used (but should preferably be > 0·02) since in determination of traces a lower precision is acceptable.

The reference (comparison) solution is usually the solvent or a reagent blank prepared strictly according to the procedure. Although the reagent blank should automatically compensate for the amount of determinand adventitiously obtained from the reagents, apparatus, and atmosphere [171–173], in trace analysis work it is better to use the solvent as reference and to run separate blank tests, since it is important to avoid the errors of reporting a substance as present when it is absent, or absent when it is present. For this it is necessary to know the signal to noise ratio at the level of the blank (i.e. the standard deviation) [174].

Lowering the blank value permits determination of smaller trace contents in the test samples. This can be done in various ways. In sampling and comminution of the sample, the equipment used should be made, if possible, from material which does not contain the element to be determined. Sometimes, however, contaminants unavoidably introduced in this way can be removed by means of a solvent which will dissolve the impurity without attacking the sample (e.g. removal of iron with hydrochloric acid from a silicon sample comminuted in a steel mortar). Instead of glassware, quartz, polyethylene, and Teflon vessels may be used (depending on the species to be determined). To prevent contact of the sample with laboratory air, certain chemical operations may be conducted in closed chambers ("dry-boxes") flushed with purified air or inert gas. For ultratrace work special "clean rooms" may be necessary [175].

The reagents are the source of the bulk of contaminant in the blank test. The water used should be distilled in a quartz apparatus after demineralization with ion-exchangers [176]. Acids (H_2SO_4, HNO_3, $HClO_4$, HCl, HBr) can be purified by slow distillation in quartz vessels. Hydrofluoric acid is purified by fractional distillation in platinum or palladium vessels. Ammonia of suitable purity is obtained by saturation of distilled water in a polyethylene or Teflon container with gaseous ammonia, appropriate precautions being taken to prevent suck-back of the solution if a cylinder of ammonia is used. Reagents so purified are stored in polyethylene bottles.

Solutions of other reagents are usually purified by solvent extraction and precipitation methods with collectors.

Some volatile reagents (HCl, HBr, ammonia) can be brought to a very high degree of purity by the isothermal (isopiestic) distillation method [177,178] or the "sub-boiling" technique [179]. Iron traces are removed from concentrated hydrochloric and hydrobromic acids by passing the acids through a strongly basic anion-exchanger [180].

Various problems connected with spectrophotometric trace analysis are discussed in detail in the literature [1, 7,181–192].

References

1. Sandell, E. B., *Colorimetric Determination of Traces of Metals*, 3rd Ed., Interscience, New York (1959).
2. Boltz, D. F. (ed.), *Colorimetric Determination of Nonmetals*, Interscience, New York (1958).
3. Snell, F. D. and Snell, C. T., *Colorimetric Methods of Analysis*, Van Nostrand, New York (1959).
4. Charlot, G., *Dosages colorimétriques des éléments minéraux*, Masson, Paris (1961).
5. Świętosławska, J. (ed.), *Spektrofotometria absorpcyjna* (Absorption Spectrophotometry), PWN, Warsaw (1962).
6. Kortüm, G., *Kolorimetrie, Photometrie und Spektrometrie*, Springer, Berlin (1962).
7. Koch, O. G. and Koch-Dedic, G. A., *Handbuch der Spurenanalyse*, 2nd Ed., Springer, Berlin (1974).
8. Peshkova, V. M. and Gromova, M. I., *Prakticheskoe rukovodstvo po spektrofotometrii i kolorimetrii* (Practical Text-Book of Spectrophotometry and Colorimetry), Izdat. Mosk. Univ., Moscow (1965).
9. Babko, A. K. and Pilipenko, A. T., *Fotometricheskii analiz. T. I. Obshchie svedeniya i apparatura* (Photometric Analysis, Part I. Principles and Apparatus), Izdat. Khimiya, Moscow (1968).
10. Bulatov, M. I. and Kalinkin, I. P., *Prakticheskoe rukovodstvo po kolorimetricheskim i spektrofotometricheskim metodam analiza* (Practical Text-Book of Colorimetric and Spectrophotometric Methods of Analysis), 3rd Ed., Izdat. Nauka, Moscow (1972).
11. Lange, B., *Kolorimetrische Analyse*, Verlag Chemie, Weinheim (1970).
11a. Edisbury, J. R., *Practical Hints on Absorption Spectrometry* 200-800 mμ, Hilger, London (1966).
12. Mellon, M. G., *Anal. Chem.* **24**, 924 (1952).
13. Peshkova, V. M. and Tsyurupa, M. G., *Vestn. Mosk. Univ. Khim.* **1958**, No. 6, 165; **1959**, No. 4, 215; No. 6, 210; **1960**, No. 6, 58.
14. Szabadváry, F., *Talanta* **5**, 109 (1960).
15. Yoe, J. H., *Anal. Chem.* **29**, 1246 (1957).
16. Trémillon, B., *Bull. Soc. Chim. France* **1960**, 775.
17. Kharlamov, I. P., *Zavodsk. Lab.* **32**, 1035 (1966).
18. Savvin, S. B., *Zh. Analit. Khim.* **22**, 1627 (1967).
19. Juster, N. J., *J. Chem. Educ.* **39**, 596 (1962).
20. Cherkesov, A. I., *Zh. Analit. Khim.* **17**, 16, 276 (1962).
21. Bartecki, A., *Spektroskopia elektronowa związków nieorganicznych i kompleksowych* (Electron Spectroscopy of Inorganic and Complex Compounds), PWN, Warsaw (1971).
22. Kęcki, Z., *Podstawy spektroskopii molekularnej* (Bases of Molecular Spectroscopy), PWN, Warsaw (1972).
23. Pfeiffer, H. G. and Liebhafsky, H. A., *J. Chem. Educ.* **28**, 123 (1951).
24. Lothian, G. F., *Analyst* **88**, 678 (1963).
25. Tsyurupa, M. G. and Peshkova, V. M., *Vestn. Mosk. Univ. Khim.* **1964**, No. 1, 60.
26. Shcherbov, D. P., *Zh. Analit. Khim.* **26**, 1013 (1971).
27. Hughes, H. K., *Anal. Chem.* **24**, 1349 (1952); **33**, 1968 (1961).
28. Cannon, C. G. and Butterworth, I. S., *ibid.* **25**, 168 (1953).
29. Agterdenbos, J. and Vink, J., *Talanta* **18**, 467 (1971).
29a. Agterdenbos, J., Vlogtman, J. and van Broekhoven, L., *ibid.* **21**, 225 (1974).
29b. Agterdenbos, J. and Vlogtman, J., *ibid.* **21**, 231 (1974).
30. Buijs, K. and Maurice, M. J., *Anal. Chim. Acta* **47**, 469 (1969).
30a. Donbrow, M., *Instrumental Methods of Analysis, Their Principles and Practice*, Vol. 2, *Optical Methods*, Chap. 2, Pitman, London (1967).

31. Fog, J. and Osnes, E., *Analyst* **87**, 760 (1962).
32. Schaffer, F. L., Fong, J. and Kirk, P. L., *Anal. Chem.* **25**, 343 (1953).
33. Pohl, F. A. and Demmel, H., *Anal. Chim. Acta* **10**, 554 (1954).
34. Wallach, D. F. and Surgenor, D. M., *Anal. Chem.* **30**, 1879 (1958).
35. Blank, A. B., *Zh. Analit. Khim.* **19**, 363 (1964).
36. Alimarin, I. P. and Petrikova, M. N., *ibid.* **23**, 161 (1968).
37. Klimeš, I., Štěpánková, K. and Janák, J., *Collection Czech. Chem. Commun.* **34**, 776 (1969).
38. Reilley, C. N. and Hildebrand, G. P., *Anal. Chem.* **31**, 1763 (1959).
39. Mahr, C., *Z. Anal. Chem.* **241**, 133 (1968).
40. Bondorf, H. and Mahr, C., *ibid.* **256**, 110 (1971).
41. Świętosławska, J., *Wiad. Chem.* **9**, 404 (1955).
42. Barkovskii, V. F. and Ganopol'skii, V. I., *Differentsial'nyi spektrofotometricheskii analiz* (Differential Spectrophotometric Analysis), Izdat. Khimiya, Moscow (1969).
43. Hiskey, C. F., *Anal. Chem.* **21**, 1440 (1949).
44. Hiskey, C. F., Rabinowitz, J. and Young, I. G., *ibid.* **22**, 1464 (1950).
45. Young, I. G. and Hiskey, C. F., *ibid.* **23**, 506, 1196 (1951).
46. Bastian, R., *ibid.* **21**, 972 (1949); **23**, 580 (1951); **25**, 259 (1953).
47. Ringbom, A. and Österholm, K., *ibid.* **25**, 1798 (1953).
48. Reilley, C. N. and Crawford, C. M., *ibid.* **27**, 716 (1955).
49. Dobkina, B. M. and Malyutina, T. M. *Zavodsk. Lab.* **24**, 1336 (1958).
50. Crawford, C. M., *Anal. Chem.* **31**, 343 (1959).
51. Ganopol'skii, V. I., *Zh. Analit. Khim.* **24**, 654 (1969).
52. Tereshin, G. S., *ibid.* **14**, 388, 516 (1959).
53. Svehla, G., Páll, A. and Erdey, L., *Talanta* **10**, 719 (1963).
54. Marczenko, Z. and Ramsza, A., *Chem. Anal. (Warsaw)* **18**, 425 (1973).
55. Ingle, J. D., Jr., *Anal. Chem.* **45**, 861 (1973).
56. Blank, A. B., *Zh. Analit. Khim.* **28**, 1435 (1973).
57. Ross, S. D. and Wilson, D. W., *Analyst* **85**, 51, 276 (1960).
58. Svehla, G., *Talanta* **13**, 641 (1966).
59. Poppe, H. and den Boef, G., *Z. Anal. Chem.* **228**, 244 (1967).
60. Banks, C. V., Grimes, P. G. and Bystroff, R. I., *Anal. Chim. Acta* **15**, 367 (1956).
61. Bastian, R., Weberling, R. and Palilla, F., *Anal. Chem.* **22**, 160 (1950).
62. Barkovskii, V. F. and Vtorygina, I. N., *Zh. Analit. Khim.* **17**, 39, 865 (1962).
63. Barkovskii, V. F. and Khurtova, L. N., *ibid.* **20**, 911 (1965).
64. Malyutina, T. M. and Dobkina, B. M., *Zavodsk. Lab.* **31**, 650 (1965).
65. Lazarev, A. I., Lazareva, V. I. and Reguzova, Z. V., *ibid.* **31**, 1064 (1965).
66. Chalmers, R. A., Personal communication.
67. Headridge, J. B., *Talanta* **1**, 293 (1958).
68. Headridge, J. B., *Photometric Titrations*, Pergamon, New York (1961).
69. Peshkova, V. M. and Efimov, I. P., *Zavodsk. Lab.* **25**, 678 (1959).
70. Flaschka, H., *Talanta* **8**, 381 (1961).
71. Flaschka, H. and Sawyer, P., *ibid.* **9**, 249 (1962).
72. Kotrlý, S., *Anal. Chim. Acta* **29**, 552 (1963).
73. Fortuin, J. M. H., Karsten, P. and Kies, H. L., *ibid.* **10**, 356 (1954).
74. Still, E. and Ringbom, A., *ibid.* **33**, 50 (1965).
75. Groeneveld, E. R. and den Boef, G., *Z. Anal. Chem.* **219**, 328 (1966).
76. Ringbom, A., Skrifvars, B. and Still, E., *Anal. Chem.* **39**, 1217 (1967).
77. Goddu, R. F. and Hume, D. N., *ibid.* **26**, 1679, 1740 (1954).
78. Flaschka, H. and Sawyer, P., *Talanta* **8**, 521 (1961).
79. Flaschka, H. and Butcher, J., *ibid.* **12**, 913 (1965).
80. Flaschka, H. and Speights, R., *ibid.* **15**, 1467 (1968).
81. Malmstadt, H. V. and Vassallo, D. A., *Anal. Chem.* **31**, 206 (1959).
82. Galík, A., *Talanta* **13**, 109 (1966); **15**, 771 (1968); **17**, 115 (1970).
83. Britt, R. D., Jr., *Anal. Chem.* **34**, 1728 (1962).

84. Lang, W., *Mikrochim. Acta* **1967**, 52.
85. Lindquist, J. E., *Anal. Chim. Acta* **41**, 158 (1968).
86. Havel, J., *Chem. Listy* **62**, 1250 (1968).
87. Foreman, J. K. and Stockwell, P. B., *Automation of Laboratory Methods*, Horwood, Chichester (1975).
88. Clevett, K. J., *Handbook of Process Stream Analysis*, Horwood, Chichester (1974).
89. Scholes, P. H. and Thulbourne, C., *Analyst* **89**, 466 (1964).
90. Sebborn, W. S., *ibid.* **94**, 324 (1969).
91. Johns, P. and Price, W. J., *ibid.* **95**, 138 (1970).
92. Meehan, E. J. and Beattie, W. H., *Anal. Chem.* **33**, 632 (1961).
93. Podobed, N. D., Kozhevnikova, E. S. and Strigina, L. V., *Izv. Vyssh. Ucheb. Zaved., Khim. Khim. Tekhnol.* **5**, 544 (1962).
94. Coleman, R. L., Shults, W. D., Kelley, M. T. and Dean, J. A., *Anal. Chem.* **44**, 1031 (1972).
95. Bobtelsky, M., *Heterometry*, Elsevier, Amsterdam (1960).
96. Blank, A. B., *Zh. Analit. Khim.* **17**, 1040 (1962).
97. Mustafin, I. S., *Zavodsk. Lab.* **28**, 664 (1962).
98. Barney, J. E., II, *Talanta* **14**, 1363 (1967).
99. Ayres, G. H. and Narang, B. D., *Anal. Chim. Acta* **24**, 241 (1961).
100. Goldring, L. S., Hawes, R. C., Hare, G. H., Beckman, A. O. and Stickney, M. E., *Anal. Chem.* **25**, 869 (1953).
101. Busev, A. I. and Ivanov, V. M., *Izv. Vyssh. Ucheb. Zaved., Khim. Khim. Tekhnol.* **4**, 914 (1961).
102. Vdovenko, V. M., *Spektroskopicheskie metody v khimii kompleksnykh soedinenii* (Spectroscopic Methods of Investigation in Complex Chemistry), pp. 102–110, Izdat. Khimiya, Moscow (1964).
103. Irving, H. M. and Butler, E. J., *Analyst* **78**, 571 (1953).
104. Marczenko, Z., *Chem. Anal. (Warsaw)* **7**, 893 (1962).
105. Püschel, R., *Mikrochim. Acta* **1968**, 82.
106. Ayres, G. H., *Anal. Chem.* **21**, 652 (1949).
107. Gridgeman, N. T., *ibid.* **24**, 445 (1952).
108. Agterdenbos, J., *Z. Anal. Chem.* **154**, 401 (1957).
109. Cook, G. B., Crespi, M. B. and Minczewski, J., *Talanta* **10**, 917 (1963).
110. Lingane, P. J., *Anal. Chim. Acta* **47**, 529 (1969).
111. Komar', N. P. and Samoilov, V. P., *Zh. Analit. Khim.* **18**, 1284 (1963); **22**, 1285 (1967).
112. Komar', N. P. and Samoilov, V. P., *ibid.* **24**, 1133, 1800 (1969); **26**, 437 (1971).
113. Mojski, M. and Marczenko, Z., *Electron Technology* **5**, 61 (1972).
114. Chalmers, R. A. and Sinclair, A. G., *Anal. Chim. Acta* **34**, 412 (1966).
115. Babko, A. K. and Gridchina, G. I., *Zavodsk. Lab.* **30**, 773 (1964).
116. Cheng, K. L., *Anal. Chem.* **33**, 783 (1961).
117. West, P. W., *Acta Chim. Acad. Sci. Hung.* **34**, 143 (1962); *Analyst* **87**, 630 (1962).
118. Hulanicki, A., *Talanta* **9**, 549 (1962).
119. Ringbom, A., *J. Chem. Educ.* **35**, 282 (1958).
120. Perrin, D. D., *Masking and Demasking of Chemical Reactions*, Interscience, New York (1970).
121. Kuznetsov, V. I. and Petrova, T. V., *Zh. Analit. Khim.* **14**, 404 (1959).
122. Petrova, T. V., *ibid.* **22**, 15 (1967).
123. Lingane, J. J. and Collat, J. W., *Anal. Chem.* **22**, 166 (1950).
124. Ciecierska-Stokłosa, D., Gorczyńska, K., Świętosławska, J. and Walędziak H., *Chem. Anal. (Warsaw)* **4**, 803, 809 (1959).
125. Kulichenko, L. B. and Espinosa, E. Z., *Z. Anal. Chem.* **206**, 248 (1964).
126. Dodin, E. I. and Kharlamov, I. P., *Zavodsk. Lab.* **35**, 1304 (1969).
127. Shibata, S., Furukawa, M. and Goto, K., *Anal. Chim. Acta* **46**, 271 (1969); **53**, 369 (1971).

128. Marczenko, Z., *Chem. Anal. (Warsaw)* **6**, 3 (1961).
129. Babko, A. K. and Pilipenko, A. T., *Zh. Analit. Khim.* **22**, 1679 (1967).
130. Slovák, Z., Fischer, J. and Borák, J., *Talanta* **16**, 215 (1969).
131. Patrovský, V., *Chem. Listy* **65**, 1121 (1971).
132. Babko, A. K., *Pure Appl. Chem.* **10**, 557 (1965); *Talanta* **15**, 721 (1968).
133. Blyum, I. A. and Oparina, L. I., *Zavodsk. Lab.* **36**, 897 (1970).
134. Blyum, I. A., *Ekstraktsionno-fotometricheskie metody analiza s primeneniem osnovnykh krasitelei* (Extractive-Photometric Methods of Analysis with Use of Basic Dyes), Izdat. Nauka, Moscow (1970).
135. Koch, S. and Ackermann, G., *Chem. Anal. (Warsaw)* **18**, 1109 (1973).
136. Pilipenko, A. T. and Tananaiko, M. M., *Zh. Analit. Khim.* **28**, 745 (1973); *Talanta* **21**, 501 (1974).
137. Jackwerth, E. and Graffmann, G., *Z. Anal. Chem.* **257**, 265 (1971).
138. Belcher, R., *Talanta* **15**, 357 (1968).
139. Weisz, H., *Mikrochim. Acta* **1970**, 1057.
140. Weisz, H. and Fritsche, U., *Mikrochim. Acta* **1973**, 361.
141. Nazarenko, V. A., *Tr. Komis. po Analit. Khim. Akad. Nauk SSSR* **14**, 3 (1963).
142. Babko, A. K. and Zharovskii, F. G., *ibid.* **14**, 218 (1963).
143. Babko, A. K. and Vasilenko, V. T., *Ukr. Khim. Zh.* **26**, 515 (1960); **27**, 396 (1961).
144. Babko, A. K. and Kish, P. P., *Zh. Analit. Khim.* **17**, 693 (1962).
145. Babko, A. K. and Shtokalo, M. I., *Ukr. Khim. Zh.* **29**, 963 (1963); **30**, 220 (1964).
146. Job, P., *Ann. Chim. (Paris)* **9**, 113 (1928).
147. Vosburgh, W. C. and Cooper, G. R., *J. Am. Chem. Soc.* **63**, 437 (1941).
148. Yoe, J. H. and Jones, A. L., *Ind. Eng. Chem., Anal. Ed.* **16**, 111 (1944).
149. Harvey, A. E. and Manning, D. L., *J. Am. Chem. Soc.* **72**, 4488 (1950).
150. Babko, A. K., Volkova, A. I. and Get'man, T. E., *Zh. Analit. Khim.* **23**, 1437 (1968).
151. Sommer, L., *Folia Prirod. Fak. Univ. J. E. Purkyne, Brno, Chimia*, **1964**, No. V, 1.
152. Babko, A. K., *Analiza fizykochemiczna związków kompleksowych w roztworach* (Physico-Chemical Analysis of Complex Compounds in Solution), PWN, Warsaw (1959).
153. McBryde, W. A. E., *Talanta* **21**, 979 (1974).
154. Schläfer, H. L., *Komplexbildung in Lösung*, Springer, Berlin (1961).
155. Asmus, E., *Z. Anal. Chem.* **178**, 104 (1960); **183**, 321, 401 (1961).
156. Sommer, L. and Jin Tsin-Jao, *Chem. Listy* **55**, 574 (1961).
157. Buděšínský, B., *Z. Anal. Chem.* **206**, 262 (1964); **207**, 178, 241, 247; **209**, 379 (1965).
158. Klausen, K. S. and Langmyhr, F. J., *Anal. Chim. Acta* **40**, 167 (1968).
159. Klausen, K. S., *ibid.* **44**, 377 (1969).
160. Likussar, W. and Boltz, D. F., *Anal. Chem.* **43**, 1265 (1971).
161. Ringbom, A. and Harju, L., *Anal. Chim. Acta* **59**, 33, 49 (1972).
162. Kirkbright, G. F., *Talanta* **13**, 1 (1966).
163. Koźlicka, M., *Chem. Anal. (Warsaw)* **15**, 683 (1970).
164. Milkey, R. G., *Anal. Chem.* **26**, 1800 (1954).
165. Beneš, P., *Chem. Listy* **60**, 153 (1966); **66**, 561 (1972).
166. West, F. K., West, P. W., and Iddings, F. A., *Anal. Chem.* **38**, 1566 (1966); *Anal. Chim. Acta* **37**, 112 (1967).
167. Struempler, A. W., *Anal. Chem.* **45**, 2251 (1973).
168. Al-Sibaai, A. A. and Fogg, A. G., *Analyst* **98**, 732 (1973).
169. Chalmers, R. A., *Standards and Standardization in Chemical Analysis*, in Wilson, C. L. and Wilson, D. W. (eds.), *Comprehensive Analytical Chemistry*, Vol. III, Elsevier, Amsterdam (1975).
170. Yamamura, S. S., *Anal. Chem.* **36**, 2515 (1964).
171. Knížek, M. and Provazník, J., *Chem. Listy* **55**, 389 (1961).
172. Robertson, D. E., *Anal. Chem.* **40**, 1067 (1968).
173. Kloster, M. B. and Hach, C. C., *ibid.* **44**, 1061 (1972).

References

174. Wilson, A. L., *Talanta* **20**, 725 (1973).
175. Tölg, G., *ibid.* **19**, 1489 (1972).
176. Hughes, R. C., Mürau, P. C. and Gunderson, G., *Anal. Chem.* **43**, 691 (1971).
177. Irving, H. M. and Cox, J. J., *Analyst* **83**, 526 (1958).
178. Kwestroo, W. and Visser, J., *ibid.* **90**, 297 (1965).
179. Kuehner, E. C., Alvarez, R., Paulsen, P. J. and Murphy, T. J., *Anal. Chem.* **44**, 2050 (1972).
180. Bradshaw, G. and Rands, J., *Analyst* **85**, 76 (1960).
181. Yoe, J. H. and Koch, H. J. (eds.), *Trace Analysis*, Wiley, New York (1957).
182. Pinta, M., *Recherche et dosage des éléments traces*, Dunod, Paris (1962).
183. Alimarin, I. P. (ed.), *Metody analiza veshchestv vysokoi chistoty* (Methods for Analysis of High Purity Materials), Izdat. Nauka, Moscow (1965).
184. Korenman, I. M., *Analiticheskaya khimiya malykh kontsentratsii* (Analytical Chemistry of Low Concentrations), Izdat. Khimiya, Moscow (1966).
185. Meinke, W. W. and Scribner, B. F. (eds.), *Trace Characterization, Chemical and Physical*, NBS, Washington (1967).
186. Monnier, D., *Chimia* **13**, 314 (1959).
187. Alimarin, I. P., *Pure Appl. Chem.* **7**, 455 (1963); *Zh. Analit. Khim.* **18**, 1412 (1963).
188. Babko, A. K., *Zavodsk. Lab.* **29**, 518 (1963).
189. Vasilevskaya, L. S., Muravenko, V. P. and Kondrashina, A. I., *Zh. Analit. Khim.* **20**, 540 (1965).
190. Blank, A. B., *ibid.* **20**, 3 (1965); **21**, 769 (1966).
191. Specker, H., *Z. Anal. Chem.* **221**, 33 (1966).
192. Hume, D. N., *Chem. Anal. (Warsaw)* **13**, 989 (1968).

Chapter **2**

SPECTROPHOTOMETRIC REAGENTS

The reagents which give the colour reactions upon which spectrophotometric methods are based are called spectrophotometric reagents. They can be grouped into two categories: organic and inorganic. Most spectrophotometric methods, especially the more sensitive ones, are based on organic reagents [1,2]. The more important reagents mentioned later when dealing with individual elements are discussed in the present section.

2.1 Dithizone

Dithizone (diphenylthiocarbazone, H_2Dz) is one of the foremost organic spectrophotometric reagents [3]. It provides the basis of sensitive methods for the determination of lead, zinc, cadmium, mercury, silver, copper, bismuth, palladium and other metals. It has often been used in the extraction of traces of many metals before their determination.

2.1.1 Reactions with Metals

Dithizone is a weak acid and is insoluble in water at pH < 7. In alkaline media it dissolves, forming an orange solution of HDz^- anions.

Dithizone dissolves, giving green solutions, in hydrocarbons, alcohols, ketones, and chlorinated hydrocarbons such as carbon tetrachloride and chloroform. The green colours of dithizone solutions differ somewhat depending on the solvent. Figure 2.1 shows the effect of pH on dithizone distribution between water and carbon tetrachloride or chloroform.

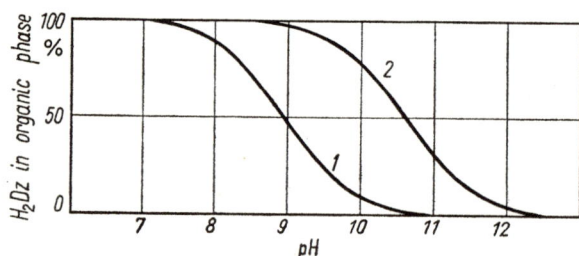

Fig. 2.1 Curves of dithizone distribution between the aqueous and the carbon tetrachloride (1) or chlorofo (2) phases

Dithizone

In organic solvents dithizone exists in the keto (I) and the enol (II) tautomeric forms. It forms coloured chelates soluble in inert organic sol-

$$S=C\begin{matrix}NH-NH-C_6H_5\\N=N-C_6H_5\end{matrix} \rightleftharpoons HS-C\begin{matrix}N-NH-C_6H_5\\N=N-C_6H_5\end{matrix}$$

$$(I) \qquad\qquad\qquad (II)$$

vents with a large group of elements adjacent in the periodic table (Table 2.1). These chelates are formed on shaking a dithizone solution (e.g. in CCl_4 or $CHCl_3$) with an aqueous solution of a given metal at a suitable pH.

Table 2.1 Elements reacting with dithizone

H																	He
Li	Be											B	C	N	O	F	Ne
Na	Mg											Al	Si	P	S	Cl	Ar
K	Ca	Sc	Ti	V	Cr	Mn	Fe	Co	Ni	Cu	Zn	Ga	Ge	As	Se	Br	Kr
Rb	Sr	Y	Zr	Nb	Mo	Tc	Ru	Rh	Pd	Ag	Cd	In	Sn	Sb	Te	I	Xe
Cs	Ba	La	Hf	Ta	W	Re	Os	Ir	Pt	Au	Hg	Tl	Pb	Bi	Po	At	Rn
Fr	Ra	Ac	Th	U													

The conditions for formation of dithizone complexes with individual metals are outlined in the sections dealing with the respective elements. Gallium dithizonate [4,5] is formed in weakly acidic medium, and iron(II) dithizonate [6] in weakly alkaline medium within narrow pH limits. Tellurium [7,8] and selenium [9] react with dithizone in strongly acidic media, although the mechanism of the selenium reaction is controversial [10,11].

A primary or a secondary dithizonate* is obtained depending on whether dithizone reacts with a metal as the anion of the monobasic (HDz^-) or the dibasic (Dz^{2-}) acid. Primary dithizonates (keto dithizonates) are formed by all the metals indicated in Table 2.1, e.g. AgHDz, $Cu(HDz)_2$, and $Bi(HDz)_3$. Secondary dithizonates (enol dithizonates) are formed by

---

* Certain authors [3, 15] consider that secondary dithizonates do not exist, and that in strongly alkaline media or if insufficient dithizone is present, mixed-ligand complexes are formed such as HgCl(HDz) or CuOH(HDz), where HDz is the anion resulting from dissociation of a proton from dithizone (H_2Dz). It is rather confusing that these authors have chosen to use the abbreviation HDz for dithizone, so that the anion is then Dz^-.

only some of those metals (Cu, Ag, Au, Pd, Pt, and Hg) [12–14]. They are represented by such formulae as Ag_2Dz and $CuDz$.

Generally speaking, acidic media and excess of dithizone favour the formation of primary dithizonates, whereas alkaline media and reagent deficiency favour secondary dithizonates. Only primary dithizonates are used in spectrophotometric analysis, because of their more intense colours and higher solubilities in organic solvents, in comparison with secondary dithizonates.

Recent studies have demonstrated that mixed secondary dithizone complexes can be formed (e.g. with silver and mercury) [16,17].

Fischer [12], who was the first to employ dithizone in analytical chemistry, believed the metal in the primary dithizonates to be bonded to two nitrogen atoms. More recent investigations [18–25] of the structures of these compounds have shown the metal to be bonded to the sulphur atom by replacement of the hydrogen in the thiol group, and also co-ordinately bonded to a nitrogen atom according to the formula:

$$\text{Ph}-\underset{H}{N}-\underset{}{N}=\underset{\underset{M/n}{|}}{\underset{S}{C}}-N=N-\text{Ph}$$

The solubility of primary metal dithizonates, like that of dithizone itself, is higher in chloroform (10^{-2}–$10^{-3}M$) than in carbon tetrachloride (10^{-3}–$10^{-4}M$). Although $Cd(HDz)_2$ and $Pb(HDz)_2$ are among the least soluble in carbon tetrachloride, they are sufficiently soluble for spectrophotometric analysis.

Solutions of metal dithizonates in organic solvents are intensely coloured, their colours differing considerably from that of dithizone. An exception is the grey-green solution of palladium dithizonate, $Pd(HDz)_2$. Figure 2.2 shows the absorption spectra of dithizone and several dithizonates dissolved in carbon tetrachloride. The molar absorptivities of the metal dithizonates are within the range $3 \cdot 0$–$9 \cdot 0 \times 10^4$. The molar absorptivity of dithizone in CCl_4 at $\lambda_{max} = 620$ nm is $3 \cdot 20 \times 10^4$ [26,27]. The presence of chloroform in the carbon tetrachloride affects the molar absorptivity of dithizone [27].

The most stable dithizonates (those of Pt, Pd, Au, Ag, Hg, and Cu) can be extracted from strongly acid solutions. Some metals (Bi, In, Zn) are extractable from weakly acid media, whereas other metals (Co, Ni, Pb, Tl, Cd) are extractable from neutral or alkaline media. The higher the excess of dithizone, the lower the pH at which the dithizonate forms. The pH at which extraction of individual metals starts is approximately 1·5 pH units higher with dithizone in chloroform than with dithizone in carbon tetrachloride.

Irrespective of their thermodynamic stability the dithizonates of certain metals (Ag, Hg, Pb, Cd) are extracted rapidly, whereas those of other metals (Pd, Cu, Zn) require prolonged shaking with the organic

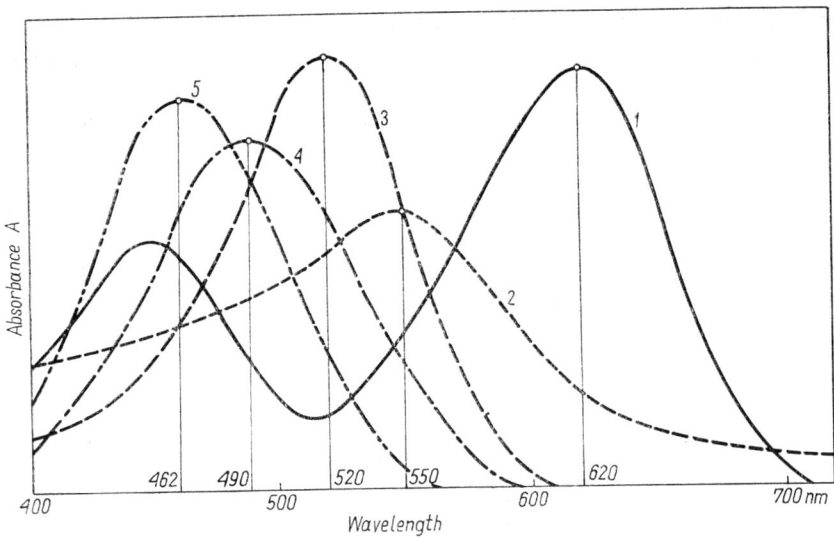

Fig. 2.2 Absorption spectra of: (1) dithizone H_2Dz, (2) $Cu(HDz)_2$, (3) $Pb(HDz)_2$, (4) $Bi(HDz)_3$, (5) AgHDz, in carbon tetrachloride

solution of dithizone. This is explicable in terms of the kinetics of dithizonate formation. It is believed that certain metal ions have difficulty in losing their hydration spheres, which inhibits the formation and extraction of the dithizonates.

Selectivity in spectrophotometric methods for determining metals with dithizone is attained by controlling the pH of the aqueous medium and using masking agents such as cyanide, EDTA, thiosulphate, and iodide.

The analogue of dithizone, di-2-naphthylthiocarbazone (dinaphthizone) is also applied to the spectrophotometric determination of some

metals. The reagent solutions in CCl_4 or $CHCl_3$ have a blue colour. The absorbances of dinaphthizone and metal dinaphthizonates are considerably higher than those of dithizone and the corresponding dithizonates [28–30]. Unfortunately, the solubility of metal dinaphthizonates is rather low.

2.1.2 Spectrophotometric Applications

Two techniques are used for the spectrophotometric determination of metal ions by dithizone: the mono-colour and mixed-colour (two-colour) procedures. The *mono-colour method* consists in extracting the metal from aqueous solution with an excess of dithizone solution in CCl_4 or $CHCl_3$, removing the free dithizone from the non-aqueous phase by shaking with an alkaline aqueous solution and measuring the absorbance of the coloured metal dithizonate solution. Alternatively, in the case of certain metals the increase in absorbance of the green dithizone solution may be measured after the metal complex has been decomposed in the organic phase by shaking the extract with dilute acid or other suitable stripping agent (Irving reversion technique).

In the *mixed-colour method*, the excess of dithizone is left in the organic phase with the metal dithizonate. Then, either the colour of the organic extract is compared with a series of standards ranging from pure dithizonate to pure dithizone, or the absorbance of the coloured solution is measured at the absorption maximum of the dithizonate or of free dithizone.

The mixed-colour method is susceptible to errors since, when the aqueous medium is alkaline, an equilibrium amount of the surplus dithizone passes to the aqueous phase and this may vary with small fluctuations in the pH of the solution (Fig. 2.1). The *Irving reversion method* [31] enables the errors from this source to be avoided. The absorbance of the mixed-colour solution is measured at the dithizone absorption maximum. Then, the extract is shaken with a suitable solution which decomposes the metal dithizonate and liberates an equivalent amount of free dithizone. The absorbance of the solution is measured a second time at the same wavelength. The amount of the metal being determined corresponds to the difference between the two absorbance measurements. Mixtures of dithizonates can be dealt with similarly by successive stripping with suitable reagents [32].

An *extractive titration method* may be used with metals which react rapidly with dithizone in an acidic or neutral medium. The sample solution is shaken with small portions of the standard dithizone solution in CCl_4 or $CHCl_3$ added from a burette until the dithizone no longer changes colour but remains green. Before addition of a subsequent portion, the organic phase is drained from the separating funnel and discarded.

Dithizone preparations are always contaminated since atmospheric oxygen and other oxidants oxidize dithizone to diphenylthiocarbadiazone

$$S=C\begin{matrix}NH-NH-C_6H_5\\N=N-C_6H_5\end{matrix} \xrightarrow{ox.} S=C\begin{matrix}N=N-C_6H_5\\N=N-C_6H_5\end{matrix}$$

which, unlike dithizone, is insoluble in aqueous alkaline solutions but soluble in carbon tetrachloride and in chloroform, giving a brown solution. This compound is chemically inactive and does not react with metal ions. Irving *et al.* [33–36] have found additional oxidation products of dithizone. The active dithizone content of a reagent preparation diminishes with time. For example, a preparation stored for 10 years was found to contain only 12% active dithizone.

Pure stable dithizone may be obtained by the following procedure which is based on the insolubility of diphenylthiocarbadiazone in aqueous ammonia. The reagent (0·5 g) is dissolved in chloroform (50 ml), and the solution filtered through a sintered glass filter and shaken in a separating funnel with two 10-ml portions of ammonia (1+100). The ammoniacal extracts are separated from the brown chloroform layer and acidified with dilute hydrochloric acid to precipitate the dithizone which is then extracted with small portions of pure chloroform. The combined extracts are shaken with two portions of water, then placed in a beaker and the solvent is evaporated off in a water-bath at about 50°C.

Carbon tetrachloride and chloroform are the normal solvents for dithizone. The solubility of dithizone is greater in $CHCl_3$ (1 g/100 ml) than in CCl_4 (0·08 g/100 ml). Unless otherwise stated, carbon tetrachloride is preferable as a solvent because of its lower volatility and greater specific gravity, which result in a more rapid phase separation on shaking with the aqueous solution. Carbon tetrachloride is less soluble in water (0·08%) than chloroform is (0·8%), and is also less toxic.

If the concentration of dithizone in the solution cannot be evaluated by weighing the purified solid preparation, it is determined by measuring the absorbance of the green organic solution [37]. The dithizone content of the solution can also be determined by the extractive titration of a standard silver solution (see p. 492).

Solutions of about 0·01% (10 mg/100 ml) can be stored for a long time if kept cool and in the dark. The dilute dithizone solutions (0·001–0·002% w/v) normally used in spectrophotometric determinations must not be stored. The stability of dithizone solutions and of metal dithizonates depends on the purity of the solvents used [27]. It is very important that the carbon tetrachloride and chloroform contain no oxidizing substances (such as chlorine from their decomposition). The procedure for purifying and recovering carbon tetrachloride and chloroform is given later (see p. 493).

2.2 Azo Reagents

Azo dyes [38,39] comprise the largest group of organic reagents used in spectrophotometric analysis. Methods using azo dyes are very sensitive. The azo reagents PAN, PAR, Arsenazo I and Arsenazo III (see below) have been thoroughly investigated and in consequence are widely used in analysis. Hückel molecular orbital theory has been used to calculate the electronic characteristics of azo compounds and their complexes with metal ions [40].

2.2.1 N-Heterocyclic Azo Compounds

Recently, pyridylazo and thiazolylazo dyes have become increasingly important spectrophotometric reagents [41–43].

1-(2-Pyridylazo)-2-naphthol (**I**) (PAN), first recommended as a spectrophotometric reagent by Cheng and Bray [44] and 4-(2-pyridylazo)-resorcinol (**II**) (PAR) are the foremost pyridylazo reagents.

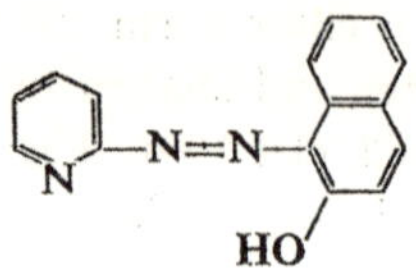

(I) and (II) structures shown.

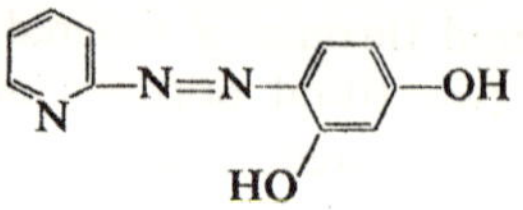

Depending on the pH, the reagent PAN exists in solution in three forms [45–47]. Acid solutions (pH ⩽ 2) contain the water-soluble yellow-green protonated H_2R^+ ion. Between pH 3 and 11, PAN occurs as the neutral HR molecule soluble in organic solvents to give a yellow colour (colloidal suspensions can be formed in aqueous systems). In alkaline solutions (pH > 11) PAN exists as the water-soluble red R^- anion.

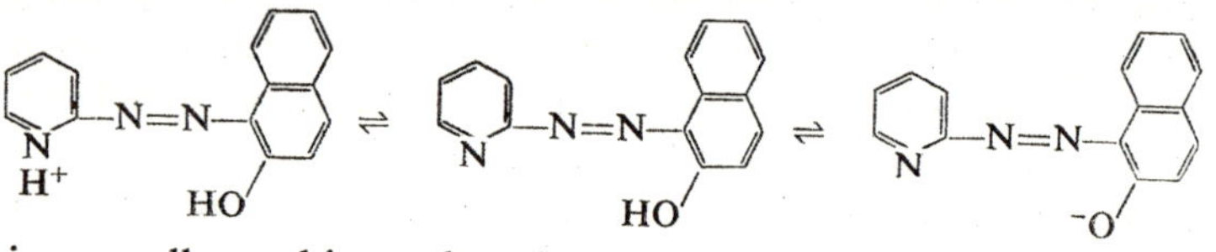

PAN is normally used in methanol or ethanol. It acts as a terdentate ligand complexing with metals through the hydroxyl oxygen, pyridine nitrogen, and one of the azo group nitrogen atoms.

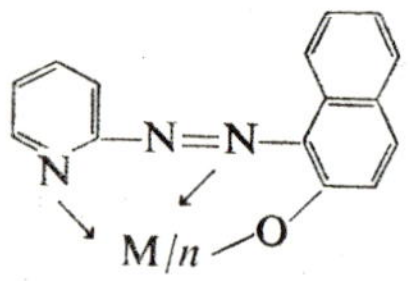

Metals which give colour reactions with PAN are shown in Table 2.2. Metal complexes with PAN are sparingly soluble in water. A considerable number of the metals listed in Table 2.2 give neutral complexes extractable with inert solvents ($CHCl_3$, C_6H_6). This permits the extractive-spectrophotometric determination of Mn, Zn, Cd, Cu, Ni, Co, In, U, Ga, Hg, Fe, Pd, and other metals [48–51]. During extraction with an organic solvent, the

Table 2.2 Metals giving colour reactions with 1-(2-pyridylazo)-2-naphthol (PAN)

H																	He
Li	Be											B	C	N	O	F	Ne
Na	Mg											Al	Si	P	S	Cl	Ar
K	Ca	Sc	Ti	V	Cr	Mn	Fe	Co	Ni	Cu	Zn	Ga	Ge	As	Se	Br	Kr
Rb	Sr	Y	Zr	Nb	Mo	Tc	Ru	Rh	Pd	Ag	Cd	In	Sn	Sb	Te	I	Xe
Cs	Ba	La	Hf	Ta	W	Re	Os	Ir	Pt	Au	Hg	Tl	Pb	Bi	Po	At	Rn
Fr	Ra	Ac	Th		U												

metal complex and the uncombined PAN pass to the extract. The absorption maxima of the complexes are usually very different from that of the reagent. The molar absorptivities of the PAN complexes lie within the range $2.0–6.0 \times 10^4$.

The selectivity of the methods using PAN is enhanced by suitable selection of pH and masking agents. Iron, cobalt, and nickel, for example, react quantitatively with PAN at pH > 4. At this pH, Mn, Hg, Zn, and Cd do not react with PAN. Cyanide enables manganese to be determined in the presence of Ni, Zn, Cd, Co, and Cu, which form stable cyanide complexes. Zinc and cadmium can be demasked from the cyanide complexes with formaldehyde. Uranium is determined specifically in the presence of EDTA. Careful choice of the extractant also improves the selectivity of PAN.

Unlike PAN, *4-(2-pyridylazo)resorcinol* (PAR) is water-soluble and forms water-soluble complexes with metals [41–43,46,52–55]. Hniličková and Sommer [56] have found that PAR can exist in 6 different forms. In 90% H_2SO_4, 50% H_2SO_4, and at pH < 2, the protonated forms H_5R^{3+}, H_4R^{2+} and H_3R^+ respectively, are present (absorption maxima at 433, 390, and 395 nm). The neutral PAR molecule exists between pH 2·1 and 4·2 (λ_{max} = 385 nm). The anion HR^- (λ_{max} = 413 nm) occurs over the pH range 4·2–7·0. In alkaline solutions (pH 10·5–13·2) both hydroxyl groups are dissociated, forming R^{2-} (λ_{max} = 490 nm).

With multivalent metal ions PAR gives 1:1 or 1:2 coloured complexes. The spectrophotometric determination of metals with PAR is performed in aqueous solution. In $0.5–0.05 M$ H_2SO_4 the reagent reacts with Cu(II), Bi, Ti, Zr, Pd, and Tl(III) ions. In acetate medium (pH 3–6) PAR gives colour reactions with Zn, Cd, Co, Ni, Hg, Mn, U, Pb, La, Ga, In, and other metals. Some metals are determined with PAR in a weakly alkaline medium. Solutions of PAR complexes have a red or violet colour.

A number of PAN and PAR derivatives have been suggested as spectrophotometric reagents for metals [57–61].

As spectrophotometric reagents, thiazole azo compounds have similar properties to PAN and PAR. Their reactions with metals are more selective, principally as a consequence of the lower stability of their complexes. The following reagents are examples of thiazole azo compounds: 1-(2-thiazolylazo)-2-naphthol (TAN) (I) [62], 4-(2-thiazolylazo)resorcinol (TAR) (II) [63–65], and 2-(2-thiazolylazo)-5-dimethylaminophenol (TAM) (formula, p. 582) [66,67].

These reagents bond to metal ions in a similar manner to PAN.

2.2.2 ARSONIC AZO COMPOUNDS

In 1941 Kuznetsov synthesized the first reagents of this group: Arsenazo I (I) and Thoron I (II) [68]. This group of reagents is characterized by the presence of an arsonic acid group ortho to the azo group. The hydroxyl group is usually also ortho to the azo group.

(I)   (II)

The presence of the arsonic acid group, $AsO(OH)_2$, causes the formation of stable complexes of some metals even in fairly strong acid solutions. The presence of the azo group ensures the colour reaction whereas the hydroxyl group enables a second ring to be formed with the metal. This not only stabilizes the complex but also involves a considerable deepening of the colour obtained. The sulphonic acid groups render these reagents and their complexes water-soluble [69].

Until Arsenazo I and related reagents were introduced into spectrophotometric analysis, there were no sensitive reagents for such elements as Th, Zr, Hf, U, and rare earths.

Arsenazo I (Neothoron, Uranon) and a number of its analogues are derivatives of chromotropic acid (formula, p. 558), a well-known reagent for titanium. Arsenazo I is a sensitive and selective reagent applied to the spectrophotometric determination of U, rare earth elements, Th, Zr, Al, Ti, Be, Nb, In, Ca, and other metals. The high selectivity of the methods using Arsenazo I is obtained by the choice of pH and suitable masking agents.

Thoron I (APANS) is less sensitive but more selective than Arsenazo I. It is most commonly applied to the determination of thorium and also U, Zr, Li, Be and rare earth elements.

Arsenazo III, suggested by Savvin in 1959 [70], is a very useful spectrophotometric reagent [71–74]. Arsenazo III is a bisazo dye based on

chromotropic acid and *o*-aminophenylarsonic acid. It is moderately soluble in neutral and acid solutions, and readily soluble in slightly alkaline solutions. The reagent is stable in the solid state and in aqueous solution although strong oxidizing (H_2O_2, Cl_2, Br_2) or reducing ($TiCl_3$, $Na_2S_2O_3$) agents cause decomposition. In acid solutions (from $10M$ HCl to pH 4)

Arsenazo III has a purplish-red colour while at higher pH values it is blue-violet.

In strongly acid solutions (1–10M HCl) Arsenazo III reacts with Th, Zr, Hf, U(IV), and Pa. The molar absorptivities of the complexes with these metals are about 10^5. At pH 1–4 Arsenazo III reacts with U(VI), Sc, Fe(III), Bi, and rare earth elements to form coloured complexes. The sensitivity of the colour reactions is lower in this case ($\varepsilon \sim 5 \times 10^4$).

The use of Arsenazo III in strongly acid medium overcomes difficulties connected with the hydrolysis or polymerization of some multivalent metals (e.g. Zr, Th, and U). In the determination of these metals, the high acidity enhances the selectivity of the reagent. Interference from sulphate and phosphate is considerably reduced in strongly acid solutions.

The absorbance of free Arsenazo III (λ_{max} = 520–530 nm) at the absorption maxima of the metal complexes (λ = 655–675 nm) is very slight. The large difference ($\Delta\lambda$) between the wavelengths of the absorption maxima of the complexes and the free reagent is important.

Arsenazo III forms 1:1 complexes with bivalent or tervalent cations (e.g. UO_2^{2+}, Pb^{2+}, La^{3+}, and Ce^{3+}) and 1:1 or 1:2 complexes with quadrivalent cations (e.g. Th^{4+} and U^{4+}) depending on the pH and the excess of reagent.

One half of the symmetrical Arsenazo III molecule complexes with a metal ion. The metal ion bonds to the nitrogen atom of the azo group, the oxygen atom of the arsonic acid group, and the oxygen atom of the hydroxyl group. The distortion of the symmetry of the reagent molecule gives rise to two neighbouring absorption maxima in the visible spectra of the Arsenazo III metal complexes (Fig. 54.2, p. 541). According to Nemodruk [75] the high value of $\Delta\lambda$ in the Arsenazo III colour reactions with metals results from the fact that the second, non-reacting, half of the reagent is converted from the azoid into the quinonohydrazone form.

It is possible to determine Arsenazo III by measuring its absorbance in concentrated H_2SO_4 solution at 675 nm. Arsenazo I does not absorb under these conditions [76]. An electrophoretic and chromatographic method has been developed for the purification and analysis of Arsenazo III samples [77,78].

A large number of other spectrophotometric reagents containing arsonic acid groups have been suggested. These reagents and their applications are reviewed at length in the monograph by Savvin [71]. Arsenazokhimdu [79,80] and Sulpharsazen [81] are new reagents of this class.

A related group of reagents comprises azo dyes containing phosphonic acid groups: for example, the bisazo dye *Chlorophosphonazo III*.

[Structure: Cl—C6H3(PO3H2)—N=N—C10H4(OH)(OH)(SO3H)(SO3H)—N=N—C6H3(H2O3P)—Cl, with HO3S and SO3H on the naphthalene]

This reagent has been recommended for the determination of many metals such as Zr, Ti, U, Sc, Ba, and Sr [71].

2.2.3 Other Azo Compounds

Many azo reagents of other classes, such as *o*-hydroxyazo and *o,o'*-dihydroxyazo compounds, have been employed in spectrophotometric methods for determining various elements. Sulphochlorophenol S is an example of the latter group of compounds.

[Structure of Sulphochlorophenol S: Cl-C6H3(HO3S)(OH)—N=N—naphthalene(OH)(OH)(HO3S)(SO3H)—N=N—C6H3(OH)(SO3H)-Cl]

This reagent has been suggested for determining such elements as Nb, Zr, Sc, Mo, Cu, V, and Al in acid media [71].

An analogue of Sulphochlorophenol S is Sulphonitrophenol M, with nitro groups substituted for the two chlorine atoms of the former compound. This reagent is of interest in the determination of Pb, Nb, Al, Ga, Zr, V, and other metals [71].

Another example of an *o,o'*-dihydroxyazo compound is Picramine-

[Structure of Picramine-epsilon: (O2N)2-C6H2(OH)—N=N—naphthalene(OH)(SO3H)(HO3S)]

epsilon, a sensitive spectrophotometric reagent for copper, zirconium and other metals [82].

Chromotrope 2B, a reagent for thorium and rare earth elements, is a suitable example of an *o*-hydroxyazo compound.

[Structure of Chromotrope 2B: naphthalene(HO)(OH)(HO3S)(SO3H)—N=N—C6H4—NO2]

In both these groups of azo reagents the oxygen atom of the *o*-hydroxyl group and the nitrogen atom of the azo group participate in complex formation with metal ions. The presence of the sulphonic acid groups in the azo reagents makes these reagents and their complexes soluble in aqueous solutions.

2.3 Triphenylmethane and Xanthene Reagents

The reagents discussed below are derived structurally from triphenylmethane (I) and xanthene (II).

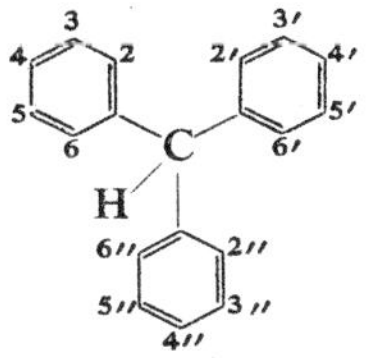

(I) 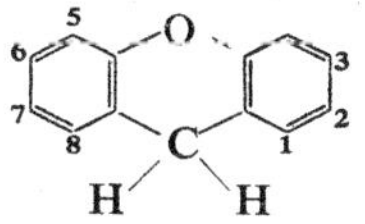   (II)

Such reagents are intensely coloured and provide the basis of sensitive spectrophotometric methods for the determination of a number of metals.

The following are common triphenylmethane reagents: Pyrocatechol Violet (III) [83,84], Eriochrome Cyanine R (Solochrome Cyanine R) (IV) [85,86], and Chrome Azurol S (Alberon) (formula, p. 144) [87].

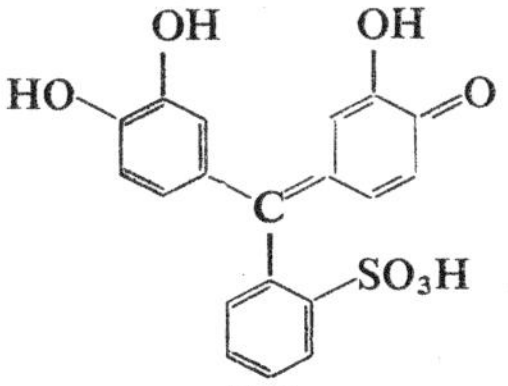

(III) 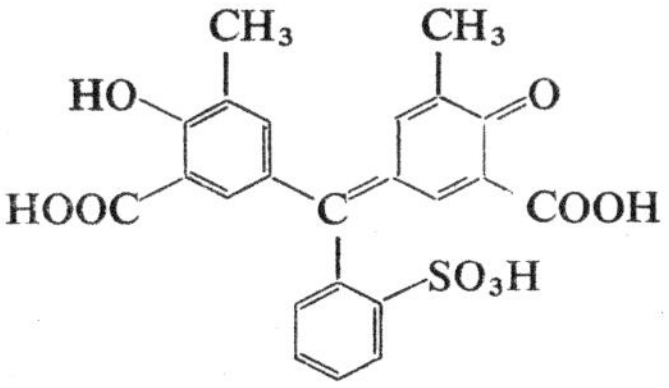   (IV)

The triphenylmethane reagents Xylenol Orange (V) [88–91]

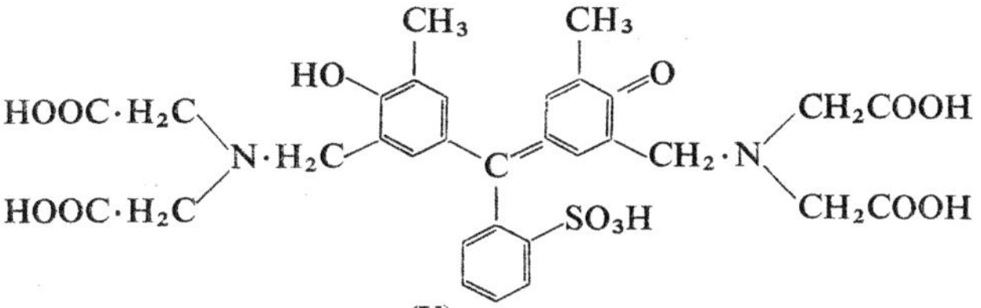

(V)

and Methylthymol Blue (VI) [92–94] are characterized by the presence

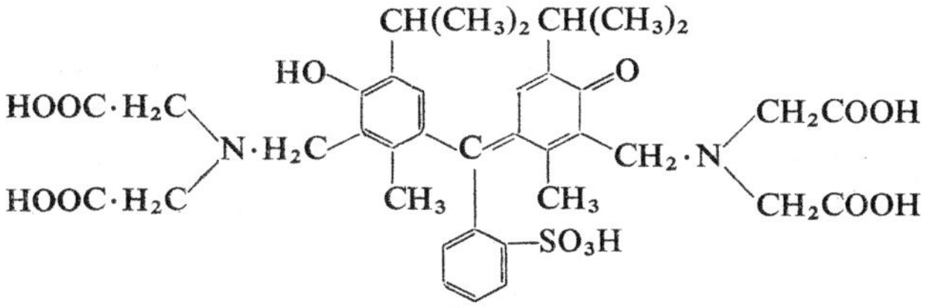

(VI)

of the iminodiacetic acid groups which occur in complexones such as EDTA.

These reagents can be purified by various methods [86,90,94]. They are usually contaminated with unchanged starting material and the product containing only one iminodiacetic acid group (e.g. Semixylenol Orange).

Common xanthene reagents are Pyrogallol Red, Bromopyrogallol Red (**VII**) [95,96], and Calcein (Fluorexone) (**VIII**).

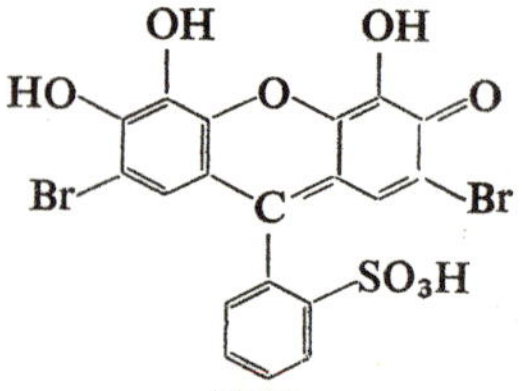

(**VII**)

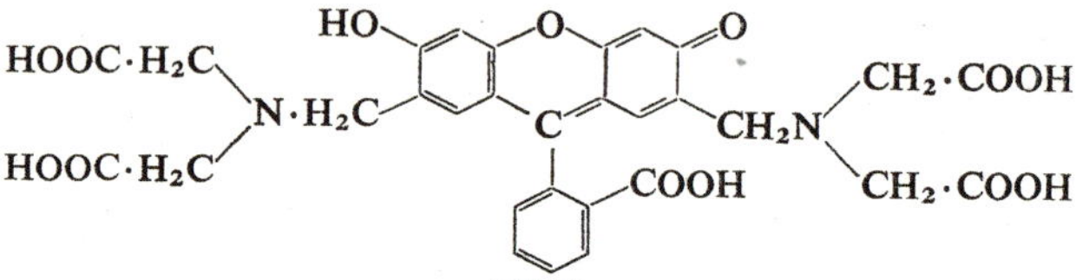

(**VIII**)

These reagents undergo stepwise dissociation and change their colours according to the pH of the medium. In the case of Xylenol Orange, 10 different ionized species were identified [88,89].

The triphenylmethane and xanthene dyes under discussion react with numerous metal ions (e.g. Al, Ga, In, Tl, Bi, Fe, Ti, Zr, V, Cu, Zn, and rare earth elements) in weakly acid or neutral solutions. Most of these reagents yield chelate complexes with metals, bonding through two oxygen atoms as ligands. In the case of reagents like Xylenol Orange, the nitrogen atom of the iminodiacetic acid group also participates in bonding to the metal ion. The metal complexes are usually 1:1 (M:R) but occasionally 1:2 [89,93].

Reagents containing sulphonic acid and iminodiacetic acid groups and their metal complexes are soluble in aqueous solutions. Reagents without such groups (e.g. Aluminon) yield with metal ions coloured suspensions stabilized by protective colloids.

Pyrocatechol Violet, Xylenol Orange, Methylthymol Blue, and Bromopyrogallol Red were used as complexometric indicators in titrations with EDTA, before they became used as spectrophotometric reagents.

Triphenylmethane basic dyes such as Brilliant Green (**IX**), Malachite Green (**X**),

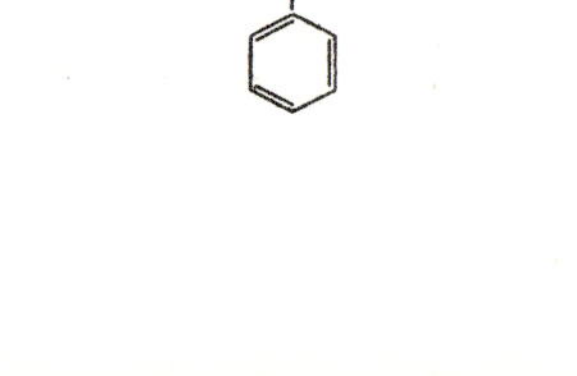

(**IX**): R=C$_2$H$_5$
(**X**): R=CH$_3$

2.3] Triphenylmethane and Xanthene Reagents

Crystal Violet (**XI**) and Methyl Violet (**XII**) serve as the basis of sensitive, extractive-spectrophotometric methods for the determination of metals.

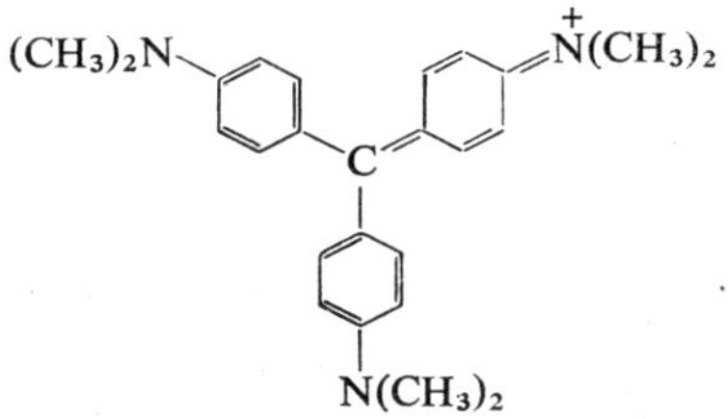

(XI) structure with $(CH_3)_2N$-, $\overset{+}{N}(CH_3)_2$, and $N(CH_3)_2$ groups

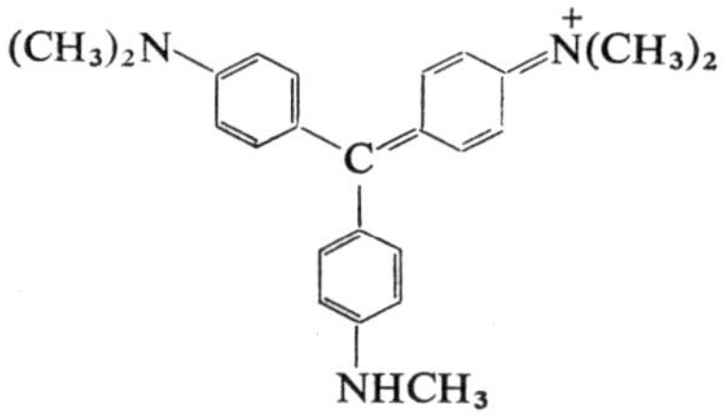

(XII) structure with $(CH_3)_2N$-, $\overset{+}{N}(CH_3)_2$, and $NHCH_3$ groups

Cations of the dyes are bonded into ion-pairs (association complexes) with anionic halide complexes (e.g. $SbCl_6^-$, $AuCl_4^-$, TaF_6^-, BF_4^-, $HgBr_3^-$, and $TeBr_5^-$) or with anions (e.g. ClO_4^- and I^-). These ion-pairs are extractable with organic solvents such as benzene, toluene, and di-isopropyl ether. An advantage is that in these methods the excess of dye does not usually pass from the aqueous to the organic phase [97–101].

The most common *xanthene basic dye* is Rhodamine B (C or S) (**XIII**) [102].

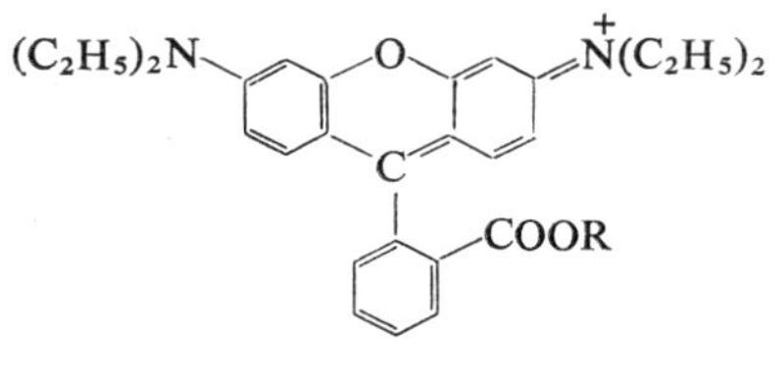

Structure with $(C_2H_5)_2N$-, O, $\overset{+}{N}(C_2H_5)_2$, and COOR groups

(XIII): R=H
(XIV): R=C_2H_5
(XV): R=C_4H_9

Further reagents in this class are Rhodamine 3B (Ethyl Rhodamine B) (**XIV**), Butylrhodamine B (C or S) (**XV**) [103], and Rhodamine 6G (**XVI**).

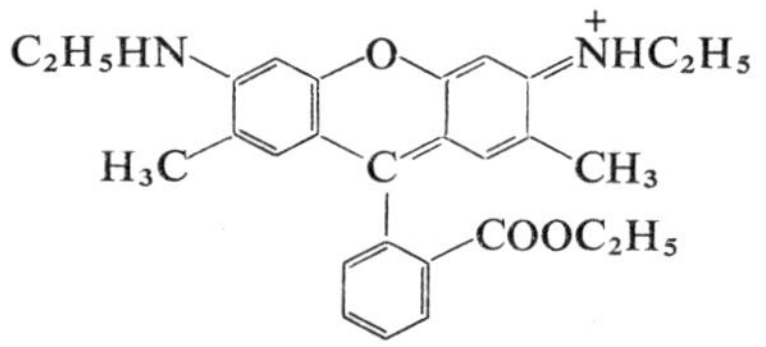

(XVI) structure with C_2H_5HN-, O, $\overset{+}{NHC_2H_5}$, H_3C, CH_3, and $COOC_2H_5$ groups

The existence of several protonated forms of Rhodamine dyes in acid solutions has been established [104]. All the above-mentioned reagents are used in extractive-spectrophotometric methods for determining antimony, gallium, thallium, gold, and other metals [99,100].

Trihydroxyfluorones [105–107] comprise another class of xanthene reagents which is applied to the spectrophotometric determination of Sn, Ge, Sb, Zr, Mo, Fe, and other metals.

2.4 Dithiocarbamates

Sodium diethyldithiocarbamate (Na–DDTC) is the dithiocarbamate most commonly used in spectrophotometric analysis [108–110].

$$\begin{array}{c} C_2H_5 \\ \diagdown \\ N-C \\ \diagup \diagdown \\ C_2H_5 S \cdot Na \end{array} \begin{array}{c} S \\ \diagup \\ \end{array}$$

Metals forming complexes with Na–DDTC are listed in Table 2.3. The metal diethyldithiocarbamates are sparingly soluble in water, but dissolve in organic solvents such as chloroform, carbon tetrachloride, diethyl ether, isoamyl acetate, pyridine, and acetone.

Table 2.3 Elements reacting with Na-DDTC. Shaded elements give colour reactions (absorption at $\lambda > 360$ nm)

H																	He
Li	Be											B	C	N	O	F	Ne
Na	Mg											Al	Si	P	S	Cl	Ar
K	Ca	Sc	Ti	V	Cr	Mn	Fe	Co	Ni	Cu	Zn	Ga	Ge	As	Se	Br	Kr
Rb	Sr	Y	Zr	Nb	Mo	Tc	Ru	Rh	Pd	Ag	Cd	In	Sn	Sb	Te	I	Xe
Cs	Ba	La	Hf	Ta	W	Re	Os	Ir	Pt	Au	Hg	Tl	Pb	Bi	Po	At	Rn
Fr	Ra	Ac	Th	U													

Spectrophotometric methods using Na–DDTC are rather insensitive since the colours of metal complexes with DDTC are not intense. Figure 2.3 shows the absorption spectra of some metal diethyldithiocarbamates.

In the reactions of diethyldithiocarbamate with metal ions, chelates with the uncommon four-membered ring are formed.

$$\begin{array}{c} C_2H_5 \\ \diagdown \\ N-C \diagdown \\ \diagup | M/n \\ C_2H_5 S \diagup \end{array} \begin{array}{c} S \\ \diagup \\ \end{array}$$

A thorough examination of the dependence of the extraction of metal dithiocarbamates on pH has been carried out by Bode [111]. Carbon tetrachloride extracts complexes of Bi, Fe, Cu, Ni, Co, Pb; Te; As, Se; and Mn over the pH ranges 4–11; 4–9; 4–6; and 6–9, respectively. Complexes of many metals can be extracted into chloroform from fairly strongly acid solutions (e.g. from $0 \cdot 1 M$ HCl) [112]. The diethyldithiocarbamates of Nb, Ru, Rh, Os, Ir and Pt are rather poorly extractable [113–116].

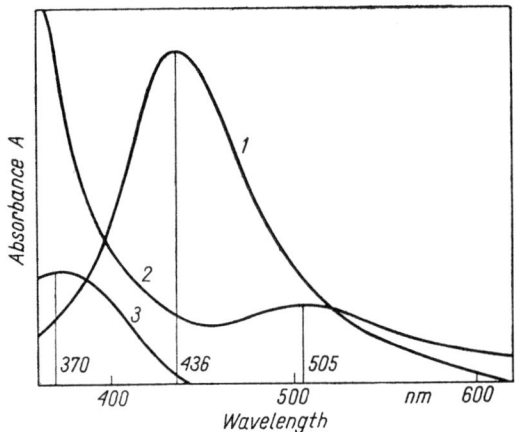

Fig. 2.3 Absorption spectra of diethyldithiocarbamates of: (1) copper, (2) manganese, (3) bismuth in carbon tetrachloride

Sodium diethyldithiocarbamate decomposes in acid solution, forming diethylamine and carbon disulphide [111].

$(C_2H_5)_2 N.CS.SNa + H^+ \rightarrow (C_2H_5)_2NH + CS_2 + Na^+$

Hence Na–DDTC solutions are stored in dilute alkali (pH ~ 9).

Figure 2.4 illustrates the reagent distribution between the aqueous and the organic phases during extraction with carbon tetrachloride.

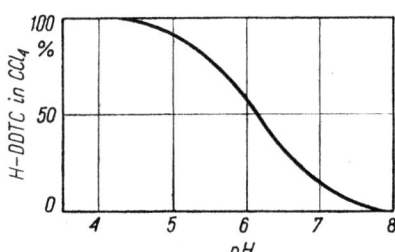

Fig. 2.4 Curve of diethyldithiocarbamate distribution between the aqueous and carbon tetrachloride phases

Diethylammonium diethyldithiocarbamate, which dissolves in chloroform and is stable in acid solution, is a more convenient reagent than Na–DDTC.

$$\begin{array}{c} C_2H_5 \\ \diagdown \\ C_2H_5 \diagup \end{array} N-\overset{\overset{\displaystyle S}{\|}}{C}-S^- \cdot {}^+NH_2 \begin{array}{c} \diagup C_2H_5 \\ \\ \diagdown C_2H_5 \end{array}$$

The acidity of the solutions being analysed may be higher than in the case of Na–DDTC [117,118].

The stability of diethyldithiocarbamate complexes with metals decreases in the series:

Hg > Ag > Co > Cu > Ni = Bi > Pb > Cd > Fe(III) > Zn > Mn

Taking advantage of the differences in stability of individual metal dithiocarbamates, it is possible to use chloroform solutions of the relatively less-stable metal dithiocarbamates for the extraction of the metals which give more stable complexes [119,120]. In determining copper, for example, the colourless chloroform solution of lead diethyldithiocarbamate, Pb(DDTC)$_2$, is used as the reagent. In this way the selectivity of metal reactions with dithiocarbamates may be enhanced.

Diethyldithiocarbamate is a group reagent. The selectivity of the methods for determining individual metals may be enhanced by using masking agents, such as KCN and EDTA, and by appropriate choice of pH.

Other dithiocarbamates which have been applied as spectrophotometric reagents are: dibenzyldithiocarbamate, pyrazolinedithiocarbamate [121], phenylhydrazinedithiocarbamate [122], glycinedithiocarbamate [123], dibutyldithiocarbamate [124], and pyrrolidinedithiocarbamate (I) [125,126].

$$\begin{array}{c} CH_2CH_2 \\ | \\ CH_2CH_2 \end{array} \!\!\!\! N\!-\!C \!\!\!\! \begin{array}{c} S \\ \\ S \cdot NH_4 \end{array}$$
(I)

Dithiocarbamates which give water-soluble metal complexes have been investigated [127].

2.5 8-Hydroxyquinoline and Derivatives

8-Hydroxyquinoline (HOx, oxine) is amphoteric [128]. It dissolves in alkaline solutions as the oxinate ion and in acid solutions as the positive oxinium ion. Oxine is soluble in chloroform, benzene, carbon tetrachloride,

ethanol, acetone, and other organic solvents [129–131]. The distribution of oxine between chloroform and water is pH-dependent (Fig. 2.5)

With certain metals, oxine forms chelates extractable into chloroform and similar solvents. These metals are indicated in Table 2.4.

The conditions of formation and extraction of particular metal hydroxyquinolinates have been thoroughly investigated [132–135]. Metals form complexes with oxine in the M:Ox ratio 1:2, 1:3, or 1:4, according to the charge on the metal ion. In the case of uranium(VI) the oxine molecule, HOx, bonds to the UO$_2$Ox$_2$ species, and uranium is extracted as H[UO$_2$Ox$_3$]. The data characterizing some metal oxinates are given in Table 2.5.

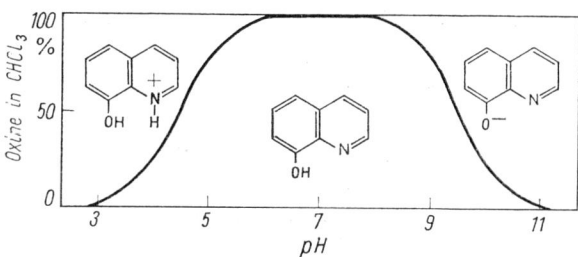

Fig. 2.5 Distribution of 8-hydroxyquinoline between chloroform and water in relation to pH

Table 2.4 Metal 8-hydroxyquinolinates extractable into chloroform or other solvents

H																	He
Li	Be											B	C	N	O	F	Ne
Na	Mg											Al	Si	P	S	Cl	Ar
K	Ca	Sc	Ti	V	Cr	Mn	Fe	Co	Ni	Cu	Zn	Ga	Ge	As	Se	Br	Kr
Rb	Sr	Y	Zr	Nb	Mo	Tc	Ru	Rh	Pd	Ag	Cd	In	Sn	Sb	Te	I	Xe
Cs	Ba	La	Hf	Ta	W	Re	Os	Ir	Pt	Au	Hg	Tl	Pb	Bi	Po	At	Rn
Fr	Ra	Ac	Th	U													

Table 2.5 Characteristics of some 8-hydroxyquinolinates extractable with chloroform

Metal	Complex	pH of extraction	Molar absorptivities (λ_{max} nm)
Copper	$CuOx_2$	5–9	5.8×10^3 (410)
Tin(II)	$SnOx_2$	6–10	4.4×10^3 (390)
Titanium	$TiOOx_2$	4–9	6.4×10^3 (385)
Molybdenum	MoO_2Ox_2	2–6	8.7×10^3 (370)
Vanadium(V)	$VO(OH)Ox_2$	3–5	3.0×10^3 (550)
Uranium(VI)	$H[UO_2Ox_3]$	7–9	7.1×10^3 (380)
Aluminium	$AlOx_3$	5–11	7.3×10^3 (390)
Gallium	$GaOx_3$	4–11	7.0×10^3 (390)
Scandium	$ScOx_3$	6–8	6.4×10^3 (378)
Iron(III)	$FeOx_3$	2–12	5.5×10^3 (470)
Cerium(IV)	$CeOx_4$	10–11	6.7×10^3 (495)
Thorium	$ThOx_4$	5–10	11.6×10^3 (378)

In chloroform solution, 8-hydroxyquinoline has an absorption maximum at 315 nm. Solutions of all the metal oxinates are coloured, yellow predominating. Extractive-spectrophotometric methods for the determination of metals with 8-hydroxyquinoline are moderately sensitive (cf. the determination of aluminium, cerium, and vanadium). The molar absorptivities of oxinates do not in general exceed 1×10^4.

8-Hydroxyquinoline is a group reagent often applied to the precipitation or extraction of a large number of elements. The selectivity of metal reactions with oxine may be enhanced by masking agents such as EDTA, tartrate, oxalate, and cyanide.

In acid medium, and with some metal ions [e.g. Fe(III), Mo(V), and Cr(III)], oxine forms coloured, water-soluble cationic complexes which can be used for the spectrophotometric determination of these metals.

Bivalent cations such as Sr^{2+}, which have a coordination number (6) too large to be satisfied by the ligand atoms available in the two oxine ions required for complexation, form dihydrates which are not extractable into chloroform unless enough oxine (or n-butylamine) is added to displace the water (synergistic effect) [136].

Oxygenated solvents extract the charged forms of the reagent as ion-pairs [137]. Some metals (Ni, Zn, U) form anionic oxine complexes giving association compounds with basic dyes which are extractable into benzene [138].

Derivatives of 8-hydroxyquinoline are also used in spectrophotometric analysis. *Chloro-oxine* (5,7-dichloro-8-hydroxyquinoline) **(I)** and *bromo-oxine* (5,7-dibromo-8-hydroxyquinoline) **(II)** react similarly to oxine [139].

(I): X = Cl
(II): X = Br

The absorption maxima of the complexes in chloroform solutions are shifted towards longer wavelengths and the sensitivity of the reactions is higher than in the case of oxine (cf. the determination of indium by bromo-oxine, p. 290).

8-Hydroxyquinoline derivatives containing sulphonic acid groups, e.g. ferron [a reagent for iron(III)], give water-soluble complexes with metal ions.

8-Mercaptoquinoline (thio-oxine), the sulphur analogue of 8-hydroxyquinoline, is often used in spectrophotometric methods [140–142]. This reagent forms sparingly soluble complexes with metal ions. Thio-oxine compounds are coloured and extractable into chloroform. The more intensely coloured complexes are those of Fe, Cu, Mn, Co, Mo, V, Ni, Pd, Os, and Pt. The molar absorptivities range from 5×10^3 to 10×10^3. 8-Mercaptoquinolinates are extracted from much more acidic solutions than are oxinates, but are susceptible to oxidation.

2.6 Formaldoxime

Formaldoxime (formaldehyde oxime) dissociates in alkaline medium, forming the CH_2NO^- anion which bonds to multivalent metals through

$$\begin{array}{c} H \\ \diagdown \\ C=NOH \\ \diagup \\ H \end{array}$$

the oxygen atom of the oxime group to form water-soluble complexes [143–147].

Formaldoxime complexes with metals having chromophoric properties (e.g. Ni, Mn, Cu, Fe, V, Co, and Ce) are coloured. Figure 2.6 shows the absorption spectra of some formaldoximates. The importance

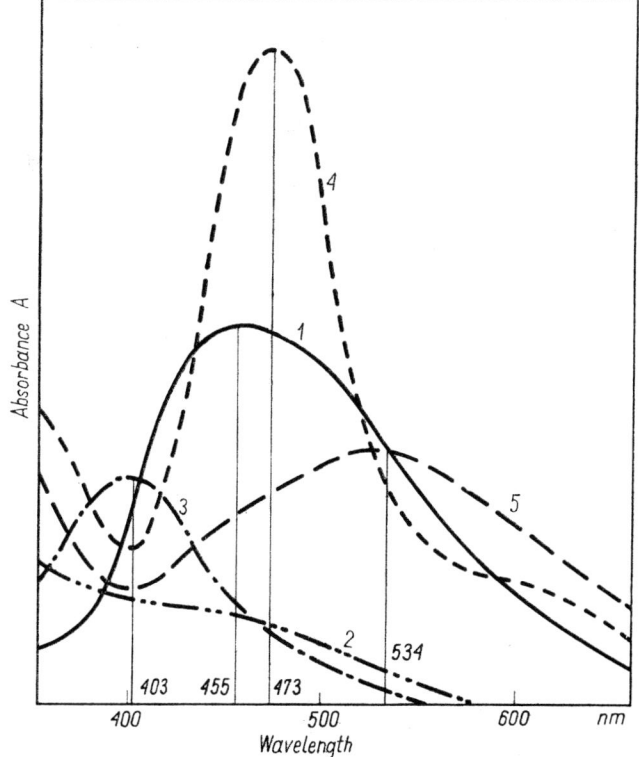

Fig. 2.6 Absorption spectra of formaldoximates of: (1) manganese, (2) cerium, (3) vanadium, (4) nickel, (5) iron

of formaldoxime as a spectrophotometric reagent lies in the determination of manganese (p. 342) and cerium (p. 200).

Atmospheric oxygen oxidizes formaldoxime complexes of metals in low oxidation states to higher oxidation states. Thus, the yellow-orange iron(II) complex, $[Fe(CH_2NO)_6]^{4-}$, is oxidized to the violet iron(III) complex, $[Fe(CH_2NO)_6]^{3-}$, and the colourless manganese(II) complex is oxidized to the brown-red manganese(IV) complex, $[Mn(CH_2NO)_6]^{2-}$.

Formaldoxime is most often bonded to metal ions in the ratio M:F = 1:6.

In alkaline medium formaldoxime undergoes auto-oxidation by reacting with oxygen to form peroxides.

Formaldoxime is employed in the form of an acid solution obtained either by mixing formaldehyde and hydroxylamine hydrochloride solutions, or by dissolving the solid hydrochloride, $(CH_2NOH)_3 \cdot HCl$ (preparation, p. 343).

Formaldoxime progressively loses its ability to react with metal ions as the solution is diluted [144]. A $1M$ solution ($\sim 5\%$) contains 55% of the "active" reagent, whereas a $0 \cdot 1M$ solution contains only 5%. Since the loss of reactivity proceeds rapidly, formaldoxime must be added in considerable excess, and the solution made alkaline immediately thereafter.

2.7 Thiocyanate

Thiocyanate (added as NH_4SCN, KSCN, or NaSCN) is one of the most important spectrophotometric reagents. The availability of the reagent and the simplicity of thiocyanate methods are responsible for its great popularity in analytical laboratories [148–151].

The metals forming thiocyanate complexes are listed in Table 2.6.

Table 2.6 Metals forming thiocyanate complexes. Complexes used in spectrophotometric methods are shaded

H																	He
Li	Be											B	C	N	O	F	Ne
Na	Mg											Al	Si	P	S	Cl	Ar
K	Ca	Sc	Ti	V	Cr	Mn	Fe	Co	Ni	Cu	Zn	Ga	Ge	As	Se	Br	Kr
Rb	Sr	Y	Zr	Nb	Mo	Tc	Ru	Rh	Pd	Ag	Cd	In	Sn	Sb	Te	I	Xe
Cs	Ba	La	Hf	Ta	W	Re	Os	Ir	Pt	Au	Hg	Tl	Pb	Bi	Po	At	Rn
Fr	Ra	Ac	Th	U													

Thiocyanate is principally used in the determination of iron(III), molybdenum, tungsten, niobium, rhenium, cobalt, uranium, and titanium.

The determination of metals by thiocyanate is carried out in aqueous or aqueous-acetone media, or after extraction with oxygenated solvents. The extractability of metal complexes depends on the acidity of the medium, the concentration of thiocyanate, and the organic solvent.

The more acidic the aqueous phase, and the higher the thiocyanate concentration, the more thiocyanic acid, HSCN, is also extracted by the organic phase [152]. From 1, 2, and $4M$ NH_4SCN (in $0.5M$ HCl), 73%, 88%, and 96%, respectively, of the thiocyanate is extracted by diethyl ether.

Stepwise formation of thiocyanate complexes gives cationic (e.g. $FeSCN^{2+}$), neutral [e.g. $Fe(SCN)_3$], and anionic [e.g. $Fe(SCN)_4^-$] species. $Fe(SCN)_4^-$ is formed at high thiocyanate concentrations.

With organic bases such as pyridine [153], tributylamine, and diantipyrylmethane (DAM) [154], anionic thiocyanate complexes form ion-pairs which can be extracted into chloroform and other inert solvents before spectrophotometric determination.

Increased selectivity in the determination of metals by thiocyanate is obtained by the choice of acidity, thiocyanate concentration, masking agent, and metal oxidation state. For example, the presence of a reducing agent is necessary for colour reactions with molybdenum, tungsten, and rhenium. The reducing medium precludes the colour reaction of thiocyanate with iron.

Thiocyanate methods differ widely in sensitivity. The methods for determining titanium, iron, and niobium are highly sensitive, whereas those for uranium and cobalt are less sensitive.

The colour stability of some thiocyanate systems is low (e.g. that with iron). This is connected with either the reducing properties of the thiocyanate ion or the slow polymerization of thiocyanic acid in acid solutions which causes yellowing. Solvents miscible with water increase the colour intensity of thiocyanate complexes in aqueous solutions. The lowered dielectric constant of the medium inhibits dissociation of the complexes.

2.8 Other Reagents

Some other frequently used organic and inorganic spectrophotometric reagents merit brief discussion.

Substituted hydroxylamines are often used in spectrophotometric determinations of metals [155–157]. Acyl hydroxylamines are called hydroxamic acids. For example, the substitution of a benzoyl group ($C_6H_5CO—$) for one of the hydrogen atoms in hydroxylamine ($H_2N—OH$) gives benzohydroxamic acid (or benzoylhydroxamic acid) (formula, p. 595), a reagent for vanadium. Other important reagents in this group are salicylhydroxamic acid, nicotinohydroxamic acid, *N*-furoylphenylhydroxylamine, and *N*-benzoyl-*N*-phenylhydroxylamine (BPHA), also called *N*-phenylbenzohydroxamic acid (formula, p. 593). These hydroxylamine derivatives form extractable chelates and are used for determining multivalent metals such as vanadium, manganese, titanium, iron(III), and niobium.

Morin (formula, p. 617) and quercetin (formula, p. 617) are commonly used hydroxy derivatives of flavone [158,159]:

Flavonols contain a highly conjugated aromatic system and consequently exhibit intense absorption spectra. In acid solutions flavonols form coloured complexes with metal ions such as Al, Fe(III), Mo(VI), U(VI), Zr, Th, and Nb. These complexes are sparingly soluble in water but dissolve in certain organic solvents.

1,10-Phenanthroline, 2,2'-bipyridyl, and related compounds [160–162] contain structural units which react with iron(II), copper(I), vanadium, cadmium, cobalt etc. They can be made specific for copper(I) by substitution of bulky groups in positions where they will cause intermolecular or intramolecular steric hindrance. Spectrophotometric reagents of this group are discussed in the sections on copper and iron.

Dioximes such as dimethylgloxime (p. 370) and α-furildioxime (p. 373) constitute a numerous group of spectrophotometric reagents used in the determination of Ni, Pd, Re, Cu, Fe and Co. Their analytical applications have been reviewed [163].

Derivatives of *anthraquinone*, e.g. Alizarin Red S, quinalizarin, and

dianthrimide, are discussed in the sections on boron, zirconium, and aluminium.

In the spectrophotometric determination of such elements as silicon, germanium, phosphorus(V), arsenic(V), and vanadium(V), the yellow *heteropoly acids* occurring in acid solutions in the presence of an excess of molybdate or tungstate are important. The blue reduction products (e.g. phosphomolybdenum blue) are the basis of sensitive spectrophotometric methods for determining these elements. The conditions of formation and extraction of these compounds have been investigated [164–169].

Other inorganic reagents utilized in spectrophotometric methods are:

hydrogen peroxide (in the determination of titanium, vanadium, and uranium), iodide (in the determination of antimony, bismuth, and palladium), and bromide (in the determination of gold). Numerous other reagents will be mentioned in Part II, where the elements are dealt with individually.

References

1. Welcher, F. J., *Organic Analytical Reagents*, Vol. I–IV, Van Nostrand, New York (1947–1948).
2. Marczenko, Z., *Odczynniki organiczne w analizie nieorganicznej* (Organic Reagents in Inorganic Analysis), PWN, Warsaw (1959).
3. Iwantscheff, G., *Das Dithizon und seine Anwendung in der Mikro- und Spurenanalyse*, 2nd Ed., Verlag Chemie, Weinheim (1972).
4. Pierce, T. B. and Peck, P. F., *Anal. Chim. Acta* **27**, 392 (1962).
5. Iwantscheff, G. and Jörrens, C., *ibid.* **38**, 470 (1967); *Z. Anal. Chem.* **264**, 131 (1973).
6. Marczenko, Z. and Mojski, M., *Chim. Anal. (Paris)* **53**, 529 (1971).
7. Mabuchi, H. and Okada, I., *Bull. Chem. Soc. Japan* **38**, 1478 (1965).
8. Marhenke, K. and Sandell, E. B., *Anal. Chim. Acta* **38**, 421 (1967).
9. Mabuchi, H. and Nakahara, H., *Bull. Chem. Soc. Japan* **36**, 151 (1963).
10. Ramakrishna, R. S. and Irving, H. M., *Anal. Chim. Acta* **49**, 9 (1970).
11. Starý, J., Marek, J., Kratzer, K. and Šebesta, F., *ibid.* **57**, 393 (1971).
12. Fischer, H., *Wiss. Veröff. Siemens-Werke* **4**, 158 (1925).
13. Nowicka-Jankowska, T. and Irving, H. M., *Anal. Chim. Acta* **54**, 489 (1971).
14. Irving, H. M. and Kiwan, A. M., *ibid.* **56**, 435 (1971).
15. Freiser, B. S. and Freiser, H., *Anal. Chem.* **42**, 305 (1970).
16. Irving, H. M. and Nowicka-Jankowska, T., *Anal. Chim. Acta* **54**, 55 (1971).
17. Kiwan, A. M. and Irving, H. M., *ibid.* **54**, 351 (1971).
18. Irving, H. M., *Anal. Chem.* **29**, 857 (1957).
19. Oh, J. S. and Freiser, H., *ibid.* **39**, 295, 1671 (1967).
20. Duncan, J. F. and Thomas, F. G., *J. Chem. Soc.* **1960**, 2814.
21. Bryan, R. F. and Knopf, P. M., *Proc. Chem. Soc.* **1961**, 203.
22. Laing, M. and Alsop, P. A., *Talanta* **17**, 242 (1970).
23. Laing, M., Sommerville, P. and Alsop, P. A., *J. Chem. Soc. (A)* **1971**, 1247.
24. Mawby, A. and Irving, H. M., *Anal. Chim. Acta* **55**, 269 (1971).
25. Kemula, W. and Gańko, T., *Roczn. Chem.* **46**, 387 (1972).
26. Geiger, R. W. and Sandell, E. B., *Anal. Chim. Acta* **8**, 197 (1953).
27. Marczenko, Z. and Mojski, M., *Chim. Anal. (Paris)* **54**, 29 (1972).
28. Grzhegorzhevskii, A. S., *Zh. Analit. Khim.* **11**, 689 (1956); *Tr. Komis. po Analit. Khim. Akad. Nauk SSSR* **11**, 165 (1960).
29. Lombardi, O. W., *Anal. Chem.* **36**, 415 (1964).
30. Dubois, R. J. and Knight, S. B., *ibid.* **36**, 1313, 1316 (1964).
31. Irving, H. M. and Butler, E. J., *Analyst* **78**, 571 (1953).
32. Chalmers, R. A. and Dick, D. M., *Anal. Chim. Acta* **32**, 117 (1965).
33. Irving, H. M., Rupainwar, D. C. and Sahota, S. S., *ibid.* **45**, 249 (1969).
34. Irving, H. M. and Rupainwar, D. C., *ibid.* **48**, 187 (1969).
35. Irving, H. M., Mahnot, U. S. and Rupainwar, D. C., *ibid.* **49**, 261 (1970).
36. Irving, H. M., Kiwan, A. M., Rupainwar, D. C. and Sahota, S. S., *ibid.* **56**, 205 (1971).
37. King, H. G. and Pruden, G., *Analyst* **96**, 146 (1971).
38. Savvin, S. B. and Kuzin, E. L., *Zh. Analit. Khim.* **22**, 1058 (1967).
39. Buděšínský, B., in Flaschka, H. A. and Barnard, A. J., *Chelates in Analytical Chemistry*, Vol. 2, Dekker, New York (1969).

40. Kuzin, E. L., Likhonina, E. A. and Savvin, S. B., *Zh. Analit. Khim.* **27**, 350 (1972).
41. Anderson, R. G. and Nickless, G., *Analyst* **92**, 207 (1967).
42. Shibata, S., in Flaschka, H. A. and Barnard, A. J., *Chelates in Analytical Chemistry*, Vol. 4, Dekker, New York (1972).
43. Busev, A. I. and Ivanov, V. M., *Zh. Analit. Khim.* **19**, 1238 (1964).
44. Cheng, K. L. and Bray, R. H., *Anal. Chem.* **27**, 782 (1955).
45. Pease, B. F. and Williams, M. B., *ibid.* **31**, 1044 (1959).
46. Pilipenko, A. T., Savranskii, L. I. and Skorokhod, E. G., *Zh. Analit. Khim.* **27**, 1080 (1972).
47. Betteridge, D. and John, D., *Analyst* **98**, 377, 390 (1973).
48. Berger, W. and Elvers, H., *Z. Anal. Chem.* **171**, 185 (1959).
49. Shibata, S., *Anal. Chim. Acta* **23**, 367 (1960); **25**, 348 (1961).
50. Betteridge, D., Fernando, Q. and Freiser, H., *Anal. Chem.* **35**, 294 (1963).
51. Püschel, R., *Z. Anal. Chem.* **221**, 132 (1966).
52. Pollard, F. H., Hanson, P. and Geary, W. J., *Anal. Chim. Acta* **20**, 26 (1959).
53. Geary, W. J., Nickless, G. and Pollard, F. H., *ibid.* **26**, 575 (1962); **27**, 71 (1962),
54. Iwamoto, T., *Bull. Chem. Soc. Japan* **34**, 605 (1961).
55. Tanaka, M., Funahashi, S. and Shirai, K., *Anal. Chim. Acta* **39**, 437 (1967).
56. Hniličková, M. and Sommer, L., *Collection Czech. Chem. Commun.* **26**, 2189 (1961).
57. Pollard, F. H., Nickless, G. and Anderson, R. G., *Talanta* **13**, 725 (1966).
58. Geary, W. J. and Bottomley, F., *ibid.* **14**, 537 (1967).
59. Anderson, R. G. and Nickless, G., *ibid.* **14**, 1221 (1967); *Anal. Chim. Acta* **39**. 469 (1967).
60. Shibata, S., Furukawa, M., Kamata, E. and Goto, K., *Anal. Chim.Acta* **50**, 439 (1970).
61. Betteridge, D. and John, D., *Analyst* **98**, 377, 390 (1973).
62. Navrátil, O., *Collection Czech. Chem. Commun.* **29**, 2490 (1964).
63. Hniličková, M. and Sommer, L., *Talanta* **13**, 667 (1966).
64. Stanley, R. W. and Cheney, G. E., *ibid.* **13**, 1619 (1966).
65. Nickless, G., Pollard, F. H. and Samuelson, T. J., *Anal. Chim. Acta* **39**, 37 (1967).
66. Minczewski, J. and Kasiura, K., *Chem. Anal. (Warsaw)* **10**, 21 (1965).
67. Kasiura, K., *ibid.* **12**, 401 (1967).
68. Kuznetsov, V. I., *Dokl. Akad. Nauk SSSR* **31**, 895 (1941); *Zh. Obshch. Khim.* **14**, 914 (1944); *Zh. Analit. Khim.* **14**, 7 (1959).
69. Savvin, S. B., *Usp. Khim.* **32**, 195 (1963).
70. Savvin, S. B., *Dokl. Akad. Nauk SSSR* **127**, 1231 (1959).
71. Savvin, S. B., *Organicheskie reagenty grupy arsenazo III* (Organic Reagents of the Arsenazo III Group), Atomizdat, Moscow (1971).
72. Savvin, S. B., *Talanta* **8**, 673 (1961); **11**, 1, 7 (1964).
73. Savvin, S. B., *Zh. Analit. Khim.* **17**, 785 (1962); *Zavodsk. Lab.* **29**, 131 (1963).
74. Savvin, S. B. and Dedkov, Yu. M., *Zh. Analit. Khim.* **19**, 21 (1964).
75. Nemodruk, A. A., *ibid.* **19**, 790 (1964).
76. Nemodruk, A. A., *ibid.* **22**, 629 (1967).
77. Savvin, S. B., Propistsova, R. F., Akimova, T. G. and Dedkova, V. P., *ibid.* **24**, 1231 (1969).
78. Savvin, S. B., Akimova, T. G., Krysin, E. P. and Davydova, M. M., *ibid.* **25**, 430 (1970).
79. Basargin, N. N., Akhmedli, M. K. and Islamov, Sh. U., *Zavodsk. Lab.* **37**, 269 (1971).
80. Yong Keun Lee, Kyn Ja Whang, and Ueno, K., *Talanta* **19**, 1665 (1972).
81. Petrova, G. S., Yagodnitsyn, M. A. and Lukin, A. M., *Zavodsk. Lab.* **36**, 776 (1970).
82. Dedkov, Yu. M. and Kotov, A. V., *Zh. Analit. Khim.* **25**, 650 (1970).
83. Ryba, O., Cífka, J., Malát, M. and Suk, V., *Collection Czech. Chem. Commun.* **21**, 349 (1956).

84. Biryuk, E. A. and Ravitskaya, R. V., *Zh. Analit. Khim.* **25**, 576 (1970).
85. Suk, V. and Mikeťuková, V., *Collection Czech. Chem. Commun.* **24**, 3629 (1959).
86. Dixon, E. J., Grisley, L. M. and Sawyer, R., *Analyst* **95**, 945 (1970).
87. Malát, M., *Anal. Chim. Acta* **25**, 289 (1961).
88. Řehák, B. and Körbl, J., *Collection Czech. Chem. Commun.* **25**, 797 (1960).
89. Murakami, M., Yoshino, T. and Harasawa, S., *Talanta* **14**, 1293 (1967).
90. Sangal, S. P., *Chim. Anal. (Paris)* **47**, 239 (1965).
91. Buděšínský, B. in Flaschka, H. A. and Barnard, A. J., *Chelates in Analytical Chemistry*, Vol. 1, Dekker, New York (1967).
92. Körbl, J. and Přibil, R., *Collection Czech. Chem. Commun.* **23**, 873 (1958).
93. Tikhonov, V. N., *Zh. Analit. Khim.* **21**, 1172 (1966); **22**, 658 (1967).
94. Yoshino, T., Imada, H., Kuwano, T. and Iwasa, K., *Talanta* **16**, 151 (1969).
95. Vodák, Z. and Leminger, O., *Collection Czech. Chem. Commun.* **21**, 1522 (1956).
96. Suk, V., Malát, M. and Jeničková, A., *ibid.* **21**, 418, 1257 (1956).
97. Blyum, I. A. and Pavlova, N. N., *Zavodsk. Lab.* **29**, 1407 (1963); *Zh. Analit. Khim.* **20**, 898 (1965).
98. Lomonosov, S. A., *Zh. Analit. Khim.* **22**, 1125 (1967).
99. Blyum, I. A., *Ekstraktsionno-fotometricheskie metody analiza s primeneniem osnovnykh krasitelei* (Extractive-Photometric Methods of Analysis with Use of Basic Dyes), Izdat. Nauka, Moscow (1970).
100. Fogg, A. G., Burgess, C. and Burns, D. T., *Talanta* **18**, 1175 (1971).
101. Lomonosov, S. A., Sorokin, G. Kh. and Shukolyukova, N. I., *Zh. Analit. Khim.* **28**, 2085 (1973).
102. Ramette, R. W. and Sandell, E. B., *J. Am. Chem. Soc.* **78**, 4872 (1956).
103. Kuznetsov, V. I. and Bol'shakova, L. I., *Zh. Analit. Khim.* **15**, 523 (1960).
104. Zorov, N. B., Golovina, A. P., Alimarin, I. P. and Khvatkova, Z. M., *ibid.* **26**, 1466 (1971).
105. Nazarenko, V. A., Shelikhina, E. I. and Antonovich, V. P., *ibid.* **27**, 642 (1972).
106. Sivanova, O. V., Ivankovich, G. S. and Gritsienko, N. N., *ibid.* **27**, 2321 (1972).
107. Nazarenko, V. A. and Antonovich, V. P., *Trioksifluorony* (Trihydroxyfluorones), Izdat. Nauka, Moscow (1973).
108. Delépine, M., *Bull. Soc. Chim. France*, **1958**, 5.
109. Halls, D. J., *Mikrochim. Acta* **1969**, 62.
110. Hulanicki, A., *Talanta* **14**, 1371 (1967).
111. Bode, H., *Z. Anal. Chem.* **142**, 414; **143**, 182 (1954); **144**, 165 (1955).
112. Malissa, H. and Gomišček, S., *Z. Anal. Chem.* **169**, 401 (1959).
113. Šedivec, V. and Flek, J., *Collection Czech. Chem. Commun.* **23**, 1977 (1958); **29**, 1310 (1964).
114. Babko, A. K., Freger, S. V., Ovrutskii, M. I. and Lisetskaya, G. S., *Zh. Analit. Khim.* **22**, 670 (1967).
115. Tulyupa, F. M., Usatenko, Yu. I. and Barkalov, V. S., *Tr. Komis. po Analit. Khim. Akad. Nauk SSSR* **17**, 315 (1969).
116. Usatenko, Yu. I., Barkalov, V. S. and Tulyupa, F. M., *Zh. Analit. Khim.* **25**, 1458 (1970).
117. Bode, H. and Neumann, F., *Z. Anal. Chem.* **169**, 410 (1959); **172**, 1 (1960).
118. Förster, H., *J. Radioanal. Chem.* **4**, 1 (1970).
119. Eckert, G., *Z. Anal. Chem.* **155**, 23 (1957).
120. Bode, H. and Tusche, K. J., *ibid.* **157**, 414 (1957).
121. Byr'ko, V. M., *Tr. Komis. po Analit. Khim. Akad. Nauk SSSR* **14**, 191 (1963).
122. Musil, A. and Haas, W., *Mikrochim. Acta* **1958**, 756.
123. Haas, W. and Winterstein, P., *ibid.* **1961**, 787.
124. Tulyupa, F. M., Usatenko, Yu. I. and Barkalov, V. S., *Zh. Analit. Khim.* **23**, 1844 (1968).
125. Malissa, H. and Gomišček, S., *Anal. Chim. Acta* **27**, 402 (1962).
126. Likussar, W. and Boltz, D. F., *Anal. Chem.* **43**, 1273 (1971).

127. Bode, H., Tusche, K. J. and Wahrhausen, H. F., *Z. Anal. Chem.* **190**, 48 (1962).
128. Hollingshead, R. G., *Oxine and its Derivatives*, Vols. I–IV, Butterworths, London (1954–56); *Anal. Chim. Acta* **19**, 447 (1958).
129. Rudenko, N. P. and Sevastyanov, A. I., *Zh. Analit. Khim.* **26**, 1994 (1971).
130. Mottola, H. A. and Freiser, H., *Talanta* **14**, 864 (1967).
131. Mason, J. G. and Lipschitz, I., *ibid.* **18**, 1111 (1971).
132. Umland, F. and Hoffmann, W., *Z. Anal. Chem.* **168**, 268 (1959).
133. Umland, F., Hoffmann, W. and Meckenstock, K. U., *ibid.* **173**, 211 (1960).
134. Umland, F., *ibid.* **190**, 186 (1962).
135. Starý, J., *Anal. Chim. Acta* **28**, 132 (1963).
136. Dyrssen, D., *Svensk Kem. Tidskr.* **67**, 311 (1955).
137. Kuz'min, N. M., Khorkina, L. S., Lebedev, A. I. and Zolotov, Yu. A., *Zh. Analit. Khim.* **25**, 1257 (1970).
138. Zolotov, Yu. A., Seryakova, I. V., Vorob'eva, G. A. and Sapragonene, M. S., *ibid.* **25**, 1845 (1970).
139. Navrátil, O. and Kotas, J., *Collection Czech. Chem. Commun.* **30**, 1824 (1965).
140. Bankovskii, Yu. A., Chera, L. M. and Ievin'sh, A. F., *Zh. Analit. Khim.* **18**, 668 (1963).
141. Corsini, A., Fernando, Q. and Freiser, H., *Anal. Chem.* **35**, 1424 (1963).
142. Veveris, O. E., Bankovskii, Yu. A., Pelekis, L. L., Ainbinder, N. G., and Shekhtmeister, L. A., *J. Radioanal. Chem.* **9**, 47 (1971).
143. Marczenko, Z. and Minczewski, J., *Chem. Anal.* (*Warsaw*) **5**, 747 (1960); *Roczniki Chem.* **35**, 1223 (1961); *Zh. Analit. Khim.* **17**, 23 (1962).
144. Marczenko, Z. and Kasiura, K., *Chem. Anal.* (*Warsaw*) **6**, 37 (1961).
145. Marczenko, Z., *Bull. Soc. Chim. France* **1964**, 939; *Anal. Chim. Acta* **31**, 224 (1964).
146. Bartušek, M. and Okač, A., *Collection Czech. Chem. Commun.* **26**, 52, 883, 2174 (1961).
147. Bečka, J. and Jokl, J., *ibid.* **36**, 2467, 3263 (1971).
148. Bock, R., *Z. Anal. Chem.* **133**, 110 (1951).
149. Różycki, C., *Chem. Anal.* (*Warsaw*) **11**, 447 (1966); **15**, 3 (1970).
150. Różycki, C., *ibid.* **14**, 755 (1969).
151. Różycki, C. and Lachowicz, E., *ibid.* **15**, 255 (1970).
152. Jurriaanse, A. and Kemp, D. M., *Talanta* **15**, 1287 (1968).
153. Ayres, G. H. and Baird, S. S., *ibid.* **7**, 237 (1961).
154. Zhivopistsev, V. P., *Zavodsk. Lab.* **31**, 1043 (1965).
155. Bass, V. C. and Yoe, J. H., *Talanta* **13**, 735 (1966).
156. Shendrikar, A. D., *ibid.* **16**, 51 (1969).
157. Majumdar, A. K., *N-Benzoylphenylhydroxylamine and Its Analogues*, Pergamon, Oxford (1972).
158. Katyal, M., *Talanta* **15**, 95 (1968).
159. Nevskaya, E. M. and Nazarenko, V. A., *Zh. Analit. Khim.* **27**, 1699 (1972).
160. Smith, G. F., *Anal. Chem.* **26**, 1534 (1954).
161. Stephen, W. I., *Talanta* **16**, 939 (1969).
162. Schilt, A. A., *Analytical Applications of 1,10-Phenanthroline and Related Compounds*, Pergamon, New York (1969).
163. Egneus, B., *Talanta* **19**, 1387 (1972).
164. Jean, M., *Chim. Anal.* (*Paris*) **44**, 195, 243 (1962).
165. Alimarin, I. P., Sudakov, F. P. and Klitina, V. I., *Usp. Khim.* **34**, 1368 (1965).
166. Boulin, R., *Chim. Anal.* (*Paris*) **51**, 369 (1969).
167. Wünsch, G. and Umland, F., *Z. Anal. Chem.* **250**, 248 (1970).
168. Halász, A. and Pungor, E., *Talanta* **18**, 557, 569 (1971).
169. Halász, A., Pungor, E. and Polyák, K., *ibid.* **18**, 577 (1971).

Chapter 3

PRECONCENTRATION AND SEPARATION OF ELEMENTS

The spectrophotometric determination of particular elements is usually preceded by their separation from the major components (matrix) of the sample, and from other elements. In the trace analysis of high-purity materials, separation from the matrix involves simultaneous concentration (enrichment) of the trace components. Separation methods often enable a particular element to be determined in a solution containing no interfering elements. General methods for preconcentrating and separating elements have been outlined in several publications [1-12].

The present section comprises a discussion of the following preconcentration and separation methods: precipitation and coprecipitation with collectors, solvent extraction, volatilization, and ion-exchange separation. These methods are also used before other analytical techniques such as emission spectrography, atomic-absorption spectrometry, and polarography.

3.1 Precipitation and Coprecipitation

Precipitation methods for the separation of elements are based on the differences in solubility of the elements (or their compounds) in aqueous solutions. Precipitation methods are used for separating trace elements alone, as well as for separating macrocomponents from the traces. Trace elements are separated quantitatively from the solution by using collectors (scavengers or carriers). When macrocomponents are precipitated, the aim is to prevent trace elements from coprecipitating with the large mass of the macrocomponent precipitate. This prerequisite restricts the application of the method to cases in which coprecipitation of trace elements with the macrocomponent precipitate is negligible. It is evident, therefore, that coprecipitation is exploited in the separation of traces with collectors, but must be minimized in the precipitation of the macrocomponents.

3.1.1 SEPARATION OF TRACES BY COPRECIPITATION

When a precipitant is added to a solution containing trace quantities of an ionic species (0·1–100 µg in 100–250 ml, 10^{-8}–$10^{-5}M$), the ionic species may be only partly precipitated (or not at all), even though the solubility product has been exceeded.

The *formation of a precipitate*, i.e. crystal growth, is a complex and slow process [13–15]. Precipitation begins with nucleation, which results from association of oppositely charged ions. When nucleation and further growth of nuclei occur slowly, a condition of permanent supersaturation exists in the solution. Nuclei are formed when the first portion of precipitant is added to the solution. If the ions to be precipitated occur in macro-amounts in the solution, further addition of the precipitant causes rapid transformation of the nuclei into crystals by the formation of more of the sparingly-soluble compound. If, however, the solution contains only trace amounts of the ions to be precipitated, then the process is virtually complete after the nucleation stage.

Nucleation, transformation of the amorphous nuclei into crystal nuclei, and growth of the crystals proceed rapidly if the quantity of the resulting solid phase is appreciable. That is why the precipitation of traces from a solution undergoes an essential change when the traces are coprecipitated with a collector [16–18].

The *mechanism of coprecipitation* depends on experimental conditions and on the chemical and physical properties of both the trace element and the collector. The ions coprecipitated from the solution accumulate with the forming particles of the collector precipitate.

Coprecipitation may involve isomorphous solid solution, mixed-crystal formation, or adsorption. Sometimes it occurs by mechanical retention of the trace substance in the collector precipitate. Many cases of coprecipitation of traces and collectors as sulphides or hydroxides are ascribed to isomorphism.

Separation by coprecipitation is not restricted by the very low concentration of the trace species [16]. Smaller traces can be separated by coprecipitation than by solvent extraction, which is limited by the stability of the complex extracted, although even here enhancement can be achieved by addition of a "coextracting" carrier [19].

The formation of a solid solution, e.g. the separation of traces of lead with lanthanum hydroxide as collector, may occur by occupation of some crystal-lattice sites by the trace ion instead of the collector ion. This mechanism of coprecipitation is possible provided the ionic radii [20] of the trace and collector are similar, as is the case with La^{3+} (1·34 Å) and Pb^{2+} (1·47 Å). However, when the collector and the trace species are chemically similar, the latter may be coprecipitated, apparently by mixed crystal formation, even though this would not be predicted on the basis of crystallography and ionic radii. This phenomenon is known as anomalous mixed crystal formation.

Coprecipitation may result in the formation of chemical compounds, provided the trace element and the collector have opposite chemical properties (acidic and basic) [17,18]. For example, traces of germanium or vanadium coprecipitated with iron, aluminium, or lanthanum hydroxide yield the respective germanates and vanadates, while traces of tungsten or

3.1] Precipitation and Coprecipitation

molybdenum coprecipitated with ferric hydroxide yield ferric tungstate or molybdate.

Elements which can be preconcentrated by precipitation with collectors are listed in Table 3.1. The most important forms for the coprecipitation of particular elements are indicated.

Table 3.1 Separation of traces by coprecipitation*

Be a																
Mg ae											Al ae	Si a	P a			Cl f
Ca ec	Sc ae	Ti ae	V ae	Cr ae	Mn ab	Fe abe	Co ab	Ni ab	Cu abd	Zn ab	Ga ae	Ge ab	As b	Se da		Br f
Sr ce	Y ae	Zr ae	Nb ae	Mo bea		Ru db	Rh db	Pd db	Ag bda	Cd ab	In abe	Sn ab	Sb abd	Te da		I f
Ba ce	La ae	Hf ae	Ta ae	W ae	Re ba	Os db	Ir db	Pt db	Au dab	Hg bd	Tl abe	Pb abc	Bi abd			
Ra c																

Ce ae	Th ae	U ae

The traces precipitated as:
a — hydroxides or acids   d — elements after reduction
b — sulphides   e — 8-hydroxyquinolinates
c — sulphates   f — silver salts

*In Tables 3.1–3.3 and 3.5–3.8 the lanthanides and actinides are represented by cerium, thorium and uranium only.

Elements which are coprecipitated as hydroxides can also be isolated by organic reagents of the R.OH type, such as 8-hydroxyquinoline, cupferron, or β-diketones. Metal ions giving sparingly soluble sulphides may be coprecipitated by organic thiols (R.SH) such as dithiocarbamate or thionalide.

Enough collector should be added to the sample solution to ensure that the precipitation is rapid, and that sufficient precipitate is formed for easy filtration or centrifugation. At the same time the quantity of the collector should be sufficiently small for adsorption of interfering ions to be negligible. The quantity of collector used depends on the volume of precipitate formed. In practice, 2–5 mg of collector are used per 50–200 ml of sample solution.

Hydroxides are frequently used for precipitation of traces with collectors [21–30]. With iron, aluminium, or lanthanum as collector, traces of most Analytical Group I–III metals are separated by using excess of ammonia. Metals forming ammine-complexes, e.g. Ag, Cu, Ni, Co, Zn, and Cd, remain in solution.

When excess of sodium hydroxide is used for precipitation, amphoteric metals such as Al, Pb, Zn, Sn, and Cr remain unprecipitated. In this case, iron, titanium, magnesium, or lanthanum may be employed as the collector. Lanthanum is convenient [22,27,30] since it usually does not have to be determined in the trace concentrate. It has no chromophoric properties, and does not interfere in most spectrophotometric methods.

Traces of readily hydrolysable metals [e.g. Sn, Sb, Tl(III), and Bi] are separated from acidic medium with hydrous MnO_2 (formed by reacting Mn^{2+} with MnO_4^-) as the collector [31,32].

The separation of traces of metals as *sulphides* is carried out with hydrogen sulphide or thioacetamide. Copper, mercury, and other metals of the hydrogen sulphide group are used as collectors [33–36]. After the collector has been added and the pH adjusted, the solution is heated with thioacetamide or the hot solution is saturated with hydrogen sulphide. All the Analytical Group I–III metals are precipitated when an ammoniacal solution (pH 8–9) is saturated with hydrogen sulphide.

To separate traces of heavy metals from macroquantities of zinc, sodium sulphide is added to the solution at pH 3·5. Some zinc sulphide is precipitated with the heavy metal sulphides and serves as the collector [36].

Further examples of inorganic trace coprecipitation methods are: coprecipitation of lead and strontium with barium sulphate as collector [35,37]; bismuth as phosphate with cobalt as collector [38]; iron as phosphate with various metals as collectors [39]; and tungsten, niobium, and tantalum with molybdophosphate as collector [40].

Besides 8-hydroxyquinoline [41–43], other organic precipitants have been employed. Tannin and thionalide [44–46] are commonly used in mixtures with 8-hydroxyquinoline, aluminium and indium being used as collectors. These reagents give concentrates of all metals, except the alkali metals.

Traces of Ag, Cu, Bi, Sn, In, and Ga in cadmium are coprecipitated with cadmium diethyldithiocarbamate after a small amount of Na–DDTC has been added to the solution at pH 4 [47].

Traces of noble and semi-noble metals (e.g. Au, Ag, Hg, and Cu) are separated electrolytically from acid medium on a small platinum or gold cathode [48]. In cementation (i.e. reduction to metal *in situ* by another metal) methods, small amounts of semi-noble metals (Cu, Bi, Sb) are deposited on less noble metals such as tin, iron, and zinc.

In addition to group systems for separating elements, there are also other specific methods for precipitation of particular elements. Palladium, for example, can be precipitated with nickel as scavenger (and vice versa) by means of dimethylglyoxime [23]. The rare earths and thorium may be separated as the sparingly-soluble oxalates or fluorides. Molybdenum and tungsten coprecipitate as benzoinoximates. Traces of sulphur are separated as sulphate with chromate as scavenger and Ba^{2+} as precipitant.

Traces of metals can be precipitated by organic collectors [49–53]. These are high molecular-weight organic compounds which are either acidic or basic in character (e.g. cinchonine and Methyl Violet). Their selection depends on the chemical nature of the traces with which they form sparingly-soluble salts, chelates, or ion-pairs. The organic collectors may be precipitated simultaneously. The anionic thiocyanate complex of a trace metal, for example, forms an insoluble ion-pair with the Methyl Violet cation, and at the same time the excess of Methyl Violet precipitates as a sparingly-soluble salt with the excess of thiocyanate. PAN has been suggested as a precipitant and collector for traces of many metals [54]. An advantage of organic collectors is their easy separation from traces simply by igniting the precipitate. There is, however, a risk of losses of non-volatile trace elements owing to the formation of light aerosols, which can be easily blown away by the slightest atmospheric motion around the crucible in which the organic collector is being burnt.

Volatile inorganic collectors such as mercury and arsenic are also used. After precipitation along with the traces, they are distilled off when the precipitate is heated. Losses of trace elements, as aerosols, are liable to occur, however.

Masking plays a fundamental role in the separation of traces from macrocomponents by precipitation. The aim is to keep the macrocomponents in solution while the traces are coprecipitated with a collector. The masking agent selected must complex the sample matrix, thus allowing the quantitative precipitation of the trace elements and the collector to proceed without interference.

The more important masking agents for particular elements are indicated in Table 3.2. Owing to their similarity in forming complexes with the same metals, cyanide and ammonia, as well as citrate and tartrate, are considered jointly (although citrate complexes are normally the more stable in more acid media). The oxo- (e.g. WO_4^{2-} and ReO_4^-) and hydroxo- (e.g. $[Al(OH)_4]^-$) complexes are also considered jointly.

Knowledge of the stability constants of complexes is not enough to predict suitable masking agents. The stability constants of complexes are apparent constants which vary with the pH and with the concentration of other species capable of complexation (competing reactions). The effect of these factors is taken into account in the conditional stability constants of complexes [55].

The prediction of conditions for forming stable complexes requires knowledge of general formation tendencies and stability variations of particular complex types. General relationships between the position of an element in the Periodic Table (and hence its electronic configuration) and

Table 3.2 Some masking (complex-forming) agents

Be											B			
eia											ga			
Mg											Al	Si	P	
de											adg	g	a	
Ca	Sc	Ti	V	Cr	Mn	Fe	Co	Ni	Cu	Zn	Ga	Ge	As	Se
de	de	egh	ahj	aej	aed	dej	cdj	cdj	cd	cad	ade	afj	abf	a
Sr	Y	Zr	Nb	Mo		Ru	Rh	Pd	Ag	Cd	In	Sn	Sb	Te
de	de	egh	egh	aeg		acf	cf	cf	c	cdf	dej	abf	abe	a
Ba	La	Hf	Ta	W	Re	Os	Ir	Pt	Au	Hg	Tl	Pb	Bi	
de	de	egh	egh	aeg	aej	acf	cf	cf	cf	fej	aed	dej		

Ce	Th	U
de	edi	ieh

Complexes:
a — oxo and hydroxo
b — sulphide
c — cyanide and ammine
d — EDTA
e — citrate and tartrate
f — halo (Cl$^-$, Br$^-$, I$^-$)
g — fluoro
h — peroxo
i — carbonate
j — oxalate

its ability to form complexes with various ligands have been formulated [55,56].

Some examples of masking macrocomponents in the coprecipitation of traces with collectors are considered below.

During the precipitation of trace metals in the analysis of high-purity tin, the matrix is masked as the soluble hydroxo-complex (stannate) [35].

In the determination of trace impurities in uranium, sodium carbonate masks uranium as the soluble carbonate complex while precipitating most trace elements as hydroxides or carbonates.

Macroquantities of tungsten are kept in acid solution by tartaric acid [34]. Heavy metals (Bi, Pb, Sn, Cd, As, Sb) are precipitated as sulphides with copper as collector.

When the trace metals contained in silver are precipitated as hydroxides, the matrix is kept in solution as the ammine complex [22].

Macroquantities of nickel, zinc, and manganese are kept in alkaline solution as the cyanide complexes while calcium and magnesium are separated as phosphates with lanthanum as the collector [57].

The presence of complexing agents in the sample solution may interfere in the separation of some metals. In the example of the trace analysis of tin, sodium hydroxide not only masks tin(IV) as the soluble complex, but also masks aluminium, arsenic, and antimony. In the trace analysis of gold, the sample is dissolved in aqua regia (HCl + HNO$_3$). Microgram quantities of silver remain in solution as the AgCl$_2^-$ complex, and cannot be separated by using a collector.

3.1.2 Separation of Macrocomponents

The precipitation of macrocomponents to separate them from trace elements is less common. In the separation of macrocomponents, their quantitative precipitation is not necessary.

The precipitation methods for individual matrix elements are shown in Table 3.3. The largest group of elements comprises those isolated from solution in the elemental form as a result of reduction (usually electro-

Table 3.3 Separation of matrix by precipitation

Li b	Be b																
Na b	Mg b											Al b	Si d				
K b	Ca cb	Sc b	Ti b		Cr f	Mn f	Fe f	Co f	Ni f	Cu fb	Zn f	Ga b	Ge f	As e	Se a		
Rb b	Sr cb	Y b	Zr b	Nb d	Mo ef		Ru a	Rh af	Pd af	Ag fab	Cd f	In b	Sn ef	Sb ef	Te a		
Cs b	Ba cb	La b	Hf b	Ta d	W de	Re f	Os a	Ir af	Pt af	Au af	Hg abf	Tl fb	Pb cbf	Bi fb			
		Ce b															

The macrocomponents precipitated as:
a — elements (chemical reduction)
b — halides (also after saturation with HCl)
c — sulphates
d — hydroxides or acids
e — sulphides
f — electrolytically

chemical). In acid solution, the electrolytic deposition of metals on a solid cathode is limited to noble and semi-noble metals. For example, anodic dissolution is used to separate macroquantities of copper in the trace analysis of copper and its compounds [58,59]. A sample in the form of a bar, plate, or wire is the anode in the electrolytic system. When current is passed through the electrolyte (nitric acid + persulphate), copper is deposited on the graphite cathode while most trace elements accumulate in the solution. Macroquantities of bismuth can be separated at a controlled-potential cathode [60].

The use of a mercury cathode [61] enables a considerable number of metals mentioned in Table 3.3 to be isolated from dilute acid solution. After electrolysis with the mercury cathode, the following metals remain in solution: Al, Ti, Zr, V, Nb, U, Th, Be, Mg, Ca, Sr, Ba, and rare earths. A mercury cathode is used to separate Fe, Ni, Cr, Mo, and Mn when steel is analysed for Al, V, and Ti.

Gold, silver, mercury, and platinum metals, as well as selenium and tellurium, can be precipitated from acid solution in the elemental form by reduction with chemical reagents such as zinc, hydroxylamine, hydrazine, sulphur dioxide, and formic acid. In the trace analysis of high-purity mercury, the sample is dissolved in nitric acid and the solution warmed in the presence of formic acid [62]. First of all nitric acid, then mercury, is reduced. The mercury forms a separate liquid phase, and the impurities remain in the aqueous solution. During the trace analysis of silver, the sample is dissolved in nitric acid, and formic acid and mercury are added. The silver liberated on reduction dissolves in the mercury to form an amalgam [62].

The belief that isolation of macrocomponents from solution as sparingly-soluble compounds is inadmissible in trace analysis because of the considerable losses of traces caused by adsorption, is not necessarily true if the precipitation is carried out in acid medium. This has been confirmed in the following examples.

In the analysis of lead for impurities, the matrix can be precipitated from nitric acid medium as lead sulphate. By using radioisotopes, it was found that none of the 24 elements investigated had coprecipitated with $PbSO_4$ [63]. Most of the lead can also be separated from nitric acid medium as $PbCl_2$ without any perceptible coprecipitation of other components [64]. The solubility of $Pb(NO_3)_2$ in water is ~ 300 times higher than that in concentrated nitric acid, permitting the concentration of traces in the analysis of high-purity lead [65].

Before the determination of trace impurities in bismuth the latter is removed from nitric acid solution as the sparingly soluble iodide [66] or basic nitrate [67,68]. Bismuth macroquantities can also be separated as BiOBr [69].

Macroquantities of molybdenum are isolated from hydrochloric acid medium as the α-benzoinoximate [70], while macroquantities of zirconium are precipitated from $2.5M$ HCl with mandelic acid [71]. The precipitate is separated by flotation with chloroform or isoamyl alcohol.

In the precipitation of tungsten macroquantities from $6M$ HCl as tungstic acid, coprecipitation of As, Sb, Cr, Ga, In, Fe, and Mn is negligible [72].

In the analysis for Cu, Zn, Cd, Ni, Pb, Mn, and Fe traces in silver, the matrix is precipitated from dilute nitric acid medium as AgCl [22,73]. The trace elements do not collect with the precipitate.

Before analysis of certain chlorides (e.g. K, Na, Ca, Al), for trace elements, the matrix may be precipitated by saturating the solution with gaseous hydrogen chloride.

All these examples of the "pure" separation of major components offer every reason to believe that other matrix elements forming sparingly-soluble compounds in acid medium are also capable of being separated by precipitation.

3.2 Solvent Extraction

The extraction process and extractive methods for preconcentration and separation of elements are described in several monographs [5,74–79] and review papers [80–83]. An exhaustive literature review of the extraction of inorganic compounds between 1945 and 1967 has been published [84], and also a book on extraction of halide complexes [85].

Solvent extraction separation is based on differences in the solubilities of elements and their compounds in two immiscible liquid phases. Usually, the initial phase is water and the second phase is an organic solvent immiscible with water.

Some properties of the more common organic solvents are listed in Table 3.4.

The species extracted must be uncharged, i.e. it must be an undissociated complex, ion-association compound (ion-pair), or a covalent compound or element.

Stripping (scrubbing, back-extraction, re-extraction) involves bringing the element from the organic extract back into the aqueous phase.

The *extraction efficiency*, i.e. the degree of transference of the species from the aqueous to the organic phase, is defined in terms of the *distribution* (or extraction) *coefficient*, D. D is the ratio of total concentration (i.e. the concentration of all the existing forms) of the element in the organic phase (ΣC_o) to the total concentration in the aqueous phase (ΣC_{aq}) at equilibrium.

$$D = \frac{\Sigma C_o}{\Sigma C_{aq}}$$

The extraction efficiency (E) is also expressed as

$$E = \frac{100 \cdot D}{D + (V_{aq}/V_o)} \%$$

where D is the distribution coefficient, and V_{aq} and V_o are the volumes of the aqueous and the organic phases, respectively, after separation.

Table 3.4 Physical properties of some organic solvents

Solvent	Density g/cm^3	b.p. °C	Dielectric constant	Water-solubility
Acetate				
amyl	0·875	149	4·8	0·2%
butyl	0·881	126·5	5·0	0·5%
ethyl	0·901	77·2	6·0	8·6%
Acetone	0·891	56·5	20·7	complete
Alcohol				
n-amyl	0·814	138·1	13·8	2·2%
n-butyl	0·813	117·7	17·1	7·9%
ethyl	0·789	78·3	24·3	complete
methyl	0.796	64·7	32·6	complete
Benzene	0·894	80·1	2·3	0·18%
Carbon tetrachloride	1·595	76·5	2·2	0·08%
Chloroform	1·498	61·3	4·8	1%
Cyclohexane	0.783	80·7	2·0	0·01%
o-Dichlorobenzene	1·300	180·5	9·9	0·01%
Dichloroethane	1·257	83·6	10·4	0·9%
Dioxan	1·034	101·3	2·2	complete
Ether				
2,2'-dichlorodiethyl (chlorex)	1·222	178	23·0	1%
diethyl	0·719	34·5	4·3	7·4%
isopropyl	0·728	67·5	3·9	0·7%
Hexane	0·660	69·0	1·9	0·02%
Ketone				
isobutyl methyl (IBMK)	0·801	115·8	13·1	2%
methyl ethyl	0·805	79·6	18·5	35%
Pyridine	0·988	115·3	12·3	complete
Tetrachloroethylene	1·631	121·2	2·3	0·02%
Tributyl phosphate (TBP)	0·973	177 (25 mm Hg)	8·0	0·6%
Trichloroethylene	1·456	87	3·4	0·1%
Toluene	0·866	110·8	2·4	0·05%

The dependence of the extraction efficiency on the pH of the aqueous solution is usually presented graphically as log D vs. pH for larger values of D, and as E vs. pH for smaller values of D. The variation of D and E with pH in the extraction of aluminium into chloroform as the 8-hydroxyquinolinate is presented in both ways in Fig. 3.1.

When the distribution coefficient of a given element in a specified system is large (e.g. 1000, i.e. $\log D = 3$), a single extraction (shaking of the aqueous solution with only one portion of the organic solvent) will suffice, even though the volume of the organic solvent is much smaller than that of the aqueous phase.

Shaking the phases in a separating funnel during an extraction or re-extraction must be continued until equilibrium is attained. The time required for the system to reach equilibrium varies from seconds to several minutes, depending on the kinetics of the extraction [86,87]. When the shaking time recommended is more than two minutes, it is advisable to use a mechanical shaker.

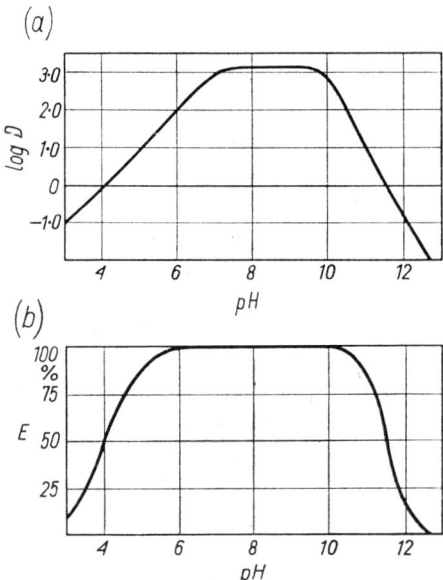

Fig. 3.1 Extraction of aluminium 8-hydroxyquinolinate into chloroform in relation to pH

Extraction is equally useful in the preconcentration and separation of small amounts of elements, and in the separation of macrocomponents from traces. Extraction methods generally require less time than precipitation methods. The former also give "purer" separations of elements owing to the small area of phase contact. Co-extraction [88–91] has been little explored in separation methods, though it may prove useful [19].

In most solvent extraction systems, the distribution coefficients vary with the concentration of the species extracted. Usually they decrease with decreasing concentrations of the elements [92,93].

Extraction systems may be divided into two classes: uncharged covalent species (simple molecules and chelates), and ion-association complexes (uncharged electrovalent species).

Simple molecules (e.g. I_2, $HgCl_2$, $AsCl_3$, BiI_3, $GeCl_4$, and OsO_4) are extracted with non-polar (inert) solvents such as benzene, chloroform, and carbon tetrachloride. The extraction of this type of compounds is comparatively selective [74,78,79,94,95].

Uncharged chelates are formed when metal ions react with bifunctional ligands [75,76,96,97], such as dithizone, 8-hydroxyquinoline, dithiocarbamates, dioximes, cupferron (ammonium salt of N-nitrosophenylhydroxylamine) (**I**), N-benzoyl-N-phenylhydroxylamine (BPHA) (p. 593) [98–100], acetylacetone (**II**) [101–103], and thenoyltrifluoroacetone (TTA) (**III**) [104,105].

78 Preconcentration and Separation [Ch. 3

(I) — N-phenyl-N-nitroso hydroxylamine ammonium salt (cupferron)

(II) — acetylacetone (H₃C-C(OH)=CH-C(=O)-CH₃)

(III) — thenoyltrifluoroacetone

Such chelates are extractable with non-polar solvents provided the ligand satisfies all the coordination sites. Synergism [106–110] is important in the extraction of some chelates, especially those in which water completes the coordination sphere, e.g. strontium oxinate dihydrate.

Increased selectivity can be obtained by using the exchange technique, in which a less-stable metal chelate is the source of the chelating agent [111].

The extraction of chelates is usually limited to the preconcentration and separation of small and trace amounts of metals. Owing to their low solubility in organic solvents, most chelates cannot be used for the extraction of macrocomponents. Cupferronates and acetylacetonates are exceptions.

A separate group of chelates comprises compounds of metal ions with alkyl- and aryl-phosphoric and thiophosphoric acids which are extractable from strongly acidic solutions into such solvents as carbon tetrachloride, chloroform, and toluene [78,112,113]. The following are examples of these chelating reagents: di-(2-ethylhexyl)phosphoric acid (HDEHP) **(IV)** and di-n-butyldithiophosphoric acid **(V)**:

(IV) — di-(2-ethylhexyl)phosphoric acid

(V) — di-n-butyldithiophosphoric acid

Ion-association complexes are formed by the electrostatic combination of oppositely-charged ions which have, in general, considerable molecular weights [77–79]. The ion-association extraction systems may be divided into the categories listed below.

Halometallic acids (e.g. $HFeCl_4$, $HSbCl_6$, $HAuBr_4$, and H_2CdI_4) are produced in the reactions of multivalent metal ions with hydrohalic acids. These compounds are extractable from acid solutions containing high concentrations of halide ions by oxygenated solvents such as ethers, higher alcohols, ketones, and esters [114–122]. The extraction of halometallic acids is made possible by the solvation of the protons with the oxygenated solvent and secondary solvation of the ion-pair formed, e.g. $[H(OBu_2)_3^+][FeCl_4^-]$. Since the solvent molecule coordinates through its oxygen atom, such systems are often called oxonium extraction systems.

Heteropoly acids [oxygen compounds of Mo(VI), W(VI), Si, P(V), As(V), V(V), Ge, and other elements] and their reduction products (molybdenum blues) are extracted with oxygenated solvents by a similar mechanism.

Another group of ion-association systems is represented by solvated salts (usually nitrates, but also halides and sulphates) [77,78]. Solutions of tri-n-butyl phosphate (TBP) (**VI**) in hexane, benzene, cyclohexane, carbon tetrachloride and solutions of tri-n-octylphosphine oxide (TOPO) (**VII**) in cyclohexane are predominantly used as the extractants.

$$\begin{array}{cc} C_4H_9O \diagdown & C_8H_{17} \diagdown \\ C_4H_9O-P{\rightarrow}O & C_8H_{17}-P{\rightarrow}O \\ C_4H_9O \diagup & C_8H_{17} \diagup \\ (\text{VI}) & (\text{VII}) \end{array}$$

Solvation with TBP or TOPO (through the strongly basic oxygen atom of the phosphoryl group) enables metal salts, such as $UO_2(NO_3)_2 \cdot 2TBP$, $Th(NO_3)_4 \cdot 2TBP$, $Zr(NO_3)_4 \cdot 2TOPO$, $UO_2Cl_2 \cdot 3TBP$, and $TiOSO_4 \cdot 2TOPO$, to be extracted with non-polar solvents.

High molecular-weight amines, including quaternary ammonium salts, are of particular importance in the extraction of ion-association compounds [123–127]. They form ion-association complexes with acids (e.g. HSCN, $HReO_4$, HNO_3, and HI), metal-complex acids [e.g. H_2PtCl_6, $HFeCl_4$, and $H_2UO_2(SO_4)_2$], and heteropoly acids. These complexes are extractable into non-polar solvents (benzene, chloroform, carbon tetrachloride) and polar solvents (methyl isobutyl ketone, amyl alcohol). Tertiary amines such as tribenzylamine (TBA), tri-iso-octylamine (TIOA), and tri-n-octylamine (TOA) are most commonly used.

Liquid amines which extract anions from aqueous solutions as ion-pairs are called liquid anion-exchangers [128]. They are marketed commercially under such trade-names as Aliquat-336 and Amberlite LA-1.

The extraction processes may be represented schematically:

$(R_3N)_o + H_{aq}^+ + A_{aq}^- \rightleftarrows \{[R_3NH^+][A^-]\}_o$ (extraction)

$\{[R_4N^+][A^-]\}_o + B_{aq}^- \rightleftarrows \{[R_4N^+][B^-]\}_o + A_{aq}^-$ (anion-exchange)

(A^- denotes the acid anion, and the subscripts o and aq denote the organic and aqueous phases, respectively).

Antipyrine and its derivatives [129–131] such as diantipyrylmethane (DAM) (**VIII**) are high molecular-weight amine extractants

$$\begin{array}{c} H_3C-C=\!\!=\!\!C-CH_2-C=\!\!=\!\!C-CH_3 \\ | \qquad | \qquad \qquad | \qquad | \\ H_3C-N\diagdown_N\diagup C{=}O \ \ O{=}C\diagdown_N\diagup N-CH_3 \\ | \qquad \qquad \qquad | \\ C_6H_5 \qquad \qquad \quad C_6H_5 \\ (\text{VIII}) \end{array}$$

In acid solutions, DAM is protonated, and the resulting cation combines with an anion to form an ion-pair (e.g. $[DAM \cdot H^+][TlBr_4^-]$ or

[DAM.H$^+$]$_2$ [CdI$_4^{2-}$]) which is extractable into inert solvents. In slightly acid or neutral media, DAM acts similarly to TBP and TOPO by solvating the metal ion, e.g. [Ti(DAM) (SCN)$_4$].

Non-solvated salts are ion-pairs formed by large unsolvated cations and unsolvated anions, and are extracted with inert solvents (e.g. carbon tetrachloride, chloroform, benzene, and toluene) [132–135]. Examples of large cations forming non-solvated salts are tetraphenylphosphonium, (C$_6$H$_5$)$_4$P$^+$, and tetraphenylarsonium, (C$_6$H$_5$)$_4$As$^+$, ions, basic dyes such as Rhodamine B and Methyl Violet, and complex cations, e.g. [Cu(neocuproine)$_2$]$^+$ and [Fe(phen)$_3$]$^{2+}$ (ferroin). These cations form ion-association complexes with anions such as ClO$_4^-$, ReO$_4^-$, SbCl$_6^-$, GaCl$_4^-$, CdI$_4^{2-}$, Fe(SCN)$_4^-$, SCN$^-$, and I$^-$.

3.2.1 Separation of Traces

The many elements which can be separated from each other and from the matrix in small and trace amounts by extraction are displayed in Table 3.5. The extractable compounds are also indicated for each element. An important role in concentrating traces by extraction is played by organic reagents which form chelates with metal ions [136].

Table 3.5 Separation of traces by solvent extraction

Be bd												B e				
Mg b												Al bd	Si ge	P g	S f	
Ca b	Sc bd	Ti bd	V bcd	Cr bdh	Mn bc	Fe bcd	Co ace	Ni acb	Cu acb	Zn acb		Ga bed	Ge eg	As ceg	Se ecf	Br f
Sr b	Y bd	Zr bd	Nb edb	Mo deb	Tc e	Ru he	Rh ech	Pd ace	Ag ac	Cd abc		In bae	Sn cde	Sb edc	Te ce	I f
Ba b	La bd	Hf bd	Ta ed	W ebg	Re eh	Os he	Ir ce	Pt ace	Au aec	Hg ace		Tl ace	Pb acb	Bi ace		
		Ce bd	Th bd	U bde												

The traces are extracted as:
a — dithizonates
b — 8-hydroxyquinolinates
c — dithiocarbamates
d — cupferronates
e — halogen compounds
f — elements
g — heteropoly acids
h — oxygen compounds

In Table 3.5, only the more important group reagents are mentioned. Metals which give sparingly-soluble sulphides can be extracted as complexes with ligands which bond through sulphur atoms (e.g. dithizone and dithiocarbamates). Ligands complexing through oxygen atoms (e.g. 8-hydroxyquinoline and cupferron) react preferentially with hydrolysable metal ions.

The use of dithizone can be illustrated by the group extraction of heavy-metal traces in the analysis of antimony [137], silicate minerals [138], various chemical reagents [139], and alkali-metal hydroxides [140].

Dithiocarbamates are applied to the separation of heavy metal traces in the analysis of biological ash [141], silicate minerals [142], chromium [143], and gallium phosphide [144].

Several trace elements in sea water can be concentrated by using 5,7-dibromo-8-hydroxyquinoline [145].

Some investigators have isolated metal traces by sequential extraction, using various reagents and suitable pH values. For example, dithizone and pyrrolidinedithiocarbamate have been used in the trace analysis of aluminium, titanium, zirconium, and their compounds, and selenium [146]; 8-hydroxyquinoline and diethyldithiocarbamate in the analysis of alkali metals [147]; 8-hydroxyquinoline and dithizone in the analysis of geological materials [148]; cupferron and pyrrolidinedithiocarbamate in the trace analysis of soils [149]. Pohl [150] has concentrated traces in the analysis of waters, plant materials, and minerals by using diethyldithiocarbamate, 8-hydroxyquinoline, and dithizone.

For the isolation and separation of several individual elements there exist highly selective, and even specific, extraction systems, such as nickel with dimethylglyoxime, and cobalt with 1-nitroso-2-naphthol.

The separation of Mo, In, Re, Ga, Fe, Tl, Au, Sb, and Sn chloride complexes from geological materials [151] is an example of the use of halide complexes in the extraction of small amounts of metals.

To prevent the co-extraction of the matrix with the minor components, the matrix is often converted into a stable complex, which remains in the aqueous solution during the extraction (the matrix is masked). The most common masking agents for individual elements are listed in Table 3.2.

Extraction is particularly useful in the separation of trace elements preconcentrated by precipitation, group extraction, or any other technique. The extraction and separation of the four chemically similar metals, Zn, Cd, Co, and Ni as dithizonates, is an example [152].

The scheme, outlined in Fig. 3.2 for the complete separation of a trace concentrate consisting of 18 metals is based on solvent extraction and coprecipitation methods. The scheme has been verified by the authors in the trace analysis of high-purity nitric and hydrofluoric acids which contained not more than $\sim 10^{-6}\%$ of any of the trace elements [153]. The procedure in the analysis performed according to this scheme is explained below.

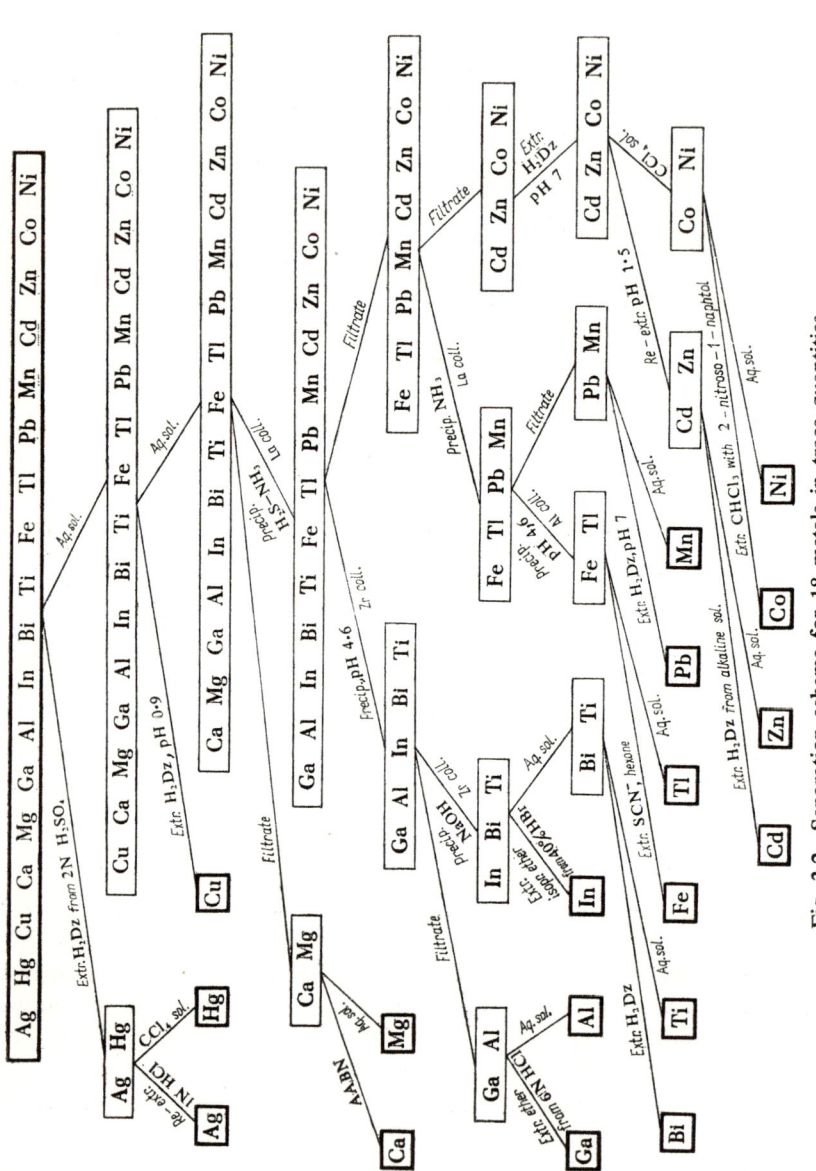

Fig. 3.2 Separation scheme for 18 metals in trace quantities

From the trace concentrate solution in $\sim 1M$ H_2SO_4, silver and mercury are co-extracted with dithizone and subsequently separated by the re-extraction of silver with dilute hydrochloric acid. Both metals are determined with dithizone. Next, from a less acidic medium (pH ~ 1), copper is extracted and determined with dithizone. Most other metals are then separated from the calcium and magnesium by coprecipitation as sulphides or hydroxides with lanthanum. Calcium is separated from magnesium by extraction with Azo-azoxy BN reagent (p. 183), and determined with glyoxal bis(2-hydroxyanil) (GBHA). Magnesium is determined with Eriochrome Black T.

From the solution, which contains the major group of metals, Ga, Al, In, Bi, and Ti are precipitated with zirconium as collector (pH 4·6). Iron and thallium, previously reduced to Fe(II) and Tl(I), remain in the filtrate. Gallium and aluminium are separated as soluble hydroxides from In, Bi, and Ti. Gallium is isolated from aluminium by extraction of the chloride complex (from $6M$ HCl) with ether. Gallium is determined with Rhodamine B, and aluminium with Eriochrome Cyanine R. Indium is isolated from Bi and Ti by extraction with di-isopropyl ether as a bromide complex, and determined with PAR. Bismuth is separated from titanium and determined with dithizone. Titanium is determined by the thiocyanate method.

With excess of ammonia, Fe(III), Tl(III), Pb, and Mn are coprecipitated as hydroxides with lanthanum as collector, and separated from Cd, Zn, Co, and Ni which form soluble ammine complexes. After the hydroxides have been dissolved in acid, iron and thallium hydroxides are reprecipitated at pH 4·6 (with aluminium as collector). Iron is separated from thallium by extraction, and determined by the thiocyanate method, while thallium is determined with Rhodamine B. Lead is separated from manganese and determined with dithizone. Manganese is determined with PAN.

After Cd, Zn, Co, and Ni have been converted into dithizonates (in CCl_4), cadmium and zinc are separated by re-extraction with dilute hydrochloric acid. Cadmium is extracted from alkaline solution and determined with dithizone, and zinc is determined in the residual solution with dithizone. Cobalt is extracted from nickel and determined as the 2-nitroso-1-naphtholate. The remaining nickel is determined with α-furildioxime.

These spectrophotometric methods are described in detail in the second part of this book, in the chapters dealing with particular elements.

The complete separation of elements before their determination makes it possible to determine many elements in a single sample. Such methods require only small samples, thereby reducing the cost of the sample required for analysis of high-purity compounds (which can be very expensive).

The procedure is simplified if fewer trace elements are to be analysed.

Later experience [153] proved that it is more convenient to separate and determine bismuth immediately after copper has been determined, by extracting bismuth with dithizone from the solution at pH 3·1 (± 0.1).

3.2.2 Separation of Macrocomponents

The elements which can be extracted in larger quantities are shown in Table 3.6.

The most numerous group comprises the halide and pseudohalide complexes listed below:

fluoride (Ta, Nb, Sn)
chloride [Fe(III), Sb, As, Ga, Ge, Au, Mo, Tl]
bromide (Au, In, Tl, Ga)
iodide (Bi, Sb, Cd, Hg, Sn)
thiocyanate [Zn, Co, Fe(III), Ti, Mo, U].

There are numerous examples of the extraction of macrocomponents as chloride complexes in the analysis of various materials. The extraction of iron (III) from hydrochloric acid medium, before the determination of trace elements, has been thoroughly investigated [154–159]. Molybdenum has been extracted from $6M$ HCl with diethyl ether [70]; antimony (V) from hydrochloric acid medium with butyl acetate [160] or di(2-chloroethyl) ether [161]; gallium macroquantities from 6–$7M$ HCl with di(2-chloroethyl) ether [161] or butyl acetate [162]; thallium with di(2-chloroethyl) ether [161]; gold with diethyl ether [163] or isoamyl acetate [164]; and indium with di-isopropyl ether or isoamyl acetate [165].

Table 3.6 Separation of the matrix by solvent extraction

	Be af															
											Al fe				S ca	
	Sc ba	Ti ae	V afe	Cr df	Mn f	Fe aef	Co ae	Ni ae	Cu ae	Zn ae	Ga af	Ge a	As a	Se ca	Br c	
	Y be	Zr ef	Nb ae	Mo ae		Ru ad	Rh a	Pd af	Ag a	Cd a	In af	Sn a	Sb ae	Te a	I c	
	La e	Hf ef	Ta ae	W ae	Re da	Os ad	Ir a	Pt a	Au ab	Hg ae	Tl af	Pb ae	Bi ae			
			Ce^{IV} bf	Th bf	U bef											

The macrocomponents are extracted as:
a — halogen compounds
b — nitrates
c — elements
d — oxygen compounds
e — cupferronates
f — acetylacetonates

An example of the application of fluoride complexes is the extraction of tantalum macroquantities into cyclohexanone [166].

As the bromide complexes, indium has been extracted with diethyl ether [144] or isopropyl ether [165], and gold (from $3M$ HBr) with di-isopropyl ether [167].

Jackwerth [168] has separated macroquantities of mercury, bismuth, and cadmium as iodide complexes by fractional extraction. The mercury, bismuth, and cadmium are extracted quantitatively from the aqueous phase with cyclohexane by adding sufficient hydriodic acid to form HgI_2 and HgI_3^-, BiI_3 and BiI_4^-, and CdI_2 and CdI_3^- complexes. The two complex species of each element are extracted.

Only those organic reagents, such as cupferron and acetylacetone, which form chelates which are highly soluble in non-polar organic solvents can be used in the extraction of matrix elements. In analysis of niobium, tantalum, molybdenum, and tungsten for manganese, for example, the matrix metals are extracted as cupferronates with chloroform [169]. Titanium macroquantities are extracted similarly [170].

Smaller groups of matrix elements may be extracted as nitrate complexes [U, Th, Ce(IV), Y] or oxides (Os, Ru). Matrix yttrium is extracted with TBP from $12M$ nitric acid [171].

Extraction methods which are specific for certain elements are also known. Beryllium, for example, may be extracted with chloroform as an acetate complex [172], niobium with TBP in benzene as a sulphate complex (from $10M\ H_2SO_4$) [173], and chromium(VI) with ethers as a peroxide compound.

Carboxylic acids are particularly useful reagents for separation of matrix elements, as they have very high capacity [174].

3.3 Volatility

These methods for separating elements are based on differences in the vapour pressures of individual elements and their compounds. Covalent compounds are generally fairly volatile whereas ionic compounds are not, though polymeric covalent species (e.g. diamond) are non-volatile. Covalently bonded compounds are also more soluble in non-polar solvents. Therefore, volatile compounds bear certain resemblances to those soluble in non-polar solvents. Examples are $AsCl_3$, $GeCl_4$, OsO_4, and certain volatile chelates (the last are just beginning to be used for separations). Gas chromatography may be regarded as a special case of volatilization separation.

3.3.1 Separation of Traces

Unlike precipitation and solvent extraction, volatilization is usually utilized for separating individual trace elements from the sample before their determination, rather than for concentrating groups of trace elements. The pertinent elements and their volatile species are shown in Table 3.7. These methods are concerned principally with non-metallic and amphoteric elements which have high vapour pressures when in the elemental form (e.g. Cl, Br, S), or in compounds with halogen (e.g. SiF_4), hydrogen, or oxygen. Other volatilization methods exist for the separation of certain elements (e.g. the distillation of boron as methyl borate [175]).

Table 3.7 Volatilization of traces

																B	C	N		F
																d	b	cb		c

Be												B	C	N		F
d												d	b	cb		c
												Al	Si	P	S	Cl
												d	d	c	cb	ac
	Ti		Cr	Mn	Fe			Zn	Ga	Ge	As	Se	Br
	d		d	b	d			a	d	d	cd	d	a
	Zr	Nb	Mo		Ru			Cd	In	Sn	Sb	Te	I
	d	d	d		b			a	a	d	cd	d	a
	Hf	Ta		Re	Os			Hg	Tl	Pb			
	d	d		bd	b			ad	a	a			

The traces are volatilized as:
a — elements
b — oxygen compounds
c — hydrogen compounds
d — halogen compounds

A method for enriching traces of the more volatile metals (e.g. Zn, Cd, Tl, In, Pb) by heating samples in quartz tubes to $\sim 1000°C$ in a stream of hydrogen has been devised [176]. The sublimed metals are collected on a cold-finger. Small amounts of beryllium are separated as the fluoride after the sample has been mixed with CeF_3 and heated in a quartz tube at 1000°C in a humid stream of nitrogen [177].

Volatile acetylacetonates and other β-diketonates can be used for the separation of metals [178–181]. Over the temperature gradient from 15 to 170°C, 21 metal acetylacetonates have been recrystallized in discrete zones.

The boiling points of anhydrous Sn, Ti, Al, Ga, Ta, Nb, Mo, Fe, Hf, and Zr chlorides range from 114 to 331°C. This permits the separation of these metals by gas chromatography [182].

The Conway micro-diffusion method [183] is well suited to the separation of e.g. HF, SiF_4, or HCN.

The separation of traces is carried out in a closed system and involves absorbing the traces in a suitable absorbent, for example hydrogen sulphide in a zinc acetate solution (in the Methylene Blue method for determining sulphur, see p. 506), ammonia in dilute HCl, and methyl borate in dilute NaOH solution.

Except in the micro-diffusion method, a carrier gas is indispensable, e.g. hydrogen, nitrogen, chlorine, or water vapour.

Selectivity in the distillation of trace elements may be enhanced by masking interfering elements as less-volatile complexes. Germanium and arsenic, for example, may be separated from tin by distillation as the chlorides after tin(IV) has been masked as the non-volatile phosphate complex.

3.3.2 SEPARATION OF MACROCOMPONENTS

If the macrocomponents can be removed by volatilization (distillation, sublimation) without introducing large quantities of reagents, the trace elements may be greatly concentrated. This favourable situation, in which relatively large samples can be used for trace analyses, arises when the sample is volatile, especially when it is a liquid such as water, an organic solvent, a volatile acid, or an ammonia solution.

During evaporation of the macrocomponent, substances are sometimes added to reduce the volatility of the trace components which are to be retained in the vessel. When volatile samples are being evaporated (e.g. water, hydrochloric acid, nitric acid [184]), a few drops of concentrated sulphuric acid are usually added, and the solution is evaporated till fuming. This technique reduces losses of volatile metal chlorides.

To minimize losses of volatile chlorides, hydrochloric acid solutions are sometimes evaporated in a closed distillation apparatus equipped with a reflux condenser, rather than in an open vessel [185].

Attention should be paid to the behaviour of the substance when present in trace amounts. For example, during the evaporation of a solution containing large quantities of boric acid, losses due to volatilization are negligible. If, however, only microgram amounts of boron are present, some is lost because of the volatility of boric acid [186]. Such a solution, therefore, should not be evaporated unless made alkaline.

In certain cases, the trace element is kept in solution as a less-volatile complex during the distillation of the sample matrix. In the determination of traces of boron in chlorosilanes, silicon is removed as the volatile fluoride complex. To prevent the formation of volatile BF_3, mannitol is added to form the non-volatile complex with boron [187].

The principal forms in which major elements can be separated by distillation or sublimation are shown in Table 3.8.

Chlorine, bromine, iodine, sulphur and mercury are distilled as the elements.

Examples of the distillation of major elements as the fluoride complexes can be found in the trace analysis of silicon [187,188] and boron [175].

Table 3.8 Volatilization of macrocomponents

1	2	3	4	5	6	7	8	9	10	11	12	13	14	15	16	17
	Be e											B db	C bc	N abc		F ac
												Al d	Si d	P d	S abc	Cl ac
			Ti d	V d	Cr d		Fe d					Ga d	Ge d	As d	Se db	Br ac
			Zr d	Nb d	Mo d		Ru b				Cd a	In d	Sn d	Sb d	Te d	I ac
			Hf d	Ta d	W d	Re b	Os b				Hg ad					

The macrocomponents are volatilized as:
a — elements
b — oxygen compounds
c — hydrogen compounds
d — halogen compounds
e — acetylacetonates

Tin can be sublimed as the iodide [189]. The matrices are removed as volatile chloride complexes in the analysis of chromium (CrO_2Cl_2) [190,191], and germanium [185], and as bromide complexes in the trace analysis of selenium [192], silicon [193], and tin [194].

The fluorides formed by heating metal oxides (at 140–400°C) in a stream of hydrogen fluoride gas can be removed by volatilization [195]. The following volatile fluorides are said to be formed under such conditions: AsF_5 (b.p. −53°C), GeF_4 (−35°C), TeF_6 (−35°C), SeF_6 (−34°C), WF_6 (19°C), MoF_6 (35°C), VF_5 (148°C)*, NbF_5 (229°C), TaF_5 (230°C), TiF_4 (284°C). Non-volatile fluoride complexes are formed by Al, Fe, Mg, and Mn.

In the trace analyses of titanium [170], and zirconium [198], volatile $TiCl_4$ and $ZrCl_4$ are sublimed after heating the samples with chlorine.

In the trace analysis of rubidium and caesium arsenates, arsenic is distilled off as arsine [199].

At a temperature > 315°C, selenium volatilizes as the oxide, SeO_2 [200]. Boron forms the volatile oxide, B_2O_3, when heated to 600°C in a stream of water vapour [201].

* The boiling point of VF_5 is variously reported as 48°C [196], 111°C [197], and 148°C [195].

To separate the matrix in the trace analysis of nickel, Shvarts [202] has converted nickel into the volatile carbonyl, $Ni(CO)_4$, by passing carbon monoxide over a finely-ground sample in an autoclave at a pressure of 150 atmospheres.

In the trace analysis of high-purity cadmium, the matrix can be separated by distillation at 630°C [203].

By heating aluminium metal with ethyl bromide, Neeb [204] has converted the aluminium into ethylaluminium bromide, a liquid which boils at 130°C under reduced pressure. Traces of Fe, Cu, Co, Ag, Pb, Mn, Cr, and Ni remain in the residue quantitatively.

Mineralization of organic samples [205–211], which precedes the determination of inorganic components, is also an example of the separation of the major components by volatilization (principally as CO_2 and H_2O). In the mineralization of organic substances by dry ashing (combustion), the mineral residue may be so small and light that considerable losses result from the formation of "volatile" aerosols. These losses are prevented by adding a mineral collector, e.g. by wetting the sample with a $Mg(NO_3)_2$, K_2SO_4, or Na_2CO_3 solution. The collector must be neither volatile nor low-melting. The temperature should not exceed 400–500°C to prevent distillation of the more volatile elements [212–214]. At 400°C As and Hg volatilize completely, and Ag, Au, Fe, Sb, Zn, and Pb partially volatilize.

Mercury is lost during wet mineralization with acids. It volatilizes on heating a sulphuric acid medium to fumes.

Smith and Diehl [215,216] have reported a number of modifications of the mineralization method for organic samples, using perchloric acid as the major oxidant. A 50% hydrogen peroxide solution has been recommended for the decomposition of difficultly-oxidizable samples (e.g. nylon and polyethylene) [217–220]. The oxygen bomb method has also been suggested for the decomposition of organic materials before trace element analysis [221]. Nitric acid vapour has also been used for decomposing plant material [222].

3.4 Ion-Exchange

Separation and concentration of trace elements can be achieved by using organic synthetic ion-exchangers. Ion-exchange processes are based on the reversible exchange of ions between the solid ion-exchange resin and a liquid eluent. Separation depends on the differences in stability of complexes, and the associated differences in distribution coefficients.

Detailed information about the ion-exchange process and about separation methods using ion-exchangers can be found in the excellent monographs by Samuelson [223], Trémillon [224], Korkisch [5], Rieman and Walton [225], Marcus and Kertes [77], and also in several review papers [226–231].

Generally, ion-exchangers consist of a matrix of cross-linked polymerized hydrocarbons which contains ionizable functional groups. The functional groups in cation-exchangers are —SO_3H, —COOH, and —OH; those in anion-exchangers are —N^+R_3, —NR_2, —NHR, and —NH_2.

A typical strongly acidic cation-exchanger is prepared by sulphonation of a copolymer consisting of polystyrene cross-linked with divinylbenzene (DVB). The DVB content of the polymerization reaction mixture determines the degree of cross-linking, and is, therefore, important in deciding the properties of the exchanger. For example, Dowex 50-X8 cation exchanger contains 8% DVB.

The hydrogen ion of acid groups is exchanged for other cations in the exchange process. The cation-exchanger may be used in the hydrogen (H^+), sodium (Na^+), or similar form, depending on the cation attached to its acid groups. Strongly acidic character is imparted to cation-exchangers by sulphonic acid groups.

In anion-exchangers, strongly basic character is imparted by quaternary amines. Anion-exchangers are most often used in the hydroxyl (OH^-) or the chloride (Cl^-) form.

Ion-exchange is usually carried out by the column technique in separation methods. An ion-exchange column fitted with a capillary to prevent column drainage is shown in Fig. 3.3. The swelled resin is placed in a glass, quartz, or plastic column, and the solution under investigation is run onto the column. In simple ion-exchange separations, some components of the

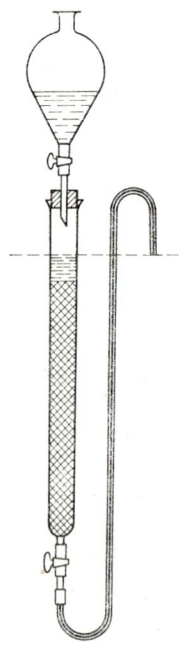

Fig. 3.3 Ion-exchange column

solution are retained in the column while others are eluted. In ion-exchange chromatography, the components retained by the ion-exchanger can often be separated by treatment with suitable eluents, the separation being achieved by pH-control and/or use of complexing agents.

The size of the column (resin bed) is selected according to the quantity of ions to be retained in the column. The resin bed depth in the column should be 10–20 times its diameter. Columns 8–10 mm in inner diameter are often used in laboratories. When the quantities of ions retained are in the microgram or milligram range, columns 3–5 mm in inner diameter are sufficient. The particle size of the resin is selected according to the column dimensions. In columns of the dimensions given above, 30–50 mesh or 100–200 mesh resins may be used (see [232] for details).

The *ion-exchange capacity* of the column depends on the quantity of resin in the column and on the ion-exchange capacity of the resin used. The latter is a measure of the ability of the exchanger to bond counter-ions, and totals 4–5 meq of ionic species per gram of dry resin for strongly-acidic cation-exchangers, and 3–4 meq per gram for strongly-basic anion-exchangers. In point of fact, it is not the overall capacity of the ion-exchange column which is important, but the break-through capacity, which is lower and depends on the column shape, resin particle size, elution rate, and other experimental conditions. The break through capacity is the actual maximum capacity for retention of the ion on the column.

The *distribution coefficient* (D) for a particular element, on a given ion-exchanger, and in a given medium, is the ratio of the element concentration in the resin bed phase, C_r (mmole/g of dry resin), to the element concentration in the solution at equilibrium with the ion-exchanger, C_{aq} (mmole/ml of solution)

$$D = \frac{C_r}{C_{aq}}$$

The ability of an ion-exchanger to retain an element from a solution of a particular complexing agent can be represented by a graph of log D vs. complexing agent concentration. As an example, the variation of the distribution coefficient of zinc on a strongly basic anion-exchanger with the concentration of hydrochloric acid is shown in Fig. 3.4. Reagents which form complexes with ions radically alter the affinity of ion-exchangers towards these ions [233].

The use of liquid ion-exchangers in separation methods is increasing [128,234].

More selective ion-exchangers incorporating chelating ligands as functional groups have been developed [235–241]. Redox ion-exchangers are also used in separation methods [242].

Recently, ion-exchange methods in mixed aqueous–organic solvents have been the subject of much research. A comprehensive survey of these methods is given by Moody and Thomas [243]. The addition of organic

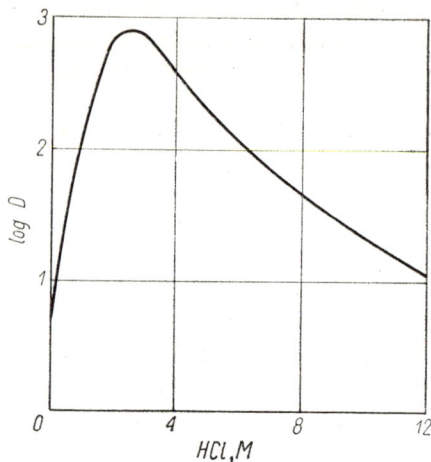

Fig. 3.4 Distribution coefficients of zinc in HCl solutions on a strongly basic anion-exchanger, Dowex 1

solvents to the aqueous system modifies the affinity of the ions for ion-exchangers by changing the solvation of the ions, decreasing the dielectric constant of the system, and stabilizing certain complexes which are very weak in aqueous solution.

3.4.1 Use of Cation-Exchangers

Strongly acidic cation-exchangers such as Dowex 50, Amberlite IR-120, and KU-2 are most often used in ion-exchange separations of elements. Detailed examples are given in the sections dealing with the individual elements.

The behaviour of elements on strongly acidic cation-exchangers in hydrohalic acid solutions has been extensively investigated.

Distribution curves (cf. Fig. 3.4) have been reported for individual elements on a strongly acidic cation-exchanger in hydrochloric acid medium [244–246]. With $0.1–0.5M$ HCl as irrigant, platinum metals, Au, Hg, As, and Se are not retained by the cation-exchanger. New possibilities for metal separations are created by mixed media, such as HCl–water–organic solvent [247–254]. The following water-miscible solvents are used in various ratios to water: acetone [247,248,253,254], methanol [249,250], ethanol [249,252], dimethylformamide [250], tetrahydrofuran [249], dimethyl sulphoxide (DMSO) [251], and other organic solvents.

The behaviour of elements on strongly acidic cation-exchangers in hydrofluoric acid medium has also been thoroughly investigated. Al, Cd, Mo(VI), Nb, Sc, Sn(IV), Ta, Ti, U(VI), W, and Zr can be eluted from a cation-exchange column with $0.1M$ HF. These metals, as well as As, Ba, Cr, Fe(III), Ga, Hg, Mn, Sb, V, and Zn, can be eluted with $1M$ HF [255,256]. Ag, Co, Cu, Mg, Mn, Ni, Pb, and Zn in a solution of steel in dilute hydrofluoric acid, are retained on Dowex 50, and eluted with $4M$ HNO_3 [257]. Korkisch and Huber [258] have investigated the cation-exchange behaviour of several elements in HF–water–organic solvent mixed media.

A detailed survey has been made of the behaviour of metals on cation-exchangers in $0.1–12M$ HBr [259]. Irrespective of the HBr concentration, platinum metals, Hg, Pb, Bi, Sb, As, Se, and Te are not retained in the column. Korkisch and Klakl [260] have investigated the cation-exchange behaviour of several elements in HBr–organic solvent media.

The following complexing agents have also been used for the separation of elements on strongly acid cation-exchangers: thiocyanate [261], thiocyanate in the presence of various organic solvents [262], sulphate [263], oxalate [264], acetate [265], nitrite [266], sulphosalicylate [267], and various complexones (EDTA, NTA, EGTA, DTPA) [268]. Perchloric acid [246,269] and nitric acid (in organic solvents) [250,270] have also been used as eluents. Of the cations retained by Dowex 50 from tetrahydrofuran–$1M$ HNO_3 (19:1) medium, a $0.01M$ dithizone solution in the same solvent elutes only Ag, Cu, and Bi [271].

Precipitation ion-exchange is an interesting variation which has been used to preconcentrate trace elements in the analysis of sodium, potassium, barium, and strontium chlorides [272]. The cation-exchanger retains both matrix and trace metals from the salt solutions. When concentrated hydrochloric acid is run onto the column, NaCl, KCl, $BaCl_2$, and $SrCl_2$ are precipitated in the resin bed, while trace metals are eluted.

3.4.2 USE OF ANION-EXCHANGERS

Many examples of the use of strongly basic anion-exchangers, e.g. Dowex 1, Amberlite IRA-400, and AV-17 can be found later, in the sections devoted to individual elements.

The behaviour of elements on strongly basic anion-exchangers in hydrochloric acid medium has been studied [273,274]. From $4–6M$ HCl medium, the following elements are retained: Au, Bi, Cd, Fe(III), Ga, Hg, Mo, Re, Sb, Sn, Tl(III), U(VI), W, Zn, and platinum metals. An example of preconcentration of microtraces is the retention by Amberlite IRA-400 of Au, Bi, and Cd from sea water acidified to be $0.1M$ in HCl [274]. The behaviour of several elements in mixed solvents has been studied [250,275,276].

Strongly basic anion-exchangers retain Al, As, B, Be, Bi, Fe(III), Hf, Mo, Nb, Re, Sc, Sb, Sn, Ta, Te, Ti, U, W, and Zr, from solutions of various hydrofluoric acid concentrations [256,277]. Schemes have been developed for the preconcentration and separation of Ti, W, Mo, Nb, Ta, V, and Zr, using HF and/or HCl as eluent [278,279]. The behaviour of a number of elements in a mixed HF–HCl medium has been examined [280].

Anion-exchange data for various elements in nitric acid and sulphuric acid media can be found in references [281–283] and [284–286], respectively. Mixed eluents containing various organic solvents are also described [283,284,286].

Comprehensive studies have been carried out of the behaviour of elements on strongly basic anion-exchangers in hydrobromic acid [287], acetic acid [288,289], oxalic acid [290,291], tartrate medium [292], nitrite medium [266], and with different complexones [268].

De Gélis [293] has devised exhaustive schemes for the isolation of iron and the separation of its alloying constituents, Al, Bi, Co, Cr, Mn, Mo, Nb, Ni, Sn, Ta, Ti, V, W, and Zr. Columns with both a strongly basic anion-exchanger and a strongly acidic cation-exchanger are used with various media, such as HF, HCl, and H_2SO_4.

Various metals have been separated on weakly basic ion-exchangers in chloride [294] or thiocyanate media [295].

References

1. De, A. K., *Separation of Heavy Metals*, Pergamon, London (1961).
2. Berg, E. W., *Physical and Chemical Methods of Separation*, McGraw-Hill, New York (1963).
3. Dean, J. A., *Chemical Separation Methods*, Van Nostrand-Reinhold, London (1969).
4. Korkisch, J., *Modern Methods for the Separation of Rarer Metal Ions*, Pergamon, London (1969).
5. Minczewski, J., Chwastowska, J. and Dybczyński, R., *Analiza śladowa. Metody rozdzielania i zagęszczania* (Trace Analysis. Separation and Enrichment Methods), WNT, Warsaw (1973).
6. Specker, H. and Hartkamp, H., *Angew. Chem.* **67**, 173 (1955).
7. West, T. S., *Anal. Chim. Acta* **25**, 405 (1961).
8. Marczenko, Z., *Chem. Anal. (Warsaw)* **11**, 347 (1966).
9. Minczewski, J., *Pure Appl. Chem.* **10**, 567 (1965); *Preconcentration in Trace Analysis in*: Meinke, W. W. and Scribner, B. F. (eds.), *Trace Characterisation, Chemical and Physical*, NBS, Washington (1967).
10. Goryushina, V. G., *Zavodsk. Lab.* **37**, 513 (1971).
11. Chalmers, R. A., *Pure Appl. Chem.* **31**, 569 (1972).
12. Irving, H. M., *Z. Anal. Chem.* **263**, 264 (1973).
13. Melikhov, I. V. and Merkulova, M. S., *Tr. Komis. po Analit. Khim. Akad. Nauk SSSR* **15**, 244 (1965).
14. Walton, A. G., *The Formation and Properties of Precipitates*, Interscience, New York (1967).
15. Lieser, K. H., *Angew. Chem., Int. Ed. Engl.* **8**, 188 (1969).
16. Chuiko, V. T., *Zh. Neorgan. Khim.* **2**, 685 (1957).
17. Rudnev, N. A. and Malofeeva, G. I., *Tr. Komis. po Analit. Khim. Akad. Nauk SSSR* **15**, 224 (1965).
18. Rudnev, N. A. and Malofeeva, G. I., *Zh. Analit. Khim.* **19**, 151, 785 (1964); *Talanta* **11**, 531 (1964).
19. Doolan, K. J. and Smythe, L. E., *Talanta* **20**, 241 (1973).
20. Shannon, R. D. and Prewitt, C. T., *Acta Cryst.*, B **25**, 925 (1969).
21. Lavrukhina, A. K., *Zh. Analit. Khim.* **10**, 203 (1955); **12**, 41 (1957).
22. Marczenko, Z. and Kasiura, K., *Chem. Anal. (Warsaw)* **9**, 87 (1964).
23. Marczenko, Z., *Chim. Anal. (Paris)* **46**, 286 (1964).

24. Püschel, R. and Lassner, E., *Mikrochim. Acta* **1965**, 751.
25. Plotnikov, V. I. and Kochetkov, V. L., *Zh. Analit. Khim.* **23**, 377 (1968).
26. Tiptsova, V. G., Dvortsan, A. G. and Golitsyna, M. I., *ibid.* **23**, 1684 (1968).
27. Ko, R. and Anderson, P., *Anal. Chem.* **41**, 177 (1969).
28. Strohal, P., Molnar, K. and Bačić, I., *Mikrochim. Acta* **1972**, 586.
29. Lebedinskaya, K. P. and Chuiko, V. T., *Zh. Analit. Khim.* **28**, 863 (1973).
30. Reichel, W. and Bleakley, B. G., *Anal. Chem.* **46**, 59 (1974).
31. Burke, K. E., *ibid.* **42**, 1536 (1970).
32. Tiptsova-Yakovleva, V. G., Dvortsan, A. G. and Semenova, I. B., *Zh. Analit. Khim.* **25**, 686 (1970).
33. Pohl, F. A., *Angew. Chem.* **66**, 603 (1954); *Z. Anal. Chem.* **142**, 19 (1954).
34. Filimonov, L. N. Makulov, N. A. and Zakharova, Z. A., *Tr. Komis. po Analit. Khim. Akad. Nauk SSSR* **12**, 227 (1960).
35. Marczenko, Z. and Kasiura, K., *Chem. Anal. (Warsaw)* **10**, 449 (1965).
36. Kasiura, K. and Marczenko, Z., *Zh. Analit. Khim.* **22**, 1398 (1967).
37. Cohen, A. I. and Gordon, L., *Talanta* **7**, 195 (1961).
38. Chuiko, V. T. and Reva, N. I., *Zavodsk. Lab.* **33**, 1503 (1967).
39. Kovaleva, N. V. and Chuiko, V. T., *Zh. Analit. Khim.* **28**, 1985 (1973).
40. Tarasevich, N. I., Khlystova, A. D. and Semenenko, K. A., *Tr. Komis. po Analit. Khim. Akad. Nauk SSSR* **15**, 263 (1965).
41. Marczenko, Z. and Stępień, A., *Chem. Anal. (Warsaw)* **5**, 247 (1960).
42. Lyle, S. J. and Southern, D. L., *Talanta* **11**, 1239 (1964).
43. Bailey, T. H. and Lyle, S. J., *ibid.* **12**, 563 (1965).
44. Dehm, R. L., Dunn, W. G. and Loder, E. R., *Anal. Chem.* **33**, 607 (1961).
45. Silvey, W. D. and Brennan, R., *ibid.* **34**, 784 (1962).
46. Cruft, E. F. and Husler, J., *ibid.* **41**, 175 (1969).
47. Grushina, N. V., Tsevun, V. I., Khrapchenkova, G. V., Erdenbaeva, M. I. and Kozin, L. F., *Zh. Analit. Khim.* **21**, 980 (1966).
48. Kozlovskii, M. T., *Tr. Komis. po Analit. Khim. Akad. Nauk SSSR* **15**, 132 (1965).
49. Kuznetsov, V. I., *Zh. Analit. Khim.* **9**, 199 (1954); *Tr. Komis. po Analit. Khim. Akad. Nauk SSSR* **15**, 279 (1965).
50. Korenman, I. M., *Usp. Khim.* **23**, 89 (1954); *Zavodsk. Lab.* **22**, 146 (1956).
51. Tappmeyer, W. P. and Pickett, E. E., *Anal. Chem.* **34**, 1709 (1962).
52. Myasoedova, G. V., *Zh. Analit. Khim.* **21**, 598 (1966).
53. Lisetskaya, G. S., Freger, S. V. and Zaionchkovskaya, E. K., *Zavodsk. Lab.* **33**, 282 (1967).
54. Püschel, R., Lassner, E., Martin, W. and Martin, D., *Mikrochim. Acta* **1969**, 145.
55. Ringbom, A., *Complexation in Analytical Chemistry*, Wiley, New York (1963).
56. Schwarzenbach, G., *Experientia*, Suppl. No. 5, 162 (1956); *Angew. Chem.* **70**, 451 (1958).
57. Marczenko, Z. and Kasiura, K., *Chem. Anal. (Warsaw)* **7**, 775 (1962).
58. Babina, F. L., Karabash, A. G., Peizulaev, Sh. I. and Semenova, E. F., *Zh. Analit. Khim.* **20**, 501 (1965).
59. Barabas, S. and Lea, S. G., *Anal. Chem.* **37**, 1132 (1965).
60. Gladyshev, V. P., Synkova, D. P., Enikeev, R. Sh., Kucherenko, N. A. and Voilokova, V. V., *Tr. Komis. po Analit. Khim. Akad. Nauk SSSR* **15**, 213 (1965).
61. Bock, R. and Hackstein, K. G., *Z. Anal. Chem.* **138**, 339 (1953).
62. Meyer, J., *ibid.* **219**, 147 (1966); **231**, 241 (1967).
63. Karabash, A. G., Bondarenko, L. S., Morozova, G. G. and Peizulaev, Sh. I., *Zh. Analit. Khim.* **15**, 623 (1960).
64. Degtyareva, O. F. and Ostrovskaya, M. F., *ibid.* **20**, 814 (1965).
65. Ustimov, A. M. and Tember, G. A., *Zavodsk. Lab.* **35**, 1440 (1969).
66. Krauz, L. S., Karabash, A. G., Peizulaev, Sh. I., Lipatova, V. M. and Moleva, V. S., *Tr. Komis. po Analit. Khim. Akad. Nauk SSSR* **12**, 175 (1960).
67. Baranova, L. L. and Solodovnik, S. M., *Zh. Analit. Khim.* **19**, 588 (1964).

68. Mizuike, A., Kawaguchi, H. and Kono, T., *Mikrochim. Acta* **1970**, 1095.
69. Chalkov, N. Ya. and Ustimov, A. M., *Zavodsk. Lab.* **37**, 149 (1971).
70. Karabash, A. G., Samsonova, Z. N., Smirnova-Averina, N. I. and Peizulaev, Sh. I., *Tr. Komis. po Analit. Khim. Akad. Nauk SSSR* **12**, 255 (1960).
71. Bondarenko, L. S., Sotnikova, N. P., Karabash, A. G. and Peizulaev, Sh. I., *Zavodsk. Lab.* **25**, 1476 (1959).
72. Gebauhr, W., *Z. Anal. Chem.* **197**, 212 (1963).
73. Jackwerth, E. and Schmidt, W., *ibid.* **225**, 352 (1967).
74. Morrison, G. H. and Freiser, H., *Solvent Extraction in Analytical Chemistry*, Wiley, New York (1957).
75. Starý, J., *The Solvent Extraction of Metal Chelates*, Pergamon, Oxford (1964).
76. Zolotov, Yu. A., *Ekstraktsiya vnutrikompleksnykh soedinenii* (Extraction of Chelate Compounds), Nauka, Moscow (1968).
77. Marcus, Y. and Kertes, A. S., *Ion Exchange and Solvent Extraction of Metal Complexes*, Wiley, London (1969).
78. De, A. K., Khopkar, S. M. and Chalmers, R. A., *Solvent Extraction of Metals*, Van Nostrand-Reinhold, London (1970).
79. Zolotov, Yu. A. and Kuz'min, N. M., *Ekstraktsionnoe kontsentrirovanie* (Concentration by Extraction), Khimiya, Moscow (1971).
80. Babko, A. K. and Zharovskii, F. G., *Zavodsk. Lab.* **25**, 42 (1959); **28**, 1287 (1962).
81. Green, H., *Metallurgia* **70**, 143, 201, 254, 299 (1964).
82. Zolotov, Yu. A., *Tr. Komis. po Analit. Khim. Akad. Nauk SSSR* **15**, 3 (1965); *Zh. Analit. Khim.* **22**, 1644 (1967).
83. Alimarin, I. P. and Zolotov, Yu. A., *Chem. Anal. (Warsaw)* **13**, 941 (1968).
84. Bagreev, V. V., Zolotov, Yu. A., Kurilina, N. A. and Kalinina, G. F., *Ekstraktsiya neorganicheskikh soedinenii. Bibliograficheskii ukazatel*: *1945–1967* (Extraction of Inorganic Compounds. Bibliographical Index: 1945–1967), Izdat. Nauka, Moscow (1971).
85. Zolotov, Yu. A., Iofa, B. Z. and Chuchalin, L. K., *Ekstraktsiya galogenidnykh kompleksov metallov* (Extraction of Halide Metal Complexes), Izdat. Nauka, Moscow (1973).
86. Zolotov, Yu. A., Alimarin, I. P. and Bodnya, V. A., *Zh. Analit. Khim.* **19**, 28 (1964).
87. Alimarin, I. P., Zolotov, Yu. A. and Bodnya, V. A., *Pure Appl. Chem.* **25**, 667 (1971).
88. Golovanov, V. I. and Zolotov, Yu. A., *Zh. Analit. Khim.* **25**, 1264 (1970).
89. Zolotov, Yu. A., Golovanov, V. I. and Vanifatova, N. G., *ibid.* **28**, 5 (1973).
90. Kiseleva, O. A., Shakhova, N. V. and Zolotov, Yu. A., *ibid.* **29**, 15 (1974).
91. Shakhova, N. V., Rybakova, E. V. and Zolotov, Yu. A., *ibid.* **29**, 682 (1974).
92. Bock, R. and Monerjan, A., *Z. Anal. Chem.* **226**, 29 (1967).
93. Bock, R. and Freitag, K. D., *ibid.* **254**, 104 (1971).
94. Študlar, K., *Collection Czech. Chem. Commun.* **31**, 1999 (1966).
95. Byrne, A. R. and Gorenc, D., *Anal. Chim. Acta* **59**, 81 (1972).
96. Schweitzer, G. K., *ibid.* **30**, 68 (1964).
97. Zolotov, Yu. A. and Bagreev, V. V., *Tr. Komis. po Analit. Khim. Akad. Nauk SSSR* **17**, 251 (1969).
98. Förster, H., *J. Radioanal. Chem.* **6**, 11 (1970).
99. Riedel, A., *ibid.* **13**, 125 (1973).
100. Chwastowska, J., Lissowska, K. and Sterlińska, E., *Chem. Anal. (Warsaw)* **19**, 671 (1974).
101. Steinbach, J. F. and Freiser, H., *Anal. Chem.* **25**, 881 (1953); **26**, 375 (1954).
102. Krishen, A. and Freiser, H., *ibid.* **31**, 923 (1959).
103. Minczewski, J. and Jaskólska, H., *Nukleonika* **11**, 521 (1966).
104. Poskanzer, A. M. and Foreman, B. M., *J. Inorg. Nucl. Chem.* **16**, 323 (1961).
105. Onishi, H. and Sekine, K., *Talanta* **19**, 473 (1972).

106. Kertes, A. S. and Marcus, Y. (eds.), *Solvent Extraction Research*, Part IV, Wiley, London (1969).
107. Zolotov, Yu. A. and Gavrilova, L. G., *Zh. Analit. Khim.* **25**, 813, 1054 (1970).
108. Marczenko, Z. and Mojski, M., *Wiad. Chem.* **25**, 677 (1971); *Anal. Chim. Acta* **54**, 469 (1971).
109. Madic, C., *Chim. Anal. (Paris)* **54**, 102 (1972).
110. Dolgashova, N. V. and Fridman, Ya. D., *Zh. Analit. Khim.* **27**, 1453 (1972).
111. Spivakov, B. Ya. and Zolotov, Yu. A., *Zh. Analit. Khim.* **25**, 616 (1970).
112. Handley, T. H. and Dean, J. A., *Anal. Chem.* **34**, 1312 (1962).
113. Goryushina, V. G. and Biryukova, E. Ya., *Zh. Analit. Khim.* **24**, 580 (1969).
114. Specker, H., Cremer, M. and Jackwerth, E., *Angew. Chem.* **71**, 492 (1959).
115. Zolotov, Yu. A., *Zh. Analit. Khim.* **26**, 20 (1971).
116. Ishimori, T., Watanabe, K. and Nakamura, E., *Bull. Chem. Soc. Japan* **33**, 636 (1960).
117. Boswell, C. R. and Brooks, R. R., *Mikrochim. Acta* **1965**, 814; *Anal. Chim. Acta* **33**, 117 (1965).
118. Gagliardi, E. and Wöss, H. P., *Anal. Chim. Acta* **48**, 107 (1969).
119. Bock, R., Kusche, H. and Bock, E., *Z. Anal. Chem.* **138**, 167 (1953).
120. Pohl, F. A. and Bonsels, W., *ibid.* **161**, 108 (1958).
121. Luke, C. L., *Anal. Chim. Acta* **39**, 447 (1967).
122. Prokoshev, A. A., Chuchalin, L. K., Iofa, B. Z. and Zolotov, Yu. A., *Zh. Analit. Khim.* **27**, 1364 (1972).
123. Maeck, W. J., Booman, G. L., Kussy, M. E. and Rein, J. E., *Anal. Chem.* **33**, 1775 (1961).
124. Olenovich, N. L., Mazurenko, E. A., Ermilova, V. N. and Rogachko, M. M., *Zavodsk. Lab.* **30**, 389 (1964).
125. Moore, F. L., *Anal. Chem.* **37**, 1235 (1965).
126. Ueno, K. and Saito, A., *Anal. Chim. Acta* **56**, 427 (1971).
127. Shmidt, V. S., *Ekstraktsiya aminami* (Extraction with Amines), Atomizdat, Moscow (1970).
128. Green, H., *Talanta* **11**, 1561 (1964); **20**, 139 (1973).
129. Zhivopistsev, V. P., *Zavodsk. Lab.* **31**, 1043 (1965); *Tr. Komis. po Analit. Khim. Akad. Nauk SSSR* **17**, 309 (1969).
130. Zhivopistsev, V. P., Petrov, B. I., Selezneva, E. A. and Sibiryakov, N. F., *ibid.* **17**, 304 (1969).
131. Akimov, V. K. and Busev, A. I., *Zh. Analit. Khim.* **26**, 134, 964 (1971).
132. Knižek, M., *Chem. Listy* **62**, 299 (1968).
133. Yamamoto, Y., Okamoto, N. and Tao, E., *Anal. Chim. Acta* **47**, 127 (1969).
134. Behrends, K., *Z. Anal. Chem.* **250**, 161 (1970).
135. Gibson, N. A. and Weatherburn, D. C., *Anal. Chim. Acta* **58**, 149, 159 (1972).
136. Kuz'min, N. M., *Zavodsk. Lab.* **34**, 395 (1968).
137. Häberli, E., *Z. Anal. Chem.* **160**, 15 (1958).
138. Minczewski, J., Karczewska, B. and Marczenko, Z., *Chem. Anal. (Warsaw)* **6**, 501 (1961).
139. Marczenko, Z., Krasiejko, M. and Chołuj, Ł., *ibid.* **8**, 375 (1963).
140. Marczenko, Z., *Mikrochim. Acta* **1965**, 281.
141. Erkelens, P. C. van, *Anal. Chim. Acta* **25**, 129 (1961).
142. Minczewski, J. Chwastowska, J. and Marczenko, Z., *Chem. Anal. (Warsaw)* **6**, 509 (1961).
143. Marchenko, P. V., *Ukr. Khim. Zh.*, **32**, 1216 (1966).
144. Brodskaya, B. D., Notkina, M. A., Korneeva, S. A. and Men'shova, N. P., *Zh. Analit. Khim.* **21**, 1447 (1966).
145. Riley, J. P. and Topping, G., *Anal. Chim. Acta* **44**, 234 (1969).
146. Koch, O. G., *Mikrochim. Acta* **1958**, 92, 151, 347, 402.

147. Chernikhov, Yu. A., Cherkashina, T. V., Notkina, M. A., Petrova, E. I., Menshova, N. P., Lugovskaya, V. I. and Goryanskaya, G. P., *Zh. Analit. Khim.* **21**, 714 (1966).
148. Brooks, R. R., *Talanta* **12**, 511 (1965).
149. Doll, W. and Specker, H., *Z. Anal. Chem.* **161**, 354 (1958).
150. Pohl, F. A., *ibid.* **139**, 241, 423 (1953); **141**, 81 (1954).
151. Brooks, R. R., *Talanta* **12**, 505 (1965).
152. Marczenko, Z., Mojski, M. and Kasiura, K., *Zh. Analit. Khim.* **22**, 1805 (1967).
153. Marczenko, Z. and Mojski, M., *Chem. Anal. (Warsaw)* **12**, 1155 (1967).
154. Bankmann, E. and Specker, H., *Z. Anal. Chem.* **162**, 18 (1958).
155. Specker, H. and Shirodker, R., *ibid.* **214**, 401 (1965).
156. Claassen, A. and Bastings, L., *ibid.* **160**, 403 (1958).
157. Dean, G. A. and Herringshaw, J. F., *Analyst* **86**, 106 (1961).
158. Morgunov, A. F. and Fomin, V. V., *Zh. Neorgan. Khim.* **8**, 508 (1963).
159. Yudelevich, I. G., Buyanova, L. M., Protopopova, N. P. and Yudina, N. G., *Zh. Analit. Khim.* **25**, 1177 (1970).
160. Lysenko, V. I., *Tr. Komis. po Analit. Khim. Akad. Nauk SSSR* **15**, 195 (1965).
161. Yudelevich, I. G., Artyukhin, P. I., Chuchalina, L. S., Protopopova, N. P., Skrebkova, L. M., Gilbert, E. N. and Pronin, V. A., *Zh. Analit. Khim.* **21**, 1457 (1966).
162. Lysenko, V. I. and Kim, A. G., *Tr. Komis. po Analit. Khim. Akad. Nauk SSSR* **15**, 200 (1965).
163. Zelle, A. and Fijałkowski, J., *Chem. Anal. (Warsaw)* **7**, 317 (1962).
164. Marczenko, Z., Kasiura, K. and Krasiejko, M., *ibid.* **14**, 1277 (1969).
165. Kasiura, K. *ibid.* **11**, 141 (1966); **13**, 849 (1968).
166. Nazarenko, V. A., Biryuk, E. A., Shustova, M. V., Shitareva, G. G., Vinkovetskaya, S. Ya. and Flyantikova, G. V., *Zavodsk. Lab.* **32**, 267 (1966).
167. Pohl, F. A. and Bonsels, W., *Mikrochim. Acta* **1961**, 314.
168. Jackwerth, E., *Z. Anal. Chem.* **202**, 81; **206**, 269 (1964); **211**, 254, (1965); **216**, 73 (1966).
169. Donaldson, E. M. and Inman, W. R., *Talanta* **13**, 489 (1966).
170. Nazarenko, V. A., Shustova, M. B., Shitareva, G. G., Yagnyatinskaya, G. Ya. and Ravitskaya, R. V., *Zavodsk. Lab.* **28**, 645 (1962).
171. Slyusareva, R. L., Kondrat'eva, L. I. and Peizulaev, Sh. I., *ibid.* **31**, 557 (1965).
172. Karabash, A. G., Peizulaev, Sh. I., Slyusareva, R. L. and Lipatova, V. M., *Tr. Komis. po Analit. Khim. Akad. Nauk SSSR* **12**, 331 (1960).
173. Kuznetsova, N. N. and Krauz, L. S., *Zh. Analit. Khim.* **18**, 1090 (1963).
174. Miller, F. I., *Talanta* **21**, 685 (1974).
175. Marczenko, Z., *Chem. Anal. (Warsaw)* **9**, 1093 (1964).
176. Geilmann, W., *Z. Anal. Chem.* **160**, 410 (1958).
177. Geilmann, W. and de Alvaro Estebaranz, A., *ibid.* **190**, 60 (1962).
178. Berg, E. W. and Hartlage, F. R., *Anal. Chim. Acta* **33**, 173 (1965).
179. Berg, E. W. and Acosta, J. J., *ibid.* **40**, 101 (1968).
180. Berg, E. W. and Reed, K. P., *ibid.* **42**, 207 (1968).
181. Berg, E. W. and Herrera, N. M., *ibid.* **60**, 117 (1972).
182. Stumpp, E., *Z. Anal. Chem.* **242**, 225 (1968).
183. Novak, J. and Tomašová, H., *Chem. Listy* **64**, 1039 (1970).
184. Marczenko, Z., *Chem. Anal. (Warsaw)* **8**, 849 (1963).
185. Wąsowicz, S. and Rutkowski, W., *ibid.* **11**, 603 (1966).
186. Korenman, I. M. and Sidorenko, L. V., *Zh. Analit. Khim.* **22**, 388 (1967).
187. Marczenko, Z., Mojski, M. and Kasiura, K., *Chem. Anal. (Warsaw)* **14**, 1331 (1969).
188. Morachevskii, Yu. V., Zil'bershtein, K. I., Piryutko, M. M. and Nikitina, O. N., *Zh. Analit. Khim.* **17**, 614 (1962).
189. Schweinsberg, D. P. and Heffernan, B. J., *Talanta* **17**, 332 (1970).

References

190. Chwastowska, J., *Chem. Anal. (Warsaw)* **7**, 731 (1962).
191. Heffelfinger, R. E., Blosser, E. R., Perkins, O. E. and Henry, W. M., *Anal. Chem.* **34**, 621 (1962).
192. Schreiber, E., *Z. Anal. Chem.* **210**, 93 (1965).
193. Pohl, F. A., Kokes, K. and Bonsels, W., *ibid.* **174**, 6 (1960).
194. Malakhov, V. V., Protopopova, N. P., Trukhacheva, V. A. and Yudelevich, I. G., *Tr. Komis. po Analit. Khim. Akad. Nauk SSSR* **16**, 89 (1968).
195. Fratkin, Z. G. and Shebunin, V. S., *ibid.* **15**, 127 (1965).
196. Clark, H. C. and Emeleus, H. J., *J. Chem. Soc.* **1957**, 2119.
197. Ruff, O. and Lickfett, H., *Chem. Ber.* **44**, 2539 (1911).
198. Sotnikova, N. P., Romanovich, L. S., Peizulaev, Sh. I. and Karabash, A. G., *Tr. Komis. po Analit. Khim. Akad. Nauk SSSR* **12**, 151 (1960).
199. Fedyashina, A. F., Yudelevich, I. G. and Strokina, T. G., *Zh. Analit. Khim.* **21**, 1232 (1966).
200. Ginzburg, V. L., Glukhovetskaya, N. P. and Danilova, N. N., *ibid.* **17**, 1096 (1962).
201. Glukhovetskaya, N. P. and Lerner, L. A., *Zavodsk. Lab.* **30**, 1082 (1964).
202. Shvarts, D. M., *ibid.* **26**, 966 (1960).
203. Rajić, S. R. and Marković, S. V., *Anal. Chim. Acta* **50**, 169 (1970).
204. Neeb, K. H., *Z. Anal. Chem.* **221**, 200 (1966).
205. Middleton, G. and Stuckey, R. E., *Analyst* **78**, 532 (1953); **79**, 138 (1954).
206. Analytical Methods Committee, *ibid.* **85**, 643 (1960).
207. Schulek, E. and Laszlovszky, J., *Mikrochim. Acta* **1960**, 485.
208. Gorsuch, T. T., *The Destruction of Organic Matter*, Pergamon, New York (1970).
209. Fabry, J. and Nangniot, P., *Analusis* **1**, 117, 177 (1972).
210. Minczewski, J., Chwastowska, J. and Hong Mai P., *Chem. Anal. (Warsaw)* **18**, 1189 (1973).
211. Adrian, W. J., *Analyst* **98**, 213 (1973).
212. Gorsuch, T. T., *ibid.* **87**, 112 (1962).
213. Hamilton, E. I., Minski, M. J. and Cleary, J. J., *ibid.* **92**, 257 (1967).
214. Strohal, P., Lulić, S. and Jelisavčić, O., *ibid.* **94**, 678 (1969).
215. Smith, G. F. and Diehl, H., *Talanta* **2**, 209; **3**, 41 (1959); **4**, 185 (1960).
216. Smith, G. F., *Anal. Chim. Acta* **8**, 397 (1953); **17**, 175 (1957); *Talanta* **11**, 633 (1964); **15**, 489 (1968).
217. Taubinger, R. P. and Wilson, J. R., *Analyst* **90**, 429 (1965).
218. Down, J. L. and Gorsuch, T. T., *ibid.* **92**, 398 (1967).
219. Analytical Methods Comittee, *ibid.* **92**, 403 (1967).
220. Denbsky, G., *Z. Anal. Chem.* **267**, 350 (1973).
221. Fujiwara, S. and Narasaki, H., *Anal. Chem.* **40**, 2031 (1968).
222. Thomas, A. D. and Smythe, L. E., *Talanta* **20**, 469 (1973).
223. Samuelson, O., *Ion Exchange Separations in Analytical Chemistry*, Wiley, New York (1963).
224. Trémillon, B., *Les séparations par les résines échangeuses d'ions* (Separations with Ion Exchange Resins), Gauthier-Villars, Paris (1965).
225. Rieman, W., III, and Walton, H. F., *Ion Exchange in Analytical Chemistry*, Pergamon, New York (1970).
226. Schindewolf, U., *Angew. Chem.* **69**, 226 (1957).
227. Nikitina, N. G., Galkina, N. K. and Senyavin, M. M., *Zh. Analit. Khim.* **21**, 1165 (1966).
228. Marhol, M., *Chem. Listy* **56**, 728 (1962).
229. Chernobrov, S. M., *Zavodsk. Lab.* **23**, 1052 (1957); **29**, 1281 (1963); **33**, 539 (1967); **37**, 1 (1971).
230. Walton, H. F., *Anal. Chem.*, **42**, 86R (1970); **44**, 256R (1972).
231. Sopkova, A. and Chomič, J., *Chem. Listy* **66**, 1150 (1972).
232. Aubouin, G. and Laverlochère, J., *J. Radioanal. Chem.* **1**, 123 (1968).
233. Trémillon, B., *Z. Anal. Chem.* **236**, 472 (1968).

234. Kaminski, E. E. and Fritz, J. S., *J. Chromatog.* **53**, 345 (1970).
235. Schmuckler, G., *Talanta* **12**, 281 (1965).
236. Bernhard, H. and Grass, F., *Mikrochim. Acta* **1966**, 426.
237. Myasoedova, G. V., Eliseeva, O. P. and Savvin, S. B., *Zh. Analit. Khim.* **26**, 2172 (1971).
238. Vernon, F. and Eccles, H., *Anal. Chim. Acta* **63**, 403 (1973).
239. Eccles, H. and Vernon, F., *ibid.* **66**, 231 (1973).
240. Parrish, J. R. and Stevenson, R., *ibid.* **70**, 189 (1974).
241. Dingman, J. F., Jr., Gloss, K. M., Milano, E. A. and Siggia, S., *Anal. Chem.* **46**, 774 (1974).
242. Sansoni, B. and Wiegand, W., *Talanta* **17**, 973, 987 (1970).
243. Moody, G. J. and Thomas, J. D., *Analyst* **93**, 557 (1968).
244. Strelow, F. W., *Anal. Chem.* **32**, 1185 (1960).
245. Mann, C. K. and Swanson, C. L., *ibid.* **33**, 459 (1961).
246. Nelson, F., Murase, T. and Kraus, K. A., *J. Chromatog.* **13**, 503 (1964).
247. Van Erkelens, P. C., *Anal. Chim. Acta* **25**, 42 (1961).
248. Fritz, J. S. and Rettig, T. A., *Anal. Chem.* **34**, 1562 (1962).
249. Korkisch, J. and Ahluwalia, S. S., *Anal. Chim. Acta* **34**, 308 (1966); *Talanta* **14**, 155 (1967).
250. Cummings, T. and Korkisch, J., *Talanta* **14**, 1185 (1967).
251. Birze, L. Marple, L. W. and Diehl, H., *ibid.* **15**, 1441 (1968).
252. Strelow, F. W., Van Zyl, C. R. and Bothma, C. J., *Anal. Chim. Acta* **45**, 81 (1969).
253. Strelow, F. W., Victor, A. H., Van Zyl, C. R. and Eloff, C., *Anal. Chem.* **43**, 870 (1971).
254. Strelow, F. W. and Victor, A. H., *Talanta* **19**, 1018 (1972).
255. Fritz, J. S., Garralda, B. B. and Karraker, S. K., *Anal. Chem.* **33**, 882 (1961).
256. Nikitin, M. K., *Dokl. Akad. Nauk SSSR* **148**, 595 (1963).
257. Danielsson, L. and Ekström, T., *Acta Chem. Scand.* **20**, 2402, 2415 (1966).
258. Korkisch, J. and Huber, A., *Talanta* **15**, 119 (1968).
259. Nelson, F. and Michelson, D. C., *J. Chromatog.* **25**, 414 (1966).
260. Korkisch, J. and Klakl, E., *Talanta* **16**, 377 (1969).
261. Majumdar, A. K. and Mitra, B. K., *Z. Anal. Chem.* **208**, 1 (1965).
262. Pietrzyk, D. J. and Kiser, D. L., *Anal. Chem.* **37**, 233 (1965).
263. Kawabuchi, K., Ito, T. and Kuroda, R., *J. Chromatog.* **39**, 61 (1969).
264. Nozaki, T., Hiraiwa, O., Henmi, C. and Koshiba, K., *Bull. Chem. Soc. Japan* **42**, 245 (1969).
265. Jha, S. K., De Corte, F. and Hoste, J., *Anal. Chim. Acta* **62**, 163 (1972).
266. Bhatnagar, R. P., Trivedi, R. G. and Bala, Y., *Talanta* **17**, 249 (1970).
267. Fritz, J. S. and Palmer, T. A., *ibid.* **9**, 393 (1962).
268. Kratochvil, V., Povondra, P. and Šulcek, Z., *Chem. Listy* **63**, 1185 (1969).
269. Strelow, F. W. and Sondorp, H., *Talanta* **19**, 1113 (1972).
270. Korkisch, J., Feik, F. and Ahluwalia, S. S., *ibid.* **14**, 1069 (1967).
271. Orlandini, K. A. and Korkisch, J., *Anal. Chim. Acta* **43**, 459 (1968).
272. Tera, F., Ruch, R. R. and Morrison, G. H., *Anal. Chem.* **37**, 358 (1965).
273. Yoshimura, J. and Waki, H., *Bull. Chem. Soc. Japan* **35**, 416 (1962).
274. Brooks, R. R., *Analyst* **85**, 745 (1960).
275. Fritz, J. S. and Gillette, M. L., *Talanta* **15**, 287 (1968).
276. Koch, W. and Korkisch, J., *Mikrochim. Acta* **1972**, 687; **1973**, 245, 263, 877.
277. Faris, J. P., *Anal. Chem.* **32**, 520 (1960).
278. Wilkins, D. H., *Talanta* **2**, 355 (1959).
279. Ferraro, T. A., *ibid.* **16**, 669 (1969).
280. Nelson, F., Rush, R. M. and Kraus, K. A., *J. Am. Chem. Soc.* **82**, 339 (1960).
281. Ichikawa, F., Uruno, S. and Imai, H., *Bull. Chem. Soc. Japan* **34**, 952 (1961).
282. Faris, J. P. and Buchanan, R. F., *Anal. Chem.* **36**, 1157 (1964).
283. Walter, C. W. and Korkisch, J., *Mikrochim. Acta* **1971**, 81, 137, 158, 181, 194.

284. Korkisch, J. and Ahluwalia, S. S., *Z. Anal. Chem.* **215**, 86 (1966).
285. Strelow, F. W. and Bothma, C. J., *Anal. Chem.* **39**, 595 (1967).
286. Lavrukhina, A. K. and Akol'zina, L. D., *Zh. Analit. Khim.* **22**, 542 (1967).
287. Klakl, E. and Korkisch, J., *Talanta* **16**, 1177 (1969).
288. Van den Winkel, P., de Corte, F. and Hoste, J., *Anal. Chim. Acta* **56**, 241 (1971).
289. De Corte, F., Van Acker, P. and Hoste, J., *ibid.* **64**, 177 (1973).
290. De Corte, F., van den Winkel, P., Speecke, A. and Hoste, J., *ibid.* **42**, 67 (1968).
291. Strelow, F. W., Weinert, C. H. and Eloff, C., *Anal. Chem.* **44**, 2352 (1972).
292. Pitstick, G. F., Sweet, T. R. and Morie, G. P., *ibid.* **35**, 995 (1963).
293. Gélis, P. de, *Chim. Anal. (Paris)* **53**, 673 (1971).
294. Kuroda, R., Ishida, K. and Kiriyama, T., *Anal. Chem.* **40**, 1502 (1968).
295. Fritz, J. S. and Kaminski, E. E., *Talanta* **18**, 541 (1971).

Part II

METHODS FOR SEPARATION AND DETERMINATION OF INDIVIDUAL ELEMENTS

Some Practical Notes

The following chapters on individual elements begin with a discussion of separation methods, followed by spectrophotometric methods for the determination of the elements. The more important analytical methods are presented in detail.

The procedures are given in a unified form. The approximate maximum quantity of the element being determined is given at the beginning, and refers to solutions in 50-ml volumetric flasks and absorbance measurements in 1-cm cells. The reagent quantities recommended should be altered according to the capacity of the flasks containing the coloured solutions (e.g. 10, 25, or 100 ml).

The preparation of standard solutions containing 1 mg of element per ml is described. Working solutions (e.g. 0·1 mg/ml, 10 µg/ml, or 1 µg/ml) are prepared by appropriate dilution of the standard solutions (1 mg/ml) with water, unless stated otherwise. Unless otherwise stated, reagent solutions are prepared with distilled water.

Calibration curves are prepared by using suitable aliquots of the working standard solution of the element being determined, instead of the sample solution.

The sensitivities of the methods are characterized by the *molar absorptivity* (ε; l.mole^{-1}.cm^{-1}), and the *specific absorptivity* ($a = \varepsilon/$at. wt. $\times 1000$).

Since most colour systems used in spectrophotometric methods obey Beer's law, attention is drawn only to those systems which deviate from it.

In the details of the preparation of reagent solutions no mention is usually made of purity. It is assumed that suitably pure reagents are used.

References are given at the end of each chapter.

Chapter 4

ALKALI METALS

Lithium (Li, at.wt. 6·94), sodium (Na, at.wt. 22·99), potassium (K, at.wt. 39·10), rubidium (Rb, at.wt. 85·47), and caesium (Cs, at.wt. 132·91) yield colourless M^+ ions. The alkali metals react violently with water, forming strong bases (MOH). Potassium, rubidium, and caesium are very similar in chemical properties. Sodium, however, has somewhat different properties. Lithium resembles magnesium more than it resembles other alkali metals. Lithium and sodium are characterized by the ability to form complexes (e.g. with pyrophosphate and EDTA).

Trace amounts of the alkali metals are best determined by flame photometry.

4.1 Isolation and Separation of the Alkali Metals

Owing to the good water-solubility of the alkali metal salts, the simplest way to isolate them from the Analytical Group I—IV metals is by leaching a carefully powdered sample with very dilute acid (e.g. HCl) or dilute alkali (e.g. ammonia), depending on the nature of the sample under examination.

Multivalent metals can be separated from the alkali metals by solvent extraction or precipitation methods. Cation-exchangers and electrolysis with a mercury cathode are also useful in separating the metals of other Groups from the alkali metals.

The separation of the alkali metals from each other by cation-exchange chromatography using strongly acidic cation-exchangers has been thoroughly investigated [1-5]. The isolation of small quantities of lithium, and of caesium and rubidium, are described in references [6,7] and [8-10], respectively.

In the separation of caesium, the following inorganic cation-exchangers are particularly useful: potassium hexacyanocobalt(II)ferrate(II) [11-13], ammonium molybdophosphate [13,14], and thallous tungstophosphate [15]. Ammonium molybdophosphate has also been used to separate potassium from sodium [16].

The alkali metal tetraphenylborates have been separated by virtue of their different distribution coefficients when extracted with nitrobenzene [17]. The alkali metals have also been separated by extraction chromatography with dipicrylamine and nitrobenzene [18]. Caesium has been isolated by extraction into nitrobenzene with TTA [19] or tetraiodobismuthite [20].

Caesium has also been isolated from fission products by double precipitation as the tetraphenylborate from a weak acid solution, and subsequent precipitation as the iodobismuthite (for separation from Rb) in an acetic acid solution [21].

Dioxan leaches only lithium chloride from a mixture of dried alkali metal chlorides [22].

4.2 Determination of Lithium

Lithium is the only alkali metal which gives colour reactions with some azo dyes. The first reagent used in the direct determination of lithium was Thoron I (formula p. 48) [23–26]. The reaction is carried out in aqueous potassium hydroxide solution or acetone (70%)–water–KOH medium ($\varepsilon \sim 6 \times 10^3$ at 486 nm). A fiftyfold amount of sodium and a tenfold amount of magnesium do not interfere.

For the spectrophotometric determination of lithium, Dziomko et al. [27, 28] have suggested Nitroanthranylazo, which they believe to be the best reagent for lithium. The colour reaction is carried out in aqueous

$$O_2N-\underset{}{\underset{}{\bigcirc}}\underset{-N=N-}{\overset{-COOH}{}} \underset{CH_3}{\overset{OH}{\underset{=N}{\overset{|}{\underset{|}{N-C_6H_5}}}}}$$

acetone made alkaline with KOH, and the absorbance is measured at 530 nm ($\varepsilon = 1{\cdot}2 \times 10^4$). The colour intensity of the complex depends on the medium (acetone–water, dioxan, methanol, dimethylformamide), on the excess of Nitroanthranylazo, and on the ionic strength. In the determination of 2 μg of lithium, there is no interference from 200 μg of Rb, 100 μg of Mg and Ca, and 50 μg of Ba, Sr, and Na. The method has been used to determinate lithium in ferrites [29].

Quinazolinazo [30], Arsenazo III, and Phosphonazo R [26] are other azo reagents recommended for the determination of Li.

In an indirect method, Li_3PO_4 is precipitated, and the phosphorus in the precipitate is determined as phosphomolybdenum blue [31].

4.3 Determination of Sodium

Sodium is separated from other alkali metals by precipitation with zinc uranyl acetate from acetic acid medium as the triple acetate $NaZn(UO_2)_3 \cdot (CH_3COO)_9 \cdot xH_2O$ [32,33], but supersaturation must be guarded against. When the potassium content exceeds the sodium content by greater than one hundred fold, it is partly coprecipitated with the sodium. In this case double precipitation is advisable. The precipitated triple acetate is then dissolved, and sodium is determined indirectly by the spectrophotometric determination of zinc with dithizone. Sodium can also be precipitated with nickel uranyl acetate. In this instance the sodium content is determined by the spectrophotometric determination of nickel with dimethylglyoxime [34].

In another indirect method, sodium is precipitated with 5-benzamidoanthraquinone-2-sulphonic acid, and the absorbance of the excess of the colour reagent in the solution is measured [35].

In water–dimethylformamide–acetone (12:2:86) solution made alkaline with tetramethylammonium hydroxide, sodium reacts with Nitroanthranylazo (a reagent for lithium). The sodium content is calculated from the decrease in the absorbance of the reagent at 540 nm [36].

The potassium salt of 1'-(3-methyl-1-phenyl-5-pyrazolone-4-azo)-4'-nitrophenyl-2'-sulphonic acid is suitable for spectrophotometric determination of sodium [36a].

4.4 Determination of Potassium

Potassium is determined spectrophotometrically by indirect methods. The method using dipicrylamine (hexanitrodiphenylamine) is often employed [37–42].

$$O_2N-\underset{\underset{NO_2}{|}}{\overset{\overset{NO_2}{|}}{\bigcirc}}-\overset{H}{N}-\underset{\underset{NO_2}{|}}{\overset{\overset{NO_2}{|}}{\bigcirc}}-NO_2$$

Potassium is precipitated from a neutral or slightly alkaline solution with sodium, lithium, or magnesium dipicrylaminate. The red precipitate is washed with ether, and dissolved in acetone. The solution is then diluted with water, the pH adjusted to ~ 10, and the absorbance of the yellow-orange solution measured at 400 nm. The sample solution must not contain Rb^+, Cs^+, NH_4^+, or Tl^+ ions which are also precipitated by dipicrylamine. Potassium has been determined in foodstuffs [39] and biological materials [40] by this method. Potassium dipicrylaminate is extractable into nitrobenzene from a solution at pH 7–10 [43,44].

In another indirect method, potassium is precipitated as the cobaltinitrite, $K_2Na[Co(NO_2)_6]$. The precipitate is filtered off, washed and dissolved, and the cobalt is determined spectrophotometrically with nitroso-R salt [45], as the coloured EDTA complex [46], or as the oxinate [47]. The potassium content can also be calculated by determining the nitrite by the Griess method [45].

Dilituric acid (5-nitrobarbituric acid) forms a sparingly soluble compound with potassium ions. The decrease in absorbance of the reagent is a measure of the potassium content in the solution [48]. Potassium ions are precipitated by tetraphenylborate. The reagent excess forms an ion-pair with the positively charged complex of copper(I) and neocuproine, and this ion-pair is extractable into ethyl acetate. The absorbance of the extract is measured at 456 nm [49].

4.5 Determination of Rubidium and Caesium

Rubidium and caesium are determined, like potassium, by indirect spectrophotometric methods, such as precipitation as the dipicrylaminate [42, 50,51], or the cobaltinitrite [50]. They can also be determined indirectly with picric acid [50,52].

Caesium is separated from small quantities of rubidium and potassium as sparingly soluble caesium tungstosilicate. The precipitate is dissolved, the tungstosilicic acid is reduced to silicotungsten blue, and the absorbance is measured at 725 nm [53] or 640 nm [54]. Caesium is also determined indirectly as phosphomolybdenum blue (absorbance measured at 805 nm) after the precipitation and isolation of caesium molybdophosphate [55].

References

1. Tsubota, H. and Kitano, Y., *Bull. Chem. Soc. Japan* **33**, 770 (1960).
2. Nelson, F., Michelson, D. C., Phillips, H. O. and Kraus, K. A., *J. Chromatog.* **20**, 107 (1965).
3. Strelow, F. W. E., Coetzee, J. H. and van Zyl, C. R., *Anal. Chem.* **40**, 196 (1968).
4. Strelow, F. W. E., Liebenberg, C. J. and Toerien, S., *Anal. Chim. Acta* **43**, 465 (1968).
5. Huber, J. F. and van Urk-Schoen, A. M., *ibid.* **58**, 395 (1972).
6. Hering, H., *ibid.* **6**, 340 (1952).
7. Ratner, R. and Ludmer, Z., *Israel J. Chem.* **2**, 21 (1964).
8. Ring, S. A., *Anal. Chem.* **28**, 1200 (1956).
9. Hahn, R. B., Johnson, J. L. and McKay, J. B., *Talanta* **13**, 1613 (1966).
10. Shestakov, G. I., *Zh. Analit. Khim.* **26**, 273 (1971).
11. Boni, A. L., *Anal. Chem.* **38**, 89 (1966).
12. Prout, W. E., Russell, E. R. and Groh, H. J., *J. Inorg. Nucl. Chem.* **27**, 473 (1965).
13. Terada, K., Hayakawa, H., Sawada, K. and Kiba, T., *Talanta* **17**, 955 (1970).
14. Oldham, G., Ware, A. R. and Sykes, D. J., *ibid.* **16**, 430 (1969).
15. Caron, H. L. and Sugihara, T. T., *Anal. Chem.* **34**, 1082 (1962).
16. Coetzee, C. J. and Rohwer, E. F., *Anal. Chim. Acta* **44**, 293 (1969).
17. Sekine, T. and Dyrssen, D., *ibid.* **45**, 433 (1969).
18. Kyrš, M. and Kadlecová, L., *J. Radioanal. Chem.* **1**, 103 (1968).
19. Crowther, P. and Moore, F. L., *Anal. Chem.* **35**, 2081 (1963).
20. Kyrš, M. and Podešva, S., *Anal. Chim. Acta* **27**, 183 (1962).
21. Fiedler, H. J. and Archer, N. P., *Z. Anal. Chem.* **226**, 114 (1967).
22. Blasius, E. and Wolf, F., *ibid.* **174**, 349 (1960).
23. Kuznetsov, V. I., *Zh. Analit. Khim.* **3**, 295 (1948).
24. Thomason, P. F., *Anal. Chem.* **28**, 1527 (1956).
25. Apple, R. F. and White, J. C., *Talanta* **13**, 43 (1966).
26. Lazarev, A. I. and Lazareva, V. I., *Zh. Analit. Khim.* **23**, 36 (1968).
27. Dziomko, V. M., Zelichenok, S. L. and Markovich, I. S., *ibid.* **23**, 170 (1968).
28. Dziomko, V. M., Zelichenok, S. L., Markovich, I. S. and Rodionov, A. N., *ibid.* **24**, 985 (1969).
29. Budyak, N. F. and Gryaznova, I. S., *Zh. Prikl. Khim.* **44**, 669 (1971).
30. Dziomko, V. M., Zelichenok, S. L. and Markovich, I. S., *Zh. Analit. Khim.* **18**, 937 (1963).

References

31. Nozaki, T., *J. Chem. Soc. Japan, Pure Chem. Sect.* **76**, 445 (1955).
32. Shell, H. R., *Anal. Chem.* **22**, 575 (1950).
33. Markova, L. V. and Kleiner, K. E., *Zavodsk. Lab.* **25**, 144 (1959).
34. Tsyvina, B. S., *ibid.* **15**, 139 (1949).
35. Fujinaga, T. and Nishida, T., *J. Chem. Soc. Japan, Pure Chem. Sect.* **85**, 547 (1964).
36. Markovich, I. S., Zelichenok, S. L., Filyagina, N. A., Chubarova, E. V. and Dziomko, V. M., *Zh. Analit. Khim.* **26**, 1097 (1971).
36a. Markovich, I. S., Zelichenok, S. L., Filyagina, N. A. and Dziomko, V. M., *ibid.* **28**, 227 (1973).
37. Faber, R. and Dirkse, T. P., *Anal. Chem.* **25**, 808 (1953).
38. Gastinger, E., *Z. Anal. Chem.* **140**, 335 (1953).
39. Bois, E. and Jean, M., *Anal. Chem.* **26**, 727 (1954).
40. Lewis, P. R., *Analyst* **80**, 768 (1955).
41. Riedler, K. and Schreiner, L., *Chem. Tech. (Berlin)* **11**, 593 (1959).
42. Gorbenko-Germanov, D. S. and Zenkova, R. A., *Zh. Analit. Khim.* **20**, 1020 (1965).
43. Iwahito, T. and Toei, K., *Bull. Chem. Soc. Japan* **37**, 1276 (1964).
44. Kyrš, M., Pivoňková, M. and Selucký, P., *Anal. Chim. Acta* **43**, 132 (1968).
45. Dupuis, T., *ibid.* **9**, 493 (1953).
46. Bultasová, H. and Konopásek, E., *Chem. Listy* **49**, 769 (1955).
47. Baar, S., *Analyst* **78**, 353 (1953).
48. Palouš, R., Pavelka, V. and Mára, M., *Collection Czech. Chem. Commun.* **24**, 3910 (1959).
49. Khreish, E. A. and Boltz, D. F., *Mikrochim. Acta* **1970**, 1174.
50. Duval, C. and Doan, M., *ibid.* **1953**, 200.
51. Kyrš, M., Rais, J. and Selucký, P., *Talanta* **16**, 1169 (1969).
52. Hejtmánek, M. and Hozmanová, E., *Mikrochim. Acta* **1966**, 97.
53. Krochta, W. G. and Mellon, M. G., *Anal. Chem.* **29**, 1181 (1957).
54. Gorenc, B. and Kosta, L., *Z. Anal. Chem.* **206**, 321 (1964).
55. Huey, F. and Hargis, L. G., *Anal. Chem.* **39**, 125 (1967).

Chapter 5

ALUMINIUM

Aluminium (Al, at.wt. 26·98) occurs in solution exclusively in the +III oxidation state. Between pH 4 and 9, the hydroxide is precipitated. Above pH 9, it is converted into the soluble tetrahydroxoaluminate anion. Aluminium forms stable complexes with fluoride, oxalate and tartrate, and weak complexes with acetate.

5.1 Methods of Separation

5.1.1 Precipitation

By precipitation of $Al(OH)_3$ with ammonia, or with an acetate buffer at pH 4·5–5·0, aluminium can be separated from metals having hydroxides precipitated at higher pH values. Traces of aluminium are separated by using titanium, lanthanum, zirconium, or iron(III) as scavenger. Iron is a rather inconvenient scavenger, because it usually has to be separated before the aluminium is determined. It is possible to precipitate $Al(OH)_3$ selectively with ammonia at pH ~ 5 in the presence of iron masked by reduction to Fe(II). Chromium can be masked by oxidation to Cr(IV), and titanium and vanadium can be masked with hydrogen peroxide.

Aluminium is separated as the soluble tetrahydroxo complex from iron, titanium, and other metals which are insoluble in excess of NaOH [1–3]. The aluminium coprecipitated with the insoluble metal hydroxides should be recovered by dissolving the precipitate in dilute hydrochloric acid, and then reprecipitating the interfering metals with excess of NaOH. A mixed solution of NaOH and Na_2S is sometimes used as precipitant, since aluminium is coprecipitated to a lesser degree in the sulphides than in the hydroxides. Water leaches aluminium (as aluminate) from an Na_2CO_3 melt, while iron, titanium, zirconium, and other metals remain in the solid.

Small amounts of aluminium may be precipitated as the 8-hydroxyquinolinate [4]. If fluoride is present, it is masked as the stable, soluble beryllium complex [5]. Iron(III), titanium, and zirconium are separated from aluminium by precipitation with cupferron from an acidic medium [6].

Mercury cathode electrolysis allows small quantities of aluminium to be separated from large amounts of Fe, Ni, Co, Cu, Zn, Mn, Cr, Mo, Pb, etc., while Be, V, Ti, Zr, Mg, Ca, and the lanthanides remain in the solution together with the aluminium [6,7].

5.1.2 Extraction

Extraction is used mainly for the preliminary separation of metals which interfere in the determination of aluminium. After Fe(III), Ti, Zr, and Cu cupferronates have been extracted from dilute HCl into chloroform, aluminium cupferronate is extracted at pH ~ 3·5 [8,9].

Aluminium has been extracted as chelates with acetylacetone [10–12], trifluoroacetylacetone [13], 8-hydroxyquinoline [4,14], 8-hydroxyquinaldine [15], and BPHA [16].

5.1.3 Ion-Exchange

Since aluminium does not form a stable chloro complex, it can be separated by passing a solution in $9M$ HCl through a strongly basic anion-exchange column. Aluminium [as well as Ni, Mn(II), Ca, Mg, Be, and Ti] is eluted, whereas Fe, Cu, Zn, Co, Cd, etc. are retained on the column [17,18]. Similarly, plutonium has been separated from aluminium by retaining the chloride complex of plutonium on an anion-exchanger [19].

From solutions of zirconium and aluminium in $0·06M$ HCl and $0·8M$ HF, only zirconium is retained by Dowex-1 anion-exchanger [20]. Iron is retained on Amberlite IRA-400 from an acidic thiocyanate solution, while aluminium is eluted [21].

Aluminium, gallium and indium can be retained on an anion-exchanger from a solution of 2-methoxyethanol and $6M$ hydrochloric acid, and can then be sequentially eluted with $1M$ HCl [22].

Aluminium and other cations are adsorbed on cation-exchangers, whereas phosphate and other interfering anions are eluted [18]. The anionic fluoride and sulphosalicylate complexes of aluminium can be separated by cation-exchangers from metals not forming corresponding complexes [23,24].

5.2 Methods of Determination

The extraction method using 8-hydroxyquinoline is not very sensitive, but it is highly selective, provided appropriate masking agents are used. The Eriochrome Cyanine R method is highly sensitive, and has the advantage that a true solution is obtained. Chrome Azurol S is another valuable reagent for aluminium.

The various spectrophotometric methods for the determination of aluminium have been reviewed by Tikhonov [25].

5.2.1 8-Hydroxyquinoline Method

Between pH 4·5 and 10, 8-hydroxyquinoline (oxine) (p. 56) forms the chelate $Al(C_9H_6ON)_3$, which is sparingly soluble in water but dissolves readily in chloroform. The yellow chloroform extract of aluminium oxinate is the basis of the spectrophotometric method [14,26,27]. Carbon tetrachloride and trichloroethylene are also used as extractants.

The absorption maximum of the chloroform extract is at 390 nm ($\varepsilon = 7\cdot3 \times 10^3$, $a = 0\cdot27$). The absorption of 8-hydroxyquinoline in chloroform rapidly increases below 390 nm.

Although 8-hydroxyquinoline is a group reagent and reacts with many metals, the use of appropriate masking agents makes the oxine method specific for aluminium.

Larger quantities of iron are usually extracted as the chloride, thiocyanate, cupferronate or TTA complexes [28,29]. Smaller quantities of iron may be masked with bipyridyl or 1,10-phenanthroline [30], or with cyanide [after reduction to Fe(II)] [31]. Middleton [32] extracted first iron(III) oxinate at pH 2·9–3·0, and then aluminium oxinate at pH 5·5. At pH 3, iron(III) is extracted quantitatively, while no aluminium is extracted. Motojima and Hashitani [33] have developed a simultaneous photometric method for the determination of aluminium and iron with oxine.

Hydrogen peroxide prevents oxine from reacting with Ti, V, Nb, U, and Ce. Cyanide masks Ni, Co, Cu, Zn, Cd, Ag, and Fe(II).

EDTA or tartrate keeps aluminium in solution at pH values at which it normally hydrolyses. The presence of EDTA does not interfere in the extraction of aluminium with oxine when the pH of the solution is higher than 8. An almost specific method for the extraction of aluminium oxinate uses EDTA as masking agent [34].

Heavy metals can be separated from aluminium by extraction as dithiocarbamates [35] and dithizonates. Titanium and zirconium can be stripped from a chloroform extract of Al, Ti, and Zr oxinates by shaking with ammoniacal solution at pH 9·2 [29]. Vanadium and titanium can be separated from aluminium by extracting the cupferronates from $1M$ H_2SO_4 medium with chloroform. Thorium is extracted as the TTA complex from a solution at pH 1·5 [28].

Aluminium can be determined indirectly by precipitation with oxine [19]. The precipitate is filtered off, washed, and dissolved in hydrochloric acid, then the absorbance of the acidic oxinate solution is measured at 360 nm.

In Pyatnitskii and Glushchenko's indirect method [36], aluminium displaces an equivalent amount of iron(III) or manganese from their oxinates. The iron is determined as the thiocyanate complex, and the manganese as permanganate.

An exchange-reaction in the organic phase is used to convert aluminium acetylacetonate into the oxinate [11].

Aluminium has been determined photometrically by the 8-hydroxyquinoline method in cast iron and steel [16,26,29,34,37,38], nickel and copper alloys [2], chromium [39], heat-resistant alloys [31], thorium [28,40], uranium [16], metallurgical products [41], beryllium [42], silicate rocks and minerals [11,30,43], rare earths [35], alkalis [27], polyethylene [44], plant materials [32,45,46], soil extracts [45], and organometallic compounds [47].

Reagents

1. 8-Hydroxyquinoline (oxine): 1% solution in chloroform.
2. Standard aluminium solution: 1 mg/ml. Dissolve 17·59 g of $KAl(SO_4)_2 \cdot 12H_2O$ in water containing 5 ml of concentrated H_2SO_4, and dilute the solution with water in a volumetric flask to 1 litre. Working solutions are obtained by suitable dilution of this stock solution with dilute ($\sim 0.01N$) sulphuric acid.

Procedure

To a solution containing not more than 60 μg of Al, add 2 ml of 5% EDTA solution and 1 ml of 3% H_2O_2 solution. After 5 minutes adjust the solution with ammonia to pH ~ 9, add ~ 0.1 g of Na_2SO_3 and ~ 0.1 g of KCN, and heat to 70–80°C. Cool the solution and readjust the pH to 9·0 (± 0.2) with dilute ($\sim 0.1M$) HCl.

Transfer the solution to a separating funnel, and extract the aluminium with two portions of the chloroform oxine solution. Wash the combined extracts with two portions of water and make the solution up to the mark with the oxine solution in a 50-ml (or smaller, depending on the amount of Al) volumetric flask. Measure the absorbance of the yellow chloroform solution at 390 nm using as reference a reagent blank solution prepared exactly as above.

Note. During the extraction and washing in the separating funnel, the liquid must not be shaken too vigorously (lest a stable emulsion be formed).

5.2.2 ERIOCHROME CYANINE R METHOD

The colour of Eriochrome Cyanine R (ER, Solochrome Cyanine R) (formula, p. 51) in aqueous solution depends on the pH:

ER^+	ER	ER^-	ER^{2-}	ER^{3-}
pH ~ 0	pH ~ 1	pH ~ 3	pH ~ 7	pH ~ 12
pink	orange	pink-red	yellow	violet

In weakly acidic medium, Eriochrome Cyanine R reacts with aluminium to form a violet-red, water-soluble complex which is used for the spectrophotometric determination of aluminium [48–51]. Hill [50] proved that the ratio Al:ER in the complex is 1:3 in the presence of an excess of Eriochrome Cyanine R. At lower Eriochrome Cyanine R concentrations, 1:1 and 1:2 complexes are formed.

The absorption spectra of Eriochrome Cyanine R and its aluminium complex at pH 6·2 are shown in Fig. 5.1. The molar absorptivity of the complex at $\lambda_{max} = 535$ nm is $\sim 6.5 \times 10^4$ ($a = 2.4$). The Eriochrome Cyanine R method is, therefore, the most sensitive spectrophotometric method for determining aluminium.

The optimum pH for the colour reaction lies between the rather narrow limits 6·1–6·2. The absorptivity of the complex drops rapidly at higher or lower pH values [51].

Eriochrome Cyanine R should be added to an acidic solution of aluminium (pH 1–2) before an acetate buffer is introduced. The aluminium complex is unstable at elevated temperatures; heating the solution (even to 40°C) probably accelerates hydrolysis of the complex.

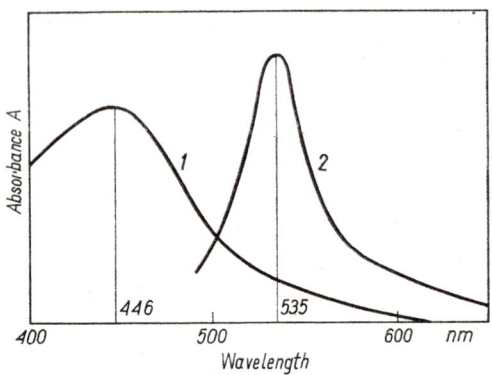

Fig. 5.1 Absorption spectra of Eriochrome Cyanine R (1) and its aluminium complex at pH 6·2 (2)

Iron(III) interferes, but if reduced to Fe(II) with ascorbic or thioglycollic acid, iron does not interfere even in hundredfold amount relative to aluminium. Titanium and vanadium give positive errors when present in larger amounts than aluminium, but they can be masked with H_2O_2. With chromium(III) present in fiftyfold amount the colour of the Al–ER system is diminished by 50%. Copper, which forms a coloured complex with Eriochrome Cyanine R, is masked with thiosulphate. Before its determination, aluminium should be separated from Be, V, and Zr. Tartrate and citrate interfere, but there is no interference from La, In, Zn, Pb, Ni, Sn, and Mn.

The chromophoric group of Eriochrome Cyanine R is reversibly reduced by such reducing agents as sodium bisulphite or hyposulphite. The reaction can be reversed by addition of hydrogen peroxide. Some other reducing agents, e.g. thioglycollic acid, ascorbic acid, hydroxylamine, and hydrogen sulphide, have no effect on Eriochrome Cyanine R.

The Eriochrome Cyanine R method has been used for determining aluminium in steel [1,7,53–56], iron ores [50], copper alloys [49,52,57], zinc [58], silicon tetrachloride [51], biological materials [9,18], food [9], and water [59].

Reagents

1. Eriochrome Cyanine R: 0·10% solution in water adjusted with hydrochloric acid to pH ~ 2·5 (1 ml of 1M HCl per 250 ml of solution).
2. Standard aluminium solution: 1 mg/ml (p. 113).

Procedure

To about 25 ml of the sample solution containing not more than 15 μg of Al, add 2 ml of 1% ascorbic acid solution. Adjust the sample solution to pH ∼ 2·0. After 5 minutes, add 5 ml of the Eriochrome Cyanine R solution and 5 ml of 50% ammonium acetate solution with stirring. Adjust the solution to pH 6·1–6·2 with dilute ammonia solution added dropwise. Dilute the solution to 50 ml in a volumetric flask with water, and measure the absorbance at 535 nm, using a blank solution as reference.

5.2.3 Chrome Azurol S Method

Chrome Azurol S (Alberon) (formula, p. 144) is a triphenylmethane dye similar to Eriochrome Cyanine R. The reagent reacts with aluminium ions to form a water-soluble blue complex which can be used for the spectrophotometric determination of aluminium [60–63].

The molar absorptivity (under the conditions specified in Procedure) is $4·9 \times 10^4$ at $\lambda_{opt} = 610$ nm (specific absorptivity $a = 1·81$). The method is thus less sensitive than the Eriochrome Cyanine R method, but the pH is less critical. Moreover, in this method there is a greater contrast between the colour of the reagent (orange) and that of the complex (blue). The absorbance of the free reagent is insignificant at 610 nm. Solutions of the Chrome Azurol S–aluminium complex do not obey Beer's law.

The intensity of the colour of the complex in solution depends on the concentration of Chrome Azurol S. The colour weakens with increasing concentration of sodium acetate buffer.

Ascorbic acid, thioglycollic acid, or hydroxylamine are used to reduce Fe(III) which interferes [64]. Copper is masked by thiosulphate.

In the presence of cetyltrimethylammonium chloride, the absorbance of the Al–Chrome Azurol S complex is less sensitive to pH changes, and the absorption maximum is shifted to 620 nm [65].

By the Chrome Azurol S method, aluminium has been determined in steels [3,60–62], various alloys [60], iron ores [62,66], tin [67], magnesium and titanium compounds [68], uranium alloys [69], magnetic alloys [70], and zircons [71].

Reagents

1. Chrome Azurol S: 0·10% solution.
2. Standard aluminium solution: 1 mg/ml (p. 113).

Procedure

To a solution in dilute hydrochloric acid, containing not more than 25 μg of Al, add 2 ml of 1% ascorbic acid solution and adjust the solution with ammonia to pH ∼ 2. After 5 minutes dilute the solution with water to ∼ 30 ml, and add 5 ml of Chrome Azurol S solution and 2 ml of 25%

sodium acetate solution. Adjust the solution with ammonia to pH 6·0 (±0·2), dilute with water to 50 ml in a volumetric flask, and measure the absorbance at 610 nm, using a reagent blank as reference.

5.2.4 Other Methods

Until recently, Aluminon (ammonium aurintricarboxylate) was an important colorimetric reagent for aluminium. This triphenylmethane reagent forms a sparingly soluble red chelate with aluminium ions in acetate buffer [72]. Protective colloids (e.g. gelatin) are used to stabilize the lake [59, 73-75a].

Other triphenylmethane dyes forming water-soluble coloured complexes are now more popular reagents for aluminium. The methods using Eriochrome Cyanine R and Chrome Azurol S have been discussed in detail above.

Recently, methods using Pyrocatechol Violet [76-80], Xylenol Orange [81-85], and Methylthymol Blue [86-89] have become popular. The reactions of these reagents with aluminium ions and their applications in the determination of aluminium are similar to those of Eriochrome Cyanine R and Chrome Azurol S, but the sensitivity is lower ($\varepsilon = 2\cdot1 \times 10^4$ for Xylenol Orange; $\varepsilon = 1\cdot9 \times 10^4$ for Methylthymol Blue). A considerable enhancement of sensitivity is obtained when the ternary complex of aluminium with Pyrocatechol Violet and the cetyltrimethylammonium ion is formed [80]. Aluminium has been determined with these three triphenylmethane reagents in uranium [82], bismuth [89], minerals [77], soil extracts [83], steel and copper alloys [78], titanium and titanium compounds [84, 86,87], and water [85].

The following triphenylmethane dyes have also been recommended for determining aluminium: Chromoxane Violet R [90,91], Chromoxane Pure Blue B [92] and Sulphochrome [93]. The structurally related reagent Pyrogallol Red is also used [79,94].

Stilbazo is an important azo reagent for aluminium [95-97]. The molar absorptivity of the aluminium complex is $1\cdot8 \times 10^4$ at 496 nm. The

reagent absorbs fairly strongly at λ_{max} for the complex. With Stilbazo, aluminium may be determined in the presence of a considerable excess of chromium(III). This method has been used for determining aluminium, *inter alia*, in steel [96] and silver [97].

A number of other azo dyes have been suggested as reagents for aluminium (e.g. Arsenazo I [6,98], Chromotrope 2C [99], SPADNS [100], Calcichrome (formula, p. 187) [101], Stilbazochrome ($\varepsilon = 5\cdot8 \times 10^4$ at 665 nm) [102], Stilbazogall I [103], Calmagite [104], Sulphochlorophenol S (formula, p. 50) [105], 5-sulpho-4'-diethylamino-2,2'-dihydroxyazobenzene

($\varepsilon = 4\cdot 1 \times 10^4$ at 540 nm) [106], chlorocyanoformazan [107], and 2-(2-carboxypyridyl-3-azo)-chromotropic acid [108]).

In slightly acidic media Alizarin S (Alizarin Red S) (formula, p. 611) forms a sparingly soluble red complex with aluminium ions which exists as a stable suspension [109–112]. According to Parker and Goddard [109], a ternary complex is formed in the presence of calcium, resulting in enhanced sensitivity. Aluminium has been determined by the Alizarin S method in, *inter alia*, biological materials [110], steel [111], and silicate minerals [112].

Biryuk and Ravitskaya [113] have studied a number of 9-substituted trihydroxyfluorones as spectrophotometric reagents for aluminium. The best was 9-(5-bromo-2-hydroxyphenyl)-2,3,7-trihydroxy-6-fluorone ($\varepsilon = 7\cdot 5 \times 10^4$ at 540 nm).

Other organic spectrophotometric reagents for the determination of aluminium are ferron (a derivative of 8-hydroxyquinoline) [114–116], 2-salicylideneaminophenol [117,118], 2-quinizarinsulphonic acid [119], 2-phenoxyquinizarin-3,4′-disulphonic acid [120], phenoxydinaphthofuchsonedicarboxylic acid [121], dibromopyrogallolsulphophthalein [122], and haematoxylin [123,124].

Alizarin Complexone has been applied to determination of aluminium [125].

References

1. Hill, U. T., *Anal. Chem.* **31**, 429 (1959).
2. Burke, K. E., *ibid.* **38**, 1608 (1966).
3. Konkin, V. D. and Kvichko, L. A., *Zavodsk. Lab.* **37**, 538 (1971).
4. Marczenko, Z., *Mikrochim. Acta* **1965**, 281.
5. Tananaev, I. V. and Vinogradova, A. D., *Zh. Analit. Khim.*, **14**, 487 (1959).
6. Kuznetsov, V. I. and Golubtsova, R. B., *Zavodsk. Lab.* **22**, 161 (1956).
7. Blair, D., Power, K. Griffiths, D. L. and Wood, J. H., *Talanta* **7**, 80 (1960).
8. Corbett, J. A., *Analyst* **78**, 20 (1953).
9. Thaler, H. and Mühlberger, F. H., *Z. Anal. Chem.* **144**, 241 (1955).
10. Steinbach, J. F. and Freiser, H., *Anal. Chem.* **26**, 375 (1954).
11. Chalmers, R. A. and Basit, M. A., *Analyst* **93**, 629 (1968).
12. Donaldson, E. M., *Talanta* **18**, 905 (1971).
13. Scribner, W. G., Treat, W. J., Weis, J. D. and Moshier, R. W., *Anal. Chem.* **37**, 1136 (1965).
14. Kambara, T. and Hashitani, H., *ibid.* **31**, 567 (1959).
15. Zolotov, Yu. A., Demina, L. A. and Petrukhin, O. M., *Zh. Analit. Khim.* **25**, 1487 (1970).
16. Villarreal, R., Krsul, J. R. and Barker, S. A., *Anal. Chem.* **41**, 1420 (1969).
17. Horton, A. D. and Thomason, P. F., *ibid.* **28**, 1326 (1956).
18. Seibold, M., *Z. Anal. Chem.* **173**, 388 (1960).
19. Evans, H. B. and Hashitani, H., *Anal. Chem.* **36**, 2032 (1964).
20. Freund, H. and Miner, F. J., *ibid.* **25**, 564 (1953).
21. Teicher, H. and Gordon, L., *ibid.* **23**, 930 (1951); *Anal. Chim. Acta* **9**, 507 (1953).
22. Korkisch, J. and Hazan, I., *Anal. Chem.* **36**, 2308 (1964).
23. Noll, C. A. and Stefanelli, L. J., *ibid.* **35**, 1914 (1963).
24. Marczenko, Z., *Chem. Anal. (Warsaw)* **4**, 437 (1959).

25. Tikhonov, V. N., *Zh. Analit. Khim.* **21**, 829 (1966).
26. Kassner, J. L. and Ozier, M. A., *Anal. Chem.* **23**, 1453 (1951).
27. Kenyon, O. A. and Bewick, H. A., *ibid.* **24**, 1826 (1952).
28. Goldstein, G., Manning, D. L. and Menis, O., *Talanta* **2**, 52 (1959).
29. Dagnall, R. M., West, T. S. and Young, P., *Analyst* **90**, 13 (1965).
30. Riley, J. P., *Anal. Chim. Acta* **19**, 413 (1958).
31. Zibulsky, H., Slowinski, M. F. and White, J. A., *Anal. Chem.* **31**, 280 (1959).
32. Middleton, K. R., *Analyst* **89**, 421 (1964).
33. Motojima, K. and Hashitani, H., *Bull. Chem. Soc. Japan* **29**, 458 (1956); *Japan Analyst* **6**, 642 (1957); **7**, 478 (1958).
34. Claassen, A., Bastings, L. and Visser, J., *Anal. Chim. Acta* **10**, 373 (1954).
35. Chernikhov, Yu. A. and Dobkina, B. M., *Zavodsk. Lab.* **25**, 131 (1959).
36. Pyatnitskii, I. V. and Glushchenko, L. M., *Zh. Analit. Khim.* **25**, 1491 (1970).
37. Specker, H., Kuchtner, M. and Hartkamp, H., *Z. Anal. Chem.* **142**, 166 (1954).
38. Blazejak-Ditges, D., *Z. Anal. Chem.* **246**, 241 (1969).
39. Tumanov, A. A. and Petukhova, V. G., *Zavodsk. Lab.* **35**, 654 (1969).
40. Margerum, D. W., Sprain, W. and Banks, C. W., *Anal. Chem.* **25**, 249 (1953).
41. Angermann, W., Kässner, B., Lorenz, G. and Geissler, M., *Chem. Anal. (Warsaw)* **16**, 261 (1971).
42. Pollock, E. N. and Zopatti, L. P., *Anal. Chim. Acta* **28**, 68 (1963).
43. Riley, J. P. and Williams, H. P., *Mikrochim. Acta* **1959**, 825.
44. Bolleter, W. T., *Anal. Chem.* **31**, 201 (1959).
45. Paul, J., *Mikrochim. Acta* **1966**, 1075.
46. Frink, C. R. and Peaslee, D. E., *Analyst* **93**, 469 (1968).
47. Belcher, R., Crossland, B. and Fennell, T. R., *Talanta* **17**, 639 (1970).
48. Hegedüs, A. J., *Mikrochim. Acta* **1963**, 831.
49. Pohl, H., *Z. Anal. Chem.* **133**, 322 (1951).
50. Hill, U. T., *Anal. Chem.* **28**, 1419 (1956); **38**, 654 (1966).
51. Marczenko, Z., Kasiura, K. and Mojski, M., *Chem. Anal. (Warsaw)* **16**, 203 (1971).
52. Mal'tsev, V. F., Luk'yanenko, L. L. and Kukui, D. M., *Zavodsk. Lab.* **27**, 807 (1961).
53. Scholes, P. H. and Smith, D. V., *Analyst* **83**, 615 (1958).
54. Lilie, H. and Rosin, H., *Z. Anal. Chem.* **160**, 261 (1958).
55. Wille, K. D., *ibid.* **250**, 23 (1970).
56. Scholes, P. H. and Thulbourne, C., *Analyst* **88**, 702 (1963).
57. Dozinel, C., *Chim. Anal. (Paris)* **38**, 244 (1956).
58. Ikenberry, L. C. and Thomas, A., *Anal. Chem.* **23**, 1806 (1951).
59. Giebler, G., *Z. Anal. Chem.* **184**, 401 (1961).
60. Kashkovskaya, E. A. and Mustafin, I. S., *Zavodsk. Lab.* **24**, 1189 (1958).
61. Brockmann, H. and Keller, H., *Arch. Eisenhüttenw.* **35**, 367 (1964).
62. Buck, L., *Chim. Anal. (Paris)* **47**, 10 (1965).
63. Pakalns, P., *Anal. Chim. Acta* **32**, 57 (1965).
64. Tikhonov, V. N., *Zh. Analit. Khim.* **26**, 65 (1971).
65. Shijo, Y. and Takeuchi, T., *Japan Analyst* **17**, 61 (1968).
66. Bhargava, O. P. and Hines, W. G., *Anal. Chem.* **40**, 413 (1968).
67. Shvaiger, M. I. and Rudenko, É. I., *Zavodsk. Lab.* **26**, 939 (1960).
68. Tikhonov, V. N., *Zh. Analit. Khim.* **19**, 1204 (1964).
69. Mal'tseva, L. S. and Kubareva, L. V., *Zavodsk. Lab.* **35**, 1299 (1969).
70. Babenko, A. S. and Volodchenko, T. T., *ibid.* **35**, 650 (1969).
71. Boix, A. and Debras-Guédon, J., *Chim. Anal. (Paris)* **53**, 459 (1971).
72. Smith, W. H., Sager, E. E. and Siewers, I. J., *Anal. Chem.* **21**, 1334 (1949).
73. Luke, C. L. and Braun, K. C., *ibid.* **24**, 1120, 1122 (1952).
74. Codell, M. and Norwitz, G., *ibid.* **25**, 1437 (1953).
75. Committee on Methods for the Analysis of Trade Effluents, *Analyst* **82**, 443 (1957).

75a. Herrmann, M. and Weber, H., *Z. Anal. Chem.* **267**, 13 (1973).
76. Anton, A., *Anal. Chem.* **32**, 725 (1960).
77. Wilson, A. D. and Sergeant, G. A., *Analyst* **88**, 109 (1963).
78. Mustafin, I. S., Molot, L. A. and Arkhangel'skaya, A. S., *Zh. Analit. Khim.* **22**, 1808 (1967).
79. Mustafin, I. S., Arkhangel'skaya, A. S. and Molot, L. A., *Tr. Komis. po Analit. Khim., Akad. Nauk SSSR* **17**, 205 (1969).
80. Chester, J. E., Dagnall, R. M. and West, T. S., *Talanta* **17**, 13 (1970).
81. Otomo, M., *Bull. Chem. Soc. Japan* **36**, 809 (1963).
82. Budeshinskii, B., *Zh. Analit. Khim.* **18**, 1071 (1963).
83. Pritchard, D. T., *Analyst* **92**, 103 (1967).
84. Tikhonov, V. N., *Zh. Analit. Khim.* **20**, 941 (1965).
84a. Tikhonov, V. N. and Petrova, L. F., *ibid.* **28**, 1413 (1973).
85. Dvořák, J. and Nývltová, E., *Mikrochim. Acta* **1966**, 1082.
86. Tikhonov, V. N., *Zh. Analit. Khim.* **21**, 275 (1966).
87. Tikhonov, V. N. and Grankina, M. Ya., *Zavodsk. Lab.* **32**, 278 (1966).
88. Nazarenko, V. A. and Nevskaya, E. M., *Zh. Analit. Khim.* **24**, 839 (1969).
89. Vinogradov, A. V. and Filippova, M. P., *Zavodsk. Lab.* **35**, 1165 (1969).
90. Mustafin, I. S. and Lisenko, N. F., *Zh. Analit. Khim.* **17**, 1052 (1962).
91. Lisenko, N. F. and Mustafin, I. S., *ibid.* **22**, 25 (1967).
92. Sheyanova, F. R. and Malenskaya, V. P., *Zavodsk. Lab.* **23**, 907 (1957).
93. Petrova, G. S., Lukin, A. M., Etingen, N. B., Molot, L. A. and Arkhangel'skaya, A. S., *Zh. Analit. Khim.* **24**, 1332 (1969).
94. Tanaka, T., Nakagawa, Y. and Honda, S., *Japan Analyst* **10**, 1148 (1961).
95. Wetlesen, C. U. and Omang, S. H., *Anal. Chim. Acta* **24**, 294 (1961).
96. Wetlesen, C. U., *ibid.* **26**, 191 (1962).
97. Kabanova, O. L. and Danyushchenkova, M. A., *Zh. Analit. Khim.* **18**, 780 (1963).
98. Kuznetsov, V. I. and Golubtsova, R. B., *Zavodsk. Lab.* **21**, 1422 (1955).
99. Majumdar, A. K. and Savariar, C. P., *Z. Anal. Chem.* **174**, 269 (1960).
100. Kotelyanskaya, L. I. and Kish, P. P., *Zavodsk. Lab.* **36**, 523 (1970).
101. Ishii, H. and Einaga, H., *Bull. Chem. Soc. Japan* **39**, 1721 (1966).
102. Cherkesov, A. I., Kazakov, B. I. and Shchepko, V. I., *Zavodsk. Lab.* **34**, 786 (1968).
103. Elinson, S. V., Pushinov, Yu. V. and Tsvetkova, V. T., *Zh. Analit. Khim.* **26**, 718 (1971).
104. Woodward, C. and Freiser, H., *Talanta* **15**, 321 (1968).
105. Romanov, P. N. and Sokolovskaya, L. A., *Tr. Komis. po Analit. Khim. Akad. Nauk SSSR* **17**, 357 (1969).
106. Florence, T. M., *Anal. Chem.* **37**, 704 (1965).
107. Malevannyi, V. A., Ermakova, M. I. and Lel'chuk, Yu. L., *Zavodsk. Lab.* **35**, 414 (1969).
108. Majumdar, A. K. and Chatterjee, A. B., *Mikrochim. Acta* **1967**, 663.
109. Parker, C. A. and Goddard, A. P., *Anal. Chim. Acta* **4**, 517 (1950).
110. Oelschläger, W., *Z. Anal. Chem.* **154**, 321 (1957).
111. Corbett, J. A. and Guerin, B. D., *Analyst* **91**, 490 (1966).
112. King, H. G. and Pruden, G., *ibid.* **93**, 601 (1968).
113. Biryuk, E. A. and Ravitskaya, R. V., *Zh. Analit. Khim.* **27**, 459 (1972).
114. Bril, J., *Mikrochim. Acta* **1958**, 212.
115. Green, H., *Metallurgia* **57**, 157 (1958).
116. Goto, K., Tamura, H., Onodera, M. and Nagayama, M., *Talanta* **21**, 183 (1974).
117. Tumanov, A. A. and Efimychev, V. S., *Zh. Analit. Khim.* **22**, 700 (1967).
118. Babko, A. K. and Lisichenok, S. L., *Ukr. Khim. Zh.* **35**, 98 (1969).
119. Owens, E. G., II, and Yoe, J. H., *Anal. Chem.* **31**, 384 (1959).
120. Owens, E. G., II, and Yoe, J. H., *Talanta* **8**, 505 (1961).

121. Adamovich, L. P., Mirnaya, A. P. and Khukhryanskaya, A. K., *Zavodsk. Lab.* **35**, 781 (1969).
122. Bokra, Y. and Luca, C., *Bull. Soc. Chim. France* **1970**, 3761.
123. Nowicka-Jankowska, T., Gołkowska, A., Pietrzak, I. and Żmijewska, W., *Chem. Anal. (Warsaw)* **3**, 977 (1958).
124. Kapitańczyk, K., Kurzawa, Z. and Miedziński, M., *Roczniki Chem.* **30**, 607 (1956).
125. Ingman, F., *Talanta* **20**, 999 (1973).

Chapter 6

ANTIMONY

Antimony (Sb, at.wt. 121·75) occurs in its compounds in the oxidation states −III (stibine, SbH_3), +III (antimonious compounds), and +V (antimonic compounds). The antimonious compounds are the most stable. The Sb^{3+} ions hydrolyse at pH values as low as ∼ 1, but the hydroxide $Sb(OH)_3$ dissolves at pH ∼ 10, forming the antimonite ion, SbO_2^-. Antimony(V) is more acidic than antimony(III). Antimony(III and V) form sulphide, halide, tartrate, and oxalate complexes, and also weak sulphate complexes.

6.1 Methods of Separation

6.1.1 Precipitation

Trace amounts of antimony are separated from an acid medium (HNO_3, H_2SO_4) as hydrated antimonic acid with hydrous manganese dioxide as the collector [1-6]. The latter is formed *in situ*, by the slow reaction of a hot solution of Mn^{2+} and MnO_4^- ions. The MnO_2 collector can also be formed by the reaction between MnO_4^- ions and ethanol [5]. Although antimony is quantitatively precipitated within the pH range 1-7, the reaction is usually carried out at pH 1-1·2 to prevent coprecipitation of other metals such as Bi, As, Au, Fe(III), Tl(III), Pb, and Cu. Both tin and antimony are precipitated quantitatively. Reagents which complex antimony (e.g. fluoride) interfere. If both lead and sulphate are present, lead sulphate must be removed before precipitation of the antimony. The MnO_2 precipitate with the collected antimony is filtered off and dissolved in hydrochloric acid (1+1) containing some H_2O_2. The filter paper can be mineralized in concentrated H_2SO_4 containing some concentrated HNO_3.

Traces of antimony are precipitated with hydrogen sulphide [7,8] from 0·5-1N HCl or H_2SO_4 in the absence of large amounts of interfering Analytical Group I and Group II metals (which precipitate as sulphides from acid media). Copper or molybdenum collectors are used. Tin does not precipitate in the presence of oxalic acid, while tungsten and vanadium are masked by tartaric acid.

Antimony can be deposited from a solution by the cementation method, e.g. on tin [9], copper, or iron. Traces of antimony in tellurium have been coprecipitated with tellurous acid and separated by precipitation of elemental tellurium [10].

6.1.2 Extraction

The antimony(V) chloride complex can be extracted into di-isopropyl ether from 1–10M hydrochloric acid [11–13]. Although the most efficient extraction (~ 99%) occurs at a concentration of 5–7M HCl, it is convenient to extract antimony(V) from 2M HCl (94% extraction) since iron(III) is not extracted under these conditions. A considerable difference in the distribution coefficients of the Sb(V) and the Sb(III) chloride complexes into di-isopropyl ether enables these two forms of antimony to be separated [11]. Iron(III) can be separated by extraction from 6–7M HCl before the Sb(III) is oxidized to Sb(V). Antimony chloride complexes can be extracted with inert solvents in the presence of amines, e.g. tribenzylamine [14] or tridodecylamine [15].

From a 10N H_2SO_4 and 0·005–0·05M KI medium, antimony(III) can be extracted into benzene as the iodide, SbI_3 [16,17]. Antimony is then re-extracted with 1N sulphuric acid. In the presence of bromide, antimony is extracted into benzene as $SbBr_3$ from H_2SO_4 medium [17].

Extraction of the antimony(III) dithiocarbamate complexes from acid media enables antimony to be separated from several metals [7,18,19]. A preliminary extraction of bismuth and copper dithiocarbamates can be carried out after oxidation of antimony to Sb(V) [18].

Antimony can also be extracted as complexes with cupferron [20,21] or BPHA [21].

6.1.3 Ion-Exchange

Antimony(III) can be separated on cation-exchangers from Cu, Ni, Co, Cd, Zn, Mn(II), Fe(III), and Bi by elution with tartrate [22–24]. These metals form weaker tartrate complexes than antimony, and are retained in the column.

Antimony and tin have been separated on an Amberlite IRA-400 anion-exchanger in the malonate form [25]. The antimony(V) chloride complex is retained by strongly basic anion-exchangers [26].

6.1.4 Distillation

Traces of antimony have been separated as the volatile hydride, SbH_3. The antimony is reduced in fairly concentrated hydrochloric acid with amalgamated zinc [27], or in a sulphuric acid solution of cadmium sulphate with zinc shot [28]. When analysing organic substances, Jeník [29] converted the antimony into magnesium antimonide, decomposed the latter with dilute sulphuric acid, and trapped the evolved SbH_3 by oxidation in a sodium nitrite scrubber.

Under suitable conditions, antimony can be separated from arsenic and tin by distillation of volatile $SbCl_3$ from 11M HCl [30,31].

6.2 Methods of Determination

The extractive-spectrophotometric methods for determining antimony, based on the ion-association complexes (ion-pairs) of $[SbCl_6]^-$ with Rhodamine B and other basic dyes, are sensitive and selective. A simple iodide method is commonly used to determine higher antimony concentrations.

6.2.1 RHODAMINE B METHOD

With Rhodamine B (p. 53), the antimony(V) chloride complex, $[SbCl_6]^-$, forms a sparingly water-soluble ion-pair $[Rhod. B]^+[SbCl_6]^-$, which is extractable into benzene or di-isopropyl ether. The violet-pink solutions of the complex in these solvents are the basis of a sensitive spectrophotometric method for antimony [8,28,32–36]. Rhodamine B dissolves in acid, giving a pink solution, but is insoluble in benzene and di-isopropyl ether. Figure 6.1 shows the absorption spectrum of the complex in di-isopropyl ether ($\varepsilon = 9\cdot7 \times 10^4$, $a = 0\cdot80$, at $\lambda_{max} = 552$ nm). The absorption maximum of the complex in benzene is at 565 nm.

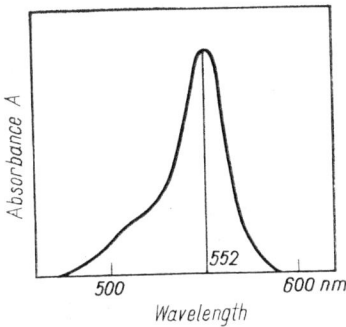

Fig. 6.1 Absorption spectrum of the ion-association complex of $SbCl_6^-$ with Rhodamine B in di-isopropyl ether

Since the solubility of $[Rhod. B]^+[SbCl_6]^-$ in benzene is rather low, the antimony concentration must not exceed 2 µg/ml of benzene. The solubility in di-isopropyl ether is much higher.

Ceric sulphate is most often used as oxidant since it oxidizes Sb(III) to Sb(V) quantitatively and quickly in $6M$ HCl, even in the cold. Since the excess of cerium(IV) could oxidize Rhodamine B, it is reduced with hydroxylamine, which does not reduce the antimony(V) under the conditions of the reaction.

That cerium(IV) is reduced by the hydrochloric acid itself is indicated by the bleaching of the yellow solution even before the addition of $NH_2OH \cdot HCl$. Hydroxylamine also reduces the chlorine formed, which, if left in the solution, would oxidize the Rhodamine B added later. As an alternative to cerium(IV), sodium nitrite can be used for oxidizing antimony(III) (the excess is reduced with urea) [37].

Metals forming chloride complexes which give the same reactions as antimony with Rhodamine B interfere in the determination. These metals are Au(III), Tl(III), Ga(III), and Fe(III). Gold can be separated after reduction to the element with sulphite. Gallium and iron can be separated by extraction as chloride complexes before the oxidation of antimony to Sb(V). Traces of iron(III) are masked with phosphoric acid.

The antimony(V) chloride complex can be readily extracted with di-isopropyl ether from 1·5–2M HCl [38]. The colourless ether extract is shaken with an aqueous solution of Rhodamine B to form the coloured complex in the ether. The iron(III) chloride complex is only slightly extracted by di-isopropyl ether from 1·5–2M HCl [38].

If antimony(V) is not first extracted into ether, Rhodamine B is added after the excess of the oxidizing agent has been reduced, and the aqueous solution is then extracted with benzene or di-isopropyl ether. Alcohol can be added to clear any turbidity in the benzene extract [33].

Rhodamine B must be added rapidly to the solution after the oxidation of antimony, since $[SbCl_6]^-$ slowly hydrolyses to $[Sb(OH)Cl_5]^-$ which does not react with Rhodamine B [35].

Antimony has been determined by the Rhodamine B method in lead [2,39,40], tin and its alloys [41], arsenic [42], copper [43], zinc [13], steel [44], germanium and silicon [45], titanium dioxide [36], soils and rocks [28,37], and sea water [5].

Reagents

1. Rhodamine B: 0·02% solution in 1M HCl.
2. Standard antimony solution: 1 mg/ml. Dissolve 2·668 g of potassium antimonyl tartrate $KSbO.C_4H_4O_6$, in hydrochloric acid (1+1) and dilute the solution accurately to 1 litre with this acid. Working solutions are obtained by suitable dilution of the stock solution with hydrochloric acid (1+1).
3. Potassium permanganate: 1% solution.
4. Manganous nitrate: 1% solution.
5. Ceric sulphate: 3% solution in 1N H_2SO_4.
6. Hydroxylamine hydrochloride: 1% solution. The solution is unstable.

Procedure

Separation of Sb by coprecipitation with MnO_2. To the chloride-free sample solution, containing 3 ml of concentrated HNO_3 per 100 ml of the solution, and heated almost to boiling, add 1 ml of $KMnO_4$ solution and 2 ml of $Mn(NO_3)_2$ solution. Keep the solution just below boiling for 30 minutes. Filter off the precipitate on a small filter paper and wash with hot dilute HNO_3 (1+100). Dry the filter paper and the precipitate and ignite; fuse the residue with 0·3 g of Na_2O_2 and a granule of NaOH in a small nickel crucible. Heat the melt till dark red. Leach the cooled melt with hot water, transfer all the contents of the crucible to a beaker, and acidify with 5 ml of concentrated HCl. Evaporate the solution to a suitable volume, add 5 ml of Rhodamine B solution, and extract with benzene or di-isopropyl ether as described below.

Separation of Sb by extraction. To the sample solution containing not more than 50 μg of Sb add sufficient concentrated HCl for its concentration in the solution to be $6M$ (the solution volume is 20–30 ml). Add 10 drops of the $Ce(SO_4)_2$ solution and stir well. After 5 minutes, add 10 drops of the $NH_2OH \cdot HCl$ solution and stir well. After 2 minutes dilute the solution with two volumes of water, transfer it to a separating funnel, and shake with two portions of di-isopropyl ether for 30 seconds. Wash the ether extract with two portions of $1M$ HCl.

Determination of Sb. To the separating funnel with the ether solution of the antimony(V) chloride complex add 5 ml of the Rhodamine B solution and shake. Transfer the coloured ether extract to a 50-ml (or smaller) volumetric flask (according to the antimony content), add di-isopropyl ether to the mark, mix, and measure the absorbance at 552 nm, using the solvent as a reference.

When using benzene as a solvent proceed as follows: 2 minutes after the addition of $NH_2OH \cdot HCl$, add 5 ml of the Rhodamine B solution, and dilute the solution with water so that the concentration of HCl is $2M$. Transfer the solution to the separating funnel and shake for 1 minute with 2 or 3 portions of benzene. Place the combined benzene extracts in a volumetric flask. If the extract is turbid (due to an emulsion), add 1 ml of ethanol. Dilute to the mark with benzene, mix, and measure the absorbance of the solution at 565 nm against benzene.

6.2.2 IODIDE METHOD

Antimony(III) forms a greenish-yellow complex with iodide in sulphuric acid medium [46,47]. The colour intensity increases with increasing iodide concentration up to 5% KI, after which it remains constant. The concentration of sulphuric acid in the final solution should be $1 \cdot 5$–$2 \cdot 5M$. When the concentration of H_2SO_4 is higher than $2 \cdot 5M$, an increase in the colour intensity is caused by fairly rapid liberation of iodide. To prevent liberation of iodine by atmospheric oxygen, ascorbic acid, hypophosphite, or sulphite are added as reducing agents.

The aqueous acid solution of the antimonious iodide complex $[SbI_4]^-$ exhibits absorption maxima at 425 nm ($\varepsilon = 4 \cdot 0 \times 10^3$, $a = 0 \cdot 033$) and 330 nm. The absorption spectra of both antimony and bismuth iodide complexes are shown in Fig. 6.2.

Bismuth also forms a yellow iodide complex, and thus interferes in the determination of antimony. In a solution containing 1% of KI, bismuth gives a colour reaction while antimony does not. By measuring the absorbances of sample solutions containing low (1%) and high (5%)

concentrations of potassium iodide, it is possible to determine both bismuth and antimony.

Antimony can be separated from bismuth by extraction into benzene from iodide medium. From $5M$ H_2SO_4 and $0.05M$ KI solution (4 ml of 10% KI solution in 50 ml of sample solution), SbI_3 is extracted in good yield, whereas the bismuth complex is not extracted at all.

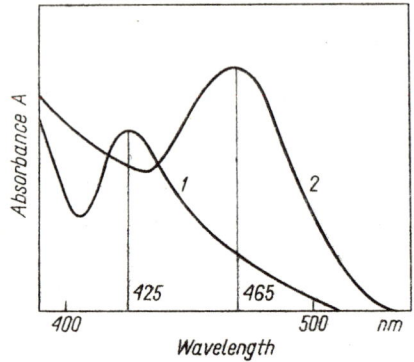

Fig. 6.2 Absorption spectra of antimony (1) and bismuth (2) iodide complexes in aqueous solutions

The colour of the benzene extract of SbI_3 is about one-fifth as intense as that of the $[SbI_4]^-$ complex in aqueous solution (under optimum conditions).

Any thallium present is precipitated as TlI [48] and separated by filtration. Copper is masked as a stable colourless complex with thiourea, which also acts as a reducing agent (instead of ascorbic acid, for example) [49].

In determination of antimony by the iodide method, As, Hg, Fe, Cu, Pb, and W do not interfere provided they are not present in greater amounts than Sb.

Antimony has been determined by the iodide method in cast iron [50,51], steels [52], lubricants [53], tin [54], lead [55], white metals [56], bronzes [49], and technical glass [57]. Antimony hydride in gaseous mixtures has been determined by the iodide method, after oxidation to Sb(III) [58].

A sensitive amplification method for determining antimony consists in extracting SbI_3 into benzene from $5M$ H_2SO_4 medium in the presence of a small excess of KI. The benzene extract is washed with sulphuric acid, and SbI_3 is re-extracted with water. In this aqueous extract, the iodide is oxidized to iodate with bromine. After removal of the excess of bromine, the iodate reacts with the potassium iodide added, to yield iodine which is determined by its colour reaction with starch (see determination of iodide, p. 296, and determination of palladium, p. 413). Thus 18 iodine atoms are formed for each antimony atom (in SbI_3). The molar absorptivity (with respect to Sb) in this method is $\sim 3\cdot 0 \times 10^5$ ($a = 2\cdot 6$).

Reagents

1. Potassium iodide: 40% solution.
2. Standard antimony solution: 1 mg/ml. The solution may be prepared either from potassium antimonyl tartrate (p. 124) or in the following manner. Dissolve 0·1000 g of powdered antimony metal in 25 ml of concentrated H_2SO_4. Dilute the cooled solution with water while stirring, cool, and make up to 100 ml in a volumetric flask. Working solutions are obtained by suitable dilutions of the stock solution with $0\cdot 5M$ H_2SO_4.

Procedure

To the sample solution in a 50-ml volumetric flask containing not more than 1 mg of Sb, add 10 ml of H_2SO_4 (1+1), 5 ml of 2% ascorbic acid solution, and 10 ml of potassium iodide solution. Dilute the solution to the mark with water and stir well. After 5 minutes, measure the absorbance of the yellow solution at 425 nm, using water as a reference.

6.2.3 OTHER METHODS

Kish and Onishchenko [59] have studied extractive–spectrophotometric determinations of antimony with xanthene dyes such as Pyronine G, Rhodamines B, G and 6G, and Butylrhodamine B (the Rhodamine B method is discussed above in detail). The molar absorptivities in these methods range from $8\cdot 5 \times 10^4$ to $1\cdot 2 \times 10^5$.

Sensitive extractive–spectrophotometric methods for determining antimony have also been based on ion-pair formation between $SbCl_6^-$ and the triphenylmethane dyes [60]: Methyl Violet [14, 20, 61–64], Crystal Violet [64–68], and Brilliant Green [9,64,69–74]. Other basic dyes used are Methylene Blue [75], Safranine T [76], Safranine O [77], antipyrine dyes [10,78], and azo dyes [79,80]. Traces of antimony have been determined with these reagents in tellurium [10], lead [20,63], various alloys [61], molybdenum and its compounds [62], germanium dioxide [67], organic matter [29], and air [68].

Another reagent recommended for spectrophotometric determination of antimony is methylfluorone [81–84]. In a weakly acidic medium (pH ~ 2),

the orange Sb(III)–methylfluorone complex, which exists as a stable sol in the presence of gelatin or poly(vinyl alcohol), is formed. The molar absorptivity is 4.0×10^4 at 530 nm. A method based on phenylfluorone is less sensitive and less selective [51,85–87].

Methods using coloured chelates of antimony(III) with azo dyes have been reported recently. These dyes include PAR [88], PAN [89], other pyridylazo compounds [90–92], thiazolylazo compounds [93], and 6-(2-quinolylazo)-3,4-dimethylphenol [94].

A number of other organic reagents have been used in spectrophotometric methods for antimony, e.g. Bromopyrogallol Red ($\varepsilon = 3.5 \times 10^4$ at 560 nm) [51,95], Pyrocatechol Violet [51,96], tetrahydroxyflavone (kaempferol) [97], 8-mercaptoquinoline [98], pyrrolidinedithiocarbamate [7], silver-DDTC (with SbH_3) [99], and the silver complex of p-sulphamoylbenzoic acid (with SbH_3) [100].

Methods based on the yellow antimony bromide complex [101] and on the reduced antimonomolybdophosphate species [102] have been described.

References

1. Babko, A. K. and Shtokalo, M. I., *Zavodsk. Lab.* **21**, 767 (1955).
2. Luke, C. L., *Anal. Chem.* **31**, 1680 (1959).
3. Ogden, D. and Reynolds, G. F., *Analyst* **89**, 538 (1964).
4. Reynolds, G. F. and Tyler, F. S., *ibid.* **89**, 579 (1964).
5. Portmann, J. E. and Riley, J. P., *Anal. Chim. Acta* **35**, 35 (1966).
6. Burke, K. E., *Anal. Chem.* **42**, 1536 (1970).
7. Kovács, E. and Guyer, H., *Z. Anal. Chem.* **186**, 267 (1962); **208**, 255 (1965).
8. Onishi, H. and Sandell, E. B., *Anal. Chim. Acta* **11**, 444 (1954).
9. Soldatova, L. A., Kilina, Z. G. and Kataev, G. A., *Zh. Analit. Khim.* **19**, 1267 (1964).
10. Busev, A. I., Tiptsova, V. G., Bogdanova, E. S. and Andreichuk, A. M., *ibid.* **20**, 812 (1965).
11. Edwards, F. C. and Voigt, A. F., *Anal. Chem.* **21**, 1204 (1949).
12. Schweitzer, G. K. and Storms, L. E., *Anal. Chim. Acta* **19**, 154 (1958).
13. Van Aman, R. E., Hollibaugh, F. D. and Kanzelmeyer, J. H., *Anal. Chem.* **31**, 1783 (1959).
14. Marchenko, P. V. and Voronina, A. I., *Zh. Analit. Khim.* **23**, 868 (1968).
15. Alian, A. and Sanad, W., *Talanta* **14**, 659 (1967).
16. Ramette, R. W., *Anal. Chem.* **30**, 1158 (1958).
17. Grimanis, A. P. and Hadzistelios, I., *Anal. Chim. Acta* **41**, 15 (1968).
18. Wyatt, P. F., *Analyst* **80**, 368 (1955).
19. Bode, H., *Z. Anal. Chem.* **144**, 165 (1955).
20. Cyrankowska, M. and Downarowicz, J., *Chem. Anal. (Warsaw)* **10**, 67 (1965).
21. Lyle, S. J. and Shendrikar, A. D., *Anal. Chim. Acta* **36**, 286 (1966).
22. Marczenko, Z., *Chem. Anal. (Warsaw)* **2**, 255 (1957).
23. Khorasani, S. S. and Khundkar, M. H., *Anal. Chim. Acta* **25**, 292 (1961).
24. Šulcek, Z., Böseová, M. and Doležal, J., *Collection Czech. Chem. Commun.* **34**, 787 (1969).
25. Dawson, J. and Magee, R. J., *Mikrochim. Acta* **1958**, 330.
26. Stroński, I. and Rybakov, W. N., *Chem. Anal. (Warsaw)* **4**, 877 (1959).
27. Zaĭkovskiĭ, F. V., *Zh. Analit. Khim.* **9**, 155 (1954).
28. Schnepfe, M. M., *Talanta* **20**, 175 (1973).
29. Jeník, J., *Collection Czech. Chem. Commun.* **23**, 1056 (1958).
30. Bruyne, P. de and Hoste, J., *Bull. Soc. Chim. Belg.* **70**, 221 (1961).

References

31. Patek, P., *Mikrochim. Acta* **1968**, 282.
32. Kuznetsov, V. I., *Dokl. Akad. Nauk SSSR* **52**, 231 (1946); *Zavodsk. Lab.* **18**, 618 (1950).
33. Nielsch, W. and Böltz, G., *Z. Anal. Chem.* **143**, 264 (1954).
34. MacNulty, B. J. and Woollard, L. D., *Anal. Chim. Acta* **13**, 64 (1955).
35. Ramette, R. W. and Sandell, E. B., *ibid.* **13**, 455 (1955); *J. Am. Chem. Soc.* **78**, 4872 (1956).
36. Jablonski, W. Z. and Watson, C. A., *Analyst* **95**, 131 (1970).
37. Greenhalgh, R. and Riley, J. P., *Anal. Chim. Acta* **27**, 305 (1962).
38. Ward, F. N. and Lakin, H. W., *Anal. Chem.* **26**, 1168 (1954).
39. Luke, C. L., *ibid.* **25**, 674 (1953).
40. Corbett, J. A., *Metallurgia* **65**, 43 (1962).
41. Coppins, W. C. and Price, J. W., *ibid.* **53**, 183 (1956).
42. de Souza, T. L. and Kerbyson, J. D., *Anal. Chem.* **40**, 1146 (1968).
43. Dozinel, C. M., *Z. Anal. Chem.* **157**, 401 (1957).
44. Kidman, L. and Waite, C. B., *Metallurgia* **66**, 143 (1962).
45. Luke, C. L. and Campbell, M. E., *Anal. Chem.* **25**, 1588 (1953).
46. Elkind, A., Gayer, K. H. and Boltz, D. F., *ibid.* **25**, 1744 (1953).
47. Washington, R. A., *Analyst* **90**, 502 (1965).
48. Maurice, M. J. and van Lingen, R. L., *Anal. Chim. Acta* **28**, 91 (1963).
49. Volodarskaya, R. S., *Zavodsk. Lab.* **25**, 143 (1959).
50. Rooney, R. C., *Analyst* **82**, 619 (1957).
51. Amsheeva, A. A., *Izv. Vyssh. Ucheb. Zaved., Khim. Khim. Tekhnol.* **13**, 339 (1970).
52. Blazejak-Ditges, D. and Klingeleers, H., *Z. Anal. Chem.* **248**, 18 (1969).
53. Norwitz, G. and Gralan, M., *Anal. Chim. Acta* **61**, 413 (1972).
54. Coppins, W. C. and Price, J. W., *Metallurgia* **46**, 52 (1952).
55. Bassett, J. and Jones, J. C., *Analyst* **91**, 176 (1966).
56. Dym, A., *ibid.* **88**, 232 (1963).
57. Völker, H. and Schwarz, G., *Silikattechnik* **22**, 44 (1971).
58. Gann, W., *Z. Anal. Chem.* **221**, 254 (1966).
59. Kish, P. P. and Onishchenko, Yu. K., *Zh. Analit. Khim.* **26**, 514 (1971).
60. Kish, P. P., Onishchenko, Yu. K. and Pogoida, I. I., *ibid.* **28**, 1746 (1973).
61. Jean, M., *Anal. Chim. Acta* **7**, 462 (1952); **11**, 82 (1954).
62. Lazarev, A. I. and Lazareva, V. I., *Zavodsk. Lab.* **25**, 405 (1959).
63. Cyrankowska, M. and Downarowicz, J., *Chem. Anal. (Warsaw)* **12**, 137 (1967).
64. Kish, P. P. and Onishchenko, Yu. K., *Zh. Analit. Khim.* **29**, 102 (1974).
65. Študlar, K. and Janoušek, I., *Collection Czech. Chem. Commun.* **25**, 1965 (1960).
66. Blyum, I. A., Solov'yan, I. T. and Shebalkova, G. N., *Zavodsk. Lab.* **27**, 950 (1961).
67. Řezáč, R. and Ditz, J., *ibid.* **29**, 1176 (1963).
68. Sorokin, G. Ch. and Lomonosov, S. A., *Zavodsk. Lab.* **40**, 23 (1974).
69. Stanton, R. E. and McDonald, A. J., *Analyst* **87**, 299 (1962).
70. Burke, R. W. and Menis, O., *Anal. Chem.* **38**, 1719 (1966).
71. Fogg, A. G., Jillings, J., Marriott, D. R. and Burns, D. T., *Analyst* **94**, 768 (1969).
72. Kerr, G. O. and Gregory, G. R., *ibid.* **94**, 1036 (1969).
73. Yadav, A. A. and Khopkar, S. M., *Bull. Chem. Soc. Japan* **44**, 693 (1971).
74. Fogg, A. G., Burgess, C. and Burns, D. T., *Analyst* **98**, 347 (1973).
75. Kish, P. P. and Onishchenko, Yu. K., *Zh. Analit. Khim.* **23**, 1651 (1968).
76. Pilipenko, A. T. and Nguyen Mong Shinh, *Ukr. Khim. Zh.* **34**, 1286 (1968).
77. Burgess, C., Fogg, A. G. and Burns, D. T., *Analyst* **98**, 605 (1973).
78. Busev, A. I., Bogdanova, E. S. and Tiptsova, V. G., *Zh. Analit. Khim.* **20**, 585 (1965).
79. Kish, P. P. and Onishchenko, Yu. K., *ibid.* **25**, 112, 500 (1970).
80. Kish, P. P. and Onishchenko, Yu. K., *Zavodsk. Lab.* **36**, 520 (1970).
81. Gillis, J., Hoste, J. and Claeys, A., *Anal. Chim. Acta* **1**, 291 (1947).
82. Meyer, S. and Koch, O. G., *Z. Anal. Chem.* **179**, 175 (1961).

83. Asmus, E. and Brandt, K., *ibid.* **208**, 189 (1965).
84. Koch, O. G., *ibid.* **265**, 29 (1973).
85. Nazarenko, V. A. and Lebedeva, N. V., *Zh. Analit. Khim.* **10**, 289 (1955); **11**, 560 (1956).
86. Biryuk, E. A., *Zavodsk. Lab.* **30**, 651 (1964).
87. Iyer, S. G., Kadam, B. V. and Venkateswarlu, C., *Indian J. Chem.* **11**, 385 (1973).
88. Talipov, Sh. T., Dzhiyanbaeva, R. Kh. and Abdisheva, A. V., *Zavodsk. Lab.* **37**, 387 (1971).
89. Rakhmatullaev, K., Rakhmatullaeva, M. A., Talipov, Sh. T. and Mamatov, A., *ibid.* **37**, 1027 (1971).
90. Gusev, S. I., Poplevina, L. V. and Pesis, A. S., *Zh. Analit. Khim.* **22**, 731 (1967).
91. Gusev, S. I. and Poplevina, L. V., *ibid.* **23**, 541 (1968).
92. Fomina, A. I., Agrinskaya, N. A., Zolotov, Yu. A. and Seryakova, I. V., *ibid.* **26**, 2376 (1971).
93. Gusev, S. I., Kurepa, G. A., Poplevina, L. V., Shalamova, G. G., Shchurowa, L. M. and Pesis, A. S., *ibid.* **24**, 1319 (1969).
94. Rakhmatullaev, K., Rakhmatullaeva, M. A., Rakhimov, Kh. R. and Talipov, Sh. T., *ibid.* **25**, 1132 (1970).
95. Christopher, D. H. and West, T. S., *Talanta* **13**, 507 (1966).
96. Bailey, B. W., Chester, J. E., Dagnall, R. M. and West, T. S., *ibid.* **15**, 1359 (1968).
97. Garg, B. S., Trikha, K. C. and Singh, R. P., *ibid.* **16**, 462 (1969).
98. Stará, V., *ibid.* **18**, 228 (1971).
99. Hulanicki, A. and Głąb, S., *Chem. Anal. (Warsaw)* **15**, 1089 (1970).
100. Ciuhandu, Gh. and Rocsin, M., *Z. Anal. Chem.* **174**, 118 (1960).
101. Nielsch, W. and Böltz, G., *Mikrochim. Acta* **1954**, 313.
102. Guyon, J. C. and Matulis, R. M., *Chemist-Analyst* **56**, 22 (1967).

Chapter 7

ARSENIC

Arsenic (As, at.wt. 74·92) occurs in its compounds in the oxidation states $-III$ (arsine, AsH_3), $+III$ (arsenite), and $+V$ (arsenate). Arsenic ($+III$) and ($+V$) are amphoteric, but with much more acidic than basic character. The sulphides are characteristically capable of yielding soluble complexes (thio-salts). Arsenic (V) forms heteropoly acids.

7.1 Methods of Separation

7.1.1 Distillation

The separation of arsenic as the toxic gas arsine, AsH_3, is the first step in both the Gutzeit method and the Ag–DDTC method for determining arsenic. Both methods are discussed in detail later. Arsenic has also been separated as AsH_3 before its determination as arsenomolybdenum blue [1–6]. Normally, arsenic is reduced to AsH_3 by treatment with zinc and acid (3–6M HCl), but it can also be reduced electrolytically [5]. The arsine generated is carried over in the hydrogen stream into an alkaline iodine absorption solution. Dyfverman and Bonnichsen [4] burned the AsH_3, separating the arsenic in the form of a mirror, and subsequently dissolved it in hypobromite. Oliver and Funnell [3] distilled the AsH_3 and absorbed it on solid mercury iodide, and then removed the arsenic with iodine solution. Arsenic may be separated as AsH_3 if the sample solutions do not contain large amounts of heavy metal compounds that can be reduced easily to the free metal.

A common method for separating arsenic involves distillation of arsenic(III) chloride, $AsCl_3$, from moderately concentrated HCl (5–7M), after arsenic(V) has been reduced to As(III) with hydrazine [7–10]. Germanium tetrachloride (b.p. 86°C) and selenium and tellurium oxychlorides distil along with $AsCl_3$ (b.p. 130°C). Addition of phosphoric acid prevents volatilization of stannic chloride (b.p. 115°C). When the sample solution contains Sb, the distillation of $AsCl_3$ should be carried out below 110°C. The distillation is more efficient when nitrogen or carbon dioxide is passed through the solution [10]. The distillate is collected either in cold water or dilute HNO_3, or in an alkaline iodine or hydrogen peroxide solution.

Arsenic(III) may also be distilled from hydrobromic acid medium as $AsBr_3$ [11].

7.1.2 Extraction

Arsenic(III) chloride can be extracted into carbon tetrachloride, chloroform or benzene from 8–12M HCl [12–15]. GeCl$_4$ is co-extracted with AsCl$_3$; other elements (e.g. Se, Bi, Sb) have very low distribution coefficients. Arsenic and germanium may be separated from each other by extraction after oxidation of the former to As(V) [12]. Arsenic is stripped from the organic solvent with dilute hydrochloric acid, water, or dilute ammonia solution. Arsenic(III) can also be extracted as AsBr$_3$ and AsI$_3$ [16–18].

A chloroform solution of diethylammonium diethyldithiocarbamate (1%) quantitatively extracts arsenic(III), and also Sn(II) and Sb(III), from 1–10N H$_2$SO$_4$ medium [2,19–21]. Copper, bismuth, and mercury, which are also extractable, can be separated by a preliminary extraction in the presence of hydrogen peroxide. Neither arsenic(V), Sn(IV), nor Sb(V) reacts with the dithiocarbamate. After Cu, Bi, and Hg have been extracted, iodide and ascorbic acid are added to the aqueous solution, thereby enabling arsenic to be extracted. Arsenic is separated from germanium by dithiocarbamate extraction in the presence of oxalic acid [19]. Lounamaa [22] has separated traces of arsenic in silicates by extraction with dithiocarbamate from hydrofluoric acid, using polyethylene apparatus.

Liedermann *et al.* [8] extracted arsenic from petroleum distillation products with a mixture of concentrated sulphuric acid and perhydrol.

Arsenic(V) has been separated from a number of elements by extraction into butanol from acidic medium as the molybdoarsenic heteropoly acid [23,24]. The extraction of arsenic as complexes with ethyl xanthate [25,26] and thionalide [27] has also been reported.

7.1.3 Precipitation

Traces of arsenic(V) are quantitatively precipitated as ferric arsenate with Fe(OH)$_3$ as the collector, by adding ammonia to an acid solution till the pH reaches 8–9 [28,29].

Arsenic has been separated from ammonium fluoride solutions by coprecipitation with hydrous MnO$_2$ [29a].

Arsenic(V) can be coprecipitated with magnesium ammonium phosphate [30,31]. Traces of arsenic have been precipitated from 3·7M HClO$_4$ as the sulphide by means of thioacetamide with molybdenum as the collector [32]. Thionalide has also been used to coprecipitate arsenic [33].

7.2 Methods of Determination

Of the methods described below, the arsenomolybdenum blue method is the most widely used. The spectrophotometric method using the reaction of arsine with Ag–DDTC is less sensitive. In certain cases, the classical Gutzeit method is convenient. In this method, the intensity of the colour produced by arsine on mercuric bromide test paper is compared.

7.2.1 Arsenomolybdenum Blue Method

After separation, arsenic is oxidized to As(V) (e.g. by evaporation to dryness with nitric acid) and reacted in a suitably acid solution with ammonium molybdate to form the practically colourless molybdoarsenic heteropoly acid. The molybdoarsenic acid is then reduced to arsenomolybdenum blue, and the absorbance measured either in the aqueous solution, or after extraction into an oxygenated organic solvent [1,9,10,34,35]. The reaction conditions are so selected that the unreacted molybdate ions are not reduced.

Hydrazine, stannous chloride, and ascorbic acid are employed as reducing agents [36]. When hydrazine is used, the reaction is carried out in a hot solution.

The best solvent for the extraction of arsenomolybdenum blue is butyl alcohol (n- or iso), although amyl alcohols, ethers, and ethyl acetate are also used.

The molar absorptivity of arsenomolybdenum blue in butanol solution is 3.0×10^4 at $\lambda_{max} = 800$ nm (specific absorptivity $a = 0.40$) (see Fig. 7.1). The molar absorptivity and λ_{max} of arsenomolybdenum blue vary slightly depending on the reducing agent and extraction solvent used.

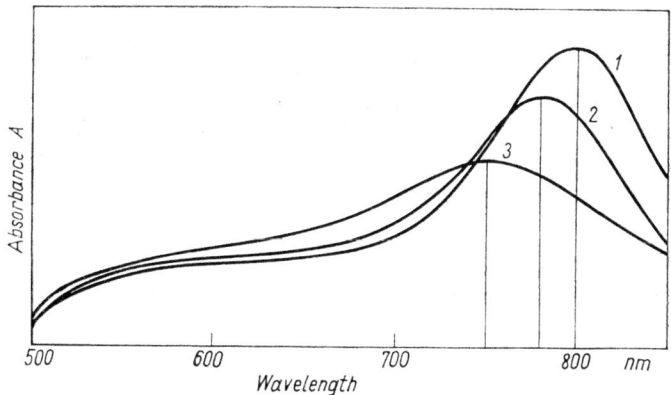

Fig. 7.1 Absorption spectra of arsenomolybdenum blue (1), phosphomolybdenum blue (2) (reduction with hydrazine and extraction into butanol), silicomolybdenum blue (3) [extraction into amyl alcohol and reduction with tin(II) chloride]

Alternatively, arsenic may be extracted as molybdoarsenic acid with the solvents mentioned, and a reducing agent may then be added to the organic phase [23].

Phosphorus(V) forms phosphomolybdenum blue under the reaction conditions used to form arsenomolybdenum blue. The corresponding silicon and germanium compounds are obtained under different reaction conditions. DeSesa and Rogers [37] have reported conditions for determining arsenic, phosphorus, and silicon present together, while Daniels [23] has discussed the conditions for determining arsenic(V) in the presence of arsenic(III).

Arsenic has been determined by the arsenomolybdenum blue method in iron and steel [27,38–41], antimony and its compounds [42–44], gallium and antimony chlorides of high purity [45], bismuth [5], tin [46], lead and its alloys [47,48], lead and zinc products [16,25], copper and its alloys [18,49], silver alloys [50], platinum and gold [51], gallium, indium, thallium, vanadium, and niobium [2], selenium [52,53], germanium and silicon [19], boron [14], coal [7], silicate minerals [1,22], ores [9], petroleum fractions and reforming catalysts [8], silicate and carbonate sediments [33], biological materials [3,4,21,54], organic compounds [55], fruits and vegetables [11], and natural water [33,56].

Reagents

1. Molybdate reagent. (*a*) Dissolve 1·0 g of ammonium molybdate in 100 ml of $4N$ H_2SO_4 (1+8); (*b*) dissolve 0·10 g of hydrazine sulphate in 100 ml of water. Immediately before use, mix 10 ml of solution (*a*) with 10 ml of solution (*b*), and dilute the solution with water to 100 ml. Solutions (*a*) and (*b*) should not be kept for longer than 3–4 days.
2. Standard arsenic(III) solution: 1 mg/ml. Dissolve 1·320 g of arsenious oxide, As_2O_3, in 20 ml of $2M$ NaOH. Dilute the solution with a little water, acidify slightly with $2M$ HCl, and dilute the solution with water in a volumetric flask to 1 litre.
3. Arsenic-free conc. hydrochloric acid. To 20 ml of conc. HCl, add 1 drop of 1% aqueous potassium iodide, allow the acid to stand for 5 min, and then shake it with three 5-ml portions of benzene.

Procedure

Extractive separation of As. Evaporate a sample solution containing not more than 150 µg of As to dryness at a temperature of $< 130°C$ with nitric or sulphuric acid (the solution must not contain halide ions). To the cooled solution, add 10 ml of As-free conc. HCl and 2 drops of 1% KI solution. Mix and allow the solution to stand for 5 minutes. Transfer the solution to a separating funnel, and shake it with three 5-ml portions of benzene. Strip the combined extracts with two 5-ml portions of water.

Determination of As. To the aqueous phase, in a beaker, add 2 ml of conc. HNO_3. Stir well, and evaporate the solution to dryness. Add 25 ml of the molybdate reagent, stir well, and place the beaker in a boiling water-bath for 10 minutes. After cooling, transfer the solution to a separating funnel, and extract the arsenomolybdenum blue with two portions (e.g. 10–15 ml) of butyl alcohol. Make up the blue extract in a volumetric flask with the solvent to 50 ml, and measure the absorbance at 800 nm, using the solvent as reference.

Note. It is also possible to measure the absorbance of the aqueous solution of arsenomolybdenum blue. In this case the cooled coloured solution is transferred to a volumetric flask, and diluted to the mark with the molybdate reagent.

7.2.2 SILVER DIETHYLDITHIOCARBAMATE (Ag–DDTC) METHOD

In the Vašák and Šedivec method [57] using Ag–DDTC, the arsine evolved by nascent hydrogen is absorbed in a pyridine solution of silver diethyldithiocarbamate. The pyridine-soluble arsenic–DDTC complex has an intense red-violet colour, whereas the pyridine Ag–DDTC solution is pale yellow [57–62].

The molar absorptivity of the coloured product is $1 \cdot 4 \times 10^4$ ($a = 0 \cdot 19$) at $\lambda_{max} \approx 535$ nm. The Ag–DDTC reagent absorbs at $\lambda < 500$ nm.

Arsenic is determined by using the simple apparatus shown in Fig. 7.2.

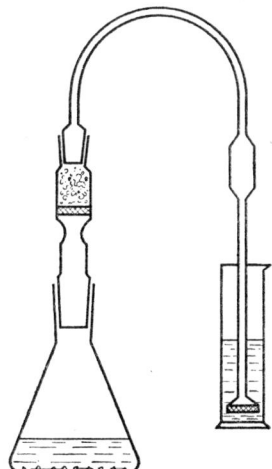

Fig. 7.2 Apparatus for determining arsenic with Ag–DDTC

The functions of the various reagents in the reduction of arsenic compounds (i.e. zinc, hydrochloric acid, nickel chloride, tin(II) chloride, and potassium iodide) are discussed below in connection with the Gutzeit method.

Hydrogen sulphide, which interferes in the reaction, is separated from arsine on cotton wool impregnated with lead acetate.

During the extraction of $AsCl_3$ with inert solvents, antimony remains in the aqueous phase. Under the conditions specified in the procedure below, antimony present in approximately the same amount as arsenic does not interfere in the Ag–DDTC method [56]. When larger amounts of antimony are present, it is advisable to increase the quantity of $SnCl_2$ added. In such conditions antimony(III) is reduced to the element, and not to SbH_3 [29].

A solution of Ag–DDTC in chloroform containing organic bases [61,63,64] (e.g. 1% of ethanolamine) has been suggested as an alternative to the obnoxious pyridine solution.

Arsenic has been determined by the silver diethyldithiocarbamate method in naphthas [65], petroleum stocks and catalysts [66], elemental sulphur [67,68], germanium dioxide [69,70], copper and its compounds [31,63], steel [63,71], various reagents [29], tungsten compounds [72], hydrofluoric acid [73], phosphoric acid [68], organic matter [74] water samples [75], and food [76].

Reagents

1. Silver diethyldithiocarbamate (Ag–DDTC) pyridine solution: Dissolve 1 g of the reagent in 100 ml of pyridine. To facilitate dissolution, the pyridine may be heated slightly (to 50–60°C) and crystalline Ag–DDTC added.

   Preparation of crystalline Ag–DDTC. Dissolve 1·8 g of silver nitrate in 20 ml of water, and 2·6 g of Na–DDTC. 3H_2O in 20 ml of water. Add the $AgNO_3$ solution slowly (during 15–20 min) with careful stirring to the Na–DDTC solution. Filter off the Ag–DDTC precipitate on a No.3 sintered-glass crucible, and wash it with water. Dry the precipitate at 100°C to constant weight. (Nearly 2·4 g of product is obtained; $\sim$ 90% yield.)
2. Standard arsenic(III) solution: 1 mg/ml (p. 134).
3. Zinc (arsenic-free), granulated. Melt down granulated zinc in a quartz crucible and pour it in a thin jet into a tall beaker filled with arsenic-free water. Dry the comminuted zinc and store it in a stoppered vessel.
4. Cotton wool impregnated with lead acetate (p. 137).

Procedure

Place in a 50-ml conical flask (Fig. 7.2) the almost neutral sample solution containing not more than 150 μg of As. Dilute to 10 ml with water, and then add successively 5 ml of conc. HCl, 5 ml of 10% KI solution, 4 drops of 10% $SnCl_2$ solution in 6M HCl, 2 drops of 10% $NiCl_2$ solution, and 1·5 g of zinc. Close the flask with a head carrying the cotton wool impregnated with lead acetate, and connected to a tube immersed in the receiver, which contains 10 ml of the Ag–DDTC pyridine solution.

After 30 minutes (even though all the zinc may not have dissolved) disconnect the receiver and rinse the delivery tube with pyridine. Dilute the coloured solution with pyridine in a 50-ml (or smaller depending on the amount of As) volumetric flask, and measure the absorbance at 535 nm, using a blank solution as reference.

7.2.3 Gutzeit Method

This method [77–79] for the determination of traces of arsenic is based on the visual comparison of coloured spots. The arsenic in the sample solution is reduced to volatile arsine with zinc and HCl. A disk of paper impregnated with mercuric bromide is placed across the flow of the arsine and hydrogen evolved. The reaction of AsH_3 with $HgBr_2$ gives coloured compounds such as yellow $H(HgBr)_2As$, brown $(HgBr)_3As$, and black Hg_3As_2. The resultant coloured spot is compared with a set of standard spots corresponding to known amounts of arsenic. The method is sensitive and permits the estimation of arsenic in the range 0·2–5 μg.

The zinc used should be arsenic-free, and suitably comminuted for fast dissolution. Nickel(II) ions added in small quantities catalyse the zinc dissolution. Stannous chloride and potassium iodide facilitate quantitative reduction of arsenic traces.

Hydrogen sulphide in the arsine evolved is removed before contact with the mercuric bromide paper by absorption by lead acetate-impregnated cotton wool and paper, since H_2S also gives coloured products with mercuric bromide. PH_3, SbH_3, and GeH_4 also interfere in the determination of arsenic. The only phosphorus compounds which can be reduced to phosphine under the reaction conditions are phosphite and hypophosphite. Phosphate and sulphate do not interfere.

Babko *et al.* [80] have reduced arsenic compounds to AsH_3 electrolytically and produced spots on silver nitrate test paper by photographic development. The enhanced sensitivity of this method allows the determination of 0·1 μg of arsenic in 25 ml of solution.

Arsenic has been determined by the Gutzeit method in germanium and its compounds [81,82], silicon [82,83], soil [84], lead [85], petroleum products [79], and foodstuffs [86].

Reagents and apparatus

1. Mercuric bromide papers. Place thin, compact filter paper disks (~ 20 mm diameter) in a freshly prepared 5% $HgBr_2$ solution in ethanol for 30 minutes. Lay the papers on a watch-glass to dry in air. The papers may be stored in an amber-glass jar for not longer than week after preparation.
2. Standard arsenic(III) solution: 1 mg/ml (p. 134).
3. Filter paper and cotton wool impregnated with lead acetate. Impregnate filter paper (2·5 × 4 cm) for 30 minutes in a 10% lead acetate solution, dry at 105°C, and pleat into narrow folds. Soak and dry the cotton wool in the same way.
4. The apparatus for determining arsenic (Fig. 7.3) consists of a 50-ml amber-glass bottle (1), fitted with a stopper (2) which carries a glass tube (3), 6–7 mm in inner diameter. The tube has a side hole (4) preventing closure of the passage by liquid carried by the gas. A neck above the stopper holds the folded filter paper (5) impregnated with lead acetate. Above the paper is placed cotton wool (6) impregnated with lead acetate. The paper disk impregnated with $HgBr_2$ is sandwiched between evenly cut tube ends held tightly together by a pair of rubber

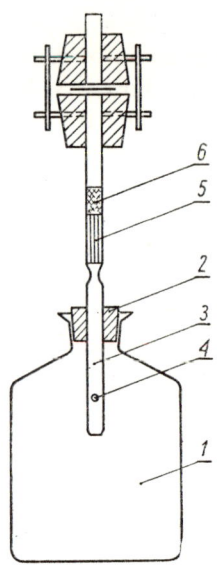

Fig. 7.3 Gutzeit apparatus

stoppers fitted with projecting glass rods for clamping by means of rubber bands.
5. Zinc (arsenic-free) (p. 136).

Procedure

Place the sample solution (at room temperature) containing 0·5–5 µg of As in the apparatus bottle, and dilute to 10 ml with water. Add to the almost neutral solution 10 ml of hydrochloric acid (1+1), then 5 ml of 10% KI solution, 4 drops of 10% $SnCl_2$ solution in $6M$ HCl, and 2 drops of 10% $NiCl_2$ solution, and stir well. Then add 1·5 g of comminuted zinc, and immediately close the bottle with a head containing the fresh disk of mercuric bromide paper.

When the evolution of hydrogen is complete (after 30–60 minutes), remove the paper disk with the coloured spot from the apparatus. Compare the colour of the spot with a set of standards, prepared at the same time and corresponding to 0, 0·5, 1, 2, 3, and 5 µg of arsenic. The zero standard is a blank test to allow for any arsenic in the reagents.

7.2.4 OTHER METHODS

Various modifications of the arsenomolybdenum blue method have been proposed for the determination of arsenic. The absorbance of molybdo-arsenic acid in n-butanol ($\varepsilon = 5\cdot1 \times 10^3$ at 370 nm) [87], and in water-acetone solution [88], has been exploited. The photometric determination of arsenic(V) as the mixed heteropoly acid with vanadium and molybdenum has been investigated [89,90]. Molybdoarsenic acid has been extracted, and the molybdenum determined by the thiocyanate method [91].

Babko *et al.* [92] have separated the molybdoarsenic acid compound with Butylrhodamine B by flotation with ether, and measured the absorbance of the coloured compound in ether–acetone solution. Arsenic has also been determined as the complex of the heteropoly acid with Crystal Violet (CV) [93]. This complex, which has an As:CV ratio of 1:3, is very intensely coloured in cyclohexanone–toluene–acetone solution ($\varepsilon = 3 \cdot 2 \times 10^5$ at 582 nm).

Archer and Doolittle [94] have determined hexafluoroarsenate by extracting the ion-association complex of $[AsF_6]^-$ with $[Fe(phen)_3]^{2+}$ into n-butyronitrile.

Růžička and Starý [95] have determined arsenic indirectly by measuring the absorbance (at 620 nm) of the dithizone liberated in a ligand-exchange reaction between As(III)–DDTC and AgHDz.

Spectrophotometric methods have been developed based on the coloured sols formed when arsenic is reduced to the element with hypophosphite [96,97], and when arsenic reacts with the silver complex of *p*-sulphamoylbenzoic acid [98,99].

Quercetin [100], 8-mercaptoquinoline (thio-oxine) [101], and curcumin [102] are other colorimetric reagents for arsenic.

References

1. Onishi, H. and Sandell, E. B., *Mikrochim. Acta* **1953**, 34.
2. Nazarenko, V. A., Flyantikova, G. V. and Lebedeva, N. V., *Zavodsk. Lab.* **23** 891 (1957).
3. Oliver, W. T. and Funnell, H. S., *Anal. Chem.* **31**, 259 (1959).
4. Dyfverman, A. and Bonnichsen, R., *Anal. Chim. Acta* **23**, 491 (1960).
5. Jackwerth, E., *Z. Anal. Chem.* **211**, 254 (1965).
6. Pavelka, F., *Mikrochim. Acta* **1965**, 117.
7. Edgcombe, L. J. and Gold, H. K., *Analyst* **80**, 155 (1955).
8. Liederman, D., Bowen, J. E. and Milner, O. I., *Anal. Chem.* **30**, 1543 (1958).
9. Finkel'shtein, D. N. and Kryuchkova, G. N., *Zh. Analit. Khim.* **12**, 196 (1957).
10. Butenko, G. A., Korzh, V. P. and Rodionova, E. M., *ibid.* **16**, 692 (1961).
11. Bartlet, J. C., Wood, M. and Chapman, R. A., *Anal. Chem.* **24**, 1821 (1952).
12. Fischer, W., Harre, W., Freese, W. and Hackstein, K. G., *Angew. Chem.* **66**, 165 (1954).
13. Beard, H. C. and Lyerly, L. A., *Anal. Chem.* **33**, 1781 (1961).
14. Marczenko, Z., *Chem. Anal. (Warsaw)* **9**, 1093 (1964).
15. Brink, G. O., Kafalas, P., Sharp, R. A., Weiss, E. L. and Irvine, J. W., Jr, *J. Am. Chem. Soc.* **79**, 1303 (1957).
16. Milaev, S. M. and Voroshnina, K. P., *Zavodsk. Lab.* **29**, 410 (1963).
17. Kolesnikova, N. M., Iofa, B. Z. and Nesmeyanov, A. N., *Izv. Vyssh. Ucheb. Zaved., Khim. Khim. Tekhnol.* **12**, 1023 (1969).
18. Nivière, P. and Winternheimer, M., *Chim. Anal. (Paris)* **47**, 448 (1965).
19. Luke, C. L. and Campbell, M. E., *Anal. Chem.* **25**, 1588 (1953).
20. Wyatt, P. F., *Analyst* **80**, 368 (1955).
21. Analytical Methods Committee, *ibid.* **85**, 629 (1960).

22. Lounamaa, K., *Z. Anal. Chem.* **146**, 422 (1955).
23. Daniels, M., *Analyst* **82**, 133 (1957).
24. Wünsch, G. and Umland, F., *Z. Anal. Chem.* **247**, 287 (1969).
25. Dreulle, N., *Chim. Anal. (Paris)* **43**, 165 (1961).
26. Chakrabarty, T. and De, A. K., *Z. Anal. Chem.* **242**, 152 (1968).
27. Nakaya, S., *Japan Analyst* **12**, 483 (1963).
28. Plotnikov, V. I. and Usatova, L. P., *Zh. Analit. Khim.* **19**, 1183 (1964).
29. Marczenko, Z. and Mojski, M., *Chem. Anal. (Warsaw)* **14**, 495 (1969).
29a. Briska, M. and Hoffmeister, W., *Z. Anal. Chem.* **268**, 347 (1974).
30. Nazarenko, V. A. and Byk, G. I., *Ukr. Khim. Zh.* **22**, 234 (1956).
31. Meyer, J., *Z. Anal. Chem.* **210**, 84 (1965).
32. Reymont, T. M. and Dubois, R. J., *Anal. Chim. Acta* **56**, 1 (1971).
33. Portmann, J. E. and Riley, J. P., *ibid.* **31**, 509 (1964).
34. Duval, Z., *Chim. Anal. (Paris)* **51**, 415 (1969).
35. Pakalns, P., *Anal. Chim. Acta* **47**, 225 (1969).
36. Morosanova, S. A., Grishko, G. A. and Shkatova, L. A., *Zh. Analit. Khim.* **29**, 529 (1974).
37. DeSesa, M. A. and Rogers, L. B., *Anal. Chem.* **26**, 1381 (1954).
38. Jean, M., *Anal. Chim. Acta* **14**, 172 (1956).
39. Bohnstedt, U. and Budenz, R., *Z. Anal. Chem.* **159**, 95 (1957).
40. Nall, W.R., *Analyst* **96**, 398 (1971).
41. Fogg, A. G., Marriott, D. R. and Burns, D. T., *ibid.* **97**, 657 (1972).
42. Norwitz, G., Cohen, J. and Everett, M. E., *Anal. Chem.* **32**, 1132 (1960).
43. Kowalczyk, M., *Chem. Anal. (Warsaw)* **9**, 331 (1964).
44. Marczenko, Z., Mojski, M. and Skibe, H., *ibid.* **17**, 881 (1972).
45. Goryushina, V. G., Romanova, E. V. and Razumova, L. S., *Zh. Analit. Khim.* **28**, 601 (1973).
46. Coppins, W. and Price, J. W., *Metallurgia* **46**, 52 (1952).
47. Pohl, H., *Z. Anal. Chem.* **134**, 177 (1951).
48. Goszczyńska, H. and Kowalczyk, M., *Chem. Anal. (Warsaw)* **12**, 1261 (1967).
49. Scholes, I. R. and Waterman, W. R., *Analyst* **88**, 374 (1963).
50. Skorko-Trybuła, Z. and Chwastowska, J., *Chem. Anal. (Warsaw)* **8**, 859 (1963).
51. Marczenko, Z. and Lenarczyk, Ł., *ibid.* **19**, 679 (1974).
52. Reed, J. F., *Anal. Chem.* **30**, 1122 (1958).
53. Ebner, E., *Z. Anal. Chem.* **206**, 106 (1964).
54. Evans, R. J. and Bandemer, S. L., *Anal. Chem.* **26**, 595 (1954).
55. Tuckerman, M. M., Hodecker, J. H., Southworth, B. C. and Fleischer, K. D., *Anal. Chim. Acta* **21**, 463 (1959).
56. Johnson, D. L. and Pilson, M. E., *ibid.* **58**, 289 (1972).
57. Vašák, V. and Šedivec, V., *Chem. Listy* **46**, 341 (1952).
58. Martin, F. and Floret, A., *Bull. Soc. Chim. France* **1965**, 404.
59. Dubois, L., Teichman, T. and Monkman, J. L., *Mikrochim. Acta* **1966**, 415.
60. Dubois, L., Teichman, T., Baker, C. J., Zdrojewski, A. and Monkman, J. L., *ibid.* **1969**, 185.
61. Bode, H. and Hachmann, K., *Z. Anal. Chem.* **229**, 261 (1967); **241**, 18 (1968).
62. Gastinger, E., *Mikrochim. Acta* **1972**, 526.
63. Hulanicki, A. and Głąb, S., *Chem. Anal. (Warsaw)* **15**, 1089 (1970).
64. Kopp, J. F., *Anal. Chem.* **45**, 1786 (1973).
65. Albert, D. K. and Granatelli, L., *ibid.* **31**, 1593 (1959).
66. Liederman, D., Bowen, J. E. and Milner, O. I., *ibid.* **31**, 2052 (1959).
67. Steinke, I., *Z. Anal. Chem.* **240**, 184 (1968).
68. Bahr, H. and Bahr, H., *Chem. Anal. (Warsaw)* **16**, 427 (1971).
69. Fowler, E. W., *Analyst* **88**, 380 (1963).
70. Řezáč, Z. and Ditz, J., *Zavodsk. Lab.* **29**, 1176 (1963).
71. Bhargava, O. P., Donovan, J. F. and Hines, W. G., *Anal. Chem.* **44**, 2402 (1972).

72. Blechta, V., *Chem. Průmysl* **14**, 373 (1964).
73. Meyer, J., *Z. Anal. Chem.* **229**, 409 (1967).
74. Jureček, M. and Jeník, J., *Collection Czech. Chem. Commun.* **20**, 550 (1955).
75. Fresenius, W. and Schneider, W., *Z. Anal. Chem.* **203**, 417 (1964).
76. Hundley, H. K. and Underwood, J. C., *J. Ass. Offic. Anal. Chem.* **53**, 1176 (1970).
77. Satterlee, H. S. and Blodgett, G., *Ind. Eng. Chem., Anal. Ed.* **16**, 400 (1944).
78. Berton, A., *Bull. Soc. Chim. France* **1945**, 296.
79. Maranowski, N. C., Snyder, R. E. and Clark, R. O., *Anal. Chem.* **29**, 353 (1957).
80. Babko, A. K., Pilipenko, A. T. and Rozenfel'd, A. L., *Zavodsk. Lab.* **30**, 1060 (1964).
81. Payne, S. T., *Analyst* **77**, 278 (1952).
82. Tumanov, A. A., Sidorenko, A. N. and Taradenkova, F. S., *Zavodsk. Lab.* **30**, 652 (1964).
83. Rigin, V. I. and Mel'nichenko, N. N., *ibid.* **32**, 394 (1966).
84. Almond, H., *Anal. Chem.* **25**, 1766 (1953).
85. Lur'e, Yu. Yu. and Minenko, A. N., *Zavodsk. Lab.* **23**, 785 (1957).
86. Tupalska, M., *Roczniki Państwowego Zakładu Hig.* **5**, 39 (1954).
87. Wadelin, C. and Mellon, M. G., *Analyst* **77**, 708 (1952).
88. Chalmers, R. A. and Sinclair, A. G., *Anal. Chim. Acta* **33**, 384 (1965).
89. Gullstrom, D. K. and Mellon, M. G., *Anal. Chem.* **25**, 1809 (1953).
90. Baghurst, H. C. and Norman, V. J., *ibid.* **29**, 778 (1957).
91. Kristalev, P. V., Kristaleva, L. B. and Shor, N. A., *Tr. Komis. po Analit. Khim. Akad. Nauk SSSR* **16**, 19 (1968).
92. Babko, A. K., Chalaya, Z. I. and Mikitchenko, V. F., *Zavodsk. Lab.* **32**, 270 (1966).
93. Babko, A. K. and Ivashkovich, E. M., *Zh. Analit. Khim.* **27**, 120 (1972).
94. Archer, V. S. and Doolittle, F. G., *Talanta* **14**, 921 (1967).
95. Růžička, J. and Starý, J., *ibid.* **14**, 909 (1967).
96. Cyrankowska, M., *Chem. Anal.* (*Warsaw*) **8**, 679 (1963).
97. Gann, W., *Z. Anal. Chem.* **221**, 254 (1966).
98. Ciuhandu, Gh. and Rocsin, M., *ibid.* **172**, 268 (1960).
99. Ciuhandu, Gh., Roscovanu, A. and Cutui, M., *Chim. Anal.* (*Paris*) **50**, 489 (1968).
100. Tanaka, T. and Hiiro, K., *Japan Analyst* **11**, 1180 (1962).
101. Stará, V. and Starý, J., *Talanta* **17**, 341 (1970).
102. Hiiro, K. and Tanaka, T., *J. Chem. Soc. Japan, Pure Chem. Sect.* **83**, 1258 (1962).

Chapter 8

BERYLLIUM

Beryllium (Be, at.wt. 9·012) forms colourless compounds. In its chemical properties, beryllium more closely resembles aluminium than magnesium. Beryllium is amphoteric; $Be(OH)_2$ is precipitated at pH $\sim$ 6, and dissolves in excess of base (pH $\sim$ 13·5). Freshly precipitated $Be(OH)_2$ dissolves in Na_2CO_3 solution owing to the formation of a rather unstable carbonate complex. Beryllium also forms weak complexes with citrate, tartrate, and fluoride. Beryllium and its compounds are highly toxic.

8.1 Methods of Separation

8.1.1 PRECIPITATION

Precipitation of $Be(OH)_2$ [1–4] with ammonia at pH $\sim$ 8, with aluminium and iron(III) as collectors, enables beryllium to be separated from Ca, Mg, Mn, and Cr(VI). Iron also remains in solution when masked with thioglycollic acid. In the presence of EDTA, the only Analytical Group III metals which ammonia precipitates are Be, Ti, U, Nb, and Ta.

When the sample containing beryllium is fused with NaOH in a nickel crucible, and the cooled melt leached with water, beryllium remains in solution, while the insoluble residue contains Fe, Mn, Cu, Ni, Mg, Ti and other metal hydroxides. In the presence of large amounts of metals having hydroxides which do not dissolve in NaOH, the precipitate may retain some beryllium. In such circumstances, the precipitate is separated and ignited, fused with NaOH, and leached once again.

Fischer and Wernet [1] have separated aluminium and titanium from traces of beryllium by saturating the solution containing NH_4Cl with gaseous hydrogen chloride. The aluminium and titanium chlorides precipitated in this way are free from beryllium.

Beryllium may be separated from many metals by precipitation as the phosphate with titanium as collector [5]. Traces of beryllium have been precipitated from a solution at pH 8–10 with Methylene Blue and tannin [6].

Beryllium remains in solution during electrolytic separation at the mercury cathode [2,7].

8.1.2 EXTRACTION

Extraction with acetylacetone in benzene or carbon tetrachloride separates beryllium from a solution of magnesium, calcium, and phosphate at pH 4–5 [7–11]. In the presence of EDTA and at the same pH, iron and aluminium also remain in the aqueous phase. Traces of Fe(III) and Al are, however, also extracted. The beryllium can be back-extracted into hydrochloric acid ($\sim 5M$).

Trifluoroacetylacetone [12], hexafluoroacetylacetone [12], and o-(2-hydroxy-5-methylphenylazo)benzoic acid [13] have also been used to extract beryllium.

8.1.3 ION-EXCHANGE

Aluminium and iron form more stable oxalate complexes than does beryllium. Ryabchikov and Bukhtiarov [14] have separated beryllium from aluminium and iron by exploiting this difference. The oxalate solution at pH 4·4 is passed through a strongly acidic cation-exchanger. The iron and aluminium complexes are eluted until the thiocyanate test for iron in the eluate is negative. The beryllium, which is retained on the column, is then eluted with 10% hydrochloric acid. Beryllium has also been separated from Fe and Al on an anion-exchanger in fluoride medium [15].

When a solution at pH 3·5 containing EDTA and H_2O_2 as masking agents is passed through an Amberlite IR-120 cation exchange column, Al, Ti, Fe, Ca, and phosphate are eluted, whereas Be is adsorbed on the column [16].

The separation of beryllium from a number of cations by using cation-exchangers has been achieved by consecutive elution of the metals with increasing acid concentrations [17].

Korkisch and Ahluwalia [18] have separated beryllium on an anion-exchanger from large quantities of uranium, using a mixture of $5M$ nitric acid and methanol to elute the beryllium.

o-(2-Hydroxy-5-dodecylphenylazo)benzoic and arsonic acids have been used as chelating liquid ion-exchangers for separation of beryllium from aluminium [19].

8.2 Methods of Determination

Triphenylmethane and azo reagents are used in most methods for determining beryllium spectrophotometrically. A method using Chrome Azurol S and a more sensitive method using Eriochrome Cyanine R are presented in detail below. Both methods are almost specific.

8.2.1 CHROME AZUROL S METHOD

Chrome Azurol S (Alberon) forms a coloured complex with beryllium which has been used for the spectrophotometric determination of this element [20–26]. Alberon is so named because of its use as a reagent for aluminium and beryllium.

The reaction of Chrome Azurol S with beryllium is carried out either at pH 4–5 [21,23], or at pH 6–7 [20,22]. The method is more specific at pH 4–5. A 1:1 complex is formed at this pH [25].

In acetate buffer at pH 4·6, and in the presence of EDTA as masking agent, the Chrome Azurol S method is highly selective for beryllium. The absorbance of the complex depends on the pH of the solution, and on the concentrations of Chrome Azurol S, EDTA, and the acetate buffer. The absorbance increases with increasing Chrome Azurol S concentration, and decreases with increasing EDTA and acetate concentrations.

The pH 4·6 medium is the most suitable. Below this pH value the absorbance of Chrome Azurol S increases considerably, and above it the absorbance of the beryllium complex is decreased more by EDTA.

Under such circumstances it is difficult to define the objective value of the molar absorptivity of the complex. The absorption maximum of the complex, vs. the reagent solution as reference, is at $\sim$ 570 nm. At this wavelength, and under the reaction conditions given in the procedure below, the molar absorptivity is $6·0 \times 10^3$ ($a = 0·66$).

At pH 4·6 the complex of beryllium with Chrome Azurol S forms instantaneously, and remains stable for at least 4 hours. The absorbance is constant between 15° and 35°C, but decreases at above 35°C.

EDTA successfully masks Cd, Co, Cu, Fe, Mn, Mo, Pb, V(IV), W, and Zn. Copper, in an amount of 2 mg, causes a positive error of 2%. Ascorbic acid reduces Fe(III) to Fe(II), and V(V) to V(IV). The EDTA complexes of Al and Cr(III) are formed slowly, and it is necessary to heat the solution after the addition of EDTA. Zirconium interferes in the determination even in the presence of EDTA, but can be masked by adding tartaric acid.

Anions which interfere are fluoride and, to a lesser degree, phosphate.

Beryllium has been determined by the Chrome Azurol S method in minerals, ores, and bronzes [21], and in various salts [23].

Reagents

1. Chrome Azurol S: 0·06% solution.
2. Standard beryllium solution: 1 mg/ml. Dissolve 1·964 g of $BeSO_4 \cdot 4H_2O$ in water containing 1 ml of conc. HCl, and dilute the solution to 100 ml in a volumetric flask with water.

3. Acetate buffer: pH 4·6. Dissolve 48 g of sodium acetate in 500 ml of water, add 20 ml of glacial acetic acid, and dilute the solution with water to 1 litre.

Procedure

To the slightly acidic solution containing not more than 50 μg of Be, add ~ 20 mg of ascorbic acid and 2 ml of 10% EDTA solution. Add dilute NaOH solution to adjust the pH to 4–5, then after 5 min add 10 ml of the acetate buffer, dilute with water to ~ 40 ml, and add 5 ml of the Chrome Azurol S solution. Dilute the solution accurately to 50 ml with water, and measure the absorbance at 570 nm, using a blank solution as reference.

Note. In the presence of aluminium and chromium (>1 mg), heat the solution to ~ 90°C, and cool before adding the Chrome Azurol S. If the solution contains zirconium, add tartaric acid as masking agent.

8.2.2 Eriochrome Cyanine R Method

Although Eriochrome Cyanine R (formula, p. 51) reacts with beryllium ions over a fairly wide pH range, ammoniacal medium (pH ~ 9·7) is recommended for the spectrophotometric determination of beryllium [27–30]. At lower pH values EDTA interferes, and at higher pH values the colour of the free reagent interferes. At pH 9·7, the reagent is orange (λ_{max} = 435 nm) while its water-soluble beryllium complex is red-purple (λ_{max} = 525 nm). The molar absorptivity at 525 nm is $1·5 \times 10^4$ (a = 1·65).

With EDTA as masking agent, this method has considerable selectivity. Iron(III), aluminium, and a great number of other metals are masked by EDTA. Tartrate [27] is not such an efficient masking agent. Cyanide is used to mask Zn, Cd, Ni, etc. [28].

Fluoride and phosphate interfere in the determination of beryllium with Eriochrome Cyanine R. When phosphate is present in ten times the concentration of beryllium, the results are reduced by 20%.

The colour of the complex slowly diminishes with time; therefore the absorbance should be measured soon after the colour reaction is complete.

Sommer and Kubáň [24] have examined in detail the equilibria in the Be^{2+} ion reaction with Eriochrome Cyanine R.

Reagents

1. Eriochrome Cyanine R: 0·10% solution. Dissolve 100 mg of the reagent in water containing 0·5 ml of 1M HCl, and dilute the solution with water to 100 ml.
2. Standard beryllium solution: 1 mg/ml (p. 144).

Procedure

To a solution in dil. HCl (pH 2–3), containing not more than 20 μg of Be, add 2 ml of 10% EDTA solution, 10 ml of 25% ammonium acetate solution, exactly 5 ml of the Eriochrome Cyanine R solution, water to ~ 40 ml, and ammonia to pH 9·7 (±0·1). Transfer the solution to a 50-ml volumetric flask, dilute with water to the mark, and mix well. Measure the absorbance of the solution within 5 minutes, at 525 nm, using a blank solution as reference.

8.2.3 Other Methods

A large number of spectrophotometric methods for determining beryllium are based on azo dyes, e.g. the Beryllon group of reagents. The most important of these is Beryllon II (I), suggested by Lukin and Zavarikhina [31]. Beryllon I, described at the same time by these authors, is similar

(I) Structure: naphthalene with HO$_3$S groups, —OH, —N=N— linked to naphthalene bearing HO$_3$S, HO, OH, and SO$_3$H groups.

in structure to Beryllon II (but with the —OH peri to the azo group replaced by —NH$_2$), but has found no practical application. Beryllon II forms a blue complex with beryllium at pH 12–13 [32,33]; the molar absorptivity of the complex is $1·2 \times 10^4$ at $\lambda_{max} = 630$ nm. The method has been used to determine beryllium in beryllium ores [5], aluminium alloys [34], and natural waters [35].

After investigating many spectrophotometric reagents, Kuznetsov et al. [36] recommended Beryllon III (II) and Beryllon IV (III) as reagents for beryllium.

(II) Structure: naphthalene with HO$_3$S groups, —OH, —N=N—, HO, and phenyl—N(C$_2$H$_5$)$_2$.

(III) Structure: phenyl with AsO$_3$H$_2$, —N=N—, naphthalene with OH, HO$_3$S, and N(CH$_2$COOH)$_2$.

Methods using these reagents [37–39] are more sensitive than the Beryllon II method.

Other azo dyes proposed as spectrophotometric reagents for beryllium are p-nitrophenylazo-orcinol [40–42], Arsenazo I [43, 44], Arsenazo III (molar absorptivity $1·8 \times 10^4$ at 580 nm) [45], Thoron I [46–48], Chlorophosphonazo R [49], Chromotrope 2C [50], Fast Sulphon Black F ($\varepsilon = 1·37 \times 10^4$ at 630 nm) [51], and p-sulphophenylazosalicylic acid [52]. In these procedures, selectivity is gained by appropriate choice of pH and masking agents.

Triphenylmethane dyes other than Chrome Azurol S and Eriochrome Cyanine R (see above) have also been used to determine beryllium, e.g. Solochrome Azurine BS [53,54], Aluminon [55–59], Xylenol Orange [60], Methylthymol Blue [61], Chromal Blue G [62], and Eriochrome Brilliant Violet B [63]. These last two reagents are more sensitive than Eriochrome Cyanine R. Beryllium has been determined with Aluminon in copper alloys [55], niobium alloys [58], and air [56,57].

Spectrophotometric methods for beryllium have also been developed based on the following organic reagents: quinalizarin [1,5], 2-phenoxyquinizarin-3,4'-disulphonic acid [64], rufigallol [65], phenoxydinaphthofuchsonedicarboxylic acid (Naphthochrome Green G) [66].

In the extractive spectrophotometric method for determining beryllium the absorbance is measured of the chloroform extract of the chelate of beryllium with 8-hydroxyquinaldine ($\varepsilon = 3.5 \times 10^3$ at 380 nm) [67,68].

In an indirect spectrophotometric method, beryllium is precipitated as beryllium ammonium phosphate, and the phosphorus subsequently determined as yellow molybdophosphoric acid [69].

References

1. Fischer, W. and Wernet, J., *Angew. Chem.* **60**, 729 (1948).
2. Toribara, T. Y. and Sherman, R. E., *Anal. Chem.* **25**, 1594 (1953).
3. Rutkowski, W., *Chem. Anal. (Warsaw)* **8**, 389 (1963).
4. Sevast'yanov, A. I. and Rudenko, N. P., *Radiokhimiya* **10**, 487 (1968).
5. Tsyvina, B. S. and Davidovich, N. K., *Zavodsk. Lab.* **23**, 280 (1957).
6. Sudhalatha, K., *Talanta* **10**, 934 (1963).
7. Toribara, T. Y. and Chen, P. S., Jr., *Anal. Chem.* **24**, 539 (1952).
8. Steinbach, J. F. and Freiser, H., *ibid.* **25**, 881 (1953).
9. Alimarin, I. P. and Gibalo, I. M., *Zh. Analit. Khim.* **11**, 389 (1956).
10. Athavale, V. T., Iyer, C. S., Tillu, M. M. and Vaidya, G. M., *Anal. Chim. Acta*, **24**, 263 (1961).
11. Merrill, J. R., Honda, M. and Arnold, J. R., *Anal. Chem.* **32**, 1420 (1960).
12. Scribner, W. G., Borchers, M. J. and Treat, W. J., *ibid.* **38**, 1779 (1966).
13. Blasius, E. and Janzen, K. P., *Z. Anal. Chem.* **255**, 10 (1971).
14. Ryabchikov, D. I. and Bukhtiarov, V. E., *Zh. Analit. Khim.* **9**, 196 (1954).
15. Eristavi, V. D., Eristavi, D. I. and Brouchek, F. I., *ibid.* **23**, 782 (1968).
16. Nadkarni, M. N., Varde, M. S. and Athavale, V. T., *Anal. Chim. Acta* **16**, 421 (1957).
17. Strelow, F. W., *Anal. Chem.* **33**, 542 (1961).
18. Korkisch, J. and Ahluwalia, S. S., *Talanta* **11**, 1623 (1964).
19. Blasius, E. and Finkenauer, H. J., *ibid.* **20**, 639 (1973).
20. Wood, J. H., *Mikrochim. Acta* **1955**, 11.
21. Mustafin, I. S. and Matveev, L. O., *Zavodsk. Lab.* **24**, 259 (1958).
22. Silverman, L. and Shideler, M. E., *Anal. Chem.* **31**, 152 (1959).
23. Pakalns, P., *Anal. Chim. Acta* **31**, 576 (1964).
24. Sommer, L. and Kubáň, V., *Collection Czech. Chem. Commun.* **32**, 4355 (1967).
25. Sommer, L. and Kubáň, V., *Anal. Chim. Acta* **44**, 333 (1969).
26. Adamovich, L. P., Morgul-Meshkova, O. H. and Yutsis, B. V., *Zh. Analit. Khim.* **17**, 678 (1962).
27. Umemoto, S., *Bull. Chem. Soc. Japan* **29**, 845 (1956).
28. Hill, U. T., *Anal. Chem.* **30**, 521 (1958).

29. Kohara, H., Ishibashi, N. and Fukamachi, K., *Japan Analyst* **17**, 1400 (1968).
30. Kasiura, K., *Chem. Anal. (Warsaw)* **16**, 407 (1971).
31. Lukin, A. M. and Zavarikhina, G. B., *Zh. Analit. Khim.* **11**, 393 (1956).
32. Karanovich, G. G., *ibid.* **11**, 400 (1956).
33. Adamovich, L. P. and Mirnaya, A. P., *ibid.* **18**, 292 (1963).
34. Budanova, L. M. and Zhukova, N. A., *Zavodsk. Lab.* **25**, 411 (1959).
35. Kornienko, T. G. and Samchuk, A. I., *Ukr. Khim. Zh.* **38**, 917 (1972).
36. Kuznetsov, V. I., Bol'shakova, L. I. and Fan Min-E, *Zh. Analit. Khim.* **18**, 160 (1963).
37. Pakalns, P. and Flynn, W. W., *Analyst* **90**, 300 (1965).
38. Budanova, L. M. and Pinaeva, S. N., *Zavodsk. Lab.* **32**, 401 (1966).
39. Shibata, S., Goto, K., Amano, T. and Miyazaki, Y., *Japan Analyst* **18**, 604 (1969).
40. Pollock, J. B., *Analyst* **81**, 45 (1956).
41. White, J. C., Meyer, A. S. and Manning, D. L., *Anal. Chem.* **28**, 956 (1956).
42. Covington, L. C. and Miles, M. J., *ibid.* **28**, 1728 (1956).
43. Shibata, S., Takeuchi, F. and Matsumae, T., *Bull. Chem. Soc. Japan* **31**, 888 (1958).
44. Tolmachev, V. N., Kvichko, L. A. and Konkin, V. D., *Zh. Analit. Khim.* **22**, 11 (1967).
45. Talipov, Sh. T., Khadeeva, L. A. and Popova, R., *Izv. Vyssh. Ucheb. Zaved., Khim. Khim. Tekhnol.* **14**, 343 (1971).
46. Adamovich, L. P. and Yutsis, B. V., *Ukr. Khim. Zh.* **22**, 523 (1956); **23**, 784 (1957).
47. Einaga, H. and Ishii, H., *Anal. Chim. Acta* **54**, 113 (1971).
48. Keil, R., *Z. Anal. Chem.* **262**, 273 (1972).
49. Luk'yanov, V. F., Lukin, A. M., Knyazeva, E. M. and Kalinina, I. D., *Zh. Analit. Khim.* **18**, 562 (1963).
50. Majumdar, A. K. and Savariar, C. P., *Z. Anal. Chem.* **176**, 170 (1960).
51. Cabrera, A. M. and West, T. S., *Anal. Chem.* **35**, 311 (1963).
52. Adamovich, L. P. and Vu Van Nyan, *Zh. Analit. Khim.* **23**, 994 (1968).
53. Katsube, Y., Uesugi, K. and Yoe, J. H., *Bull. Chem. Soc. Japan* **34**, 72 (1961).
54. Sharma, C. L. and Tandon, S. N., *Z. Anal. Chem.* **250**, 383 (1970).
55. Luke, C. L. and Campbell, M. E., *Anal. Chem.* **24**, 1056 (1952).
56. Crawley, R. H., *Anal. Chim. Acta* **22**, 413 (1960).
57. McCloskey, J. P., *Microchem. J.* **12**, 32, 40 (1967).
58. Tsyvina, B. S. and Ogareva, M. B., *Zavodsk. Lab.* **28**, 917 (1962).
59. Dhond, P. V., and Khopkar, S. M., *Anal. Chem.* **45**, 1937 (1973).
60. Otomo, M., *Bull. Chem. Soc. Japan* **38**, 730 (1965).
61. Pilipenko, A. T. and Belyaev, Yu. D., *Ukr. Khim. Zh.* **37**, 193 (1971).
62. Uesugi, K., *Bull. Chem. Soc. Japan* **42**, 2998 (1969).
63. Uesugi, K., *Anal. Chim. Acta* **49**, 89 (1970).
64. Owens, E. G., II and Yoe, J. H., *Anal. Chem.* **32**, 1345 (1960); *Talanta* **8**, 505 (1961).
65. Azim, M. A. and Ayaz, A. A., *Mikrochim. Acta* **1969**, 153.
66. Adamovich, L. P., Mirnaya, A. P. and Khukhryanskaya, A. K., *Zh. Analit. Khim.* **24**, 1816 (1969).
67. Motojima, K., *Bull. Chem. Soc. Japan* **29**, 71 (1956).
68. Keil, R., *Mikrochim. Acta* **1973**, 919.
69. Sunderasan, M. and Sankar Das, M., *Analyst* **80**, 697 (1955).

Chapter 9

BISMUTH

Bismuth (Bi, at.wt. 208·98) occurs in its compounds in the +III and +V oxidation states. Bismuth(V) (bismuthate) exists only in solids (e.g. $NaBiO_3$, a powerful oxidant). In solution only compounds of bismuth(III) are found. Bismuth(III) hydrolyses at pH 1–2, and shows no amphoteric properties. Bismuth(III) forms citrate, oxalate, iodide, thiosulphate, and EDTA complexes.

9.1 Methods of Separation

9.1.1 EXTRACTION

Extraction (in the presence of masking agents) with dithizone in $CHCl_3$ or CCl_4 is a selective method for separating traces of bismuth. The bismuth extracted may be determined directly as the dithizonate (see below), or after reaction with other spectrophotometric reagents [1]. In the absence of masking agents, dithizone extracts bismuth together with other heavy metals [2].

Bismuth diethyldithiocarbamate, $Bi(DDTC)_3$, is extracted into CCl_4 or $CHCl_3$ from alkaline solution containing tartrate, cyanide, and EDTA. No other metal is extracted under such conditions. Once extracted, bismuth may be determined either directly as the coloured complex $Bi(DDTC)_3$ (see below), or by other methods [3–5].

Bismuth cupferronate [6,7] is extracted with chloroform from a solution at pH 1, thereby separating bismuth from major quantities of lead, and also from Sn(IV), Hg, Mo, Co, and other metals. A cupferron analogue, BPHA, has been used to extract bismuth from a solution at pH 2 [8].

Bismuth can be separated from strongly acidic media (1–5N H_2SO_4 or 2M $HClO_4$) as the iodide complex, with isoamyl acetate–isoamyl alcohol mixture as extractant [9]. The extraction of bismuth as the bromide and the chloride complexes has also been investigated [10,11].

Bismuth has also been extracted as complexes with di-(2-ethylhexyl) phosphoric acid (HDEHP) [12,13], and with 4-methoxybenzothiohydroxamic acid [14].

9.1.2 PRECIPITATION

When precipitated as the sulphide from fairly acidic medium (2M HCl) with copper as scavenger [15], bismuth is separated from Pb, Sn, and Cd:

other Analytical Group I and II metals are coprecipitated with bismuth from 0·2–0·3M HCl. Bismuth has been preconcentrated from niobium and chromium alloys by precipitation as the sulphide [16].

When traces of bismuth are precipitated as the hydroxide with ammonia, iron [4,17], aluminium [16], and lanthanum [18] can be used as collectors. Traces of bismuth can be coprecipitated with manganese dioxide from a nitric acid solution at pH 1–2·5 [19,20]. Antimony and tin are precipitated quantitatively together with the bismuth. Traces of bismuth have also been coprecipitated as the phosphate with cobalt as the carrier [21].

Organic precipitants have also been used to concentrate traces of bismuth. For example, the $[BiI_4]^-$–Methyl Violet ion-association compound is quantitatively coprecipitated with the sparingly soluble Methyl Violet–iodide complex [22]. A similar precipitate of $[Bi(SCN)_4]^-$ and Methyl Violet has been separated by flotation with toluene [23].

Small amounts of bismuth have been separated by electrolysis [24], and by cementation with metallic zinc from a slightly acidic medium (H_2SO_4) [25].

9.1.3 Ion-Exchange

Strongly basic anion-exchangers retain bismuth from dilute hydrochloric acid (0·1–1M), allowing the separation of bismuth from iron and very many other metals [26,27]. Bismuth is eluted from the column with thiourea in dilute sulphuric acid, or with dilute nitric acid.

From a solution of nitric acid in n-propanol [28], or propylene glycol [29], Dowex-1 adsorbs only Bi, U, Th, and La. It is possible to separate the bismuth from the other metals by using appropriate eluents.

Bismuth has been separated from most other metals by selective elution from cation-exchange columns with varying concentrations of hydrobromic [30] or sulphuric [31] acid. Separation of traces of bismuth in platinum metal alloys has been achieved by adsorption of bismuth on a cation-exchanger, and elution of the platinum metal anionic chloride complexes [32].

9.2 Methods of Determination

The conventional iodide and thiourea methods are rapid and simple, but lack sensitivity. The extractive spectrophotometric dithiocarbamate method is also rather insensitive, but is important because of its specificity. The Xylenol Orange method is more sensitive. The high sensitivity of the dithizone method makes this method suitable for the determination of traces of bismuth.

9.2.1 Dithizone Method

With dithizone (formula, p. 41), bismuth(III) ions form an orange-brown dithizonate, $Bi(HDz)_3$, which is soluble in CCl_4 and $CHCl_3$, and stable over the pH range 3–9·5. This dithizonate is the basis of the spectrophotometric method [33–36].

The absorption spectrum of bismuth dithizonate in CCl_4 is shown in Fig. 2.2 (p. 43) ($\varepsilon = 7.92 \times 10^4$; $a = 0.38$ at $\lambda_{max} = 490$ nm).

When cyanide and tartrate (or citrate) are present as masking agents, only lead, thallium(I), and tin(II) dithizonates are co-extracted with bismuth from slightly alkaline medium (pH 8–9·5). Tin(IV) does not react with dithizone, so Sn(II) interference can be avoided by preliminary oxidation. Lead and thallium can easily be separated from bismuth since their dithizonates are unstable in slightly acidic medium (pH 3·0–3·5). After Bi, Pb, and Tl have been extracted into carbon tetrachloride, the lead and thallium are stripped into an aqueous solution at pH 3·3, while bismuth dithizonate remains in the organic phase.

If the solution being analysed contains considerably more lead than bismuth, it is more convenient to prevent the extraction of lead. In this case, the noble metals (Pt, Pd, Au, Ag, and Hg) and the copper in the sample are quantitatively extracted with dithizone at pH 0·5–1·0, after which the pH of the aqueous solution is readjusted to 3·1, and the bismuth is then extracted with dithizone.

If the amounts of zinc, cadmium, and lead in the aqueous solution considerably exceed that of bismuth (e.g. hundredfold), then traces of Zn, Cd, and Pb dithizonates are also extracted. The cadmium and lead are quantitatively removed by stripping with an aqueous solution at pH 3·3. Zinc dithizonate, however, is not decomposed at pH 3·3, and below this pH bismuth dithizonate is partly decomposed. The traces of zinc are stripped from the extract with a dilute ($\sim 0.05\%$) KCN solution buffered at pH 9·5 (an unbuffered cyanide solution is sufficiently alkaline to partly decompose bismuth dithizonate).

At higher concentrations, halide ions inhibit the extraction of bismuth from acid medium. The effect is most severe for iodide. Tartrate, citrate, and acetate similarly inhibit the extraction.

Busev and Bazhanova [37] have determined bismuth in the presence of Hg, Ag, and Cu by the dithizone reversion method, using EDTA and KCl solutions at pH 2 as reversion agents. Chalmers and Dick used $0.2M$ nitric acid as reversion agent to determine Bi in presence of Ag and Hg [37a].

Bismuth has been determined by the dithizone method in platinum [32], gold [35], silver [18], silver alloys [38], tin [39], boron [40], alkalis [2], natural water [27], and biological material [33].

Di-2-naphthylthiocarbazone forms a complex with the bismuth ion, the molar absorptivity of which is 1.7×10^5 [41].

Reagents

1. Dithizone: 0·001% solution in CCl_4. Preparation, see p. 492.
2. Standard bismuth solution: 1 mg/ml. Dissolve 2·321 g of $Bi(NO_3)_3$.

5H$_2$O in 100 ml of HNO$_3$ (1+3), and dilute the solution with water in a volumetric flask to 1 litre. Working solutions are obtained by diluting the stock solution with 0·01M HNO$_3$.

3. Potassium cyanide: 10% solution. Preparation, see p. 325.
4. Wash solution (pH 3·3 ±0·1). Adjust an approximately 0·1% NH$_4$Cl solution with dilute HCl to pH 3·3 ±0·1.
5. Buffer solution (pH 9·5). Dissolve 60 g of NH$_4$Cl in water, add 120 ml of conc. ammonia (25%), and dilute the solution with water to 1 litre.

Procedure

Any Cu, Ag, Au, Pt, Pd, or Hg in the sample should be removed (e.g. by extraction with dithizone at pH 0·5–1·0). Adjust the pH of the sample solution (containing not more than 50 μg of Bi) cautiously with ammonia to pH 3·1 ±0·1. Extract bismuth with several aliquots of the dithizone solution in CCl$_4$ (1 ml of 0·001% H$_2$Dz solution is equivalent to 2·7 μg of Bi). Wash the combined extracts by shaking successively with the wash solution, water, and then with a solution containing 3 drops of the buffer solution (pH 9·5) and 3 drops of 10% KCN solution per 10 ml of water. Finally wash the bismuth dithizonate extract with water, and dilute in a 50 ml or smaller volumetric flask (depending on the colour intensity) with carbon tetrachloride. Measure the absorbance of the extract at 490 nm, using the solvent or a blank solution as reference.

Notes: (1) If other metals (e.g. Zn, Pb and Cd) are to be determined in the aqueous solution after extraction of Bi, then add a few drops of conc. HCl and H$_2$O$_2$ (30%) to the combined acidic and alkaline aqueous washings, evaporate the solution to 3–4 ml, and combine the residue with the aqueous solution from which bismuth was extracted.

(2) Both bismuth and lead dithizonates are extracted from cyanide solution at pH 9·5. To remove the lead, shake the extract with two portions of the pH 3·3 wash solution, and then with the pH 9·5 buffer solution to remove the free dithizone liberated by the decomposition of Pb(HDz)$_2$.

(3) Before extracting bismuth and lead from ammoniacal cyanide medium, add purified potassium sodium tartrate (see p. 325) to the initial acidic solution to prevent precipitation of hydrolysable metals [e.g. Al, Fe(III), Ti and Zr].

9.2.2 Iodide Method

In acid medium (0·4–4N H$_2$SO$_4$), and in the presence of an excess of iodide, bismuth forms the orange-yellow complex [BiI$_4$]$^-$, the basis of this spectrophotometric method [42–44]. Up to a concentration of 3% KI in the solution, the absorbance increases; above this concentration, the absorbance remains constant. Liberation of free iodine, due to oxidation of the iodide by atmospheric oxygen or oxidizing substances present in the sample solution, is prevented by the addition of reducing agents, such as ascorbic acid, sulphite, or hypophosphite.

The absorption spectrum of the bismuth iodide complex (Fig. 6.2) exhibits an intense maximum at 337 nm, and a less intense maximum in the visible spectrum at 465 nm [43]. The molar absorptivity of the complex at 465 nm is $9·1 \times 10^3$ ($a = 0·044$).

Interference in the iodide method by antimony may be overcome by preliminary separation, or by a slight modification of the reaction conditions. At 1·5% KI concentration, antimony does not form a coloured complex, while the intensity of the colour of the bismuth complex is reduced only by $\sim 10\%$.

Besides antimony and bismuth, platinum(IV), palladium(II), and tin(IV) also form coloured iodide complexes. Silver, thallium, copper, and lead are precipitated as their sparingly soluble iodides. Chloride and fluoride weaken the colour of the bismuth iodide complex.

The sensitivity of the iodide method can be enhanced by extracting the ion-pairs formed by $[BiI_4]^-$ with brucine [45], caprolactam [46] or Methyl Green [47].

Bismuth has been determined by the iodide method in steel [48], lead [6,44], lead–antimony alloys [24], non-ferrous metals [4], copper [38], chromium and its alloys [16], chromium–nickel steel [22], cast iron and steel [26,49,50], niobium and its alloys [16,51], and vanadium [51].

The extraction of BiI_3 with benzene from H_2SO_4–KI solution forms the basis of the very sensitive indirect starch–iodine amplification method for determining bismuth [52]. The BiI_3 is re-extracted with dilute H_2SO_4. Iodide ions in the aqueous solution are oxidized to IO_3^-, which react with added potassium iodide to release iodine which is determined by the starch–iodine method (see determination of iodide, p. 296), and determination of palladium, p. 413). For each atom of bismuth (in BiI_3), 18 atoms of iodine are released. The molar absorptivity (relative to Bi) is $3·2 \times 10^5$ ($a = 1·5$).

Reagents

1. Potassium iodide: 20% solution.
2. Standard bismuth solution: 1 mg/ml (p. 151).

Procedure

To a solution in a volumetric flask, containing not more than 0·7 mg of Bi, add 5 ml of sulphuric acid (1+1), 2 ml of 2% ascorbic acid, and 10 ml of potassium iodide solution. Dilute the solution to the mark, and mix well. After 5 minutes, measure the absorbance of the coloured solution at 465 nm against water.

9.2.3 DIETHYLDITHIOCARBAMATE METHOD

With dithiocarbamates (especially with sodium diethyldithiocarbamate, Na–DDTC, p. 54), bismuth forms coloured chelates, which are soluble in organic solvents (such as carbon tetrachloride and isoamyl alcohol) [3].

The maximum absorbance of $Bi(DDTC)_3$ in CCl_4 is at 370 nm ($\varepsilon = 8·6 \times 10^3$). At 400 nm, the molar absorptivity is $6·7 \times 10^3$ (specific absorptivity 0·033). The colorimetric method for determining bismuth as the yellow Bi(DDTC)$_3$ complex [53–55] is not very sensitive, but it is important because it is specific.

In ammoniacal medium (pH 9–11) containing EDTA and cyanide as masking agents, bismuth is the only metal which forms a coloured complex with DDTC. Mercury(II) diethyldithiocarbamate is also extractable under these conditions, but it does not absorb at 400 nm [although it does absorb slightly at 370 nm, λ_{max} for Bi(DDTC)$_3$].

The extraction of bismuth diethyldithiocarbamate is not affected by the presence of tartrate or citrate as masking agents for aluminium, titanium, and similar elements.

Kovács and Guyer [55], as well as Lau *et al.* [56], have used pyrrolidinedithiocarbamate for the extractive spectrophotometric determination of bismuth. Bismuth has also been determined with pyrazolinedithiocarbamate [57], glycinedithiocarbamate [58], and dibenzyldithiocarbamate [59,60].

The dithiocarbamate methods for determining bismuth have been used in the analysis of steel [55], non-ferrous metals and alloys [55], niobium compounds [57], lead alloys [54,56] and tin alloys [56].

Reagents

1. Sodium diethyldithiocarbamate (Na–DDTC): 0·2% aqueous solution adjusted with ammonia to pH 8–9.
2. Standard bismuth solution: 1 mg/ml. (p. 151).

Procedure

To a solution containing not more than 1·0 mg of Bi add 5 ml of 10% EDTA solution, neutralize the solution with conc. ammonia solution, and add 5 ml excess. Next add 5 ml of 10% KCN solution and 2 ml of Na–DDTC solution. Transfer the solution to a separating funnel and shake with two portions of CCl$_4$. Dilute the extracts with CCl$_4$ to the mark in a suitable volumetric flask (e.g. 50 ml). Mix the solution, and measure its absorbance at 400 nm, using the solvent as a reference.

Note. The amounts of the EDTA and KCN solutions added depend on the amount of metals in the solution which have to be masked before the extraction of bismuth diethyldithiocarbamate.

9.2.4 Xylenol Orange Method

Xylenol Orange (p. 51) forms a water-soluble red-violet 1:1 complex with bismuth ions in acid medium [1,61–64].

The absorption maximum of the bismuth–Xylenol Orange complex is at 540 nm, whereas that of the reagent is at 440 nm (Fig. 9.1). Maximum colour intensity of the complex is obtained in 0·05–0·1M H$_2$SO$_4$ or 0·08–0·12M HNO$_3$. According to Cheng [61], the molar absorptivity of the complex in 0·05M H$_2$SO$_4$ is $2·4 \times 10^4$ ($a = 0·11$). The colour of the complex is stable.

Iron(III), which interferes in the determination of bismuth, is reduced to Fe(II) with ascorbic acid. By means of citric or tartaric acid, it is possible to mask Zr, Hf, Sn(II), and Sb(III), which also form coloured complexes with Xylenol Orange. Tin can also be masked conveniently with fluoride, which at low concentration does not affect the reaction between Xylenol Orange and bismuth.

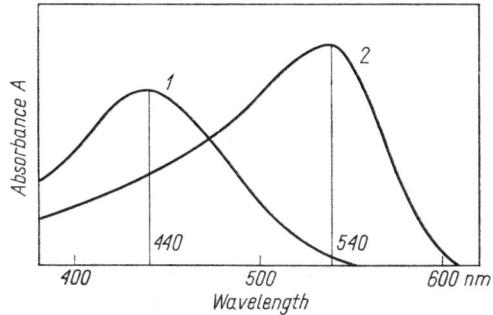

Fig. 9.1 Absorption spectra of Xylenol Orange (1) and its bismuth complex (2) at pH ~ 1

Chloride, bromide, and iodide decompose the bismuth–Xylenol Orange complex. Since chloride exerts no influence on the coloured complexes of Xylenol Orange with Zr, Hf, Fe, and Sn, this specific action on the bismuth complex can be used for determining bismuth in the presence of these metals [58].

Bismuth has been determined by the Xylenol Orange method in bismuth–manganese thin films [61], lead [65], copper alloys [65], cast iron [19], copper [66,67], nickel [66], and silver [67].

Reagents

1. Xylenol Orange: 0·05% solution in $1N$ sulphuric acid.
2. Standard bismuth solution: 1 mg/ml (p. 151).

Procedure

Place a sample solution at pH ~ 1 containing not more than 300 μg of Bi in a 50-ml volumetric flask. Add 2 ml of 2% ascorbic acid solution, 1 ml of 10% citric acid solution, and (after 2–3 minutes) 5 ml of the Xylenol Orange solution. Dilute with $0·1N$ sulphuric acid to the mark, and mix well. After 10 minutes, measure the absorbance at 545 nm, using a blank solution as reference.

9.2.5 OTHER METHODS

The cationic yellow bismuth–thiourea complex, stable in acid medium, is the basis of the conventional, rather insensitive method for determining bismuth [68–70]. At 470 nm, the molar absorptivity is ~ $9·0 \times 10^3$. The colour of the complex varies slightly depending on the acid present in the solution (usually ~ 1N nitric acid is used). The concentration of thiourea in the solution must be high (~ 6%). Coloured complexes are also formed by thiourea with Sb, Te, Pd, Fe(III), Os, and Ru. At higher concentrations of some metals (Ag, Hg, Cu, Pb, Cd, Sn, Ti), precipitates are formed with thiourea. The thiourea method has been used to determine bismuth in, *inter alia*, lead [69-71], tin [70,72], and ferrous and other metals [25].

Aoki and Tomioka [73] have extracted the bismuth–thiourea complex from perchloric acid medium with TBP, and Rudnev *et al.* [74] with alkylphosphoric acids.

Busev *et al.* [75] have suggested extractive–photometric methods for the determination of bismuth by means of its ternary complexes with *N*-phenyl derivatives of thiourea and thiocyanate or perchlorate. The molar absorptivity is ~ 1.0×10^4.

Other extractive–spectrophotometric methods are based on the ion-pairs formed by bismuth halide complexes with thiolactams [76,77], diantipyrylmethylmethane (DAMM) and diantipyrylpropylmethane (DAPM) [78], 1,10-phenanthroline [79], chlorpromazine hydrochloride [80], and tri-n-octylamine [81].

More sensitive methods for determining bismuth utilize the azo dyes Thoron I [82], PAR [83], and Arsenazo III (molar absorptivity 2.7×10^4 at 610 nm) [84], and the triphenylmethane dyes Pyrocatechol Violet [85] and Methylthymol Blue [86] (the Xylenol Orange method has been discussed in detail above).

Further organic reagents recommended for the spectrophotometric determination of bismuth are Bromopyrogallol Red [87] 2,6-dimercapto-3,5-diphenylthiopyran-4-one ($\varepsilon = 3.2 \times 10^4$ at 410 nm) [5,88], 5-mercapto-3-(4-bromophenyl)-1,3,4-thiadiazole-2-thione (a derivative of Bismuthol II) [89], 1,3-bis[8'-mercaptotheophyllinyl-(7')]-propane [90], CMAB-oxine [91], quinoxaline-2,3-dithiol [92], and dithiopyrylmethane [93].

Lastly, the less sensitive methods based on bismuth complexes with thiocyanate [17] and bromide [94] should be mentioned.

References

1. Onishi, H. and Ishiwatari, N., *Talanta* **8**, 753 (1961).
2. Marczenko, Z., *Mikrochim. Acta* **1965**, 281.
3. Bode, H., *Z. Anal. Chem.* **144**, 165 (1955).
4. Strel'nikova, N. P. and Lystsova, G. G., *Zavodsk. Lab.* **28**, 659 (1962).
5. Usatenko, Yu. I., Arishkevich, A. M. and Akhmetshin, A. G., *ibid.* **31**, 788 (1965).
6. Bode, H. and Henrich, G., *Z. Anal. Chem.* **135**, 98 (1952).
7. Ishihara, Y., Shibata, K., Kishi, H. and Hori, T., *Japan Analyst* **11**, 91 (1962).
8. Chwastowska, J. and Bragińska, J., *Chem. Anal.* (*Warsaw*) **11**, 169 (1966).
9. Mottola, H. A. and Sandell, E. B., *Anal. Chim. Acta* **24**, 301; **25**, 520 (1961).
10. Shevchuk, I. A. and Degtyarenko, L. I., *Ukr. Khim. Zh.* **28**, 1112 (1962).
11. Kimura, K., *Bull. Chem. Soc. Japan* **33**, 1038 (1960); **34**, 63 (1961).
12. Neirinckx, R. D., *Anal. Chim. Acta* **54**, 357 (1971).
13. Levin, I. S., Yukhin, Yu. M. and Zelinskii, A. G., *Zh. Analit. Khim.* **27**, 1976 (1972).

14. Skorko-Trybuła, Z. and Polanowska, J., *Chem. Anal. (Warsaw)* **15**, 635 (1970).
15. Moore, V. J., *Analyst* **81**, 553 (1956).
16. Mukhina, Z. S., Tikhonova, A. A. and Zhemchuzhnaya, I. A., *Tr. Komis. po Analit. Khim. Akad. Nauk SSSR* **12**, 298 (1960).
17. Różycki, C. and Maksjan, J., *Chem. Anal. (Warsaw)* **15**, 391 (1970).
18. Marczenko, Z. and Kasiura, K., *ibid.* **9**, 87 (1964).
19. Amsheeva, A. A. and Bezuglyi, D. V., *Zh. Analit. Khim.* **19**, 97 (1964).
20. Blakeley, S. J., Manson, A. and Zatka, V. J., *Anal. Chem.* **45**, 1941 (1973).
21. Chuiko, V. T. and Reva, N. I., *Zavodsk. Lab.* **33**, 1503 (1967).
22. Kuznetsov, V. I. and Papushina, L. I., *Zh. Analit. Khim.* **11**, 686 (1956).
23. Babko, A. K. and Marchenko, P. V., *Zavodsk. Lab.* **25**, 1047 (1959).
24. Andreev, A. S. and Kopets, N. P., *ibid.* **22**, 538 (1956).
25. Kovalenko, P. N., Bagdasarov, K. N., Kasatkina, S. K. and Malygina, S. V., *Zh. Analit. Khim.* **23**, 1173 (1968).
26. Léontovitch, N., *Chim. Anal. (Paris)* **47**, 458 (1965).
27. Portmann, J. E. and Riley, J. P., *Anal. Chim. Acta* **34**, 201 (1966).
28. Korkisch, J. and Tera, F., *Z. Anal. Chem.* **186**, 290 (1962).
29. Feik, F. and Korkisch, J., *Talanta* **11**, 1585 (1964).
30. Fritz, J. S. and Garralda, B. B., *Anal. Chem.* **34**, 102 (1962).
31. Akki, S. B. and Khopkar, S. M., *Sepn. Sci.* **5**, 707 (1970).
32. Marczenko, Z., Kasiura, K. and Krasiejko, M., *Mikrochim. Acta* **1969**, 625.
33. Hubbard, D. M., *Anal. Chem.* **20**, 363 (1948).
34. Barcza, L., *Acta Chim. Acad. Sci. Hung.* **28**, 143 (1961).
35. Marczenko, Z., Kasiura, K. and Krasiejko, M., *Chem. Anal. (Warsaw)* **14**, 1277 (1969).
36. Bidleman, T. F., *Anal. Chim. Acta* **56**, 221 (1971).
37. Busev, A. I. and Bazhanova, L. A., *Vestn. Mosk. Univ. Khim., Ser.* II **16**, No. 6, 47 (1961).
37a. Chalmers, R. A. and Dick, D. M., *Anal. Chim. Acta* **32**, 117 (1965).
38. Skorko-Trybuła, Z. and Chwastowska, J., *Chem. Anal. (Warsaw)* **8**, 859 (1963).
39. Marczenko, Z. and Kasiura, K., *ibid.* **10**, 449 (1965).
40. Marczenko, Z., *ibid.* **9**, 1093 (1964).
41. Grzhegorzhevskii, A. S., *Zh. Analit. Khim.* **11**, 689 (1956).
42. Lur'e, Yu. Yu. and Ginzburg, L. V., *Zavodsk. Lab.* **15**, 21 (1949).
43. Lisicki, N. M. and Boltz, D. F., *Anal. Chem.* **27**, 1722 (1955).
44. Englis, D. T. and Burnett, B. B., *Anal. Chim. Acta* **13**, 574 (1955).
45. Oosting, M., *Mikrochim. Acta* **1956**, 528.
46. Sikorska-Tomicka, H., *Z. Anal. Chem.* **187**, 258 (1962).
47. Shestidesyatnaya, N. L., Kish, P. P. and Merenich, A. V., *Zh. Analit. Khim.* **25**, 1547 (1970).
48. Koch, O. G., *Z. Anal. Chem.* **255**, 269 (1971).
49. Siekierska, J. and Piotrowski, A., *Chem. Anal. (Warsaw)* **11**, 845 (1966).
50. Lazareva, V. I. and Lazarev, A. I., *Zavodsk. Lab.* **31**, 1437 (1965).
51. Nazarenko, V. A. and Biryuk, E. A., *ibid.* **25**, 28 (1959).
52. Marczenko, Z., Żołądek, I. and Limbach, A., *Chem. Anal. (Warsaw)* **14**, 741 (1969).
53. Lacoste, R. J., Earing, M. H. and Wiberley, S. E., *Anal. Chem.* **23**, 871 (1951).
54. Cheng, K. L., Bray, R. H. and Melsted, S. W., *ibid.* **27**, 24 (1955).
55. Kovács, E. and Guyer, H., *Chimia* **13**, 164 (1959); *Z. Anal. Chem.* **186**, 267; **187**, 188 (1962).
56. Lau, H. K., Droll, H. A. and Lott, P. F., *Anal. Chim. Acta* **56**, 7 (1971).
57. Busev, A. I., Byr'ko, V. M. and Zhukova, R. G., *Vestn. Mosk. Univ. Khim., Ser.* **21**, No. 6, 72 (1966).
58. Haas, W. and Winterstein, P., *Mikrochim. Acta* **1961**, 787.
59. Yamane, T., Suzuki, T. and Mukoyama, T., *Anal. Chim. Acta* **62**, 137 (1972).
60. Yamane, T., Mukoyama, T. and Sasamoto, T., *ibid.* **69**, 347 (1974).

61. Cheng, K. L., *Talanta* **5**, 254 (1960).
62. Onishi, H. and Ishiwatari, N., *Bull. Chem. Soc. Japan* **33**, 1581 (1960).
63. Bagdasarov, K. N., Kovalenko, P. N. and Shemyakina, M. A., *Zh. Analit. Khim.* **23**, 515 (1968).
64. Kantcheva, D., Nenova, P. and Karadakov, B., *Talanta* **19**, 1450 (1972).
65. Danilova, V. N. and Marchenko, P. V., *Zavodsk. Lab.* **28**, 654 (1962).
66. Adamiec, I., *Chem. Anal. (Warsaw)* **13**, 147 (1968).
67. Bagdasarov, K. N., Vladimirova, V. F. and Shemyakina, M. A., *Zavodsk. Lab.* **34**, 1306 (1968).
68. Nielsch, W. and Böltz, G., *Z. Anal. Chem.* **142**, 321; **143**, 13, 168 (1954).
69. Pohl, H., *Angew. Chem.* **64**, 608 (1952).
70. Asmus, E., *Z. Anal. Chem.* **142**, 255 (1954).
71. Karanov, R. A. and Karolev, A. N., *Zavodsk. Lab.* **26**, 48 (1960).
72. Shvaiger, M. I., Paklina, V. P. and Medvedeva, A. S., *ibid.* **24**, 16 (1958).
73. Aoki, F. and Tomioka, H., *Bull. Chem. Soc. Japan* **38**, 1557 (1965).
74. Rudnev, V. V., Buzina, N. I. and Kosychenko, L. I., *Zh. Analit. Khim.* **28**, 1351 (1973).
75. Busev, A. I , Shvedova, N. V., Akimov, V. K. and Fursova, E. G., *ibid.* **24**, 1679, 1833 (1969).
76. Sikorska-Tomicka, H., *Chem. Anal. (Warsaw)* **12**, 1291 (1967); **13**, 341 (1968); **14**, 97 (1969).
77. Sikorska-Tomicka, H., *Mikrochim. Acta* **1968**, 1106.
78. Busev, A. I., Akimov, V. K. and Said Alisha Saber, *Zh. Analit. Khim.* **25**, 918 (1970).
79. Buhl, F. and Skibe, H., *Chem. Anal. (Warsaw)* **17**, 285 (1972).
80. Basińska, H. and Tarasiewicz, M., *ibid.* **13**, 1287 (1968).
81. Tsukahara, I. and Yamamoto, T., *Anal. Chim. Acta* **64**, 337 (1973).
82. Mottola, H. A., *ibid.* **27**, 136 (1962); **29**, 261 (1963).
83. Tomioka, H. and Terashima, K., *Japan Analyst* **16**, 698 (1967).
84. Barkovskii, V. F. and Povet'eva, Z. N., *Zavodsk. Lab.* **35**, 555 (1969).
85. Malát, M., *Z. Anal. Chem.* **186**, 418 (1962).
86. Enoki, T., Mori, I. and Izumi, Y., *Japan Analyst* **18**, 963 (1969).
87. Suk, V. and Smetanová, M., *Collection Czech. Chem. Commun.* **30**, 2532 (1965).
88. Usatenko, Yu. I., Arishkevich, A. M. and Akhmetshin, A. G., *Zh. Analit. Khim.* **20**, 462 (1965).
89. Busev, A. I., Simonova, L. N. and Gaponyuk, E. I., *ibid.* **23**, 59 (1968).
90. Asmus, E. and Marsen, G., *Z. Anal. Chem.* **225**, 252 (1967).
91. Röbisch, G., *Anal. Chim. Acta* **48**, 161 (1969).
92. Chernomorchenko, L. I. and Butenko, G. A., *Zavodsk. Lab.* **39**, 1448 (1973).
93. Dolgorev, A. V., Lysak, Ya. G. and Lukoyanov, A. P., *ibid.* **40**, 247 (1974).
94. Nielsch, W. and Böltz, G., *Anal. Chim. Acta* **11**, 438 (1954).

Chapter 10
BORON

Boron (B, at.wt. 10·81) is a non-metal with properties somewhat similar to those of silicon. In chemical analysis, only boron(III) compounds are of importance. Boron forms complexes with fluoride and polyalcohols (e.g. mannitol and glycerol). In anhydrous medium, boric acid reacts with methanol to form volatile trimethyl borate.

10.1 Methods of Separation

10.1.1 Volatilization as Trimethyl Borate

Distillation of boron as volatile trimethyl borate (b.p. 65°C) is the most common method of isolating boron before its spectrophotometric determination [1–5]. When separating small amounts of boron, a quartz distillation apparatus should be used since laboratory glassware contains boron. An anhydrous medium promotes the quantitative formation and distillation of methyl borate (water hydrolyses the ester). The usual procedure is to add methanol and concentrated sulphuric acid to a dried sample, and heat the still in a glycerol- or oil-bath, gradually raising the temperature to 120°C. The distillate is collected in a quartz or platinum receiver containing dilute NaOH solution (see Procedure below). If the sample solution contains fluoride, boron partly distils as volatile BF_3. This is prevented by masking fluoride as the stable aluminium complex [6]. Colloidal silica partially traps boron, thus inhibiting its quantitative distillation.

Traces of boron have been separated as trimethyl borate by microdiffusion methods [7,8].

10.1.2 Extraction and Leaching

With the tetraphenylarsonium cation $[(C_6H_5)_4As]^+$, the fluoroborate anion $[BF_4]^-$ forms an ion-pair which is extracted with chloroform. Quantitative conversion of boron into $[BF_4]^-$ and subsequent quantitative extraction are ensured by using an excess of fluoride (at pH 2–3), and by leaving the solution to react for sufficient time before extraction [9,10]. Substitution of cationic dyes for the tetraphenylarsonium ion enables boron to be separated and determined spectrophotometrically with the same reagent.

Traces of boron can be extracted from liquid chlorosilanes with concentrated sulphuric acid. If a suitable spectrophotometric reagent for boron (e.g. quinalizarin) is dissolved in the sulphuric acid, the boron can be determined from the absorbance of the acid extract [11].

To determine boron in fluoride salts, Ross *et al.* [12] dissolved the sample in an $AlCl_3$ solution (in H_2O–HCl–ethanol), and extracted the boron with diethyl ether.

Boron can sometimes be leached from solid samples. For example, Rynasiewicz *et al.* [13] leached boric acid with ethanol from sodium chloride.

10.1.3 Ion-Exchange

Strongly acidic cation-exchangers retain metal ions from acid solutions, whereas boric acid is eluted [14]. Large quantities of titanium have been separated from traces of boron by adsorption of the peroxy-titanium(IV) complex on a cation-exchanger [15,16].

From a weakly acid solution AlF_6^{3-}, SO_4^{2-}, PO_4^{3-}, and NO_3^- can be retained on anion-exchangers, while boric acid, being only very slightly dissociated, is eluted. Borate ions can be adsorbed on anion-exchangers only in neutral or alkaline media [17,18]. The boric acid-mannitol complex can also be retained on an anion-exchanger [19,20].

Mixed resin beds of cation- and anion-exchangers have been used to preconcentrate boron [21].

10.1.4 Other Methods

The major components of certain samples (e.g. silicon and germanium compounds) containing traces of boron may be separated by volatilization. However, boron is relatively volatile in acid media, escaping almost completely from a hydrochloric acid solution, and partially from solutions of other acids, during evaporation. This loss of boron can be prevented by masking boron as the non-volatile boric acid-mannitol complex [22].

In the determination of traces of boron in chlorosilanes, some chlorotriphenylmethane [23–25] or N,N-dimethylaniline [26] is added to form non-volatile complexes with boron, thereby preventing its loss during the volatilization of the sample matrix.

10.2 Methods of Determination

The most important methods for the determination of boron have been compared in several reviews [27–30]. The method using curcumin, which has exceptionally high sensitivity but presents some difficulties in operation, is the foremost spectrophotometric method. Methods using carmine, quinalizarin, and dianthrimide in concentrated sulphuric acid media are easier to use but less sensitive. Colorimetric methods in which the ion-pairs formed between $[BF_4]^-$ and organic dyes are extracted are also important.

10.2.1 CURCUMIN METHOD

The curcumin method is both the most sensitive spectrophotometric method for the determination of boron, and the most sensitive of all the known direct spectrophotometric methods for the determination of any element.

Curcumin is a natural compound extracted from the curcuma root, and purified by crystallization. It has the formula:

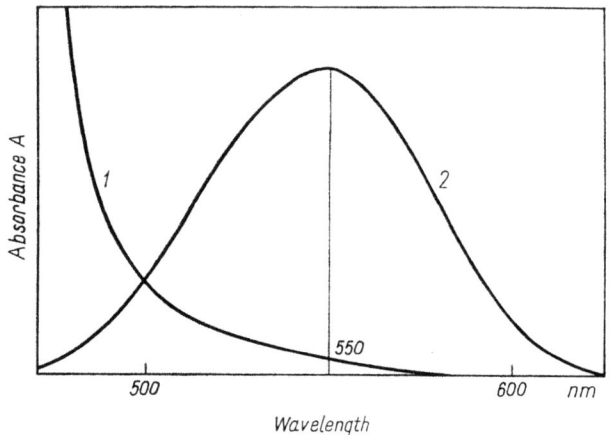

The reagent is insoluble in water, but dissolves, giving a yellow colour, in methanol, ethanol, acetone, and glacial acetic acid. In acid media and in the absence of water, curcumin and boron form a violet-red 2:1 complex called rosocyanin [31,32]. Figure 10.1 shows absorption spectra of ethanol solutions of curcumin and of its complex with boron (rosocyanin).

Fig. 10.1 Absorption spectra of curcumin (1) (against water as reference), and its boron complex (2) (against curcumin solution as reference)

The sensitivity of the method and the reproducibility of the results obtained depend on the quality of the curcumin reagent, and on rigorous observance of the reaction conditions (temperature, time, reagent quantities, solvents—see Procedure below) [33–39]. Commercial curcumin samples differ considerably in quality. Under the most favourable conditions, the molar absorptivity of rosocyanin is 1.8×10^5 at $\lambda_{max} = 550$ nm ($a = 16.6$). In practice, lower molar absorptivities are obtained, e.g. $1.6–1.7 \times 10^5$.

A modification of the curcumin method for determining boron, in which the ternary complex between curcumin, boron, and oxalic acid is formed, is more rapid, but only about half as sensitive [40–41a]. This ternary complex (rubrocurcumin) contains curcumin, boron, and oxalate in the ratio 1:1:1 [31].

Thierig and Umland [42] have extracted rosocyanin into a solution of cyclohexanone and phenol from H_2SO_4–CH_3COOH medium, after the addition of water. Vrchlabský [42a] has extracted the rosocyanin into dichloromethane after the addition of an acetate buffer to the reaction medium.

Numerous elements (e.g. Fe, Mo, Ti, W, Ge, Be, and Ta) form coloured complexes with curcumin, and interfere in the determination of boron with this reagent. Oxidants (e.g. HNO_3), and substances forming stable complexes with boron (e.g. HF), also interfere. In general, therefore, boron is first separated by distillation as trimethyl borate, and trapped in dilute alkali. The trimethyl borate is hydrolysed by the water, and glycerol or mannitol is added to prevent loss of boron during evaporation of the distillate.

In certain cases the distillation separation of boron before its determination is not necessary. When determining traces of boron, for example, in high-purity semiconductor-grade silicon, the latter is expelled as SiF_4 while the elements remaining in the residue containing the boric acid–mannitol complex are present in such small quantities that they do not interfere [38].

The exceptionally high sensitivity of the curcumin method has enabled traces of boron to be determined in silicon [2,43–45], chlorosilanes [26, 38,46], germanium [2], uranium [47–49], zirconium and its alloys [50–52], hafnium and titanium [52], nickel [53–55], copper alloys [55], steels [5,49, 56,57], metallic sodium [57a], beryllium and magnesium [49], silicates [58], phosphates [59], soils [60], animal tissue [60a], plant material [60], chemical reagents [61–63], sea water [64,65] and fibre building boards [65a].

The curcumin method has been adapted for the continuous, automatic determination of boron in sea water [65] and aqueous solutions [66].

Reagents and Apparatus

1. Curcumin; 0·1% solution in glacial acetic acid. The solution is prepared on the day of use.
2. Standard boron solution: 1 mg/ml. Dissolve 0·5716 g of H_3BO_3 in water, and dilute the solution with water to 100 ml.
3. Conc. H_2SO_4 and glacial CH_3COOH reagent (1+1). Mix equal volumes of the two acids immediately before use.
4. Alkaline solution. Dissolve 1 g of NaOH in water, add 3 g of glycerol, and dilute with water to 100 ml. Store the solution in a polyethylene bottle.
5. Methanol. Purify by distillation from solid NaOH in a quartz still. Store in a polyethylene bottle.
6. Quartz distillation apparatus (50–75 ml capacity). The distance between

the bulb of the distillation flask and the side-arm to the condenser should be at least 10 cm.

Procedure

Separation of boron by distillation. Evaporate an alkaline solution containing microgram quantities of B to dryness. If it is necessary to ignite and fuse the residue (e.g. if mannitol is present), a platinum vessel should be used. Add to the residue 1–2 ml of conc. H_2SO_4, stir with a plastic rod, and wash the contents of the vessel with 25 ml of methanol into a still fitted with a simple condenser. Immerse the end of the condenser in the trapping solution (2 ml of alkaline solution +18 ml of water) contained in a platinum vessel. Distil the contents of the still by heating in an oil-bath, the temperature of which should be 120°C at the end of distillation. After all the methanol has been distilled off, cool the still, add 10 ml of fresh methanol, and repeat the distillation.

Determination of boron. Take an aliquot, containing not more than 1·5 μg of B, from the distillate, and evaporate it to dryness in a platinum crucible. Ignite the residue to burn off the organics. Place the vessel in a water-bath at 60°C, add from a pipette exactly 5 ml of the curcumin solution, and keep the vessel on the bath for about 3 minutes with occasional stirring. To the cooled vessel, add 2 ml of the H_2SO_4–CH_3COOH mixture, and mix thoroughly by swirling the vessel. After 20 minutes wash the contents with ethanol (70%) into a 50-ml volumetric flask and dilute to the mark with ethanol. Mix the coloured solution, and measure its absorbance at 550 nm, using the blank solution as a reference.

Notes. (1) The final dilution with ethanol causes the deprotonation of the excess of curcumin, which would otherwise cause spectral interference. An ethanolic acetate buffer has also been used as diluent [36].

(2) It is imperative that the blank solution is prepared under exactly the same conditions as the sample solutions.

(3) If boron is determined without distillation as trimethyl borate, any traces of HF or HNO_3 must be carefully removed before the addition of curcumin (e.g. by evaporating the solution 2 or 3 times with dilute HCl in the presence of mannitol).

10.2.2 CARMINE METHOD

Carmine belongs to the group of boron reagents derived from α-hydroxyanthraquinone. These reagents are characterized by the ability to give coloured complexes with boron in concentrated sulphuric acid medium [67–70]. Boron occurs as the B^{3+} cation in concentrated H_2SO_4, and as BO^+ in less concentrated H_2SO_4.

Carmine (Carmine Red), a glycoside derivative of α-hydroxyanthraquinone, is a natural product obtained from cochineal. In concentrated H_2SO_4, carmine reacts with boron to form a complex according to the following equation:

164   **Boron**   [Ch. 10

$$\text{carmine structure} \xrightarrow[\text{H}_2\text{SO}_4 \text{ conc.}]{\text{B}^{3+}} \text{boron complex}$$

(Structure 1: anthraquinone with CH$_3$O, OH, CO(CHOH)$_4$CH$_3$, HO, HOOC, O, OH substituents)

(Structure 2: boron complex with B^{2+} bridging two O atoms on the anthraquinone with CH$_3$O, O, CO(CHOH)$_4$CH$_3$, HO, HOOC, O, OH)

The reagent is red (λ_{max} = 520 nm), whereas the boron complex is violet-blue. The molar absorptivity of the complex at 615 nm (against the reagent solution as the reference) is $5 \cdot 5 \times 10^3$ ($a = 0 \cdot 51$).

Boron reacts slowly with carmine. The reaction may be accelerated by diluting the H$_2$SO$_4$ (e.g. to ~ 92%), but the acid concentration should not be lower than 90% [71–73]. The absorbance reaches a maximum within 45–60 minutes, after which it remains constant for a few hours. Oxidizing agents and fluoride interfere in the carmine method.

Before its determination, boron is normally separated either as volatile trimethyl borate [4,74] or by ion-exchange [15,18].

The carmine method is more suitable than the curcumin method for determining larger amounts of boron. Boron has been determined by the carmine method in steel [75], molybdenum alloys [73], zirconium and its alloys [74], titanium and its alloys [15,76], cobalt and nickel alloys [76], uranium alloys with aluminium [77], uranyl nitrate [78,79], silicon [80], glass [4], rocks and minerals [81,82], fertilizers [18,83,84], fluorides [12,85], sewage effluents and water [81,86], coal by-products [87], and biological materials [71,88].

Reagents

1. Carmine solution: Dissolve 25 mg of carmine in 100 ml of conc. sulphuric acid.
2. Standard boron solution: 1 mg/ml (p. 162).

Procedure

Add 5 ml of solution containing not more than 20 μg of B to a 50-ml volumetric flask. Add 5 drops of conc. HCl, 25 ml of conc. H$_2$SO$_4$, cool the solution, and add 20 ml of the carmine solution. Mix the solution in the flask thoroughly, and set aside for 1 hour. Measure the absorbance of the coloured solution at 615 nm, using a blank solution as reference.

Note. If boron is separated by distillation (see p. 163), the distillate should be evaporated to dryness, and the residue dissolved in 5 ml of water.

10.2.3 OTHER METHODS

Like carmine, quinalizarin (1,2,5,8-tetrahydroxyanthraquinone) is a hydroxyanthraquinone reagent which reacts with boron in concentrated sulphuric acid [89,90] or concentrated sulphuric–acetic acid medium [91].

[Structure of quinalizarin]

The sensitivity of the quinalizarin method is lower ($\varepsilon = 3.5 \times 10^3$ at 600 nm), and the difference between the absorption maxima of the reagent and the complex is smaller than in the carmine method. With acetylquinalizarin [92], there is higher contrast between the colours of the reagent and complex.

The quinalizarin method has been used to determine boron in steel [93], high-temperature alloys [94], zirconium and its alloys with niobium [95], titanium alloys [16], aluminium alloys [96], nickel alloys [97], chlorosilanes [10], and plants [98].

Dianthrimide (1,1'-dianthraquinonylamine) also reacts with boron in a concentrated H_2SO_4 medium, but only after heating at 70–90°C for 2–4 hours. A complex with the following structure is obtained:

[Structure of dianthrimide–boron complex]

The colour of dianthrimide in concentrated H_2SO_4 is olive-green, whereas that of the boron complex is dark blue. The absorption spectrum is shown in Fig. 10.2 ($\varepsilon = 1.9 \times 10^4$ at $\lambda_{max} = 630$ nm). The reaction is very de-

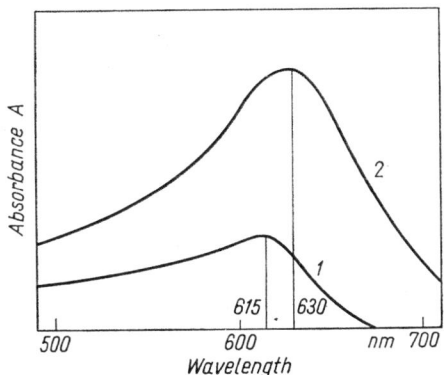

Fig. 10.2 Absorption spectra of boron complexes with carmine (1) and with 1,1'-dianthrimide (2)

pendent upon the time and temperature of heating, and the concentration of sulphuric acid. The dianthrimide method is more sensitive than the carmine and quinalizarin methods, but considerably less sensitive than the curcumin method [99–102a].

Boron has been determined by the dianthrimide method in cast iron and steel [100], nickel alloys [103], aluminium and its alloys [104], titanium alloys [105], minerals [1,106], fertilizers [1], plant materials [1,99], foods [107], and biological materials [108].

Other anthraquinone derivatives have been suggested as spectrophotometric reagents for boron, e.g. alizarin and Alizarin Red S [30], chrysazin [109], tribromoanthrarufin [110], diaminochrysazin [111], tetrabromochrysazin [112], 1-hydroxy-4-*p*-toluidinoanthraquinone [113], and 1,1-bis(6-chloroanthraquinonyl)amine [114].

Ducret [115] has extracted the Methylene Blue–tetrafluoroborate ion-association complex (molar absorptivity 6.5×10^4 at 660 nm) into dichloroethane. The method has been investigated by various authors [116,117] and has been applied especially in the analysis of steel [116,118–121], and also in the analysis of uranium fluoride [122], and soils, sediments, and rocks [123]. Pasztor and Bode [124] have found the related dye, Azure C (monomethylthionine), to be superior to Methylene Blue.

Boron has been determined as the tetrafluoroborate ion-pair with many other basic dyes, e.g. Brilliant Green [125,126], Crystal Violet [127], Methyl Violet [128], Nile Blue A [129–132], ferroin [133], Capri Blue [129], Rhodamine B [134], and various antipyrine dyes [135]. The borosalicylate–ferroin ion-association complex has been similarly exploited [136]. Boron has also been determined as the borosalicylate–Crystal Violet complex [19,137], and the borosalicylate–Rhodamine 6G complex [138].

Azo reagents which have been proposed for the spectrophotometric determination of boron are Arsenazo I [30,139], SPADNS [30,140], Chromotrope 2B [76], Stilbazo [141], Beryllon II [30,141], Victoria Violet [142], and H-resorcinol [143]. Other organic reagents used include Azomethine H [14,144–147], haematoxylin [148], morin [149], quercetin [150], Pyrocatechol Violet [151], Phthalein Violet [152], and nitropyrocatechol [153].

Indirect methods for the determination of boron have been based on the liberation of the intensely coloured chloranilate ion from its insoluble barium salt [154,155], on the formation of molybdenum blue [156], and on the yellow titanium–peroxide complex [157].

References

1. Roth, H. and Beck, W., *Z. Anal. Chem.* **141**, 404, 414 (1954).
2. Luke, C. L., *Anal. Chem.* **27**, 1150 (1955).
3. Spicer, G. S. and Strickland, J. D., *Anal. Chim. Acta* **18**, 523 (1958).
4. Ehrlich, P. and Keil, T., *Z. Anal. Chem.* **165**, 188 (1959).
5. Borrowdale, J., Jenkins, R. H. and Shanahan, C. E., *Analyst* **84**, 426 (1959).
6. Gaestel, C. and Huré, J., *Bull. Soc. Chim. France* **1949**, 830.

7. Umland, F. and Janssen, A., *Z. Anal. Chem.* **219**, 121 (1966).
8. Landry, J. C., Landry, M. F. and Monnier, D., *Anal. Chim. Acta* **62**, 177 (1972).
9. Coursier, J., Huré, J. and Platzer, R., *ibid.* **13**, 379 (1955).
10. Ducret, L. and Seguin, P., *ibid.* **17**, 207 (1957).
11. Haas, C. S., Pellin, R. A. and Everingham, M. R., *Anal. Chem.* **36**, 245 (1964).
12. Ross, W. J., Meyer, A. S., Jr. and White, J. C., *ibid.* **29**, 810 (1957).
13. Rynasiewicz, J., Sleeper, M. P. and Ryan, J. W., *ibid.* **26**, 935 (1954).
14. Capelle, R., *Anal. Chim. Acta* **24**, 555; **25**, 59 (1961).
15. Calkins, R. C. and Stenger, V. A., *Anal. Chem.* **28**, 399 (1956).
16. Newstead, E. G. and Gulbierz, J. E., *ibid.* **29**, 1673 (1957).
17. Ryabchikov, D. I. and Kuril'chikova, G. E., *Zh. Analit. Khim.* **19**, 1495 (1964).
18. Lang, K., *Z. Anal. Chem.* **163**, 241 (1958).
19. Vinkovetskaya, S. Ya. and Nazarenko, V. A., *Zavodsk. Lab.* **32**, 1202 (1966).
20. Barbier, Y. and Rosset, R., *Bull. Soc. Chim. France* **1968**, 5072.
21. Callicoat, D. L., Wolszon, J. D. and Hayes, J. R., *Anal. Chem.* **31**, 1437 (1959).
22. Feldman, C., *Anal. Chem.* **33**, 1916 (1961).
23. Pchelintseva, A. F., Rakov, N. A. and Slyusareva, L. P., *Zavodsk. Lab.* **28**, 677 (1962).
24. Vecsernyés, L. and Hangos, I., *Z. Anal. Chem.* **208**, 407 (1965).
25. Kawasaki, K. and Higo, M., *Anal. Chim. Acta* **33**, 497 (1965).
26. Miyamoto, M., *Japan Analyst* **12**, 233 (1963).
27. Capelle, R., *Chim. Anal. (Paris)* **45**, 303 (1963).
28. Goward, G. W. and Wiederkehr, V. R., *Anal. Chem.* **35**, 1542 (1963).
29. Vlačil, F. and Drbal, K., *Chem. Listy* **62**, 1371 (1968).
30. Kasiura, K., *Chem. Anal. (Warsaw)* **16**, 219 (1971).
31. Spicer, G. S., Strickland, J. D. H. and Bellamy, L. J., *J. Chem. Soc.* **1952**, 4644, 4650, 4653.
32. Dyrssen, D. W., Novikov, Y. P. and Uppström, L. R., *Anal. Chim. Acta* **60**, 139 (1972).
33. Spicer, G. S. and Strickland, J. D., *ibid.* **18**, 231 (1958).
34. Lima, F. W., Pagano, C. and Schneiderman, B., *Analyst* **85**, 909 (1960).
35. Silverman, L. and Trego, K., *Anal. Chem.* **25**, 1264 (1953).
36. Uppström, L. R., *Anal. Chim. Acta* **43**, 475 (1968).
37. Grinstead, R. R. and Snider, S., *Analyst* **92**, 532 (1967).
38. Marczenko, Z., Mojski, M. and Kasiura, K., *Chem. Anal. (Warsaw)* **14**, 1331 (1969).
39. Umland, F. and Janssen, A., *Z. Anal. Chem.* **249**, 186 (1970).
40. Miyamoto, M., *Bull. Chem. Soc. Japan* **36**, 1208 (1963).
41. Williams, D. E. and Vlamis, J., *Anal. Chem.* **33**, 1098 (1961).
41a. Monnier, D., Kapetanidis, I. and Wenger, P. E., *Helv. Chim. Acta* **44**, 1856 (1961).
42. Thierig, D. and Umland, F., *Z. Anal. Chem.* **211**, 161 (1965).
42a. Vrchlabský, M., *Collection Czech. Chem. Commun.* **37**, 3590 (1972).
43. Luke, C. L. and Flaschen, S. S., *Anal. Chem.* **30**, 1406 (1958).
44. Pohl, F. A., Kokes, K. and Bonsels, W., *Z. Anal. Chem.* **174**, 6 (1960).
45. Berthel, K. H., Döge, H. G., Ehrlich, G., Köthe, A. and Schmidt, A., *Mikrochim. Acta* **1963**, 702.
46. Schneer, A., Halmos, T. and Székely, T., *Z. Anal. Chem.* **182**, 178 (1961).
47. Silverman, L. and Trego, K., *Anal. Chim. Acta* **15**, 439 (1956).
48. Onishi, H., Ishiwatari, N. and Nagai, H., *Bull. Chem. Soc. Japan* **33**, 830 (1960).
49. Hayes, M. R. and Metcalfe, J., *Analyst* **87**, 956 (1962).
50. Freegarde, M. and Cartwright, J., *ibid.* **87**, 214 (1962).
51. Hayes, M. R. and Metcalfe, J., *ibid.* **88**, 471 (1963).
52. Elwell, W. T. and Wood, D. F., *ibid.* **88**, 475 (1963).
53. Luke, C. L., *Anal. Chem.* **30**, 1405 (1958).
54. Andrew, T. R. and Nichols, P. N., *Analyst* **91**, 664 (1966).

55. Pakalns, P., *ibid.* **94**, 1130 (1969).
56. Harrison, T. S. and Cobb, W. D., *ibid.* **91**, 576 (1966).
57. Tolk, A., Tap, W. A. and Lingerak, W. A., *Talanta* **16**, 111 (1969).
57a. Pohl, F. A., *Z. Anal. Chem.* **157**, 6 (1957).
58. Lechner, A., Toman, E. and Ferenczy, Z., *Acta Chim. Acad. Sci. Hung.* **28**, 231 (1961).
59. Kocher, J., *Bull. Soc. Chim. France* **1962**, 1247.
60. Dible, W. T., Truog, E. and Berger, K. C., *Anal. Chem.* **26**, 418 (1954).
60a. Mair, J. W., Jr. and Day, H. G., *ibid.* **44**, 2015 (1972).
61. Kemula, W., Brzozowski, S. and Janowski, A., *Chem. Anal. (Warsaw)* **3**, 905 (1958).
62. Gallus-Olender, J., *ibid.* **10**, 1039 (1965).
63. Ciecierska-Tworek, Z., Walędziak, H., Gorczyńska, K. and Ciecierska-Stokłosa D., *ibid.* **13**, 773 (1968).
64. Greenhalgh, R. and Riley, J. P., *Analyst* **87**, 970 (1962).
65. Hulthe, P., Uppström, L. and Östling, G., *Anal. Chim. Acta* **51**, 31 (1970).
65a. Bethge, P. O. and Hakanen, O., *Svensk Papperstidn.* **74**, 517 (1971).
66. Crawley, R. H., *Analyst* **89**, 749 (1964).
67. Sommer, L. and Hniličková, M., *Collection Czech. Chem. Commun.* **22**, 1432 (1957).
68. Ráb, F., *ibid.* **24**, 3654 (1959).
69. Nazarchuk, T. N., *Ukr. Khim. Zh.* **28**, 233 (1962).
70. Brown, R. S., *Anal. Chim. Acta* **50**, 157 (1970).
71. Hatcher, J. T. and Wilcox, L. V., *Anal. Chem.* **22**, 567 (1950).
72. Callicoat, D. L. and Wolszon, J. D., *ibid.* **31**, 1434 (1959).
73. Higgs, D. G., *Analyst* **85**, 897 (1960).
74. Pollock, E. N. and Zopatti, L. P., *Talanta* **10**, 118 (1963).
75. Martynchenko, I. U. and Bondarenko, A. M., *Zh. Analit. Khim.* **12**, 495 (1957).
76. Golubtsova, R. B., *ibid.* **15**, 481 (1960).
77. Puphal, K. W., Merrill, J. A., Booman, G. L. and Rein, J. E., *Anal. Chem.* **30**, 1612 (1958).
78. Nowicka-Jankowska, T. and Szyszko, H., *Chem. Anal. (Warsaw)* **1**, 285 (1956).
79. Ross, W. J. and White, J. C., *Talanta* **3**, 311 (1960).
80. Marczenko, Z. and Kasiura, K., *Chem. Anal. (Warsaw)* **8**, 185 (1963).
81. Malyuga, D. P., *Zavodsk. Lab.* **35**, 279 (1969).
82. Fleet, M. E., *Anal. Chem.* **39**, 253 (1967).
83. Borland, H., Brownlie, I. A. and Godden, P. T., *Analyst* **92**, 47 (1967).
84. Peterson, H. P. and Zoromski, D. W., *Anal. Chem.* **44**, 1291 (1972).
85. Nowicka-Jankowska, T. and Szyszko, H., *Chem. Anal. (Warsaw)* **3**, 969 (1958).
86. Lionnel, L. J., *Analyst* **95**, 194 (1970).
87. Wnękowska, L., *Chem. Anal. (Warsaw)* **1**, 301 (1956).
88. Smith, W. C., Jr., Goudie, A. J. and Sivertson, J. N., *Anal. Chem.* **27**, 295 (1955).
89. Johnson, E. A. and Toogood, M. J., *Analyst* **79**, 493 (1954).
90. Langmyhr, F. J. and Holme, A., *Anal. Chim. Acta* **35**, 220 (1966).
91. Gupta, H. K. and Boltz, D. F., *Mikrochim. Acta* **1971**, 577.
92. Nemodruk, A. A., Palei, P. N. and Hun-i Ho, *Zavodsk. Lab.* **28**, 406 (1962).
93. Kysil, B. and Vobora, J., *Collection Czech. Chem. Commun.* **24**, 3893 (1959).
94. Jones, A. H., *Anal. Chem.* **29**, 1101 (1957).
95. Palei, P. N., Nemodruk, A. A. and Pyzhova, Z. I., *Tr. Komis. po Analit. Khim. Akad. Nauk SSSR* **11**, 223 (1960).
96. Scharnbeck, C., *Chem. Techn.* **9**, 416 (1957).
97. Wójtowicz, M. and Kubica, M., *Chem. Anal. (Warsaw)* **13**, 65 (1968).
98. Bardzicka, B. and Krauze, A., *ibid.* **5**, 791 (1960).
99. Ellis, G. H., Zook, E. G. and Baudisch, O., *Anal. Chem.* **21**, 1345 (1949).
100. Danielsson, L., *Talanta* **3**, 138, 203 (1959).

References

101. Langmyhr, F. J. and Skaar, O. B., *Acta Chem. Scand.* **13**, 2107 (1959); *Anal. Chim. Acta* **25**, 262 (1961).
102. Skaar, O. B. and Langmyhr, F. J., *Acta Chem. Scand.* **14**, 550 (1960).
102a. Gupta, H. K. and Boltz, D. F., *Anal. Lett.* **4**, 161 (1971).
103. Burke, K. E. and Albright, C. H., *Talanta* **13**, 49 (1966).
104. Kerin, D., *Mikrochim. Acta* **1964**, 670.
105. Codell, M. and Norwitz, G., *Anal. Chem.* **25**, 1446 (1953).
106. Werner, H., *Z. Anal. Chem.* **168**, 266 (1959).
107. Raber, H. and Likussar, W., *Mikrochim. Acta* **1970**, 577.
108. Kaczmarczyk, A., Messer, J. R. and Peirce, C. E., *Anal. Chem.* **43**, 271 (1971).
109. Ruggieri, R., *Anal. Chim. Acta* **25**, 145 (1961).
110. Cogbill, E. C. and Yoe, J. H., *ibid.* **12**, 455 (1955); *Anal. Chem.* **29**, 1251 (1957).
111. Eberle, A. R. and Lerner, M. W., *ibid.* **32**, 146 (1960).
112. Karpen, W. L., *ibid.* **33**, 738 (1961).
113. Bell, D. and McArthur, K., *Analyst* **93**, 298 (1968).
114. Grob, R. L., Cogan, J., Mathias, J. J., Mazza, S. M. and Piechowski, A. P., *Anal. Chim. Acta* **39**, 115 (1967).
115. Ducret, L., *ibid.* **17**, 213 (1957).
116. Pasztor, L., Bode, J. D. and Fernando, Q., *Anal. Chem.* **32**, 277 (1960).
117. Skaar, O. B., *Anal. Chim. Acta* **28**, 200 (1963).
118. Rosotte, R., *Chim. Anal. (Paris)* **44**, 208 (1962).
119. Bhargava, O. P. and Hines, W. G., *Talanta* **17**, 61 (1970).
120. Vernon, F. and Williams, J. M., *Anal. Chim. Acta* **51**, 533 (1970).
121. Blazejak-Ditges, D., *Z. Anal. Chem.* **247**, 20 (1969).
122. Karalova, Z. K. and Nemodruk, A. A., *Zh. Analit. Khim.* **18**, 615 (1963).
123. Stanton, R. E. and McDonald, A. J., *Analyst* **91**, 775 (1966).
124. Pásztor, L. and Bode, J. D., *Anal. Chem.* **32**, 1530 (1960); *Anal. Chim. Acta* **24**, 467 (1961).
125. Babko, A. K. and Marchenko, P. V., *Zavodsk. Lab.* **26**, 1202 (1960); **27**, 801 (1961).
126. Karalova, Z. K. and Nemodruk, A. A., *Zh. Analit. Khim.* **17**, 985 (1962).
127. Blyum, I. A., Dushina, T. K., Semenova, T. V. and Shcherba, I. Ya., *Zavodsk. Lab.* **27**, 644 (1961).
128. Poluéktov, N. S., Kononenko, L. I. and Lauer, R. S., *Zh. Analit. Khim.* **13**, 396 (1958).
129. Skaar, O. B., *Anal. Chim. Acta* **32**, 508 (1965).
130. Gagliardi, E. and Wolf, E., *Mikrochim. Acta* **1968**, 140.
131. Nicholson, R. A., *Anal. Chim. Acta* **56**, 147 (1971).
132. Gagliardi, E. and Höllinger, W., *Mikrochim. Acta* **1972**, 136.
133. Archer, V. S., Doolittle, F. G. and Young, L. M., *Talanta* **15**, 864 (1968).
134. Onishi, H. and Nagai, H., *Japan Analyst* **17**, 345 (1968).
135. Busev, A. I., Yakovlev, P. Ya. and Kozina, G. V., *Zh. Analit. Khim.* **22**, 1227 (1967).
136. Bassett, J. and Matthews, P. J., *Analyst* **99**, 1 (1974).
137. Vasilevskaya, A. E. and Lenskaya, L. K., *Zh. Analit. Khim.* **20**, 747 (1965).
138. Vasilevskaya, A. E., *Nauchn. Trudy Vses. Inst. Min. Resursov*, **1971**, 22.
139. Hiiro, K., *Japan Analyst* **11**, 223 (1962).
140. Truhaut, R., Boudene, C. and Nguyen Phu Lich, *Bull. Soc. Chim. France* **1966**, 2551.
141. Hiiro, K., *Japan Analyst* **10**, 1281 (1961); **11**, 337 (1962).
142. Reynolds, C. A., *Anal. Chem.* **31**, 1102 (1959).
143. Grizo, V. A. and Poluektova, E. N., *Zh. Analit. Khim.* **13**, 434 (1958).
144. Shanina, T. M., Gel'man, N. E. and Mikhailovskaya, V. S., *ibid.* **22**, 782 (1967).
145. Basson, W. D., Böhmer, R. G. and Stanton, D. A., *Analyst* **94**, 1135 (1969).
146. Hofer, A., Brosche, E. and Heidinger, R., *Z. Anal. Chem.* **253**, 117 (1971).

147. Basson, W. D., Pille, P. P. and Du Preez, A. L., *Analyst* **99**, 168 (1974).
148. Monnier, D., Marcantonatos, N., Feraud, R. and Haerdi, W., *Helv. Chim. Acta* **46**, 1047 (1963).
149. Murata, A. and Yamauchi, F., *J. Chem. Soc. Japan, Pure Chem. Sect.* **79**, 1454 (1954).
150. Hiiro, K., *Bull. Chem. Soc. Japan* **34**, 1748 (1961).
151. Hiiro, K., *ibid.* **34**, 1743 (1961).
152. Patrovsky, V., *Talanta* **10**, 175 (1963).
153. Hakoila, E. J., Kankare, J. J. and Skarp, T., *Anal. Chem.* **44**, 1857 (1972).
154. Srivastava, R. D., van Buren, P. R. and Gesser, H., *ibid.* **34**, 209 (1962).
155. Peterson, D. R. and Hayes, J. R., *ibid.* **37**, 305 (1965).
156. Campbell, R. H. and Mellon, M. G., *ibid.* **32**, 50 (1960).
157. Fukamauchi, H., *Z. Anal. Chem.* **229**, 413 (1967).

Chapter 11

BROMINE

Bromine (Br, at.wt. 79·90) is a dark amber-red liquid. Saturated bromine water contains 3·6% (w/v) of bromine (at 20°C). Bromine forms bromide (Br^-) and hypobromite (BrO^-) in alkaline solution. The most stable forms of bromine are bromide and bromate (BrO_3^-). Bromide has reducing properties, whereas bromine (Br_2), hypobromite, and bromate are oxidants. Many bromides are sparingly soluble compounds, and soluble bromide complexes are formed with the same metals as form soluble chloride complexes.

11.1 Separation of Bromine and Bromide

Bromine is volatile and can be distilled from acidified solutions.

Bromide is most often separated by distillation after oxidation to bromine. Distillation is carried out in a stream of gas such as air, nitrogen, or carbon dioxide. It is possible to separate iodide, bromide, and chloride from each other by selective oxidation [1,2]. First, the iodine produced by oxidation of iodide with hydrogen peroxide in phosphoric acid medium (pH 1) is distilled. Then dilute nitric acid (2·5–4M) is used to oxidize bromide to bromine. Iodide in the presence of bromide can also be oxidized with nitrite in acetic acid medium. The iodine (and subsequently the bromine) liberated can be separated by extraction into chloroform, carbon tetrachloride, or other solvents [3].

Bromide may also be separated by conversion into volatile cyanogen bromide, CNBr, in dilute H_2SO_4 containing cyanide and chromic acid [4,5].

De Guiso et al. [6] separated chloride, bromide, and iodide on a strongly basic anion-exchanger, using sodium nitrate solutions as eluents.

Other methods of separating halide ions are mentioned in Chapter 16 on chlorine (p. 204).

Precipitation of traces of bromide as silver bromide, using chloride as collector, is sometimes a suitable method of preconcentration and separation [7,8].

11.2 Determination of Bromine and Bromide

The determination of bromide is, in general, based either on indirect methods, or on methods involving preliminary oxidation to bromine. The bromine formed undergoes subsequent bromination or oxidation reactions to give coloured products. The bromination of Phenol Red is described below.

11.2.1 Phenol Red Method

The reaction of bromine with Phenol Red (i.e. the triphenylmethane dye, phenolsulphonephthalein) to form tetrabromophenolsulphonephthalein (Bromophenol Blue) is the basis of a sensitive spectrophotometric method [9,10].

The change in colour of a solution at pH 5·5 from yellow to violet is very pronounced. The molar absorptivity of the violet product at $\lambda_{max} = 580$ nm is $1 \cdot 14 \times 10^4$ ($a = 0 \cdot 14$).

In Larsen and Ingber's method [9], the oxidation of bromide to bromine, and the bromination of Phenol Red are carried out in a weakly alkaline medium. Calcium hypochlorite can be used as oxidizing agent, and the excess of hypochlorite reduced with arsenite. The periods of time specified for oxidation (2 min) and for bromination (4 min) must be adhered to strictly (see procedure below). With shorter times, oxidation and bromination are incomplete: with longer oxidation, bromine is oxidized to bromate.

Traces of bromide have been determined by this method in uranium fluorides and oxides [9], wines [10], food [11], and water samples [12].

Before the distillation, bromides are oxidized to bromine with a mixture of chromic ($0 \cdot 8M$) and sulphuric ($7M$) acids [9]. This mixture does not oxidize chloride. During the distillation, nitrogen is used to carry the bromine into a trapping solution of $0 \cdot 15N$ sodium sulphite, which is a better scrubber than NaOH solution. Since traces of chromium(VI) are also carried into the receiver, an additional extraction of bromine is performed. Bromides in the receiver are oxidized with a cold CrO_3–H_2SO_4 mixture, and the bromine formed is extracted into carbon tetrachloride. The bromine is stripped from the organic layer with $2M$ ammonia, the ammonia is driven off, and the bromide determined as detailed below.

Reagents

1. Phenol Red: 0·01% solution. Dissolve 10·0 mg of the reagent in 1 ml of $0 \cdot 1M$ NaOH and dilute the solution with water to 100 ml.
2. Standard bromide solution: 1 mg/ml. Dissolve in water 1·490 g of KBr dried at 110°C, and dilute the solution with water in a volumetric flask to 1 litre.

3. Borate buffer: pH 8·7–8·8. Saturated borax ($Na_2B_4O_7 \cdot 10H_2O$) solution.
4. Calcium hypochlorite: 0·4% solution, filtered.
5. Sodium arsenite: $\sim 0 \cdot 1N$ solution (13 g of salt per litre of solution).
6. Acetate buffer: pH 4·6–4·7. Dissolve 68 g of sodium acetate trihydrate in water, add 30 ml of glacial acetic acid, and dilute the solution with water to 1 litre.

Procedure

Place an approximately neutral sample solution ($\sim$ 25 ml), containing not more than 150 µg of Br^-, in a 50-ml volumetric flask, add 10 ml of the borate buffer, 1 ml of the hypochlorite solution, and shake for 2 min. Add 2·0 ml of the Phenol Red solution, stir well, and let the solution stand for 4 min. Add 2·5 ml of the arsenite solution and 7·5 ml of the acetate buffer, and dilute to the mark with water. Measure the absorbance of the solution at 580 nm against a blank solution.

11.2.2 OTHER METHODS

Several other methods for determining bromide are also based on the coloured products formed by bromination of various organic compounds. The reaction with rosaniline gives tetrabromorosaniline [13–16], and fluorescein yields tetrabromofluorescein (eosin) [7,17]. The product of the reaction of bromine with fuchsine can be extracted into chloroform [18]. Bromine has been determined by such methods in selenium [7,18], body fluids [14], water [15], and alkali metal chlorides [16].

As with chlorine, it is possible to determine bromine from its oxidizing effect on *o*-tolidine [19,20]. Bromine obtained by oxidation of bromide has also been used to bleach Chromotrope 2B [21] and Methyl Orange [22,23].

A less sensitive method is based on measuring the absorbance at 417 nm of a yellow bromine solution in carbon tetrachloride, after oxidation of bromide with calcium hypochlorite [3].

Bromide has also been determined by the benzidine–pyridine method more often used for determining chloride [4].

When a yellow solution of silver dithizonate in chloroform is shaken with a sample solution containing bromide, an amount of dithizone equivalent to the bromide present is liberated in the $CHCl_3$ phase, and its absorbance is measured at 598 nm [24]. In another method employing dithizone, AgBr is precipitated, separated, washed, and reduced by hydrazine. The silver is then dissolved in nitric acid, and determined with dithizone [8].

Other indirect methods involve the reaction with the mercury(II) complexes of diphenylcarbazone [25,26] or Methylthymol Blue [27], or the displacement of thiocyanate ion from AgSCN by bromide, followed by the colorimetric determination of thiocyanate with ferric ions [28].

The association complex of Br^- and the oxazine dye Nile Blue has been extracted into chloroform, and measured spectrophotometrically at

626 nm [29]. Bromide ions can be selectively extracted as bromosulphinate into 1,2-dichloroethane with triphenylbenzylphosphonium ion and sulphite, and the absorbance measured at 359 nm [30].

A suspension of silver bromide can be used for the turbidimetric determination of bromide [1,19].

11.3 Determination of Bromate and Perbromate

The spectrophotometric determination of bromate is based on reactions of BrO_3^- ions with *o*-arsanilic acid [31], *o*-aminobenzoic acid [32], antipyrine [33], and tris(2-diethylaminoethoxy)benzene [34]. Bromate may be reduced to bromine, which reacts with fuchsine, forming a coloured product extractable into chloroform [35].

Brown and Boyd [36] have extracted perbromate ions into chlorobenzene as an ion-association complex with Crystal Violet, and determined the perbromate from the absorbance at 596 nm.

References

1. Murphy, T. J., Clabaugh, W. S. and Gilchrist, R., *J. Res. Nat. Bur. Stds* **53**, 13 (1954).
2. Kahane, E. and Kahane, M., *Bull. Soc. Chim. France* **1954**, 396.
3. Collins, A. G. and Watkins, J. W., *Anal. Chem.* **31**, 1182 (1959).
4. Van Pinxteren, J. A., *Analyst* **77**, 367 (1952).
5. Winefordner, J. D. and Tin Maung, *Anal. Chem.* **35**, 382 (1963).
6. De Guiso, R. C., Rieman III, W. and Lindenbaum, S., *ibid.* **26**, 1840 (1954).
7. Pohl, F. A., *Mikrochim. Acta* **1956**, 414.
8. Koch, H. and Schulze, K., *Z. Anal. Chem.* **210**, 90 (1965).
9. Larsen, R. P. and Ingber, N. M., *Anal. Chem.* **31**, 1084 (1959).
10. Jaulmes, P., Brun, S. and Cabanis, J. C., *Chim. Anal. (Paris)* **44**, 327 (1962).
11. Kretzschmann, F. and Engst, R., *Mikrochim. Acta* **1970**, 270.
12. Archimbaud, M. and Bertrand, M. R., *Chim. Anal. (Paris)* **52**, 531 (1970).
13. Hunter, G. and Goldspink, A. A., *Analyst* **79**, 467 (1954).
14. Hunter, G., *Biochem. J.* **60**, 261 (1955).
15. Moldán, B. and Zýka, J., *Mikrochem. J.* **13**, 357 (1968).
16. Joy, E. F., Bonn, J. D. and Barnard, A. J., Jr., *Anal. Chem.* **45**, 856 (1973).
17. Pohl, F. A., *Z. Anal. Chem.* **149**, 68 (1956).
18. Siwecka, J., *Chem. Anal. (Warsaw)* **3**, 1001 (1958).
19. Creitz, E. C., *Anal. Chem.* **37**, 1690 (1965).
20. Scheubeck, E. and Ernst, O., *Z. Anal. Chem.* **249**, 370 (1970).
21. Elbeih, I. I. and El-Sirafy, A. A., *Chemist-Analyst* **54**, 8 (1965).
22. Tamarchenko, L. M. and Toropova, V. F., *Zh. Analit. Khim.* **23**, 1028 (1968).
23. Laitinen, H. A. and Boyer, K. W., *Anal. Chem.* **44**, 920 (1972).
24. Kirsten, W. J., *Mikrochim. Acta* **1955**, 1086.
25. Tomonari, A., *J. Chem. Soc. Japan, Pure Chem. Sect.* **83**, 459 (1962).
26. Okutani, T., *ibid.* **88**, 737 (1967).
27. Nomura, T. and Komatsu, S., *ibid.* **90**, 168 (1969).
28. Utsumi, S., *ibid.* **74**, 32 (1953).
29. Likussar, W., Pokorny, G. and Zechman, H., *Anal. Chim. Acta* **49**, 97 (1970).
30. Behrends, K. and Klein, H., *Z. Anal. Chem.* **249**, 165 (1970).
31. MacDonald, J. C. and Yoe, J. H., *Anal. Chim. Acta* **28**, 383 (1963).

32. Hashmi, M. H., Ahmad, H., Rashid, A. and Ayaz, A. A., *Anal. Chem.* **36**, 2028 (1964).
33. Qureshi, M., Qureshi, S. Z. and Zehra, N., *Mikrochim. Acta* **1970**, 831.
34. Odler, I., *Anal. Chem.* **41**, 1116 (1969).
35. Dangoumau, A. and Ducos, R., *Chim. Anal. (Paris)* **44**, 292 (1962).
36. Brown, L. C. and Boyd, G. E., *Anal. Chem.* **42**, 291 (1970).

Chapter 12

CADMIUM

Cadmium (Cd, at.wt. 112·40) occurs in its compounds exclusively in the +II oxidation state. Unlike $Zn(OH)_2$ and $Pb(OH)_2$, $Cd(OH)_2$ shows no amphoteric properties and is insoluble in excess of NaOH. Cadmium forms ammine, cyanide, halide, and EDTA complexes.

12.1 Methods of Separation

12.1.1 PRECIPITATION

Precipitation of cadmium as the sulphide at low acidity (pH ~ 1·5) enables traces of cadmium to be separated from large quantities of zinc; a small amount of sodium sulphide is added to the solution to precipitate cadmium, other Group II metals, and some of the zinc, which acts as a collector. Further separation from zinc is achieved by a dithizone extraction. Precipitation of cadmium in the presence of KCN allows its separation from copper. Traces of cadmium have been coprecipitated with various collectors [1–4].

Cadmium can be separated as the sparingly soluble compounds formed by cadmium halide complexes with basic dyes such as Methylene Blue, Methyl Violet and Malachite Green [5].

Using suitable carriers, DeVoe and Meinke [6] isolated cadmium by precipitation with Reinecke's salt $(NH_4[Cr(NH_3)_2(SCN)_4]\cdot H_2O)$, or with 2-(o-hydroxyphenyl)benzoxazole, while Hibbits et al. [7] used benzotriazole as precipitant.

12.1.2 EXTRACTION

Before its spectrophotometric determination, cadmium can be extracted into chloroform or carbon tetrachloride as the diethyldithiocarbamate [8,9]. In the presence of tartrate and cyanide at pH 11–12, only bismuth, lead, and thallium(III) are co-extracted with cadmium.

Schweitzer and Randolph [10] have determined the conditions required for chloroform extraction of cadmium chelates with 8-hydroxyquinoline, TTA, acetylacetone, anthranilic acid, and various other reagents.

Iodide and thiocyanate complexes of cadmium are extractable with oxygenated solvents [11–14]. In iodide medium, cadmium is separated from zinc by extraction with cyclohexanone and tetrahydrofuran (5+1) [12].

Anionic cadmium iodide, bromide, chloride, and thiocyanate complexes form ion-pairs with basic compounds which are extractable into non-polar solvents. For this purpose, DAM [15,16], tribenzylamine [16,17], and liquid anion-exchangers [18–20] have been used.

Cadmium is often separated together with traces of other heavy metals by extraction with dithizone.

12.1.3 Ion-Exchange

Kallmann et al. [21] have separated cadmium and zinc on a Dowex 1 anion-exchanger. In a $0.12M$ HCl solution (containing 100 g of NaCl per litre), zinc and cadmium chloride complexes are retained in the column, while other metals are eluted. First, zinc is eluted from the column with $2M$ NaOH (containing 20 g of NaCl per litre); then cadmium is eluted with $1M$ HNO_3.

Berg and Truemper [22] have separated cadmium, zinc, and mercury as their chloride complexes by anion-exchange. Brooks [23] has concentrated microtraces of cadmium from sea water (acidified to be $0.1M$ in HCl) on an anion-exchanger.

Since the stability of the cadmium iodide complex is greater than that of the corresponding zinc complex, the two metals can be separated on ion-exchangers [24–26]. Cadmium has also been separated on anion-exchangers from a large number of metals in hydrobromic acid solution [27].

When metals dissolved in $0.5M$ HCl are passed through a strongly acidic cation-exchanger, cadmium and tin(IV) are eluted, whereas Zn, Cu, Mn(II), Ni, Co, U(VI), and Ti are retained on the column [28].

Cadmium, Sn(IV), Bi, and Hg have been separated from most other metals by elution from a cation-exchanger with $0.4M$ HBr [29].

Akki and Khopkar [30] have investigated the effect of various eluents on the separation of cadmium from other metals on Dowex 50 cation-exchange resin.

Korkisch and Feik [31] have separated cadmium from zinc on an anion-exchanger, using an ethanol–nitric acid medium (90% ethanol + 10% $5M$ HNO_3).

12.2 Methods of Determination

The extraction and spectrophotometric determination of cadmium with dithizone is the most sensitive and selective method. None of the more recently published methods has been generally accepted as an improvement.

12.2.1 Dithizone Method

The cadmium ion reacts with dithizone (p. 40) in neutral to strongly alkaline media to give a pink cadmium dithizonate, which is moderately soluble in carbon tetrachloride and more readily soluble in chloroform.

The stability of Cd(HDz)$_2$ in strongly alkaline media (5–20% NaOH) allows cadmium to be extracted from Pb, Bi, Sn(II), and Zn, the dithizonates of which cannot exist under such conditions. Dimethylglyoxime is added to mask nickel and cobalt. Tartrate prevents the precipitation of metals as hydroxides. Since the noble metals (Au, Pt, Pd, Ag, Hg) and copper form secondary dithizonates in alkaline media, these metals must be removed before cadmium is extracted. They are most simply pre-extracted with dithizone from acidic medium.

The dithizone method for determining cadmium [32–35] is very sensitive. The molar absorptivity is 8.8×10^4 ($a = 0.78$) at $\lambda_{max} = 520$ nm.

Only the monocolour method is used in the determination of cadmium with dithizone since, during the extraction from aqueous solutions at high pH values, the excess of dithizone passes completely to the aqueous phase.

Saltzman [33] does not employ a preliminary extraction from acidic medium when determining cadmium with dithizone, but uses instead a double extraction of cadmium from alkaline solutions containing cyanide; first, from a solution containing 6–7% NaOH and 0.2% KCN, and secondly from a solution containing 6–7% NaOH and 0.01% KCN. Tartaric acid is used as stripping agent after the first extraction.

Although a high concentration of cyanide prevents extraction of cadmium, a relatively low one does not, provided sufficient dithizone is used. Saltzman's conditions allow separation of the cadmium from all but traces of lead and zinc in the first extraction, and the stripping separates cadmium from silver and mercury.

Dilute hydrochloric acid (pH 2) readily decomposes cadmium and zinc dithizonates, whereas nickel and cobalt dithizonates are unaffected. Thus, cadmium (together with zinc) may be separated from nickel and cobalt [2]. Cadmium is separated from zinc by extraction from $2M$ KOH into a CCl$_4$ solution of dithizone: zinc is not extracted from such an alkaline solution [36].

To prevent oxidation of dithizone by oxygen in extraction of strongly alkaline solutions (especially those containing manganese), hydroxylamine should be added to the aqueous solution.

Cadmium has been determined by the dithizone method in the following materials: zinc [16], zinc ore concentrates [19,25], zinc sulphide [3], lead [32], tin [2,37], silver [38], boron [39], aluminium [40], nickel and cobalt [17], chromium and its alloys [1], beryllium [7], vanadium [8], niobium and its alloys [8,41], bismuth [42], zirconium alloys [20], uranyl nitrate [43], chemical reagents [44,45], urine [46], biological materials [33,47], organic matter [48], foods [47], waste waters [49], sea water [50,51], and air [52].

Free cadmium and cadmium oxide in cadmium sulphide have also been determined with dithizone after distillation of the metallic Cd and extraction of the CdO from the residue with acetic acid [53].

Reagents

1. Dithizone: 0·002% solution in CCl_4 (p. 492).
2. Standard cadmium solution: 1 mg/ml. Dissolve 1·631 g of cadmium chloride, dried at 110°C, in water containing 2 ml of conc. HCl, and dilute the solution with water to 1 litre.
3. Potassium sodium tartrate: 20% solution (p. 325).
4. Hydroxylamine hydrochloride: 10% solution (p. 325).
5. Dimethylglyoxime: 1% solution in ethyl alcohol.
6. Sodium hydroxide: 20% solution. Kept in a polyethylene bottle.

Procedure

Acidify a solution containing not more than 40 μg of Cd to pH ~ 2, and shake it with portions of the dithizone solution in CCl_4 until the colour of the organic phase no longer changes. Discard the organic extracts. To the aqueous solution, add tartrate solution (the amount depending on the quantity of metals present), 0·5 ml of dimethylglyoxime solution, and ammonia till neutral. Allow to stand for 1 min, then add 1 ml of hydroxylamine solution, and sufficient 20% NaOH to give a minimum final NaOH concentration of 5%. Extract the cadmium with portions of dithizone in CCl_4 (1 ml of 0·002% H_2Dz corresponds to 4·4 μg of Cd) until the extract is no longer pink with $Cd(HDz)_2$. Wash the combined organic extracts with 0·5% aqueous NaOH solution, and water. Dilute the pink solution with the solvent to the mark in a 50-ml (or smaller) volumetric flask, and measure its absorbance at 520 nm against carbon tetrachloride.

Note. When determining cadmium in a solution containing a considerable excess of zinc, the two-stage extraction with KCN added should be used. After tartrate, hydroxylamine, and dimethylglyoxime have been added to the sample solution, add NaOH and KCN solutions to give concentrations of ~ 10% of the first and 0·2% of the second. Extract with carbon tetrachloride, and then strip the cadmium with 0·02M hydrochloric acid. To the aqueous back-extract, add a little tartrate, hydroxylamine, sodium hydroxide (to 10% concentration), and KCN (to 0·05% concentration). Re-extract with CCl_4, wash this extract with dilute NaOH and water, and measure its absorbance as above.

12.2.2 OTHER METHODS

Many azo dyes are used for the spectrophotometric determination of cadmium, e.g. PAN [54], 1-[(5-chloro-5-pyridyl)azo]-2-naphthol (5-Cl-β-PAN) [55], PAR [56], Cadion (I) [57], Cadion 2B [58], Bromobenzothiazo (II) [59,60], bromobenzothiazolylazocresol [61], and Arsenazo (III) [62].

(I)   (II)

Several spectrophotometric methods are based on the formation of ion-pairs between basic dyes and the stable iodide complex $[CdI_4]^{2-}$. In the Courtot-Coupez and Guerder [63] method, di-isopropyl ether is added to a suspension of the Crystal Violet–$[CdI_4]^{2-}$ complex in water. This complex, which is also insoluble in ether, collects at the interface on gentle shaking. The aqueous solution is decanted, and acetone is added to obtain a clear and intensely coloured ether–acetone solution ($\varepsilon \sim 1\cdot3 \times 10^5$).

Rhodamine B [9] and antipyrine derivatives [64] are other basic dyes used in similar methods.

The cadmium iodide complex also forms a coloured ion-pair with the cationic complex between 2,2'-bipyridyl and iron(II). This ion-pair can be extracted into 1,2-dichloroethane [65].

Other organic reagents which have found application in spectrophotometric methods for determining cadmium are 8-hydroxyquinoline [66], Xylenol Orange [67,68], thiothenoyltrifluoroacetone [69], thiodibenzoylmethane [70], and glyoxal bis(2-hydroxyanil) (GBHA) [71], a well-known reagent for calcium.

References

1. Mukhina, Z. S., Tikhonova, A. A. and Zhemchuzhnaya, I. A., *Tr. Komis. po Analit. Khim. Akad. Nauk SSSR* **12**, 298 (1960).
2. Marczenko, Z. and Kasiura, K., *Chem. Anal. (Warsaw)* **10**, 449 (1965).
3. Kasiura, K. and Marczenko, Z., *Zh. Analit. Khim.* **22**, 1398 (1967).
4. Nielsch, W. and Böltz, G., *Chemiker Ztg.* **79**, 364 (1955).
5. Babko, A. K. and Marchenko, P. V., *Tr. Komis. po Analit. Khim. Akad. Nauk SSSR* **9**, 65 (1958).
6. DeVoe, J. R. and Meinke, W. W., *Anal. Chem.* **31**, 1428 (1959).
7. Hibbits, J. O., Kallmann, S., Oberthin, H. and Oberthin, J., *Talanta* **8**, 104 (1961).
8. Nazarenko, V. A. and Biryuk, E. A., *Zavodsk. Lab.* **25**, 28 (1959).
9. Lazarev, A. I. and Lazareva, V. I., *ibid.* **25**, 783 (1959).
10. Schweitzer, G. K. and Randolph, D. R., *Anal. Chim. Acta* **26**, 567 (1962).
11. Kinnunen, J. and Wennerstrand, B., *Chemist-Analyst* **43**, 34 (1954).
12. Hartkamp, H. and Specker, H., *Naturwissenschaft.* **43**, 421 (1956).
13. Kataev, G. A. and Shpaer, I. S., *Izv. Vyssh. Ucheb. Zaved., Khim. Khim. Tekhnol.* **7**, 891 (1964).
14. Gagliardi, E. and Tümmler, P., *Talanta* **17**, 93 (1970).
15. Babko, A. K. and Danilova, V. N., *Ukr. Khim. Zh.* **34**, 394 (1968).
16. Marchenko, P. V. and Voronina, A. I., *ibid.* **35**, 652 (1969).
17. Vasyutinskii, A. I., Kisel', N. A. and Matveeva, E. N., *Zh. Analit. Khim.* **23**, 1847 (1968).
18. Watanabe, H. and Akatsuka, K., *Bull. Chem. Soc. Japan* **41**, 620 (1968).
19. McDonald, C. W. and Moore, F. L., *Anal. Chem.* **45**, 983 (1973).
20. Ghersini, G. and Mariottini, S., *Talanta* **18**, 442 (1971).
21. Kallmann, S., Steele, C. G. and Chu, N. Y., *Anal. Chem.* **28**, 230 (1956).
22. Berg, E. W. and Truemper, J. T., *ibid.* **30**, 1827 (1958).
23. Brooks, R. R., *Analyst* **85**, 745 (1960).
24. Baggott, E. R. and Willcocks, R. G., *ibid.* **80**, 53 (1955).
25. Kallmann, S., Oberthin, H. and Liu, R., *Anal. Chem.* **30**, 1846 (1958); **32**, 58 (1960).
26. Razumova, V. P., *Izv. Vyssh. Ucheb. Zaved., Khim. Khim. Tekhnol.* **5**, 709 (1962); **8**, 192 (1965).

References

27. Strelow, F. W., Louw, W. J. and Weinhert, C. H., *Anal. Chem.* **40**, 2021 (1968).
28. Strelow, F. W., *ibid.* **32**, 363 (1960).
29. Fritz, J. S. and Garralda, B. B., *ibid.* **34**, 102 (1962).
30. Akki, S. B. and Khopkar, S. M., *Z. Anal. Chem.* **249**, 228 (1970).
31. Korkisch, J. and Feik, F., *Anal. Chim. Acta* **32**, 110 (1965).
32. Silverman, L. and Trego, K., *Analyst* **77**, 143 (1952).
33. Saltzman, B. E., *Anal. Chem.* **25**, 493 (1953).
34. Petzold, A. and Lange, I., *Z. Anal. Chem.* **146**, 1 (1955).
35. Schweitzer, G. K. and Dyer, F. F., *Anal. Chim. Acta* **22**, 172 (1960).
36. Marczenko, Z., Mojski, M. and Kasiura, K., *Zh. Analit. Khim.* **22**, 1805 (1967).
37. Coppins, W. C. and Price, J. W., *Metallurgia* **48**, 149 (1953).
38. Marczenko, Z. and Kasiura, K., *Chem. Anal.* (*Warsaw*) **9**, 87 (1964).
39. Marczenko, Z., *ibid.* **9**, 1093 (1964).
40. Hashimoto, S. and Tanaka, R., *Japan Analyst* **8**, 564 (1959).
41. Mukhina, Z. S., Tikhonova, A. A. and Zhemchuzhnaya, I. A., *Tr. Komis. po Analit. Khim. Akad. Nauk SSSR* **12**, 71 (1960).
42. Sinyakova, S. I. and Tsvetkova, L. A., *ibid.* **12**, 191 (1960).
43. Gołkowska, A., *Chem. Anal.* (*Warsaw*) **5**, 389 (1960).
44. Kemula, W., Brachaczek, W. and Hulanicki, A., *ibid.* **3**, 923 (1958).
45. Marczenko, Z., *Mikrochim. Acta* **1965**, 281.
46. Smith, J. C., Kench, J. E. and Lane, R. E., *Biochem. J.* **61**, 698 (1955).
47. Shirley, R. L., Benne, E. J. and Miller, E. J., *Anal. Chem.* **21**, 300 (1949).
48. Analytical Methods Committee, *Analyst* **94**, 1153 (1969).
49. Committee on Analysis of Trade Effluents, *ibid.* **82**, 764 (1957).
50. Mullin, J. B. and Riley, J. P., *Nature* **174**, 42 (1954).
51. Korkisch, J. and Dimitriades, D., *Talanta* **20**, 1295 (1973).
52. Pines, I., *Chem. Anal.* (*Warsaw*) **15**, 103 (1970).
53. Tumanov, A. A. and Shakhverdi, N. M., *Zavodsk. Lab.* **36**, 918 (1970).
54. Berger, W. and Elvers, H., *Z. Anal. Chem.* **171**, 255 (1959); **199**, 166 (1964).
55. Shibata, S., Furukawa, M. and Ishiguro, Y., *Mikrochim. Acta* **1972**, 721.
56. Kitano, M. and Ueda, J., *J. Chem. Soc. Japan, Pure Chem. Sect.* **91**, 760 (1970).
57. Chavanne, P. and Geronimi, C., *Anal. Chim. Acta* **19**, 377 (1958).
58. Hong-Kang Dao, Shih-Hsuan Jen and Nai-Ching Wang, *Acta Chim. Sinica* **29**, 344 (1963).
59. Drapkina, D. A., Brudz', V. G., Smirnova, K. A. and Doroshina, N. I., *Zh. Analit. Khim.* **17**, 940 (1962).
60. Shkrobot, E. P. and Bakinovskaya, L. M., *Zavodsk. Lab.* **32**, 1452 (1966).
61. Gusev, S. I., Zhvakina, M. V. and Kozhevnikova, I. A., *Zh. Analit. Khim.* **26**, 1493 (1971).
62. Michaylova, V. and Yuroukova, L., *Anal. Chim. Acta* **68**, 73 (1974).
63. Courtot-Coupez, J. and Guerder, P., *Bull. Soc. Chim. France* **1961**, 1942.
64. Zhivopistsev, V. P. and Chelnokova, M. N., *Zh. Analit. Khim.* **18**, 717 (1963).
65. Kotsuji, K., *Bull. Chem. Soc. Japan* **38**, 988 (1965).
66. Umland, F. and Hoffmann, W., *Z. Anal. Chem.* **168**, 268 (1959).
67. Otomo, M., *Bull. Chem. Soc. Japan* **37**, 504 (1964).
68. Pasechnova, R. A. and Mokhov, A. A., *Zh. Analit. Khim.* **27**, 2146 (1972).
69. Solanke, K. R. and Khopkar, S. M., *Sep. Sci.* **8**, 511 (1973).
70. Schuknecht, B., Röbisch, G. and Uhlemann, E., *Anal. Chim. Acta* **69**, 329 (1974).
71. Oi, N., *Japan Analyst* **9**, 770 (1960).

Chapter 13

CALCIUM

Calcium (Ca, at.wt. 40·08) occurs in aqueous solutions in the +II oxidation state. The hydroxide, $Ca(OH)_2$ (solubility 1·3 g/l.), is a strong base. Calcium ions form sparingly soluble compounds with oxalate and carbonate, and also form weak complexes with EDTA and hydroxyl compounds such as carbohydrates and tartrate.

13.1 Methods of Separation

13.1.1 Precipitation

A preliminary separation of the Analytical Group I–III metals which interfere in the spectrophotometric determination of calcium is often accomplished by precipitation of their sulphides or hydroxides (pH ~ 9), or of their sparingly soluble compounds with organic reagents (e.g. oxine and DDTC). Macroquantities of a large number of metals are separated from calcium by electrolysis at a mercury cathode.

Precipitation with oxalate at pH 3–4 separates calcium from metals giving soluble oxalate complexes (e.g. Fe, Al, and Ti), and also from phosphate. Calcium oxalate can also be precipitated from homogeneous solution at pH 6–8 by releasing calcium from its EDTA complex by oxidation with hydrogen peroxide [1].

Traces of calcium and magnesium may be separated from matrices of metals forming cyanide complexes (e.g. Ni, Zn, and Mn) by precipitation as phosphate with a lanthanum collector in an alkaline cyanide medium [2].

Small amounts of calcium can be separated from magnesium by coprecipitation with strontium sulphate from aqueous ethanol [3]. Traces of calcium have been separated from macroamounts of barium and strontium by precipitation of barium with a stoichiometric amount of CrO_4^{2-}, and precipitation of strontium with a slight excess of ammonium sulphate in the presence of a small amount of EDTA (pH ~ 5) [4].

When Fe(III), Cr, Mn, and phosphate are precipitated in the presence of a suspension of $PbCO_3$, calcium and magnesium are not coprecipitated [5].

13.1.2 Extraction

Liquid–liquid extraction has often been used to separate the Group I–III metals from calcium, which remains in the aqueous phase.

Gorbenko et al. [6–8] have extracted calcium (barium and strontium are not extractable) with a solution of carbon tetrachloride and TBP (20%) from an alkaline medium (0·1–1M NaOH) as a complex with Azo-azoxy BN (I)

$$\underset{\text{(I)}}{\text{[structure: naphthol–OH with N=N linkage to phenyl ring bearing N=N(→O) group, connected to another phenyl ring with OH and CH}_3\text{ substituents]}}$$

Solvation with TBP of two of the Ca^{2+} coordination sites renders the calcium Azo-azoxy BN complex soluble in the organic phase. During the extraction the colour of the CCl_4–TBP phase changes from the orange-red of the Azo-azoxy BN to the red of the calcium Azo-azoxy BN complex. n-Butylamine or "butylcellosolve" may be used instead of TBP. Calcium is stripped from the organic phase with water or dilute hydrochloric acid. After this stripping the carbon tetrachloride solution of Azo-azoxy BN may be reused.

Calcium may also be separated by extraction of its complexes with 1-phenyl-3-methyl-4-benzoyl-5-pyrazolone [9], or with TTA and TBP [10], into amyl alcohol or carbon tetrachloride, respectively.

13.1.3 Ion-Exchange

Calcium is often separated from Mg, Sr, Ba, and other metals by cation-exchange chromatography in the presence of EDTA [11,12], EGTA [13], chloride [14–17], formate [18,19], malonate [20], or glycollate [21] as complexants. Strelow and Weinert [22] have examined the cation-exchange behaviour of the alkaline earths with an extensive range of complexing agents.

Anion-exchange separation of calcium from magnesium has been carried out in oxalate [23] and citrate media [24]. Nitric acid–methanol or propanol systems facilitate the separation of calcium from magnesium [25] and from strontium [26].

13.2 Methods of Determination

One of the best spectrophotometric reagents for determining calcium is glyoxal bis(2-hydroxyanil). This reagent has high sensitivity and selectivity, and can be used in water or in organic solvents. The specific preseparation as the Azo-azoxy BN complex enables calcium to be determined with less selective reagents.

13.2.1 Glyoxal bis(2-hydroxyanil) Method

Glyoxal bis(2-hydroxyanil) (GBHA) forms a sparingly soluble red complex with calcium in alkaline media [27–32].

GBHA structure reacting with Ca^{2+} to form calcium complex (structure shown at top of page).

The compound dissolves in 50% aqueous methanol by the substitution of two molecules of methanol for the two water molecules coordinatively bonded to the calcium.

The methanol solution of GBHA is colourless. As sodium hydroxide is added and the hydroxyl groups dissociate, the solution becomes more and more intensely yellow. The optimum alkalinity of the solution ($\sim 0.04M$ NaOH) is that at which the reaction of GBHA with calcium is complete, but the colour of the reagent is still weak.

The absorption maximum of the red-violet complex is at 516 nm. (The molar absorptivity of the compound is 1.8×10^4; specific absorptivity 0.45.)

Figure 13.1 shows the absorption spectra of both GBHA and its calcium complex.

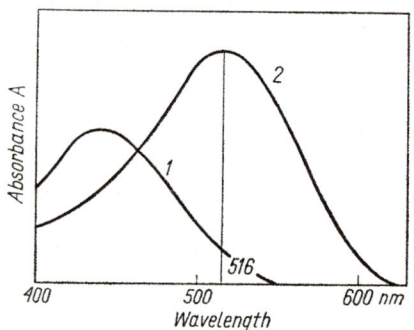

Fig. 13.1 Absorption spectra of glyoxal bis(2-hydroxyanil) (GBHA) (1) and its calcium complex (2) (alkaline methanol–water solutions)

The absorbance of the solution should be measured within 10 minutes of the development of the colour, otherwise a slow fading of the colour will be observed (2–3% in 30 minutes).

GBHA reacts with a number of bivalent metals other than calcium (e.g. barium and strontium). Magnesium and the alkali metals do not interfere. The Analytical Group I–III metals interfere and should be separated: small quantities may, however, be masked by adding a little cyanide and sodium sulphide (and some hydroxylamine when iron is present). These masking agents do not affect the reaction between calcium and GBHA. To mask barium and strontium a small amount of sulphate and carbonate is added to the sample solution. Fluoride, oxalate, tartrate, citrate, and EDTA interfere in the reaction of calcium with GBHA.

The GBHA complex of calcium can be extracted into chloroform from aqueous methanol or aqueous ethanol [27,31,33]. The presence of hexanol aids the extraction. Keil [34] has extracted the complex into octanol.

The complex may also be extracted with butanol, but the colour of the extract fades rapidly. Twenty minutes after the extraction the absorbance of the solution has decreased by $\sim 20\%$.

When the red complex of calcium with GBHA is extracted with 1,2-dichloroethane in the presence of benzyldimethyltetradecylammonium chloride, the absorbance of the organic extract remains unchanged for 6 hours. The organic phase is dried over Na_2SO_4 and the absorbance is measured at 530 nm against water [35].

The GBHA method has been used to determine calcium in milk [36], uranium [37], indium [38], potassium chloride [33], silicate minerals [39], amphoteric metals [40], alkali metal compounds [41,42], and barium compounds [42].

Calcium may be extracted with a 0·01% solution of Azo-azoxy BN in CCl_4 containing 20% TBP, and then stripped into very dilute HCl, before its determination with GBHA [40,42].

Lindstrom and Milligan [43,44] have examined the analytical properties of various derivatives of GBHA. Some methyl derivatives offered slightly better sensitivity or colour stability, but no dramatically superior reagent was found.

Reagents

1. Glyoxal bis(2-hydroxyanil) (GBHA): 0·05% solution in methanol. Solutions more than a week old must be discarded.
2. Standard calcium solution: 1 mg/ml. Dissolve 2·498 g of $CaCO_3$, dried at 110°C, in 40 ml of $2M$ HCl. Remove CO_2 by boiling, and dilute the solution with water in a volumetric flask to 1 litre.
3. Sodium hydroxide: $1M$ and $5M$ solutions.
4. Azo-azoxy BN: 0·01% solution. Dissolve 12 mg of reagent in 100 ml of carbon tetrachloride. Add 2 ml of TBP to 10 ml of this CCl_4 solution before it is used for the extraction.

Procedure

Extraction of Ca. Add sufficient $5M$ NaOH to the sample solution to make the NaOH concentration 0·5–1M. Shake the solution for 1 minute with the solution of Azo-azoxy BN in CCl_4 containing 20% TBP. Strip calcium from the extract with a small volume of $1M$ hydrochloric acid. (The Azo-azoxy BN solution is reusable.)

Determination of Ca. The sample solution should contain not more than 50 μg of Ca, and be free from ammonium salts and Analytical Group I–III metals (to mask small amounts of these metals, add ~ 10 mg each of KCN and Na_2S). Neutralize the sample solution (to pH 7–8) with NaOH or HCl, and add ~ 20 mg of Na_2SO_4. Add 25 ml of the GBHA solution and 2 ml of 1M NaOH, and dilute the solution with water to the mark in a 50-ml volumetric flask. After 10 minutes, measure the absorbance of the coloured solution at 516 nm against a blank solution.

Note. If the Ca is first extracted by the Azo-azoxy BN method, neutralize the acidic back-extract, add the GBHA solution, and proceed as described above.

13.2.2 Murexide Method

Murexide (ammonium purpurate, **II**) forms a 1:1 complex with calcium ions in alkaline solution. The change in colour from violet to pink when

$$\left[O = \underset{\underset{H}{N}-\underset{O}{\overset{H}{\underset{|}{N}}}}{\overset{H}{\underset{|}{N}}-\overset{O}{\underset{|}{\overset{|}{N}}}}-N=\underset{\underset{O}{\overset{|}{N}}-\underset{H}{\overset{|}{N}}}{\overset{O}{\underset{|}{\overset{H}{N}}}}=O \right]^{-} NH_4^+ \quad \textbf{(II)}$$

the reagent complexes calcium is the basis of this spectrophotometric method [45–47].

The optimum pH for the reaction is ~ 12·5 (2·5 ml of 1M NaOH per 50 ml of the final solution volume). In this medium, magnesium (in fivefold amount relative to calcium) does not interfere: at lower pH values magnesium causes positive errors. Since the absorbance of the calcium murexide complex decreases gradually with increasing pH, care should be taken to maintain the sample and the standard solutions at the same pH.

Under the conditions given in the procedure below, the molar absorptivity of the calcium murexide complex is ~ $1·4 \times 10^4$ ($a = 0·35$).

Figure 13.2 shows the absorption spectra of both murexide and its complex with calcium in ~ 0·05M NaOH. The absorption maximum of the reagent is at 552 nm, while that of the complex is at 514 nm. In the colorimetric determination of calcium, the absorbance is measured at 500–505 nm, where the absorbance of the free reagent interferes least. This system does not obey Beer's law.

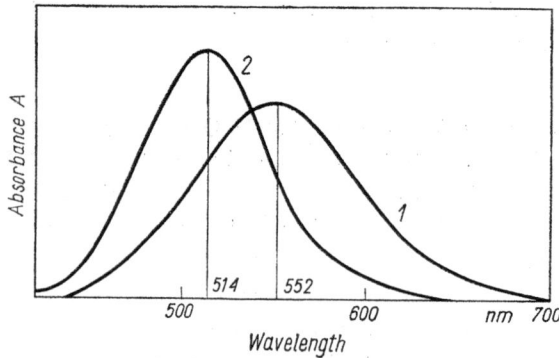

Fig. 13.2 Absorption spectra of murexide (1) and its calcium complex (2) in 0·05M NaOH

Murexide reacts with ions of numerous metals. Therefore, the Analytical Group I–III metals should be separated before the determination of calcium. Trace amounts of some metal ions may be masked with cyanide. Barium and strontium react like calcium with murexide. Alkali metals do not interfere.

Since murexide is readily oxidized, a reducing agent is sometimes added to the sample solution. Some commercial murexide preparations are very impure.

Calcium has been determined by the murexide method in zirconium powder [48], and nickel, zinc, and manganese salts [2].

Reagents

1. Murexide: 0·02% solution. Dissolve 20 mg of the reagent in 35 ml of water and dilute the solution with ethanol to 100 ml. If the solution is stored in the dark, it may be used for one week.
2. Standard calcium solution: 1 mg/ml (p. 185).

Procedure

The sample solution must be neutral (pH 6–8) and free from ammonium salts, and contain not more than 70 µg of Ca. To the sample solution in a 50-ml volumetric flask, add ∼ 10 mg of ascorbic acid and ∼ 10 mg of potassium cyanide. After 2–3 minutes, add 5 ml of the murexide solution and 2·5 ml of $1M$ NaOH, and dilute with water to the mark. Measure the absorbance of the coloured solution at 500 nm against a blank solution.

Note. KCN is added to mask possible traces of Group III metals. Ascorbic acid reduces any Fe(III) present to Fe(II).

13.2.3 OTHER METHODS

West *et al.* have found Calcichrome (**III**) to be a highly selective, stable reagent for calcium [49–51]. Its molar absorptivity is $7·6 \times 10^3$ at $\lambda_{max} = 615$ nm. Although this reagent has also been studied by other authors, its structure has not been conclusively established, and it is probable that it is identical to the reagent Calcion [52–54]. Calcichrome has been used for determining calcium in soil extracts [55].

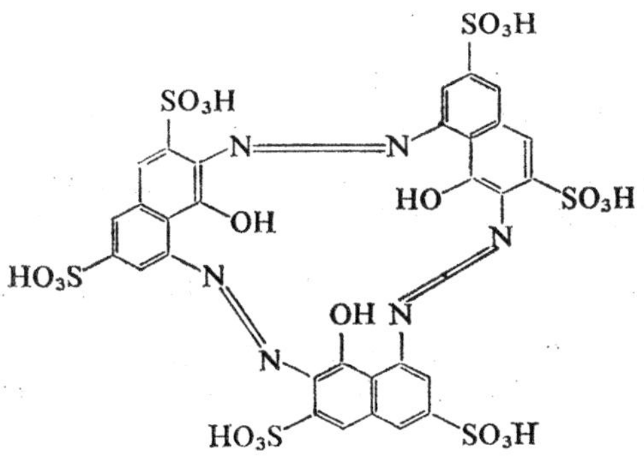

(III) (after Close and West [50])

Chlorophosphonazo III (formula, p. 50) permits the determination of calcium in a neutral medium ($\varepsilon = 6\cdot4 \times 10^4$, at 669 nm, pH 7) or, with less sensitivity ($\varepsilon = 1\cdot5 \times 10^4$), in a slightly acid medium (pH 2·2) [56,57]. With this promising reagent, calcium has been determined in steels [58], cobalt [59], molybdenum alloys [60], aluminium alloys [61], and biological fluids [62].

Other azo reagents for calcium include: Arsenazo I [63,64], Arsenazo III [61,62], Calmagite [67], Eriochrome Black T [68–70], Erio SE [71], Calcon [72], and Acid Chrome Blue K [73].

Calcium may also be determined spectrophotometrically with Calcein (from the decrease in absorbance at 506 nm of a Calcein solution when calcium is added) [74], Metalphthalein (o-Cresolphthalein Complexone; formula, p. 50) [15,47], Thymolphthalexone [75], and di(o-hydroxyphenylimino)ethane [76].

The extractive photometric methods for determining calcium are not numerous. Calcium oxinate is extracted into chloroform in the presence of n-butylamine [77] or 2-butoxyethanol ("butylcellosolve") [78]. Complexes of calcium with bromo-oxine [79] or with TTA [80] have been extracted into benzene as ion-association compounds with Rhodamine B.

Azo-azoxy BN (see p. 183), synthesized by Dziomko and Dunaevskaya [81], is an excellent reagent for the extraction and spectrophotometric determination of calcium.

There exist indirect methods for determining calcium with chloranilic acid [82], 2-chloro-5-cyano-3,6-dihydroxybenzoquinone (an analogue of chloranilic acid) [83], picrolonic acid [84,85] rhodizonic acid [86], or naphthalhydroxamic acid [87] as precipitants. Either the decrease in absorbance of the coloured precipitant solution is measured, or the precipitate is separated and the organic component determined by colour-forming reactions. After precipitation of calcium as $CaMoO_4$, the molybdenum can be determined by the thiocyanate method [88]. Calcium has also been determined indirectly by exchange reactions with EGTA complexes [89].

Lastly, calcium has been determined turbidimetrically as a suspension of calcium oxalate [90].

References

1. Grzeskowiak, R. and Turner, T. A., *Talanta* **20**, 351 (1973).
2. Marczenko, Z. and Kasiura, K., *Chem. Anal. (Warsaw)* **7**, 775 (1962).
3. Kar, K. R. and Singh, G., *Anal. Chim. Acta* **35**, 259 (1966).
4. Tarkovskaya, I. A. and Gorbenko, F. P., *Zh. Analit. Khim.* **20**, 1185 (1965).
5. Bezuglyi, D. V., Bazalei, N. V., Kleopova, E. S. and Petrusevich, I. A., *ibid.* **26**, 1010 (1971).
6. Gorbenko, F. P. and Sachko, V. V., *ibid.* **18**, 1198 (1963); **24**, 15 (1969); **25**, 1884 (1970).
7. Gorbenko, F. P. and Sachko, V. V., *Ukr. Khim. Zh.* **30**, 402 (1964).
8. Gorbenko, F. P., Shevchuk, I. A., Tselinskii, Yu. K. and Sachko, V. V., *Zh. Analit. Khim.* **18**, 1397 (1963).
9. Zolotov, Yu. A. and Lambrev, V. G., *ibid.* **20**, 659 (1965).

10. Sekine, T. and Dyrssen, D., *Anal. Chim. Acta* **37**, 217 (1967).
11. Wade, M. A. and Seim, H. J., *Anal. Chem.* **33**, 793 (1961).
12. Chwastowska, J. and Szymczak, S., *Chem. Anal. (Warsaw)* **14**, 1161 (1969).
13. Povondra, P. and Přibil, R., *Talanta* **10**, 713 (1963).
14. Nelson, F., Holloway, J. H. and Kraus, K. A., *J. Chromatogr.* **11**, 258 (1963).
15. Bosholm, J., *Anal. Chim. Acta* **34**, 71 (1966).
16. Holzapfel, H. and Tischer, W., *J. Prakt. Chem.* **27**, 91 (1965).
17. Strelow, F. W. E. and van Zyl, C. R., *Anal. Chim. Acta* **41**, 529 (1968).
18. Tsubota, H. and Kitano, Y., *Bull. Chem. Soc. Japan* **33**, 770 (1960).
19. Tsubota, H., *ibid.* **38**, 159 (1965).
20. Strelow, F. W. E., van Zyl, C. R. and Nolte, C. R., *Anal. Chim. Acta* **40**, 145 (1968).
21. Senegačnik, M., Paljk, Š. and Korošin, J., *Z. Anal. Chem.* **244**, 365 (1969).
22. Strelow, F. W. E. and Weinert, C. H., *Talanta* **17**, 1 (1970).
23. Winowski, Z., *Chem. Anal. (Warsaw)* **12**, 1271 (1967).
24. Nelson, F. and Kraus, K. A., *J. Am. Chem. Soc.* **77**, 801 (1955).
25. Fritz, J. S. and Waki, H., *Anal. Chem.* **35**, 1079 (1963).
26. Fritz, J. S., Waki, H. and Garralda, B. B., *ibid.* **36**, 900 (1964).
27. Umland, F. and Meckenstock, K. U., *Z. Anal. Chem.* **176**, 96 (1960).
28. Kerr, J. R., *Analyst* **85**, 867 (1960).
29. Williams, K. T. and Wilson, J. R., *Anal. Chem.* **33**, 244 (1961).
30. Lindstrom, F. and Milligan, C. W., *ibid.* **39**, 132 (1967).
31. Kuczerpa, A. V., *ibid.* **40**, 581 (1968).
32. Hunter, G., *Analyst* **97**, 233 (1972).
33. Glasner, A. and Skurnik, S., *Israel J. Chem.* **2**, 363 (1965).
34. Keil, R., *Z. Anal. Chem.* **253**, 15 (1971).
35. Nishimura, M. and Noriki, S., *Japan Analyst* **21**, 640 (1972).
36. Nickerson, T. A., Moore, E. E. and Zimmer, A. A., *Anal. Chem.* **36**, 1676 (1964).
37. Abrao, A., *ibid.* **37**, 437 (1965).
38. Kasiura, K., *Chem. Anal. (Warsaw)* **11**, 141 (1966).
39. King, H. G. and Pruden, G., *Analyst* **94**, 39 (1969).
40. Gorbenko, F. P. and Degtyarenko, L. I., *Zavodsk. Lab.* **31**, 1309 (1965).
41. Gorbenko, F. P., Tarkovskaya, I. A. and Olevinskii, M. I., *Ukr. Khim. Zh.* **30**, 640 (1964).
42. Gorbenko, F. P. and Sachko, V. V., *Zh. Analit. Khim.* **18**, 1497 (1963); **20**, 309 (1965).
43. Lindstrom, F. and Milligan, C. W., *Anal. Chem.* **36**, 1334 (1964).
44. Milligan, C. W. and Lindstrom, F., *ibid.* **44**, 1822 (1972).
45. Tammelin, L. E. and Mogensen, S., *Acta Chem. Scand.* **6**, 988 (1952).
46. Williams, M. B. and Moser, J. H., *Anal. Chem.* **25**, 1414 (1953).
47. Pollard, F. H. and Martin, J. V., *Analyst* **81**, 348 (1956).
48. Gordon, H. and Norwitz, G., *Talanta* **19**, 7 (1972).
49. West, T. S., *Analyst* **87**, 630 (1962).
50. Close, R. A. and West, T. S., *Talanta* **5**, 221 (1960).
51. Herrero-Lancina, M. and West, T. S., *Anal. Chem.* **35**, 2131 (1963).
52. Bezděková, A. and Buděšínský, B., *Collection Czech. Chem. Commun.* **30**, 811 (1965).
53. Mendes-Bezerra, A. E. and Stephen, W. I., *Analyst* **94**, 1117 (1969).
54. Lukin, A. M., Smirnova, K. A. and Zavarikhina, G. B., *Zh. Analit. Khim.* **18**, 444 (1963).
55. Pakalns, P. and Florence, T. M., *Anal. Chim. Acta* **30**, 353 (1964).
56. Ferguson, J. W., Richard, J. J., O'Laughlin, J. W. and Banks, C. V. *Anal. Chem.* **36**, 796 (1964).
57. Lukin, A. M., Smirnova, K. A. and Vysokova, N. N., *Zavodsk. Lab.* **34**, 1436 (1968).
58. Yakovlev, P. Ya. and Zhukova, M. P., *ibid.* **36**, 1169 (1970); **37**, 1292 (1971).

59. Zhukova, M. P. and Yakovlev, P. Ya., *ibid.* **39**, 661 (1973).
60. Zhukova, M. P., Shemyakin, F. M. and Yakovlev, P. Ya., *Zh. Analit. Khim.* **28**, 1705 (1973).
61. Zhukova, M. P., Matrosova, T. V. and Yakovlev, P. Ya., *ibid.* **26**, 2231 (1971).
62. Howell, D. S., Pita, J. C. and Marquez, J. F., *Anal. Chem.* **38**, 434 (1966).
63. Polyak, L. Ya., *Zavodsk. Lab.* **27**, 803 (1961).
64. Vlasov, N. A. and Morgen, E. A., *Zh. Prikl. Khim.* **38**, 998 (1965).
65. Michaylova, V. and Ilkova, P., *Anal. Chim. Acta* **53**, 194 (1971).
66. Michaylova, V. and Kouleva, N., *Talanta* **21**, 523 (1974).
67. Ingman, F. and Ringbom, A., *Microchem. J.* **10**, 545 (1966).
68. Young, A., Sweet, T. R. and Baker, B. B., *Anal. Chem.* **27**, 356, 418 (1955).
69. Menon, V. P. and Das, M. S., *Analyst* **83**, 434 (1958).
70. Impedovo, S., Traini, A. and Papoff, P., *Talanta* **18**, 97 (1971).
71. Brush, J. S., *Anal. Chem.* **33**, 798 (1961).
72. Reilley, C. N. and Hildebrand, G. P., *ibid.* **31**, 1763 (1959).
73. Goryushina, V. G. and Archakova, T. A., *Zavodsk. Lab.* **28**, 796 (1962).
74. Robinson, C. and Weatherell, J. A., *Analyst* **93**, 722 (1968).
75. Bezdeková, A. and Budešinský, B., *Collection Czech. Chem. Commun.* **30**, 818 (1965).
76. Fifield, J. A. and Blezard, R. G., *Analyst* **94**, 503 (1969).
77. Umland, F. and Meckenstock, K. U., *Z. Anal. Chem.* **165**, 161 (1959).
78. Luke, C. L., *Anal. Chim. Acta* **32**, 221 (1965).
79. Bel'tyukova, S. V. and Poluektov, N. S., *Zh. Analit. Khim.* **25**, 1714 (1970).
80. Poluetkov, N. S. and Bel'tyukova, S. V., *ibid.* **25**, 2106 (1970).
81. Dziomko, V. M. and Dunaevskaya, K. A., *ibid.* **15**, 661 (1960).
82. Frost-Jones, R. E. and Yardley, J. T., *Analyst* **77**, 468 (1952).
83. Rehwoldt, R. E., Chasen, B. L. and Li, J. B., *Anal. Chem.* **38**, 1018 (1966).
84. Mackereth, F. J., *ibid.* **76**, 482 (1951).
85. Nonowa, D. C., *Mikrochim. Acta* **1958**, 111.
86. Prokopov, T. S., *ibid.* **1973**, 429.
87. Trinder, P., *Analyst* **85**, 889 (1960).
88. Harrison, G. E. and Raymond, W. H., *ibid.* **78**, 528 (1953).
89. Nakagawa, G., Namiki, I. and Tanaka, M., *Talanta* **13**, 1135 (1966).
90. Podobed, N. D. and Malyshkina, E. A., *Zh. Analit. Khim.* **21**, 269 (1966).

Chapter 14
CARBON

Carbon (C, at.wt. 12·01) occurs in compounds principally in the +IV oxidation state. Carbonate (CO_3^{2-}) gives sparingly soluble salts with all cations except Group I metals and the ammonium ion, as well as soluble complexes with uranium and thorium ions. Of the carbon compounds generally considered inorganic, the cyanides (CN^-) and thiocyanates (SCN^-) are of great importance in analysis. The analytical reactions of cyanide and thiocyanate are similar to those of the halogens. Cyanide gives stable complexes with such metals as Hg(II), Ag, Cu(I), Fe(II and III) (ferro- and ferricyanide), Ni, and Co(III). Coloured thiocyanate complexes [e.g. with Fe(III), Co, Mo, Re, and Nb] are of major importance in spectrophotometric analysis.

14.1 Separation of Cyanides

Since cyanides are very poisonous, suitable precautions should be taken during their separation and determination.

Hydrogen cyanide is such a weak acid that it volatilizes from cyanide solutions at pH < 9. Usually hydrogen cyanide is separated by distillation from solutions acidified with tartaric acid to pH ~ 3. Nitrogen or air is passed through the heated solution to accelerate the separation of HCN, which is then absorbed in dilute NaOH or Na_2CO_3 solutions [1-3].

Any sulphide in the sample solution is precipitated with zinc or cadmium ions before the distillation of HCN [4-6]. Complex cyanides are precipitated with zinc ions if it is necessary to separate HCN derived only from simple cyanides (e.g. KCN) [7,8].

To determine total cyanide in solutions containing both simple cyanides and also metal cyanide complexes, more drastic decomposition conditions are necessary before the HCN is distilled off. Decomposition of complex cyanides occurs on heating with non-volatile mineral acids (H_2SO_4, H_3PO_4) in the presence of complexing agents such as EDTA, citric acid, and tartaric acid [9]. The cyanide complexes of Zn, Cd, Ni, and Fe(III) are fairly rapidly decomposed. On the other hand, Co(III), Fe(II), Cu, Hg, and Pd complexes are decomposed only with difficulty. Decomposition of this latter group requires several hours of heating. Care must be taken, however, not to use too concentrated acid, or complete decomposition may take place, with production of CO [10].

Hilbert and Darwish [11] have separated and determined cyanide, complex cyanides, cyanates, and ammonia, using selective steam distillation at various pH values.

Traces of hydrogen cyanide have been conveniently separated by the microdiffusion method [5,12].

14.2 Determination of Cyanides

The outstanding spectrophotometric methods for determining cyanides are based on the formation of polymethine dyes. These methods are highly sensitive and almost specific. The benzidine–pyridine method is used most frequently. Lower colour stability is a drawback of the pyrazolone method. The barbituric acid method can also be recommended.

Spectrophotometric (and other) methods for determining cyanides are the subject of two review papers [13,14].

14.2.1 BENZIDINE–PYRIDINE METHOD

In this method, developed by Aldridge [15], cyanide reacts with bromine water to form cyanogen bromide, CNBr, which then reacts with pyridine to yield glutaconaldehyde. This aldehyde is condensed with an aromatic amine (benzidine) to form a red polymethine dye. The excess of bromine is reduced with arsenite. The colour intensity reaches a maximum after 15 minutes, and remains constant for a further 30 minutes.

The method is highly sensitive. The molar absorptivity (with respect to CN^-) at $\lambda_{max} = 520$–530 nm is $\sim 6 \cdot 0 \times 10^4$ ($a = 2 \cdot 3$). The colour obtained depends on the reaction conditions [16,17].

When the sample solution is turbid or is itself coloured, the coloured polymethine dye may be extracted into butanol.

Any thiocyanate present in the sample solution also reacts with bromine to form cyanogen bromide. Since thiocyanate is non-volatile, cyanide can be conveniently separated by distillation as HCN.

The benzidine–pyridine method is used principally for determining cyanide in effluents and water samples [2,7,17–20], but has also been used for the determination in food [21,22].

Bark and Higson [23] recommend the use of p-phenylenediamine instead of benzidine, which is an active carcinogen. The resulting dye is more intensely coloured than the corresponding benzidine dye. This modified method has been applied in the analysis of river water and effluents [24,25].

Asmus and Garschagen [26] have utilized the polymethine dye which is formed in the reaction of cyanogen chloride (from cyanide and chloramine-T) with pyridine and barbituric acid. With cyanogen bromide as the intermediate, greater colour stability is obtained [27]. The barbituric acid method has been used for determining cyanide in plant materials [3] and wine [28,29].

In Epstein's method [30] for determining cyanide, cyanogen chloride reacts with a pyridine-pyrazolone reagent to form a blue dye ($\lambda_{max} = 630$ nm). This method has been applied to the determination of cyanide in wastes [1], biological materials [31,32], water [33], organic compounds [34]; free cyanide in ferro- and ferricyanides [35], and cyanide in its platinum and palladium complexes [36].

The violet polymethine dye obtained from the reaction of dimedone with glutaconaldehyde (from CNCl and pyridine) is the basis of another modification of this method for cyanide [37].

Reagents

1. Benzidine hydrochloride: 1% solution. Add 0·5 g of the reagent to 50 ml of 0·5M HCl, heat to boiling, cool, and filter. Keep the filtrate in an amber-glass bottle.
2. Benzidine–pyridine reagent. Mix 18 ml of redistilled pyridine, 12 ml of water, and 3 ml of conc. HCl. Add 10 ml of benzidine solution (1), and shake until a clear solution is obtained. Prepare this solution fresh daily.
3. Standard cyanide solution: 1 mg/ml. Dissolve 0·2503 g of KCN in cold water which has been previously boiled, and dilute the solution with similar water to 100 ml in a volumetric flask. Store the solution in a polyethylene bottle. This solution is unstable.

Procedure

Place the alkaline sample solution (~ 20 ml) containing not more than 20 μg of CN^- in a 50-ml volumetric flask. Acidify the solution with conc. acetic acid and add 1 ml in excess. Immediately add 2 ml of bromine water. Mix the solution thoroughly, and let it stand for 10 minutes with occasional shaking. Add a 1·5% solution of sodium arsenite dropwise to reduce the excess of bromine, then 2–4 drops more. Add 10 ml of the benzidine-pyridine reagent, and stir the solution. After 30 seconds add 10 ml of ethanol, and make up to the mark with water. After 15 minutes, measure the absorbance of the solution at 530 nm against a reagent blank.

14.2.2 Other Methods

In an extractive spectrophotometric method for determining cyanide, a coloured ion-pair formed between cyanide and the iron(II)–1,10-phenanthroline complex is extracted into chloroform [38,39].

A specific field method for determining cyanide depends on the formation of a coloured stain of Prussian blue on a filter paper impregnated with $FeSO_4$ and NaOH [40].

There are several indirect methods for determination of cyanide, which involve the displacement of metals from their complexes with organic reagents by cyanide, and the subsequent change in colour of the solutions. On this principle are based the methods using mercury complexes with diphenylcarbazone [41], *p*-dimethylaminobenzylidenerhodanine [5,12,42], chloranilate [43], Methylthymol Blue [44], and Metalphthalein [45]; silver complexes with dithizone [46] and thiofluorescein [47]; the palladium complex with α-furildioxime [4,48]; and the copper complex with DDTC [49]. Further indirect methods utilize the association complexes of silver with 1,10-phenanthroline and Bromopyrogallol Red [50,51], and of iron(II) with 1,10-phenanthroline and iodide [52]. Methods using haemoglobin [53] and organic disulphides [54,55] also deserve mention.

Cyanide enhances the redox potential of copper(II). Colorimetric methods for determining cyanide based on this property consist of oxidizing phenolphthalin to phenolphthalein (coloured in alkaline medium) with Cu(II) [56], and of oxidizing Variamine Blue [6].

14.3 Separation and Determination of Other Carbon Compounds

Thiocyanate is separated from cyanide by distilling off HCN from a feebly acid medium. With bromine and chloramine-T, thiocyanate is converted into cyanogen bromide and cyanogen chloride, respectively, and determined as a polymethine dye by the benzidine–pyridine method [15,19].

In the absence of cyanide, thiocyanate can be determined as the thiocyanate complex $[FeSCN]^{2+}$, which is formed in acid medium with excess of iron(III) [18,57]. The colour obtained fades in light owing to the oxidation of SCN^- by Fe(III).

Thiocyanate may be extracted and determined as the association complexes formed by SCN^- with Methylene Blue [58], with Cu^{2+} and pyridine [1,59], with the tris(1,10-phenanthroline)iron(II) chelate [60], and with Rhodamine B [61].

Thiocyanate forms with Hg(II) in the presence of quinoline a mixed-ligand mercury complex and is extracted into chloroform (pH 5·1–6·5). This complex is treated with dithizone to form the $Hg(HDz)_2$ complex [62].

Methods using dithiofluorescein [63], Variamine Blue [64], and the rhenium complex with SCN^- [65] are also worth noting.

Cyanate, CNO^-, is determined by an extractive spectrophotometric method as the coloured $CHCl_3$-soluble ion-pair formed with copper(II) and pyridine [66]. The Nessler method (see p. 394) is also applicable after cyanate ions have been converted into ammonium ions by passing an alkaline cyanate solution through a cation-exchanger (in the Na^+ form) to separate cyanate from cations (except Na^+), and acidifying the eluate with sulphuric acid [67].

Ferrocyanide, $[Fe(CN)_6]^{4-}$, is determined spectrophotometrically by adding a solution of Fe(III) in HCl to the sample solution containing citric acid, thereby forming Prussian blue [68,69]. Prussian blue can be extracted into chloroform in presence of ajatin (a quaternary amine) [70].

Asmus and Seifert [71] have converted ferrocyanide into HCN (by heating with tartaric acid and EDTA). The HCN was then distilled off, and determined by the pyridine–barbituric acid method.

Ferricyanide, $[Fe(CN)_6]^{3-}$, oxidizes o-dianisidine, producing a blue-green colour which changes to red on acidification of the solution [72].

Carbon in steels and cast-iron alloys is determined turbidimetrically when present in amounts lower than 0·05%. Carbon dioxide produced by burning a sample in a furnace is trapped in a $Ba(OH)_2$ solution and the resulting turbidity is measured [73]. For determination of carbon in titanium the sample is dissolved in H_2SO_4–HF–H_3BO_3 and HNO_3 is added to dissolve the remaining titanium carbide. A yellow organic nitro compound is then formed, and the absorbance of the solution measured [74].

Free carbon in tungsten and molybdenum carbides remains as amorphous carbon after dissolution of the sample and is determined from the decrease in absorbance of dye solutions added (e.g. Bromothymol Blue) caused by adsorption of the dye on the amorphous carbon [75].

Ducret and Cornet [76] have reported an interesting method for determining traces of carbon in silicon and germanium. The sample is heated to 1100°C with sulphur in a sealed quartz tube, and the resulting CS_2 is made to react with diethylamine to yield diethyldithiocarbamate, which is determined spectrophotometrically as its complex with copper(II).

The reducing properties of carbon monoxide are made use of in spectrophotometric methods for its determination. In the reaction of CO with iodine pentoxide, iodine is released and absorbed in KI solution, and the absorbance of the resulting brown solution is measured [77]. Carbon monoxide displaces silver from an alkaline solution of silver p-sulphamoylbenzoate. The absorbance of the coloured silver sol is measured [78–83]. Traces of CO in air may be determined by using silica gel saturated with palladium and molybdenum salts. Under the influence of CO, the silica gel changes colour from yellow to green or blue [84]. Carbon monoxide forms a red-violet product in the reaction with the palladium-1,10-phenanthroline complex [85].

A continuous spectrophotometric method for determining carbon dioxide in gases takes advantage of the acidic properties of CO_2 when the gas is passed through a solution of the acid-base indicator Phenol Red [86]. Nazarenko *et al.* [87] have determined traces of carbon in titanium, zirconium, and other metals by oxidizing them with a $PbCrO_4$–V_2O_5 mixture at 1200°C, and absorbing the resulting CO_2 in a Thymol Blue solution.

Carbon disulphide in air can be determined by absorption in an ethanolic solution containing copper(II) acetate, diethylamine and triethanolamine. The yellow colour produced is compared visually with standard colours or measured spectrophotometrically [88].

References

1. Kruse, J. M. and Mellon, M. G., *Anal. Chem.* **25**, 446 (1953).
2. Ludzack, F. J., Moore, W. A. and Ruchhoft, C. C., *ibid.* **26**, 1784 (1954).

3. Pulss, G., *Z. Anal. Chem.* **190**, 402 (1962).
4. Brooke, M., *Anal. Chem.* **24**, 583 (1952).
5. Ohlweiler, O. A. and Meditsch, J. O., *ibid.* **30**, 450 (1958).
6. Gregorowicz, Z., Buhl, F. and Śliwa, E., *Z. Anal. Chem.* **186**, 407 (1962).
7. Russell, F. R. and Wilkinson, N. T., *Analyst* **84**, 751 (1959).
8. Roberts, R. F. and Jackson, B., *ibid.* **96**, 209 (1971).
9. Leschber, R. and Schlichting, H., *Z. Anal. Chem.* **245**, 300 (1969).
10. Williams, H. E., *Cyanogen Compounds*, 2nd ed., p. 168, Arnold, London (1948).
11. Hilbert, F. and Darwish, N. A., *ibid.* **255**, 357 (1971).
12. Ohlweiler, O. A. and Meditsch, J. O., *Anal. Chim. Acta* **11**, 111 (1954).
13. Bark, L. S. and Higson, H. G., *Analyst* **88**, 751 (1963).
14. Gregorowicz, Z. and Górka, P., *Chem. Anal. (Warsaw)* **16**, 703 (1971).
15. Aldridge, W. N., *Analyst* **69**, 262 (1944).
16. Baker, M. O., Foster, R. A., Post, B. G. and Hiett, T. A., *Anal. Chem.* **27**, 448 (1955).
17. Higson, H. G. and Bark, L. S., *Analyst* **89**, 338 (1964).
18. Recommended Methods for Trade Effluents, *ibid.* **83**, 230 (1958).
19. Wagner, F., *Z. Anal. Chem.* **162**, 106 (1958).
20. Royer, J. L., Twichell, J. E. and Muir, S. M., *Anal. Lett.* **6**, 619 (1973).
21. Piekacz, H. and Mazur, H., *Roczniki Państw. Zakł. Hig.* **12**, 481 (1961).
22. Mazur, H. and Piekacz, H., *ibid.* **12**, 523 (1961).
23. Bark, L. S. and Higson, H. G., *Talanta* **11**, 471, 621 (1964).
24. Montgomery, H. A., Gardiner, D. K. and Gregory, J. G., *Analyst* **94**, 284 (1969).
25. Casapieri, P., Scott, R. and Simpson, E. A., *Anal. Chim. Acta* **49**, 188 (1970).
26. Asmus, E. and Garschagen, H., *Z. Anal. Chem.* **138**, 414 (1953).
27. Murty, G. V. and Viswanathan, T. S., *Anal. Chim. Acta* **25**, 293 (1961).
28. Deibner, L. and Bardou, P., *Chim. Anal. (Paris)* **48**, 278 (1966).
29. Deibner, L. and Hérédia, N., *ibid.* **49**, 90 (1967).
30. Epstein, J., *Anal. Chem.* **19**, 272 (1947).
31. Marsden, K., *Analyst* **84**, 746 (1959).
32. Baar, S., *ibid.* **91**, 268 (1966).
33. Goulden, P. D., Afghan, B. K. and Brooksbank, P., *Anal. Chem.* **44**, 1845 (1972).
34. Pauer, K., *Chem. Anal. (Warsaw)* **18**, 1069 (1973).
35. Kruse, J. M. and Thibault, L. E., *Anal. Chem.* **45**, 2260 (1973).
36. Gilbert, B. L., Olson, B. L. and Reuter, W., *ibid.* **46**, 170 (1974).
37. Kratochvil, V., *Collection Czech. Chem. Commun.* **25**, 299 (1960).
38. Schilt, A. A., *Anal. Chem.* **30**, 1409 (1958).
39. Kodura, I. and Łada, Z., *Chem. Anal. (Warsaw)* **17**, 871 (1972).
40. Dixon, B. E., Hands, G. C. and Bartlett, A. F., *Analyst* **83**, 199 (1958).
41. Okutani, T. and Utsumi, S., *J. Chem. Soc. Japan, Pure Chem. Sect.* **87**, 444 (1966).
42. Tanaka, Y., *Japan Analyst* **21**, 767 (1972).
43. Humphrey, R. E. and Hinze, W., *Anal. Chem.* **43**, 1100 (1971).
44. Nomura, T., *Bull. Chem. Soc. Japan* **41**, 1619 (1968).
45. Nomura, T., *J. Chem. Soc. Japan, Pure Chem. Sect.* **88**, 635 (1967).
46. Miller, A. D. and Aranovich, M. I., *Zavodsk. Lab.* **26**, 426 (1960).
47. Wroński, M., *Chem. Anal. (Warsaw)* **5**, 457 (1960).
48. Yamasaki, K. and Ito, R., *J. Chem. Soc. Japan, Pure Chem. Sect.* **79**, 914 (1958); **80**, 271 (1959).
49. Komatsu, S., Nomura, T. and Mochizuki, F., *ibid.* **90**, 944 (1969).
50. Dagnall, R. M. and West, T. S., *Talanta* **11**, 1627 (1964).
51. Dagnall, R. M., El-Ghamry, M. T. and West, T. S., *ibid.* **15**, 107 (1968).
52. Lambert, J. L. and Manzo, D. J., *Anal. Chem.* **40**, 1354 (1968).
53. Baumeister, R. and Schievelbein, H., *Z. Anal. Chem.* **255**, 362 (1971).
54. Humphrey, R. E. and Hinze, W., *Talanta* **18**, 491 (1971).
55. Humphrey, R. E. and Alvarez, J. J., *Microchem. J.* **16**, 652 (1971).

56. Maute, R. L. and Owens, M. L., *Anal. Chem.* **26**, 1723 (1954).
57. Whiston, T. G. and Cherry, G. W., *Analyst* **87**, 819 (1962).
58. Koh, T. and Iwasaki, I., *Bull. Chem. Soc. Japan* **40**, 569 (1967).
59. Danchik, R. S. and Boltz, D. F., *Anal. Chem.* **40**, 2215 (1968).
60. Yamamoto, Y., Tarumoto, T. and Hanamoto, Y., *Bull. Chem. Soc. Japan* **42**, 268 (1969).
61. Guerrero, A. H. and Roig, A. M., *Anal. Chem.* **45**, 1943 (1973).
62. Einaga, H. and Ishii, H., *Talanta* **20**, 1017 (1973).
63. Wroński, M., *Chem. Anal. (Warsaw)* **14**, 1183 (1969).
64. Gregorowicz, Z. and Buhl, F., *Z. Anal. Chem.* **187**, 1 (1962).
65. Neas, R. E. and Guyon, J. C., *Anal. Chem.* **41**, 1470 (1969).
66. Martin, E. L. and McClelland, J., *ibid.* **23**, 1519 (1951).
67. Shaw, W. H. and Bordeaux, J. J., *ibid.* **27**, 136 (1955).
68. Marier, J. R. and Clark, D. S., *Analyst* **85**, 574 (1960).
69. Roberts, R. F. and Wilson, R. H., *ibid.* **93**, 237 (1968).
70. Galík, A. and Vopravilová, J., *Talanta* **21**, 307 (1974).
71. Asmus, E. and Seifert, B., *Z. Anal. Chem.* **239**, 99 (1968).
72. Buscaróns, F. and Artigas, J., *Anal. Chim. Acta* **19**, 434 (1958).
73. Agassant, P. and Andrieux, J. L., *Bull. Soc. Chim. France* **1950**, 253.
74. Codell, M., Norwitz, G. and Simmons, O. W., *Anal. Chim. Acta* **9**, 555 (1953).
75. Nazarchuk, T. N. and Pechentkovskaya, L. E., *Zavodsk. Lab.* **27**, 256 (1961).
76. Ducret, L. and Cornet, C., *Anal. Chim. Acta* **25**, 542 (1961).
77. Nelson, K. H., Grimes, M. D., Smith, D. E. and Heinrich, B. J., *Anal. Chem.* **29**, 180 (1957).
78. Ciuhandu, G., *Z. Anal. Chem.* **155**, 321 (1957); **161**, 345 (1958).
79. Ciuhandu, G. and Rusu, V., *ibid.* **222**, 393 (1966).
80. Ciuhandu, G. and Krall, G., *ibid.* **172**, 81 (1960).
81. Ciuhandu, G. and Bockel, B., *Chim. Anal. (Paris)* **52**, 525 (1970).
82. Ciuhandu, G. and Chicu, A., *Z. Anal. Chem.* **255**, 35 (1971).
83. Bock, R. and Bockholt, B., *ibid.* **260**, 274 (1972).
84. Shepherd, M., Schuhmann, S. and Kilday, M. V., *Anal. Chem.* **27**, 380 (1955).
85. Burianec, Z. and Burianová, J., *Collection Czech. Chem. Commun.* **28**, 2895 (1963).
86. Maxon, W. D. and Johnson, M. J., *Anal. Chem.* **24**, 1541 (1952).
87. Nazarenko, V. A., Biryuk, E. A. and Antonovich, V. P., *Zavodsk. Lab.* **33**, 22 (1967).
88. Hunt, E. C., McNally, W. A. and Smith, A. F., *Analyst* **98**, 585 (1973).

Chapter 15

CERIUM

Cerium (Ce, at.wt. 140·12) is the commonest of the rare earth elements (see p. 438). It occurs in its compounds in the +III and +IV oxidation states. Cerium(III) has the same chemical properties as all the other rare earth elements (REE). Cerium(IV) resembles thorium, zirconium, and uranium(IV) in many of its chemical properties but is also a powerful oxidant. White $Ce(OH)_3$ precipitates at pH 7·5, whereas yellow $Ce(OH)_4$ precipitates at pH ~ 1. Neither hydroxide is amphoteric. The stability of ceric compounds increases in alkaline medium. Cerium(III) can be oxidized in acidic medium with bismuthate, silver(II) oxide, persulphate (in the presence of Ag^+), or bromate (in $9M$ HNO_3).

15.1 Methods of Separation

15.1.1 Precipitation

Like other rare earth elements cerium(III) can be separated as the sparingly soluble oxalate [1–4] or fluoride [5]. Calcium, barium, or any of the other rare earth elements (e.g. lanthanum) can be used as the collector.

The precipitation of cerium(IV) hydroxide (pH ~ 1) enables cerium to be isolated from the remaining REE, the hydroxides of which start to precipitate between pH 6·3 and 7·8. Titanium, zirconium, or iron(III) may be used as the collector.

Cerium(IV) iodate has been precipitated from nitric acid medium to separate cerium from other REE [6]. Cerium(III) has been precipitated as the phosphate at pH 5·5 with beryllium as scavenger, and EDTA as masking agent for Fe, Cr, Ni, and V [7].

15.1.2 Extraction

Cerium(IV) has been extracted from nitric acid solutions with MIBK [8], TBP [2,9], or nitroethane [10]. Moore [11] has separated cerium(IV) from other REE by extraction from HNO_3 medium with Aliquat 336 in xylene. Suzuki and Oki [12,13] have separated cerium(IV) by extraction of its acetylacetonate into benzene.

Other methods of separating cerium as one of the rare earth elements are given on p. 438.

15.2 Methods of Determination

Spectrophotometric methods for the determination of cerium can be divided into three groups. (1) Methods for cerium(III) which are common for all the REE (discussed in Chapter 41, p. 440). (2) Methods based on coloured cerium(IV) complexes or the colour of cerium(IV) ions. (3) Indirect methods, in which the colour change is produced by cerium(IV) oxidizing various organic reagents. In methods from groups 2 and 3, the separation of cerium from the remaining rare earth elements is unnecessary.

Two methods from the second group, the formaldoxime method and the extractive colorimetric method using 8-hydroxyquinoline, are described in detail below. Both methods are less sensitive than those (e.g. with Arsenazo I or Arsenazo III) which do not differentiate between the lanthanides.

15.2.1 8-Hydroxyquinoline Method

8-Hydroxyquinoline (oxine) (formula, p. 56) forms a sparingly soluble chelate with cerium ions in ammoniacal media. The red-brown chelate can be extracted into chloroform and other organic solvents, thereby allowing the spectrophotometric determination of cerium [14–16].

First reports of the 8-hydroxyquinoline method stated mistakenly that the coloured extract contains cerium(III) oxinate. Alimarin *et al.* [15] have found that cerium in the red-brown oxinate is in the +IV oxidation state. In the presence of reducing agents a pale yellow cerium(III) oxinate is obtained. When reducing agents are absent, the yellow cerium(III) oxinate is rapidly oxidized to the more intensely coloured cerium(IV) oxinate by atmospheric oxygen.

The absorption maximum of a chloroform solution of the complex is at 495 nm ($\varepsilon = 6 \cdot 7 \times 10^3$; $a = 0 \cdot 048$). The chloroform extracts are stable with respect to time.

Other rare earth elements and thorium form, like cerium(III), pale yellow oxinates (these elements accompany cerium during its separation as the oxalate or fluoride). Since such oxinates absorb weakly at $\lambda = 495$ nm, spectral interference from REE and thorium in the determination of cerium is eliminated by measuring the cerium(IV) oxinate absorbance at ~ 530 nm (which, of course, results in lower sensitivity than measurement at 495 nm).

Chloroform extracts cerium(IV) oxinate almost completely from ammoniacal aqueous solution at pH 9·9–10·6. The absorption spectra of cerium(IV) oxinate in dichloroethane, carbon tetrachloride, and benzene differ slightly from that in chloroform. The chloroform solution exhibits the greatest ε value.

The presence of citrate in the aqueous solution prevents the quantitative extraction of cerium oxinate. The sample solution must not contain other Group I–IV metals, except REE and thorium. In the presence of

EDTA, titanium is not extracted from aqueous solution at pH 9·9–10·5 [15].

Cerium has been determined by the 8-hydroxyquinoline method in cast iron and steel [14,16].

Misumi and Nagano [17] have applied 2-methyl-8-hydroxyquinoline to the determination of cerium. The absorption maximum of the complex in CCl_4 solution is at 485 nm.

Reagents

1. 8-Hydroxyquinoline: 1% solution in ethanol.
2. Standard cerium solution: 1 mg/ml. Dissolve 3·100 g of $Ce(NO_3)_3 \cdot 6H_2O$ in water containing 2 ml of conc. HNO_3, and dilute the solution with water in a volumetric flask to 1 litre.

Procedure

Separation of Ce as the oxalate. To an acidic sample solution add 5–20 ml (depending on the concentration of other metals in the solution) of 8% oxalic acid solution, and 3–6 mg of lanthanum (in the form of a salt solution). Adjust the pH of the solution to 2–3, heat the solution to 70–80°C, and keep it at this temperature for 1 hour. After 2–3 hours filter off the precipitate, and wash it thoroughly with 1% oxalic acid solution and then water. Ignite the precipitate to the oxide, and dissolve it in a small amount of hot $4M$ HCl.

Determination of Ce. Dilute the acid solution thus obtained, containing not more than 0·5 mg of Ce, with water to 10–15 ml, and add 1 ml of the 8-hydroxyquinoline solution, 2 drops of 1% solution of phenolphthalein in ethanol, and ammonia solution (1+1) until a red colour appears. Add 1 ml more of the ammonia solution (1+1) (the pH of the solution should be within the range 9·9–10·6), and quantitatively transfer the solution to a separating funnel. Extract the cerium oxinate with two portions of chloroform, shaking with each portion for 5 minutes. Make up the combined extracts with chloroform to the mark in a 50-ml (or smaller depending on the amount of cerium) volumetric flask and measure the absorbance of the solution at 530 nm against the solvent.

15.2.2 Formaldoxime Method

Addition of alkali to a solution containing cerium ions and excess of formaldoxime (see p. 59) produces, under the influence of atmospheric oxygen, an orange colour, due principally to the complex $[Ce(CH_2NO)_6]^{2-}$, but also to small amounts of cerium(IV) hydroxo-formaldoxime complexes [18,19]. The optimum concentration of NaOH in the solution is 0·05–0·1M. The colour is stable over a period of 5–15 minutes after development. Afterwards, the absorbance of the solution decreases. The absorption maximum lies in the near ultraviolet at 340 nm ($\varepsilon = 4 \cdot 7 \times 10^3$). At 400 nm ε is $3 \cdot 2 \times 10^3$ ($a = 0 \cdot 023$).

Citrate, tartrate, oxalate, and sulphosalicylate do not interfere in the reaction of formaldoxime with cerium. On the other hand, the solution must not contain fluoride, phosphate, or EDTA which form more stable complexes with cerium than does formaldoxime. Owing to its reducing properties, ascorbic acid interferes in the reaction by preventing the oxidation of the colourless cerium(III) formaldoxime complex to the coloured cerium(IV) complex.

Cyanide masks Ni, Co, Cu, and Fe(II) present in the solution without affecting the reaction of cerium with formaldoxime. Other REE and thorium form colourless complexes with formaldoxime and do not interfere. These other elements in the sample solution can act as collectors during the separation of traces of cerium as the oxalate or fluoride. Before its determination by formaldoxime, cerium should be carefully separated from manganese.

Cerium has been determined by the formaldoxime method in apatite concentrates [18].

Reagents

1. Formaldoxime: $1M$ solution (p. 343).
2. Standard cerium solution: 1 mg/ml (p. 200).

Procedure

To the sample solution containing not more than 1·0 mg of Ce in a 50-ml volumetric flask, add 5 ml of formaldoxime solution, and immediately make the solution alkaline with $1M$ NaOH by adding 4 ml more than required to neutralize the solution. Dilute the solution to the mark with water, and mix well. After 10 minutes, measure the absorbance of the coloured solution at 400 nm, using water as the reference.

Notes. (1) If cerium is first separated as the oxalate, the precipitate should be carefully washed, then dissolved in hot dilute hydrochloric acid. After cooling, the colour reaction may be carried out. To obtain complete separation of cerium from manganese by the oxalate method, the oxalate should be precipitated twice.

(2) If cerium is separated as the fluoride, the precipitate should be washed and heated with H_2SO_4 (or $HClO_4$) to fumes to remove HF completely. After the cooled residue has been diluted, the reaction with formaldoxime may be carried out.

(3) Addition of KCN (10–20 mg) before formaldoxime is added prevents the reaction of the reagent with any traces of Ni, Co, Cu, and Fe present. When traces of Fe are present, it is advisable to add some NH_2OH in addition to KCN.

15.2.3 OTHER METHODS

Cerium(IV) ions in H_2SO_4 or HNO_3 medium are yellow. The molar absorptivity at $\lambda_{max} = 320$ nm is $5·6 \times 10^3$, whereas it is only one-sixth as great at ~ 400 nm [20–24]. This method is suitable for determining larger amounts of cerium. Persulphate in sulphuric acid (> $0·1N$) containing silver ions as catalyst is usually used to oxidize Ce(III). Interference from other species which also absorb in the near ultraviolet is minimized if the absorbance is measured *vs.* an aliquot of sample solution with the cerium(IV) reduced by oxalic acid [22] or azide [5] [this reagent does not reduce chromium(VI)].

The addition of potassium carbonate in excess and also of hydrogen peroxide to a cerium(III) solution produces a yellow colloidal solution ($\lambda_{max} = 304$ nm) [25,26]. A true solution of cerium(IV) peroxide complex is obtained in alkaline medium containing citrate [27–29] or EDTA [30–32]. Methods based on the colour of the cerium(IV) peroxide complex exhibit low sensitivity.

Cerium(IV) gives a coloured complex ($\lambda_{max} = 450$ nm) with TTA which is extractable into benzene [33] or xylene [34–35a]. Holland et al. [36] have used 3-thianapthenoyltrifluoroacetone (the molar absorptivity at 425 nm is $5 \cdot 51 \times 10^3$).

Furthermore, cerium(IV) has been determined by colour reactions with BPHA [37] and Tiron [38], and as the molybdoceric heteropoly acid [39] or molybdocerophosphoric acid [39a].

The popular method using ferroin [the red iron(II) 1,10-phenanthroline complex] is based on the oxidizing properties of Ce(IV). The cerium is determined from the bleaching of the ferroin [40–42]. In related methods, a known amount of iron(II) is added to the solution containing cerium(IV), which oxidizes a stoichiometric amount of the iron(II) to iron(III). Either the iron(III) is determined with thiocyanate, or the iron(II) is determined with 1,10-phenanthroline [43,44].

A number of other methods for determining cerium are based on colour effects resulting from the oxidation of the following compounds by cerium(IV): *o*-dianisidine [45,46], benzidine [47], *o*-dimethoxybenzene [48], *o*-tolidine [49–51], DAM and other antipyrine derivatives [52], Methylene Blue [2,53,54], sulphanilic acid [55,56], phenylanthranilic acid [57], Tetron (*N,N'*-tetramethyl-*o*-tolidine) ($\varepsilon = 2 \cdot 53 \times 10^4$ at $\lambda_{max} = 470$ nm) [58], Methyl Red [59], and *N,N'*-bis-(2-hydroxypropyl)-*o*-phenylenediamine [60].

In the indirect permanganate method [4], cerium is determined from the bleaching of the $KMnO_4$ solution used to oxidize cerium(III) to cerium(IV).

References

1. Rudenko, E. I. and Shvaiger, M. I., *Zavodsk. Lab.* **30**, 400 (1964).
2. Eremin, Yu. G., Raevskaya, V. V. and Romanov, P. N., *Zh. Analit. Khim.* **21**, 1303 (1966).
3. Bykhovskii, D. N. and Petrova, I. K., *Radiokhimiya* **10**, 520 (1968).
4. Lev, I. E. and Kovtun, M. S., *Zavodsk. Lab.* **28**, 273 (1962).
5. Gräbner, H. J., *Z. Anal. Chem.* **201**, 401 (1964).
6. Willard, H. H. and T'sai Yu, S., *Anal. Chem.* **25**, 1754 (1953).
7. Shakurov, V. G. and Kharlamov, I. P., *Zavodsk. Lab.* **36**, 925 (1970).
8. Glendenin, L. E., Flynn, K. F., Buchanan, R. F. and Steinberg, E. P., *Anal. Chem.* **27**, 59 (1955).
9. Nikolaev, A. V., Ryabinin, A. I. and Afanas'ev, Yu. A., *Dokl. Akad. Nauk SSSR* **150**, 820 (1963).
10. Marsh, S. F., Maeck, W. J., Booman, G. L. and Rein, J. E., *Anal. Chem.* **34**, 1406 (1962).

References

11. Moore, F. L., *ibid.* **41**, 1658 (1969).
12. Suzuki, N. and Oki, S., *Bull. Chem. Soc. Japan* **35**, 233, 237 (1962).
13. Oki, S., *ibid.* **38**, 522 (1965).
14. Westwood, W. and Mayer, A., *Analyst* **73**, 275 (1948).
15. Alimarin, I. P., Przheval'skii, E. S., Puzdrenkova, I. V. and Golovina, A. P., *Tr. Komis. po Analit. Khim. Akad. Nauk SSSR* **8**, 152 (1958).
16. Roberts, J. E. and Ryterband, M. J., *Anal. Chem.* **37**, 1585 (1965).
17. Misumi, S. and Nagano, N., *ibid.* **34**, 1723 (1962).
18. Marczenko, Z. and Minczewski, J., *Chem. Anal. (Warsaw)* **5**, 903 (1960).
19. Marczenko, Z., *Acta Chim. Acad. Sci. Hung.* **26**, 347 (1961); *Anal. Chim. Acta* **31**, 224 (1964).
20. Freedman, A. J. and Hume, D. N., *Anal. Chem.* **22**, 932 (1950).
21. Medalia, A. I. and Byrne, B. J., *ibid.* **23**, 453 (1951).
22. Huré, J. and St. James-Schonberg, R., *Anal. Chim. Acta* **9**, 415 (1953).
23. Gottschalk. G. and Bartsch, H., *Z. Anal. Chem.* **174**, 423 (1960).
24. Blatz, L. A., *Anal. Chem.* **33**, 249 (1961).
25. Telep, G. and Boltz, D. F., *ibid.* **25**, 971 (1953).
26. Conca, N. and Merritt, C., Jr., *ibid.* **28**, 1264 (1956).
27. Ryabchikov, D. I. and Strelkova, Z. G., *Zh. Analit. Khim.* **3**, 226 (1948).
28. Babko, A. K. and Volkova, A. I., *Ukr. Khim. Zh.* **20**, 211 (1954).
29. Federov, A. A. and Ozerskaya, F. A., *Zavodsk. Lab.* **27**, 139 (1961).
30. Malinek, M. and Klir, L., *Chem. Listy* **50**, 1317 (1956).
31. Babko, A. K. and Eremenko, O. M., *Zh. Analit. Khim.* **13**, 206 (1958).
32. Tikhonov, V. N., *ibid.* **22**, 886 (1967).
33. Khopkar, S. M. and De, A. K., *Anal. Chem.* **32**, 478 (1960).
34. Onishi, H. and Banks, C. V., *ibid.* **35**, 1887 (1963).
35. Onishi, H. and Toita, Y., *ibid.* **36**, 1867 (1964).
35a. Onishi, H. and Toita, Y., *Japan Analyst*, **21**, 756 (1972).
36. Holland, W. J., Veel, A. E. and Gerrard, J., *Mikrochim. Acta* **1970**, 297.
37. Murugaiyan, P. and Sankar Das, M., *Anal. Chim. Acta* **48**, 155 (1969).
38. Püschel, R., *Mikrochim. Acta* **1960**, 344.
39. Shakhova, Z. F. and Gavrilova, S. A., *Zh. Analit. Khim.* **13**, 211 (1958).
39a. Johnson, H. N., Kirkbright, G. F. and Whitehouse, R. J., *Anal. Chem.* **45**, 1603 (1973).
40. Korabel'nik, R. K., *Zh. Analit. Khim.* **11**, 419 (1956).
41. Culkin, F. and Riley, J. P., *Anal. Chim. Acta* **24**, 167 (1961).
42. Culkin, F. and Riley, J. P., *ibid.* **32**, 197 (1965).
43. Verbeek, F., *Bull. Soc. Chim. Belg.* **70**, 415 (1961).
44. Gordon, L. and Feibush, A. M., *Anal. Chem.* **27**, 1050 (1955).
45. Popa, G., Negoiu, D. and Baiulescu, G., *Z. Anal. Chem.* **167**, 329 (1959).
46. Thomann, H. J. and Junghans, U., *Neue Hütte* **7**, 421 (1962).
47. Murthy, T. K. and Rao, B. S. R., *J. Indian Chem. Soc.* **27**, 383 (1950).
48. Antoniades, H. N., *Chemist-Analyst* **44**, 34 (1955).
49. Iordanov, N. and Daiev, Kh., *Zh. Analit. Khim.* **15**, 443 (1960).
50. Lev, I. E., *Izv. Vyssh. Ucheb. Zaved. Khim. Khim. Tekhnol.* **8**, 698 (1965).
51. Blazejak-Ditges, D., *Z. Anal. Chem.* **251**, 11 (1970).
52. Kanaev, N. A., *Zh. Analit. Khim.* **18**, 575 (1963).
53. Gotó, H. and Kakita, Y., *J. Chem. Soc. Japan Pure Chem. Sect.* **79**, 1524 (1958).
54. Vernon, F., *Anal. Chim. Acta* **48**, 425 (1969).
55. Sarma, P. L. and Dieter, L. H., *Talanta* **13**, 347 (1966).
56. Pollock, E. N., *ibid.* **16**, 1323 (1969).
57. Bondareva, T. N., Shvarev, V. S. and Perkina, V. P., *Zavodsk. Lab.* **32**, 907 (1966).
58. Jordanov, N., Antonova, N. and Daiev, C., *Talanta* **13**, 1459 (1966).
59. Cherkesov, A. I. and Zhigalkina, T. S., *Zh. Analit. Khim.* **16**, 364 (1961).
60. Kasterka, B. and Ostrowski, S., *Chem. Anal. (Warsaw)* **11**, 1135 (1966).

Chapter 16

CHLORINE

Chlorine (Cl, at.wt. 35·45) is a gas (Cl_2) which has oxidizing properties. It occurs in several oxidation states: chloride $-I$, hypochlorite $+I$, chlorite $+III$, chlorate $+V$ and perchlorate $+VII$. Chloride exhibits reducing properties towards such powerful oxidants as Mn(VII) and Ce(IV). In its other oxidation states, chlorine has oxidizing properties. Hypochlorite and chlorite are rather unstable, and are subject to gradual disproportionation into chloride and chlorate. Of the chloro-anions, only chloride reveals strong complex-forming capacity.

16.1 Separation of Chloride and Chlorine

Small amounts of chloride can be separated from many other elements by precipitation as silver chloride from dilute nitric acid [1]. Bromide, iodide, and thiocyanate are also precipitated. Microgram amounts of chloride have been separated as AgCl by using $BaSO_4$ as the collector [2], or as $PbCl_2$ with lead phosphate [2a].

Chloride is often oxidized to chlorine and separated by distillation. The chlorine may be spectrophotometrically determined directly (e.g. by the Methyl Red method given below). Chloride is oxidized with potassium periodate or permanganate in a sulphuric acid medium, after separation of bromide and iodide by oxidation to bromine and iodine with potassium iodate in dilute nitric acid.

Chloride may also be separated by distilling volatile hydrogen chloride, which is collected in a solution containing excess of silver ions [3].

When stable complexes containing Cl^-, Br^-, or I^- ligands are heated with $(NH_4)_2HPO_4$, the appropriate ammonium halides are formed. They are separated from the melt by sublimation in a sealed glass-tube [4].

Successful separations of mixtures of halide ions have been achieved by using strongly basic anion-exchange resins [5,6] or chromatography [6a].

16.2 Determination of Chloride and Chlorine

Small amounts of chloride are usually determined by the simple and rapid silver chloride turbidimetric method. Spectrophotometric methods are more accurate, and often more sensitive. Several such methods are based on the oxidation of chloride to chlorine, which undergoes a subsequent redox reaction resulting in either the appearance or disappearance of colour (e.g. the Methyl Red method described below). In methods directly involving chloride ions, advantage is taken of the higher stability of the colourless mercury(II) chloride complex in comparison with that of coloured mercury(II) complexes with organic reagents.

16.2.1 TURBIDIMETRIC (AgCl) METHOD

The method involves comparing the turbidity formed when silver nitrate is added to an acidified (with HNO_3) sample solution containing chloride, with the turbidity formed in standard solutions. The method is simple but has rather low precision. Instead of visual comparison of the turbidities in colorimetric cylinders, the absorbance can be measured with a photometer [2, 7–9].

A time lapse of 15–20 minutes after the addition of silver nitrate is necessary to allow the AgCl precipitate to become stabilized. Variations of temperature within the range 20–30°C, and of acidity between 0·01 and 0·1M HNO_3 have no effect on the determination.

An aqueous acetone medium stabilizes the suspension [9]. However, commercial acetone may contain chloride, and should, therefore, be purified by distillation in the presence of sodium hydroxide.

Higher concentrations of electrolytes and organic compounds have adverse effects on the formation of the AgCl sol. Before the determination of traces of chloride in materials such as mineral salts, it is advisable to separate the chloride by coprecipitation of AgCl with $BaSO_4$ [2]. The silver chloride is readily leached from the precipitate with dilute ammonia solution.

Ions which form precipitates with silver nitrate in acid medium, i.e. bromide, iodide, thiocyanate, interfere in the turbidimetric determination of chloride.

Chloride has also been determined indirectly as the Ag_2S sol produced when the AgCl is separated and dissolved in ammonia solution, and Na_2S is added [10].

Reagents

1. Silver nitrate: 0·1M solution ($\sim$ 2%).
2. Standard chloride solution: 1 mg/ml. Dissolve in water 1·649 g of sodium chloride previously ignited at 400–500°C, and dilute the solution with water to 1 litre in a volumetric flask.
3. Potassium sulphate: 2% solution (chloride-free).
4. Barium nitrate: 2% solution (chloride-free).

Procedure

Separation of chloride. Acidify the sample solution (100–200 ml) with nitric acid, add 5 ml of the K_2SO_4 solution, and heat to $\sim$ 80°C. Add dropwise a mixture of 6 ml of the $Ba(NO_3)_2$ solution, and 2 ml of the $AgNO_3$ solution with stirring. Keep the solution at $\sim$ 60°C for 1 hour.

Then cool and filter, using paper washed free from chloride, and wash the precipitate three times with ~ $0.01M$ HNO_3 by decantation. Discard the filtrate, and add 10 ml of $2M$ NH_3 to a beaker containing the precipitate of $BaSO_4$ and AgCl. Thoroughly mix the liquid and the precipitate, re-filter (using the same filter paper), and wash the filter paper and the precipitate with dilute ammonia solution. The filtrate contains the separated chloride.

Determination of chloride. Place the sample solution (or the ammoniacal filtrate obtained as described above) containing not more than 40 μg of chloride in a Nessler cylinder, acidify with dilute nitric acid (to make the final solution $0.05M$ HNO_3), dilute with water to 40 ml, add 2 ml of $AgNO_3$ solution, and stir well. Prepare at the same time a series of standards covering the range 0–40 μg of Cl^- in exactly equivalent Nessler cylinders. Let the cylinders stand for 15 minutes in the dark, and then compare the turbidity obtained in the sample solution with the standards. The cylinders should be observed from above, against a black background, in a brightly and uniformly lit location.

16.2.2 METHYL RED METHOD

The chloride is oxidized to chlorine, which is steam-distilled and trapped in an acidified Methyl Red (**I**) solution. The oxidation of Methyl Red by

$$(CH_3)_2N-\langle\bigcirc\rangle-N=N-\langle\bigcirc\rangle\text{-COOH} \quad \textbf{(I)}$$

chlorine results in a partial bleaching of the red solution, thus providing the basis of an indirect spectrophotometric method for determining chloride [11].

One molecule of the dye reacts with two molecules of chlorine (Cl_2). The absorption spectra of the Methyl Red solution before and after reaction with a definite amount of chlorine are shown in Fig. 16.1. A measure of the sensitivity of the method is obtained from the molar absorptivity calculated from the difference in the absorbances ($A_1 - A_2$) of the coloured solutions ($\varepsilon = 1.17 \times 10^4$ at $\lambda_{max} = 515$ nm; $a = 0.33$).

For the oxidation of chloride, potassium periodate in a sulphuric acid solution gives more reproducible results than permanganate. The solution of KIO_4 in H_2SO_4 is first heated to distil off any chlorine formed from chloride impurities in the reagents. The optimum acid concentration for the Methyl Red colour reaction with chlorine is $1M$ H_2SO_4.

Bromide and iodide interfere in this method for determining chloride. The presence of 5 μg of Br^- enhances the results by 60–70% in determination of 5 μg of Cl^-. Positive errors due to iodide are smaller and less reproducible than those due to bromide.

The Methyl Red method has been employed for determining traces of chloride in chemical reagents (sulphates, phosphates, oxides, and hydroxides) [11]. This method cannot be used to determine traces of chloride in nitrates, since Methyl Red is oxidized and discoloured by nitric acid which distils into the receiver under the conditions of the method.

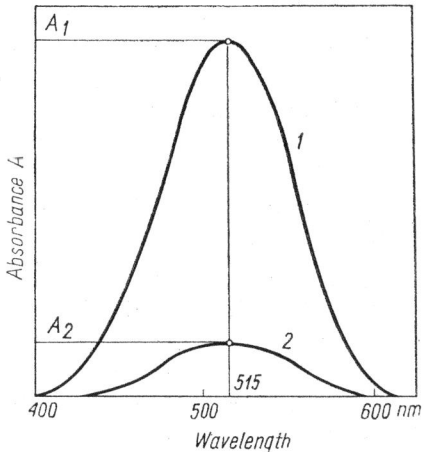

Fig. 16.1 Absorption spectra of Methyl Red in $1M$ H_2SO_4 before (1) and after (2) reaction with chlorine

In the determination of free chlorine in tap water, the sample is placed in a distillation flask (without the oxidizing agent), and the chlorine is steam-distilled and trapped in the Methyl Red solution. Experience has shown that distillation is, in this case, unnecessary, the water sample being added directly to a fixed amount of the Methyl Red solution. Owing to the high sensitivity of this method, the water samples may require suitable dilution with distilled water.

Reagents

1. Methyl Red: 0·0005% solution in $1M$ H_2SO_4. Dissolve 50·0 mg of the dye in ~ $1M$ H_2SO_4 and dilute the solution with the acid in a volumetric flask to 1 litre. Dilute 25 ml of this solution with $1M$ H_2SO_4 in a volumetric flask to 250 ml.
2. Standard chloride solution: 1 mg/ml (p. 205).
3. Periodate reagent. Add 200 ml of conc. H_2SO_4 to 300 ml of water, and mix. Place the cooled solution in a 750-ml distillation flask, and add 12 g of analytically pure KIO_4. When the salt is dissolved, add a few fragments of porous porcelain, connect the flask to the condenser, and distil off and discard the first 200 ml of water. To the cooled solution in the distillation flask, add 200 ml of water with stirring. Add a fragment of porcelain, and again distil 200 ml of water. Transfer the cooled

chlorine-free reagent solution from the distillation flask to a glass-stoppered bottle.

Procedure

Preliminary preparation. Place 50 ml of the periodate reagent in the distillation flask, add 20 ml of water, and stir well. Add a few fragments of porous porcelain and distil 20 ml of water into a receiver containing 10 ml of the Methyl Red solution. Transfer the solution from the receiver to a 50-ml volumetric flask, make up to the mark with water, and measure the absorbance of the solution against water. The absorbance measured should not be lower than the absorbance of the solution obtained by diluting 10 ml of the Methyl Red solution with water in a 50-ml volumetric flask. If the absorbance is lower, add 20 ml of water to the distillation flask, and repeat the procedure described above.

Determination of Cl. Introduce 20 ml of the sample solution containing not more than 50 µg of Cl (chloride) into the distillation flask containing the periodate reagent prepared as described above. Mix the solution thoroughly, add a fragment of porcelain, and distil 20 ml of water together with the separated chlorine into the receiver containing exactly 10 ml of the Methyl Red solution. Transfer the solution from the receiver to a 50-ml volumetric flask, and dilute to the mark with water. Measure the absorbance of the partly decolourized red solution at 515 nm against water.

Notes. (1) If a weighed portion of the sample is added to the distillation flask containing the periodate reagent, 20 ml of water are then added.

(2) Complete decolorization of the Methyl Red solution implies that too large a sample was taken.

(3) The weight of a sample containing traces of chloride is limited by the solubility of the sample in the periodate reagent. During distillation, the solution in the flask should remain clear.

(4) The same aliquot of periodate reagent in the distillation flask can be used for several successive determinations of chloride (e.g. when preparing a calibration curve).

16.2.3 Other Methods for Determining Chloride and Chlorine

Several spectrophotometric methods for determining small amounts of chlorine (and chloride after its separation by oxidation to chlorine and subsequent distillation) are based on the oxidation of organic reagents such as Methyl Red (discussed above in detail), *o*-tolidine [1,12–14], dimethylnaphthidine [15], Methyl Orange [16,16a], and Fast Green FCF [17].

In the sensitive benzidine–pyridine method [18,19], chlorine reacts with cyanide ions to form cyanogen chloride. The product of the reaction of CNCl with pyridine (i.e. glutaconaldehyde) is condensed with primary amines to form polymethine dyes. To improve the sensitivity of the method, Asmus and Garschagen [20] recommend condensing barbituric acid with the glutaconaldehyde. Chlorine has been determined by this method in water [21,22], and in selenium [23].

Kul'berg and Borzova [24] have determined chlorine on the basis of the indophenol reaction which takes place when sodium phenate and aniline react in the presence of chlorine.

Chlorine oxidizes iodide to iodine in dilute H_2SO_4. The iodine formed can be easily determined by the starch–iodine method (see p. 296).

When a solution containing chloride is passed through a column of granular silver iodate, the chloride displaces an equivalent amount of IO_3^-, which reacts with iodide to give iodine (3 molecules of I_2 per Cl^- ion), which can be determined by the starch–iodine method (see p. 296) [25].

The indirect thiocyanate method for determining chloride [26–29] is based on the displacement of SCN^- ions from the mercury(II) thiocyanate complex [a saturated $Hg(SCN)_2$ solution in water is $\sim 0.07\%$ w/v] by chloride ions to give a stable mercury chloride complex. After the addition of iron(III), the red $[FeSCN]^{2+}$ complex is formed, and the absorbance is measured at 480 nm. Chloride has been determined by this method in water [2a,30,30a], petroleum [31], perhydrol [32], polybutene polymers [33], biological materials [34], glycols [35], and silicate rocks [36]. Alternatively, the SCN^- ions liberated have been extracted into nitrobenzene as the ion-pair with tris(1,10-phenanthroline)iron(II) (ferroin) [37].

There are several modifications of the spectrophotometric method for determining chloride which exploits the fact that the mercury(II) chloride complex is more stable than the violet mercury(II) diphenylcarbazone complex [38–44]. The violet colour of the solution is a function of the chloride concentration. The reaction is carried out in dilute acid.

Chloride ions can be determined by their reaction with sparingly soluble mercuric chloranilate to liberate soluble reddish purple chloranilic acid [45].

Dithizone has also been applied in the indirect spectrophotometric determination of chloride. First, chloride is precipitated as AgCl, and the silver in the precipitate is then reduced to the element, separated from chloride, dissolved in nitric acid, and determined with dithizone (see p. 491) [46,47]. In another method [48], a yellow solution of silver dithizonate in chloroform is shaken in a separating funnel with the sample solution containing chloride, which displaces an equivalent amount of dithizone from the silver complex. The absorbance of the dithizone released is measured at 598 nm (the excess of AgHDz does not absorb at this wavelength).

A less sensitive method for determining chloride is based on the formation of the yellow chloro-complex of iron(III) [49]. It is temperature-sensitive and dependent on the acidity and iron(III) concentration.

Continuous spectrophotometric methods for determining small amounts of chlorine in gases include those utilizing the bleaching of Methyl Red [50], and the displacement of iodine from potassium iodide by chlorine to form a yellow solution (the iodine is then removed with activated charcoal) [51].

16.3 Determination of Other Chlorine Compounds

Perchlorate ions form extractable coloured ion-pairs with basic dyes such as Methylene Blue [52,53], Brilliant Green [54–56], Crystal Violet [57], and Neutral Red [58]. Chloroform, dichloroethane, benzene, chlorobenzene, toluene, xylene, and nitrobenzene have been used as extractants. Perchlorate impurities in potassium chlorate have been determined by these sensitive extractive spectrophotometric methods [54,56]. Several ions interfere, including nitrate.

Gregorowicz *et al.* [59] have proposed a method for the determination of perchlorate based on the extraction of the ClO_4^--ferroin ion-pair into n-amyl alcohol. In n-butyronitrile solution, the molar absorptivity of the ion-association complex $[Fe(phen)_3^{2+}][ClO_4^-]_2$ is $1·08 \times 10^4$ at 510 nm [60], and in 1,2-dichloroethane it is $1·24 \times 10^4$ at 512 nm [61]. Methods based on the solvent extraction of ClO_4^- as ion-association complexes with bipyridyl-iron(II) [62] and neocuproine–copper(I) [63,63a] have also been reported.

Fogg *et al.* [64] have determined perchlorate by extraction with the tetrabutylphosphonium ion into *o*-dichlorobenzene, followed by displacement of the ClO_4^- by the iron(III)–thiocyanate complex anion to form an intensely-coloured complex.

The oxidizing properties of chlorate are utilized in its spectrophotometric determination. Chlorate in ammonium perchlorate has been determined by its colour reactions with brucine [65], or after reduction to chlorine, with benzidine [66]. Chlorate in water has been determined with *o*-tolidine [67]. Chloride and nitrate do not interfere.

In the presence of perchlorate, chlorate may be selectively reduced to chloride, and determined as the AgCl sol [68].

In an indirect method [69], advantage is taken of the interference by chlorate in the formation of the rhenium complex with α-furildioxime.

Chlorine dioxide is determined either from its absorbance in carbon tetrachloride [70], or as the coloured product ($\lambda_{max} = 490$ nm) of its reaction with tyrosine. In the latter method, hypochlorite and chloramines do not interfere [71,72]. Masschelein [73] has determined chlorine dioxide spectrophotometrically with Acid Chrome Violet K.

Prince [74] has devised a scheme for analysing mixtures of chloride, hypochlorite, chlorite, chlorate, perchlorate, and chlorine dioxide by using spectrophotometric and other methods.

References

1. Scheubeck, E. and Ernst, O., *Z. Anal. Chem.* **254**, 185 (1971).
2. Chwastowska, J., Marczenko, Z. and Stolarczyk, U., *Chem. Anal. (Warsaw)* **8**, 517 (1963).
2a. Radabaugh, R. D. and Upperman, G. T., *Anal. Chim. Acta* **60**, 434 (1972).
3. Reichel, W. and Acs, L., *Anal. Chem.* **41**, 1886 (1969).
4. Preetz, W. and Homborg, H., *Z. Anal. Chem.* **251**, 98 (1970).

5. Holzapfel, H. and Gürtler, O., *J. Prakt. Chem.* **35**, 113 (1967).
6. Zalevskaya, T. L. and Starobinets, G. L., *Zh. Analit. Khim.* **24**, 721 (1969).
6a. Michal, J., *Inorganic Chromatographic Analysis*. Van Nostrand Reinhold, London (1974).
7. Challis, H. J. and Jones, J. T., *Analyst* **81**, 703 (1956).
8. Blanc, P., Bertrand, P. and Liandier, L., *Chim. Anal. (Paris)* **38**, 156 (1956).
9. Palei, P. N. and Udal'tsova, N. I., *Zavodsk. Lab.* **30**, 151 (1964).
10. Kuroda, P. K. and Sandell, E. B., *Anal. Chem.* **22**, 1144 (1950).
11. Marczenko, Z. and Chołuj-Lenarczyk, Ł., *Chem. Anal. (Warsaw)* **11**, 1221 (1966).
12. Kaszper, W. and Kęsy-Dąbrowska, I., *ibid.* **9**, 1063 (1964).
13. Johnson, J. D. and Overby, R., *Anal. Chem.* **41**, 1744 (1969).
14. Scheubeck, E. and Ernst, O., *Z. Anal. Chem.* **249**, 370 (1970).
15. Belcher, R., Nutten, A. J. and Stephen, W. I., *Anal. Chem.* **26**, 772 (1954).
16. Athavale, V. T., Krishnan, C. V. and Subramanian, A. R., *Analyst* **87**, 707 (1962).
16a. Laitinen, H. A. and Boyer, K. W., *Anal. Chem.* **44**, 920 (1972).
17. Gordon, H. T., *ibid.* **24**, 857 (1952).
18. Nusbaum, I. and Skupeko, P., *ibid.* **23**, 1881 (1951).
19. Asmus, E. and Garschagen, H., *Z. Anal. Chem.* **136**, 269 (1952).
20. Asmus, E. and Garschagen, H., *ibid.* **138**, 404 (1953).
21. Webber, H. M. and Wheeler, E. A., *Analyst* **90**, 372 (1965).
22. Nicolson, N. J., *ibid.* **90**, 187 (1965).
23. Baresel, D. and Jaetsch, K., *Z. Anal. Chem.* **249**, 234 (1970).
24. Kul'berg, L. M. and Borzova, L. D., *Ukr. Khim. Zh.* **22**, 100 (1956); *Zh. Analit. Khim.* **11**, 470 (1956).
25. Lambert, J. L. and Yasuda, S. K., *Anal. Chem.* **27**, 444 (1955).
26. Iwasaki, I., Utsumi, S. and Ozawa, T., *Bull. Chem. Soc. Japan* **25**, 226 (1952).
26a. Iwasaki, I., Utsumi, S., Hagino, K. and Ozawa, T., *ibid.* **29**, 860 (1956).
27. Elsheimer, H. N. and Kochen, R. L., *Anal. Chem.* **38**, 145 (1966).
28. Elsheimer, H. N., Johnston, A. L. and Kochen, R. L., *ibid.* **38**, 1684 (1966).
29. Florence, T. M. and Farrar, Y. J., *Anal. Chim. Acta* **54**, 373 (1971).
30. Zall, D. M., Fisher, D. and Garner, M. Q., *Anal. Chem.* **28**, 1665 (1956).
30a. Rodabaugh, R. D. and Upperman, G. T., *Anal. Chim. Acta* **60**, 434 (1972).
31. Bergmann, J. G. and Sanik, J., Jr., *Anal. Chem.* **29**, 241 (1957).
32. Geld, I. and Sternman, I., *ibid.* **31**, 1662 (1959).
33. Rowe, R. D., *ibid.* **37**, 368 (1965).
34. Kulhánek, V. and Fišer, C., *Collection Czech. Chem. Commun.* **31**, 1890 (1966).
35. Marquardt, R. P., *Anal. Chem.* **43**, 277 (1971).
36. Huang, W. H. and Johns, W. D., *Anal. Chim. Acta* **37**, 508 (1967).
37. Yamamoto, Y., Kumamaru, T., Tatehata, A. and Yamada, N., *ibid.* **50**, 433 (1970).
38. Clarke, F. E., *Anal. Chem.* **22**, 553 (1950).
39. Gerlach, J. L. and Frazier, R. G., *ibid.* **30**, 1142 (1958).
40. Kemula, W., Hulanicki, A. and Janowski, A., *Chem. Anal. (Warsaw)* **3**, 581 (1958); *Talanta* **7**, 65 (1960).
41. Tomonari, A., *J. Chem. Soc. Japan, Pure Chem. Sect.* **82**, 864 (1961).
42. Utsumi, S. and Okutani, T., *ibid.* **85**, 543 (1964).
43. Novák, J. and Hauptman, Z., *Z. Anal. Chem.* **217**, 340 (1966).
44. Miller, G., Grabiec-Koska, W., Paterok, N. and Jaki, J., *Chem. Anal. (Warsaw)* **15**, 833 (1970).
45. Barney II, J. E. and Bartolacini, R. J., *Anal. Chem.* **29**, 1187 (1957); **30**, 202 (1958).
46. Iwantscheff, G., *Angew. Chem.* **62**, 361 (1950).
47. Suter, H., and Hadorn, H., *Z. Anal. Chem.* **160**, 335 (1958).
48. Kirsten, W. J., *Mikrochim. Acta* **1955**, 1086.
49. West, P. W. and Coll, H., *Anal. Chem.* **28**, 1834 (1956).
50. Enders, D., *Chem. Techn.* **8**, 67 (1956).
51. Waszak, S., *Chem. Anal. (Warsaw)* **4**, 351 (1959).

52. Nabar, G. M. and Ramachandran, C. R., *Anal. Chem.* **31**, 263 (1959).
53. Iwasaki, I., Utsumi, S. and Kang, C., *Bull. Chem. Soc. Japan* **36**, 325 (1963).
54. Golosnitskaya, V. A. and Petrashen', V. I., *Zh. Analit. Khim.* **17**, 878 (1962).
55. Reusmann, G., *Z. Anal. Chem.* **226**, 346 (1967).
56. Fogg, A. G., Burgess, C. and Burns, D. T., *Analyst* **96**, 854 (1971).
57. Uchikawa, S., *Bull. Chem. Soc. Japan* **40**, 798 (1967).
58. Tsubouchi, M., *Anal. Chim. Acta* **54**, 143 (1971).
59. Gregorowicz, Z., Buhl, F. and Klima, Z., *Mikrochim. Acta* **1963**, 116.
60. Fritz, J. S., Abbink, J. E. and Campbell, P. A., *Anal. Chem.* **36**, 2123 (1964).
61. Knížek, M. and Musilová, M., *Collection Czech. Chem. Commun.* **33**, 757 (1968).
62. Yamamoto, Y. and Kotsuji, K., *Bull. Chem. Soc. Japan* **37**, 785 (1964).
63. Collinson, W. J. and Boltz, D. F., *Anal. Chem.* **40**, 1896 (1968).
63a. Weiss, J. A. and Staubury, J. B., *ibid.* **44**, 619 (1972).
64. Fogg, A. G., Burns, D. T. and Yeowart, E. H., *Mikrochim. Acta* **1970**, 974.
65. Eger, C., *Anal. Chem.* **27**, 1199 (1955).
66. Burns, E. A., *ibid.* **32**, 1800 (1960).
67. Urone, P. and Bonde, E., *ibid.* **32**, 1666 (1960).
68. Forster, C. F., *Analyst* **79**, 90 (1954).
69. Trautwein, N. L. and Guyon, J. C., *Anal. Chim. Acta* **41**, 275 (1968).
70. Sherman, M. I. and Strickland, J. D., *Anal. Chem.* **27**, 1778 (1955).
71. Hodgden, H. W. and Ingols, R. S., *ibid.* **26**, 1224 (1954).
72. Tumanova, T. A., Pakhomova, L. N. and Maiorova, L. P., *Zavodsk. Lab.* **36**, 1036 (1970).
73. Masschelein, W., *Anal. Chem.* **38**, 1839 (1966).
74. Prince, L. A., *ibid.* **36**, 613 (1964).

Chapter 17
CHROMIUM

Chromium (Cr, at.wt. 52·00) forms a series of colour species corresponding to the +II, +III and +VI oxidation states. In spectrophotometric analysis, chromic (Cr^{3+}), chromate (CrO_4^{2-}), and dichromate ($Cr_2O_7^{2-}$) ions are important. Chromic hydroxide, which precipitates at pH ~ 5, is amphoteric, dissolving in alkali (pH ~ 13). Chromium(III) forms stable oxalate, tartrate, and EDTA complexes. Chromium(II) is a strong reducing agent, whereas chromium(VI) has oxidizing properties.

17.1 Methods of Separation

17.1.1 EXTRACTION

Comprehensive studies of the behaviour of traces of chromium (microgram and even nanogram quantities), and of methods of separation, have been carried out by Beyermann [1].

A conventional, selective, and fairly simple method for separating chromium is the extraction of chromium(VI) with MIBK from 1–3M hydrochloric acid [1–5]. The method enables chromium to be completely separated from most elements (e.g. V, Fe, Mn, and Ni): only In, Tl, Sb, Hg, W, and Re are co-extracted in significant quantities. The recommended procedure involves extracting chromium from 2M HCl with two portions of MIBK, washing the combined extracts once with 2M HCl, and stripping the chromium with two portions of water. According to Specker and Arend [3], chromium is extracted as $HCrO_3Cl$ solvated with two molecules of MIBK. Shinde and Khopkar [6] extracted chromium(VI) from a 1M HCl and 2·5M KCl medium with mesityl oxide, and stripped the chromium with dilute ammonia.

Inert solvents (e.g. chloroform, benzene, and dichloroethane) extract association complexes of chromium(VI) with high molecular-weight amines from acid media (HCl, H_2SO_4, $HClO_4$, HNO_3). Amines used include trioctylamine [7–9], tribenzylamine [10–12], and tetrabutylammonium hydroxide [13]. Mann and White [14] have extracted chromium(VI) with a benzene solution of TOPO.

In a slightly acidic medium (optimum pH ~ 1·7) and at a temperature not higher than 10°C, chromium(VI) reacts with hydrogen peroxide (the final H_2O_2 concentration is ~ 0·02M) to form blue perchromic acid which can be extracted into ethyl acetate, isoamyl alcohol, ether, or similar solvents. This method permits the separation of chromium from V, Fe, and most other metals [15–17]. Sastri and Sundar have extracted perchromic acid both with a mixture of TBP and benzene (1+3) [18], and with tertiary and quaternary amines [19]. Perchromic acid, generally formulated as CrO_5, readily forms anionic species of the type $[CrO(O_2)_2X]^-$ where X^- is (for example) chloride or hydroxide [19,20]. In acid solution, the neutral extractable species H_2CrO_6 is formed by protonation of the anionic hydroxide complex.

Since the rate of formation of the complex of acetylacetone with chromium(III) is much slower than with other metals (Fe, Al, V, Mo, Ti), extraction with a mixture of acetylacetone and chloroform (1+1) separates chromium(III) from these metals [1,21]. If, however, a chromium(III) solution is boiled with acetylacetone, chromium acetylacetonate is formed, and may be extracted into chloroform [22]. The chromium(III)–DCTA complex can be extracted into chloroform with Aliquat-336 [22a,22b].

17.1.2 Precipitation

Traces of chromium(III) are precipitated as the hydroxide with NaOH (in moderate excess), or with ammonia. Fe(III), Al, La, or Zn are used as collectors [23–28].

When a sample is fused with Na_2O_2 or Na_2CO_3 (with or without KNO_3), chromium(III) is oxidized to chromate, which is leached from the melt with water, whereas Fe, Mn (reduced with ethanol from MnO_4^{2-} to hydrous MnO_2), Cu, Ni, Co, Ti, and most other metals remain in the solid phase. Losses of chromium due to retention in the precipitate are insignificant. Such elements as V, Mo, As, and Al pass into the alkaline solution together with Cr(VI) when the melt is leached with water. Alternatively, chromium(III) can be oxidized in a hot alkaline solution (NaOH) to soluble CrO_4^{2-} with hydrogen peroxide or bromine.

Traces of chromium(VI) can be separated by coprecipitation with barium or lead sulphate.

17.1.3 Other Methods

Since chromium(III) is not retained on anion-exchangers when 0·02–12M HCl is used as eluent, it can be separated from a number of metals [1].

Strongly basic anion-exchangers retain V, Cr, and Mo from acetate solution at pH 2·5–3. Vanadium, chromium, and molybdenum are washed out with 0·6M NaOH, 8M HCl, and 1M HCl, respectively [29]. Mulokozi [30] has separated chromium(VI) from aluminium by sorption on a strongly basic anion-exchanger and sequential elution of aluminium with NaOH solution, and chromium with carbonate solution. He has also separated the chromate from the peroxide used for the oxidation [30].

Water samples containing chromate are acidified to pH 5 and passed through an anion-exchange resin bed (Cl^-) in ascending flow. The chromate is eluted with small volumes of an acidic reductant solution [30a].

Chromium traces can be distilled as CrO_2Cl_2 from a perchloric acid medium at 200–210°C, through which CO_2 and HCl are bubbled [1].

17.2 Methods of Determination

Chromium is usually determined by the highly sensitive and selective diphenylcarbazide method. This method, which is more than one hundred times as sensitive as those based on the colours of the chromate or dichromate ions, is particularly useful for determining traces of chromium. Larger amounts of chromium are determined either by the chromate method, or by the method based on the coloured chromium(III)–EDTA complex.

17.2.1 Diphenylcarbazide Method

Diphenylcarbazide (*sym*-diphenylcarbazide, diphenylcarbohydrazide) reacts in acid medium with chromium(VI) ions to give a violet solution which is the basis of this sensitive method.

Many investigators have studied the reaction [31–42], offering rather divergent explanations of its mechanism. Pflaum and Howick [32], among others [35,38,40], have shown that the cationic chromium(III)-diphenylcarbazone complex is formed by oxidation of diphenylcarbazide with chromium(VI). Direct mixing of solutions of chromium(III) and diphenylcarbazone fails to yield a violet colour. In all probability, the reaction involves unhydrated chromium(III) ions formed during the oxidation of diphenylcarbazide to diphenylcarbazone.

$$O=C\begin{matrix}NH-NH-C_6H_5\\NH-NH-C_6H_5\end{matrix} \xrightarrow{ox.} O=C\begin{matrix}NH-NH-C_6H_5\\N=N-C_6H_5\end{matrix}$$

This explanation is, however, incomplete since, when the coloured reaction product is extracted into isoamyl alcohol or chloroform in the presence of perchlorate, the remaining colourless aqueous phase contains half of the chromium [33, 36].

When studying the reactions of diphenylcarbazide and diphenylcarbazone with various metal cations, Balt and van Dalen [43] found that diphenylcarbazide only forms metal chelates after its oxidation to diphenylcarbazone.

According to Allen [44], the molar absorptivity of the coloured product of the chromium(VI) reaction with diphenylcarbazide is $4 \cdot 17 \times 10^4$ (specific absorptivity 0·80) at $\lambda_{max} = 546$ nm. The colour intensity obtained in the reaction is affected by the quality of the diphenylcarbazide reagent used [45,46].

The oxidation of Cr^{3+} to $Cr_2O_7^{2-}$ is normally carried out in acid medium with $KMnO_4$ or with $(NH_4)_2S_2O_8$ in the presence of silver ions

[47,48]. The excess of permanganate is either decomposed by azide or precipitated as hydrous MnO_2 in the presence of Mn^{2+} ions. The excess of persulphate is either decomposed by boiling the solution or reduced by azide. Other oxidizing agents used to oxidize chromium are perchloric acid [49], bromine [24], silver(II) oxide [50], sodium bismuthate, and sodium and hydrogen peroxides [51]. Fusing solid samples with sodium carbonate or sodium peroxide converts chromium into Cr(VI). Oelschläger [52] has emphasized that the quantitative oxidation of traces of chromium(III) is rather difficult.

Since the absorbance of the solution varies with acidity, the pH should be kept constant at the optimum value of pH $\sim$ 1 (0·05–0·1M H_2SO_4). The presence of hydrochloric acid should be avoided.

The diphenylcarbazide method is almost specific for chromium(VI). Interferences in the determination of chromium are produced only by Fe, V, Mo, Cu, and Hg present in much higher concentrations than the chromium. Larger amounts of iron(III) are masked by phosphoric acid or EDTA. Iron(III) can also be separated as the hydroxide after chromium has been oxidized to Cr(VI), or by extraction.

Vanadium is separated from chromium(VI) by extraction as the oxinate at pH $\sim$ 4 [53]. Molybdenum is masked with oxalic acid, while mercury(II) is converted into the chloride complex.

When small amounts of chromium(III) are to be determined in the presence of chromium(VI), the chromic ions are first separated by precipitation of the hydroxide, with aluminium or iron as the collector and ammonia as the precipitant. The precipitate is then dissolved, and the Cr(III) is oxidized to Cr(VI) and determined with diphenylcarbazide [25].

The preliminary extraction of Cr(VI) before its determination with diphenylcarbazide has been discussed at length elsewhere [2,6,15,54].

The cationic coloured product of the reaction of chromium(VI) with diphenylcarbazide can be extracted into isoamyl alcohol from aqueous solutions containing large amounts of chloride [55], sulphate [56], or naphthalene-2-sulphonate [57].

Chromium has been determined by the diphenylcarbazide method in iron and steel [49,56], ferro-alloys [58], nickel [53,55,59], vanadium [53], tantalum [22], antimony [60], tin [61], aluminium [62], beryllium [63], titanium alloys [64], rhenium [65], uranyl nitrate [66], cement [67], glass [68], ilmenite [69], sapphire and ruby [70,71], minerals [51], sewage and industrial wastes [72], wood [73], collagen sutures [74], biological materials [47,75,76], water [24,28,47,53a], and air [47,77].

Reagents

1. Diphenylcarbazide: 0·25% solution in acetone. Dissolve 0·25 g of the reagent in 100 ml of acetone containing 1 ml of H_2SO_4 (1+9). Keep the solution in an amber-glass bottle. The solution should not be stored for too long.

2. Standard chromium(VI) solution: 1 mg/ml. Dissolve in water 2·830 g of $K_2Cr_2O_7$ previously dried at 140°C, and dilute the solution with water in a volumetric flask to 1 litre.
3. Potassium permanganate: $\sim 0.1N$ ($0.02M$) solution.
4. Sodium azide NaN_3: 2·5% solution.

Procedure

Place in a beaker a solution containing not more than 40 μg of Cr. If the solution contains chloride add a little sulphuric acid, and evaporate to fumes. Cool the residue, add ~ 25 ml of water and 3–5 drops of $KMnO_4$ solution, cover the beaker with a watch glass, and heat without boiling for ~ 15 minutes. The acidity of the solution should at this point be ~ 0.05–$0.1M$ H_2SO_4. If in the course of heating the pink colour disappears, add more $KMnO_4$ solution dropwise. Reduce (decolorize the pink solution) the excess of oxidant (a hydrous MnO_2 suspension is liable to form) by adding sodium azide solution dropwise, waiting a few seconds after the addition of each drop. Avoid introducing a large excess of the reducing agent.

Transfer the cooled solution to a 50-ml volumetric flask, add 1 ml of the diphenylcarbazide solution, dilute the solution with water to the mark, and mix thoroughly. Measure the absorbance at 546 nm using water as the reference.

17.2.2 Chromate Method

The spectrophotometric method for the determination of chromium, based on the colour of $Cr_2O_7^{2-}$ or CrO_4^{2-} ions, is a good example of a precise but rather insensitive method. It may be based either on the yellow colour of chromate (CrO_4^{2-}) ions present in alkaline solution, or on the orange colour of dichromate ($Cr_2O_7^{2-}$) ions, formed from CrO_4^{2-} ions by acidification of the solution [78].

Figure 17.1 shows the absorption spectra of dichromate (1) and chromate (2). The absorption peaks lie in the near ultraviolet at 350 nm ($\varepsilon = 7.5 \times 10^2$) and at 373 nm ($\varepsilon = 1.4 \times 10^3$), respectively. The sensitivity of the method decreases considerably as the wavelength increases, e.g. for dichromate at 400 nm, ε is only 1.6×10^2 ($a = 0.003$). When the absorbance is measured with a spectrophotometer, greater sensitivity is achieved by using alkaline solutions (CrO_4^{2-}). When working with a photoelectric colorimeter equipped with filters, it is best to measure the absorbance of acid solutions, thus gaining closer conformity to Beer's law.

The concentration of the acid in dichromate solutions affects the colour. An approximately $1M$ H_2SO_4 medium is most suitable.

If chromium is present as Cr(III), it must first be oxidized to Cr(VI). In acid media (H_2SO_4, $HClO_4$), chromium can be oxidized with permanganate, persulphate in the presence of Ag^+ ions as catalyst, bismuthate, or periodate. Perchloric acid oxidizes chromium(III) when heated to boiling ($\sim 200°C$).

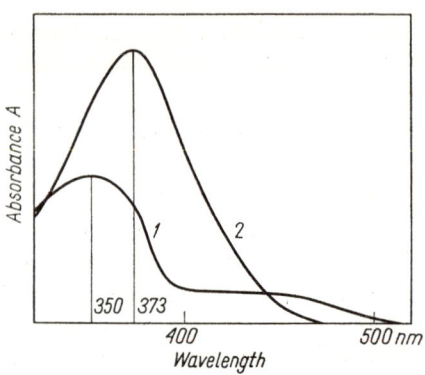

Fig. 17.1 Absorption spectra of dichromate (1) (in $1M$ H_2SO_4) and of chromate (2) (in ammoniacal medium)

In alkaline media, chromium(III) can be oxidized with bromine, H_2O_2, or Na_2O_2. Fusion with an alkaline flux (Na_2CO_3, Na_2O_2) also converts chromium into Cr(VI). The alkaline medium enables chromium(VI) to be easily separated from the majority of metals which form coloured solutions and give sparingly soluble hydroxides [Fe(III), U(VI), Ce(IV), Cu, Co, Ni]. Any coloured MnO_4^- ions formed during the oxidation are reduced with sodium azide or oxalic acid.

In the presence of phosphoric acid [79,80] which may be added to mask Fe(III), the colour of the solution is slightly changed owing to the formation of the mixed chromate–phosphate ions, $HCrPO_7^{2-}$ and $H_2CrPO_7^-$.

It is possible to determine $Cr_2O_7^{2-}$ ions in the presence of MnO_4^- ions provided the absorbance is measured at two appropriate wavelengths [81,82].

The chromate method has been used for determining chromium in, among other things, steels [82,83], aluminium and bauxites [81], and the raw material used in match production [84].

Savichev *et al.* [85] have determined chromium(VI) as its association complex with the basic dye, Methyl Violet, after extraction with benzene. Ziegler and Pohl [86] have extracted dichromate ions into methylene chloride as the ion-association complex with triphenylselenonium ion, $[(C_6H_5)_3Se]_2[Cr_2O_7]$. Other extractants used were listed on p. 213 [7–14].

Reagents

1. Potassium permanganate: $\sim 0.1N$ ($0.02M$) solution.
2. Sodium azide NaN_3: 2.5% solution.
3. Standard chromium(VI) solution: 1 mg/ml (p. 217).

Procedure

Oxidize the chromium in a solution containing not more than 5 mg of Cr as described in the diphenylcarbazide method (p. 217). More $KMnO_4$ will be needed in this case if the Cr(III) content is greater.

Neutralize the chromium(VI) solution and then add 10 ml of H_2SO_4 (1+3). Make the solution up to the mark with water in a 50-ml volumetric flask, and measure the absorbance at 400 nm against a water blank.

Notes. (1) Chromium(VI) standard solution is used to construct the calibration curve. The procedure is thereby simplified, the oxidation of chromium is unnecessary.

(2) For measurement of chromate an alkaline medium is required; $0.1M$ sodium or potassium carbonate or a 2% sodium metaborate solution is suitable.

17.2.3 EDTA Method

Ethylenediaminetetra-acetic acid (EDTA) forms coloured complexes with cations which have chromophoric properties (e.g. Fe, Cr, Cu, Co, and Ni).

$$\begin{array}{c} HOOC-CH_2 \\ \diagdown \\ HOOC-CH_2 \end{array} N-CH_2-CH_2-N \begin{array}{c} CH_2-COOH \\ \diagup \\ CH_2-COOH \end{array}$$

These complexes, which are not very intensely coloured, form the basis of several less sensitive spectrophotometric methods, such as that for chromium(III) [87–90].

With chromium(III), EDTA forms a violet complex in slightly acidic medium. The complex is formed slowly in the cold, but more rapidly if the solution is heated [91]. The sensitivity of the method is not high. The molar absorptivity is 1.4×10^2 at 540 nm ($a \sim 0.003$). The colour intensity diminishes as the pH is reduced. In a hot solution, EDTA reduces chromium(VI) to Cr(III). This reaction is catalysed by traces of manganese(II).

Coloured ions and those giving coloured complexes with EDTA interfere in the spectrophotometric determination of chromium as its EDTA complex. If metal ions which form colourless complexes with EDTA are present in the solution, more EDTA must be added. Oxalate and citrate interfere in the colour reaction; tartrate does not.

The chromium(III)–EDTA complex has been extracted by solutions of Aliquat-336 in 1,2-dichloroethane [92].

The EDTA method has been applied in the spectrophotometric determination of chromium in bronze [93], steel and aluminium alloys [94], iron ores [95], and chromium ores and ceramic wares [96,97].

Similarly to EDTA (Complexone III), 1,2-diaminocyclohexanetetra-acetic acid (DCTA, Complexone IV) has also been used as a spectrophotometric reagent for chromium [98–99a]. The applicability of other complexones has also been investigated [100].

Reagents

1. EDTA: 2.5% solution.
2. Standard chromium(III) solution: 1 mg/ml. Dissolve 9.197 g of alum $CrNH_4(SO_4)_2 \cdot 12H_2O$ in water containing 2 ml of conc. H_2SO_4, and dilute the solution with water to 1 litre. Working solutions are obtained

by suitable dilution of the standard solution with dilute H_2SO_4 (e.g. $0.01M$).

Procedure

To the solution (20–30 ml) containing not more than 5 mg of Cr, add a little $NH_2OH.HCl$ and 5 ml of the EDTA solution, adjust the solution with ammonia to pH 4–5, and heat to boiling. Continue gentle boiling for 2 minutes, then cool and dilute the solution with water in a 50-ml volumetric flask. Measure the absorbance at 540 nm, using water as the reference.

17.2.4 OTHER METHODS

There are many spectrophotometric methods for the determination of chromium, based on triphenylmethane dyes such as Pyrocatechol Violet [101,102], Chrome Azurol S [103], Xylenol Orange [104,105], and Eriochrome Cyanine R [106]. In their review of reagents for chromium Tataev and Abdulaev [107] additionally mention Methylthymol Blue, Glycinethymol Blue and Glycinecresol Red. The most sensitive method uses Eriochrome Cyanine R (molar absorptivity 2.6×10^4 at 540 nm).

The following other organic reagents have been recommended for determining chromium: PAR [108,109], oxine [110,111], 8-hydroxyquinaldine [110,112], Alizarin Red S [113,114], TTA [115], 3-thianaphthenoyltrifluoroacetone [116], pyridine-2,6-dicarboxylic acid [117,118], o-aminophenyldithiocarbamic acid [119], and thioglycollic acid [120].

The colorimetric methods using o-dianisidine [23,121,122], and 3,3'-diaminobenzidine [123] are based on the oxidation of these reagents by chromium(VI).

Chromium has also been determined as blue perchromic acid [16], and as the green chromium pyrophosphate complex [124].

References

1. Beyermann, K., *Z. Anal. Chem.* **190**, 4; **191**, 346 (1962).
2. Blundy, P. D., *Analyst* **83**, 555 (1958).
3. Specker, H. and Arend, A., *Naturwissenschaft.* **48**, 524 (1961).
4. Katz, S. A., McNabb, W. M. and Hazel, J. F., *Anal. Chim. Acta* **25**, 193 (1961); **27**, 405 (1962).
5. Dinstl, G. and Hecht, F., *Mikrochim. Acta* **1962**, 321.
6. Shinde, V. M. and Khopkar, S. M., *Z. Anal. Chem.* **249**, 239 (1970).
7. Deptuła, C., *Roczniki Chem.* **41**, 3 (1967).
8. Deptuła, C. and Moszyńska, K., *Chem. Anal. (Warsaw)* **13**, 211 (1968).
9. Adam, J., *Talanta* **18**, 91 (1971).
10. Fasolo, G. B., Malvano, R. and Massaglia, A., *Anal. Chim. Acta* **29**, 569 (1963).
11. Shevchuk, I. A. and Simonova, T. N., *Ukr. Khim. Zh.* **30**, 983 (1964).
12. Tserkovnitskaya, I. A., Il'inskaya, G. I. and Belyaev, V. P., *Zh. Analit. Khim.* **24**, 1357 (1969).
13. Maeck, W. J., Kussy, M. E. and Rein, J. E., *Anal. Chem.* **34**, 1602 (1962).
14. Mann, C. K. and White, J. C., *ibid.* **30**, 989 (1958).

References

15. Brookshier, R. K. and Freund, H., *ibid.* **23**, 1110 (1951).
16. Glasner, A. and Steinberg, M., *ibid.* **27**, 2008 (1955).
17. Tuck, D., *Anal. Chim. Acta* **27**, 296 (1962).
18. Sastri, M. N. and Sundar, D. S., *Z. Anal. Chem.* **195**, 343 (1963).
19. Sastri, M. N. and Sundar, D. S., *Anal. Chim. Acta* **33**, 340 (1965).
20. Tuck, D. G. and Faithful, B. D., *J. Chem. Soc.* **1965**, 5753.
21. McKaveney, J. P. and Freiser, H., *Anal. Chem.* **29**, 290 (1957); **30**, 1965 (1958).
22. Hofer, A. and Heidinger, R., *Z. Anal. Chem.* **233**, 415 (1968).
22a. Kinhikar, G. M. and Dara, S. S., *Talanta* **21**, 1208 (1974).
22b. Adam, J. and Přibil, R., *ibid.* **21**, 1205 (1974).
23. Sunderman, D. N. and Meinke, W. W., *Anal. Chem.* **29**, 1578 (1957).
24. Cline, R. W., Simmons, R. E. and Rossmassler, W. R., *ibid.* **30**, 1117 (1958).
25. Novikov, A. I., *Zh. Analit. Khim.* **17**, 1076 (1962).
26. Plotnikov, W. I., Kochetkov, V. L. and Gibova, E. G., *ibid.* **22**, 86 (1967).
27. Chuecas, L. and Riley, J. P., *Anal. Chim. Acta* **35**, 240 (1966).
28. Fuhrman, D. L. and Latimer, G. W., Jr., *Talanta* **14**, 1199 (1967).
29. Hall, F. M. and Bryson, A., *Anal. Chim. Acta* **24**, 138 (1961).
30. Mulokozi, A. M., *Analyst* **97**, 820 (1972); *Talanta* **20**, 1341 (1973).
30a. Pankow, J. F. and Janauer, G. E., *Anal. Chim. Acta* **69**, 97 (1974).
31. Bose, M., *ibid.* **10**, 201, 209 (1954).
32. Pflaum, R. T. and Howick, L. C., *J. Am. Chem. Soc.* **78**, 4862 (1956).
33. Lichtenstein, I. E. and Allen, T. L., *ibid.* **81**, 1040 (1959); *J. Phys. Chem.* **65**, 1238, (1961).
34. Babko, A. K. and Get'man, T. E., *Zh. Obshch. Khim.* **29**, 2416 (1959).
35. Minczewski, J. and Żmijewska, W., *Roczniki Chem.* **34**, 1559 (1960); *Chem. Anal. (Warsaw)* **5**, 429 (1960).
36. Sano, H., *Anal. Chim. Acta* **27**, 398 (1962).
37. Kemula, W., Kublik, Z. and Najdeker, E., *Roczniki Chem.* **36**, 937 (1962).
38. Kovalenko, E. V. and Petrashen', V. I., *Zh. Analit. Khim.* **18**, 743 (1963).
39. Zittel, H. E., *Anal. Chem.* **35**, 329 (1963).
40. Marchart, H., *Anal. Chim. Acta* **30**, 11 (1964).
41. Szczepaniak, W., *Chem. Anal. (Warsaw)* **9**, 1115 (1964).
42. Kemula, W. and Najdeker, E., *Roczniki Chem.* **44**, 2243, 2289 (1970).
43. Balt, S. and van Dalen, E., *Anal. Chim. Acta* **25**, 507 (1961); **27**, 188 (1962); **29**, 466 (1963); **30**, 434 (1964).
44. Allen, T. L., *Anal. Chem.* **30**, 447 (1958).
45. Urone, P. F., *ibid.* **27**, 1355 (1955).
46. Willems, G. J., Lontie, R. A. and Seth-Paul, W. A., *Anal. Chim. Acta* **51**, 544 (1970).
47. Saltzman, B. E., *Anal. Chem.* **24**, 1016 (1952).
48. Oelschläger, W., *Z. Anal. Chem.* **144**, 27 (1955).
49. Scholes, P. H. and Smith, D. V., *Metallurgia* **67**, 153 (1963).
50. Appelbaum, J. and Mashall, J., *Anal. Chim. Acta* **35**, 409 (1966).
51. Fröhlich, F., *Z. Anal. Chem.* **170**, 383 (1959).
52. Oelschläger, W., *ibid.* **145**, 81 (1955).
53. McAloren, J. T. and Reynolds, G. F., *Metallurgia* **57**, 52 (1958).
53a. Cline, R. W., Simmons, R. E. and Rossmassler, W. R., *Anal. Chem.* **30**, 1117 (1958).
54. Dean, J. A. and Beverly, M. L., *ibid.* **30**, 977 (1958).
55. Babko, A. K. and Get'man, T. E., *Zavodsk. Lab.* **25**, 1429 (1959).
56. Kammori, O. and Ono, A., *Japan Analyst* **14**, 1137 (1965).
57. Kovalenko, E. V. and Petrashen', V. I., *Tr. Komis. po Analit. Khim. Akad. Nauk SSSR* **15**, 101 (1965).
58. Imai, T. and Nagumo, S., *J. Chem. Soc. Japan, Ind. Chem. Sect.* **61**, 53 (1958).
59. Luke, C. L., *Anal. Chem.* **30**, 359 (1958).
60. Nazarenko, V. A., Shustova, M. B., Ravitskaya, R. V. and Nikonova, M. P., *Zavodsk. Lab.* **28**, 537 (1962).

61. Pilipenko, A. T., Voronina, A. I. and Nabivanets, B. I., *ibid.* **36**, 273 (1970).
62. Erdey, L. and Inczédy, J., *Acta Chim. Acad. Sci. Hung.* **4**, 289 (1954).
63. Pollock, E. N. and Zopatti, L. P., *Anal. Chim. Acta* **28**, 68 (1963).
64. Norwitz, G. and Codell, M., *ibid.* **9**, 546 (1953).
65. Ryabchikov, D. I., Lazarev, A. I. and Lazareva, V. I., *Zh. Analit. Khim.* **19**, 1110 (1964).
66. Żmijewska, W., *Chem. Anal. (Warsaw)* **3**, 994 (1958).
67. Flatt, R. and Cusani, P., *Anal. Chim. Acta* **21**, 181 (1959).
68. Gottlieb, A. and Hecht, F., *Mikrochemie* **35**, 523 (1950).
69. Pilkington, E. S. and Smith, P. R., *Anal. Chim. Acta* **39**, 321 (1967).
70. Dodson, E. M., *Anal. Chem.* **34**, 966 (1962).
71. Chirnside, R. C., Cluley, H. J., Powell, R. J. and Proffitt, P. M., *Analyst* **88**, 851 (1963).
72. Christe, A. A., Kerr, J. R., Knowles, G. and Lowden, G. F., *ibid.* **82**, 336 (1957).
73. Williams, A. I., *ibid.* **93**, 611 (1968).
74. Hoffmann, E. R. and Comfort, M. G., *Microchem. J.* **2**, 263 (1959).
75. Grogan, C. H., Cahnmann, H. J. and Lethco, E., *Anal. Chem.* **27**, 983 (1955).
76. Miller, D. O. and Yoe, J. H., *Clin. Chim. Acta* **4**, 378 (1959).
77. Pilz, W., *Z. Anal. Chem.* **219**, 350 (1966).
78. Tong, J. Y. and King, E. L., *J. Am. Chem. Soc.* **75**, 6180 (1953).
79. Cardone, M. J. and Compton, J., *Anal. Chem.* **24**, 1903 (1952).
80. Holloway, F., *J. Am. Chem. Soc.* **74**, 224 (1952).
81. Lacroix, S. and Labalade, M., *Anal. Chim. Acta* **3**, 262 (1949).
82. Lingane, J. J. and Collat, J. W., *Anal. Chem.* **22**, 166 (1950).
83. Wood, A. A., *Analyst* **78**, 54 (1953).
84. Marczenko, Z. and Skorko-Trybuła, Z., *Chem. Anal. (Warsaw)* **5**, 71 (1960).
85. Savichev, E. I., Iskhakova, E. I. and Flazhnikova, L. F., *Zavodsk. Lab.* **28**, 412 (1962).
86. Ziegler, M. and Pohl, K. D., *Z. Anal. Chem.* **204**, 413 (1964).
87. Přibil, R. and Klubalová, J., *Collection Czech. Chem. Commun.* **15**, 42 (1950).
88. Nielsch, W. and Böltz, G., *Metall* **10**, 916 (1956).
89. Den Boef, G., De Jong, W. J., Krijn, G. C. and Poppe, H., *Anal. Chim. Acta* **23**, 557 (1960).
90. Stolarov, K. P and Agrest, F B., *Zh. Analit. Khim.* **19**, 457 (1964).
91. Hamm, R. E., *J. Am. Chem. Soc.* **75**, 5670 (1953).
92. Irving, H.M.N.H. and Al-Jarrah, R.H., *Anal. Chim. Acta* **74**, 321 (1975).
93. Goryushina, V. G. and Gailis, E. Ya., *Zavodsk. Lab.* **21**, 642 (1955).
94. Nordling, W. D., *Chemist-Analyst* **49**, 78 (1960).
95. Usatenko, Yu. I. and Klimkovich, E. A. *Zavodsk. Lab.* **22**, 279 (1956); *Tr. Komis. po Analit. Khim. Akad. Nauk SSSR* **8**, 169 (1958).
96. Pinus, A. M., *Zavodsk. Lab.* **23**, 662 (1957).
97. Bennett, H. and Marshall, K., *Analyst* **88**, 877 (1963).
98. Khalifa, H., Roberts, J. E. and Khater, M. M., *Z. Anal. Chem.* **188**, 428 (1962).
99. Selmer-Olsen, A. R., *Anal. Chim. Acta* **26**, 482 (1962).
99a. Fuhrman, D. L. and Latimer, G. W., Jr., *Talanta* **14**, 1199 (1967).
100. Den Boef, G. and Poeder, B. C., *Anal. Chim. Acta* **30**, 261 (1964).
101. De Angelis, G. and Chiacchierini, E., *Ricerca Scient.* **36**, 53 (1966).
102. Golubtsova, R. B. and Yaroshenko, A. D., *Zavodsk. Lab.* **36**, 147 (1970).
103. Malát, M. and Hrachovcová, M., *Collection Czech. Chem. Commun.* **29**, 2484 (1964).
104. Tonosaki, K., Otomo, M. and Tanaka, K., *Japan Analyst* **15**, 683 (1966).
105. Tataev, O. A. and Abdulaev, R. R., *Zavodsk. Lab.* **36**, 1173 (1970).
106. Abdulaev, R. R., Tataev, O. A. and Busev, A. I., *ibid.* **37**, 389 (1971).
107. Tataev, O. A. and Abdulaev, R. R., *Zh. Analit. Khim.* **25**, 930 (1970).
108. Akhmedov, S. A., Tataev, O. A. and Abdulaev, R. R., *Zavodsk. Lab.* **37**, 756 (1971).

109. Yotsuyanagi, T., Takeda, Y., Yamashita, R. and Aomura, K., *Anal. Chim. Acta* **67**, 297 (1973).
110. Tandon, J. P. and Mehrotra, R. C., *Z. Anal. Chem.* **176**, 87 (1960).
111. Gillet, A. C., Jr., *ibid.* **247**, 163 (1969).
112. Motojima K. and Hashitani, H., *Anal. Chem.* **33**, 239 (1961).
113. Malik, W. U., Haque, R. and Pratap, S., *Bull. Chem. Soc. Japan* **36**, 744 (1963).
114. Sangal, S. P., *Chim. Anal.* **46**, 492 (1964).
115. Majumdar, S. K. and De, A. K., *Anal. Chem.* **32**, 1337 (1960).
116. Johnston, J. R. and Holland, W. J., *Mikrochim. Acta* **1972**, 321.
117. Hartkamp, H., *Z. Anal. Chem.* **187**, 16 (1962).
118. Den Boef, G. and Poeder, B. C., *ibid.* **199**, 348 (1964).
119. Gagliardi, E. and Haas, W., *ibid.* **147**, 321 (1955).
120. Jacobsen, E. and Lund, W., *Anal. Chim. Acta* **36**, 135 (1966).
121. Buscaróns, F. and Artigas, J., *ibid.* **16**, 452 (1957).
122. Ariel, M. and Manka, J., *ibid.* **25**, 248 (1961).
123. Cheng, K. L. and Goydish, B. L., *Chemist-Analyst* **52**, 73 (1963).
124. Songina, O. A. and Dausheva, M. R., *Zavodsk. Lab.* **32**, 910 (1966).

Chapter 18

COBALT

Cobalt (Co, at.wt. 58·93) occurs predominantly in the +II oxidation state. In some complexes it is readily oxidizable to Co(III). The hydroxide Co(OH)$_2$ is precipitated at pH ~ 7·5 and is insoluble in excess of NaOH. Cobalt forms ammine, cyanide, halide, tartrate and EDTA complexes. Blue chloride complexes are formed in fairly concentrated chloride solutions.

18.1 Methods of Separation

18.1.1 EXTRACTION

1-Nitroso-2-naphthol [1-3], a popular colorimetric reagent for cobalt, is sometimes used in the preliminary chloroform extraction of cobalt from other metals before its determination with (for example) nitroso-R salt.

Very small amounts of cobalt can be separated together with traces of other heavy metals by extraction with dithizone [4-5]. Within the pH range from 6 to 10 the extraction of cobalt is quantitative. The stability of cobalt dithizonate to dilute HCl permits the separation of cobalt from other metals which form easily decomposed dithizonates (e.g. Cd, Zn, and Pb) [6].

Other organic reagents for the extraction of cobalt include diethyldithiocarbamate [7] and 8-hydroxyquinoline [8]. Schweitzer and Howe [9] have studied the extraction of 25 cobalt chelates into several organic solvents.

Hicks [10] has separated Co(II) from Co(III) by exploiting the fact that only the cobalt(III) forms a benzene-soluble acetylacetonate.

Cobalt forms stable halide and pseudohalide complexes which are the basis for its extractive separation from other metals, in particular from nickel [11-13]. Specker and Werding [14] have investigated the extraction of cobalt complexes with thiocyanate, chloride, bromide, and iodide using TBP, cyclohexanone, and MIBK as solvents. From a solution of nickel in HCl (1+1), cobalt can be extracted with a 10% solution of trioctylamine in CCl$_4$ [15].

The extracted cobalt thiocyanate complexes play an important role in the spectrophotometric determination of cobalt.

18.1.2 Ion-Exchange

When a solution of metals in $9M$ HCl is passed through a strongly basic anion-exchanger column, the chloride complexes of Co, Cu, Zn, and Fe(III) are retained by the resin, whereas Ni, Mn, and Cr are eluted since they do not form stable chloride complexes. Cobalt is eluted from the column with $4M$ HCl. More dilute hydrochloric acid (e.g. $0.01M$) elutes Cu, Zn, and Fe(III) [16–19]. Anion-exchange separation of Co and Ni is also possible in a malonic acid–sodium nitrite medium [20].

Hazan and Korkisch [21] have separated Co, Ni, and Fe as chloride complexes on a Dowex 1 anion-exchanger, using water–acetone solutions.

Cation-exchange separation of cobalt from other metals with various eluents has been discussed [22,23]. Orlova [24] has separated cobalt and manganese sorbed on a cation-exchange column by selective elution of the cobalt with a solution of nitroso-R salt.

18.1.3 Other Methods

Luke [25] has separated traces of cobalt from larger amounts of nickel by masking Co as the stable Co(III) ammine complex, and precipitating nickel as hexa-amminonickel perchlorate. A similar technique (but with the bromide) was used by Richards to purify the nickel salts used in his atomic-weight work [25a]. Traces of cobalt can be precipitated with lanthanum [26] or aluminium [27] hydroxides. Gorshkov et al. [28,29] have precipitated traces of cobalt with various organic collectors.

Magnesium hydroxide selectively sorbs cobalt(II) from an ammoniacal solution containing Co(II) and Co(III) [30].

18.2 Methods of Determination

Three highly sensitive organic reagents for the spectrophotometric determination of cobalt are 1-nitroso-2-naphthol, 2-nitroso-1-naphthol, and nitroso-R salt. Thanks to their characteristic nitroso-hydroxyl group, they are specific for cobalt under set conditions. The first two reagents are used in the extractive spectrophotometric methods, while the third enables cobalt to be determined in aqueous media. The well-known thiocyanate method, either with or without an extraction step, is suitable for the determination of relatively large amounts of cobalt.

Besides these methods which are discussed below in detail, there exist a large variety of other spectrophotometric methods for cobalt.

18.2.1 1-Nitroso-2-naphthol and 2-Nitroso-1-naphthol Methods

1-Nitroso-2-naphthol (α-nitroso-β-naphthol, Cobaltone) (**I**) and its isomer, 2-nitroso-1-naphthol (β-nitroso-α-naphthol) (**II**) react with cobalt ions in a similar, highly selective manner to form chelates soluble in non-polar solvents.

1-Nitroso-2-naphthol, one of the first organic analytical reagents known (Ilinski, 1885), has long been used as a reagent for the gravimetric determination of cobalt. In the reaction with nitrosonaphthols (in the tautomeric quinone oxime form), cobalt is oxidized by atmospheric oxygen to the +III oxidation state. The complex of cobalt with 1-nitroso-2-naphthol is orange, whereas that with 2-nitroso-1-naphthol is brown-pink. The absorption spectra of the two complexes in chloroform solutions are shown in Fig. 18.1.

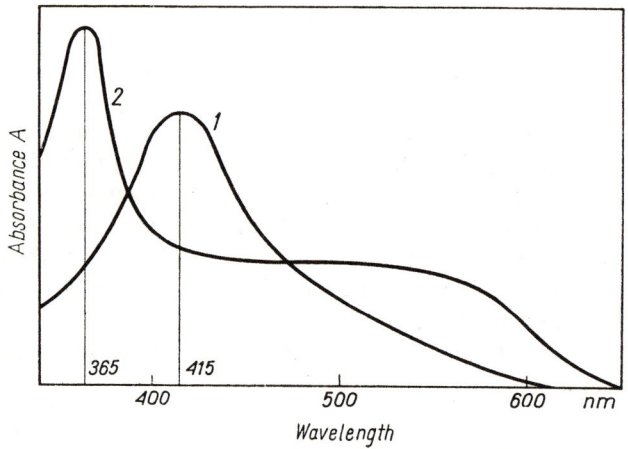

Fig. 18.1 Absorption spectra of cobalt complexes with 1-nitroso-2-naphthol (1) and with 2-nitroso-1-naphthol (2) in chloroform solutions

The molar absorptivity of the complex with 1-nitroso-2-naphthol at $\lambda_{max} = 415$ nm is $2 \cdot 9 \times 10^4$ ($a = 0 \cdot 49$), and that of the complex with 2-nitroso-1-naphthol at $\lambda_{max} = 365$ nm is $3 \cdot 7 \times 10^4$ ($a = 0 \cdot 63$). 1-Nitroso-2-naphthol [31–33] and 2-nitroso-1-naphthol [33–39] are used equally often for the spectrophotometric determination of cobalt. Chloroform is the most popular solvent, although carbon tetrachloride, benzene, toluene, and isoamyl acetate are also used.

The reaction between cobalt ions and nitrosonaphthols takes place in weakly acidic medium and proceeds rather slowly. For this reason, the sample solution is allowed to stand for ∼ 30 minutes after the reagent is added, before the complex is extracted.

In the case of 1-nitroso-2-naphthol, the reaction with cobalt is carried out at pH ⩾ 3, and at pH ⩾ 4 in the case of 2-nitroso-1-naphthol.

The cobalt complex formed is stable and does not decompose even when treated with $2M$ hydrochloric acid, which decomposes other complexes formed in a weakly acidic medium, e.g. those of Ni, Cu, Fe, and Cr. Uncomplexed nitrosonaphthol is scrubbed from the organic extract with a solution of sodium hydroxide.

Larger amounts of iron(III) are either extracted first or masked with fluoride. Addition of citrate prevents the precipitation of the hydroxides of metals which hydrolyse at the pH of the reaction of cobalt with the nitrosonaphthol.

1-Nitroso-2-naphthol has been used for determining cobalt in cast iron and steel [32], tin [26], silver [40], salts of various metals [41], alkalis [42], plant material [43], and sea water [44].

2-Nitroso-1-naphthol has been used for determining cobalt in cast iron and steel [36,45,46], nickel [36,37], sodium [47], soils and rocks [35,38], and plant material [48,49].

Reagents

1. 1-Nitroso-2-naphthol: 0·5% solution in glacial acetic acid. Before use, the solution is purified by shaking with activated carbon.
2. 2-Nitroso-1-naphthol: 0·5% solution in glacial acetic acid. Purify the solution as for 1-nitroso-2-naphthol.
3. Standard cobalt solution: 1 mg/ml. Dissolve 4·780 g of $CoSO_4 \cdot 7H_2O$ in water containing 2 ml of conc. H_2SO_4, and dilute the solution with water to 1 litre. (It is also possible to weigh 2·630 g of anhydrous cobalt sulphate prepared by ignition of the hydrous salt at $\sim 400°C$. The anhydrous salt dissolves more slowly).

Procedure

To a solution containing not more than 50 µg of Co, add 1 ml of the 1-nitroso-2-naphthol (or 2-nitroso-1-naphthol) solution with stirring, adjust the pH of the solution with ammonia to 4 (or 5 in the case of 2-nitroso-1-naphthol), and allow to stand for 30 minutes. Transfer the solution to a separating funnel and extract with two portions of chloroform. Shake the combined extracts with $2M$ HCl, followed by two portions of $2M$ NaOH, and wash finally with water. Transfer the coloured organic extract to a 50-ml or smaller volumetric flask, make up to the mark with chloroform, and measure the absorbance of the solution at 415 nm (in the case of 2-nitroso-1-naphthol, at 365 nm), using the solvent as the reference.

18.2.2 Nitroso-R Salt Method

Nitroso-R salt (disodium 1-nitroso-2-hydroxynaphthalene-3,6-disulphonate) is a derivative of 1-nitroso-2-naphthol. Both reagents are specific for cobalt. The sulphonate groups in the molecule of nitroso-R salt render this reagent and its cobalt complex soluble in water but insoluble in nonpolar solvents (in contrast to 1-nitroso-2-naphthol). Hence, nitroso-R salt is used to determine cobalt spectrophotometrically in aqueous medium.

According to Malyuga [50], the quinone oxime tautomer of nitroso-R salt takes part in the reaction with cobalt.

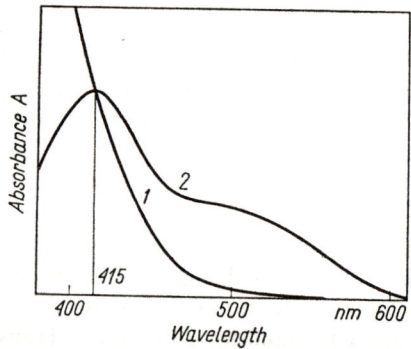

In acidic solution (pH $\sim$ 4), cobalt(II) is oxidized to Co(III) by the uncomplexed reagent; in alkaline solution (pH $\sim$ 9), aerial oxidation of Co(II) to Co(III) predominates [50a].

In acid solution, the reagent is yellow whereas the complex is red. Figure 18.2 shows the absorption spectra. The molar absorptivity of the cobalt nitroso-R salt complex at λ_{max} = 415 nm is $3 \cdot 5 \times 10^4$ (a = 0·60). At

Fig. 18.2 Absorption spectra of nitroso-R salt (1) and its cobalt complex (2) (reagent solution as reference) in aqueous solutions

500 nm, where interference from the colour of the reagent is negligible, ε is $\sim 1 \cdot 5 \times 10^4$.

The absorbance of the cobalt complex solution may be measured with higher sensitivity at 415–425 nm (reagent blank solution as reference), or with lower sensitivity at 500–520 nm (a reagent blank or water as reference).

Removal of the excess of nitroso-R salt with oxidizing agents such as bromide, bromate, or hydrogen peroxide has been recommended by some authors [51,52].

The reaction of cobalt with nitroso-R salt is usually carried out in a hot, weakly acidic medium buffered with sodium acetate. The solution is then made sufficiently acidic with hydrochloric or nitric acid to decompose nitroso-R salt complexes of other metals (e.g. Cu, Ni, Fe, and Mn), which are less stable than the cobalt(III) complex. Phosphate or fluoride masks iron(III) which has a yellow colour in hydrochloric acid medium [53–56].

Preliminary separation of cobalt from large quantities of other metals is carried out by the methods outlined above. Alternatively, large quantities of metals, e.g. iron(III), are separated from cobalt by extraction of the chloride complex of the metal concerned.

When the colour reaction takes place in the presence of relatively large amounts of metals which also react in weakly acidic medium with nitroso-R salt, a considerable amount of the reagent should be added to the solution [56].

Adam and Přibil [57] extract the complex of cobalt and nitroso-R salt with chloroform in the presence of trioctylmethylammonium chloride.

Wise and Brandt [58] use the related reagent, 2-nitroso-1-naphthol-4-sulphonic acid, to determine cobalt.

Cobalt has been determined by the spectrophotometric nitroso-R salt method in cast iron and steel [59,60], cements [59], ores and minerals [3,61,62], nickel [13,25,63], tungsten [64], zirconium and titanium [65,66], uranium [7], bismuth [2], beryllium compounds [67], aluminium alloys [68], plant materials [51,69], biological materials [1,70,71], and natural waters [5,71a].

Reagents

1. Nitroso-R salt: 0·1% solution. Store the solution in an amber-glass bottle.
2. Standard cobalt solution: 1 mg/ml (p. 227).

Procedure

Neutralize a solution containing less than 50 µg of Co with dilute ammonia solution, and then acidify with 2 ml of $1M$ HCl. Add 5 ml of the nitroso-R salt solution, 5 ml of 25% sodium acetate solution, heat to boiling, and boil gently for 1 minute. To the slightly cooled solution, add 1 ml of conc. phosphoric acid and 3 ml of hydrochloric acid (1+1). Boil the solution for a further minute. Cool the solution, transfer it to a 50-ml volumetric flask, and make it up to the mark with water. Measure the absorbance of the solution at 415 nm against a reagent blank solution.

18.2.3 THIOCYANATE METHOD

In concentrated solutions of potassium, sodium, or ammonium thiocyanate, cobaltous ions produce a blue colour which fades when the solution is diluted with water, owing to dissociation of the complex. The addition of acetone or any other water-miscible organic solvent (e.g. ethanol) which lowers the dielectric constant and thereby suppresses the dissociation of the complex, restores the blue colour to the solution. Babko and Drako [72], as well as other authors [73,74], have shown that a complex with the formula $[Co(SCN)_4]^{2-}$ is predominant in the blue solution. In solutions with lower SCN^- concentrations and in the absence of acetone, pale pink $[Co(SCN)]^+$ exists.

The colour intensity depends on the concentration of SCN^- and acetone. It has been established that in 50% acetone solution increasing the KSCN concentration above 10% produces no further increase in the colour intensity. The acidity of the solution also affects the absorbance and should therefore be kept constant (within the limits from 0·1 to $1M$ HCl in the sample and the reference solutions).

At $\lambda_{max} = 620$ nm the molar absorptivity (of a solution containing 10% of KSCN and 50% of acetone) is 1.9×10^3 ($a = 0.032$).

Since iron(III) forms coloured complexes with SCN^- ions, it interferes in the colorimetric determination of cobalt by the thiocyanate method. Larger quantities must be separated (e.g. by extraction or the zinc oxide method); smaller amounts can be masked with fluoride, phosphate, or pyrophosphate, or reduced to Fe(II) with ascorbic acid or stannous chloride.

Interfering elements include other metals which form coloured thiocyanate complexes (e.g. V, Bi, U, Cu, Mo, and W), and metals which give precipitates [Ag, Tl(I), Cu(I)] or consume thiocyanate ions to form colourless complexes (e.g. Hg). Finally, ions which are themselves coloured (e.g. Ni and Cr) interfere when present in high concentrations. Copper(II) can be masked by means of tartaric acid (in acetate-buffered media).

Using the thiocyanate method, Kitson [75] has determined cobalt, iron, and copper simultaneously by measuring the absorbances of the solution at 625, 480, and 380 nm, respectively.

The cobalt(II) thiocyanate complex can be extracted with oxygenated solvents such as a mixture of diethyl ether with isoamyl alcohol (1+1), methyl isopropyl ketone [76], MIBK, acetylacetone [77,78], cyclohexanone and TBP [79]. The molar absorptivity of the cobaltous complex in the ether–isoamyl alcohol (1+1) solution is about 30% lower than that in acetone–water medium.

The anionic cobaltous thiocyanate complex reacts with some organic bases to form ion-association complexes which can be extracted into chloroform or other non-polar solvents. Organic reagents used for this purpose include DAM [80–82], the tetraphenylarsonium [83,84], triphenylmethylarsonium [85], diphenyliodonium [86], and tricaprylmethylammonium ions [87], tri-n-butylammonium acetate [88], tri-iso-octylamine [89], tri-n-octylamine [90], hyamine [91], and Malachite Green [92].

Cobalt has been determined by various modifications of the thiocyanate method in steel and its alloys [77,83,84,86], nickel and its compounds [79–81,83], beryllium [78], manganese ores [93], and lead alloys [94].

Reagents

1. Potassium or ammonium thiocyanate: 50% solution.
2. Standard cobalt solution: 1 mg/ml (p. 227).
3. Ascorbic acid: 2% solution, freshly prepared.
4. Thiosulphate–phosphate solution. Dissolve in water 40 g of $Na_2S_2O_3$. $5H_2O$ and 10 g of $Na_3PO_4.12H_2O$; dilute the solution with water to 250 ml.

Procedure

Acetone variant. To the sample solution in a 50-ml volumetric flask containing less than 1 mg of Co, add hydrochloric acid until $0.5M$. Add 10 ml of the thiocyanate solution. If a red colour appears (indicative of iron), add the ascorbic acid solution dropwise till decolorization, and add 1 ml more. Add 25 ml of acetone, and dilute the solution with water to the mark. Measure the absorbance of the blue solution at 620 nm, using water as the reference.

Extraction variant. Slightly acidify the sample solution containing not more than 1 mg of Co. Add 5 ml of the thiosulphate–phosphate solution and 10 ml of the thiocyanate solution, and adjust to pH 3·5–4·0. Transfer the solution to a separating funnel, and extract the cobaltous thiocyanate complex with 2 or 3 portions of a mixture of ether and isoamyl alcohol $(1+1)$. Make up the combined extracts to the mark with the solvent in a 50-ml or smaller volumetric flask, and measure the absorbance at 620 nm, using the solvent as reference.

18.2.4 OTHER METHODS

Among other organic reagents for the determination of cobalt, there is a large group of nitroso compounds, related to the nitrosonaphthols, comprising: 3-nitrososalicylic acid [95], isonitrosodimedone [96], isonitrosomalonylguanidine [97,98], 5-dimethylamino-2-nitrosophenol [99,100], 5-diethylamino-2-nitrosophenol [101], nitrosoresorcinol monomethyl ether (presumably 3-methoxy-2-nitrosophenol) [102], 6-amino-4-hydroxy-2-mercapto-5-nitrosopyrimidine ($\varepsilon = 2.6 \times 10^4$ at $\lambda_{max} = 396$ nm) [103], and 3-nitroso-2,6-pyridinediol [104].

It has been established for a long time that dimethylglyoxime, a well-known reagent for nickel, reacts with cobalt in an ammoniacal or pyridine medium to give a water-soluble brown complex. Cobalt has recently been determined by the extraction of this complex in the presence of hydrophobic anions [105,106]. A number of other oximes have also been recommended as spectrophotometric reagents for cobalt, namely: α-furilmonoxime [107], α-furildioxime [108], oxamidoxime [109], pyridyl-2-aldoxime [110], 2,2'-dipyridylketoxime [111,112], p-tolueneamidoxime [113], nicotinamidoxime [114], 2,4-dimethylbenzamidoxime [115], 2,2'-diquinolylketoxime (**III**) ($\varepsilon = 5\cdot3 \times 10^4$ at 365 nm, after extraction of the complex into benzene) [116], phenanthrenequinone monoxime [117], the p-nitrophenylhydrazone of diacetylmonoxime [118], 1,2,3-cyclohexanetrione trioxime [119], dimedone dioxime [120], and 2-pyridyl-2-thienyl-β-ketoxime [121].

In recent years there has been a remarkable growth in the number of highly sensitive methods for cobalt determination based on azo compounds. With PAN as the reagent, traces of cobalt in nickel and other metals have been determined after extraction into chloroform [122–126]. PAR reacts similarly with cobalt [127–130]. PAN can be used in aqueous medium in the presence of surfactants [131].

A number of other pyridine azo compounds have been recommended [132–141], since their cobalt complexes have very high molar absorptivities. For example, the reagents 2-[(5-bromo-2-pyridyl)azo]-5-diethylaminophenol (5-Br-PAAP) (**IV**) and 4-[(5-chloro-2-pyridyl)azo]-1,3-diaminobenzene (5-Cl-PADAB) (**V**) form cobalt complexes with $\varepsilon = 9\cdot9 \times 10^4$ at 580 nm [133] and $\varepsilon = 1\cdot13 \times 10^5$ at 570 nm [139], respectively.

Thiazole azo reagents which have also been suggested include 4-(2-thiazolylazo)-resorcinol (TAR) ($\varepsilon = 5\cdot6 \times 10^4$ at 510 nm) [142], a bromo derivative of TAR [143], and 1-(2-thiazolylazo)-2-naphthol (TAN) [144]. Some further azo compounds which have been used for the determination of cobalt are Calcichrome [145], Sulpharsazen [146], Quinolinazo R [147], Acid Monochrome Green S [148], Eriochrome Black T [149], 3-methyl-5-propylpyrrole-(2-azo-2')-phenol [150], and Azonol A [151].

Methods based on reagents containing sulphur as a ligand atom are, in the main, less sensitive than those discussed above. Many authors have recommended Na–DDTC [152–154] or other dithiocarbamates [155,156] for the determination of cobalt. Furthermore, mention can be made of quinoxaline-2,3-dithiol [157–160], dithio-oxamide (rubeanic acid) and its N,N'-bis(3-dimethylaminopropyl) derivative [161], thioglycollic acid [162] and its p-toluidine derivative [163], thiotropolone [164], thiodibenzoylmethane [165], thiothenoyltrifluoroacetone [166], 1-phenylthiosemicarbazide [167], and N,N'-ethylene-bis(o-mercaptobenzamide) [168].

In a solution at pH 2–4 containing the 1,10-phenanthroline iron(III) complex, cobalt(II) reduces the iron(III), thereby giving rise to red ferroin in an amount equivalent to the cobalt. The method based on this principle has been applied in the determination of cobalt in metallic sodium [169] and in nickel [170].

Less sensitive photometric methods for cobalt determination utilize EDTA (Complexone III) [171,172] and DCTA (Complexone IV) [173].

Other organic reagents used in various spectrophotometric methods for determining cobalt are 2,2',2''-terpyridine [174], 2,4,6-tri(2'-pyridyl)-s-triazine (TPTZ) [175], 8-hydroxyquinoline [176,177], Alizarin Red S [178], TTA [179–181], Chrome Azurol S [182], pyridine-2-aldehyde-2-quinolylhydrazone [183], benzil mono-(2-pyridyl)hydrazone ($\varepsilon = 2\cdot74 \times 10^4$ at 535 nm) [184], N,N'-bis(o-aminobenzylidene)ethylenediamine [185], biuret [186], and glycerol [187]. The last two reagents react with cobalt in strongly alkaline solutions.

Among inorganic complexes other than the thiocyanate, the blue chloride complex [188–190] and the cyanate complex [191] are used for cobalt determination. Molybdophosphoric acid is reduced by cobalt(II) in the presence of EDTA to phosphomolybdenum blue, which has been used for the indirect determination of cobalt [192].

Larger amounts of cobalt can be determined as Co^{2+} ions. In $HClO_4$ medium, cobalt has been determined by differential spectrophotometry [193]. In $3M\ H_2SO_4$ medium, the coloured solutions of Co(II) obey Beer's law well ($\varepsilon = 4\cdot9$ at 510 nm) [194].

References

1. Saltzman, B. E., *Anal. Chem.* **27**, 284 (1955).
2. Nazarenko, V. A. and Shitareva, G. G., *Zavodsk. Lab.* **24**, 932 (1958).
3. Cogan, E., *Anal. Chem.* **32**, 973 (1960).
4. Monnier, D., Haerdi, W., Vogel, J. and Wenger, P. E., *Helv. Chim. Acta* **42**, 1846 (1959).
5. Forster, W. and Zeitlin, H., *Anal. Chim. Acta* **34**, 211 (1966).
6. Marczenko, Z., Mojski, M. and Kasiura, K., *Zh. Analit. Khim.* **22**, 1805 (1967).
7. Suzuki, M. and Takeuchi, T., *Japan Analyst* **9**, 179 (1960).
8. Duffield, W. D., *Analyst* **84**, 455 (1959).
9. Schweitzer, G. K. and Howe, (III), L. H., *Anal. Chim. Acta* **37**, 316 (1967).
10. Hicks, J. E., *ibid.* **45**, 101 (1969).
11. Forsythe, J. H., Magee, R. J. and Wilson, C. L., *Talanta* **1**, 249 (1958).
12. Musil, A. and Weidmann, G., *Mikrochim. Acta* **1959**, 476.
13. Athavale, V. T., Gulavane, S. V. and Tillu, M. M., *Anal. Chim. Acta* **23**, 487 (1960).
14. Specker, H. and Werding, G., *Z. Anal. Chem.* **200**, 337 (1964).
15. Ioffe, E. Sh. and Karaseva, A. D., *Zavodsk. Lab.* **33**, 1502 (1967).
16. Moore, G. E. and Kraus, K. A., *J. Am. Chem. Soc.* **74**, 843 (1952); **75**, 1460 (1953).
17. Thiers, R. E., Williams, J. F. and Yoe, J. H., *Anal. Chem.* **27**, 1725 (1955).
18. Ryabchikov, D. I. and Borisova, L. V., *Zh. Analit. Khim.* **13**, 340 (1958).
19. Vogel, J., Monnier, D., Haerdi, W. and Wenger, P. E., *Helv. Chim. Acta* **43**, 217 (1960).
20. Nambiar, O. G. B. and Subbaraman, P. R., *Talanta* **14**, 785 (1967).
21. Hazan, I. and Korkisch, J., *Anal. Chim. Acta* **32**, 46 (1965).
22. Akki, S. R. and Khopkar, S. M., *ibid.* **52**, 393 (1970).
23. Inczédy, J., Klatsmanyi-Gabor, P. and Erdey, L., *Acta Chim. Acad. Sci. Hung.* **61**, 261; **62**, 1 (1969).

24. Orlova, L. M., *Zavodsk. Lab.* **25**, 417 (1959).
25. Luke, C. L., *Anal. Chem.* **32**, 836 (1960).
25a. Richards, J. W. and Cushman, A. S., *Proc. Am. Acad. Arts. Sci.* **33**, 95 (1897).
26. Marczenko, Z. and Kasiura, K., *Chem. Anal. (Warsaw)* **10**, 449 (1965).
27. Ivanov, V. M., Busev, A. I., Smirnova, L. I. and Nemtseva, Zh. I., *Zh. Analit. Khim.* **25**, 1149 (1970).
28. Gorshkov, V. V., Mekhryusheva, L. I. and Smakhtin, L. A., *Zavodsk. Lab.* **37**, 396 (1971).
29. Gorshkov, V. V. and Mekhryusheva, L. I., *Zh. Analit. Khim.* **26**, 342 (1971).
30. Melikhov, I. V., Belousova, M. Ya. and Peshkova, V. M., *ibid.* **25**, 1144 (1970).
31. Komar', N. P. and Tolmachev, V. N., *ibid.* **5**, 21 (1950).
32. Oi, N., *J. Chem. Soc. Japan, Pure Chem. Sect.* **76**, 413 (1955).
33. Peshkova, V. M. and Bochkova, V. M., *Tr. Komis. po Analit. Khim. Akad. Nauk SSSR* **8**, 125 (1958).
34. Baron, H., *Z. Anal. Chem.* **140**, 173 (1953).
35. Almond, H., *Anal. Chem.* **25**, 166 (1953).
36. Claassen, A. and Daamen, A., *Anal. Chim. Acta* **12**, 547 (1955).
37. Pontet, M., *Chim. Anal. (Paris)* **37**, 372 (1955).
38. Clark, L. J., *Anal. Chem.* **30**, 1153 (1958).
39. Nielsch, W., *Mikrochim. Acta* **1959**, 725.
40. Marczenko, Z. and Kasiura, K., *Chem. Anal. (Warsaw)* **9**, 87 (1964).
41. Różycki, C., *ibid.* **12**, 131 (1967).
42. Marczenko, Z., *Mikrochim. Acta* **1965**, 281.
43. Krauze, A., *Chem. Anal. (Warsaw)* **6**, 711 (1961).
44. Kentner, E. and Zeitlin, H., *Anal. Chim. Acta* **49**, 587 (1970).
45. Needleman, M., *Anal. Chem.* **38**, 915 (1966).
46. Rooney, R. C., *Metallurgia* **58**, 205 (1958); **62**, 175 (1960).
47. Silverman, L. and Seitz, R. L., *Anal. Chim. Acta* **20**, 340 (1959).
48. Schüller, H., *Mikrochim. Acta* **1959**, 107.
49. Ssekaalo, H., *Anal. Chim. Acta* **51**, 503 (1970).
50. Malyuga, D. P., *Zh. Analit. Khim.* **1**, 176 (1946); **2**, 323 (1947).
50a. Lalor, G. C., *J. Inorg. Nucl. Chem.* **31**, 1783 (1969).
51. Gallego, R., Deijs, W. B. and Feldmeijer, J. H., *Rec. Trav. Chim.* **71**, 987 (1952).
52. Shipman, W. H. and Lai, J. R., *Anal. Chem.* **28**, 1151 (1956).
53. Haerdi, W., Vogel, J., Monnier, D. and Wenger, P. E., *Helv. Chim. Acta* **42**, 2334 (1959).
54. Pascual, J. N., Shipman, W. H. and Simon, W., *Anal. Chem.* **25**, 1830 (1953).
55. Shipman, W. H., Foti, S. C. and Simon, W., *ibid.* **27**, 1240 (1955).
56. Finkel'shtein, D. N., *Zavodsk. Lab.* **22**, 648 (1956).
57. Adam, J. and Přibil, R., *Talanta* **18**, 733 (1971).
58. Wise, W. M. and Brandt, W. W., *Anal. Chem.* **26**, 693 (1954).
59. Monnier, D., Haerdi, W. and Vogel, J., *Anal. Chim. Acta* **23**, 577 (1960); **24**, 365 (1961).
60. Koch, K. H., Ohls, K., Sebastiani, E. and Riemer, G., *Z. Anal. Chem.* **249**, 307 (1970).
61. Guerin, B. D., *Analyst* **81**, 409 (1956).
62. Volkova, L. P. and Pakhomova, K. S., *Zavodsk. Lab.* **33**, 414 (1967).
63. Mizuike, A., Iida, Y. and Hirano, S., *J. Chem. Soc. Japan, Ind. Chem. Sect.* **61**, 1459 (1958).
64. Norwitz, G. and Gordon, H., *Anal. Chem.* **37**, 417 (1965).
65. Wood, D. F. and Clark, R. T., *Talanta* **2**, 1 (1959).
66. Vogel, J., Monnier, D. and Haerdi, W., *Anal. Chim. Acta* **24**, 55 (1961).
67. Ovenston, T. C. and Parker, C. A., *ibid.* **4**, 142 (1950).
68. Jean, M., *ibid.* **7**, 523 (1952).

References

69. Jensen, E., *ibid.* **7**, 561 (1952).
70. Haerdi, W., Vogel, J., Monnier, D. and Wenger, P. E., *Helv. Chim. Acta* **43**, 869 (1960).
71. Dewey, D. W. and Marston, H. R., *Anal. Chim. Acta* **57**, 45 (1971).
71a. Korkisch, J. and Dimitriadis, D., *Talanta* **20**, 1287 (1973).
72. Babko, A. K. and Drako, O. F., *Zh. Obshch. Khim.* **19**, 1809 (1949); *Zavodsk. Lab.* **16**, 1162 (1950).
73. Katzin, L. I. and Gebert, E., *J. Am. Chem. Soc.* **72**, 5659 (1950).
74. West, P. W. and De Vries, C. G., *Anal. Chem.* **23**, 334 (1951).
75. Kitson, R. E., *ibid.* **22**, 664 (1950).
76. Kinnunen, J., Merikanto, B. and Wennerstrand, B., *Chemist-Analyst* **43**, 21 (1954).
77. Brown, W. B. and Steinbach, J. F., *Anal. Chem.* **31**, 1805 (1959).
78. Hibbits, J. O., Rosenberg, A. F. and Williams, R. T., *Talanta* **5**, 250 (1960).
79. Jackwerth, E. and Schneider, E. L., *Z. Anal. Chem.* **207**, 188 (1965).
80. Babko, A. K. and Danilova, V. N., *Zavodsk. Lab.* **30**, 1198 (1964); *Zh. Analit. Khim.* **20**, 1341 (1965).
81. Adamiec, I., *Chem. Anal. (Warsaw)* **14**, 115 (1969).
82. Donaldson, E. M., Charette, D. J. and Rolko, V. H., *Talanta* **16**, 1305 (1969).
83. Affsprung, H. E., Barnes, N. A. and Potratz, H. A., *Anal. Chem.* **23**, 1680 (1951).
84. Pepkowitz, L. P. and Marley, J. L., *ibid.* **27**, 1330 (1955).
85. Ellis, K. W. and Gibson, N. A., *Anal. Chim. Acta* **9**, 275 (1953).
86. Fogg, A. G., Higgens, C. T. and Burns, D. T., *Mikrochim. Acta* **1969**, 546.
87. Wilson, A. M. and McFarland, O. K., *Anal. Chem.* **35**, 302 (1963).
88. Ziegler, M., Glemser, O. and Preisler, E., *Angew. Chem.* **68**, 411 (1956).
89. Selmer-Olsen, A. R., *Anal. Chim. Acta* **31**, 33 (1964).
90. Watanabe, H. and Akatsuka, K., *ibid.* **38**, 547 (1967).
91. Gundersen, N. and Jacobsen, E., *ibid.* **42**, 330 (1968).
92. Kotelyanskaya, L. I. and Kish, P. P. *Zh. Analit. Khim.* **28**, 1999 (1973).
93. Cyrankowska, M. and Downarowicz, J., *Chem. Anal. (Warsaw)* **11**, 727 (1966).
94. Alexandrow, A., *Mikrochim. Acta* **1972**, 664.
95. Perry, M. H. and Serfass, E. J., *Anal. Chem.* **22**, 565 (1950).
96. Van den Bossche, W. and Hoste, J., *Anal. Chim. Acta* **18**, 564 (1958).
97. Jean, M., *ibid.* **6**, 278 (1952).
98. Boulin, R., Leblond, A. M. and Jean, M., *ibid.* **56**, 45 (1971).
99. Motomizu, S., *ibid.* **56**, 415 (1971).
100. Motomizu, S., *Analyst* **97**, 986 (1972).
101. Motomizu, S., *Anal. Chim. Acta* **64**, 217 (1973).
102. Peach, S. M., *Analyst* **81**, 371 (1956).
103. Waksmundzki, A. and Przeszlakowski, S., *Chem. Anal. (Warsaw)* **9**, 69 (1964).
104. McDonald, C. W., Rhodes, T. and Bedenbaugh, J. H., *Mikrochim. Acta* **1972**, 298.
105. Zolotov, Yu. A. and Vlasova, G. E., *Zh. Analit. Khim.* **24**, 1542 (1969).
106. Pyatnitskii, I. V., Mikhelson, P. B. and Moshkovskaya, L. T., *ibid.* **28**, 2227 (1973).
107. Martinek, J. and Hovorka, V., *Collection Czech. Chem. Commun.* **22**, 246 (1957).
108. Jones, J. L. and Gastfield, J., *Anal. Chim. Acta* **51**, 130 (1970).
109. Pearse, G. A., Jr. and Pflaum, R. T., *Anal. Chem.* **32**, 213 (1960).
110. Gagliardi, E. and Presinger, P., *Mikrochim. Acta* **1965**, 791.
111. Holland, W. J. and Bozic, J., *Talanta* **15**, 843 (1968).
112. Holland, W. J., De Pooter, M. and Bozic, J., *Mikrochim. Acta* **1974**, 99.
113. Busev, A. I., Zholondkovskaya, T. N. and Teplova, G. N., *Zh. Analit. Khim.* **26**, 1133 (1971).
114. Tripathi, K. K. and Banerjea, D., *Z. Anal. Chem.* **168**, 407 (1959).
115. Manolov, K. and Motekov, N., *Mikrochim. Acta* **1974**, 231.
116. Stupavsky, S. and Holland, W. J., *ibid.* **1971**, 559.

117. Trikha, K. C., Katyal, M. and Singh, R. P., *Talanta* **14**, 977 (1967).
118. Deshmukh, G. S., Anand, V. D. and Pandey, C. M., *Z. Anal. Chem.* **182**, 170 (1961).
119. Frierson, W. J., Patterson, N., Harrill, H. and Marable, N., *Anal. Chem.* **33**, 1096 (1961).
120. Belcher, R., Ghonaim, S. and Townshend, A., *Talanta* **21**, 191 (1974).
121. Notenboom, H. R., Holland, W. J. and Billinghurst, R. C., *Mikrochim. Acta* **1973**, 467.
122. Goldstein, G., Manning, D. L. and Menis, O., *Anal. Chem.* **31**, 192 (1959).
123. Püschel, R., Lassner, E. and Katzengruber, K., *Z. Anal. Chem.* **223**, 414 (1966).
124. Flaschka, H. and Garrett, J., *Talanta* **15**, 595 (1968).
125. Flaschka, H. and Speights, R. M., *Microchem. J.* **14**, 490 (1969).
126. Püschel, R., Lassner, E., and Illaszewicz, A., *Chemist-Analyst* **55**, 40 (1966).
127. Pollard, F. H., Hanson, P. and Geary, W. J., *Anal. Chim. Acta* **20**, 26 (1959).
128. Busev, A. I. and Ivanov, V. M., *Zh. Analit. Khim.* **18**, 208 (1963).
129. Zaboeva, M. I., Zus', G. N. and Dorofeeva, M. P., *Zavodsk. Lab.* **35**, 1158 (1969).
130. Akhmedov, S. A., Tataev, O. A. and Abdullaev, R. R. *ibid.* **37**, 756 (1971).
131. Watanabe, H., *Talanta* **21**, 295 (1974).
132. Gusev, S. I., Kiryukhina, N. N. and Bitovt, Z. A., *Zh. Analit. Khim.* **23**, 889 (1968).
133. Gusev, S. I. and Kiryukhina, N. N., *ibid.* **24**, 210 (1969).
134. Talipov, Sh. T., Podgornova, V. S. and Kosolapova, S. N., *ibid.* **24**, 409, 880 (1969).
135. Podgornova, V. S., Kosolapova, S. N. and Talipov, Sh. T., *ibid.* **24**, 945 (1969).
136. Epshova, N. S., Ivanov, V. M. and Busev, A. I., *ibid.* **28**, 2220 (1973).
137. Gusev, S. I. and Dazhina, L. G., *ibid.* **29**, 813 (1974).
138. Kiss, E., *Anal. Chim. Acta* **66**, 385 (1973).
139. Shibata, S., Furukawa, M., Ishiguro, Y. and Sasaki, S., *ibid.* **55**, 231 (1971).
140. Majumdar, A. K. and Chatterjee, A. B., *Talanta* **13**, 821 (1966).
141. Shibata, S., Furukawa, M. and Goto, K., *ibid.* **20**, 426 (1973).
142. Busev, A. I., Ivanov, V. M. and Nemtseva, Zh. I., *Zh. Analit. Khim.* **24**, 414 (1969).
143. Ivanov, V. M., Busev, A. I., Nemtseva, Zh. I. and Smirnova, L. I., *Zavodsk. Lab.* **35**, 1042 (1969).
144. Navratil, O. and Frei, R. W., *Anal. Chim. Acta* **52**, 221 (1970).
145. Ishii, H. and Einaga, H., *Japan Analyst* **16**, 328 (1967).
146. Yagodnitsyn, M. A., *Zavodsk. Lab.* **35**, 788 (1969).
147. Basargin, N. N., Kadomtseva, A. V. and Petrashen', V. I., *ibid.* **35**, 16 (1969).
148. Bagreev, V. V. and Zolotov, Yu. A., *Talanta* **15**, 988 (1968).
149. Kodama, M., *Bull. Chem. Soc. Japan* **40**, 2575 (1967); **42**, 555 (1969).
150. Cherepakhin, A. I., *Zh. Analit. Khim.* **21**, 502 (1966).
151. Budešinský, B. and Svecová, J., *Anal. Chim. Acta* **49**, 231 (1970).
152. Přibil, R., Jenik, J. and Kobrová, M., *Collection Czech. Chem. Commun.* **19**, 470 (1954).
153. Stolyarov, K. P., *Zh. Analit. Khim.* **16**, 452 (1961).
154. Motojima, K., Bando, S. and Tamura, N., *Talanta* **14**, 1179 (1967).
155. Tulyupa, F. M., Usatenko, Yu. I. and Pavlichenko, V. A., *Zavodsk. Lab.* **34**, 14 (1968).
156. Motojima, K. and Tamura, N., *Anal. Chim. Acta* **45**, 327 (1969).
157. Burke, R. W. and Yoe, J. H., *Anal. Chem.* **34**, 1378 (1962).
158. Dalziel, J. A. and Slawinski, A. K., *Talanta* **15**, 367 (1968).
159. Burke, R. W. and Deardorff, E. R., *ibid.* **17**, 255 (1970).
160. Chernomorchenko, L. I., Chuiko, T. V. and Akhmetshin, A. G., *Zh. Analit. Khim.* **27**, 2265 (1972).
161. Jacobs, W. D. and Yoe, J. H., *Anal. Chim. Acta* **20**, 332, 435 (1959).
162. Anand, V. D., Deshmukh, G. S. and Pandey, C. M., *Anal. Chem.* **33**, 1933 (1961).
163. Busev, A. I. and Naku, A., *Zh. Analit. Khim.* **19**, 475 (1964).
164. Srivastava, J. N. and Singh, R. P., *Talanta* **20**, 1210 (1973).

165. Uhlemann, E. and Müller, H., *Anal. Chim. Acta* **48**, 115 (1969).
166. Mulye, R. R. and Khopkar, S. M., *Mikrochim. Acta* **1973**, 55.
167. Koshkin, N. V. and Shreiner, N. M., *Zh. Analit. Khim.* **18**, 757 (1963).
168. Busev, A. I. and Vin', D. K., *ibid.* **21**, 327 (1966).
169. Vydra, F. and Přibil, R., *Collection Czech. Chem. Commun.* **26**, 3081 (1961); *Z. Anal. Chem.* **188**, 273 (1962).
170. Ganzburg, G. M. and Mal'tseva, G. V., *Zavodsk. Lab.* **31**, 406 (1965).
171. Přibil, R. and Malik, J., *Chem. Listy* **45**, 237 (1951).
172. Silverstone, N. M. and Bach, B. B., *Metallurgia* **63**, 205 (1961).
173. Jacobsen, E. and Selmer-Olsen, A. R., *Anal. Chim. Acta* **25**, 476 (1961).
174. Miller, R. R. and Brandt, W. W., *Anal. Chem.* **26**, 1968 (1954).
175. Janmohamed, M. J. and Ayres, G. H., *ibid.* **44**, 2263 (1972).
176. Mukhedkar, A. J. and Deshpande, N. V., *ibid.* **35**, 47 (1963).
177. Oki, S., *Anal. Chim. Acta* **50**, 465 (1970).
178. Mukherji, A. K. and Dey, A. K., *Bull. Chem. Soc. Japan* **31**, 521 (1958).
179. Majumdar, S. K. and De, A. K., *Z. Anal. Chem.* **177**, 97 (1960); *Anal. Chim. Acta* **27**, 153 (1962).
180. De, A. K. and Rahaman, M. S., *ibid.* **34**, 233 (1966).
181. Rahaman, M. S. and Finston, H. L., *Anal. Chem.* **41**, 2023 (1969).
182. Horiuchi, Y. and Nishida, H., *Japan Analyst* **16**, 576 (1967).
183. Singhal, S. P. and Ryan, D. E., *Anal. Chim. Acta* **37**, 91 (1967).
184. Pflaum, R. T. and Tucker, E. S., (III), *Anal. Chem.* **43**, 458 (1971).
185. Berge, H. and Mennenga, H., *Z. Anal. Chem.* **213**, 346 (1965).
186. Gustin, V. K. and Sweet, T. R., *Anal. Chem.* **33**, 1942 (1961).
187. Nessonova, G. D., Pogosyants, E. K. and Lishevskaya, M. O., *Zavodsk. Lab.* **25**, 786 (1959).
188. Sherwood, A. E., *Metallurgia* **64**, 47 (1961).
189. Pike, L. and Yoe, J. H., *Talanta* **12**, 657 (1965).
190. Walker, C. R. and Vita, O. A., *Anal. Chim. Acta* **47**, 9 (1969).
191. Ziegler, M. and Rittner, W., *Z. Anal. Chem.* **165**, 197 (1959).
192. Klochkovskii, S. P. and Chistota, V. D., *Zavodsk. Lab.* **36**, 911 (1970).
193. Páll, A., Svehla, G. and Erdey, L., *Talanta* **11**, 1383 (1964).
194. Komar', N. P. and Palagina, L. S., *Zh. Analit. Khim.* **23**, 75 (1968).

Chapter 19

COPPER

Copper (Cu, at.wt. 63·54) occurs in its compounds mostly in the +II, and less often in the +I, oxidation state. A few copper(III) compounds exist. The properties of copper(I) are similar to those of Ag, Au(I), and Tl(I). Copper(I) forms sparingly soluble compounds with the halogens. In solution, copper(I) exists only in complexes, e.g. $[Cu(CN)_2]^-$, $[CuCl_2]^-$, and $[Cu(NH_3)_2]^+$. Cupric hydroxide, $Cu(OH)_2$, begins to precipitate at pH $\sim$ 5 and shows no amphoteric properties. Copper(II) forms ammine, chloride, tartrate, and EDTA complexes. Because of their greater stabilities, the copper(I) cyanide complex is formed from Cu(II) and cyanide, and sparingly soluble CuSCN is precipitated from Cu(II) by thiocyanate, and CuI from Cu(II) by iodide.

19.1 Methods of Separation

19.1.1 Extraction

There are many methods for determining copper which use the same reagent for the solvent extraction as for the spectrophotometric determination, e.g. those using dithizone, dithiocarbamates, and cuproine. These spectrophotometric methods are discussed later. Sometimes, copper is extracted as cupric dithizonate [1,2] or dithiocarbamate [3] before its determination with a different colorimetric reagent.

The following methods for extracting copper with non-colorimetric reagents are worthy of note: extraction of copper (I) from hydrohalic acid solutions with triphenyl phosphite in carbon tetrachloride [4], of copper(II) as the salicylate [5] or trifluoroacetylacetonate [6], and of copper(II) with phenylacetic acid [6a].

19.1.2 Precipitation and Other Methods

Copper can be conveniently separated by precipitation as the sulphide from either an acidic or a neutral (pH $\sim$ 8) medium, depending on the metals to be precipitated along with it. Metals yielding sparingly soluble sulphides (e.g. Hg, Cd, Pb, and Zn) can be used as collectors. Traces of copper are also isolated quantitatively as the sulphide when a hydroxide collector, e.g. lanthanum hydroxide, is used [7]. Zinc phosphate may also be used as a carrier for copper [8].

Since it forms soluble ammine complexes, copper may be separated from relatively small amounts of metals which are precipitated as hydroxides with ammonia (e.g. Fe, Al, and Ti). Reprecipitation is advisable when a more quantitative separation is required.

Small amounts of copper are quantitatively separated from less noble metals by electrodeposition on a platinum cathode in acidic medium [9].

By exploiting the differences in stability of various complexes of copper and other metals, it is possible to achieve separations on ion-exchange columns. For example, copper has been separated in an anion-exchange column from gallium, iron, and zinc by using various concentrations of hydrochloric acid [5]. Cation-exchange chromatography in acetone-hydrobromic acid media has also been employed [9a].

19.2 Methods of Determination

The copper(II) ion itself gives rise to colour, hence some of the great variety of spectrophotometric methods for its determination are based on colourless reagents, others on reagents with chromophoric groups. Many of these methods are highly selective, e.g. the dithizone, dithiocarbamate, cuproine, and cuprizone methods described below in more detail. The dithizone method is the most sensitive. Cuprizone is used to determine copper in the aqueous phase, whereas the other methods mentioned involve extraction procedures.

19.2.1 Dithizone Method

In acidic medium containing an excess of dithizone (formula, p. 41), copper(II) forms the violet dithizonate, $Cu(HDz)_2$, which is soluble in non-polar solvents (CCl_4, $CHCl_3$), and is the basis of a sensitive spectrophotometric method for the determination of copper [10,11]. In alkaline medium, the less-intensely coloured, yellow-brown secondary cupric dithizonate, $CuDz$, is formed. Secondary cupric dithizonate can also be formed in a neutral or acidic medium containing a deficiency of dithizone, and, like $Cu(HDz)_2$, is soluble in CCl_4 and $CHCl_3$.

The molar absorptivity of $Cu(HDz)_2$ in CCl_4 is $4 \cdot 52 \times 10^4$ at $\lambda_{max} = 550$ nm ($a = 0 \cdot 71$). The absorption spectrum of copper dithizonate is shown in Fig. 2.2 (p. 43).

Although primary copper dithizonate is very stable, it is formed rather slowly, necessitating relatively prolonged shaking during the extraction. The use of a mechanical shaker is, therefore, recommended.

Higher concentrations of dithizone in the organic phase and lower acidity of the aqueous medium promote extraction. At pH 1, the rate of extraction is highest, and the solution is too acidic for bismuth and zinc to be extracted with a 0·002% solution of dithizone. Noble metals which react with dithizone (Pt, Pd, Au, Ag, and Hg) are extracted together with copper, but can be stripped from the organic extract as their stable iodide complexes with a 1% solution of potassium iodide [12]. Extraction of silver is prevented by chloride. Citrate and tartrate slightly inhibit the extraction of copper with dithizone.

The interferences in the determination of copper due to noble metals are most conveniently eliminated by extracting these metals at the start with dithizone from a 1M mineral acid solution. Noble metals (except palladium) form yellow-orange dithizonates, and their rates of extraction are much higher than that for copper. The solution is shaken with small portions of dithizone in CCl_4 until the organic layer no longer rapidly becomes yellow. (If shaking is continued beyond this, some copper will be extracted and the solution will turn violet).

Both the mono- and two-colour dithizone methods can be applied to the determination of copper. In the monocolour method, the free dithizone should be stripped from the extract with a very dilute ammonia solution, and the shaking should be continued for only a short time, to prevent partial conversion of the primary dithizonate into the secondary dithizonate.

The dithizone method has been used to determine copper in various materials, e.g. tin [7], titanium and its alloys [13], uranium compounds [14], and biological materials [15,16].

Reagents

1. Dithizone: 0·002% solution in CCl_4 (p. 492).
2. Standard copper solution: 1 mg/ml. Dissolve 3·928 g of cupric sulphate pentahydrate, $CuSO_4.5H_2O$, in water containing 1 ml of concentrated H_2SO_4, and dilute the solution with water to 1 litre.
3. Ammonia, dilute solution: 1 drop of conc. ammonia in 25 ml of water.

Procedure

Adjust a solution containing not more than 40 μg of copper, from which the noble metals (Pt, Pd, Au, Ag, Hg) have been removed, to pH ⩽ 1 with ammonia. Transfer the solution to a separating funnel, and extract with portions of dithizone in CCl_4 (1 ml of 0·002% H_2Dz corresponds to 2·5 μg of Cu) until the last portion does not change its green colour. With each extraction, separate the organic phase before the green dithizone reagent is completely complexed as the violet $Cu(HDz)_2$. Shake the aqueous sample with the final portion of the dithizone solution for not less than 3 minutes. Remove free dithizone by shaking the combined extracts with dilute ammonia. Dilute the violet dithizonate solution with carbon tetrachloride in a 50-ml (or smaller, depending on the amount of copper) volumetric flask, and measure the absorbance at 550 nm, using the solvent as reference.

Note. For prior separation of the noble metals, adjust the concentration of the inorganic acid in the sample solution to about 1M, and shake the solution in a separating funnel with 0·1–0·2-ml portions of a 0·001% dithizone solution in CCl_4 until the green colour changes to yellow. Discontinue the extraction when the last portion of the carbon tetrachloride solution turns violet, and add this portion to the subsequent (*vide supra*) $Cu(HDz)_2$ extract.

19.2.2 Dithiocarbamate Methods

The addition of an aqueous solution of sodium diethyldithiocarbamate (Na–DDTC, cupral) (formula, p. 54) to a solution (at pH 4–11) containing small amounts of copper(II) ions produces a yellow-brown colour due to a colloidal suspension of the sparingly-soluble copper 1:2 chelate with DDTC. The reagent co-ordinates with copper through the two sulphur atoms to form a chelate with four-membered rings, which is a rather rare configuration. Protective colloids (e.g. gum arabic) stabilize the pseudosolution, and permit the colorimetric determination of copper in the aqueous phase.

Greater accuracy and sensitivity are attained by extracting the copper diethyldithiocarbamate into organic solvents such as CCl_4, $CHCl_3$, trichloroethylene, amyl acetate, and isoamyl alcohol [17–21]. These solutions are fairly stable.

The molar absorptivity of a carbon tetrachloride solution of the complex is 1.4×10^4 at $\lambda_{max} = 436$ nm (specific absorptivity 0.22).

The main interfering metals in the copper determination are Fe, Bi, Mn, Ni, Co, Cr, Mo, and U, which form coloured complexes. The selectivity of the method is considerably enhanced by the use of EDTA as a masking agent. In a tartrate or citrate medium at pH 8–9, Fe, Mn, Ni, and Co are masked by EDTA, as are Cd, Pb, Zn, and In, which form colourless complexes with DDTC. Of the metals forming coloured compounds with DDTC, only Bi, Tl(III), and Cu are not masked. Thallium, when reduced to Tl(I), does not interfere. Bismuth can be stripped from the organic extract containing copper and bismuth dithiocarbamates with $5M$ hydrochloric acid (shaking time 30 seconds). Copper dithiocarbamate is decomposed by cyanide, whereas the bismuth complex remains unaffected.

Apart from cyanide, interferences in the determination of copper are also caused by thiosulphate and species which reduce Cu(II) to Cu(I) or oxidize DDTC.

The disadvantage of sodium diethyldithiocarbamate lies in its insolubility in organic solvents. It is also quite readily decomposed in acidic solutions into diethylamine and carbon disulphide.

Sometimes diethylammonium diethyldithiocarbamate (formula, p. 55), which is soluble in chloroform and stable in acid, is used instead of Na–DDTC [22–24]. A chloroform solution of the reagent is shaken with the sample solution to extract Cu(II).

Šedivec and Vašák [25] have overcome the interference of metals such as Fe, Mn, and Zn in the determination of copper by using lead diethyldithiocarbamate instead of Na–DDTC. When a chloroform solution of $Pb(DDTC)_2$ is shaken with an aqueous solution containing copper, a displacement reaction occurs:

$$Cu^{2+} + Pb(DDTC)_2 \rightarrow Cu(DDTC)_2 + Pb^{2+}$$

Metals, the DDTC-complexes of which are more stable than the lead complex [namely Hg, Ag, Ti(III), and, to some extent, Bi] interfere by consuming reagent, but do not interfere in the determination if they are colourless. Kreimer and Lomekhov [26] have studied the kinetics of the displacement reaction.

Zinc dibenzyldithiocarbamate, which is soluble in CCl_4 and $CHCl_3$, is less selective than lead diethyldithiocarbamate but more resistant to highly acidic media [27–29].

$$\begin{array}{c} \text{Ph–CH}_2 \\ \text{Ph–CH}_2 \end{array} \!\! N\text{–C} \!\! \begin{array}{c} S \\ S \end{array} \!\! Zn/2$$

With this reagent, copper can be extracted from $1-2N$ HCl or H_2SO_4. Lead dibenzyldithiocarbamate can be used similarly [30].

Other dithiocarbamates which have been suggested for spectrophotometric determination of copper include sodium or ammonium pyrrolidinedithiocarbamate [31], sodium piperazine-bis-dithiocarbamate [32], potassium diethanoldithiocarbamate [33], and morpholinium morpholine-N-dithiocarbamate [34].

With Na–DDTC, copper has been determined in various metals [18,19], zinc [35,36], cadmium and its compounds [35–37], lead [36], antimony [38], titanium and zirconium [39], graphite [40], petroleum [41], waste water [42], water [43], and soil and plant material [44].

With $Pb(DDTC)_2$, copper has been determined in various metals [45], nickel and cobalt [46], gallium arsenide [47], and water [48].

Zinc dibenzyldithiocarbamate has been used to determine copper in foodstuffs [28,49], plant material [29], preserved timber [50], phosphate rocks [51], steel [52], and water [53,54].

Carter and Nickless [55] have developed an automated procedure for the determination of copper with zinc dibenzyldithiocarbamate.

Reagents

1. Sodium diethyldithiocarbamate: 0·1% solution adjusted with ammonia to pH $\sim$ 8·5.
2. Standard copper solution: 1 mg/ml (p. 240).
3. EDTA (Complexone III): $0\cdot1M$ ($\sim$ 3·7%) solution.
4. Potassium sodium tartrate: 20% solution. Purified by the extraction of copper traces with Na–DDTC.

Procedure

To a solution containing not more than 150 μg of Cu, add 1–5 ml of the tartrate solution and 1–5 ml of the EDTA solution, adjust the solution with ammonia to pH $\sim$ 8·5, and add 5 ml of the Na–DDTC solution. Shake the solution in a separating funnel for about 1 minute with each of two portions of CCl_4. Make up the combined extracts with the solvent to the mark in a 50-ml (or smaller) volumetric flask, and measure the absorbance at 436 nm against CCl_4.

Notes. (1) Sufficient EDTA and tartrate must be added to mask metals which would react with DDTC, or which would be precipitated after the adjustment of pH.

(2) If bismuth is absent, shaking the carbon tetrachloride extract with an aqueous 1% solution of KCN will decolorize the organic phase.

(3) Copper may be determined without extraction by adding 5 ml of 1% gum arabic solution (per 50 ml final volume) to the sample solution before the Na–DDTC is added.

19.2.3 CUPROINE METHODS

Cuproine (2,2'-biquinolyl) (I), neocuproine (2,9-dimethyl-1,10-phenanthroline) (II), and bathocuproine (2,9-dimethyl-4,7-diphenyl-1,10-phenanthroline) (III) react with copper(I) to form coloured cationic complexes, the structure of which is illustrated by the formula of the neocuproine complex:

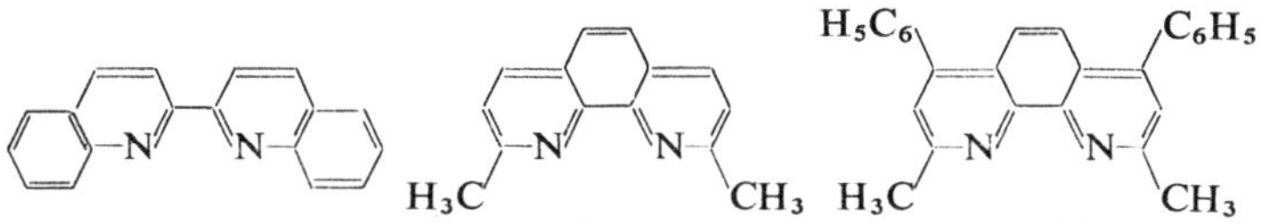

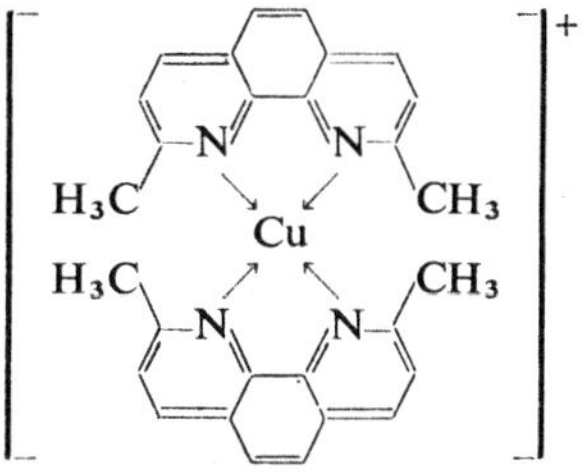

The specific structural moiety for copper(I) is the group which gives intermolecular steric-hindrance for octahedral, but not tetrahedral co-ordination:

$$X-C=N-C-C-N=C-X$$

The methods using these reagents are specific for copper, but are of relatively low sensitivity (especially in the case of cuproine).

Hoste *et al.* [56,57] have thoroughly investigated the conditions for the extractive spectrophotometric determination of copper with cuproine. Amyl alcohols and n-hexanol are most suitable for the extraction of the copper complex from aqueous solutions. A reducing agent (usually $NH_2OH.HCl$), is added to the sample solution, the pH is adjusted to the optimum range of 4–7, and the sample shaken with a colourless isoamyl alcohol solution of cuproine. Since the distribution coefficient is high (> 1000), extraction with one portion of cuproine solution is usually sufficient.

At $\lambda_{max} = 546$ nm, the molar absorptivity of the copper(I)–cuproine complex in isoamyl alcohol is 6.4×10^3 (specific absorptivity 0.10). The coloured solutions are stable.

Figure 19.1 shows the absorption spectra of the copper(I)–cuproine, –neocuproine, and –bathocuproine complexes in isoamyl alcohol.

Of the other cations, only Ti^{3+} reacts with cuproine, giving a greenish colour which is much weaker than the purple of the cuprous complex. In the presence of citrate, chromium(III) interferes due to the formation of a stable ternary complex with Cr(III), Cu(II), and citrate [58]. Of the anions, cyanide, thiosulphate, oxalate, and EDTA interfere in the determination of copper.

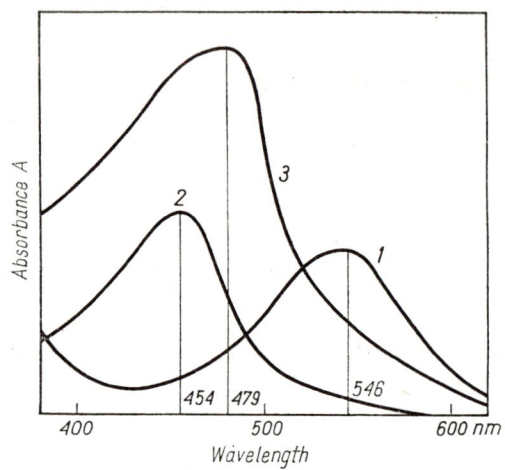

Fig. 19.1 Absorption spectra of copper(I) complexes with cuproine (1), neocuproine (2), and bathocuproine (2) in isoamyl alcohol

Hackett [59] has extracted the cuproine complex from aqueous methanol solutions into chloroform. Since cuproine and its complex with copper(I) are readily soluble in dimethylformamide, it is possible to determine copper in water–dimethylformamide (about 1:1 v/v) without solvent extraction [60]. The value of λ_{max} and the molar absorptivity are the same in this medium as in isoamyl alcohol.

The cuprous complex has a higher absorption in the near ultraviolet ($\varepsilon = 5\cdot 2 \times 10^4$ at $\lambda = 358$ nm), but the cuproine reagent also absorbs slightly at this wavelength [61].

With cuproine, copper has been determined in steel [62–65], nickel and cobalt [63,64], chromium and manganese [64], aluminium and its alloys [64], sodium chloride [66], soil and plant materials [44], silicate minerals [67], foodstuffs [49,68], biological materials [67], water [43,67], mineral oils [60], and acrylonitrile [69].

Neocuproine, recommended for the determination of copper(I) by Smith and McCurdy [70], reacts with copper(I) within the pH range from 5 to 10 to form an orange complex which is extractable with n-amyl and isoamyl alcohols, n-hexanol, and MIBK. In the presence of ethanol, the Cu(I) neocuproine complex can also be extracted into chloroform [71].

The molar absorptivity of the complex in isoamyl alcohol at λ_{max} = 454 nm is $7 \cdot 9 \times 10^3$ ($a = 0 \cdot 12$).

Neocuproine (2,9-dimethyl-1,10-phenanthroline) is as specific for copper(I) as cuproine is. Substitution of methyl groups at the 2- and 9-carbon atoms in 1,10-phenanthroline renders the reagent unreactive towards iron(II). After extraction of the copper(I) neocuproine complex with isoamyl alcohol, iron(II) can be determined in the aqueous phase by the addition of 1,10-phenanthroline [72].

Higher sensitivity is obtained by extracting with chloroform the ion-pair formed by the copper(I) neocuproine complex and the anionic dye Rose Bengal (tetrachlorotetraiodofluorescein). The molar absorptivity of the ion-pair in CHCl$_3$ is $6 \cdot 25 \times 10^4$ at 570 nm [73].

Copper has been determined by the neocuproine method in beryllium [74,75], titanium [76], arsenic and gallium [77], germanium and silicon [78], tungsten [79], aluminium and lead–tin alloys [80], mineral oils [41,81], foodstuffs [49,82], biological materials [16], water [43], plutonium [83], and tellurium [84,84a].

The bathocuproine method for the determination of copper(I) is about twice as sensitive as those based on cuproine and neocuproine. The molar absorptivity of the copper bathocuproine complex in isoamyl alcohol at λ_{max} = 479 nm is $1 \cdot 42 \times 10^4$ ($a = 0 \cdot 22$).

The Cu(I) bathocuproine complex, which is formed between pH 4 and 8, can be extracted with amyl alcohols or n-hexanol.

Bathocuproine has been applied in the determination of copper in cast iron [85]; niobium, tantalum, molybdenum and tungsten [86]; lead and nickel [87]; and products of the paper-making industry [88].

Blair and Diehl [89] have recommended the use of bathocuproine-disulphonic acid for the determination of copper, because the reagent and its Cu(I) complex are soluble in water. The sensitivity of this method is the same as in the case of bathocuproine.

The following cuproine analogues have also been recommended as spectrophotometric reagents for the determination of copper: 2,3,8,9-dibenzo-4,7-dimethyl-5,6-dihydro-1,10-phenanthroline (Cuprotest) [90], 4,4'-dihydroxy-2,2'-biquinoline [91], 3,3'-dimethylene-4,4'-diphenylbiquinolyl [92], and 2-pyridyl-substituted quinoxalines [93,94]. Dunbar and Schilt [95] have shown 2,9-dimethyl-4,7-dihydroxy-1,10-phenanthroline to be a sensitive reagent which is applicable in highly alkaline solutions.

Reagents

1. Cuproine: 0·03% solution in isoamyl alcohol. Dissolve 75 mg of reagent in 250 ml of alcohol.
2. Standard copper solution: 1 mg/ml (p. 240).
3. Acetate buffer: pH $\sim$ 4·5. Dissolve in water 80 g of sodium acetate and 60 ml of conc. acetic acid; dilute the solution with water to 500 ml.

Procedure

To an acidic solution containing not more than 300 μg of Cu, add about 0·2 g of $NH_2OH.HCl$, ammonia to pH ~ 3, and 10 ml of acetate buffer. Transfer the solution to a separating funnel, and shake for 2 minutes with each of 2 or 3 portions of the cuproine solution. The last portion should remain colourless. Dilute the combined extracts with isoamyl alcohol to the mark in a 50-ml (or smaller) volumetric flask (according to the amount of copper), and measure the absorbance at 546 nm, using the solvent as reference.

Note. If the isoamyl alcohol extract is turbid, clear it by adding a small amount of methanol, or by passing the solution through a dry filter.

19.2.4 Cuprizone Method

Colourless cuprizone (bis-cyclohexanone-oxalyldihydrazone) (**IV**) reacts with copper(II) in slightly alkaline medium (optimum pH 8–9) to form a blue water-soluble complex which constitutes the basis of a very selective

$$\underset{H_2}{\underset{H_2}{H_2}}\underset{H_2}{\overset{H_2}{\bigcirc}}=N-NH-\overset{O}{\overset{\|}{C}}-\overset{O}{\overset{\|}{C}}-NH-N=\underset{H_2}{\underset{H_2}{\overset{H_2}{\bigcirc}}}\overset{H_2}{\underset{H_2}{H_2}} \quad \text{(IV)}$$

spectrophotometric method for the determination of copper in the aqueous phase [96–100]. The complex is not formed below pH 6·5, and the colour fades above pH 12. The molar absorptivity at 595–600 nm is $1·6 \times 10^4$ ($a = 0·25$).

Copper is usually determined in an ammoniacal citrate medium which keeps most metals in solution. The presence of citrate somewhat retards the development of the colour, but once the copper cuprizone complex is formed, the colour does not fade.

Interferences in the determination of copper by this method include major quantities (about 10 mg) of Ni, Co, Fe, Cr, and U. Iron in 500-fold amount relative to copper does not interfere when citrate is present. Citrate, tartrate, and oxalate do not interfere, but cyanide does.

Copper has been determined with cuprizone in cast iron and steel [1,97–99,101], lead and its alloys [3,102], tin, aluminium, zinc and their alloys [102], titanium and zirconium [39], platinum alloys [103], cadmium sulphide [104], borate glass [104a], petroleum distillates [105], plant materials [106,107], biological materials [16], and water [43].

The spectrophotometric method for copper using bis-acetaldehyde-oxalyldihydrazone, a related reagent, is more sensitive ($\varepsilon = 2·36 \times 10^4$) [99,100,108–110]. This reagent has been applied to the determination of copper in uranium [109], steel [99], and water [43,110].

Other reagents which have been similarly recommended include quinoline-2-aldehyde-2-quinolylhydrazone (QAQH) (molar absorptivity $5·8 \times 10^4$ at 540 nm) [111–113] and bis(ethylacetoacetate)oxalyldihydrazone (neocuprizone) ($\varepsilon = 1·39 \times 10^4$ at 585 nm) [114].

Reagents

1. Cuprizone: 0·1% solution. Dissolve 100 mg of reagent in 20 ml of hot 50% ethanol, and dilute the solution with cold 50% ethanol to 100 ml.
2. Standard copper solution: 1 mg/ml (p. 240).

Procedure

To a solution in a 50-ml volumetric flask, containing not more than 150 μg of Cu, add 1–5 ml of 10% ammonium citrate solution (depending on the amount of metals in the solution which would precipitate as the pH is increased), and adjust the solution with ammonia to pH 8–9. Add 5 ml of the cuprizone solution, dilute the solution to the mark with water, and after 10 minutes measure the absorbance of the blue solution at 600 nm against water.

19.2.5 OTHER METHODS

Diphenylcarbazide, a well-known reagent for chromium(VI), is also a highly sensitive reagent for the determination of copper. The colour in an aqueous alkaline solution increases uniformly for about 8 minutes and thereafter increases erratically [115]. However, when copper(II) is extracted with a benzene solution of diphenylcarbazide, the colour of the complex (λ_{max} = 545 nm) is stable for at least 24 hours [116]. The molar absorptivity of the complex in benzene is $> 8·0 \times 10^4$. The maximum colour obtained in aqueous alkaline solution is less than that attainable in benzene. A related reagent, diphenylcarbazone, is also adopted [117–118a].

Several sensitive methods for the determination of copper are based on azo reagents, such as PAN [119], 2-(5-nitro-2-pyridylazo)-1-naphthol [120], 2-(2-pyridylazo)-1-naphthol (o-α-PAN) [120a], PAR [121], 2-(2-thiazolylazo)-5-dimethylaminophenol (TAM) ($\varepsilon = 3·94 \times 10^4$ at 570 nm) [122], 1-(2-thiazolylazo)-2-naphthol (TAN) [123], Picramine R [124], Picramine-epsilon ($\varepsilon = 3·0 \times 10^4$ at 550 nm) [125], Aminomethylazo III [126], other bisazo derivatives of chromotropic acid [127], Acid Alizarin Black SN [128], Zincon [129], Calcichrome [130], Azo-azoxy BN [131], and Magneson KhS [132].

Many methods are based on oximes, e.g. α-furildioxime [133], α-benzoinoxime (cupron) [134], salicylaldoxime [135], acetylacetone dioxime [136], pyridine-2,6-diacetoxime [137], methyl-2-pyridylketoxime [138], n-amyl-2-pyridylketoxime [139], 6-methylpyridyl-2-aldoxime [140], phenyl-2-(6-methylpyridyl)ketoxime [141], and dimethylglyoxime (+pyridine) [142]. The coloured copper complexes formed with most of these reagents can be extracted into chloroform or isoamyl alcohol.

A numerous group of methods has been developed based on organic reagents having sulphur as the ligand atom (e.g. the dithizone and Na-DDTC methods already described). Tetraethylthiuram disulphide (dicupral)

[143,144] and its derivatives [145,146] are highly selective and sensitive reagents ($\varepsilon = 2.3 \times 10^4$ with dicupral).

Rubeanic acid (dithio-oxamide), which exists in amino and imino tautomers,

$$\begin{array}{c} H_2N-C=S \\ | \\ H_2N-C=S \end{array} \rightleftharpoons \begin{array}{c} HN=C-SH \\ | \\ HN=C-SH \end{array}$$

reacts with copper ions in a weakly acidic medium to yield a sparingly soluble compound which, at low concentrations of copper, forms an olive-green pseudosolution stabilized with a protective colloid. Rubeanic acid is fairly often used for the colorimetric determination of copper [147–149].

The following other organic reagents complexing through sulphur have been proposed: 2,3-dimercaptopropanol [150], mercaptoquinoline (thio-oxine) [151], 8,8'-diquinolyl disulphide [152], dimercaptothiopyrones [153], 6-methylpicolinic acid thioamide [154], derivatives of thiazolidine-2-thione [155,156], and 2,3-quinoxalinedithiol [157].

Some β-diketones have also been used in the determination of copper, e.g. acetylacetone [158], 2-furoyltrifluoroacetone [159], TTA [160,161], thiothenoyltrifluoroacetone [162], thiodibenzoylmethane [162a], and 3-thianaphthenoyltrifluoroacetone [162b].

Diverse organic spectrophotometric reagents for the determination of copper include Aluminon [163], Chrome Azurol S [164], 2-nitroso-1-naphthol-4-sulphonic acid [165], Tiron [166], dicinchoninic acid [167, 168], 3-(2-pyridyl)-5,6-diphenyl-1,2,4-triazine (PDT) [169], bis(2-methyl-2-pyridyl)glyoxal dihydrazone [170], 2-(4-toluenesulphonamido)aniline (TSA) [171], N-8-quinolyl-4-toluenesulphonamide (QTS) [171a], 2,3-bis(salicylideneamino)benzofuran [172], 1,3-dimethyl-4-imino-5-oximinoalloxan (DAXIM) [173], and Variamine Blue [redox reaction with Cu(II)] [174].

Coloured copper complexes with EDTA (Complexone III) [175,176], DCTA (Complexone IV) [177], ethylenediaminedipropionic acid [178], and diethylenetriaminepenta-acetic acid [179] are useful in less sensitive methods for the determination of copper.

Relatively higher concentrations of copper are also determined as the blue ammine complex (molar absorptivity 120) [180], or similar copper complexes with ethylenediamine [181] and triethylenetetramine (TRIEN) [182,183].

Other less-sensitive methods are based on copper complexes with chloride (in water–acetone) [184], bromide [185], azide [186], and pyrophosphate [187]. Azide [188] and thiocyanate [189] copper ternary complexes with pyridine are extractable into chloroform. The coloured complex formed with Cu(II), thiocyanate, and 3,5-dimethylpyrazole can also be extracted with chloroform [190].

When copper(II) ions are added to a solution of molybdophosphoric acid in the presence of cyanide ions, the phosphomolybdenum blue formed can be used for the indirect photometric determination of copper [191].

Rereferences

1. Elliott, C. R., Preston, P. F. and Thompson, J. H., *Analyst* **84**, 237 (1959).
2. Marczenko, Z., *Mikrochim. Acta* **1965**, 281.
3. Thompson, J. H. and Ravenscroft, M. J., *Analyst* **85**, 735 (1960).
4. Handley, T. H. and Dean, J. A., *Anal. Chem.* **33**, 1087 (1961).
5. Schwedow, W. P. and Klug, O. N., *Chem. Anal. (Warsaw)* **11**, 237, 831 (1966).
6. Scribner, W. G., Treat, W. J., Weis, J. D. and Moshier, R. W., *Anal. Chem.* **37**, 1136 (1965).
6a. Adam, T. and Přibil, R., *Talanta* **19**, 1105 (1972).
7. Marczenko, Z. and Kasiura, K., *Chem. Anal. (Warsaw)* **10**, 449 (1965).
8. Chuiko, V. T. and Reva, N. I., *Ukr. Khim. Zh.* **34**, 193 (1968).
9. Malissa, H. and Marr, I. L., *Mikrochim. Acta* **1971**, 241.
9a. Strelow, F. W., Victor, A. H. and Weinert, C. H., *Anal. Chim. Acta* **69**, 105 (1974).
10. Geiger, R. W. and Sandell, E. B., *Anal. Chim. Acta* **8**, 197 (1953).
11. Trémillon, B., *Bull. Soc. Chim. France* **1954**, 1156, 1160.
12. Friedeberg, H., *Anal. Chem.* **27**, 305 (1955).
13. Pender, H. W., *ibid.* **30**, 1915 (1958).
14. Mareček, J. and Singer, E., *Z. Anal. Chem.* **203**, 336 (1964).
15. Masiak, M., Gradziński, A. and Sierawski, S., *Chem. Anal. (Warsaw)* **5**, 763 (1960).
16. Butler, E. J. and Forbes, D. H., *Anal. Chim. Acta* **33**, 59 (1965).
17. Jewsbury, A., *Analyst* **78**, 363 (1953).
18. Cluley, H. J., *ibid.* **79**, 561 (1954).
19. Pohl, H., *Anal. Chim. Acta* **12**, 54 (1955).
20. Claassen, A. and Bastings, L., *Z. Anal. Chem.* **153**, 30 (1956).
21. Gottschalk, G., *ibid.* **194**, 321 (1963).
22. Wyatt, P. F., *Analyst* **78**, 656 (1953).
23. Abson, D. and Lipscomb, A. G., *ibid.* **82**, 152 (1957).
24. Analytical Methods Committee, *ibid.* **88**, 253 (1963).
25. Šedivec, V. and Vašák, V., *Chem. Listy* **45**, 435 (1951).
26. Kreimer, S. E. and Lomekhov, A. S., *Zh. Analit. Khim.* **18**, 567 (1963).
27. Martens, R. I. and Githens, R. E., Sr., *Anal. Chem.* **24**, 991 (1952).
28. Abbott, D. C. and Polhill, R. D. A., *Analyst* **79**, 547 (1954).
29. Andrus, S., *ibid.* **80**, 514 (1955).
30. Dittel, F., *Z. Anal. Chem.* **229**, 188 (1967).
31. Kovács, E. and Guyer, H., *ibid.* **186**, 267 (1962); **209**, 388 (1965).
32. Hulanicki, A. and Shishkova, L., *Chem. Anal. (Warsaw)* **10**, 837 (1965).
33. Tulyupa, F. M., Bekleshova, G. E. and Vitkina, M. A., *Zh. Analit. Khim.* **21**, 783 (1966).
34. Beyer, W. and Likussar, W., *Mikrochim. Acta* **1967**, 721; **1971**, 610.
35. Kladnitskaya, K. B. and Grisevich, A. N., *Ukr. Khim. Zh.* **27**, 803 (1961).
36. Cyrankowska, M. and Downarowicz, J., *Chem. Anal. (Warsaw)* **10**, 1015 (1965).
37. Podchainova, V. N., Liplavk, I. L. and Ushkova, L. N., *Zavodsk. Lab.* **38**, 411 (1972).
38. Provaznik, J. and Knižek, M., *Chem. Listy* **55**, 79 (1961).
39. Wood, D. F. and Clark, R. T., *Analyst* **83**, 509 (1958).
40. Gorczyńska, K., Ciecierska, D. and Walędziak, H., *Chem. Anal. (Warsaw)* **2**, 52 (1957).
41. Howard, J. M. and Spauschus, H. O., *Anal. Chem.* **35**, 1016 (1963).
42. Committee on Methods for the Analysis of Trade Effluents, *Analyst* **81**, 59 (1956).
43. Tuck, B. and Osborn, E. M., *ibid.* **85**, 105 (1960).
44. Cheng, K. L. and Bray, R. H., *Anal. Chem.* **25**, 655 (1953).
45. Adamiec, I., *Rudy Metale* **5**, 409 (1960).
46. Kalinkin, I. P. and Semikozov, G. S., *Zavodsk. Lab.* **27**, 17 (1961).

47. Goryushina, V. G., Biryukova, E. Ya. and Razumova, L. S., *Zh. Analit. Khim.* **23**, 1044 (1968).
48. Kovařík, M. and Vinš, V., *Z. Anal. Chem.* **147**, 401 (1955).
49. Russell, G. and Hart, P. J., *Analyst* **83**, 202 (1958).
50. Williams, A. I., *ibid.* **93**, 611 (1968).
51. Kocher, J., *Chim. Anal. (Paris)* **44**, 161 (1962).
52. Bhargava, O. P., *Talanta* **16**, 743 (1969).
53. Abbott, D. C. and Harris, J. R., *Analyst* **87**, 497 (1962).
54. Wilson, A. L., *ibid.* **87**, 884 (1962).
55. Carter, J. M. and Nickless, G., *ibid.* **95**, 148 (1970).
56. Hoste, J., Heiremans, A. and Gillis, J., *Mikrochemie* **36/37**, 349 (1951).
57. Hoste, J., Eeckhout, J. and Gillis, J., *Anal. Chim. Acta* **9**, 263 (1953).
58. Irving, H. M. and Tomlinson, W. R., *Talanta* **15**, 1267 (1968).
59. Hackett, C. E., *Anal. Chim. Acta* **12**, 358 (1955).
60. Pflaum, R. T., Popov, A. I. and Goodspeed, N. C., *Anal. Chem.* **27**, 253 (1955)
61. Gershuns, A. L. and Grineva, L. G., *Zh. Analit. Khim.* **26**, 1485 (1971).
62. Elwell, W. T., *Analyst* **80**, 508 (1955).
63. Gräbner, H. J., *Z. Anal. Chem.* **182**, 401 (1961).
64. Leeb, A. J. and Hecht, F., *ibid.* **168**, 101 (1959).
65. Shanahan, C. E. and Jenkins, R. H., *Analyst* **86**, 166 (1961).
66. Bahr, H. and Lipińska, I., *Chem. Anal. (Warsaw)* **13**, 399 (1968).
67. Riley, J. P. and Sinhaseni, P., *Analyst* **83**, 299 (1958).
68. Peynaud, E., *Chim. Anal. (Paris)* **36**, 187 (1954).
69. Maute, R. L., Owens, M. L., Jr., and Slate, J. L., *Anal. Chem.* **27**, 1614 (1955).
70. Smith, G. F. and McCurdy, W. H., Jr., *ibid.* **24**, 371 (1952).
71. Gahler, A. R., *ibid.* **26**, 577 (1954).
72. Zak, B. and Ressler, N., *ibid.* **28**, 1158 (1956).
73. Bailey, B. W., Dagnall, R. M. and West, T. S., *Talanta* **13**, 753 (1966).
74. Hibbits, J. O., Davis, W. F. and Menke, M. R., *ibid.* **4**, 101 (1960).
75. Pollock, E. N. and Zopatti, L. P., *Anal. Chim. Acta* **28**, 68 (1963).
76. Frank, A. J., Goulston, A. B. and Deacutis, A. A., *Anal. Chem.* **29**, 750 (1957).
77. Knižek, M. and Pečenkova, V., *Zh. Analit. Khim.* **21**, 260 (1966).
78. Luke, C. L. and Campbell, M. E., *Anal. Chem.* **25**, 1588 (1953).
79. Norwitz, G. and Gordon, H., *ibid.* **37**, 417 (1965).
80. Fulton, J. W. and Hastings, J., *ibid.* **28**, 174 (1956).
81. Zall, D. M., McMichael, R. E. and Fisher, D. W., *ibid.* **29**, 88 (1957).
82. Jones, P. D. and Newman, E. J., *Analyst* **87**, 637 (1962).
83. Lindsay, J. W. and Plock, C. E., *Talanta* **16**, 414 (1969).
84. Nebesar, B., *Anal. Chem.* **36**, 1961 (1964).
84a. Kasterka, B. and Dobrowolski, J., *Chem. Anal. (Warsaw)* **16**, 619 (1971).
85. Smith, G. F. and Wilkins, D. H., *Anal. Chem.* **25**, 510 (1953).
86. Penner, E. M. and Inman, W. R., *Talanta* **10**, 407 (1963).
87. Jackwerth, E. and Döring, E., *Z. Anal. Chem.* **255**, 194 (1971).
88. Borchardt, L. G. and Butler, J. P., *Anal. Chem.* **29**, 414 (1957).
89. Blair, D. and Diehl, H., *Talanta* **7**, 163 (1961).
90. Ackermann, G. and Angermann, W., *ibid.* **15**, 79 (1968).
91. Schilt, A. A. and Hoyle, W. C., *Anal. Chem.* **41**, 344 (1969).
92. Uhlemann, E. and Waiblinger, K., *Anal. Chim. Acta* **41**, 161 (1968).
93. Stephen, W. I. and Uden, P. C., *ibid.* **39**, 357 (1967).
94. Foster, D. and Trusell, F. C., *ibid.* **47**, 154 (1969).
95. Dunbar, W. E. and Schilt, A. A., *Talanta* **19**, 1025 (1972).
96. Peterson, R. E. and Bollier, M. E., *Anal. Chem.* **27**, 1195 (1955).
97. Haywood, L. J. and Sutcliffe, P., *Analyst* **81**, 651 (1956).
98. Wetlesen, C. U., *Anal. Chim. Acta* **16**, 268 (1957).

99. Capelle, R., *Chim. Anal. (Paris)* **42**, 69, 127, 181 (1960).
100. Jacobsen, E., Langmyhr, F. J. and Selmer-Olsen, A. R., *Anal. Chim. Acta* **24**, 579 (1961).
101. Meyer, S. and Koch, O. G., *Arch. Eisenhüttenw.* **32**, 67 (1961).
102. Rohde, R. K., *Anal. Chem.* **38**, 911 (1966).
103. Forbes, J. S. and Dalladay, D. B., *Analyst* **86**, 418 (1961).
104. Cabane-Brouty, F., *Anal. Chim. Acta* **47**, 511 (1969).
104a. Banerjee, S. and Paul, A., *ibid.* **68**, 226 (1974).
105. Lambdin, C. E. and Taylor, W. V., *Anal. Chem.* **40**, 2196 (1968).
106. Middleton, K. R., *Analyst* **90**, 234 (1965).
107. Sharp, R. B., *ibid.* **91**, 212 (1966).
108. Gran, G., *Anal. Chim. Acta* **14**, 150 (1956).
109. Stevančević, D. B., *Z. Anal. Chem.* **165**, 348 (1959).
110. Capelle, R., *Chim. Anal. (Paris)* **43**, 280 (1961); **48**, 498 (1966).
111. Jensen, R. E., Bergman, N. C. and Helvig, R. J., *Anal. Chem.* **40**, 624 (1968).
112. Sims, G. G. and Ryan, D. E., *Anal. Chim. Acta* **44**, 139 (1969).
113. Abraham, J., Winpe, M. and Ryan, D. E., *ibid.* **48**, 431 (1969).
114. Ackermann, G. and Kaden, W., *Z. Anal. Chem.* **234**, 409 (1968).
115. Turkington, R. W. and Tracy, F. M., *Anal. Chem.* **30**, 1699 (1958).
116. Stoner, R. E. and Dasler, W., *ibid.* **32**, 1207 (1960).
117. Lapin, L. N. and Reis, N. V., *Zh. Analit. Khim.* **13**, 426 (1958).
118. Geering, H. R. and Hodgson, J. F., *Anal. Chim. Acta* **36**, 537 (1966).
118a. Einaga, H. and Ishii, H., *Analyst* **98**, 82 (1973).
119. Pease, B. F. and Williams, M. B., *Anal. Chem.* **31**, 1044 (1959).
120. Dahl, I., *Anal. Chim. Acta* **62**, 145 (1972).
120a. Betteridge, D., John, D. and Snape, F., *Analyst* **98**, 512 (1973).
121. Tataev, O. A., Akhmedov, S. A. and Akhmedova, Kh. A., *Zh. Analit. Khim.* **24**, 834 (1969).
122. Minczewski, J. and Kasiura, K., *Chem. Anal. (Warsaw)* **10**, 719 (1965).
123. Gusev, S. I., Ketova, L. A. and Glushkova, I. N., *Zh. Analit. Khim.* **25**, 2099 (1970).
124. Bogdanova, V. I. and Dedkov, Yu. M., *ibid.* **23**, 1046 (1968); *Zavodsk. Lab.* **34**, 688 (1968).
125. Dedkov, Yu. M., Koluzanova, V. P. and Kirakosyan, A. K., *Zh. Analit. Khim.* **25**, 1482 (1970).
126. Buděšinský, B. and Haas, K., *Z. Anal. Chem.* **214**, 325 (1965).
127. Okhanova, L. A., Bol'shakova, L. I. and Savvin, S. B., *Zh. Analit. Khim.* **23**, 1562 (1968).
128. Hosain, M. and West, T. S., *Anal. Chim. Acta* **33**, 164 (1965).
129. Rush, R. M. and Yoe, J. H., *Anal. Chem.* **26**, 1345 (1954).
130. Ishii, H. and Einaga, H., *Bull. Chem. Soc. Japan* **38**, 1416 (1965); **39**, 1154 (1966).
131. Minczewski, J. and Wieteska, E., *Chem. Anal. (Warsaw)* **9**, 365 (1964).
132. Lukin, A. M. and Vysokova, N. N., *Zavodsk. Lab.* **37**, 28 (1971).
133. Benediktova-Lodochnikova, N. V., *Zh. Analit. Khim.* **18**, 1322 (1963).
134. Madera, J., *Anal. Chem.* **27**, 2003 (1955).
135. Simonsen, S. H. and Burnett, H. M., *ibid.* **27**, 1336 (1955).
136. Ben-Bassat, A. H., Sa'at, Y. and Sarel, S., *Bull. Soc. Chim. France* **1960**, 948.
137. Gagliardi, E. and Presinger, P., *Mikrochim. Acta* **1965**, 1047.
138. Banerjea, D. K., and Tripathi, K. K., *Anal. Chem.* **32**, 1196 (1960).
139. Trusell, F. and Lieberman, K., *Anal. Chim. Acta* **30**, 269 (1964).
140. Hartkamp, H., *Z. Anal. Chem.* **176**, 185 (1960).
141. Pemberton, J. R. and Diehl, H., *Talanta* **16**, 393 (1969).
142. Peshkova, V. M., Levontin, M. E. and Litvin, K. I., *Zh. Analit. Khim.* **3**, 161 (1948).

143. Michal, J. and Zýka, J., *Collection Czech. Chem. Commun.* **20**, 305 (1955); *Zh. Analit. Khim.* **14**, 422 (1959).
144. Bilíková, A. and Zýka, J., *Chem. Listy* **59**, 91 (1965).
145. Musil, A., Wawschinek, O. and Leitner, J., *Mikrochim. Acta* **1963**, 355.
146. Wawschinek, O. and Tagger, H. H., *Anal. Chim. Acta* **35**, 109 (1966).
147. Jacobs, W. D. and Yoe, J. H., *ibid.* **20**, 332, 435 (1959).
148. McCann, D. S., Burcar, P. and Boyle, A. J., *Anal. Chem.* **32**, 547 (1960).
149. Paul, A., *Anal. Chem.* **35**, 2119 (1963).
150. Casassas, E., *Chim. Anal. (Paris)* **47**, 419 (1965).
151. Bankovskii, Yu. A. and Ievin'sh, A. F., *Zh. Analit. Khim.* **13**, 643 (1958).
152. Bankovskii, Yu. A., Ievin'sh, A. F., Luksha, E. O. and Bochkans, P. Ya., *ibid.* **16**, 150 (1961).
153. Usatenko, Yu. I., Arishkevich, A. M. and Danilevskaya, A. I., *Tr. Komis. po Analit. Khim. Akad. Nauk SSSR* **15**, 88 (1965).
154. Wawschinek, O., *Mikrochim. Acta* **1965**, 860.
155. Corbett, J. A., *Talanta* **13**, 1089 (1966).
156. Stiff, M. J., *Analyst* **97**, 146 (1972).
157. Burke, R. W. and Deardorff, E. R., *Talanta* **17**, 255 (1970).
158. Ben-Bassat, A. H. and Frydman-Kupfer, G., *Chemist-Analyst* **51**, 44 (1962).
159. Berg, E. W. and Day, M. C., *Anal. Chim. Acta* **18**, 578 (1958).
160. Khopkar, S. M. and De, A. K., *Z. Anal. Chem.* **171**, 241 (1959).
161. Akaiwa, H., Kawamoto, H. and Abe, M., *Bull. Chem. Soc. Japan* **44**, 117 (1971).
162. Shinde, V. M. and Khopkar, S. M., *Anal. Chem.* **41**, 342 (1969).
162a. Uhlemann, E., Schuknecht, B., Busse, K. D. and Pohl, V., *Anal. Chim. Acta* **56**, 185 (1971).
162b. Johnston, J. R. and Holland, W. J., *Mikrochim. Acta* **1972**, 126.
163. Mukherji, A. K. and Dey, A. K., *Anal. Chim. Acta* **18**, 546 (1958).
164. Ishida, R. and Sawaguchi, T., *Japan Analyst* **16**, 590 (1967).
165. Tolmachev, V. N. and Tul'chinskaya, A. Ya., *Zh. Analit. Khim.* **14**, 272 (1959).
166. Majumdar, A. K. and Savariar, C. P., *Anal. Chim. Acta* **21**, 53 (1959).
167. Tikhonov, V. N., Mustafin, I. S. and Grankina, M. Ya., *Zh. Analit. Khim.* **20**, 390 (1965); **21**, 1016 (1966).
168. Gregorowicz, Z., Kwapulińska, G. and Piwowarska, B., *Chem. Anal. (Warsaw)* **13**, 887 (1968).
169. Schilt, A. A. and Taylor, P. J., *Anal. Chem.* **42**, 220 (1970).
170. Valcarcel, M. and Pino, F., *Analyst* **98**, 246 (1973).
171. Betteridge, D. and Rangaswamy, R., *Anal. Chim. Acta* **42**, 293 (1968).
171a. Haworth, D. T. and Munroe, J. H., *Anal. Chem.* **41**, 529 (1969).
172. Ishii, H. and Einaga, H., *ibid.* **94**, 1038 (1969).
173. Burger, K., *Talanta* **8**, 77 (1961).
174. Gregorowicz, Z., *Z. Anal. Chem.* **171**, 246 (1959).
175. Nielsch, W. and Böltz, G., *ibid.* **143**, 1 (1954).
176. Gottschalk, G., *ibid.* **193**, 1 (1963).
177. Nielsch, W., *Mikrochim. Acta* **1959**, 419.
178. Da Silva, J. J., Calado, J. C. and De Moura, M. L., *Talanta* **12**, 467 (1965).
179. Bermejo-Martinez, F. and Rodriguez-Campos, J. A., *Microchem. J.* **11**, 331 (1966).
180. Malkina, T. G. and Podchainova, V. N., *Zh. Analit. Khim.* **19**, 668 (1964).
181. Tomic, E. A. and Bernard, J. L., *Anal. Chem.* **34**, 632 (1962).
182. Cheng, K. L., *ibid.* **34**, 1392 (1962).
183. Goydish, B. L., *Mikrochim. Acta* **1971**, 675.
184. Walker, C. R. and Vita, O. A., *Anal. Chim. Acta* **47**, 9 (1969).
185. Nielsch, W. and Böltz, G., *Z. Anal. Chem.* **142**, 94, 427 (1954).
186. Kapitańczyk, K., Kurzawa, Z. and Pryminski, Z., *Chem. Anal. (Warsaw)* **6**, 23 (1961).
187. Keattch, C. J., *Talanta* **3**, 351 (1960).

188. Clem, R. G. and Huffman, E. H., *Anal. Chem.* **38**, 926 (1966).
189. Ayres, G. H. and Baird, S. S., *Talanta* **7**, 237 (1961).
190. Janik, B., Gawron, H. and Weyers, J., *Mikrochim. Acta* **1965**, 1142.
191. Tobia, S. K., Gawargious, Y. A. and El-Shahaṭ, M. F., *Anal. Chim. Acta* **39**, 392 (1967).

Chapter 20

FLUORINE

Fluorine (F, at.wt. 19·00), exists as the diatomic gas, F_2, which is the strongest oxidizing agent known. Water reduces fluorine to hydrogen fluoride, HF. Fluoride ions form sparingly soluble compounds with a number of metals (e.g. Ba, Mg, Ce, and Th). Among the most commonly encountered fluoride complexes are those with Si, B, Al, Fe(III), Ti, Zr, Ta, and Be.

20.1 Methods of Separation

20.1.1 DISTILLATION

The conventional method for separating fluorine (fluoride) from inorganic and ashed organic samples consists in distilling it from a solution made strongly acidic with a non-volatile mineral acid. For a long time it was believed that the fluorine was distilled only as H_2SiF_6 or SiF_4. However, physico-chemical analysis of the system [1,1a] has shown that if steam-distillation is used, practically no silicon compounds are volatilized until the composition of the constant-boiling ternary mixture of HF, H_2SiF_6, and H_2O is reached. On the other hand, if a stream of nitrogen is used instead of steam as carrier-gas, then HF is volatilized if the amount of fluorine in the still is less than about 0·1 mg but H_2SiF_6 is distilled if the amount of fluorine exceeds about 3 mg, a mixture of the two being distilled for intermediate amounts of fluorine [2].

Before distillation of HF, organic samples are mineralized under alkaline conditions, usually in the presence of CaO. Organic substances with covalently bonded fluorine cannot be ashed in the usual way, but must be destroyed by special methods, e.g. by oxygen-flask combustion [3,4].

Most investigators [5–9] have distilled HF with nitrogen or steam as carrier gas, but this is not essential.

For determining microgram quantities of fluorine, 50- or 100-ml stills are suitable. Before the solution under investigation is placed in the still, it can be concentrated by evaporation in a platinum vessel (after the sample has been made slightly alkaline). A smaller volume of solution in the still makes it possible to reduce the volume of distillate collected, this being essential in the spectrophotometric determination of traces of fluoride.

Methods of Separation

Perchloric or sulphuric acid is usually used in the distillation. Perchloric acid is more suitable since it neither complexes nor precipitates metal ions used in the photometric determination of fluoride. However, perchloric acid must not be used in the presence of organic substances for fear of explosion. When HF is distilled from sulphuric acid medium, minute amounts of H_2SO_4 carried over to the distillate can interfere in the determination of fluoride by complexing zirconium or thorium. More rarely, distillation from phosphoric acid medium is employed [10,11].

The still should be fitted with a thermometer reaching to the bottom of the liquid. Towards the end of the distillation, the temperature of the solution in the still attains 140–150°C and 150–160°C with $HClO_4$ and H_2SO_4, respectively. The final temperature depends on the still design and, in particular, on the distance between the liquid level and the side-arm.

Samples containing silicate minerals which are sparingly soluble in acids are fused with Na_2CO_3 or NaOH before fluorine is distilled off [12]. Larger quantities of silica precipitate as the gel in the still and interfere in the distillation. If the silicate sample is fused with Na_2CO_3 and ZnO, silica remains in the solid phase during subsequent leaching with water. Fluorine is distilled from the silica-free filtrate [13].

In the presence of metals forming strong fluoride complexes (e.g. zirconium), distillation from phosphoric acid is advisable.

To prevent distillation of hydrogen chloride, silver sulphate or silver nitrate is added to the still.

20.1.2 Microdiffusion and Pyrohydrolysis

Traces of hydrogen fluoride are conveniently separated by microdiffusion [9,14–19a]. The sample is placed in a polyethylene, polypropylene, or polytetrafluoroethylene container which is then closed. The HF distils isothermally from conc. $HClO_4$ over several hours (e.g. 20), and is absorbed in dilute NaOH solution, or in filter paper treated with NaOH or magnesium succinate [16].

When superheated steam and oxygen are passed through a quartz, platinum, or nickel tube containing the sample at $\sim$ 1000°C, metal fluorides present are pyrolysed, releasing hydrogen fluoride which is absorbed in dilute NaOH after the gases have been cooled. This method [20–24a] is useful for materials containing stable fluorine compounds from which it is difficult to release HF by distillation.

20.1.3 Ion-Exchange and Other Methods

When phosphate, sulphate, and fluoride are sorbed on an anion-exchange column, the fluoride is readily eluted with ammonium chloride [25], sodium acetate [26], or dilute (0.25M) NaOH [27,28] solutions.

Fluoride has also been eluted from an anion-exchanger with 10M hydrochloric acid, leaving the anionic chloride complexes of multivalent metals in the column [29]. A Dowex-2 anion-exchanger has been used for the separation of fluoride in drinking water (containing ~ 1 mg of F^- per litre) [30].

Coursier and Saulnier [31] have quantitatively separated fluoride from aluminium on an anion-exchanger. After sorption on Amberlite IRA-400 as $Al(OH)_4^-$ and F^-, aluminium and fluoride are sequentially eluted with $0.2M$ NaOH and $1M$ NaOH, respectively.

Kempf [32] has separated interfering metals on a cation-exchanger before the determination of fluoride in water.

Fluoride has been separated by precipitation from natural water samples with an $Mg(OH)_2$ suspension [33], and in the analysis of minerals by coprecipitation of interfering elements with $Fe(OH)_3$ [34].

Fluoride ions may be extracted into carbon tetrachloride as tetraphenylstibonium fluoride [35], or as triphenylstibonium hydroxyfluoride [35a]. The latter extraction is subject to fewer interferences.

20.2 Methods of Determination

Fluoride ions form stable complexes with some multivalent metals, namely Zr, Th, Fe(III), Al, and Ti. The colour changes resulting from the reactions of fluoride with coloured complexes of these metals provide indirect methods for the determination of fluoride. Two examples of these methods given below are the more sensitive Eriochrome Cyanine R–zirconium method and the less sensitive sulphosalicylic acid–iron(III) method.

More recently, the direct method for determining traces of fluoride as the coloured ternary complex formed by fluoride with Alizarin Complexone and cerium(III) (or lanthanum) has become increasingly important.

Spectrophotometric methods for determining fluoride have been reviewed elsewhere [36,37].

20.2.1 Eriochrome Cyanine R–Zirconium Method

Zirconium ions (in dilute HCl) react with Eriochrome Cyanine R (ER) (formula, p. 51) to form red 1:1 and 1:2 complexes. Formation of the 1:1 complex ($\lambda_{max} = 515$ nm) is favoured by a more acidic medium (pH 0–1) and a deficiency of Eriochrome Cyanine R. Conversely the 1:2 complex ($\lambda_{max} = 540$ nm) is formed at lower acidities (pH 1–2) and in the presence of excess of Eriochrome Cyanine R. The absorption maximum of Eriochrome Cyanine R in dilute HCl is at 475 nm. Figure 20.1 shows the absorption spectra of both the reagent and the complex in a solution at pH 1.

Addition of fluoride to a solution of zirconium–Eriochrome Cyanine R complex results in a colour change due to partial decomposition of the Zr–ER complex and formation of the more stable, colourless zirconium fluoride complex. This reaction constitutes the basis of a sensitive spectrophotometric method for determining fluoride [2,38–43].

Obtaining reproducible results necessitates rigorous observance of the conditions specified for this method. Zirconium can occur in solution in various forms, owing to its well-known tendency to polymerize and hydrolyse.

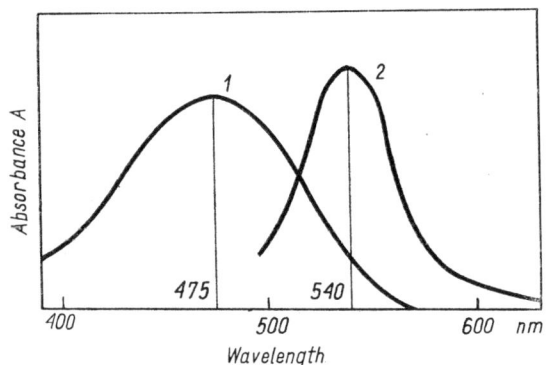

Fig. 20.1 Absorption spectra of Eriochrome Cyanine R (1) and its zirconium complex (2) at pH 1

To avoid interference by foreign ions in the reaction of fluoride with the Zr–ER complex, preliminary distillation of fluoride is recommended. Sulphate, which complexes zirconium, and metal ions forming stable fluoride complexes interfere in the determination. The most suitable molar ratio of Zr: ER in the zirconium–Eriochrome Cyanine R reagent is 1:4.

Megregian [38] has determined fluoride in a fairly acidic medium ($\sim 0.8 M$ HCl), whereas Valach [40] has recommended a less acidic solution ($\sim 0.2 M$ HCl) (pH ~ 0.7). The highest sensitivity is obtained at pH 1 (± 0.1).

The effective "molar absorptivity" calculated from the change in absorbance produced by a specific amount of fluoride (F^-) under the conditions given below in the procedure is 2.7×10^4 ($a = 1.4$) at 540 nm. At this wavelength, the difference in the absorbance before and after reaction of the reagent with fluoride is greatest.

It is recommended to use the ER solution as the reference when measuring the absorbance [38]. The cuvette holding the sample solution is placed in the spectrophotometer at the site of the reference solution.

Fluorides have been determined by the Zr–ER method in iron ore and apatite [29], rocks and minerals [44,45], various reagents [2], organic compounds [46,47], plants [48], blood serum [5], mineral waters [40], and air [49,50].

Reagents

1. Eriochrome Cyanine R: $0.004M$ solution. Dissolve 0.5364 g of the reagent in water containing 2.5 ml of $1M$ HCl, and dilute the solution with water to 250 ml.

2a. Zirconium solution: $0.005M$ in $4M$ HCl. Evaporate a weighed portion of zirconium chloride or zirconium nitrate dissolved in 25 ml of HCl (1+1) until salt crystals appear, and dilute the solution with $4M$ HCl to 1 litre. Determine the zirconium in the solution gravimetrically as ZrO_2, and dilute with $4M$ HCl to give a zirconium concentration of exactly $0.005M$.

2b. Zirconium solution $0.001M$ in $2M$ HCl. Dilute 50 ml of the $0.005M$ zirconium solution with exactly $1.5M$ HCl to 250 ml.

3. Zirconium–Eriochrome Cyanine R reagent (Zr:ER molar ratio = 1:4). Add 25 ml of $0.001M$ zirconium solution to 25 ml of Eriochrome Cyanine R solution with stirring.

4. Standard fluoride solution: 1 mg/ml. Dissolve in water 2.210 g of sodium fluoride previously ignited at $\sim 400°C$, and dilute the solution with water to 1 litre.

Procedure

Preparation for determination of F. Place 15–25 ml of conc. $HClO_4$ in a 50–100-ml still, add 25 ml of water, stir the solution well, and add a few fragments of porous porcelain. Connect the still to a condenser, and distil ~ 25 ml. Stop heating when the temperature of the solution in the still reaches $\sim 150°C$. During the distillation, the end of the condenser should be immersed in water (5 ml made slightly alkaline to phenolphthalein with ammonia) in a measuring cylinder serving as a distillation receiver. If the solution in the receiver becomes decolorized, add ammonia until the pink colour just reappears. Place exactly 5 ml of the Zr–ER reagent in a 50-ml volumetric flask. Add the distillate to the reagent. Dilute the solution with water to the mark and mix thoroughly. If the distillation system contained no fluoride, the absorbance of this solution at 540 nm should not differ from the absorbance of 5 ml of the Zr–ER reagent diluted with water in a volumetric flask to 50 ml.

Distillation and determination of F. Having conducted the test above, add to the still a solution (~ 25 ml) containing not more than 25 µg of F^-. Distil ~ 25 ml as before. Place 5 ml of the Zr–ER reagent in a 50-ml volumetric flask, add the distillate with swirling, dilute with water to the mark, and mix well. Prepare a reference solution from 5 ml of Zr–ER reagent diluted to volume with water in a 50-ml standard flask and measure the absorbance of this solution against the sample solution, at 540 nm. (Place the cuvette with the sample solution where the reference solution is usually positioned.)

Notes. (1) This method of measuring the absorbance gives a calibration curve as for direct spectrophotometric methods, i.e. zero absorbance corresponds to absence of fluoride.

(2) In the preparation of the calibration curve, the standard solution of fluoride is added to the Zr–ER reagent solution (5 ml) in a 50-ml volumetric flask, the solution is diluted to the mark with water, and the absorbance is measured as stated above.

20.2.2 Sulphosalicylic Acid–Iron(III) Method

In a slightly acidic medium (pH 2·5–3), iron(III) and sulphosalicylic acid form a 1:1 red-violet complex (in neutral and ammoniacal solutions, 1:2 and 1:3 complexes are formed).

(structure: sulphosalicylate–Fe^{2+} complex, HO$_3$S– substituent on benzene ring, with –O–Fe^{2+} and C=O–O chelation)

Addition of fluoride results in partial bleaching of the red-violet solution owing to the formation of a stable colourless iron–fluoride complex. The appropriate pH for the solution is obtained with a chloroacetate buffer. The method under discussion is rather insensitive and is suitable for determining larger amounts of fluoride [10,51].

The effective "molar absorptivity" (calculated from the decrease in absorbance at $\lambda_{max} = 500$ nm of the sulphosalicylate–iron(III) complex caused by a known amount of fluoride) is $2 \cdot 0 \times 10^2$ ($a = 0 \cdot 01$). This system does not obey Beer's law.

Interference from metals which form complexes with sulphosalicylic acid or fluoride, as well as from anions capable of reacting with iron(III), is prevented by separating the fluoride by distillation as HF.

Salicylic acid may be substituted for sulphosalicylic acid [52,53]. Fluoride has been determined in organic compounds by the salicylate–Fe(III) method [3,52–54].

Reagents

1. Sulphosalicylic acid solution. Dissolve in water 0·95 g of the dihydrate (or 0·82 g of the anhydrous acid) and dilute with water to 100 ml.
2. Iron(III) solution. Dissolve 0·30 g of Fe(NO$_3$)$_3$ in water containing 0·5 ml of conc. HClO$_4$, and dilute the solution with water to 100 ml, with stirring.
3. Chloroacetate buffer: pH 2·85–2·9. Dissolve 18·90 g of monochloroacetic acid in 100 ml of $1M$ NaOH solution, and dilute with water to 1 litre.
4. Sulphosalicylic acid–iron(III) reagent. Mix 20 ml of solution 1, 40 ml of solution 2, and 2·8 ml of $1M$ NaOH, dilute to 100 ml with the chloroacetate buffer, and allow to stand for 5 hours. The solution should not be kept longer than 10 days.
5. Standard fluoride solution: 1 mg/ml (p. 258).

Procedure

Place 25 ml of the chloroacetate buffer in a 50-ml volumetric flask, add 5 ml of reagent 4 followed by the sample solution (or distillate) containing not more than 1·5 mg of F^-, and dilute with water to the mark. Measure the absorbance of the partly bleached solution at 500 nm against water.

20.2.3 ALIZARIN COMPLEXONE METHOD

Belcher, Leonard, and West [55–57] have developed a direct spectrophotometric method for determining fluoride, which has aroused widespread interest among analysts [58–60].

Yellow Alizarin Complexone (AC) reacts with cerium(III) (or lanthanum) ions to form a red chelate, which, in turn, reacts with fluoride ions to give a blue ternary complex with the ratio AC:Ce:F = 1:1:1.

In the reaction a fluoride ion displaces one of the two molecules of water coordinately bonded to the metal ion. An equimolar mixture of AC and Ce(III) is used as the reagent. There are only small differences in the sensitivities of the methods using cerium(III) or lanthanum. The sensitivity varies with the pH used, because of the effect of pH on the spectra of both the reagent and the complex. A pH range of 5·0–5·2 and choice of wavelength from 610 to 620 nm seem optimum [57].

In the presence of certain water-miscible organic solvents (e.g. acetone, acetonitrile, and dimethylsulphoxide), equilibrium is attained more rapidly (15–20 minutes instead of ~2 hours), and the sensitivity of the reaction and the stability of the blue ternary complex are enhanced [59,60]. The optimum pH may also be shifted. The molar absorptivity in the presence of 20% acetone is $1·4 \times 10^4$ ($a = 0·74$) at 620 nm. Figure 20.2 shows the absorption spectra of yellow Alizarin Complexone, the red AC-Ce complex, and the blue AC–Ce–F complex.

In the AC method, it is absolutely essential that the pH values of the sample and the standard solutions are identical. The maximum difference in absorbance between the AC–Ce–F and AC–Ce complex occurs at ~ 620 nm.

Ions of metals giving stable fluoride compounds, namely Al, Fe(III), Sn, Ca, and Mg, interfere seriously in the determination of fluoride. Phosphate, sulphate, and oxalate which compete with fluoride in the reaction

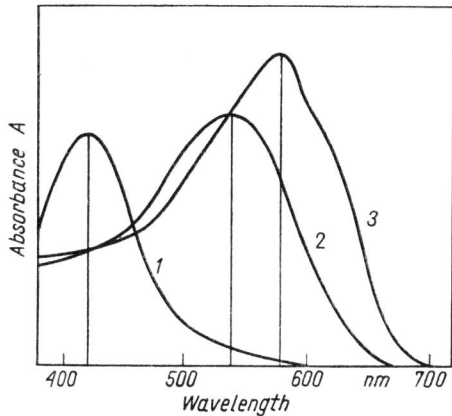

Fig. 20.2 Absorption spectra of Alizarin Complexone (AC) (1), Ce–AC complex (2) and Ce–AC–F complex (3) at pH 4·5

with the AC–Ce(III) complex, or which complex cerium(III) (or lanthanum), also interfere. Finally, certain oxidants, e.g. nitrites, interfere. As a result of these many interferences, fluoride must usually be separated first by distillation [61], ion-exchange [62], or pyrohydrolysis [63].

The anionic AC–Ce–F complex is extractable with a 5% solution of triethylamine in n-pentanol from solution at pH 9 [64]. Both the AC–Ce–F complex and the excess of AC–Ce are extracted from the weakly acidic solution (pH ~ 4·5) with a 5% solution of dioctylamine in isobutanol, the free reagent being subsequently stripped with $0.25M$ Na_2CO_3 or $0.5M$ ammonia [64a].

Fluorides have been determined by the cerium(III)- or lanthanum-Alizarin Complexone method in silicate rocks and minerals [65], phosphate rocks [62], tungsten [63], biological materials [16,66,67], soils [66], plants and wood [68,69], organic compounds [4,70,70a], and water [8,32,58,71,72].

This method has also been automated for the determination of fluoride in urine [67] and water [71].

Reagents

1. Alizarin Complexone (AC): $0.0015M$ solution. Dissolve 0·1445 g of the reagent in 50 ml of water containing 0·5 ml of conc. ammonia. Add 0·5 ml of glacial acetic acid, and dilute with water to exactly 250 ml. Store the solution in a brown bottle.

2. Cerium(III) nitrate: $0.0015M$ solution. Dissolve 0.1629 g of $Ce(NO_3)_3 \cdot 6H_2O$ in water, and dilute with water to exactly 250 ml.
3. Standard fluoride solution: 1 mg/ml (p. 258).
4. Acetate buffer: pH 4·0. Dissolve 30 g of $CH_3COONa \cdot 3H_2O$ in water, add 57·5 ml of glacial CH_3COOH and dilute the solution with water to 500 ml.

Procedure

To a 50-ml volumetric flask, add 2 ml of buffer, 5 ml of cerium(III) solution, 10 ml of acetone, and 5 ml of Alizarin Complexone solution, and mix thoroughly. To this solution, add the sample solution (distillate) containing not more than 60 μg of F^- and neutralized with dilute NaOH or HCl in the presence of phenolphthalein. The solution should be at pH ~ 4·5. Dilute the solution to 50 ml, mix well, and allow to stand for 20 minutes. Measure the absorbance at 620 nm against a reagent blank solution.

20.2.4 OTHER METHODS

As well as the zirconium–Eriochrome Cyanine R complex, the coloured systems formed by zirconium with alizarin and Alizarin Red S [73–78] are often used for the indirect photometric determination of fluoride, e.g. in water [75,76,78a], phosphates [73,77], and air [78].

Many investigators [30,79–83] have determined fluoride with the zirconium–SPADNS complex as reagent. The method has been applied in the analysis of beryllium [80], silicate rocks [81], and organic compounds [82]. Zirconium complexes with Xylenol Orange [84–86], Pyrocatechol Violet [87], and *p*-dimethylaminoazobenzenearsonic acid [88] have been used similarly.

Numerous indirect spectrophotometric methods for determining fluoride are also based on coloured thorium complexes with organic reagents such as alizarin and Alizarin Red S [6,89–92], Xylenol Orange [83,93], Methylthymol Blue [7,94], Chrome Azurol S [95], Thoron I [13,96], Arsenazo I [97,98], Amaranth [99], 2-(1,8-dihydroxy-3,6-disulpho-2-naphthylazo)-phenoxyacetic acid [100], phenylfluorone [101], and chloranilic acid [102,103]. With these reagents, fluoride has been determined in water [95,101], rocks [13], fertilizers [92], and organic compounds [98].

Aluminium complexes with organic complexants such as haematoxylin [104,105], Eriochrome Cyanine R [106], Chrome Azurol S [107,108], Arsenazo I [97,109], and morin [110] are used as reagents in numerous methods, the most sensitive of which is that with Eriochrome Cyanine R [106].

Methods based on the bleaching of iron(III) complexes include the sulphosalicylate method discussed earlier, and those with iron(III) complexes with thiocyanate [78,111,112], ferron [113], and acetylacetone [114].

According to Nichols and Condo [115], the complexes with 5-phenylsalicylic acid and 2,4-dihydroxyacetophenone are the most suitable of 18 organoferric coloured systems they investigated as reagents for fluoride.

Further complex reagents proposed for determining fluoride are the titanium(IV) peroxide complex (rather insensitive) [116,117]; the titanium chromotropic acid complex [118]; scandium complexes with Xylenol Orange [119], Pyrocatechol Violet [120], and Methylthymol Blue [121]; the beryllium complex with Chrome Azurol S [122]; the niobium complex with catechol + EDTA [123], and lanthanum chloranilate [124].

Dahl [125] has determined fluoride by extraction into benzene of the TaF_6^- ion-pair with Malachite Green.

Fluoride distils as SiF_4 from concentrated H_2SO_4 medium (containing no more than 10% of water) when silica is present. The SiF_4 hydrolyses in the distillation receiver, and the fluoride can be determined indirectly by conversion of the silicon into silicomolybdenum blue [126]. Small amounts of hydrogen fluoride in the air have been determined by this method [127].

Small amounts of fluorine in gas mixtures have been continuously determined spectrophotometrically by passing the gas through a reaction tube containing hot solid NaCl, and measuring the absorbance (at 360 nm) of the liberated chlorine in a pressure-controlled flow cell [128].

References

1. Fox, E. J. and Jackson, W. A., *Anal. Chem.* **31**, 1657 (1959).
1a. Langmyhr, F. J. and Graff, P. R., *Anal. Chim. Acta* **21**, 334 (1959).
2. Marczenko, Z. and Lenarczyk, Ł., *Chem. Anal. (Warsaw)* **13**, 405 (1968).
3. Levy, R. and Debal, E., *Mikrochim. Acta* **1962**, 224.
4. Fernandopulle, M. E. and Macdonald, A. M. G., *Microchem. J.* **11**, 41 (1966).
5. Singer, L. and Armstrong, W. D., *Anal. Chem.* **31**, 105 (1959).
6. Guntz, A. A. and Arène, M., *Chim. Anal. (Paris)* **39**, 260 (1957); **40**, 39, 453 (1958).
7. Hlucháň, E. and Mayer, J., *Chem. Zvesti* **17**, 569 (1963).
8. Quentin, K. E. and Rosopulo, A., *Z. Anal. Chem.* **241**, 241 (1968).
9. Tušl, J., *Chem. Listy* **62**, 839 (1968); **63**, 777 (1969); *Collection Czech. Chem. Commun.* **35**, 1001 (1970).
10. Lacroix, S. and Labalade, M., *Anal. Chim. Acta* **4**, 68 (1950).
11. Wade, M. A. and Yamamura, S. S., *Anal. Chem.* **37**, 1276 (1965).
12. Remmert, L. F., Parks, T. D., Lawrence, A. M. and McBurney, E. H., *ibid.* **25**, 450 (1953).
13. Grimaldi, F. S., Ingram, B., and Cuttitta, F., *ibid.* **27**, 918 (1955).
14. Singer, L. and Armstrong, W. D., *ibid.* **26**, 904 (1954).
15. Greenland, L., *Anal. Chim. Acta* **27**, 386 (1962).
16. Hall, R. J., *Analyst* **85**, 560 (1960); **88**, 76 (1963).
17. Taves, D. R., *Talanta* **15**, 969 (1968).
17a. Alcock, N. W., *Anal. Chem.* **40**, 1397 (1968).
18. Marshall, B. S. and Wood, R., *Analyst* **94**, 493 (1969).
19. Stuart, J. L., *ibid.* **95**, 1032 (1970).
19a. Hanocq, M., *Mikrochim. Acta* **1973**, 729.
20. Powell, R. H. and Menis, O., *Anal. Chem.* **30**, 1546 (1958).
21. Surak, J. G., Fisher, D. J., Burros, C. L. and Bate, L. C., *ibid.* **32**, 117 (1960).

22. Nardozzi, M. J. and Lewis, L. L., *ibid.* **33**, 1261 (1961).
23. Newman, A. C., *Analyst* **93**, 827 (1968).
24. Clements, R. L., Sergeant, G. A. and Webb, P. J., *ibid.* **96**, 51 (1971).
24a. Berns, E. G. and Van der Zwaan, P. W., *Anal. Chim. Acta* **59**, 293 (1972).
25. Newman, A. C., *ibid.* **19**, 471 (1958).
26. Nielsen, H. M., *Anal. Chem.* **30**, 1009 (1958).
27. Eger, C. and Lipke, J., *Anal. Chim. Acta* **20**, 548 (1959).
28. Zipkin, I., Armstrong, W. D. and Singer, L., *Anal. Chem.* **29**, 310 (1957).
29. Glasö, Ö. S., *Anal. Chim. Acta* **28**, 543 (1963).
30. Kelso, F. S., Matthews, J. M. and Kramer, H. P., *Anal. Chem.* **36**, 577 (1964).
31. Coursier, J. and Saulnier, J., *Anal. Chim. Acta* **14**, 62 (1956).
32. Kempf, T., *Z. Anal. Chem.* **244**, 113 (1969).
33. Shapiro, M. Ya. and Kolesnikova, V. G., *Zh. Analit. Khim.* **18**, 507 (1963).
34. Kuril'chikova, G. E., *ibid.* **25**, 563 (1970).
35. Moffett, K. D., Simmler, J. R. and Potratz, H. A., *Anal. Chem.* **28**, 1356 (1956).
35a. Chermette, H., Martelet, C., Sandino, D. and Tousset, J., *ibid.* **44**, 857 (1972).
36. Valach, R., *Talanta* **8**, 629 (1961).
37. Crosby, N. T., Dennis, A. L. and Stevens, J. G., *Analyst* **93**, 643 (1968).
38. Megregian, S., *Anal. Chem.* **26**, 1161 (1954).
39. Thatcher, L. L., *ibid.* **29**, 1709 (1957).
40. Valach, R., *Talanta* **9**, 341 (1962).
41. Oelschläger, W., *Z. Anal. Chem.* **191**, 408 (1962).
42. Sarma, P. L., *Anal. Chem.* **36**, 1684 (1964).
43. Dixon, E. J., *Analyst* **95**, 272 (1970).
44. Evans, W. H. and Sergeant, G. A., *ibid.* **92**, 690 (1967).
45. Huang, W. H. and Johns, W. D., *Anal. Chim. Acta* **37**, 508 (1967).
46. Senkowski, B. Z., Wollish, E. G. and Shafer, E. G., *Anal. Chem.* **31**, 1574 (1959).
47. Tölg, G., *Z. Anal. Chem.* **194**, 20 (1963).
48. Debiard, R. and Dupraz, M. L., *Chim. Anal. (Paris)* **47**, 657 (1965); **48**, 384 (1966).
49. Adams, D. F., Koppe, R. K. and Matzek, N. E., *Anal. Chem.* **33**, 117 (1961).
50. Marshall, B. S. and Wood, R., *Analyst* **93**, 821 (1968).
51. Sen, B., *Z. Anal. Chem.* **153**, 168 (1956)
52. Rickard, R. R., Ball, F. L. and Harris, W. W., *Anal. Chem.* **23**, 919 (1951).
53. Rogers, R. N. and Yasuda, S. K., *ibid.* **31**, 616 (1959).
54. Lewandowska, B., *Chem. Anal. (Warsaw)* **10**, 1353 (1965).
55. Belcher, R., Leonard, M. A. and West, T. S., *J. Chem. Soc.* **1959**, 3577.
56. Leonard, M. A. and West, T. S., *ibid.* **1960**, 4477.
57. Belcher, R. and West, T. S., *Talanta* **8**, 853, 863 (1961).
58. Greenhalgh, R. and Riley, J. P., *Anal. Chim. Acta* **25**, 179 (1961).
59. Yamamura, S. S., Wade, M. A. and Sikes, J. H., *Anal. Chem.* **34**, 1308 (1962).
60. Hanocq, M. and Molle, L., *Anal. Chim. Acta* **40**, 13; **42**, 349 (1968).
61. Kubota, H., *Microchem. J.* **12**, 525 (1967).
62. Schafer, H. N., *Anal. Chem.* **35**, 53 (1963).
63. Mosen, A. W. and Bevege, E., *Chemist-Analyst* **55**, 71 (1966).
64. Haarsma, J. P. and Agterdenbos, J., *Talanta* **18**, 747 (1971).
64a. Marczenko, Z. and Lenarczyk, Ł., *Chem. Anal. (Warsaw)* **20** (in press).
65. Hall, A. and Walsh, J. N., *Anal. Chim. Acta* **45**, 341 (1969).
66. Hall, R. J., *Analyst* **93**, 461 (1968).
67. Hargreaves, J. A., Ingram, G. S. and Cox, D. L., *ibid.* **95**, 177 (1970).
68. Buck, M., *Z. Anal. Chem.* **193**, 101 (1963).
69. Williams, A. I., *Analyst* **94**, 300 (1969).
70. Johnson, C. A. and Leonard, M. A., *ibid.* **86**, 101 (1961).
70a. Hanocq, M., *Mikrochim. Acta* **1972**, 707.
71. Chan, K. M. and Riley, J. P., *Anal. Chim. Acta* **35**, 365 (1966).
72. Kletsch, R. A. and Richards, F. A., *Anal. Chem.* **42**, 1435 (1970).

73. Vigier, R., *Bull. Soc. Chim. France* **1957**, 160.
74. Ashley, R. P., *Anal. Chem.* **32**, 834 (1960).
75. Lim, C. K., *Analyst* **87**, 197 (1962).
76. Meyling, A. H. and Meyling, J., *ibid.* **88**, 84 (1963).
77. Kost, T., *Z. Anal. Chem.* **203**, 260 (1964).
78. Zawadzka, E. and Pampuch-Karska, K., *Chem. Anal. (Warsaw)* **11**, 261 (1966).
78a. Wierzbicki, T., Pawlita, W. and Pietrzyk, H., *ibid.* **16**, 1079 (1971).
79. Wharton, H. W., *Anal. Chem.* **34**, 1296 (1962).
80. Smith, J. F. and Crawley, R. H., *Anal. Chim. Acta* **23**, 345 (1960).
81. Peck, L. C. and Smith, V. C., *Talanta* **11**, 1343 (1964).
82. Willis, D. E. and Cave, W. T., *Anal. Chem.* **36**, 1821 (1964).
83. Tušl, J., *ibid.* **41**, 352 (1969).
84. Růžička, J. A., Jakschová, H. and Mrklas, L., *Talanta* **13**, 1341 (1966).
85. Dobiášová, L., Hlasivcová, N. and Novák, J., *Chem. Listy* **61**, 1658 (1967).
86. Kukisheva, T. N., Sinitsyna, E. S. and Efimova, N. S., *Zh. Analit. Khim.* **26**, 953 (1971).
87. Růžička, J. A., *Collection Czech. Chem. Commun.* **30**, 2717 (1965).
88. Kamada, M., Onishi, T. and Ota, M., *Bull. Chem. Soc. Japan* **28**, 148 (1955).
89. Icken, J. M. and Blank, B. M., *Anal. Chem.* **25**, 1741 (1953).
90. Lothe, J. J., *ibid.* **28**, 949 (1956).
91. Yasuda, S. K. and Lambert, J. L., *ibid.* **30**, 1485 (1958).
92. Angot, J. and Mevel, N., *Chim. Anal. (Paris)* **45**, 111 (1963).
93. Rezáč, Z. and Ditz, J., *Z. Anal. Chem.* **186**, 424 (1962).
94. Uhlíř, Z., *Chem. Zvesti* **18**, 756 (1964).
95. Mayer, J. and Hlucháň, E., *ibid.* **12**, 143 (1958).
96. Horton, A. D., Thomason, P. F. and Miller, F. J., *Anal. Chem.* **24**, 548 (1952).
97. Kuteinikov, A. F., *Zh. Analit. Khim.* **16**, 327 (1961).
98. Larina, N. I. and Gel'man, N. E., *ibid.* **22**, 582 (1967).
99. Lambert, J. L., *Anal. Chem.* **26**, 558 (1954).
100. Shimoishi, Y. and Hayami, T., *Bull. Chem. Soc. Japan* **40**, 1139 (1967).
101. Brownley, F. I., Jr. and Howle, C. W., Jr., *Anal. Chem.* **32**, 1330 (1960).
102. Hensley, A. L., and Barney, J. E., II, *ibid.* **32**, 828 (1960).
103. Papay, M. K., Mazor, L. and Takacs, J., *Acta Chim. Acad. Sci. Hung.* **66**, 13 (1970).
104. Price, M. J. and Walker, O. J., *Anal. Chem.* **24**, 1593 (1952).
105. Hunter, G. J., MacNulty, B. J., Terry, E. A. and Beveridge, J. S., *Anal. Chim. Acta* **8**, 351; **9**, 330, 425 (1953).
106. MacNulty, B. J., Hunter, G. J. and Barret, D. G., *ibid.* **14**, 368 (1956).
107. MacNulty, B. J. and Woollard, L. D., *ibid.* **14**, 452 (1956).
108. Ballczo, H., Doppler, G. and Lanik, A., *Mikrochim. Acta* **1957**, 809.
109. Kuteinikov, A. F., Brodskaya, V. M. and Lanskoi, G. A., *Zh. Analit. Khim.* **17**, 87 (1962).
110. Beck, M. T., *Acta Chim. Acad. Sci. Hung.* **4**, 223 (1954).
111. Babko, A. K. and Kleiner, K. E., *Zh. Analit. Khim.* **1**, 106 (1946).
112. Szabó, Z. G., Beck, M. T. and Tóth, K., *Naturwissenschaft.* **43**, 156 (1956).
113. Adams, D. F., *Anal. Chem.* **32**, 1312 (1960).
114. McKaveney, J. P., *ibid.* **40**, 1276 (1968).
115. Nichols, M. L. and Condo, A. C., Jr., *ibid.* **26**, 703 (1954).
116. Monnier, D., Vaucher, R. and Wenger, P., *Helv. Chim. Acta* **33**, 1 (1950).
117. Rink, M. and Twarock, H., *Z. Anal. Chem.* **213**, 31 (1965).
118. Novak, V. P., Bogovina, V. I. and Mal'tsev, V. F., *Zavodsk. Lab.* **31**, 278 (1965).
119. Hung, S.-C., Teng, H.-C. and Liang, S.-C., *Acta Chim. Sin.* **30**, 452 (1964).
120. Marczenko, Z. and Lenarczyk, Ł., *Chem. Anal. (Warsaw)* **15**, 607 (1970).
121. Marczenko, Z. and Lenarczyk, Ł., *ibid.* **15**, 753 (1970).
122. Silverman, L. and Shideler, M. E., *Anal. Chem.* **31**, 152 (1959).

123. Nabivanets, B. I. and Lukachina, V. V., *Zavodsk. Lab.* **33**, 145 (1967).
124. Hayashi, K., Danzuka, T. and Ueno, K., *Talanta* **4**, 126 (1960).
125. Dahl, W. E., *Anal. Chem.* **40**, 416 (1968).
126. Curry, R. P. and Mellon, M. G., *ibid.* **28**, 1567 (1956).
127. Peregud, E. A. and Boikina, B. S., *Zh. Analit. Khim.* **17**, 611 (1962).
128. Weber, C. W. and Howard, O. H., *Anal. Chem.* **35**, 1002 (1963).

Chapter 21
GALLIUM

Gallium (Ga, at.wt. 69·72) forms colourless Ga^{3+} ions; it occurs in solution exclusively in the +III oxidation state. Gallium resembles aluminium and zinc in its properties. The hydroxide, $Ga(OH)_3$, precipitates at pH ~ 3, but dissolves in weakly alkaline media (pH 8-9). Gallium forms halide, oxalate, tartrate, and EDTA complexes.

21.1 Methods of Separation

21.1.1 EXTRACTION

Gallium is extracted from 5·5–6·5M HCl with diethyl ether [1,2]. In a single extraction, ~ 95% of the gallium passes into the organic phase. When the extractant is di-isopropyl ether [3], more than 99% of the gallium is extracted in one step (the optimum HCl concentration is 7–8M). The following species are also wholly or partly extracted: Fe(III), Au(III), Tl(III), Ge, In, Mo, Re, As, Sb, Sn, and Te. Many ions [e.g. Al, Ti, Fe(II), and Zn] remain quantitatively in the aqueous phase. Before gallium is extracted, iron(III) is reduced with $TiCl_3$ or ascorbic acid. The gallium chloride complex may be extracted with MIBK [4], mesityl oxide in MIBK [5], methyl ethyl ketone [6], and TBP [4,7,8].

Gallium has been extracted from hydrochloric acid media into chloroform in the presence of N-benzylaniline [8a], acetone [9], or antipyrine bases (e.g. DAM) [10]; into toluene in the presence of the quaternary base, Adogen-364 [11]; and into o-dichlorobenzene in the presence of the tetraphenylarsonium ion [12]. Gallium has also been separated by extraction from thiocyanate media [13,14].

Gallium can be extracted in the form of chelates with cupferron [15,16], TTA (and tetraphenylarsonium chloride) [17], acetylacetone [18], and alkylphosphoric acids [19].

21.1.2 ION-EXCHANGE

Alimarin et al. [20] have separated gallium from aluminium and zinc on ion-exchange columns irrigated with sulphosalicylate, oxalate, tartrate, EDTA, or carbonate solutions. In an ammonium carbonate medium, gallium exists as the anionic carbonate complex, whereas zinc exists as the cationic ammine complex.

Gallium is separated from lead and cadmium by cation-exchange chromatography of the metal sulphosalicylate, tartrate, and oxalate complexes [21].

Indium and gallium are separated by sequential elution from a Dowex 50 cation-exchanger with $0{\cdot}4M$ and $1{\cdot}3M$ hydrochloric acid, respectively [22]. Cation-exchange in acetone–aqueous HCl gives a very clean separation of Al, Ga, In, and Tl [23].

The differences in stability of the thiocyanate [24] and the chloride [25] complexes have permitted anion-exchange separation of gallium from iron, indium, and aluminium.

21.1.3 Precipitation and Other Methods

When the sample is fused with Na_2CO_3 or NaOH, the gallium is leached almost quantitatively from the melt with water, whereas iron and most of the indium remain in the solid phase. Similarly, gallium remains in solution during the precipitation of indium and iron with a sodium hydroxide solution.

Traces of gallium have been coprecipitated with $Al(OH)_3$ [26], hydrous MnO_2 [27], and Ag_2S [28] as collectors.

When an alkaline gallium solution is treated with a 0·5% sodium amalgam (60 minutes, 55°C), the gallium is reduced and collected in the amalgam. The gallium is stripped into $1M$ H_2SO_4 by electrolytic decomposition of the amalgam. This method has been applied to separate traces of gallium in aluminium [29].

21.2 Methods of Determination

The most sensitive and selective spectrophotometric methods involve extraction of ion-association complexes formed between the $GaCl_4^-$ ion and basic dyes (e.g. Rhodamine B). Methods based on coloured gallium chelates formed with triphenylmethane or azo reagents are also important.

Comparative studies have been made of spectrophotometric methods for determining gallium [30,31].

21.2.1 Rhodamine B Method

The gallium chloride complex, $[GaCl_4^-]$, reacts with the basic xanthene dye, Rhodamine B (formula, p. 53), to form a pink ion-pair which can be extracted from hydrochloric acid medium into benzene and other solvents, thereby providing the basis of a sensitive and selective spectrophotometric method for the determination of gallium [3,32–38]. Onishi and Sandell [3] used benzene, whereas Culkin and Riley [32] preferred a mixture of chlorobenzene and carbon tetrachloride (3+1) (which has the advantage of being denser than the aqueous phase). A mixture of benzene and diethyl ether has also been recommended as extractant [33].

Extraction of the Rhodamine B chlorogallate complex into the mixed $C_6H_5Cl + CCl_4$ solvent is ~ 97% (for the procedure described below) [33a]. Only a minute amount of free Rhodamine B is co-extracted.

The molar absorptivity appears to depend on the conditions used and is 5.0×10^4 ($a = 0.71$) at 560 nm for the procedure described below. The molar absorptivity is increased to ~ 9×10^4 when o-dichlorobenzene is used as extractant, but the distribution coefficient is only 1·7, which corresponds to ~ 60% extraction [37]. Since Rhodamine B is only slightly soluble in o-dichlorobenzene, but dissolves readily in the *meta*-isomer, the o-dichlorobenzene must not be contaminated with the *meta*-isomer.

The Rhodamine B complexes with gallium, iron(III) gold(III), antimony(V), and thallium(III) are all extracted from $6M$ HCl. In the presence of reducing agents, such as $TiCl_3$, ascorbic acid, $SnCl_2$, or NH_2OH, the interfering metals are reduced to their lower oxidation states and do not react with Rhodamine B. In the presence of larger quantities of these metals, it is advisable to separate gallium first by extraction from hydrochloric acid medium with di-isopropyl or diethyl ether (in the presence of a reducing agent).

The Rhodamine B method has been applied to the determination of gallium in rocks and minerals [32,33,37], bauxite [33,38], lead and zinc [34], aluminium [35,36], tungsten metal and tungstates [36], and ores [36a].

Reagents

1. Rhodamine B: 0·5% solution in hydrochloric acid (1+1).
2. Standard gallium solution: 1 mg/ml. Dissolve 0·1000 g of gallium metal in 5 ml of HCl (1+1), and dilute the solution with HCl (1+1) to exactly 100 ml. Working solutions are obtained by appropriate dilution of the stock solution with hydrochloric acid (1+1).
3. Hydrochloric acid, 6·5M, containing 1% of titanium(III) chloride, stored in a closed bottle.

Procedure

Extractive separation of Ga. To the sample solution (in 7–8M HCl) containing not more than 50 μg of Ga, add 0·05–0·5 g of ascorbic acid [depending on the amount of iron(III) present]. Extract the gallium with two portions of di-isopropyl ether (shake with each portion for about 1 minute).

Determination of Ga. Place in a beaker the ethereal extract obtained as above, add ~ 50 mg of sodium chloride, and evaporate to dryness on a water-bath. Dissolve the residue in 10 ml of hydrochloric acid (3), transfer the solution to a separating funnel, add 0·5 ml of the Rhodamine B solution, and extract the complex with two 10-ml portions of chlorobenzene–carbon tetrachloride (3+1) solution. Place the clear extract in a 50-ml volumetric flask (or smaller depending on the gallium content), dilute to the mark with the solvent, and measure the absorbance of the solution at 560 nm against a reagent blank solution.

21.2.2 OTHER METHODS

Butylrhodamine B [39] and Rhodamine 6G [30] are other xanthene dyes which have been used as spectrophotometric reagents for gallium.

Armeanu and Costinescu [40] have compared spectrophotometric methods based on extraction of the ion-pairs formed between tetrachlorogallate and basic triphenylmethane or diphenylnaphthylmethane dyes. Such systems, like those with xanthene dyes, suffer from rather low distribution coefficients. The best reagent of this class for determining gallium appears to be Crystal Violet [$\varepsilon = 4 \cdot 9 \times 10^4$ at 589 nm in chloroform–acetylacetone (6:1)] [9,40,41], but Malachite Green [42], Brilliant Green [43], Methyl Green [44], Victoria Blue 4R [45], and Astrazon Blue B [41] can also be used.

Oxazine and thiazine dyes suitable for extractive spectrophotometric methods include Methylene Blue [46], Capri Blue [47], and Meldola's Blue [48]. Some basic antipyrine dyes have been used similarly [49].

Gallium ions form coloured products with a number of triphenylmethane reagents, the most important of which is Xylenol Orange ($\bar{\varepsilon} = 2 \cdot 54 \times 10^4$ at 550 nm) [31, 50–53]. In the presence of oxine, a ternary complex which is extractable into n-butanol is formed by Xylenol Orange, oxine, and gallium [54]. Pyrocatechol Violet ($\varepsilon = 3 \cdot 2 \times 10^4$ at 620 nm, $\Delta \lambda = 160$ nm) is also a suitable reagent [31,55,56]: the sensitivity and stability can be increased by extraction of the ternary complex formed with diphenylguanidine into a solution of n-butanol and chloroform (1+1) [57]. Other triphenylmethane reagents for gallium are Methylthymol Blue [57,58], Chrome Azurol S [60,61], Eriochrome Cyanine R [62,63], and Glycinecresol Red [64].

PAR [65–67] and PAN (formulae, p. 46) [68,69] are the foremost azo reagents for gallium. Other such azo reagents include Gallion ($\varepsilon = 2 \cdot 1 \times 10^4$) [30,53,70,71], Lumogallion (I) [72], Sulphonazo [73], TAR [74],

$$HO_3S\text{–}\underset{Cl}{\underset{|}{\bigcirc}}(OH)\text{–}N=N\text{–}\bigcirc(HO)\text{–}OH \quad (I)$$

3,6-disulpho-TAN [75], Sulphochlorophenol R [76], Sulphonitrophenol R [77], Sulphonitrophenolazurin-1,4 [78], Picramine RG ($\varepsilon = 1 \cdot 96 \times 10^4$ at 560 nm) [79], Acid Chrome Dark-green C [80], 1-(2,4-dihydroxyphenylazo)-2-naphthol-4-sulphonic acid [81], 3,4-dihydroxyazobenzene-2′-carboxylic acid [82], 2-bromo-4,5-dihydroxybenzene-(1-azo-1′)-benzene-4′-sulphonic acid [83], and ethylsulphonaphtholazoaminocresol and sulphonaphtholazoaminophenol [84].

A long established method for determining gallium consists of extracting the oxinate complex into chloroform and measuring the absorbance at 393 nm ($\varepsilon = 7 \cdot 1 \times 10^3$) [1,85,86]. 5,7-Dibromo-8-hydroxyquinoline (bromo-oxine) has been used similarly [85,87].

Diverse organic reagents suggested for the spectrophotometric determination of gallium include diphenylcarbazone [31], morin [88], quercetin [89,90], 3,5,7,4′-tetrahydroxyflavone (kaempferol) [91], haematein ($\varepsilon = 4\cdot 0 \times 10^4$) [92], 3,4-dihydroxyanthraquinone-2-sulphonic acid [93], and C-cyano-N,N'-di(2-hydroxyphenyl)formazan [94].

References

1. Luke, C. L. and Campbell, M. E., *Anal. Chem.* **28**, 1340 (1956).
2. Geilmann, W., Bode, H. and Kunkel, E., *Z. Anal. Chem.* **148**, 161 (1955).
3. Onishi, H. and Sandell, E. B., *Anal. Chim. Acta* **13**, 159 (1955).
4. Henning, K. and Specker, H., *Z. Anal. Chem.* **241**, 81 (1968).
5. Chavan, M. B. and Shinde, V. M., *Anal. Chim. Acta* **59**, 165 (1972).
6. Rafaeloff, R., *Anal. Chem.* **43**, 272 (1971).
7. De, A. K. and Sen, A. K., *Talanta* **14**, 629 (1967).
8. Lanfranco, G., *Anal. Chim. Acta* **38**, 523 (1967).
8a. Khosla, M. M. and Rao, S. P., *ibid.* **61**, 156 (1972).
9. Kuznetsova, V. K., *Zh. Analit. Khim.* **18**, 1326 (1963).
10. Busev, A. I. and Skrebkova, L. M., *ibid.* **17**, 56 (1962).
11. Sherif, Sh. H., Abdel-Gawad, A. S. and El-Wakil, A. M., *Talanta* **17**, 137 (1970).
12. Finston, H. L. and Rahaman, M. S., *Mikrochim. Acta* **1969**, 78.
13. Specker, H. and Bankmann, E., *Z. Anal. Chem.* **149**, 97 (1956).
14. Alimarin, I. P., Bol'shova, T. A. and Ershova, N. I., *Vestn. Mosk. Univ. Khim.* **1970**, 568.
15. Khamid, Sh. A., Alimarin, I. P. and Puzdrenkova, I. V., *Zh. Analit. Khim.* **19**, 195 (1964); **20**, 894 (1965).
16. Vadasdi, K. G., *Anal. Chim. Acta* **44**, 471 (1969).
17. Rahaman, M. S. and Finston, H. L., *Anal. Chem.* **40**, 1709 (1968).
18. Steinbach, J. F. and Freiser, H., *ibid.* **26**, 375 (1954).
19. Levin, I. S. and Balakireva, N. A., *Zh. Analit. Khim.* **22**, 1475 (1967); *Talanta* **17**, 915 (1970).
20. Alimarin, I. P., Tsintsevich, E. P. and Gorokhova, A. N., *Zavodsk. Lab.* **22**, 1276 (1956); **26**, 144 (1960).
21. Tsintsevich, E. P. and Nazarova, G. E., *ibid.* **23**, 1068 (1957).
22. Klement, R. and Sandmann, H., *Z. Anal. Chem.* **145**, 325 (1955).
23. Strelow, F. W. E. and Victor, A. H., *Talanta* **19**, 1019 (1972).
24. Korkisch, J. and Hecht, F., *Mikrochim. Acta* **1956**, 1230.
25. Korkisch, J. and Hazan, I., *Anal. Chem.* **36**, 2309 (1964).
26. Buxbaum, P. and Vadasdi, K. G., *Chem. Anal.* (*Warsaw*) **14**, 429 (1969).
27. Biskupsky, V. S., *Anal. Chim. Acta* **46**, 149 (1969).
28. Rudnev, N. A., Tuzova, A. M. and Malofeeva, G. I., *Zh. Analit. Khim.* **26**, 886 (1971).
29. Lysenko, V. I. and Lisitsyna, E. V., *Zavodsk. Lab.* **26**, 145 (1960).
30. Shcherbov, D. P. and Ivankova, A. I., *ibid.* **24**, 667 (1958).
31. Akhmedli, M. K. and Glushchenko, E. L., *Zh. Analit. Khim.* **19**, 556 (1964).
32. Culkin, F. and Riley, J. P., *Analyst* **83**, 208 (1958); *Anal. Chim. Acta* **24**, 413 (1961).
33. Saltykova, V. S. and Fabrikova, E. A., *Zh. Analit. Khim.* **13**, 63 (1958).
33a. Marczenko, Z. and Krasiejko, M., unpublished work.
34. Mityureva, T. T. and Nizhnik, A. T., *Ukr. Khim. Zh.* **24**, 650 (1958).
35. Parissakis, G. and Issopoulos, P. B., *Mikrochim. Acta* **1965**, 28.
36. Szücs, A. I. and Klug, O. N., *Chem. Anal.* (*Warsaw*) **11**, 665 (1966); **12**, 939 (1967).
36a. Lypka, G. N. and Chow, A., *Anal. Chim. Acta* **60**, 65 (1972).
37. Rutkowski, W., and Basińska, M., *Chem. Anal.* (*Warsaw*) **13**, 641 (1968).
38. Šoljić, Z. and Marjanović, V., *Chim. Anal.* (*Paris*) **51**, 121 (1969); **52**, 285 (1970).

39. Skrebkova, L. M., *Zh. Analit. Khim.* **16**, 422 (1961).
40. Armeanu, V. and Costinescu, P., *Talanta* **14**, 699 (1967).
41. Drăgulescu, C. and Costinescu, P., *Rev. Roum. Chim.* **10**, 67, 1267 (1965).
42. Jankovský, J., *Talanta* **2**, 29 (1959).
43. Kuznetsova, V. K., *Tr. Komis. po Analit. Khim., Akad. Nauk SSSR* **16**, 33 (1968).
44. Tarayan, V. M., Ovsepyan, E. N. and Pogosyan, A. N., *Zavodsk. Lab.* **36**, 656 (1970).
45. Kish, P. P. and Bukovich, A. M., *Ukr. Khim. Zh.* **35**, 1290 (1969).
46. Kish, P. P. and Bukovich, A. M., *Zh. Analit. Khim.* **24**, 1653 (1969).
47. Matsuo, T., Funada, S. and Suzuki, M., *Bull. Chem. Soc. Japan* **38**, 326 (1965).
48. Pilipenko, A. T. and Nguyen Dyk Tu, *Ukr. Khim. Zh.* **35**, 200 (1969).
49. Busev, A. I., Skrebkova, L. M. and Zhivopistsev, V. P., *Zh. Analit. Khim.* **17**, 685 (1962).
50. Otomo, M., *Bull. Chem. Soc. Japan* **38**, 624 (1965).
51. Kish, P. P. and Golovei, M. I., *Zh. Analit. Khim.* **20**, 794 (1965).
52. Doicheva, R., Popova, S. and Mitropolitska, E., *Talanta* **13**, 1345 (1966).
53. Sotnikov, V. S., Kononova, A. M., Kononova, L. I. and Yaskevich, M. E., *Zavodsk. Lab.* **33**, 10 (1967).
54. Glushchenko, E. L., Akhmedli, M. K. and Kyazimova, A. K., *Zh. Analit. Khim.* **26**, 75 (1971).
55. Akhmedli, M. K., Bashirov, E. A., Glushchenko, E. L. and Zykova, L. I., *ibid.* **21**, 1022 (1966).
56. Ishito, T., *Japan Analyst* **21**, 752 (1972).
57. Akhmedli, M. K., Glushchenko, E. L. and Gasanova, Z. L., *Zh. Analit. Khim.* **26**, 1947 (1971).
58. Nazarenko, V. A. and Nevskaya, E. M., *ibid.* **24**, 839 (1969).
59. Lukomskaya, N. D., *Izv. Vyssh. Ucheb. Zaved., Khim. Khim. Tekhnol.* **14**, 1431 (1971).
60. Horiuchi, Y. and Nishida, H., *Japan Analyst* **16**, 1146 (1967).
61. Evtimova, B. and Nonova, D., *Anal. Chim. Acta* **67**, 107 (1973).
62. Joshi, A. P. and Munshi, K. N., *Microchem. J.* **12**, 447 (1967).
63. Garg, V. C., Shrivastawa, S. C. and Dey, A. K., *Mikrochim. Acta* **1969**, 668.
64. Kish, P. P. and Onishchenko, Yu. K., *Zh. Analit. Khim.* **21**, 944 (1966).
65. Hniličkovà, M. and Sommer, L., *Z. Anal. Chem.* **193**, 171 (1963).
66. Bansho, K. and Umezaki, Y., *Bull. Chem. Soc. Japan* **40**, 326 (1967).
67. Vadasdi, K. G., *Chem. Anal. (Warsaw)* **14**, 733 (1969).
68. Cheng, K. L. and Goydish, B. L., *Anal. Chim. Acta* **34**, 154 (1966).
69. Dobeš, I. and Salamon, M., *Chem. Listy* **60**, 68 (1966).
70. Lukin, A. M. and Zavarikhina, G. B., *Zh. Analit. Khim.* **13**, 66 (1958).
71. Karanovich, G. G., Ionova, L. A. and Podol'skaya, B. L., *ibid.* **13**, 439 (1958).
72. Salikhov, V. D. and Yampol'skii, M. Z., *ibid.* **20**, 1299 (1965).
73. Shkrobot, E. P., *ibid.* **17**, 311 (1962).
74. Langová-Hniličková, M. and Sommer, M., *Talanta* **16**, 681 (1969).
75. Busev, A. I., Zholondkovskaya, T. N., Krysina, L. S. and Golubkova, N. A., *Zh. Analit. Khim.* **27**, 2165 (1972).
76. Salikhov, V. D. and Yampol'skii, M. Z., *ibid.* **23**, 189 (1968).
77. Salikhov, V. D., Dedkov, Yu. M. and Yampol'skii, M. Z., *ibid.* **23**, 529 (1968).
78. Dedkov, Yu. M., Khokhlov, L. M. and Salikhov, V. D., *ibid.* **26**, 2350 (1971).
79. Salikhov, V. D., Khokhlov, L. M. and Dedkov, Yu. M., *ibid.* **26**, 69 (1971).
80. Devyatova, T. M. and Yampol'skii, M. Z., *ibid.* **23**, 1468 (1968).
81. Chang, T.-L. and Yoe, J. H., *Anal. Chim. Acta* **29**, 344 (1963).
82. Oskotskaya, E. P. and Yampol'skii, M. Z., *Zh. Analit. Khim.* **23**, 1307 (1968).
83. Basargin, N.N., Akhmedli, M. K. and Kafarova, A. A., *ibid.* **25**, 1497 (1970).
84. Gusev, S. I. and Dazhina, L. G., *ibid.* **24**, 1506 (1969).

85. Moeller, T. and Cohen, A. J., *Anal. Chim. Acta* **4**, 316 (1950); *Anal. Chem.* **22**, 686 (1950).
86. Keil, R., *Z. Anal. Chem.* **249**, 172 (1970).
87. Ngueyn Shi Zung and Zharovskii, F. G., *Ukr. Khim. Zh.* **36**, 1273 (1970).
88. Busev, A. I. and Shkrobot, E. P., *Vestn. Mosk. Univ. Khim.* **1959**, No. 4, 199.
89. Nazarenko, V. A., Biryuk, E. A., Antonovich, V. P. and Ravitskaya, R. V., *Ukr. Khim. Zh.* **34**, 504 (1968).
90. Olenovich, N. L. and Koval'chuk L. I., *Zh. Analit. Khim.* **28**, 2162 (1973).
91. Garg, B. S. and Singh, R. P., *Talanta* **18**, 761 (1971).
92. Graffmann, G. and Jackwerth, E., *Z. Anal. Chem.* **246**, 12 (1969).
93. Dwivedi, C. D., Munshi, K. N. and Dey, A. K., *Microchem. J.* **9**, 218 (1965).
94. Vasil'eva, N. L. and Ermakova, M. I., *Zh. Analit. Khim.* **18**, 43 (1963).

Chapter 22

GERMANIUM

Germanium (Ge, at.wt. 72·59) resembles tin and arsenic in its chemical properties. It occurs in the +IV and +II (and −IV in GeH_4) oxidation states. Germanium(II) compounds are unstable. Germanium(IV) compounds are amphoteric with the acidic properties predominating. The white sulphide, GeS_2, precipitates from strongly acidic media but dissolves in alkali. Germanium(IV) forms halide and oxalate complexes and heteropoly acids.

22.1 Methods of Separation

22.1.1 Extraction

A rapid and very selective separation of germanium from other elements is achieved by extracting $GeCl_4$ from $9M$ hydrochloric acid into carbon tetrachloride (or other inert solvents such as $CHCl_3$ or C_6H_6) [1–5]. With $9M$ HCl, the distribution coefficient for $GeCl_4$ is 500; with $8M$ it is only 50. Only $AsCl_3$ and traces of some other metals are also extracted. Extraction of arsenic is prevented by oxidation to As(V) with $KClO_3$. Water or an oxalic acid solution strips germanium from the CCl_4 phase.

A considerable difference exists between the distribution coefficients for $GeCl_4$ and $AsCl_3$ in extractions with hydrocarbons [6]. Extraction of $GeCl_4$ with MIBK [7] or with mesityl oxide [7a] is not as selective as with benzene [7]. Germanium can also be extracted as $GeBr_4$ with diethyl ether [8].

A germanium complex with tannin has been extracted with tri-n-octylamine (TOA) in n-butanol [9]. Germanium may be separated from gallium and from titanium and zirconium [10] by extraction of the Ge–BPHA chelate into chloroform or benzene.

22.1.2 Distillation

Distillation of volatile $GeCl_4$ (b.p. 84°C) from ~ $6M$ HCl quantitatively separates germanium [11–14]. Part or all of any As(III), Sn(IV), and Sb(III) distils with the $GeCl_4$. Distillation of arsenic is prevented by oxidation with $KMnO_4$ to non-volatile As(V), or reduction to the element with powdered copper. Any fluoride in the sample solution is distilled off first (as hydrogen fluoride) at 160°C from sulphuric acid medium (HCl-free).

After removal of the fluoride, NaCl is added to the still, and GeCl$_4$ is distilled.

Comparative tests have shown [14] the solvent extraction and the distillation of germanium as GeCl$_4$ to give equally good results. The extraction method is more rapid, and stripping yields germanium in aqueous solution, whereas the distillation yields germanium in moderately concentrated acid (6M HCl).

22.1.3 Other Methods

Koch and Korkisch [15] have separated germanium by solvent extraction with TBP, followed by ion-exchange removal of interfering coextracted elements. Dowex-50 cation-exchanger retains most metal ions from dilute HCl, while germanium is eluted as neutral GeCl$_4$ [16]. Cabbell *et al.* [17] separated GeCl$_4$ from both cations and anions by passing the sample solution at pH 2 through a mixed-resin bed (strongly acidic cation-exchanger + weakly basic anion-exchanger).

Strongly basic anion-exchangers retain germanium and arsenic(V) from a solution at pH 6–9. First, germanium is eluted with 0·2M CH$_3$COOH, and then arsenic is eluted with 5% H$_2$SO$_4$ [18]. Separation of germanium from boron was achieved on AV-17 anion-exchanger [19].

Germanium can be separated from zinc sulphate solutions with the aid of polyphenol exchange resins [19a].

Traces of germanium are preconcentrated by coprecipitation with Fe(OH)$_3$ [20] or ammonium molybdophosphate [21]. Germanium sulphide has been precipitated from 6M HCl with mercury or arsenic as collectors [22].

22.2 Methods of Determination

Phenylfluorone is commonly used as a sensitive spectrophotometric reagent for determining germanium. Other notable methods involve germanium heteropoly acids, and Pyrocatechol Violet. Shcherbov and Plotnikova [22a] have reviewed photometric methods of determining germanium.

22.2.1 Phenylfluorone Method

Germanium(IV) reacts with phenylfluorone (I) to yield a red-orange complex which is sparingly soluble in water [1,4,7a,12,23–26]. The reaction proceeds slowly in acid medium (pH 0–1), but fairly rapidly (1–2 minutes) in acetate-buffered solution (pH 4–5). Although a greater excess of reagent accelerates the reaction rate, it is most convenient to carry out the reaction at pH 4–5 with a small excess of phenylfluorone, and then to acidify the solution to pH < 1 before measuring the absorbance. In order to stabilize the colloidal suspension of the complex, a protective colloid is added [e.g. gum arabic, gelatin, or poly(vinyl alcohol)]. The presence of methanol (~ 40%) helps to maintain a stable, clear pseudosolution.

Figure 22.1 shows the absorption spectra of phenylfluorone (1) and its germanium complex (2). Interference due to the absorption of the reagent is slight. The molar absorptivity of the solution of germanium phenylfluorone complex, under the conditions given in the procedure below, is 5.3×10^4 at 510 nm ($a = 0.73$).

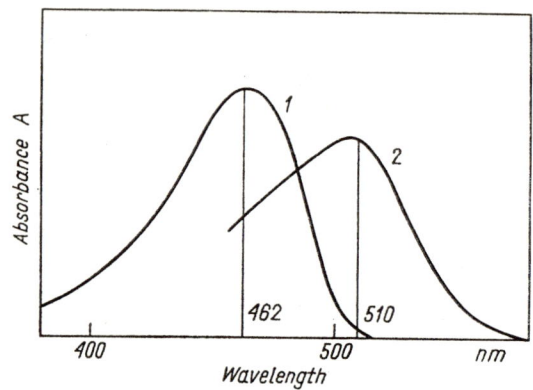

Fig. 22.1 Absorption spectra of phenylfluorone (1) and its germanium complex (2) at pH ~0.5 (in 40% methanol)

Phenylfluorone also forms coloured complexes with other multivalent metals [e.g. Sn, Sb, Ti, Fe(III), Nb, and Ta]. Low concentrations of arsenic, silicon, and fluoride do not interfere in the formation of the germanium complex. Citric and oxalic acids are used to mask Mo, V, Sn, and Ti [27,28]. Preliminary separation of germanium as $GeCl_4$ by extraction or distillation renders the phenylfluorone method specific for germanium.

When the colloidal solution of germanium phenylfluoronate is shaken with carbon tetrachloride, the precipitate agglomerates at the interface. After removal of nearly all the CCl_4 and the aqueous solution (containing the excess of phenylfluorone), the precipitate is dissolved in acetone and the absorbance measured [29]. Hillebrant and Hoste [30] have extracted germanium phenylfluoronate into benzyl alcohol, thereby increasing the sensitivity twofold.

Phenylfluorone has been used to determine germanium in coal [12,15, 31–34], coke [12], flue-dust [12], silicate minerals [15], ores [31,35], organogermanium compounds [36], and air [37].

p-Dimethylaminophenylfluorone forms a water-soluble complex with germanium, thus removing the necessity for protective colloids [38,39].

Nitrophenylfluorone and disulphophenylfluorone (in weak acid media) are suitable alternatives to phenylfluorone [40]. Nazarenko *et al.* [41] have shown that certain trihydroxyfluorones react with germanium in the presence of antipyrine and bromide (or I^-, SCN^-, ClO_4^-) to form complex species extractable into chloroform and certain other solvents.

Reagents

1. Phenylfluorone: 0·01% solution in methanol. Dissolve 25·0 mg of the reagent in methanol containing 2·5 ml of conc. HCl, and dilute the solution with methanol to 250 ml in a volumetric flask.
2. Standard germanium solution: 1 mg/ml. Dissolve 0·1000 g of germanium powder in 5 ml of $1M$ NaOH, acidify the solution slightly with hydrochloric acid, and dilute to the mark in a 100-ml volumetric flask with water.
3. Acetate buffer (pH ~ 5). Dissolve in water 450 g of sodium acetate trihydrate and 240 ml of glacial acetic acid, and dilute the solution with water to 1 litre.

Procedure

Extractive separation of Ge. Add sufficient conc. hydrochloric acid to the sample solution to give a final acid concentration of at least $9M$. (The sample solution can be concentrated by evaporation after being made slightly alkaline with NaOH). Shake the acidic solution for 2 minutes in a separating funnel with two portions of CCl_4, and wash the combined organic extracts with $9M$ HCl. Strip the germanium by shaking the CCl_4 solution with 10 ml of water followed by 5 ml of water containing 1 drop of $1M$ NaOH (shaking time 1 minute).

Determination of Ge. To a 50-ml volumetric flask, add the solution obtained (or an aliquot thereof) containing not more than 50 µg of Ge. Add water to 20 ml, 2 ml of 1% gum arabic solution, 10 ml of phenylfluorone solution, 10 ml of methanol, and 2 ml of acetate buffer, mixing the solution after the addition of each reagent. After 5 minutes, dilute the solution with hydrochloric acid (1+4) to 50 ml, and measure the absorbance at 510 nm against a blank solution.

22.2.2 OTHER METHODS

Germanium is one of the group of elements which forms yellow heteropoly acids with molybdate and other ions. The method for determining germanium based on yellow molybdogermanic acid [42,43] is insensitive (molar absorptivity ~ $2·0 \times 10^3$ at 440 nm), but reduction of the heteropoly acid to germanomolybdenum blue [44–46] considerably increases the sensitivity (ε ~ $10·0 \times 10^3$ at 800 nm). Germanium has also been determined as molybdovanadogermanic acid [47]. The β-form of molybdogermanic acid is unstable, being transformed spontaneously into the α-form, unless stabilized by addition of acetone [43].

High sensitivity characterizes methods based on the formation of sparingly water-soluble compounds of molybdogermanic acid with xanthene dyes (e.g. Rhodamine B, Rhodamine 6G) [48,49] and triphenylmethane dyes (e.g. Brillant Green) [50]. These compounds can be dissolved in ethanol or acetone after flotation with toluene or butyl acetate [49,50].

Busev et al. [51] have precipitated the ternary complexes formed with 2,3,4- or 3,4,5-trihydroxybenzoic acid and Brilliant Green, and dissolved the precipitate in aqueous acetone.

Several authors recommend Pyrocatechol Violet [52–55] as a colorimetric reagent for germanium. The determination is carried out in the aqueous phase, or directly in the organic phase after extraction of germanium as $GeCl_4$. The molar absorptivity of the complex is $4·32 \times 10^4$ at 610 nm [54].

Nazarenko et al. [56] have proposed a method based on extraction of the dinitropyrocatechol-germanic acid association complex with Brilliant Green, Nile Blue A or Methylene Blue. The molar absorptivity with Brilliant Green is $1·41 \times 10^5$.

The following organic reagents have also been proposed for the spectrophotometric determination of germanium: Rezarson [57,58], Bromopyrogallol Red [59,60], o-dihydroxychromenols [61,62], oxidized haematoxylin [63], quinalizarin [3], Purpurogallin [64], Purpurin [65], and 1,1′-dianthrimide [66].

Extraction of the yellow germanium iodide complex into cyclohexane provides a less sensitive method suitable for determining larger quantities of germanium [67].

References

1. Schneider, W. A., Jr. and Sandell, E. B., *Mikrochim. Acta* **1954**, 263.
2. Fischer, W., Harre, W., Freese, W. and Hackstein, K. G., *Angew. Chem.* **66**, 165 (1954).
3. Strickland, E. H., *Analyst* **80**, 548 (1955).
4. Luke, C. L. and Campbell, M. E., *Anal. Chem.* **28**, 1273 (1956).
5. Brink, G. O., Kafalas, P., Sharp, R. A., Weiss, E. L. and Irvine, J. W., Jr., *J. Am. Chem. Soc.* **79**, 1303 (1957).
6. Murach, N. N., Krapukhin, V. V., Kulikov, F. S., Chernyaev, V. N. and Nekhamkin, L. G., *Zh. Prikl. Khim.* **34**, 2188 (1961).
7. Senise, P. and Sant'Agostino, L., *Mikrochim. Acta* **1956**, 1445; **1959**, 572.
7a. Dhond, P. V. and Khopkar, S. M., *Anal. Chim. Acta* **59**, 161 (1972).
8. Ladenbauer, I., Slama, O. and Hecht, F., *Mikrochim. Acta* **1955**, 118.
9. Andrianov, A. M. and Koryukova, V. P., *Zh. Analit. Khim.* **24**, 1117 (1969).
10. Alimarin, I. P., Smolina, E. V., Sokolova, I. V. and Firsova, T. V., *ibid.* **25**, 2287 (1970).
11. Cluley, H. J., *Analyst* **76**, 523, 530 (1951).
12. Frederick, W. J., White, J. A. and Biber, H. E., *Anal. Chem.* **26**, 1328 (1954).
13. Wunderlich, E. and Göhring, E., *Z. Anal. Chem.* **169**, 346 (1959).
14. Basińska, M. and Rutkowski, W., *Chem. Anal. (Warsaw)* **8**, 353 (1963).
15. Koch, W. and Korkisch, J., *Mikrochim. Acta* **1973**, 101.
16. Klement, R. and Sandmann, H., *Z. Anal. Chem.* **145**, 325 (1955).

17. Cabbell, T. R., Orr, A. A. and Hayes, J. R., *Anal. Chem.* **32**, 1602 (1960).
18. Dranitskaya, R. M., Gavril'chenko, A. I. and Morozov, A. A., *Ukr. Khim. Zh.* **28**, 866 (1962).
19. Dranitskaya, R. M., Tsybul'kova, L. P. and Gavril'chenko, A. I., *Zh. Analit. Khim.* **22**, 448 (1967).
19a. Kraft, G., Dosch, H. and Gabbert, K., *Z. Anal. Chem.* **267**, 106 (1973).
20. Kuus, K. Ya., *ibid.* **16**, 166 (1961).
21. Tarasevich, N. I., Khlystova, A. D. and Okuneva, G. A., *Vestn. Mosk. Univ. Khim.* **1968**, No. 1, 90.
22. Bartelmus, G. and Hecht, F., *Mikrochim. Acta* **1954**, 148.
22a. Shcherbov, D. P. and Plotnikova, R. N., *Zh. Analit. Khim.* **27**, 740 (1972).
23. Gillis, J., Hoste, J. and Claeys, A., *Anal. Chim. Acta* **1**, 302 (1947).
24. Zharovskii, F. G. and Pilipenko, A. T., *Zavodsk. Lab.* **24**, 1192 (1958).
25. Burton, J. D. and Riley, J. P., *Mikrochim. Acta* **1959**, 586.
26. Oshman, V. A. and Volkov, V. M., *Zavodsk. Lab.* **27**, 1341 (1961).
27. Dranitskaya, R. M., Gavril'chenko, A. I. and Okhitina, L. A., *Zh. Analit. Khim.* **25**, 1740 (1970).
28. Dranitskaya, R. M., Gavril'chenko, A. I., Karpii, N. L. and Palei, L. N., *ibid.* **26**, 2137 (1971).
29. Luke, C. L., *Chemist-Analyst* **54**, 109 (1965).
30. Hillebrant, A. and Hoste, J., *Anal. Chim. Acta* **18**, 569 (1958).
31. Nazarenko, V. A., Lebedeva, N. V. and Ravitskaya, R. V., *Zavodsk. Lab.* **24**, 9 (1958).
32. Menkovskii, M. A. and Aleksandrova, A. N., *ibid.* **29**, 797 (1963).
33. Sendul'skaya, T. I., Shrirt, M. Ya. and Yurovskii, A. Z., *Zh. Analit. Khim.* **22**, 445 (1967).
34. Salikova, G. E. and Sevryukov, N. N., *Zavodsk. Lab.* **36**, 25 (1970).
35. Gregorowicz, Z., *Hutnik* **25**, 477 (1958); *Chem. Anal. (Warsaw)* **4**, 829 (1959).
36. Obtemperanskaya, S. I., Dudova, I. V. and Dikaya, G. F., *Zh. Analit. Khim.* **23**, 784 (1968).
37. Babina, M. D., *ibid.* **17**, 252 (1962).
38. Kimura, K., Saito K. and Asada, M., *Bull. Chem. Soc. Japan* **29**, 635 (1956).
39. Campe, A. and Hoste, J., *Talanta* **8**, 453 (1961).
40. Nazarenko, V. A. and Lebedeva, N. V., *Zavodsk. Lab.* **25**, 899 (1959).
41. Nazarenko, V. A., Makrinich, N. I. and Shustova, M. B., *Zh. Analit. Khim.* **25**, 1595 (1970).
42. Shakhova, Z. F., Motorkina, R. K. and Mal'tseva, N. N., *ibid.* **12**, 95 (1957).
43. Chalmers, R. A. and Sinclair, A. G., *Anal. Chim. Acta* **33**, 384 (1965).
44. Shaw, E. R. and Corvin, J. F., *Anal. Chem.* **30**, 1314 (1958).
45. Zhukovskii, Yu. G., *Zh. Analit. Khim.* **19**, 1361 (1964).
46. Paul, J., *Anal. Chim. Acta* **35**, 200 (1966).
47. Shakhova, Z. F. and Motorkina, R. K., *Zh. Analit. Khim.* **11**, 698 (1956).
48. Popa, Gr. and Paralescu, I., *Talanta* **16**, 315 (1969).
49. Ganago, L. I. and Prostak, I. A., *Izv. Vyssh. Ucheb. Zaved., Khim. Khim. Tekhnol.* **14**, 1165 (1971).
50. Ganago, L. I. and Prostak, I. A., *Zh. Analit. Khim.* **26**, 104 (1971).
51. Busev, A. I., Dzotsenidze, N. E. and Akimov, V. K., *ibid.* **24**, 556 (1969).
52. Nazarenko, V. A. and Vinarova, L. I., *ibid.* **18**, 1217 (1963).
53. Saginashvili, R. M. and Petrashen', V. I., *Zavodsk. Lab.* **32**, 661 (1966).
54. Agrinskaya, N. A., Gołosnitskaya, V. A. and Kovalenko, E. V., *ibid.* **33**, 923 (1967).
55. Leong, C. L., *Talanta* **18**, 845 (1971).
56. Nazarenko, V. A., Lebedeva, N. V. and Vinarova, L. I., *Zh. Analit. Khim.* **27**, 128 (1972).
57. Lukin, A. M., Efremenko, O. A. and Podol'skaya, B. L., *ibid.* **21**, 970 (1966).
58. Lukin, A. M., Efremenko, O. A. and Petrova, G. S., *ibid.* **22**, 1234 (1967).

59. Popa, Gr. and Paralescu, I., *Talanta* **15**, 272 (1968).
60. Nazarenko, V. A. and Makrinich, N. I., *Zh. Analit. Khim.* **24**, 1694 (1969).
61. Kononenko, L. I. and Poluektov, N. S., *ibid.* **15**, 61 (1960).
62. Nazarenko, V. A. and Makrinich, N. I., *ibid.* **25**, 719 (1970).
63. Newcombe, H., McBryde, W. A. E., Bartlett, J. and Beamish, F. E., *Anal. Chem.* **23**, 1023 (1951).
64. Nazarenko, V. A. and Poluektova, E. N., *Zh. Analit. Khim.* **19**, 1459 (1964).
65. Nazarenko, V. A., Flyantikova, G. V. and Tetereva, A. M., *ibid.* **29**, 284 (1974).
66. Skaar, O. B. and Langmyhr, F. J., *Anal. Chim. Acta* **21**, 370 (1959).
67. Tanaka, K. and Takagi, N., *ibid.* **48**, 357 (1969).

Chapter 23
GOLD

Gold (Au, at.wt. 196·97) occurs in the +I and +III oxidation states. Gold(I) compounds resemble the corresponding Ag, Cu(I) and Hg(I) compounds. Gold(I) gives stable cyanide and thiosulphate complexes. Gold(III) compounds are more stable. The hydroxide, $Au(OH)_3$, is amphoteric. Gold(III) yields stable halide complexes. Gold compounds are readily reduced to the metal, which dissolves in aqua regia to give $AuCl_4^-$.

23.1 Methods of Separation

23.1.1 Extraction

Gold(III) is usually separated from platinum metals by extraction of the chloride complex (from 4–8M HCl) [1–5] or as the bromide complex [6] into oxygenated solvents, such as amyl acetate, TBP, MIBK, mesityl oxide, and di-isopropyl ether. Iron(III) is masked with phosphoric acid before the extraction of gold.

Gold has also been extracted as the acetylacetonate [7], dithizonate [8], diethyldithiocarbamate [9], and thio-oxinate [10].

23.1.2 Precipitation

Gold is often precipitated by reduction to the element with tellurium as collector [11,12]. The separated precipitate is then dissolved in a few drops of aqua regia. Stannous chloride, zinc, magnesium, and hydrazine which are used as reductants also release Pt, Pd, Hg, and Ag. In the presence of palladium or platinum, gold may be separated by reduction with hydroquinone or oxalic acid.

Mizuike [13] separated traces of gold in copper by reduction and amalgamation with mercury in ammoniacal solution. He then distilled off the mercury from the amalgam at 350°C in a stream of nitrogen, leaving the gold in the residue.

Alternatively, traces of gold have been precipitated with 2-mercaptobenzimidazole [14] and 2-mercaptobenzothiazole [15], and with hydrogen sulphide in the presence of lead as collector. Gold has been coprecipitated with lanthanum hydroxide [16], and, as metallic gold, with $Fe(OH)_3$ in the presence of $SnCl_2$ as reductant [17].

23.1.3 Ion-Exchange and Other Methods

Traces of gold are isolated from a number of metals (e.g. from macro-quantities of copper) as the anionic chloride complex when a solution in dilute hydrochloric acid is passed through a strongly-basic anion-exchange column [18,19]. The gold is not eluted: instead, the resin containing the sorbed traces is ignited. Brooks [19] has isolated gold from sea water (9 μg of Au per ton) by passing 250 litres of water acidified with HCl to $0.1 M$ through a small Amberlite IRA-400 column at a rate of 2 ml/minute.

Dybczyński and Maleszewska [20] have separated gold from platinum metals on cation-exchangers in HBr medium. The gold retained by the resin is eluted with acetylacetone. Cation-exchange separation of gold is also possible in dilute HCl medium (pH 1–1.5) [21].

Koch and Korkisch [21a] have separated many metals from gold by sorption on Dowex 1 anion-exchanger from TBP solution.

Shashkin [22] has adsorbed on charcoal the colloidal gold released from cyanide solution by the exchange reaction with metallic zinc at pH 1–2 [$2Au(CN)_2^- + Zn \rightarrow 2Au + Zn(CN)_4^{2-}$].

Gold in high-purity mercury can be enriched by partially dissolving the sample (e.g. 10 g) in nitric acid. Almost all the gold is collected in the residue (which comprises a mercury drop of ~ 100 mg) [23].

The separation of gold and platinum metals from ores and concentrates with the classical fire assay and cupellation method is discussed in the chapter dealing with platinum (p. 431). Faye and Inman [24] have found tin to be a suitable alternative to lead as collector. The fire assay and cupellation method for gold has been critically evaluated by Chow and Beamish [25] and Trokowicz [26]. The determination of losses in the fire assay of gold is the subject of paper [27].

When sodium nitrite is added to a hot solution of gold and platinum metals (in dilute HCl), the gold is reduced to the element, while soluble complexes of the platinum metals are produced.

23.2 Methods of Determination

Many methods for determining gold are known. A sensitive and selective method using Rhodamine B and a simple bromide method suitable for determining large amounts of gold are discussed later. Among other methods, those with Methyl Violet and rhodanine as reagents merit attention.

Spectrophotometric methods for determining gold have been reviewed by Beamish [28].

23.2.1 Rhodamine B Method

In hydrochloric acid medium, the basic dye, Rhodamine B (formula, p. 53), forms an ion-association complex with the gold chloride complex,

[AuCl$_4$]$^-$, which is extractable into di-isopropyl ether [29] or benzene [30]. No free reagent is coextracted. This sensitive and selective method for determining gold was devised by MacNulty and Woolard [29]. The molar absorptivity of a benzene solution of the complex is 9.7×10^4 ($a = 0.49$) at $\lambda_{max} = 565$ nm.

Since the efficiency of extraction of the gold(III) complex depends on the concentration of hydrochloric acid in the solution, the concentration in both sample and standard solutions must be kept constant (at $\sim 0.5M$ HCl).

Interference in the determination of gold with Rhodamine B by Sb(V), Tl(III), Fe(III), Ga, and Hg(II) is overcome by coprecipitation of the gold traces with tellurium. Combined with this separation, the method is specific for gold. Antimony and thallium can be removed by coprecipitation with hydrous MnO$_2$. Iron(III) may be masked with fluoride.

Since the benzene extracts are sometimes turbid and rather difficult to clarify, it is preferable to extract the gold into di-isopropyl ether. The ethereal extracts are stable for at least 30 minutes.

Gold in silicate ores [30], copper concentrates [31], and mercury [23] has been determined by the Rhodamine B method.

Reagents

1. Rhodamine B: 0·04% solution in 1M HCl.
2. Standard gold solution: 1 mg/ml. Dissolve 0·1000 g of suitably pure gold in 4 ml of aqua regia (3 ml of conc. HCl + 1 ml of conc. HNO$_3$). Evaporate the solution nearly to dryness. Add 2 ml of conc. HCl, evaporate to half the volume, and dilute the solution with water to 100 ml in a volumetric flask. The stability of gold solutions has been discussed by Chow [32].
3. Tellurium solution: $\sim$ 1 mg/ml. Dissolve 0·10 g of tellurium in 2 ml of conc. HNO$_3$, and evaporate the solution to dryness. Dissolve the residue in 10 ml of conc. HCl, and dilute the solution with water to 100 ml.

Procedure

Separation of Au with Te as collector. To the sample solution (10–20 ml), add 1 mg of tellurium (as a solution in dilute HCl), and add hydrochloric acid till the acid concentration is 1–2M. Heat the solution nearly to boiling, and add hydrazine sulphate in small portions (0·2 g in total). Continue heating after the appearance of a turbidity, until the precipitate coagulates. Filter off the precipitate, wash it with water, and dissolve it in a few drops of conc. HCl and 1–2 drops of conc. HNO$_3$. Evaporate the solution nearly to dryness, and dilute with water.

Determination of Au. To the sample solution containing not more than 80 μg of Au, add 2·5 ml of HCl (1+1), dilute the solution with water to $\sim$ 15 ml, and add 5 ml of the Rhodamine B solution. Extract the gold with two portions of di-isopropyl ether (shaking time 30 seconds). Dilute the combined extracts with the solvent to the mark in a 50-ml or smaller volumetric flask (according to the amount of gold), and measure the absorbance at 555 nm against water.

23.2.2 Bromide Method

Gold is readily determined from the absorbance of an organic extract of the Au(III) bromide complex [33–35].

The gold bromide complex has an orange-yellow colour, the absorption maximum being at 380 nm ($\varepsilon = 4\cdot8 \times 10^3$ in di-isopropyl ether; $a = 0\cdot024$). Since the method is selective and simple, but rather insensitive, it is suitable for determining larger amounts of gold.

Potassium bromide and sulphuric acid can be used instead of hydrobromic acid [33]. Ethyl acetate is a suitable alternative to di-isopropyl ether as the solvent.

Large quantities of chloride should be avoided since they result in the formation of mixed chloride-bromide complexes which are less intensely coloured.

Iron(III) interferes in the determination of gold by forming an orange bromide complex, but it can be masked with phosphate or fluoride. In larger quantities, Cu, Ni, and Cr(III) also interfere. Of the platinum metals, only osmium interferes.

The gold(III) bromide complex ($[AuBr_4]^-$) is also extractable as coloured association complexes with polyethylene glycol [36,37], pyridine-N-oxide [38], TOPO [39], DAPM [40], and ferroin [the complex of iron(II) with 1,10-phenanthroline] [41].

Reagents

1. Hydrobromic acid: conc. (40%, $\sim 5M$) solution.
2. Standard gold solution: 1 mg/ml (p. 283).

Procedure

To 20–25 ml of the sample solution (in dilute sulphuric or hydrochloric acid) containing not more than 1·5 mg of Au, add 20 ml of conc. HBr, 1 ml of conc. H_3PO_4, and extract the complex of gold(III) with two portions of di-isopropyl ether. Wash the combined extracts with 2·5M hydrobromic acid. Dilute the ethereal solution with the solvent to the mark in a volumetric flask of suitable capacity, and measure its absorbance at 380 nm against the solvent.

23.2.3 Other Methods

One of the most widely used spectrophotometric reagents for gold is p-dimethylaminobenzylidenerhodanine (rhodanine; formula, p. 494) [42,43]; the diethyl analogue is also used [11]. The reagent reacts with gold ions in weakly acidic media (e.g. 0·1M HCl) to form a pink-violet complex, which is either stabilized in the aqueous phase, or extracted into a mixture of chloroform and benzene (3+1) or isoamyl acetate. This method is only one-third as sensitive as the Rhodamine B method. Gold has been determined with rhodanine in biological materials [11], copper [13,18], silver alloys [44], minerals [8], and sea water [14].

Dithizone (formula, p. 41) forms a yellow-brown gold complex which is soluble in chloroform or in carbon tetrachloride containing ethanol (10%) [16,45–50]. The molar absorptivity of gold dithizonate in CCl_4 (+10% ethanol) is $2 \cdot 8 \times 10^4$ at 420 nm [49].

A large group of sensitive spectrophotometric methods, similar to the Rhodamine B method, is based on extraction of ion-association complexes of $AuCl_4^-$ with various basic dyes, such as Methyl Violet ($\varepsilon = 1 \cdot 15 \times 10^5$ at 600 nm in trichloroethylene or chloroform) [51], Crystal Violet [52], Brilliant Green [53,54], Butylrhodamine B [55], Methylene Blue [56,56a], Victoria Blue 4R ($\varepsilon = 5 \cdot 5 \times 10^4$) [57], Chrompyrazole I (an antipyrine dye) ($\varepsilon = 6 \cdot 5 \times 10^4$ at $\lambda_{max} = 580$ nm in toluene) [58,59], and ferroin [60]. Nabivanets et al. [61] have compared the sensitivities of twenty extraction photometric methods based on the ion-association complexes with basic dyes. Association complexes with DAPM [62] or the tetrabutylammonium ion [63] give less sensitive methods.

Various organic reagents recommended for the spectrophotometric determination of gold include phenyl-α-pyridylketoxime [64], 2,2'-dipyridylketoxime [65], 2,2'-dipyridyl-α-glyoxime [66], pyridyl-2-aldoxime [67], diphenylcarbazide [68], sulphochlorophenolazorhodanine ($\varepsilon = 5 \cdot 2 \times 10^4$) [69], thio-Michler's ketone [70], TTA [71], formic acid hydrazide [72], and thiosalicylamide [73].

Many methods are based on the colour effects produced by the oxidation by gold(III) of the following organic compounds: o-tolidine [74], N,N'-tetramethyl-o-tolidine (Tetron) [75,76], Variamine Blue [77], Chlorpromazine [78], o-phenylenediamine [79], and leuco-Malachite Green [80].

Gold may also be determined as a coloured sol after reduction to the metal by, for example, tin(II) [81] or anthranilic acid [82].

References

1. Ishimori, T., Watanabe, K. and Nakamura, E., *Bull. Chem. Soc. Japan* **33**, 636 (1960).
2. Shinde, V. M. and Khopkar, S. M., *Anal. Chim. Acta* **43**, 146 (1968).
3. Jordanov, N. and Havesov, I., *Z. Anal. Chem.* **244**, 176 (1969).
4. Yadav, A. A. and Khopkar, S. M., *Sepn. Sci.* **5**, 637 (1970).
5. Ichinose, N., *Talanta* **18**, 105 (1971).
6. McBryde, W. A. E. and Yoe, J. H., *Anal. Chem.* **20**, 1094 (1948).
7. Wódkiewicz, L. and Jaskólska, H., *Chem. Anal. (Warsaw)* **6**, 1071 (1961).
8. Thilliez, G., *Acta Chim. Acad. Sci. Hung.* **32**, 315 (1962).
9. Kreimer, S. E., Mikhailov, P. M. and Lomekhov, A. S., *Zh. Analit. Khim.* **22**, 1105 (1967).

10. Demina, L. A., Petrukhin, O. M. and Zolotov, Yu. A., *ibid.* **25**, 1704 (1970); **27**, 593 (1972).
11. Sandell, E. B., *Anal. Chem.* **20**, 253 (1948).
12. Voskresenskaya, N. T., Zvereva, N. F. and Rivkina, L. L., *Zh. Analit. Khim.* **20**, 1288 (1965).
13. Mizuike, A., *Talanta* **9**, 948 (1962).
14. Weiss, H. V. and Lai, M. G., *Anal. Chim. Acta* **28**, 242 (1963).
15. Ujihira, Y., *J. Chem. Soc. Japan, Pure Chem. Sect.* **84**, 642 (1963).
16. Marczenko, Z. and Kasiura, K., *Chem. Anal. (Warsaw)* **9**, 87 (1964).
17. Polikarpochkin, V. V., Korotaeva, I. Ya. and Sarapulova, V. N., *Zavodsk. Lab.* **33**, 441 (1967).
18. Mizuike, A., Iida, Y., Yamada, K. and Hirano, S., *Anal. Chim. Acta* **32**, 428 (1965).
19. Brooks, R. R., *Analyst* **85**, 745 (1960).
20. Dybczyński, R. and Maleszewska, H., *ibid.* **94**, 527 (1969).
21. Pitts, A. E. and Beamish, F. E., *Anal. Chem.* **41**, 1107 (1969).
21a. Koch, W. and Korkisch, J., *Mikrochim. Acta* **1973**, 117.
22. Shashkin, M. A., *Zavodsk. Lab.* **27**, 145 (1961).
23. Jackwerth, E. and Kulok, A., *Z. Anal. Chem.* **257**, 28 (1971).
24. Faye, G. H. and Inman, W. R., *Anal. Chem.* **33**, 1914 (1961).
25. Chow, A. and Beamish, F. E., *Talanta* **14**, 219 (1967).
26. Trokowicz, J., *Chem. Anal. (Warsaw)* **15**, 1147 (1970).
27. Wall, S. G. and Chow, A., *Anal. Chim. Acta* **69**, 439 (1974).
28. Beamish, F. E., *Anal. Chem.* **33**, 1059 (1961); *Talanta* **12**, 789 (1965).
29. MacNulty, B. J. and Woolard, L. D., *Anal. Chim. Acta* **13**, 154 (1955).
30. Onishi, H., *Mikrochim. Acta* **1959**, 9.
31. Pompowski, T. and Trokowicz, J., *Chem. Anal. (Warsaw)* **10**, 1211 (1965).
32. Chow, A., *Talanta* **18**, 453 (1971).
33. Patrovský, V., *Collection Czech. Chem. Commun.* **27**, 1705 (1962).
34. Chow, A. and Beamish, F. E., *Talanta* **10**, 883 (1963).
35. Shkrobot, E. P., Shebarshina, N. I. and Bakinovskaya, L. M., *Zavodsk. Lab.* **37**, 409 (1971).
36. Ziegler, M. and Matschke, H. D., *Z. Anal. Chem.* **184**, 166 (1961).
37. Chapman, A. H. and Price, J. W., *Metallurgia* **78**, 217 (1968).
38. Ziegler, M. and Stephan, G., *Mikrochim. Acta* **1970**, 628.
39. Holbrook, W. B. and Rein, J. E., *Anal. Chem.* **36**, 2451 (1964).
40. Akimov, V. K., Busev, A. I., Shubashvili, L. V. and Shvedova, N. V., *Zavodsk. Lab.* **38**, 146 (1972).
41. Nasouri, F. G., Shahine, S. A. and Magee, R. J., *Anal. Chim. Acta* **36**, 346 (1966).
42. Hara, S., *Japan Analyst* **7**, 147 (1958).
43. Cotton, T. M. and Woolf, A. A., *Anal. Chim. Acta* **22**, 192 (1960).
44. Chwastowska, J. and Skorko-Trybuła, Z., *Chem. Anal. (Warsaw)* **9**, 123 (1964).
45. Young, R. S., *Analyst* **76**, 49 (1951).
46. Titley, A. W., *ibid.* **87**, 349 (1962).
47. Beardsley, D. A., Briscoe, G. B., Růžička, J. and Williams, M., *Talanta* **13**, 328 (1966).
48. Zolotov, Yu. A., Demina, L. A. and Petrukhin, O. M., *Zh. Analit. Khim.* **25**, 2315 (1970).
49. Marczenko, Z. and Krasiejko, M., *Chem. Anal. (Warsaw)* **17**, 1201 (1972).
50. Cox, J. J. and Servant, D. M., *Anal. Chim. Acta* **66**, 123 (1973).
51. Ducret, L. and Maurel, H., *Anal. Chim. Acta* **21**, 74 (1959).
52. Kothny, E. L., *Analyst* **94**, 198 (1969).
53. Stanton, R. E. and McDonald, A. J., *ibid.* **89**, 767 (1964).
54. Fogg, A. G., Burgess, C. and Burns, D. T., *ibid.* **95**, 1012 (1970).
55. Blyum, I. A., Pavlova, N. N. and Kalupina, F. P., *Zh. Analit. Khim.* **26**, 55 (1971).
56. Gantchev, N. and Dimitrova, A., *Mikrochim. Acta* **1969**, 1257.

56a. Ganchev, N. and Atanasova, B., *Zh. Analit. Khim.* **22**, 274 (1967).
57. Kish, P. P., Shestidesyatnaya, N. L. and Pitsur, F. F., *ibid.* **24**, 1501 (1969).
58. Gorbunova, N. N., Busev, A. I. and Ivanov, V. M., *ibid.* **25**, 461, 1471 (1970).
59. Busev, A. I., Gorbunova, N. N. and Ivanov, V. M., *Zavodsk. Lab.* **37**, 26 (1971).
60. Yamamoto, Y., Tsubouchi, M. and Okimura, I., *Japan Analyst* **16**, 1176 (1967).
61. Nabivanets, B. I., Zadorozhnaya, E. M. and Maslei, N. N., *Zh. Analit. Khim.* **28**, 1901 (1973).
62. Busev, A. I. and Babenko, N. L., *ibid.* **19**, 926 (1964).
63. Bravo, O. and Iwamoto, R. T., *Anal. Chim. Acta* **47**, 209 (1969).
64. Sen, B., *ibid.* **21**, 35 (1959).
65. Holland, W. J. and Bozic, J., *Anal. Chem.* **39**, 109 (1967).
66. Soules, D. and Holland, W. J., *Mikrochim. Acta* **1971**, 565.
67. Gagliardi, E. and Presinger, P., *ibid.* **1965**, 791.
68. Adam, J. and Přibil, R., *Talanta* **18**, 405 (1971).
69. Propistsova, R. F., Savvin, S. B. and Rozovskii, Yu. G., *Zh. Analit. Khim.* **26**, 2424 (1971).
70. Christopher, A. J., *Analyst* **94**, 397 (1969).
71. Rangnekar, A. V. and Khopkar, S. M., *Z. Anal. Chem.* **230**, 425 (1967).
72. Nashmi, M. H., Rashid, A., Umar, M. and Azam, F., *Anal. Chem.* **38**, 439 (1966).
73. Sur, K., Mazumdar, M. and Shome, S. C., *Anal. Chim. Acta* **59**, 306 (1972).
74. Schreiner, H., Brantner, H. and Hecht, F., *Mikrochemie* **36/37**, 1056 (1951).
75. Daiev, Ch. and Jordanov, N., *Talanta* **11**, 501 (1964).
76. Jordanov, N., Mareva, St., Krasnobaeva, N. and Nedyalkova, N., *ibid.* **15**, 963 (1968).
77. Mustafin, I. S., Frumina, N. S. and Agranovskaya, L. A., *Zh. Analit. Khim.* **18**, 1054 (1963).
78. Lee, K.-T., *Anal. Chim. Acta* **26**, 478 (1962).
79. Shkrobot, E. P., Shebarshina, N. I. and Bakinovskaya, L. M., *Zavodsk. Lab.* **37**, 1296 (1971).
80. Pilipenko, A. T., Pavlova, V. K. and Voevutskaya, R. N., *ibid.* **38**, 257 (1972).
81. Pantani, F. and Piccardi, G., *Anal. Chim. Acta* **22**, 231 (1960).
82. Macovschi, M. E., *Talanta* **16**, 443 (1969).

Chapter 24

INDIUM

Indium (In, at.wt. 114·82) is similar to gallium and cadmium in its chemical properties. It occurs in aqueous solution exclusively in the +III oxidation state. The hydroxide, $In(OH)_3$, precipitates above pH 3–4, but, when freshly precipitated, redissolves in fairly concentrated alkali, thereby displaying weakly amphoteric properties. Yellow In_2S_3 is precipitated at pH 2–3. Indium forms strong halide, oxalate, tartrate, and EDTA complexes.

24.1 Methods of Separation

24.1.1 Extraction

Among the methods for the separation of small amounts of indium before its determination, the solvent extraction methods in general, and extraction of indium iodide and bromide complexes in particular, are of the utmost importance.

The indium iodide complex [1–3a] is > 99% extracted into diethyl ether or other oxygenated solvents (e.g. cyclohexanone and MIBK) from 0·5–2·5M hydriodic acid (6–30%). Gallium is not extracted under these conditions, but the reverse is true when the extraction is from 6M HCl. The hydriodic acid can be made by adding 15–20% of potassium iodide to 0·5–3M sulphuric acid. Chloride, bromide, cyanide, fluoride, phosphate, and citrate do not interfere in the extraction of indium from iodide media. Under the optimum conditions for the extraction of indium, Tl, Cd, and Sn (and some Bi, Zn, Hg, and Sb) are extracted. Aluminium and iron(II), like gallium, are not extracted. The indium iodide complex has also been extracted into chloroform containing N-benzylaniline [4].

The indium bromide complex is extractable into diethyl ether from 4·5–5·5M hydrobromic acid (∼ 40%), or di-isopropyl ether from 6M HBr [5–7]. The extraction of indium from bromide solution is not as selective as that from iodide solution. Ga, Fe(III), Sb(V), Au(III), Tl(III), Sn, and Mo are extracted with the indium, which is then stripped from the ethereal solution with water. The indium bromide complex has been extracted into benzene containing tri-n-butylphosphine [8], and into toluene containing the quaternary amine Adogen-364 [9]. Indium has also been extracted with TBP from HCl medium [10].

When a solution (pH ~ 9) containing fairly large amounts of cyanide and citrate is shaken with dithizone in chloroform, indium passes into the chloroform layer accompanied by only Pb, Bi, Sn(II), and Tl [11–13]. Bismuth may be isolated first by dithizone extraction at pH 3–4. Dithizone in carbon tetrachloride extracts indium, but not thallium, from a solution at pH 5–6 [14].

Other methods for the extractive separation of indium are based on its complexes with 8-hydroxyquinoline [15,16], acetylacetone [17], and DDTC [18]. During the extraction of the indium DDTC complex at pH 3–5, extraction of gallium is prevented by the addition of oxalate.

24.1.2 Ion-Exchange

Klement and Sandmann [19] have separated indium, gallium, and germanium on Dowex-50 cation-exchanger. The indium and gallium are sorbed on the column from a chloride solution, whereas the germanium is eluted. Indium is then eluted with $0.4M$ HCl, whereas gallium is not eluted unless the acid concentration is raised to $1.3M$ HCl.

In an ammonium carbonate medium, indium forms an anionic carbonate complex, while zinc and cadmium form cationic ammine complexes, which can be separated from the indium complex on cation- or anion-exchangers [20].

Indium and tin(IV) are retained from $5M$ HCl by an anion-exchanger, in contrast to Al, Mn, Cu, and As. Indium is eluted from the column with $0.1M$ HCl [21]. Mixed media, such as $1M$ HCl and 2-methoxyethanol or $1M$ HCl and acetone, are also suitable for the anion-exchange separation of indium from Al, Ga, and Tl [22,22a].

24.1.3 Precipitation

Indium may be separated from the Analytical Group III and IV metals by precipitation as the sulphide from $0.1M$ HCl medium with tin(IV) as collector [23].

Separation from metals yielding soluble ammine complexes (Ag, Cu, Ni, Co, Zn, Cd), is achieved by precipitation as $In(OH)_3$ with ammonia. Lanthanum [14], iron(III) [24], or aluminium are suitable collectors.

Traces of indium(InI_4^-) have been coprecipitated with Methylene Blue, Nile Blue A or tannin (pH ~ 7) [25].

24.2 Methods of Determination

Since methods for determining indium are rather unselective, the separation methods are very important. The extractive spectrophotometric method involving bromo-oxine (or oxine), and the much more sensitive method based on the water-soluble coloured complex of indium with 4-(2-pyridylazo)-resorcinol (PAR) are discussed below in detail.

Babko and Kish [26] have compared 16 of the more important spectrophotometric reagents for indium.

24.2.1 Bromo-oxine Method

8-Hydroxyquinoline (oxine) [11,12] and 5,7-dibromo-8-hydroxyquinoline (bromo-oxine) [27–29] (formulae, p. 58) react with indium ions in weak acid medium (pH 3·5–4·0) to form yellow, chloroform-soluble chelates, on which is based this spectrophotometric method for determining indium.

The absorption maximum of an indium bromo-oxinate solution in chloroform is at 415 nm (the molar absorptivity being then $8·8 \times 10^3$; specific absorptivity 0·08). The corresponding values for indium oxinate are 395 nm, $6·7 \times 10^3$, 0·06. The absorption peaks of the indium complexes with the two reagents are wide, facilitating the selection of a suitable filter for measuring the absorption in a photocolorimeter.

The aqueous phase is adjusted to the appropriate pH with a phthalate buffer before the extraction of the indium chelate into chloroform.

The bromo-oxine and oxine methods for determining indium have poor selectivity. Under the optimum conditions (pH 3·5) the following metals also react: Al, Ga, Tl(III), Sn(II), Bi, Fe(III), V(V), Mo(VI), Cu, and Ni. Small amounts of lead, zinc, and thallium(I) do not interfere. As a general rule, the preliminary separation of indium, usually by extraction as either the dithizone [11,28] or the iodide [12] complex, is indispensable.

Bankovskii *et al.* [30] have determined indium with thio-oxine (in toluene) as extractant and spectrophotometric reagent.

Indium has been determined by the bromo-oxine or oxine methods in minerals [12,28], cadmium sulphide [31], germanium dioxide [11], tin [14], and zinc and lead ores [24].

Reagents

1. 5,7-Dibromo-8-hydroxyquinoline (bromo-oxine): 0·1% solution in chloroform.
2. Standard indium solution: 1 mg/ml. Dissolve 0·1000 g of metallic indium in 5 ml of HCl (1+1), and dilute the solution with water in a volumetric flask to 100 ml. Working solutions are obtained by suitable dilutions of the stock solution with $\sim 0·01 M$ HCl.
3. Phthalate buffer (pH 3·5). Dissolve 1 g of potassium hydrogen phthalate in water and adjust the solution to pH 3·5, using a pH-meter.

Procedure

To the acidic sample solution (pH 1–3) containing not more than 250 µg of In, add 1 ml of the bromo-oxine solution, a drop of 0·1% Methyl Orange solution, and ammonia till the indicator turns orange-yellow. Make up to ~ 25 ml with water, add 5 ml of the phthalate buffer, and transfer the solution to a separating funnel. Extract the indium with two portions of bromo-oxine solution in chloroform by shaking each portion for 1 minute. If the combined extracts are not clear, filter through a dry filter paper before diluting to the mark in a 50-ml or smaller volumetric flask (depending on the amount of indium) with the chloroform solution of bromo-oxine. Measure the absorbance of the solution at 415 nm against the bromo-oxine solution.

24.2.2 Pyridylazoresorcinol (PAR) Method

4-(2-Pyridylazo)-resorcinol (PAR) (formula, p. 46) reacts with indium to form two water-soluble complexes [32–36]. The 1:1 and 1:2 In:PAR complexes are formed at pH $\sim$ 3 and pH $\sim$ 6, respectively. Both complexes are present at intermediate pH values, but within the pH range from 6 to 8, only the 1:2 complex exists. The yellow colour of the reagent is constant between pH 2 and 10.

Figure 24.1 shows the absorption spectra of the reagent (λ_{max} = 425 nm), and of its pink 1:2 complex with indium (λ_{max} = 510 nm). The molar absorptivity of this complex is $\sim 4 \cdot 3 \times 10^4$ ($a = 0 \cdot 37$). The 1:1 complex which is formed at pH $\sim$ 3 also has its absorption maximum at $\sim$ 510 nm, but its colour is much weaker ($\varepsilon = 1 \cdot 8 \times 10^4$).

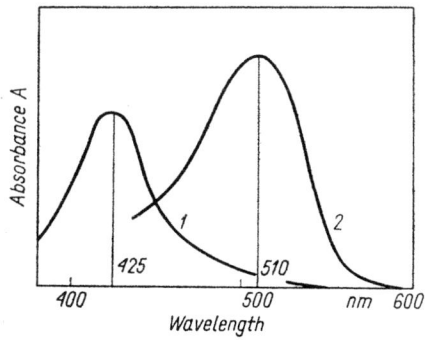

Fig. 24.1 Absorption spectra of 4-(2-pyridylazo)resorcinol (PAR) (1) and its indium complex (2) at pH 6

The PAR method for determining indium at pH 6 is sensitive but not selective. At pH 3, the selectivity of the method increases, but the sensitivity decreases [33,34]. When the reaction is carried out at pH 6, indium should be separated from Zn, Pb, Cr, Al, Sn(IV), Cd, Cu, and Mn. However, when ~ 10 μg of indium are determined at pH 3, these elements can be tolerated in the following maximum quantities: 20 μg of Zn, 40 μg of Pb and Cr, 300 μg of Al and Sn(IV), and 1000 μg of Cd, Cu, and Mn. Further interference is caused by Fe(III and II), Co, Ni, V, Zr, Bi, fluoride, oxalate, and EDTA. Tin(II) and other strong reductants decolorize the reagent irreversibly.

To determine indium in mineral concentrates containing, on average, 0·1% of this element, Kish and Orlovskii [33] have recommended the following procedure. A weighed sample (0·1–1 g) is decomposed in a mixture of HCl and HNO_3, tin is expelled as volatile $SnBr_4$, and lead is precipitated as $PbSO_4$. The hydroxides of metals which, like indium, form sparingly soluble hydroxides are then twice precipitated with excess of ammonia. After the precipitated hydroxides have been dissolved in hydrochloric acid, potassium iodide is added, and indium is extracted as the iodide complex into diethyl ether, then stripped into aqueous solution and determined spectrophotometrically with PAR.

1-(2-Pyridylazo)-2-naphthol (PAN) (formula, p. 46), is also a suitable spectrophotometric reagent for indium [35–39]. This reagent is equally as unselective as PAR, and is somewhat less sensitive ($\varepsilon = 1.9 \times 10^4$). PAN is added as a solution in methanol, and the water-insoluble complex with indium is extracted into chloroform from the aqueous solution at pH ~ 6.

Derivatives of PAR which have also been recommended as reagents for determining indium include 5-(2-pyridylazo)-2-monoethylamino-*p*-cresol (PAAC), 5-(2-pyridylazo)-4-ethoxy-2-monoethylamino-1-methyl-benzene (PAMB), and various bromine derivatives of PAAC [40].

Reagents

1. 4-(2-Pyridylazo)-resorcinol: 0·01% solution. Dissolve 10 mg of the reagent in 100 ml of water.
2. Standard indium solution: 1 mg/ml (p. 290).

Procedure

To the weakly acidic sample solution containing not more than 70 μg of In, add 5 ml of the PAR solution and 10 ml of 40% ammonium acetate solution. Dilute the solution with water to ~ 40 ml, and adjust the pH of the solution with ammonia to 6·0–6·2. Transfer the solution to a 50-ml volumetric flask, make up to the mark with water, and mix well. Measure the absorbance of the coloured solution at 510 nm, using a reagent blank solution as reference.

24.2.3 OTHER METHODS

Besides the pyridylazo dyes discussed in detail above, several other azo dyes have been recommended for determining indium, namely: TAR [41], 3,6-disulpho-TAN [42], 1-(4-carboxy-2-thiazolylazo)-2-naphthol [43], Gallion ($\varepsilon = 2.2 \times 10^4$ at 610 nm) [44], Lumogallion (formula, p. 270) [45], Sulphonazo [46], Sulphochlorophenol R [47], Stilbazo [48,49], Arsenazo I [50], and Thoron I [51].

Pyrocatechol Violet is a popular and remarkably sensitive ($\varepsilon = 3.5 \times 10^4$ at 630 nm) triphenylmethane reagent for indium [52–56]. Other reagents of this type used similarly are Xylenol Orange [57–60], Methylthymol Blue [61], Eriochrome Cyanine R [62], and Chrome Azurol S [63,64].

Owing to their high sensitivity and selectivity, methods based on extractable ion-association complexes of $InBr_4^-$ with the basic xanthene dyes Rhodamine B ($\varepsilon = 1 \cdot 1 \times 10^5$ at 557 nm) [65–67], and Rhodamine 6G [68,69] are also worthy of notice. The tetraiodoindate-Malachite Green ion-association complex is also extractable ($\varepsilon = 1 \cdot 06 \times 10^5$) [70]. Zolotov et al. [71] have extracted Brilliant Green tetrabromoindate.

Dithizone is more often used as an extractant than as a spectrophotometric reagent for indium [7,72–74]. Indium may be extracted with dithizone in carbon tetrachloride at pH 5–6, or with a chloroform solution at pH ~ 9. The absorption maximum of the pink $In(HDz)_3$ solution in $CHCl_3$ or in CCl_4 is at ~ 510 nm. The carbon tetrachloride solutions are stable for longer.

Gallein [75], phenylfluorone [13,76], salicylfluorone and other fluorones [77], morin [78,79], quercetin [79,80], and kaempferol [81] are fluorone and flavone derivatives which are also sensitive spectrophotometric reagents for indium.

The following organic reagents have also been recommended for the determination of indium: diphenylcarbazone [82], glyoxal bis(2-hydroxyanil) (a favourite reagent for calcium) [83], thiothenoyltrifluoroacetone [84], and Alizarin Red S [85]. The diphenylguanidinium salt of the indium Alizarin Red S chelate can be extracted with n-butyl acetate ($\varepsilon = 2 \cdot 65 \times 10^4$ at 525 nm).

References

1. Irving, H. M. and Rossotti, F. J. C., *Analyst* **77**, 801 (1952).
2. Hartkamp, H. and Specker, H., *Talanta* **2**, 67 (1959).
3. Gagliardi, E. and Tümmler, P., *ibid.* **17**, 93 (1970).
3a. Hasegawa, Y., Takeuchi, H. and Sekine, T., *Bull. Chem. Soc. Japan* **45**, 1388 (1972).
4. Khosla, M. M. and Rao, S. P., *Anal. Chim. Acta* **58**, 389 (1972).
5. Hudgens, J. E. and Nelson, L. C., *Anal. Chem.* **24**, 1472 (1952).
6. Kosta, L. and Hoste, J., *Mikrochim. Acta* **1956**, 790.
7. Collins, T. A., Jr. and Kanzelmeyer, J. H., *Anal. Chem.* **33**, 245 (1961).
8. Mieczkowska, E., *Chem. Anal. (Warsaw)* **14**, 683 (1969); **15**, 509 (1970).
9. Sherif, S. A., Abdel-Gawad, A. S. and El-Wakil, A. M., *Talanta* **17**, 137 (1970).
10. De, A. K. and Sen, A. K., *ibid.* **14**, 629 (1967).
11. Luke, C. L. and Campbell, M. E., *Anal. Chem.* **28**, 1340 (1956).
12. Irving, H. M., Smit, J. R. and Salmon, L., *Analyst* **82**, 549 (1957).
13. Stolarczyk, U., *Chem. Anal. (Warsaw)* **9**, 161 (1964).
14. Marczenko, Z. and Kasiura, K., *ibid.* **10**, 449 (1965).
15. Schweitzer, G. K. and Coe, G. R., *Anal. Chim. Acta* **24**, 311 (1961).
16. Zolotov, Yu. A. and Lambrev, V. G., *Zh. Analit. Khim.* **20**, 1153 (1965).
17. Steinbach, J. F. and Freiser, H., *Anal. Chem.* **26**, 375 (1954).
18. Busev, A. I., Zholondkovskaya, T. N. and Kuznetsova, Z. M., *Zh. Analit. Khim.* **15**, 49 (1960).
19. Klement, R. and Sandmann, H., *Z. Anal. Chem.* **145**, 325 (1955).
20. Alimarin, I. P., Tsintsevich, E. P. and Burlaka, V. P., *Zavodsk. Lab.* **25**, 1287 (1959).
21. Jentzsch, D., Frotscher, I., Schwerdtfeger, G. and Sarfert, G., *Z. Anal. Chem.* **144**, 8 (1955).
22. Korkisch, J. and Hazan, I., *Anal. Chem.* **36**, 2308 (1964).

22a. Strelow, F. W. and Victor, A. H., *Talanta* **19**, 1019 (1972).
23. Rudnev, N. A. and Dzhumaev, R. M., *Zh. Analit. Khim.* **19**, 443 (1964).
24. Gregorowicz, Z. and Marczak, M., *Chem. Anal. (Warsaw)* **14**, 159 (1969).
25. Kral, J., Jambor, and Sommer, L., *Chem. Listy* **63**, 1036 (1969).
26. Babko, A. K. and Kish, P. P., *Zh. Analit. Khim.* **17**, 693 (1962).
27. Johnson, J. E., Lavine, M. C. and Rosenberg, A. J., *Anal. Chem.* **30**, 2055 (1958).
28. Minczewski, J., Stolarczyk, U. and Marczenko, Z., *Chem. Anal. (Warsaw)* **6**, 51 (1961).
29. Nguyen Shi Zuong and Zharovskii, F. G., *Ukr. Khim. Zh.* **36**, 1159, 1273 (1970).
30. Bankovskii, Yu. A., Tsirule, Ya. A. and Ievin'sh, A. F., *Zh. Analit. Khim.* **16**, 562 (1961).
31. Kulik, O. P. and Mizetskaya, I. B., *Zavodsk. Lab.* **31**, 150 (1965).
32. Hagiwara, K. and Muraki, I., *Japan Analyst* **10**, 1022 (1961).
33. Kish, P. P. and Orlovskii, S. T., *Zh. Analit. Khim.* **17**, 1057 (1962).
34. Hniličková, M., *Collection Czech. Chem. Commun.* **29**, 1424 (1964).
35. Busev, A. I., Ivanov, V. M. and Khlybova, N. S., *Zh. Analit. Khim.* **22**, 547 (1967).
36. Biryuk, E. A. and Ravitskaya, R. V., *ibid.* **26**, 735 (1971).
37. Shibata, S., *Anal. Chim. Acta* **23**, 367, 434 (1960).
38. Cheng, K. L. and Goydish, B. L., *ibid.* **34**, 154 (1966).
39. Zolotov, Yu. A., Seryakova, I. V. and Vorobyeva, G. A., *Talanta* **14**, 737 (1967).
40. Gusev, S. I. and Nikolaeva, E. M., *Zh. Analit. Khim.* **21**, 166, 281, 1183 (1966).
41. Langová-Hniličková, M. and Sommer, L., *Talanta* **16**, 681 (1969).
42. Busev, A. I., Zholondkovskaya, T. N., Krysina, L. S. and Golubkova, N. A., *Zh. Analit. Khim.* **27**, 2165 (1972).
43. Drozdova, S. N., Momsenko, A. P. and Yampol'skii, M. Z., *ibid.* **26**, 291 (1971).
44. Joshi, A. P. and Munshi, K. N., *Mikrochim. Acta* **1971**, 526.
45. Salikhov, V. D. and Yampol'skii, M. Z., *Zh. Analit. Khim.* **22**, 998 (1967).
46. Shkrobot, E. P., *ibid.* **17**, 311 (1962).
47. Salikhov, V. D., Dedkov, Yu. M. and Yampol'skii, M. Z., *ibid.* **24**, 368 (1969).
48. Yampol'skii, M. Z., *Tr. Komis. po Analit. Khim. Akad. Nauk SSSR* **8**, 141 (1958); **11**, 261 (1960).
49. Ozawa, T., *Japan Analyst* **19**, 1389 (1970).
50. Matsumae, T., *ibid.* **8**, 167 (1959).
51. Mottola, H. A., *Talanta* **11**, 715 (1964).
52. Starościk, R. and Terpiłowski, J., *Chem. Anal. (Warsaw)* **7**, 803 (1962).
53. Orlovskii, S. T. and Kish, P. P., *Izv. Vyssh. Ucheb. Zaved., Khim. Khim. Tekhnol.* **5**, 892 (1962).
54. Malát, M. and Hrachovcová, M., *Collection Czech. Chem. Commun.* **29**, 1503 (1964).
55. Malát, M., and Zelinka, J., *Mikrochim. Acta* **1966**, 228.
56. Trochinskaya, G. N. and Knizhko, P. O., *Ukr. Khim. Zh.* **36**, 950 (1970).
57. Orlovskii, S. T. and Kish, P. P., *ibid.* **29**, 209 (1963).
58. Dwivedi, C. D. and Dey, A. K., *Mikrochim. Acta* **1968**, 708.
59. Beschetnova, E. T., Malinovskaya, L. N., Golovina, A. P. and Zorov, N. B., *Zh. Analit. Khim.* **27**, 2152 (1972).
60. Pyatnitskii, I. V. and Pinaeva, S. G., *ibid.* **28**, 671 (1973).
61. Popova, S. A., Karadakov, B. P., Kovalenko, P. N. and Bagdasarov, K. N., *ibid.* **24**, 682 (1969).
62. Joshi, A. P. and Munshi, K. N., *Microchem. J.* **12**, 447 (1967).
63. Starościk, R. and Ładogórski, P., *Chem. Anal. (Warsaw)* **9**, 97 (1964).
64. Evtimova, B. and Nonova, D., *Anal. Chim. Acta* **67**, 107 (1973).
65. Poluektov, I. S., Kononenko, L. I. and Lauer, R. S., *Zh. Analit. Khim.* **13**, 396 (1958).
66. Jankovský, J., *Hutn. Listy* **21**, 274 (1966).
67. Garčic, A. and Sommer, L., *Collection Czech. Chem. Commun.* **35**, 1047 (1970).
68. Blyum, I. A., Solov'yan, I. T. and Shebalkova, G. N., *Zavodsk. Lab.* **27**, 950 (1961).

69. Levin, I. S. and Azarenko, T. G., *ibid.* **28**, 1313 (1962).
70. Kish, P. P. and Pogoida, I. I., *Zh. Analit. Khim.* **29**, 52 (1974).
71. Zolotov, Yu. A., Kish, P. P., Bagreev, V. V. and Pogoida, I. I., *ibid.* **29**, 221 (1974).
72. May, I. and Hoffmann, J. I., *J. Wash. Acad. Sci.* **38**, 329 (1948).
73. Kleiner, K. E. and Markova, L. V., *Zh. Analit. Khim.* **8**, 279 (1953).
74. Athavale, V. T., Ramachandran, T. P., Tillu, M. M. and Vaidya, G. M., *Anal. Chim. Acta* **22**, 56 (1960).
75. Orlovskii, S. T. and Kish, P. P. *Ukr. Khim. Zh.* **27**, 687 (1961).
76. Stolarczyk, U. and Minczewski, J., *Chem. Anal. (Warsaw)* **9**, 151 (1964).
77. Nazarenko, V. A. and Ravitskaya, R. V., *Ukr. Khim. Zh.* **30**, 625 (1964); *Zavodsk. Lab.* **31**, 1301 (1965).
78. Busev, A. I. and Shkrobot, E. P., *Vestn. Mosk. Univ. Khim.* **1959**, No. 4, 199.
79. Olenovich, N. L., Koval'chuk, L. I. and Lozitskaya, E. P., *Zh. Analit. Khim.* **29**, 47 (1974).
80. Alimarin, I. P., Golovina, A. P. and Torgov, V. G., *Zavodsk. Lab.* **26**, 709 (1960).
81. Garg, B. S. and Singh, R. P., *Talanta* **18**, 761 (1971).
82. Nazarenko, V. A., Biryuk, E. A. and Ravitskaya, R. V., *Zh. Analit. Khim.* **13**, 445 (1958).
83. Boček, P. and Vrchlabský, M., *Chem. Prumysl* **16**, 625 (1966).
84. Solanke, K. R. and Khopkar, S. M., *Anal. Chim. Acta* **66**, 307 (1973).
85. Otomo, M. and Tonosaki, K., *Talanta* **18**, 438 (1971).

Chapter 25

IODINE

Iodine (I, at.wt. 126·90) is a solid non-metal which is fairly volatile at room temperature and easily sublimed. Iodine dissolves readily in aqueous KI solutions to yield the I_3^- complex, and is also soluble in organic solvents ($CHCl_3$, CCl_4, C_6H_6). It occurs mainly in the $-I$, $+V$, and $+VII$ oxidation states, in iodide, iodate, and periodate, respectively. Iodide reveals reducing properties, in contrast to iodine, iodate, and periodate which have oxidizing properties.

25.1 Separation of Iodine and Iodide

Elementary iodine is volatile, and small quantities are commonly separated by distillation [1–3]. In organoiodine compounds, iodine is first oxidized to iodate with chromic acid in a concentrated H_2SO_4 medium, then reduced with phosphorous acid to iodine, which is steam-distilled and collected in an alkaline trapping solution.

The oxidation of iodide to iodine in aqueous solution (e.g. with nitrite), and the subsequent extraction of the iodine with inert organic solvents, are discussed later in more detail.

Traces of iodine can be separated as AgI by coprecipitation with silver chloride [4–7].

Methods for the separation of the halide ions are described in the chapters on chlorine and bromine.

25.2 Determination of Iodine and Iodide

The colour reaction of iodine with starch is the conventional sensitive spectrophotometric method for determining iodide. Extraction of iodine into chloroform, benzene, and other solvents provides a less sensitive colorimetric method.

Ultratraces of iodine can be determined by means of the catalytic action of iodine on the reduction of cerium(IV) by arsenic(III). This is not discussed further here, since such methods are outwith the scope of this book.

25.2.1 Starch–Iodine and Extractive Methods

Iodide can be determined colorimetrically (after oxidation to iodine) either as the blue iodine–starch adsorption compound, or as coloured organic extracts containing iodine. Iodide is oxidized to iodine with nitrite or iron(III).

The sensitivity of the iodide determination is enhanced sixfold if iodide ions are first oxidized to iodate ions which, in turn, are reacted with added potassium iodide in an acidic medium.

$$IO_3^- + 5I^- + 6H^+ \rightarrow 3I_2 + 3H_2O$$

Bromine water is usually used for the oxidation of iodide to iodate [5,6]. The excess of bromine is removed by boiling, by brominating phenol, or by means of formic acid.

Permanganate oxidizes iodide to iodate in alkaline media. The excess of permanganate is reduced with nitrite, the residual nitrite being reduced with urea.

The molar absorptivity (with respect to I^-) of the starch–iodine complex (after amplification via iodate) is 1.08×10^5 at 590 nm (specific absorptivity $= 0.85$) (see Fig. 25.1).

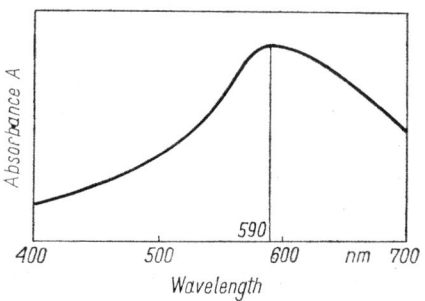

Fig. 25.1 Absorption spectrum of the starch-iodine complex in aqueous solution

In this method involving oxidation of iodide to iodate, any chloride or bromide in the sample solution does not interfere.

The starch–iodine method has been used to determine iodide (or iodine) in plant materials [1,2], natural waters [4,7,8], and silicate rocks [6].

Lambert [9] has used a starch–iodate reagent in the direct determination of iodide ions. Herbo and Sigalla [10] have investigated the effect of oxidation of iodide by dissolved oxygen.

Iodine dissolves in benzene, toluene [11], xylene [12], chloroform, and carbon tetrachloride [13,14] to form violet solutions. Higher distribution coefficients and rather more intense colours are obtained with the first three solvents; chloroform and carbon tetrachloride, however, are more convenient since they are denser than water.

The sensitivity of these extraction methods is much lower than that of the starch–iodine method. The molar absorptivity of the chloroform solution of iodine (after the reaction $I^- \rightarrow IO_3^- \rightarrow 3I_2$) is $\sim 3 \times 10^3$ (relative to I, not I_2).

The extractive spectrophotometric method has been used to determine iodine in lead telluride [12], silicon [15], rocks [14], oil-field brines [16], and water [17].

The sensitivity of the extractive method for iodine is considerably enhanced by adding alcoholic potassium iodide to the $CHCl_3$ or CCl_4 extract, and measuring the absorbance of the tri-iodide at 360 nm [18].

Reagents

1. Starch: 1% solution. Mix 1 g of starch with 5 ml of water, and add the suspension obtained, slowly with stirring, to 50 ml of boiling water. Add 50 ml of glycerol, and gently boil the solution for 5 minutes. This solution is stable for several weeks.
2. Standard iodide solution: 1 mg/ml. Dissolve 1·308 g of potassium iodide (iodate-free) in water, and dilute the solution with water in a volumetric flask to 1 litre. Store the solution in the dark.
3. Phenol: 10% solution in glacial acetic acid.
4. Potassium iodide: 0·5% solution, freshly prepared.

Procedure

Starch–iodine method. Acidify 30–40 ml of an approximately neutral sample solution, containing less than 50 μg of I (as iodide), with 2 ml of $0·5M\ H_2SO_4$, add 3 drops of bromine water, stir well, and allow to stand for one minute. Next add 3 drops of the phenol solution, stir well, and after a minute add 2 ml of the KI solution and 2 ml of the starch solution, and dilute with water to 50 ml in a volumetric flask. Measure the absorbance of the blue solution at 590 nm, using a reagent blank or water as reference.

Extractive method. Acidify a sample solution containing not more than 2 mg of I (as iodide) with 4 ml of $0·5M\ H_2SO_4$, add 10 drops of bromine water, stir well, and allow to stand for one minute. An excess of free bromine should be present in the solution. Remove this excess by heating the solution. Allow the solution to cool, and add 1 drop of the phenol reagent. After a minute, add 4 ml of the KI solution and extract iodine with two portions of chloroform, shaking for 1 minute with each portion. Make up the coloured extracts to the mark with the solvent in a 50-ml or smaller volumetric flask (depending on the amount of iodine) and measure the absorbance at 510 nm against chloroform.

25.2.2 Other Methods

Iodine may be extracted into toluene and shaken with an aqueous solution of Methyl Violet to form a highly-coloured species (which is probably a charge-transfer complex) in the organic phase [19]. Iodide (or tri-iodide) can be extracted as the coloured ion-pairs formed with ferroin (extraction into nitrobenzene) [20,21], Neutral Red [22], Crystal Violet [23], bis(neocuproine)copper(I) [24], Methylene Blue [25], or Nitron [26]. In corresponding methods, the extracted species are the ion-pairs formed by Brilliant Green [17] or Crystal Violet with I_2Cl^- [27], as well as by Basic Blue K with I_2Br^- [28].

If iodine cyanide is added to the sample solution containing I^-, then, after acidification, iodine is released by the reaction:

$$I^- + ICN + H^+ \rightarrow HCN + I_2$$

The concentration of iodine can be determined from the absorbance of either the aqueous solution or a CCl_4 extract [29].

Indirect methods have been suggested for the spectrophotometric determination of iodide ions, exploiting exchange reactions between iodide and $Hg(SCN)_2$ or AgSCN [in the presence of iron(III)] [30], the mercury (II)-diphenylcarbazone complex [31], mercury(II) dithizonate [32], and silver diethyldithiocarbamate [+copper(II) ions] [33].

The methods for determining iodide with fluorescein [34], mercaptobenzothiazole [35], and a triphenylbenzylphosphonium salt [36] are worth mentioning.

Erdey and Szabadváry [37] have determined iodine on the basis of the blue colour formed with Variamine Blue. Although o-tolidine is not oxidized by iodine in the presence of iodide ions, it does react when the iodide is complexed with mercury(II) [38]. No colour reaction is given by iodate with o-tolidine. The products of the oxidation of N,N'-bis-(β-hydroxypropyl)-o-phenylenediamine and its 4-bromo derivative with iodine are also coloured [39].

Cheng and Goydish [3] determined iodide turbidimetrically as silver iodide precipitated from ammoniacal medium.

Iodine forms a characteristic red complex with poly(vinyl acetate) or poly(vinyl alcohol) [40] but this has so far only been applied to determination of the polymers.

25.3 Determination of Iodate and Periodate

Iodate can be determined spectrophotometrically after conversion into iodine by reaction with iodide in acidic medium, and the subsequent colour reaction with starch [41,42]. Nitrite, which interferes in the determination of iodate, is removed by diazotization of sulphanilic acid.

p-Aminophenol has been oxidized by IO_3^- to quinonoimine, which immediately condenses with p-aminophenol to give a blue indamine dye [43].

$$O=\!\!\langle\bigcirc\rangle\!\!=NH + HO\!\!-\!\!\langle\bigcirc\rangle\!\!-NH_2 \rightarrow O=\!\!\langle\bigcirc\rangle\!\!=N\!\!-\!\!\langle\bigcirc\rangle\!\!-NH_2 + H_2O$$

Chlorate and bromate do not form coloured products with p-aminophenol.

An equimolar mixture of isonicotinic acid hydrazide and 2,3,5-triphenyltetrazolium chloride in dilute hydrochloric acid reacts in the presence of IO_3^- to form a pink colour which can be used to determine iodate [44]. Bromate gives a similar reaction only if the solution is heated.

Periodate can be determined spectrophotometrically by the reaction of IO_4^- with o-dianisidine [45] or with benzhydrazide [46]. Coprecipitation of periodate with ferric hydroxide [47] or aluminium hydroxide [48] allows its separation from iodate.

References

1. Gross, W. G., Wood, L. K. and McHargue, J. S., *Anal. Chem.* **20**, 900 (1948).
2. Houston, F. G., *ibid.* **22**, 493 (1950).
3. Cheng, K. L. and Goydish, B. L., *ibid.* **35**, 1965 (1963).
4. Sugawara, K., Koyama, T. and Terada, K., *Bull. Chem. Soc. Japan* **28**, 494 (1955).
5. Matthews, A. D. and Riley, J. P., *Anal. Chim. Acta* **51**, 295 (1970).
6. Crouch, W. H., Jr., *Anal. Chem.* **34**, 1698 (1962).
7. Tsunogai, S., *Anal. Chim. Acta* **55**, 444 (1971).
8. Schnepfe, M. M., *ibid.* **58**, 83 (1972).
9. Lambert, J. L., *Anal. Chem.* **23**, 1251 (1951).
10. Herbo, C. and Sigalla, J., *Anal. Chim. Acta* **17**, 199 (1957).
11. Custer, J. J. and Natelson, S., *Anal. Chem.* **21**, 1005 (1949).
12. Silverman, L., *ibid.* **34**, 701 (1962).
13. Raben, M. S., *ibid.* **22**, 480 (1950).
14. Grimaldi, F. S. and Schnepfe, M. M., *Anal. Chim. Acta* **53**, 181 (1971).
15. Nazarenko, V. A. and Shustova, M. B., *Zavodsk. Lab.* **27**, 15 (1961).
16. Collins, A. G. and Watkins, J. W., *Anal. Chem.* **31**, 1182 (1959).
17. Proskuryakova, G. F., Shveikina, R. V. and Chernavina, M. S., *Izv. Vyssh. Ucheb. Zaved., Khim. Khim. Tekhnol.* **6**, 729 (1963).
18. Ovenston, T. C. and Rees, W. T., *Anal. Chim. Acta* **5**, 123 (1951).
19. Savichev, E. I., Vasil'eva, I. G. and Golovin, E. I., *Zavodsk. Lab.* **29**, 1433 (1963).
20. Yamamoto, Y. and Kinuwaki, S., *Bull. Chem. Soc. Japan* **37**, 434 (1964).
21. Yamamoto, Y., Tarumoto, T. and Tsubouchi, M., *ibid.* **44**, 2124 (1971).
22. Tsubouchi, M., ibid. **44**, 554 (1971); *Anal. Chim. Acta* **54**, 143 (1971).
23. Lomonosov, S. A., Shukolyukova, N. I. and Chernoukhova, V. I., *Zh. Analit. Khim.* **28**, 2389 (1973).
24. Yamamoto, Y., Kumamaru, T., Hayashi, Y. and Yamamoto, M., *Anal. Chim. Acta* **69**, 321 (1974).
25. Gantschewa, A., *Mikrochim. Acta* **1967**, 601.
26. Gantschev, N. and Kiréva, Ant., *ibid.* **1972**, 889.
27. Morsches, B. and Tölg, G., *Z. Anal. Chem.* **200**, 20 (1964).
28. Podberezskaya, N. K. and Shilenko, E. A., *Zavodsk. Lab.* **32**, 918 (1966).
29. Herries, D. G. and Richards, F. M., *Anal. Chem.* **36**, 1155 (1964).
30. Utsumi, S., *J. Chem. Soc. Japan, Pure Chem. Sect.* **74**, 32, 35 (1953).
31. Okutani, T., *ibid.* **88**, 737 (1967).
32. Agterdenbos, J., Jütte, B. A. and Elberse, P. A., *Talanta* **17**, 1085 (1970).
33. Komatsu, S., Nomura, T. and Usui, Y., *J. Chem. Soc. Japan, Pure Chem. Sect.* **88**, 1164 (1967).
34. Braun, D. E. and Wadman, W. H., *Anal. Chem.* **39**, 840 (1967).
35. Kujawa, R., *Mikrochim. Acta* **1969**, 193.
36. Behrends, K. and Klein, H., *Z. Anal. Chem.* **249**, 165 (1970).
37. Erdey, L. and Szabadváry, F., *Acta Chim. Acad. Sci. Hung.* **8**, 191 (1955).
38. Johannesson, J. K., *Anal. Chem.* **28**, 1475 (1956).
39. Pasławska, S., *Chem. Anal. (Warsaw)* **16**, 951 (1971); **17**, 1267 (1972).
40. Pritchard, J. G. and Serra, F. T., *Talanta* **20**, 541 (1973).

41. Lambert, J. L., *Anal. Chem.* **23**, 1247 (1951).
42. Bahr, H. and Bahr, H., *Chem. Anal. (Warsaw)* **8**, 347 (1963).
43. Fuchs, J., Jungreis, E. and Ben-Dor, L., *Anal. Chim. Acta* **31**, 187 (1964).
44. Hashmi, M. H., Ahmad, H., Rashid, A. and Azam, F., *Anal. Chem.* **36**, 2471 (1964).
45. Guernet, M., *Bull. Soc. Chim. France* **1964**, 478.
46. Escarrilla, A. M., Maloney, P. F. and Maloney, P. M., *Anal. Chim. Acta* **45**, 199 (1969).
47. Novikov, A. I. and Finkel'shtein, E. I., *Zh. Analit. Khim.* **19**, 541 (1964).
48. Bhattacharyya, S. N. and Chetia, P. K., *Anal. Chem.* **39**, 369 (1967).

Chapter 26

IRIDIUM

Iridium (Ir, at.wt. 192·22) is one of the platinum metals. Ir(III) and Ir(IV) are the most stable species. Iridium(III) is oxidized to Ir(IV) with MnO_4^-, IO_3^-, or cerium(IV): iridium(IV) is reduced to Ir(III) with iron(II) or ascorbic acid. The reduction of iridium(III and IV) to the metal offers some difficulties. Zinc does not liberate iridium quantitatively from acid solution. Iridium (III) and (IV) form strong complexes with halides [1], ammonia, and several other ligands. Metallic iridium is insoluble in aqua regia; when fused with alkali, it is converted into soluble iridate, IrO_4^{2-}.

26.1 Methods of Separation

Methods for isolating platinum metals are discussed in the chapter on platinum (p. 431). Before its determination iridium must be separated from rhodium. This separation is discussed in the chapter on rhodium (p. 457).

Beamish *et al.* [2–4] have developed fire assay methods for separating iridium. Distillation of iridium from concentrated sulphuric or perchloric acid, at 200°C in a stream of chlorine, gives a specific separation for this element if osmium and ruthenium are absent [5].

Iridium can be separated by extraction with DAPM [6], diethyldithiocarbamate [7], or thio-oxine [8].

26.2 Methods of Determination

Since spectrophotometric methods for determining iridium are non-selective, separation of this element from other platinum metals is essential. The stannous bromide method is given in detail later. Other methods for determining iridium are discussed in several review papers [9,10].

26.2.1 TIN(II) BROMIDE METHOD

When tin(II) in hydrobromic acid solution is added to an acidic solution of the iridium bromide complex formed by heating the sample solution with HBr, the yellow ternary iridium–tin(II)–bromide complex is produced on heating. The absorbance of this complex is the basis of the spectrophotometric determination of iridium [11–15]. Maximum colour intensity is obtained by heating the solution for 1 minute in a bath of boiling water. If the heating is continued for longer than 2 minutes the colour intensity decreases. The complex may be extracted into isoamyl alcohol [12].

The molar absorptivity of the complex in aqueous solution at $\lambda_{max} = 402$ nm is $4\cdot96 \times 10^4$ ($a = 0\cdot26$).

Before its determination by this method, iridium must be separated from Rh, Pt, and Pd, which produce a brown colour under the conditions of the reaction. Other metals (e.g. Co, Ni, Cu, Fe, Sb, and Ti) interfere slightly. Oxidants must not be present. Small amounts of HCl ($\leqslant 0\cdot5M$) and higher amounts of H_2SO_4 in the sample solution may be tolerated.

The analogous method for determining iridium with stannous chloride in hydrochloric acid medium is less sensitive than the bromide method and the stannous iodide method is even less sensitive [16].

The iridium halide complexes may be extracted as coloured ion-pairs with DAM [17–19], DAPM [17,20], diphenylguanidine or tribenzylamine [19], and the tetraphenylphosphonium ion [21].

Reagents

1. Stannous bromide: 25% solution of $SnBr_2.2H_2O$ in 40% HBr.
2. Standard iridium solution: 1 mg/ml. Fuse $0\cdot1000$ g of suitably pure iridium powder with 2 g of sodium peroxide in a silver crucible. Dissolve the melt in water, acidify the solution with nitric acid, and heat to boiling. Adjust the solution to pH 6 with $NaHCO_3$. Filter off the precipitated hydrous IrO_2 and wash with water. Dissolve the precipitate in ~ 10 ml of HCl (1+3). After dilution with water, filter off any AgCl present, and wash the filter with dilute hydrochloric acid. Dilute the iridium solution with water to 100 ml in a volumetric flask.

Note. Faye and Inman [22] have developed a simple method for the preparation of iridium solutions in which iridium is alloyed with a large excess of tin, and the alloy dissolved in a mixture of HCl and H_2O_2. The bulk of the tin is distilled from the solution as a mixed halide after the addition of HCl and HBr.

Procedure

Place ~ 10 ml of sample solution, the HCl concentration of which is not higher than $0\cdot5M$, and which contains less than 150 µg of Ir, in a 50-ml volumetric flask. Add 10 ml of conc. HBr, and place in a boiling water-bath. After 10 minutes, add 10 ml of the stannous bromide solution, and mix the solution well. Within 2 minutes of the addition of Sn(II), take the flask out of the boiling water, and cool quickly under cold water. Dilute the solution in the flask to the mark with water, and measure the absorbance at 402 nm against water.

26.2.2 OTHER METHODS

In neutral media, iridium gives a cherry-red colour with *p*-nitrosodimethylaniline [23], a reagent normally used for determining palladium and platinum.

Other organic reagents which are suitable for the spectrophotometric determination of iridium include PAN [24], TTA [25], oximidobenzotetronic acid (3-nitroso-4-hydroxycoumarin) [26], 8-hydroxyquinoline-N-oxide [27], and diphenylcarbazone [28].

Several sensitive spectrophotometric methods take advantage of the oxidation by iridium(IV) of o-dianisidine [29,30], N,N'-di(2-naphthyl)-p-phenylenediamine [31,32], and leuco-Crystal Violet (molar absorptivity $4 \cdot 8 \times 10^4$ at 590 nm, pH 3·5–4·7) [33].

The coloured solutions formed by iridium(IV) in H_2SO_4 [34], H_3PO_4 + H_2SO_4 [35], and H_3PO_4 + $HClO_4$ + HNO_3 [36] are the basis of selective but insensitive methods for determining iridium.

References

1. Bus'ko, E.A., Burkov, K.A. Kalinin, S.K., *Zh. Analit. Khim.* **29**, 340 (1974).
2. Barefoot, R.R. and Beamish, F.E., *Anal. Chem.* **24**, 840 (1952).
3. Tertipis, G.G. and Beamish, F.E., *ibid.* **34**, 108 (1962).
4. Agrawal, K.C. and Beamish, F.E., *Z. Anal. Chem.* **211**, 265 (1965).
5. Gijbels, R. and Hoste, J., *Anal. Chim. Acta* **36**, 230 (1966).
6. Rudenko, N.P. and Kordyukevich, V.O., *Zh. Analit. Khim.* **23**, 1061 (1968).
7. Rakovskii, E.E. and Baevskaya, G.M., *ibid.* **26**, 1796 (1971).
8. Kordyukevich, V.O. and Rudenko, N.P., *Radiochem. Radioanalyt. Lett.* **13**, 259 (1973).
9. Beamish, F.E. and McBryde, W.A., *Anal. Chim. Acta* **9**, 349 (1953); **18**, 551 (1958).
10. Beamish, F.E., *Talanta* **12**, 789 (1965).
11. Berman, S.S. and McBryde, W.A.E., *Analyst* **81**, 566 (1956).
12. Pantani, F. and Piccardi, G., *Anal. Chim. Acta* **22**, 231 (1960).
13. Tertipis, G.G. and Beamish, F.E., *Anal. Chem.* **34**, 623 (1962).
14. Cerceo, E.C. and Markham, J.J., *ibid.* **38**, 1426 (1966).
15. Bartscher, W. and Kramer, W., *Anal. Chim. Acta* **63**, 216 (1973).
16. Berg, E.W. and Youmans, H.L., *ibid.* **25**, 470 (1961).
17. Busev, A.I. and Akimov, V.K., *Zh. Neorgan. Khim.* **8**, 302 (1963).
18. Danilova, V.N. and Lisichenok, S.L., *Zavodsk. Lab.* **34**, 1284 (1968).
19. Pilipenko, A.T., Danilova, V.N. and Lisichenok, S.L., *Zh. Analit. Khim.* **25**, 1154 (1970).
20. Busev, A.I. and Akimov, V.K., *Talanta* **11**, 1657 (1964).
21. Neeb, R., *Z. Anal. Chem.* **154**, 17 (1957).
22. Faye, G.H. and Inman, W.R., *Talanta* **3**, 277 (1960).
23. Westland, A.D. and Beamish, F.E., *Anal. Chem.* **27**, 1776 (1955).
24. Stokely, J.R. and Jacobs, W.D., *ibid.* **35**, 149 (1963).
25. Rangnekar, A.V. and Khopkar, S.M., *Chemist-Analyst* **56**, 84 (1967).
26. Manku, G.S., Bhat, A.N. and Jain, B.D., *Talanta* **16**, 1421 (1969).
27. Gupta, R.D., Manku, G.S., Bhat, A.N. and Jain, B.D., *ibid.* **17**, 772 (1970).
28. Manku, G.S., *Mikrochim. Acta* **1973**, 341.
29. Cluett, M.L., Berman, S.S. and McBryde, W.A., *Analyst* **80**, 204 (1955).
30. Berman, S.S., Beamish, F.E. and McBryde, W.A., *Anal. Chim. Acta* **15**, 363 (1956).
31. Nasouri, F.G. and Witwit, A.S., *ibid.* **50**, 163 (1970).
32. Booth, M.D., *ibid.* **59**, 304 (1972).
33. Ayres, G.H. and Bolleter, W.T., *Anal. Chem.* **29**, 72 (1957).
34. Maynes, A.D. and McBryde, W.A. *Analyst* **79**, 230 (1954).
35. Pshenitsyn, N.K., Ginzburg, S.I. and Sal'skaya, L.G., *Zh. Neorgan. Khim.* **4**, 301 (1959).
36. Ayres, G.H. and Quick, Q., *Anal. Chem.* **22**, 1403 (1950).

Chapter 27

IRON

Iron (Fe, at.wt. 55·85) occurs in solution in the +II and +III oxidation states. Compounds of Fe(III) are generally the more stable. Fe(OH)$_2$ is precipitated at above pH ~ 7·5, Fe(OH)$_3$ at above pH 2-3. Neither of the hydroxides shows acidic properties. Iron(II) exhibits properties similar to those of Ni(II) and Zn(II), forming a stable cyanide complex [Fe(CN)$_6$]$^{4-}$, ferrocyanide ion]. Iron(III) forms fluoride, chloride, cyanide, EDTA, tartrate, and oxalate complexes. In acid media, iron(III) acts as an oxidant.

27.1 Methods of Separation

27.1.1 Extraction

Although diethyl ether is used to extract macroquantities of iron(III) from hydrochloric acid media, it is replaced in the extraction of small amounts of iron by di-isopropyl ether, 2,2'-dichlorodiethyl ether, methyl ethyl ketone, and MIBK, since these solvents give much higher distribution coefficients [1-4]. For MIBK, the optimum concentration is 6-7M HCl, while in the case of di-isopropyl ether it is 7-8M HCl. During the extraction of the iron(III) chloride complex, Ga, Tl(III), Au(III), Ge, As, Sb, and Mo are also extracted. Iron(III) can also be extracted from hydrochloric acid medium with TBP, TOPO, or triphenylarsine oxide in chloroform [5-7].

From acidic solutions, iron(III) is extractable as the acetylacetonate [8,9], trifluoroacetylacetonate [10], and 2-thenoyltrifluoroacetonate [11-12a].

Well-known methods for separating iron(III) involve extracting with chloroform solutions of cupferron [13,14] or oxine [15,16]. The optimum conditions for the extraction of iron(III) oxinate with chloroform and various other solvents have been described by Zolotov and Kuz'min [16].

Traces of iron can be separated from large quantities of gallium by extraction from a chloride medium at pH 2 with a solution of PAN in isoamyl alcohol [17].

Iron is often determined directly from the absorbance of the organic phase after a solvent extraction separation, e.g. in the thiocyanate and the bathophenanthroline methods.

27.1.2 Precipitation

Iron is usually precipitated as the hydroxide, Fe(OH)$_3$. When it is precipitated with ammonia, Al(OH)$_3$ can be used as collector [18], whereas with an alkali metal hydroxide as the precipitant, hydrous MnO_2 is a suitable collector [19,20]. Ferric and aluminium hydroxides can be coprecipitated either with ammonia from a weakly acidic medium (pH 4–5) or by adding excess of ammonia (pH 9–10). Since some of the aluminium hydroxide dissolves in the latter case, lanthanum hydroxide is a better collector. When the manganese collector is used, iron can be precipitated with ammonia provided that Mn(II) and MnO_4^- are added before, and a few drops of ethanol are added after the precipitation.

Traces of iron can be separated from Al, Ti, V, U, and phosphate, by precipitation as the sulphide from tartrate medium [21]. Cadmium, lead, or any other metal giving a sparingly soluble sulphide can be used as carrier. Oxine and cupferron are also useful precipitants for the separation of iron [15].

27.1.3 Ion-Exchange

Ion-exchangers are of little importance in the separation of small amounts of iron from other metals. Iron(III) differs from bivalent cations (Cu, Ni, Mn, Mg, etc.) by forming stable anionic sulphosalicylate, tartrate, and chloride complexes. This permits the separation of iron on cation- [22,23] or anion-exchange columns [24,25].

The separation of Fe(III) from Fe(II) is achieved by sorption of both species on a cation-exchanger column, and sequential elution of the anionic iron(III) oxalate complex with $0.25M$ oxalic acid, and of iron(II) with $2M$ hydrochloric acid [26].

Kemula *et al.* [27] have isolated small amounts of iron(III) from Cu, Co, Cd, and Zn salts by passing an ammoniacal solution of the salt through an anion-exchanger saturated with ammonia solution; iron is retained in the column as the hydroxide, whereas the Cu, Co, Cd, and Zn ammine complexes are eluted. After the column has been washed with ammonia solution, iron is eluted with $0.25M$ sulphuric acid.

27.2 Methods of Determination

There are many well-known spectrophotometric methods for the determination of iron(II) and iron(III). Since the Fe^{2+} and Fe^{3+} ions have chromophoric properties, many methods utilize reagents without chromophoric groups.

The following selective and sensitive methods are considered below in more detail: the classical thiocyanate method coupled with extraction; methods using 1,10-phenanthroline and 2,2'-bipyridyl which are usually applied in aqueous media; the bathophenanthroline extraction method which has similar sensitivity to that of the thiocyanate method (extraction with MIBK), and is twice as sensitive as the 1,10-phenanthroline method. Furthermore, the simple and often convenient sulphosalicylic acid method is described, although it is less sensitive than the other methods mentioned.

27.2.1 THIOCYANATE METHOD

Thiocyanate ions react in a moderately acidic medium with ferric ions to yield a red colour which, for a long time, was the basis for the determination of iron(III), or total iron after oxidation of Fe(II) to Fe(III) [28–31]. Owing to stepwise complex formation in solution, $[Fe(SCN)]^{2+}$, $[Fe(SCN)_2]^+$, and further complexes up to $[Fe(SCN)_6]^{3-}$ are said to be formed. The concentrations of the reagents and the pH of the medium determine which complexes are prevalent. In general, only $[Fe(SCN)]^{2+}$ is formed in significant quantity at moderate concentration of SCN^- ($\log\beta_1 \approx 2\cdot1$, $\log\beta_2 \approx 3\cdot4$, $\log\beta_3 \approx 3\cdot9$, $\log\beta_6 \approx 3\cdot7$). The higher complexes are more intensely coloured.

Iron can be determined by the thiocyanate method in aqueous [28] and acetone–aqueous [29] media, or after extraction of the coloured complex with a suitable organic solvent [15,30].

When iron is determined in aqueous media or in the presence of acetone, care should be taken that the concentration of thiocyanate (ammonium, potassium, or sodium salt) is the same in the sample and the standard solutions.

The aqueous solution must be sufficiently acid to prevent hydrolysis of iron(III) which begins even below pH ~ 3. The solution should not, however, be too acidic otherwise the concentration of thiocyanate may be reduced by polymerization. Iron(III) thiocyanate complexes are not very stable, and can persist only at a relatively high concentration of SCN^-. The optimum acidity of the solution with hydrochloric, sulphuric, nitric, or perchloric acid lies within the concentration range of $0\cdot05-0\cdot2N$.

The colour of aqueous solutions of Fe(III) thiocyanate complexes is unstable owing to reduction of Fe(III) by SCN^-, fading by a few per cent in 30 minutes, and by 50% in 6 hours.

Lowering the dielectric constant of the medium by addition of solvents such as acetone or dioxan intensifies the colour. The molar absorptivity increases from $8\cdot5\times10^3$ in aqueous solution to $1\cdot8\times10^4$ in 50% (v/v) aqueous acetone. Agreement with Beer's law is not observed.

The iron(III) thiocyanate complexes can be extracted with oxygenated solvents such as ethers, higher alcohols, esters, and ketones. Certain mixed solvents are also used, e.g. diethyl ether with isoamyl alcohol $(1+1)$, or TBP with carbon tetrachloride. Depending on the solvent used, different species are extracted [32,32a]: the 1:4 Fe:SCN complex is extracted with diethyl ether, while the 1:3 complex is extracted with TBP, or with dibenzylsulphoxide in methylene chloride [33].

Extraction increases the sensitivity of the thiocyanate method twofold or more. The molar absorptivity of the Fe(III) thiocyanate complex solution in MIBK (see "Procedure" below) is $2\cdot4\times10^4$ ($a=0\cdot43$) at $\lambda_{max}=495$ nm. The position of the absorption maximum for the complex varies between 470 and 530 nm, depending on the medium.

Sensitivity as high as with MIBK can be obtained with TBP in CCl_4 (1+1), but the MIBK extraction is more selective.

The iron(III) thiocyanate complexes react with organic bases to form ion-association complexes which can be extracted into non-polar solvents. Suitable reagents include the triphenylmethylarsonium ion [34], pyridine [35], tributylamine [36,37], dihexylamine [37], tri-n-octylamine [38], and DAM [39,40]. The molar absorptivity with dihexylamine is 3.0×10^4 [37], and with DAM is 3.5×10^4 [39].

Anions which form stable complexes with iron(III) (i.e. fluoride, phosphate, citrate, and oxalate) interfere in the determination of iron by the thiocyanate method. High concentrations of chloride, sulphate, and acetate also interfere, but to a lesser degree. Interference is also caused by other metals which form coloured thiocyanate complexes under the same conditions (Co, Mo, Bi, Ti), by metals giving sparingly soluble thiocyanates, and by coloured ions. These interferences may be avoided by selective extraction.

Luke [41] separates iron preliminarily by MIBK extraction of the chloride complex from $7M$ HCl or LiCl. He then shakes the extract with solid NH_4SCN, the resulting colour intensity being proportional to the amount of iron present.

McCown and Kudera [42] shake the extract of the chloride complex in butyl acetate and MIBK with a 20% aqueous solution of NH_4SCN. Cerrai and Ghersini [43] extract iron with di-(2-ethylhexyl)phosphoric acid (HDEHP) in cyclohexane, and develop the colour by adding KSCN in ethanol. Jackwerth and Schneider [44] determine iron and cobalt simultaneously by measuring the absorbance of the TBP extract at two wavelengths.

The thiocyanate method has been used for the determination of iron in gallium [2,17], tin [21], aluminium [29], nickel and its salts [44,45], uranium alloys [42], Cu, Co, Cd, and Zn salts [27], alkalis [15], tellurium [46], titanium tetrachloride [47], and organic materials [48].

Reagents

1. Potassium thiocyanate: 20% solution. Acidify the solution with hydrochloric acid to pH $\sim$ 2. If colour due to traces of iron present in the thiocyanate is observed, shake the solution with 2 or 3 small portions of MIBK.
2. Standard iron(III) solution: 1 mg/ml. Weigh out 8.635 g of ferric alum, $FeNH_4(SO_4)_2 \cdot 12H_2O$, dissolve it in water containing 5 ml of conc. H_2SO_4, and dilute with water to 1 litre in a volumetric flask. Working solutions are obtained by suitable dilution of the stock solution with dilute sulphuric acid (e.g. $0.005M$).

Procedure

Separation of Fe(III) with the collector. If the solution contains no Ti, Bi, Pb, Al or other metals precipitated by ammonia as hydroxides, add 2 mg of lanthanum (as a nitrate solution). Heat the solution to about 60°C, and add excess of ammonia with stirring to dissolve the hydroxides of the metals which form ammine complexes. Allow the solution containing the precipitate to stand for a few minutes at 60–70°C, then filter off the coagulated precipitate on paper, and wash it with hot dilute ammonia solution. Dissolve the precipitate in a small amount of hot $2M$ HCl.

Determination of Fe(III). Adjust the solution containing not more than 70 μg of Fe(III) to pH ~ 1 by adding hydrochloric acid or ammonia. Transfer the solution to a separating funnel, add 10 ml of thiocyanate solution, and extract the iron with two portions of MIBK. Make up the combined extracts with the solvent to the mark in a 50-ml or smaller volumetric flask (according to the amount of iron), mix well, and measure the absorbance of the solution at 495 nm against the solvent.

Note. If the total iron content of a solution containing both Fe(III) and Fe(II) is required, the latter must be oxidized by heating with a small amount of ammonium persulphate or hydrogen peroxide. Before the addition of thiocyanate, the solution must be cooled.

27.2.2 1,10-PHENANTHROLINE AND 2,2'-BIPYRIDYL METHODS

1,10-Phenanthroline (*o*-phenanthroline) (**I**) and 2,2'-bipyridyl (α,α'-bipyridyl; 2,2'-bipyridine) (**II**) are organic bases with very similar chemical properties. Their hydrochlorides yield colourless aqueous solutions. Phenanthroline and bipyridyl contain the iron(II)-specific group:

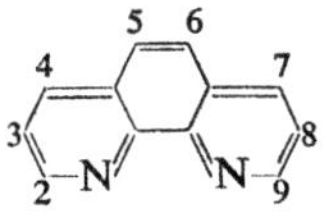

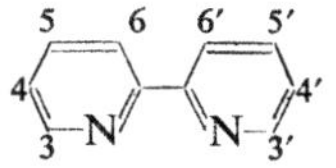

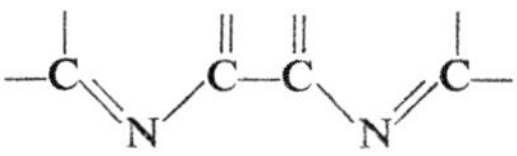

Both reagents react rapidly with Fe^{2+} ions over the wide range of pH 2–9 (to give orange-red or pink complexes, respectively). The cationic complexes formed are the basis of convenient methods for determining iron in aqueous solution [49–53]. The absorption spectra of both complexes are shown in Fig. 27.1.

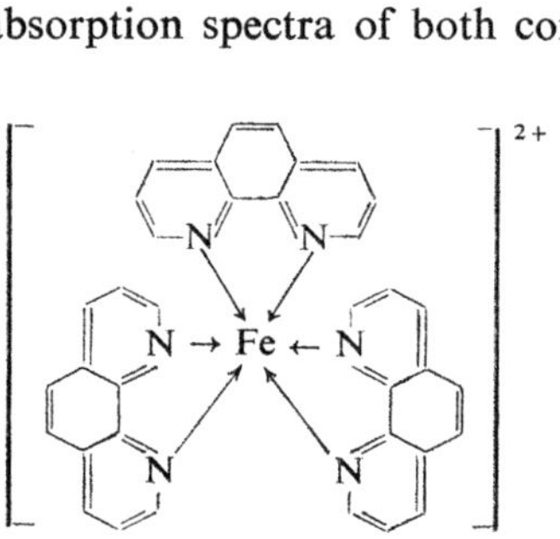

The formula above shows the structure of the iron(II) phenanthroline complex. Three molecules of the reagent are coordinatively bonded via their nitrogen atoms with one iron(II) ion to satisfy its coordination number (6). Stable 5-membered chelate rings are thus formed. If there is a deficiency of the reagent, yellow complexes are formed in which the ratio of iron to phenanthroline or bipyridyl is 1:1 [54].

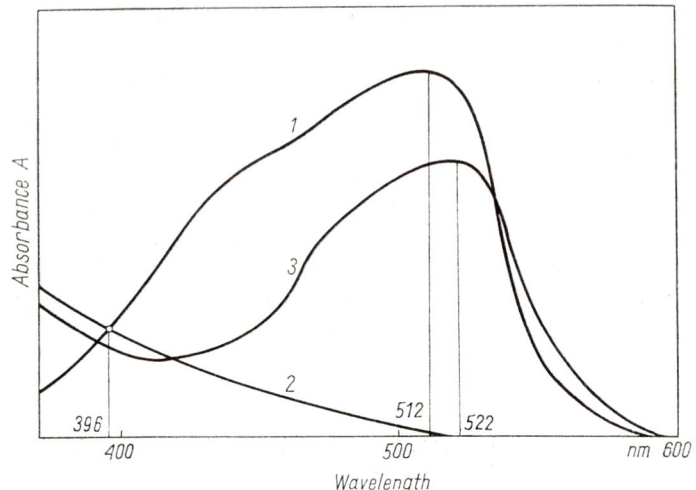

Fig. 27.1 Absorption spectra of the orange-red iron(II) 1,10-phenanthroline complex (1), the yellow iron(III) 1,10-phenanthroline complex (2), and the pink iron(II) 2,2'-bipyridyl complex (3)

The phenanthroline and bipyridyl methods have similar sensitivity. The molar absorptivity of the iron(II) phenanthroline complex is 1.1×10^4 ($a = 0.20$) at $\lambda_{max} = 512$ nm, and that of the bipyridyl complex 8.7×10^3 ($a = 0.16$) at $\lambda_{max} = 522$ nm. Solutions of the complexes with phenanthroline and bipyridyl are stable, and the complexed iron is resistant to oxidation.

Iron(II) and total iron can be determined with phenanthroline or bipyridyl [after reduction of Fe(III) to Fe(II)]. Hydroxylamine reduces iron(III) within a few minutes in a weak acid medium (pH 3–4); ascorbic acid is a better reductant in a fairly acidic solution (pH 0–1). Other reducing agents used are sulphite, dithionite [55], and hypophosphite [56].

The colour reactions are usually carried out in acetate or citrate buffers. The presence of citrate or tartrate is desirable as it prevents the precipitation of certain cations which hydrolyse in weak acid media (e.g. Ti, Al, and Bi).

Oxidation of the iron(II)–phenanthroline complex yields a pale blue complex ($\lambda_{max} = 585$ nm) which was used by Watts [57] to determine relatively high concentrations of iron. The blue complex is slowly converted into a stable yellow complex with $\lambda_{max} = 360$ nm.

The molar absorptivities of the red iron(II)–phenanthroline complex and of the yellow iron(III)–phenanthroline complex are identical at 396 nm. Harvey et al. [50] simultaneously determined iron(II) from the absorbance at 512 nm, and Fe(II) + Fe(III) from the absorbance of the same solution at 396 nm. The absorption spectra of both complexes are shown in Fig. 27.1.

Phenanthroline and bipyridyl form weakly coloured complexes with ruthenium, osmium, and copper. The Cu(I)–phenanthroline complex, but not the iron(II) complex, can be extracted with n-octanol [58]. A number of bivalent metals react with phenanthroline and bipyridyl to form colourless complexes which, in the case of Zn and Cd, are more stable than the iron(II) complex. Hence, EDTA should be used to mask Zn and Cd [59,60]. Copper can be masked with thioglycollic acid [61]. Phosphate, oxalate, and fluoride do not interfere when the pH of the solution is less than 4.

The cationic iron(II) phenanthroline or bipyridyl complexes can form coloured ion-association complexes which are extractable into various solvents. The ion-association complexes with perchlorate [2,62,63], iodide [59,61], and azide [64] ions are extractable with nitrobenzene or chloroform. Ketones are used to extract the thiocyanate association complexes [65,66].

Methods based on ion-association complexes with acidic azo and sulphophthalein dyes reveal high sensitivity [67–69a]. The molar absorptivity for the 1,10-phenanthroline iron(II) complex with Methyl Orange is 5.0×10^4 [68], and with Bromophenol Blue it is 5.9×10^4 [69a].

Jackson and Phillips [4] recommend shaking the MIBK extract of the iron(III) chloride complex with an aqueous solution of phenanthroline and hydroxylamine. The iron(III) is thus reduced, and stripped into the aqueous phase as the red phenanthroline complex.

When determining iron(III) in a nickel salt solution, Ganzburg and Mal'tseva [70] add phenanthroline and cobalt(II), thus reducing the iron(III) and forming the red phenanthroline complex.

The complex of phenanthroline with iron(II) is called ferroin and has found widespread application in titrimetric analysis as a redox indicator.

Iron has been determined with 1,10-phenanthroline or 2,2'-bipyridyl in a large variety of materials such as copper [61], various metals [62], nickel and its alloys [70,71], aluminium and its alloys [4,72], indium [2], zinc and cadmium [60], bismuth [73], plutonium [63], titanium alloys [74,74a], phosphorus and its compounds [55,75,76], uranyl nitrate [77], phosphors [25], calcium and magnesium carbonates [25,68], synthetic ruby and sapphire [78], rock salt [79], silicate minerals [79a], silicone polymers [80], selenium [80a], foodstuff [19,81,82], water [83], and iron(III) oxide [84]. Large amounts of Fe(III) can be masked with fluoride [84a].

Reagents

1. 1,10-Phenanthroline: 0·25% solution of the hydrochloride or hydrate in $\sim 0.1M$ HCl.
2. Standard iron solution: 1 mg/ml (p. 308).
3. Hydroxylamine hydrochloride: 10% solution, freshly prepared.

Procedure

To a slightly acidic solution, containing not more than 150 μg of Fe(III) or Fe(II), add 2 ml of the hydroxylamine solution, and 10% sodium citrate solution till the pH is 3–4. Transfer the solution to a 50-ml volumetric flask, add 5 ml of the 1,10-phenanthroline solution, dilute to the mark with water, and mix thoroughly. After 5 minutes, measure the absorbance of the coloured solution at 512 nm against water.

27.2.3 BATHOPHENANTHROLINE METHOD

Bathophenanthroline (4,7-diphenyl-1,10-phenanthroline) reacts with iron(II) ions very similarly to 1,10-phenanthroline. Both methods are of similar selectivity, but the bathophenanthroline method is twice as sensitive, and

the resulting iron(II) complex can be extracted with numerous organic solvents. Bathophenanthroline is soluble in ethanol [85–89].

The iron(II) bathophenanthroline complex is extractable with amyl alcohol [85–87], chloroform [86], n-hexanol [90], nitrobenzene [91], and amyl acetate [88]. The highest distribution coefficients are obtained with n-hexanol and chloroform, chloroform being the more convenient solvent owing to its greater density.

For the quantitative extraction of iron into chloroform, at least 10% of ethanol in the initial aqueous solution is necessary. Free bathophenanthroline is completely extracted. The chloroform extracts are diluted with ethanol in volumetric flasks. The red-violet extracts of the complex are stable and independent of the chloroform:ethanol ratio within the range from 5:1 to 1:5.

The molar absorptivity of the iron(II) bathophenanthroline complex in isoamyl alcohol or in chloroform–ethanol is 2.2×10^4 (specific absorptivity 0.40) at $\lambda_{max} = 533$ nm.

Iron(III) is reduced with hydroxylamine [85], hydrazine, dithionite [86], stannous chloride [90], or ascorbic acid [87] before determination with bathophenanthroline.

The optimum pH for the colour reaction is 4–7. Acetate buffer is usually used, and citrate, tartrate, or EDTA is added to keep readily hydrolysable metals in solution [87,90].

Copper interferes in the determination of iron(II) with bathophenanthroline and should, therefore, be separated, e.g. by precipitation as cuprous thiocyanate [92], or masked, e.g. with thiourea [87]. The interference is due to the preferential formation of a very stable, almost colourless, complex between copper(I) and bathophenanthroline [87].

Higher concentrations of Co, Ni, Zn, and Cd interfere slightly. Phosphate and fluoride do not interfere.

The bathophenanthroline reagent can be recovered after the determination of iron. The organic solvent is evaporated under reduced pressure, and the residue warmed with $10M$ NaOH after which the bathophenanthroline is extracted with hot benzene and purified by recrystallization [90].

Blair and Diehl [93] have proposed bathophenanthrolinedisulphonic acid as a reagent for iron. It has similar sensitivity to bathophenanthroline, but is water-soluble. Iron in chromic acid [94] and plant materials [95] has been determined with this reagent.

Iron has been determined by the bathophenanthroline method in the following materials: niobium, tantalum, and molybdenum [87,96], tungsten [86,87,96], vanadium, chromium, titanium, and uranium [96], molybdenum compounds [97], cobalt [98], beryllium [99], gallium and arsenic [100], synthetic ruby and sapphire [101], aluminium alloys [102], silicon tetrachloride [103], petroleum oils [104,105], biological materials [106,107] plant materials [108], and boiler feed water [109,110].

Bathophenanthroline is suitable for the determination of iron(II) in the presence of a large excess of iron(III) [88,111,111a].

Reagents

1. Bathophenanthroline: 0·02% solution in ethanol. Dissolve 20 mg of the reagent in 100 ml of solvent.
2. Standard iron solution: 1 mg/ml (p. 308).
3. Sodium acetate: 50% solution.

Procedure

To a solution (pH $\sim$ 1) containing less than 70 µg of Fe, add 0·1 g of $NH_2OH \cdot HCl$ and 2 ml of sodium acetate solution. Heat to boiling, transfer the cooled solution to a separating funnel, add 10 ml of bathophenanthroline solution, and extract with two portions of chloroform. Dilute the coloured extract with anhydrous ethanol in a 50-ml or smaller volumetric flask, depending on the content of iron, and measure the absorbance at 533 nm, using ethanol or water as reference.

27.2.4 SULPHOSALICYLATE METHOD

With ferric ions, sulphosalicylic acid forms three coloured complexes of different composition. At pH 2–3, a red-violet 1:1 (Fe:R) complex is

$$HO_3S-C_6H_3(OH)(COOH)$$

formed; at pH 4–7, a brown-orange 1:2 complex, and at pH 8–10, a yellow 1:3 complex [112,113].

The violet complex, which is stable in acid medium, is useful in the photometric determination of iron(III) in the presence of iron(II), and in the indirect determination of fluoride (cf. p. 259), but the method is not very sensitive (the molar absorptivity is 2.6×10^3 at $\lambda_{max} = 490$ nm).

More often, total iron [after oxidation of Fe(II)] is determined as the yellow sulphosalicylate complex which is stable in alkaline medium [114–117]. The absorption maximum of this complex is at 425 nm ($\varepsilon = 5.8 \times 10^3$: $a = 0.14$). Figure 27.2 shows the absorption spectra of sulphosalicylate complexes of iron(III) in solutions at pH 2 and 9.

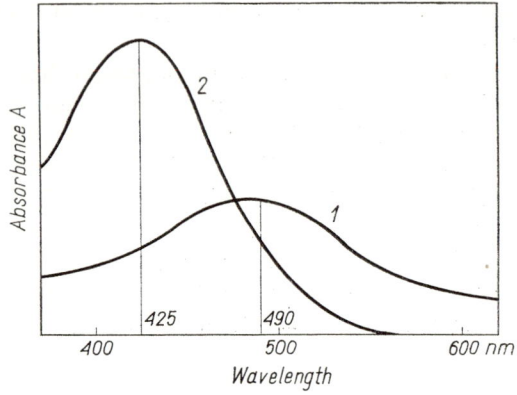

Fig. 27.2 Absorption spectra of iron(III) sulphosalicylate complexes at pH 2 (1) and at pH 9 (2)

Salicylic acid, which reacts similarly, but is less soluble in aqueous solutions, may be used in place of sulphosalicylic acid.

Sulphosalicylic acid forms water-soluble complexes with most multivalent metal ions. Coloured complexes are obtained with metals having chromophoric properties.

Coloured ions interfere with the determination of iron by the sulphosalicylate method. Copper ions can be removed by precipitation with thiocyanate [117]. If manganese is present, the addition of hydroxylamine is recommended to prevent possible precipitation of brown manganese hydrous oxides after the solution is made alkaline.

Iron has also been determined by the sulphosalicylate method in zinc and aluminium [114], noble metals [115], copper alloys [117], aluminium alloys [118], and ilmenite [119].

Reagents

1. Sulphosalicylic acid: 10% solution. This solution is stable.
2. Standard iron solution: 1 mg/ml (p. 308).

Procedure

Adjust an acid solution containing less than 300 μg of Fe(III) to pH 2–3 with ammonia, and transfer it to a 50-ml volumetric flask. Add 5 ml of sulphosalicylic acid solution, and adjust the solution to pH 9 with dilute ammonia. Dilute the yellow solution to the mark and measure the absorbance at 425 nm against water.

27.2.5 OTHER METHODS

EDTA (Complexone III) forms a yellow-brown complex with Fe(III) ($\varepsilon = 20$), which reacts with hydrogen peroxide to give a violet ternary complex ($\varepsilon = 500$). The reaction is useful in the spectrophotometric determination of larger quantities of iron [89,120–124].

Ferron (7-iodo-8-hydroxyquinoline-5-sulphonic acid) forms with iron(III) a green, water-soluble complex which can be used for the determination of iron (molar absorptivity is $4 \cdot 0 \times 10^3$ at 610 nm) [116,125].

Ziegler et al. [126,127] have extracted the ferron complex from a slightly acidic medium with amyl alcohol containing tri-n-butylamine. The amine neutralizes the sulphonic acid group and makes it possible to extract the complex. Middleton [128] has determined iron by extracting the 8-hydroxyquinoline complex into chloroform.

Collins et al. [129,130] and other authors [102,131] recommend 2,4,6-tri(2'-pyridyl)-s-triazine (TPTZ) as a spectrophotometric reagent for iron(II).

In aqueous medium the molar absorptivity is $2 \cdot 26 \times 10^4$ at 593 nm; in nitrobenzene, ε is $2 \cdot 41 \times 10^4$ at 595 nm. Other related reagents for Fe(II) include 2,2',2''-terpyridyl ($\varepsilon = 1 \cdot 15 \times 10^4$ at 522 nm) [132], ferrozine [132a], 4,7-dihydroxy-1,10-phenanthroline [133], 6-hydroxy-1,7-phenanthroline [134]; and several more analogues [135–142a].

Several oximes have been recommended as reagents, e.g. phenyl-2-

pyridylketoxime ($\varepsilon = 1.52 \times 10^4$) [89,143,144], di-2-pyridylketoxime [145], methyl-2-pyridylketoxime [146], pyridine-2-aldoxime [147], 2,6-pyridinediamidoxime [148], quinisatinoxime [149], dimethylglyoxime [150], 2,2′-dipyridyl-α-glyoxime [150a], 2,2′-dipyridyl-β-glyoxime [150b], and formaldoxime [151]. These reagents form coloured iron complexes in alkaline media.

High sensitivity is displayed by two-colour methods using the following triphenylmethane dyes: Xylenol Orange [152,153], Pyrocatechol Violet [154], Pontachrome Azure Blue B [155], Chrome Azurol S [156,157], Aluminon [158], Eriochrome Cyanine R [159], Methylthymol Blue [160–162], and Chromoxane Violet R [163]. According to Cheng [152], the molar absorptivity of the iron(III)-Xylenol Orange complex in 0·04–0·06M perchloric acid is 2.66×10^4 at $\lambda_{max} = 550$ nm.

The β-diketones acetylacetone [164,165], TTA [12,166, 167] 3-thianaphthenoyltrifluoroacetone [168], and dibenzoylmethane [169] give less sensitive extraction methods for the determination of iron. The molar absorptivity of a xylene solution of the iron(III)-TTA complex is 4.9×10^3 at 510 nm.

Sulphur-containing organic spectrophotometric reagents for iron include thioglycollic acid (the molar absorptivity is 4×10^3 at 540 nm) [116,170], 2-mercaptopyridine-N-oxide [171], unithiol [172], bithionol [173], thiovioluric acid [174], and 3-hydroxypyridine-2-thiol [174a].

The following reagents have nitrogen as a ligand atom: pyridine-2,6-dicarboxylic acid [175], 8-quinoline carboxylic acid [176], DAM [177], 1-nitroso-2-naphthol [178], nitroso-R salt [116,179,180], PAN [181,182], and PAR [182a].

Further organic reagents suggested for use in the spectrophotometric determination of iron are: chromotropic acid [183], 2,7-dichlorochromotropic acid [184], phenylfluorone [185], cacotheline [186], tartaric acid [187], meconic acid [188], 2-hydroxy-3-naphthoic acid [189], BPHA [190], Tiron [191], quercetin and morin [192], Indoferron [193], β-isopropyltropolone [194], metamizol [195], and ethyl-4,6-dihydroxy-5-nitrosonicotinate [196].

Apart from thiocyanate, inorganic ligands giving coloured species include azide ($\varepsilon = 3.6 \times 10^3$ at 460 nm) [197,198], bromide [199], and chloride [200]. The colours of chloride and bromide complexes of iron(III) are relatively weak. However, the anionic iron(III)–chloride complex can be extracted as highly coloured ion-association complexes with Crystal Violet and Rhodamine B [201,202].

Korenaga *et al.* [203] recommend sensitive methods based on the extractable ternary complexes of iron(II) with Rhodamine B and various nitrosophenols.

Of the spectrophotometric methods based on the redox reactions of iron(III), the *o*-dianisidine method [204] and the phosphomolybdenum blue method [205] are worth noting.

References

1. Specker, H. and Doll, W., *Z. Anal. Chem.* **152**, 178 (1956).
2. Nazarenko, V. A. and Flyantikova, G. V., *Zavodsk. Lab.* **27**, 1339 (1961).
3. Gagliardi, E. and Wöss, H. P., *Z. Anal. Chem.* **248**, 302 (1969).
4. Jackson, H. and Phillips, D. S., *Analyst* **87**, 712, 718 (1962).
5. Majumdar, S. K. and De, A. K., *Talanta* **7**, 1 (1960).
6. Hibbits, J. O., Davis, W. F. and Menke, M. R., *ibid.* **4**, 61 (1960).
7. Pietsch, R., *Mikrochim. Acta* **1967**, 708.
8. McKaveney, J. P. and Freiser, H., *Anal. Chem.* **29**, 290 (1957).
9. Wilkins, D. H. and Smith, G. E., *Talanta* **13**, 1049 (1966).
10. Scribner, W. G., Treat, W. J., Weis, J. D. and Moshier, R. W., *Anal. Chem.* **37**, 1136 (1965).
11. Moore, F. L., Fairman, W. D., Ganchoff, J. G. and Surak, J. G., *ibid.* **31**, 1148 (1959).
12. Testa, C., *Anal. Chim. Acta* **25**, 525 (1961).
12a. Murata, K., Yokoyama, Y. and Ikeda, S., *Anal. Chem.* **44**, 805 (1972).
13. Peterson, R. E., *ibid.* **24**, 1850 (1952).
14. Wyatt, P. F., *Analyst* **78**, 656 (1953).
15. Marczenko, Z., *Mikrochim. Acta* **1965**, 281.
16. Zolotov, Yu. A. and Kuz'min, N. M., *Zh. Analit. Khim.* **20**, 476 (1965).
17. Schwedow, W. P. and Klug, O. N., *Chem. Anal. (Warsaw)* **11**, 237 (1966).
18. Gottlieb, A., *Mikrochemie* **39**, 176 (1952).
19. Otozai, K. and Mizumoto, K., *Mikrochim. Acta* **1961**, 217.
20. Miyamoto, M., *Bull. Chem. Soc. Japan* **34**, 1435 (1961).
21. Marczenko, Z. and Kasiura, K., *Chem. Anal. (Warsaw)* **10**, 449 (1965).
22. Marczenko, Z., *ibid.* **4**, 437 (1959).
23. Mann, C. K. and Swanson, C. L., *Anal. Chem.* **33**, 459 (1961).
24. Morie, G. P. and Sweet, T. R., *ibid.* **36**, 140 (1964).
25. Lel'chuk, Yu. L. and Glukhovskaya, R. D., *Tr. Komis. po Analit. Khim. Akad. Nauk SSSR* **16**, 15 (1968).
26. Sagortschew, B., *Acta Chim. Acad. Sci. Hung.* **26**, 289 (1961); *Chem. Anal. (Warsaw)* **17**, 973 (1972).
27. Kemula, W., Brajter, K. and Kostrowicka, H., *Chem. Anal. (Warsaw)* **6**, 463 (1961).
28. Ovenston, T. C. and Parker, C. A., *Anal. Chim. Acta* **3**, 277 (1949).
29. Parissakis, G. and Issopoulos, P. B., *Mikrochim. Acta* **1965**, 28.
30. Melnick, L., Freiser, H. and Beeghly, H. F., *Anal. Chem.* **25**, 856 (1953).
31. Baily, P., *ibid.* **29**, 1534 (1957).
32. Specker, H., Jackwerth, E. and Hövermann, G., *Z. Anal. Chem.* **177**, 10 (1960).
32a. Antonova-Karataeva, I. I., Sultanova, Z. Kh. and Zolotov, Yu. A., *Zh. Analit. Khim.* **28**, 1124 (1973).
33. Ziegler, M. and Stephan, G., *Mikrochim. Acta* **1970**, 1270.
34. Dwyer, F. P. and Gibson, N. A., *Analyst* **76**, 548 (1951).
35. Ayres, G. H. and Baird, S. S., *Talanta* **7**, 237 (1961).
36. Ziegler, M., Glemser, O. and Petri, N., *Z. Anal. Chem.* **154**, 81; **157**, 19 (1957).
37. Ivanova, E. K., Parmenova, V. M. and Peshkova, V. M., *Vestn. Mosk. Univ. Khim.* **1968**, No 4, 69.
38. Watanabe, H. and Murozumi, M., *Bull. Chem. Soc. Japan* **40**, 1006 (1967).
39. Tananaiko, M. M., *Ukr. Khim. Zh.* **28**, 446 (1962).
40. Babko, A. K. and Tananaiko, M. M., *Zh. Neorgan. Khim.* **11**, 827 (1966).
41. Luke, C. L., *Anal. Chim. Acta* **36**, 122 (1966).
42. McCown, J. J. and Kudera, D. E., *Anal. Chem.* **34**, 870 (1962).
43. Cerrai, E. and Ghersini, G., *Analyst* **91**, 662 (1966).
44. Jackwerth, E. and Schneider, E. L., *Z. Anal. Chem.* **207**, 188 (1965).

45. Jackwerth, E., *ibid.* **206**, 335 (1964).
46. Dobrowolski, J. and Wilczewski, T., *Chem. Anal.* (*Warsaw*) **15**, 839 (1970).
47. Malyutina, T. M. and Orlova, V. A., *Zavodsk. Lab.* **34**, 277 (1968).
48. Marschall, R. A., *Analyst* **96**, 675 (1971).
49. Vydra, F. and Kopanica, M., *Chemist-Analyst* **52**, 88 (1963).
50. Harvey, A. E., Jr., Smart, J. A. and Amis, E. S., *Anal. Chem.* **27**, 26 (1955).
51. Hibbits, J. O., Davis, W. F. and Menke, M. R., *Talanta* **8**, 163 (1961).
52. Yamamura, S. S. and Sikes, J. H., *Anal. Chem.* **38**, 793 (1966).
53. Struszyński, M., Marczenko, Z. and Nowicka, T., *Przemysł Chem.* **32**, 293 (1953).
54. Kolthoff, I. M., Leussing, D. L. and Lee, T. S., *J. Am. Chem. Soc.* **72**, 2173 (1950).
55. Grat-Cabanac, M., *Anal. Chim. Acta* **17**, 588 (1957).
56. Somidevamma, G. and Rao, G. G., *Z. Anal. Chem.* **187**, 183 (1962).
57. Watts, H. L., *Anal. Chem.* **36**, 364 (1964).
58. Wilkins, D. H. and Smith, G. F., *Anal. Chim. Acta* **9**, 538 (1953).
59. Vydra, F. and Přibil, R., *Talanta* **3**, 72 (1959).
60. Vydra, F. and Marková, V., *Z. Anal. Chem.* **192**, 347 (1963).
61. Vydra, F. and Přibil, R., *ibid.* **186**, 295 (1962).
62. Margerum, D. W. and Banks, C. V., *Anal. Chem.* **26**, 200 (1954).
63. Plock, C. E. and Caldwell, C. E., *ibid.* **39**, 1472 (1967).
64. Rao, V. P. and Sarma, P. V., *Mikrochim. Acta* **1970**, 783.
65. Bhadra, A. K. and Banerjee, S., *Indian J. Chem.* **7**, 936 (1969).
66. Rao, V. P., Rao, K. V. and Sarma, P. V., *Talanta* **16**, 277 (1969).
67. Knižek, M. and Musilová, M., *ibid.* **15**, 479 (1968).
68. Hulanicki, A., Galus, M., Jędral, W., Karwowska, R. and Trojanowicz, M., *Chem. Anal.* (*Warsaw*) **16**, 1011 (1971).
69. Kobyakova, S. O., Savostina, V. M. and Dobychina, N. L., *Zh. Analit. Khim.* **25**, 1348 (1970).
69a. Kobyakova, S. O., Savostina, V. M. and Peshkova, V. M., *ibid.* **29**, 300 (1974).
70. Ganzburg, G. M. and Mal'tseva, G. V., *Zavodsk. Lab.* **31**, 406 (1965).
71. Riedel, K., *Z. Anal. Chem.* **159**, 110 (1957).
72. Jackson, H., Bailey, E. E. and Williams, L. H., *Metallurgia* **51**, 309 (1955).
73. Holmes, D. G., *Analyst* **82**, 528 (1957).
74. Norwitz, G. and Codell, M., *Anal. Chim. Acta* **11**, 350 (1954).
74a. Naumann, H. C., *Metall.* **27**, 247 (1973).
75. Norwitz, G., Cohen, J. and Everett, M. E., *Anal. Chem.* **36**, 142 (1964).
76. Masalovich, V. M., Agasyan, P. K. and Nikolaeva, E. R., *Zavodsk. Lab.* **32**, 914 (1966).
77. Nowicka-Jankowska, T., Gołkowska, A., Pietrzak, I. and Żmijewska, W., *Chem. Anal.* (*Warsaw*) **3**, 997 (1958).
78. Dodson, E. M., *Anal. Chem.* **34**, 966 (1962).
79. Marczenko, Z. and Stępień, A., *Chem. Anal.* (*Warsaw*) **5**, 247 (1960).
79a. French, W. J. and Adams, S. J., *Analyst* **97**, 828 (1972).
80. Fujiwara, S. and Narasaki, H., *Anal. Chem.* **36**, 206 (1964).
80a. Kasterka, B. and Weppe, J., *Chem. Anal.* (*Warsaw*) **18**, 879 (1973).
81. Salomon, R. E. and Livingston, E. M., *Microchem. J.* **2**, 109 (1958).
82. Genevois, L. and Larrouquère, J., *Bull. Soc. Chim. France* **1961**, 1905.
83. Fresenius, W. and Schneider, W., *Z. Anal. Chem.* **209**, 340 (1965).
84. Novák, J., *Chem. Zvesti* **20**, 545 (1966).
84a. Tamura, H., Goto, K., Yotsuyanagi, T. and Nagayama, M., *Talanta* **21**, 314 (1974).
85. Smith, G. F., McCurdy, W. H., Jr. and Diehl, H., *Analyst* **77**, 418 (1952).
86. Crawley, R. H. and Aspinal, M. L., *Anal. Chim. Acta* **13**, 376 (1955).
87. Penner, E. M. and Inman, W. R., *Talanta* **9**, 1027 (1962).
88. Clark, L. J., *Anal. Chem.* **34**, 348 (1962).
88a. Levillain, P. and Bourdon, R., *Bull. Soc. Chim. France* **1972**, 371, 3309.
89. Cluley, H. J. and Newman, E. J., *Analyst* **88**, 3 (1963).

90. Booth, E. and Evett, T. W., *ibid.* **83**, 80 (1958).
91. Collins, P. and Diehl, H., *Anal. Chem.* **31**, 1692 (1959).
92. Hair R. P. and Newman, E. J., *Analyst* **89**, 42 (1964).
93. Blair, D. and Diehl, H., *Talanta* **7**, 163 (1961).
94. Fuhrman, D. L. and Latimer, G. W., Jr., *ibid.* **14**, 1199 (1967).
95. Quarmby, C. and Grimshaw, H. M., *Analyst* **92**, 305 (1967).
96. Gahler, A. R., Hamner, R. M. and Shubert, R. C., *Anal. Chem.* **33**, 1937 (1961).
97. Galliford, D. J. and Newman, E. J., *Analyst* **87**, 68 (1962).
98. Uny, G., Mathien, C., Targif, J. P. and Tran Van Danh, *Anal. Chim. Acta* **53**, 109 (1971).
99. Pollock, E. N. and Zopatti, L. P., *ibid.* **28**, 68 (1963).
100. Knížek, M. and Galík, A., *Z. Anal. Chem.* **213**, 254 (1965).
101. Chirnside, R. C., Cluley, H. J., Powell, R. J. and Proffitt, P. M., *Analyst* **88**, 851 (1963).
102. Stephens, B. G. and Suddeth, H. A., *Anal. Chem.* **39**, 1478 (1967).
103. Marczenko, Z., Kasiura, K. and Mojski, M., *Chem. Anal. (Warsaw)* **16**, 203 (1971).
104. Forrester, J. S. and Jones, J. L., *Anal. Chem.* **32**, 1443 (1960).
105. Short, F. R., Eyster, H. C. and Scribner, W. G., *ibid.* **39**, 251 (1967).
106. Seven, M. J. and Peterson, R. E., *ibid.* **30**, 2016 (1958).
107. Hankiewicz, J., *Chem. Anal. (Warsaw)* **16**, 115 (1971).
108. Quinsland, D. E. and Jones, D. C., *Talanta* **16**, 282 (1969).
109. Wilson, A. L., *Analyst* **89**, 389, 402 (1964).
110. Tetlow, J. A. and Wilson, A. L., *ibid.* **89**, 442 (1964).
111. Pollock, E. N. and Miguel, A. N., *Anal. Chem.* **39**, 272 (1967).
111a. Mizuno, T., *Talanta* **19**, 369 (1972).
112. Ågren, A., *Acta Chem. Scand.* **8**, 266 (1954).
113. Ogawa, K. and Tobe, N., *Bull. Chem. Soc. Japan* **39**, 223, 227 (1966).
114. Eberius, E., *Angew. Chem.* **63**, 513 (1951).
115. Erdey, L. and Bányai, É., *Acta Chim. Acad. Sci. Hung.* **4**, 315 (1954).
116. Quast, R., *Acta Chem. Scand.* **21**, 873 (1967).
117. Hoffman, M., *Chem. Anal. (Warsaw)* **9**, 495 (1964).
118. Kemula, W., Rubel, S. and Stefańska, W., *ibid.* **15**, 361 (1970).
119. Lunina, G. E. and Romanenko, E. G., *Zavodsk. Lab.* **34**, 538 (1968).
120. Nielsch, W. and Böltz, G., *Mikrochim. Acta* **1954**, 481.
121. Lott, P. F. and Cheng, K. L., *Anal. Chem.* **29**, 1777 (1957).
122. Ringbom, A., Siitonen, S. and Saxén, B., *Anal. Chim. Acta* **16**, 541 (1957).
123. Poeder, B. C., Den Boef, G. and Franswa, C. E., *ibid.* **27**, 339 (1962).
124. Babko, A. K. and Loriya, N. V., *Zavodsk. Lab.* **34**, 1305 (1968).
125. Roubert, J., *Chim. Anal. (Paris)* **38**, 134 (1956).
126. Ziegler, M., Glemser, O. and Petri, N., *Z. Anal. Chem.* **154**, 170 (1957); *Mikrochim. Acta* **1957**, 215.
127. Ziegler, M., Glemser, O. and Petri, N., *Angew. Chem.* **69**, 174 (1957).
128. Middleton, K. R., *Analyst* **89**, 421 (1964).
129. Collins, P. F., Diehl, H. and Smith, G. F., *Anal. Chem.* **31**, 1862 (1959).
130. Collins, P. and Diehl, H., *Anal. Chim. Acta* **22**, 125 (1960).
131. Buchanan, E. B., Jr., Crichton, D. and Bacon, J. R., *Talanta* **13**, 903 (1966).
132. Morris, R. L., *Anal. Chem.* **24**, 1376 (1952).
132a. Carter, P., *Mikrochim. Acta* **1972**, 410.
133. Schilt, A. A., Smith, G. F. and Heimbuch, A., *Anal. Chem.* **28**, 809 (1956).
134. Duswalt, J. M. and Mellon, M. G., *ibid.* **33**, 1782 (1961).
135. Jensen, R. E. and Pflaum, R. T., *Anal. Chim. Acta* **32**, 235 (1965).
136. Pflaum, R. T., Smith, C. J., Jr., Buchanan, E. B., Jr. and Jensen, R. E., *ibid.* **31**, 341 (1964).
137. Walter, J. L. and Freiser, H., *Anal. Chem.* **26**, 217 (1954).
138. Smith, G. F. and Banick, W. M., *Anal. Chim. Acta* **18**, 269 (1958).

139. Stantscheff, P., *Chem. Anal. (Warsaw)* **16**, 243 (1971).
140. Stookey, L. L., *Anal. Chem.* **42**, 779 (1970).
141. Schilt, A. A. and Hoyle, W. C., *ibid.* **39**, 114 (1967).
142. Sabatino, J. D., Weber, O. W., Padmanabhan, G. R. and Senkowski, B. Z., *ibid.* **41**, 905 (1969).
142a. Schilt, A. A. *et al.*, *Talanta* **15**, 475 (1968); **16**, 519 (1969); **17**, 649 (1970); **21**, 831 (1974).
143. Trusell, F. and Diehl, H., *Anal. Chem.* **31**, 1978 (1959).
144. Chernin, R. and Simonsen, E. R., *ibid.* **36**, 1093 (1964).
145. Holland, W. J., Bozic, J. and Gerard, J. T., *Anal. Chim. Acta* **43**, 417 (1968).
146. Banerjea, D. K. and Tripathi, K. K., *Anal. Chem.* **32**, 1196 (1960).
147. Hartkamp, H., *Naturwissenschaft.* **45**, 211 (1958); *Z. Anal. Chem.* **170**, 399 (1959).
148. Wehking, M. W., Pflaum, R. T. and Tucker, E. S., III, *Anal. Chem.* **38**, 1950 (1966).
149. Ayres, G. H. and Roach, M. K., *Anal. Chim. Acta* **26**, 332 (1962).
150. Mehlig, J. P. and Robertson, D. M., *Chemist-Analyst* **43**, 32 (1954).
150a. Soules, D. and Holland, W. J., *Mikrochim. Acta* **1972**, 247.
150b. Notenboom, H. R., Holland, W. J. and Soules, D., *ibid.* **1973**, 187.
151. Marczenko, Z. and Kasiura, K., *Chem. Anal. (Warsaw)* **6**, 37 (1961).
152. Cheng, K. L., *Talanta* **3**, 147 (1959).
153. Buděšínský, B., *Z. Anal. Chem.* **188**, 266 (1962).
154. Birmantas, I. I. and Yasinskene, E. I., *Zh. Analit. Khim.* **20**, 811 (1965).
155. Katsube, Y., Uesugi, K. and Yoe, J. H., *Bull. Chem. Soc. Japan* **34**, 72 (1961).
156. Langmyhr, F. J. and Klausen, K. S., *Anal. Chim. Acta* **29**, 149 (1963).
157. Malyutina, T. M., Orlova, V. A. and Spivakov, B. Ya., *Zh. Analit. Khim.* **29**, 790 (1974).
158. Sangal, S. P., *Chim. Anal. (Paris)* **47**, 288 (1965).
159. Langmyhr, F. J. and Stumpe, T., *Anal. Chim. Acta* **32**, 535 (1965).
160. Tonosaki, K., *Bull. Chem. Soc. Japan* **39**, 425 (1966).
161. Karadakov, B., Kantcheva, D. and Nenova, P., *Talanta* **15**, 525 (1968).
162. Srivastava, K. C. and Banerji, S. K., *Mikrochem. J.* **13**, 621 (1968).
163. Mustafin, I. S. and Lisenko, N. F., *Zh. Analit. Khim.* **17**, 1052 (1962).
164. Lieser, K. H. and Schroeder, H., *Z. Anal. Chem.* **174**, 174 (1960).
165. Martinet, B., *Chim. Anal. (Paris)* **44**, 64 (1962).
166. Khopkar, S. M. and De, A. K., *Anal. Chim. Acta* **22**, 223 (1960).
167. Akaiwa, H., Kawamoto, H. and Hara, M., *ibid.* **43**, 297 (1968).
168. Gerard, J., Holland, W. J., Veel, A. E. and Bozic, J., *Mikrochim. Acta* **1969**, 724.
169. Shigematsu, T. and Tabushi, M., *J. Chem. Soc. Japan, Pure Chem. Sect.* **81**, 262 (1960).
170. Kemula, W., Hulanicki, A. and Rubel, S., *Przemysł Chem.* **34**, 99 (1955).
171. Dalziel, J. A. and Thompson, M., *Analyst* **91**, 98 (1966).
172. Ospanov, Kh. K., Makletsova, N. E. and Tember, N. I., *Zh. Analit. Khim.* **22**, 444 (1967).
173. Fogg, A. G., Gray, A. and Burns, D. T., *Anal. Chim. Acta* **45**, 196 (1969); **47**, 151 (1969).
174. Toropova, V. F., Timofeeva, O. Yu. and Evteeva, Z. P., *Zh. Analit. Khim.* **26**, 1545 (1971).
174a. Katyal, M., Kushwaha, V. and Singh, R. P., *Analyst* **98**, 659 (1973).
175. Hartkamp, H., *Z. Anal. Chem.* **190**, 66 (1962).
176. Zehner, J. M. and Sweet, T. R., *Anal. Chim. Acta* **35**, 135 (1966).
177. Polyak, L. Ya., *Zavodsk. Lab.* **27**, 388 (1961).
178. Blank, A. B., Fedorova, I. I. and Tete, L. E., *Zh. Analit. Khim.* **24**, 978, 1367 (1969).
179. Dean, J. A. and Lady, J. H., *Anal. Chem.* **25**, 947 (1953).
180. Barkovskii, V. F. and Solonenko, V. G., *Zh. Analit. Khim.* **25**, 128 (1970).

181. Püschel, R., Lassner, E. and Katzengruber, K., *Z. Anal. Chem.* **223**, 414 (1966); *Chemist-Analyst* **56**, 63 (1967).
182. Shibata, S., Goto, K. and Nakashima, R., *Anal. Chim. Acta* **46**, 146 (1969).
182a. Yotsuyanagi, T., Yamashita, R. and Aomura, K., *Anal. Chem.* **44**, 1091 (1972).
183. Sommer, L., *Chem. Listy* **52**, 1485 (1958); *Collection Czech. Chem. Commun.* **24**, 1649 (1959).
184. Basargin, N. N. and Nemtseva, Zh. I., *Zh. Analit. Khim.* **20**, 966 (1965).
185. Minczewski, J. and Stolarczyk, U., *Chem. Anal. (Warsaw)* **11**, 531, 853 (1966); **12**, 1113 (1967).
186. Rao, G. G. and Rao, V. N., *Talanta* **1**, 169 (1958).
187. Nielsch, W. and Böltz, G., *Z. Anal. Chem.* **141**, 247 (1954).
188. Mannelli, G. and Biffoli, R., *Anal. Chim. Acta* **11**, 168 (1954).
189. Majumdar, A. K. and Savariar, C. P., *ibid.* **21**, 47 (1959).
190. Per'kov, I. G., Komar', N. P. and Mel'nik, V. V., *Zh. Analit. Khim.* **22**, 485, 653 (1967).
191. Busev, A. I., Rudzit, G. P. and Tsurika, I. A., *ibid.* **25**, 2151 (1970).
192. Paletskite, V. Yu. and Finkel'shteinaite, M. L., *ibid.* **24**, 1550 (1969).
193. Sakai, T., *Bull. Chem. Soc. Japan* **43**, 3171 (1970).
194. Menis, O. and Iyer, C. S., *Anal. Chim. Acta* **55**, 89 (1971).
195. Guirgis, F. K. and Habib, Y. A., *Analyst* **95**, 614 (1970).
196. McDonald, C. W. and Bedenbaugh, J. H., *Anal. Chem.* **39**, 1476 (1967).
197. Kapitańczyk, K., Kurzawa, Z. and Prymiński, Z., *Chem. Anal. (Warsaw)* **5**, 413 (1960).
198. Dukes, E. K. and Wallace, R. M., *Anal. Chem.* **33**, 242 (1961).
199. Nielsch, W. and Böltz, G., *Z. Anal. Chem.* **142**, 102 (1954).
200. Wolf, R. H. and Orhanović, M., *ibid.* **216**, 405 (1966).
201. Likussar, W., Wawschinek, O. and Beyer, W., *Anal. Chim. Acta* **40**, 538 (1968).
202. Imai, H. and Yamada, T., *Bull. Chem. Soc. Japan* **42**, 237 (1969).
203. Korenaga, T., Motomizu, S. and Toei, K., *Anal. Chim. Acta* **65**, 335 (1973).
204. Vassiliades, C. and Manoussakis, G., *Bull. Soc. Chim. France* **1960**, 390.
205. Zaki, B. M. and Khalil, S. S., *Microchem. J.* **8**, 6 (1964).

Chapter 28

LEAD

Lead (Pb, at.wt. 207·19) occurs in its compounds in the +II and +IV oxidation states. Compounds of lead(IV) have acidic properties. Lead hydroxide, $Pb(OH)_2$, is amphoteric, precipitating within the pH range 7–13. Lead(II) forms strong tartrate, acetate, thiosulphate, and EDTA complexes; the chloride complexes, however, are relatively weak.

28.1. Methods of Separation

28.1.1 Extraction

The foremost extraction method for separating lead from other elements involves extraction with dithizone (from a neutral or slightly alkaline medium), and is described below in detail. In the separation of larger quantities of lead (e.g. 1–5 mg), a chloroform solution of dithizone is more suitable due to the higher solubility of lead dithizonate in $CHCl_3$ than in CCl_4 [1].

Diethyldithiocarbamate extracts lead from slightly acidic media [2–4] (but dibenzyldithiocarbamate is the better extractant under acidic conditions [5]), from neutral media containing citrate, and from alkaline media containing cyanide and tartrate.

The lead iodide complex is extractable into MIBK or methyl isopropyl ketone [6], or dichloromethane containing tributylamine [7]. A solution of TBP in MIBK is a suitable extractant for the lead chloride complex [8].

28.1.2 Precipitation

Microgram quantities of lead are separated from slightly acidic solutions (pH $\geqslant$ 2) as the sparingly-soluble sulphide [9–11]. Silver, copper, or mercury serve as carriers, and the presence of citrate prevents precipitation of iron. Traces of lead have been isolated quantitatively as the hydroxide with ammonia as precipitant and lanthanum as collector [12]. Lead has also been separated by coprecipitation with strontium sulphate [13], barium sulphate [11,14,15], barium chromate [16], and calcium carbonate [17].

28.1.3 Ion-Exchange and Other Methods

The anionic lead chloride complex can be separated from complexes of other metals on strongly basic anion-exchangers [18–20a]. From 1M hydrochloric acid, Pb and Bi are retained on the column, whereas Ca, Ba, Cu, Fe, Sn, and Tl are eluted. The lead is then washed from the column with $0.01M$ HCl while the bismuth is retained.

Strelow and Toerien [21] use samples in 0.1–$4M$ HBr and a strongly basic anion-exchanger. Most of the metals retained are eluted with $0.1M$ HBr, after which lead is selectively eluted with a $0.3M$ HNO_3 + $0.025M$ HBr solution while Bi, Cd, Tl(III), Hg(II), Au(III), Pt(IV), and Pd(II) remain in the column. Traces of lead in cobalt have been retained from $0.3M$ HBr solution passed through Dowex-1 [22].

Korkisch and Feik [23] have used mixed eluents comprising tetrahydrofuran and 5 or $2.5M$ HNO_3 to separate lead from many other metals on Dowex-1.

Lead sorbed on a cation-exchanger can be eluted with $1M$ ammonium acetate, and thus separated from Ba, Sr, and Al [24].

Umland and Kirchner [25] recommend a 5–9% nitric acid medium for the electrolytic anodic separation of lead as PbO_2. Piryutko and Tsvetkovskii have found that more dilute HNO_3 is suitable for the anodic deposition of small amounts of lead as PbO_2 [25a]. Microamounts can be deposited on a stainless-steel cathode [25b].

Since lead is relatively volatile, it can be separated from rocks by evaporation at 1400°C, and condensation on a water-cooled quartz surface [26].

28.2. Methods of Determination

The well-known dithizone method is surely the best extractive spectrophotometric method for determining lead. Other methods (briefly discussed below) are not numerous and are of limited value in lead spectrophotometry.

28.2.1 DITHIZONE METHOD

Shaking a neutral or slightly alkaline solution of lead with dithizone (formula, p. 41) results in the formation of the pinkish-red dithizonate, $Pb(HDz)_2$, which is soluble in CCl_4, $CHCl_3$, and C_6H_6. The optimum pH range for the extraction of lead is from 7 to 10 [27–30].

The molar absorptivity of the complex at λ_{max} = 520 nm is 6.86×10^4 (a = 0.33). $Pb(HDz)_2$ solutions are stable if shielded from direct sunlight. Curve 3 in Fig. 2.2 (p. 43) shows the absorption spectrum of lead dithizonate in CCl_4 solution.

Of the various modifications of the dithizone method for the determination of lead, the monocolour technique is the most commonly used [31]. However, other methods such as the mixed colour [32] and reversion [33] procedures, as well as extractive [34] and spectrophotometric titrations [35], have also found favour.

In the monocolour method, alkaline aqueous solutions are extracted with portions of dithizone solution. Under these conditions, some of the excess of dithizone passes into the aqueous phase, which becomes brown. The excess of dithizone remaining in the organic phase is stripped with dilute potassium cyanide solution. If the KCN concentration, and accordingly the pH, are too high, some loss of lead is to be expected owing to partial decomposition of the dithizonate. The lead content is calculated from the absorbance of the pink $Pb(HDz)_2$ solution, or of the green dithizone solution (λ_{max} = 620 nm) produced by decomposing the lead dithizonate with dilute (e.g. $0.1M$) hydrochloric or sulphuric acid.

The principal masking agent used in the spectrophotometric determination of lead with dithizone is cyanide, which forms stable complexes with Ag, Hg, Pd, Au, Cu, Zn, Cd, Ni, and Co, thus preventing their reactions with dithizone. The first five of these metals can be separated from lead by preliminary extraction with dithizone from an acidic medium (pH < 1).

In the presence of cyanide, dithizone extracts lead together with Bi, Tl(I), and In. However, indium ions are extracted by dithizone in CCl_4 from only weakly alkaline media, and if the pH of the solution is ~ 10 (as is commonly the case in the determination of lead), indium remains in the aqueous phase. Thallium(I) extracted along with lead from the ammoniacal cyanide medium is stripped from the extract together with the free dithizone by shaking with dilute KCN solution.

The most serious interference in the determination is caused by bismuth, which is quantitatively extracted at pH 3–10 by dithizone in CCl_4. The bismuth may be preliminarily extracted with dithizone at pH 3. Alternatively, the lead is stripped with $\sim 0.001M$ HCl (optimum pH 3.3 ± 0.1) from the CCl_4 extract containing the Bi and Pb dithizonates, and re-extracted with dithizone in CCl_4 after the pH of the aqueous phase has been raised [36].

Addition of tartrate or citrate to the aqueous solution before the extraction of lead prevents the precipitation of readily hydrolysed metals such as Al, Fe, and Ti. In the presence of tartrate and cyanide, small amounts of iron do not interfere in the extraction; larger quantities of iron(III) should, however, be separated preliminarily. Hydroxylamine is added to prevent oxidation of dithizone in the alkaline solution.

Bloch and Lazare [37] have drawn attention to the interference of vanadium (V) in the determination of lead with dithizone by formation of a coloured vanadium complex.

Diaper and Kuksis [38] have determined lead with dithizone in a homogeneous acetone–water system.

Henderson and Snyder [39] have developed a method for determining diethyl–lead(II) and triethyl–lead(II) in the presence of inorganic lead. With these species, dithizone forms three differently-coloured complexes which have absorption maxima at 487, 435, and 520 nm, respectively.

Measurement of the absorbances at these wavelengths allows these compounds to be distinguished. Tetraethyl-lead does not react with dithizone.

Lead has been determined by the dithizone method in cast iron and steel [5,10,40,41], chromium and its alloys [42], molybdenum and tungsten [43], vanadium [44], niobium and its alloys [44,45], silver [12], cadmium [46, 46a], tin [14], indium [16], boron [36], silicate rocks [3,11,47], monazite [48], telluric acid [49], petroleum products [50], foodstuffs [34, 51–53], organic matter [2,31,32], urine [6,54], biological substances [55], human tissue and excreta [4], plant material [56,57], water [58,59], and air [60–64].

Browett and Moss [54,62] and Bano and Crossland [65] have developed automated procedures for determining lead with dithizone.

Reagents

1. Dithizone, H_2Dz: 0·001% solution in CCl_4 (p. 492).
2. Standard lead solution: 1 mg/ml. Dissolve 1·598 g of lead nitrate (dried at 110°C) in water containing 1 ml of conc. HNO_3, and dilute the solution with water to 1 litre.
3. Potassium sodium tartrate (Seignette salt): 20% solution.
   Purification of the solution: add ammonia to make the pH ~ 8 and shake the solution in a separating funnel with small portions of dithizone in CCl_4 until the organic phase no longer turns pink. Store the solution in a polyethylene bottle.
4. Potassium cyanide: 10% solution. Prepare a 50% solution of KCN, and shake it with small portions of dithizone in CCl_4, until the organic phase no longer turns pink. Shake the aqueous solution with 2 or 3 portions of chloroform to remove the free dithizone dissolved in the alkaline aqueous solution. Dilute the clear, purified, aqueous KCN solution fivefold with water and store in a polyethylene bottle.
5. Hydroxylamine hydrochloride: 20% solution. Adjust aqueous $NH_2OH \cdot HCl$ with ammonia to pH ~ 8, and shake with dithizone as described for the tartrate solution. This solution is unstable.
6. Wash solution. Dissolve 0·25 g of NH_4Cl in water, add 0·5 ml of conc. ammonia solution, 1 ml of the 10% KCN solution, and dilute the solution with water to 250 ml. This solution is also unstable.

Procedure

To the solution containing less than 70 μg of Pb, add 5 ml of the tartrate solution, adjust with ammonia (1+1) to pH ~ 8, and transfer quantitatively to a separating funnel. Add 1 ml of the hydroxylamine solution and 1–3 ml of the KCN solution (depending on the quantities of metals to be masked), and extract with portions of the dithizone solution (1 ml of 0·001% H_2Dz solution is equivalent to 4·0 μg of Pb) until the last portion of dithizone added shows no pink colour. Shake the combined extracts in the separating funnel with two portions of the wash solution. Wash the pink extract by shaking with water, transfer it to a 50-ml or smaller volumetric flask (depending on the amount of Pb), and dilute to the mark with carbon tetrachloride. Measure the absorbance at 520 nm, using CCl_4 as reference.

Note. To separate lead from any bismuth present in the extract, shake the carbon tetrachloride extract with two portions of dilute ($\sim 0.001M$) HCl (pH 3.0–3.5), causing the lead to pass quantitatively into the aqueous phase. Adjust the aqueous solution with ammonia to pH ~ 8, add some KCN solution, and extract the lead with dithizone as stated above.

28.2.2. OTHER METHODS

With Na–DDTC, lead forms the colourless diethyldithiocarbamate, $Pb(DDTC)_2$, which is soluble in CCl_4 and $CHCl_3$. The spectrophotometric determination of lead exploits the displacement of lead from the complex by copper ions, to give the more stable and coloured $Cu(DDTC)_2$ complex:

$$Cu^{2+} + Pb(DDTC)_2 \rightarrow Pb^{2+} + Cu(DDTC)_2$$

The colourless CCl_4 extract of $Pb(DDTC)_2$ is shaken with an aqueous $CuSO_4$ solution, and the absorbance of the resulting brown $Cu(DDTC)_2$ solution is measured at 435 nm [66–68].

4-(2-Pyridylazo)resorcinol (PAR) (formula, p. 46) reacts with lead ions in weakly ammoniacal medium to form a red complex ($\varepsilon = 4.0 \times 10^4$ at $\lambda_{max} = 520$ nm) [8, 69], which has been used to determine lead in steel, brass, and bronze [17].

Lukin *et al.* recommend the azo dye Arsazen [70] and its water-soluble sulphonic acid derivative Sulpharsazen [71] as spectrophotometric reagents for lead. Gusev *et al.* [72, 73] suggest other azo reagents; 2-(5-bromopyridylazo)-5-diethylaminophenol (5-bromo-PAAP) ($\varepsilon = 4.9 \times 10^4$ at 575 nm) seems to be particularly promising.

Triphenylmethane dyes used include Xylenol Orange (molar absorptivity 1.94×10^4 at 580 nm) [74, 75], and Methylthymol Blue [76].

High sensitivity is shown by the method for determining lead with diphenylcarbazone ($\varepsilon = 7.2 \times 10^4$ at 525 nm) [77].

Other organic reagents include a derivative of 8-hydroxyquinoline, CMAB-oxine [78,79], chloranilic acid [20a], 2-(*o*-hydroxyphenyl)benzothiazoline [80], and thio derivative of TTA [81]. The colours of the lead iodide [82, 83] and lead sulphide [84] suspensions have also formed the basis of colorimetric methods for lead.

References

1. Norwitz, G., Cohen, J. and Everett, M. E., *Anal Chem.* **32**, 1132 (1960).
2. Gage, J. C., *Analyst* **83**, 672 (1958).
3. Maynes, A. D. and McBryde, W. A. E., *Anal. Chem.* **29**, 1259 (1957).
4. Tompsett, S. L., *Analyst* **81**, 330 (1956).
5. Stobart, J. A., *ibid.* **90**, 278 (1965).
6. McCord, W. M. and Zemp, J. W., *Anal. Chem.* **27**, 1171 (1955).
7. Ziegler, M., *Z. Anal. Chem.* **180**, 351 (1961).

References

8. Yadav, A. A. and Khopkar, S. M., *Talanta* **18**, 833 (1971).
9. Willmer, T. K., *Arch. Eisenhüttenw.* **29**, 159 (1958).
10. Claassen, A. and Bastings, L., *Analyst* **88**, 67 (1963).
11. Baskova, Z. A., *Zh. Analit. Khim.* **14**, 75 (1959).
12. Marczenko, Z. and Kasiura, K., *Chem. Anal. (Warsaw)* **9**, 87 (1964).
13. Nazarenko, V. A., Fuga, N. A., Flyantikova, G. V. and Esterlis, K. A., *Zavodsk. Lab.* **26**, 131 (1960).
14. Marczenko, Z. and Kasiura, K., *Chem. Anal. (Warsaw)* **10**, 449 (1965).
15. Tiptsova, V. G., Dvortsan, A. G., Golitsyna, M. I. and Kopnina, O. I., *Zh. Analit. Khim.* **23**, 1065 (1968).
16. Volkova, A. I. and Zakharova, N. N., *Ukr. Khim. Zh.* **23**, 530 (1957).
17. Dagnall, R. M., West, T. S. and Young, P., *Talanta* **12**, 583, 589 (1965).
18. Nelson, F. and Kraus, K. A., *J. Am. Chem. Soc.* **76**, 5916 (1954).
19. Morachevskii, Yu. V., Zvereva, M. N. and Rabinovich, R. Sh., *Zavodsk. Lab.* **22**, 541 (1956).
20. Johnson, E. I. and Polhill, R. D., *Analyst* **82**, 238 (1957).
20a. Wynne, E. A., Burdick, R. D. and Fine, L. H., *Anal. Chem.* **33**, 807 (1961).
21. Strelow, F. W. E. and Toerien, F. von, *ibid.* **38**, 545 (1966).
22. Uny, G., Mathien, C., Tardif, J. P. and Tran Van Danh, *Anal. Chim. Acta* **53**, 109 (1971).
23. Korkisch, J. and Feik, F., *Anal. Chem.* **36**, 1793 (1964).
24. Khopkar, S. M. and De, A. K., *Talanta* **7**, 7 (1960).
25. Umland, F. and Kirchner, K., *Z. Anal. Chem.* **143**, 259 (1954).
25a. Piryutko, M. M. and Tsvetkovskii, I. B., *Zh. Analit. Khim.* **27**, 1536 (1972).
25b. Zykora, I. S., Kashtan, M. S. and Bulatov, V. V., *ibid.* **27**, 793 (1972).
26. Marshall, R. R. and Hess, D. C., *Anal. Chem.* **32**, 960 (1960).
27. Mayer, F. X. and Schweda, P., *Mikrochim. Acta* **1956**, 485.
28. Weber, O A. and Vouk, V. B., *Analyst* **85**, 40, 46 (1960).
29. Lur'e, Yu. Yu. and Nikolaeva, Z. V., *Zavodsk. Lab.* **23**, 652 (1957).
30. Mathre, O. B. and Sandell, E. B., *Talanta* **11**, 295 (1964).
31. Gage, J. C., *Analyst* **80**, 789 (1955); **82**, 453 (1957).
32. Neumann, F., *Z. Anal. Chem.* **155**, 340 (1957).
33. Irving, H. M. and Butler, E. J., *Analyst* **78**, 571 (1953).
34. Lockwood, H. C., *ibid.* **79**, 143 (1954).
35. Jones, R. A. and Szutka, A., *Anal. Chem.* **38**, 779 (1966).
36. Marczenko, Z., *Chem. Anal. (Warsaw)* **9**, 1093 (1964).
37. Bloch, J. M. and Lazare, J., *Bull. Soc. Chim. France* **1960**, 1148.
38. Diaper, D. G. and Kuksis, A., *Can. J. Chem.* **35**, 1278 (1957).
39. Henderson, S. R. and Snyder, L. J., *Anal. Chem.* **33**, 1172 (1961).
40. Milner, G. W. and Nall, W. R., *Anal. Chim. Acta* **6**, 420 (1952).
41. Meyer, S. and Koch, O. G., *Arch. Eisenhüttenw.* **31**, 711 (1960).
42. Mukhina, Z. S., Tikhonova, A. A. and Zhemchuzhnaya, I. A., *Tr. Komis. po Analit. Khim. Akad. Nauk SSSR* **12**, 298 (1960).
43. Püschel, R. and Lassner, E., *Mikrochim. Acta* **1965**, 751.
44. Nazarenko, V. A. and Biryuk, E. A., *Zavodsk. Lab.* **25**, 28 (1959).
45. Mukhina, Z. S., Tikhonova, A. A. and Zhemchuzhnaya, I. A., *Tr. Komis. po Analit. Khim. Akad. Nauk SSSR* **12**, 71 (1960).
46. Vdovenko, M. E. and Spivakovşkaya, N. E., *Zavodsk. Lab.* **27**, 963 (1961).
46a. Krasiejko, M. and Marczenko, Z., *Mikrochim. Acta* **1975 I**, 585.
47. Stanton, R. E., McDonald, A. J. and Carmichael, I., *Analyst* **87**, 134 (1962).
48. Powell, R. A. and Kinser, C. A., *Anal. Chem.* **30**, 1139 (1958).
49. Veale, C. R. and Wood, R. G., *Analyst* **85**, 371 (1960).
50. Griffing, M. E., Rozek, A., Snyder, L. J. and Henderson, S. R., *Anal. Chem.* **29**, 190 (1957).
51. Analytical Methods Committee, *Analyst* **79**, 397 (1954).

52. Bonastre, J., *Chim. Anal. (Paris)* **39**, 104 (1957).
53. Abson, D. and Lipscomb, A. G., *Analyst* **82**, 152 (1957).
54. Browett, E. V. and Moss, R., *ibid.* **90**, 715 (1965).
55. Oelschläger, W. and Schwarz, E., *Z. Anal. Chem.* **258**, 203 (1972).
56. Riebartsch, K. and Gottschalk, G., *ibid.* **214**, 179 (1965).
57. Kerin, Ž., *Mikrochim. Acta* **1968**, 927.
58. Miller, A. D. and Libina, R. I., *Zh. Analit. Khim.* **13**, 664 (1958).
59. Abbott, D. C. and Harris, J. R., *Analyst* **87**, 387 (1962).
60. Dixon, B. E. and Metson, P., *ibid.* **85**, 122 (1960).
61. Snyder, L. J. and Henderson, S. R., *Anal. Chem.* **33**, 1175 (1961).
62. Moss, R. and Browett, E. V., *Analyst* **91**, 428 (1966).
63. Snyder, L. J., *Anal. Chem.* **39**, 591 (1967).
64. Groffman, D. M. and Wood, R., *Analyst* **96**, 140 (1971).
65. Bano, F. J. and Crossland, R. J., *ibid.* **97**, 823 (1972).
66. Šedivec, V. and Vašak, V., *Chem. Listy* **46**, 607 (1952).
67. Tertoolen, J. F., Detmar, D. A. and Buijze, C., *Z. Anal. Chem.* **167**, 401 (1959).
68. Bulgakova, A. M. and Volkova, A. M., *Zh. Analit. Khim.* **15**, 591 (1960).
69. Pollard, F. H., Hanson, P. and Geary, W. J., *Anal. Chim. Acta* **20**, 26 (1959).
70. Lukin, A. M., Chernaya, L. S., Petrova, G. S. and Sosnina, A. I., *Zavodsk. Lab.* **28**, 398 (1962).
71. Lukin, A. M. and Petrova, G. S., *Zh. Analit. Khim.* **15**, 295 (1960).
72. Gusev, S. I., Kozhevnikova, I. A., Mal'tseva, L. S. and Shchurova, L. M., *ibid.* **22**, 1191 (1967).
73. Gusev, S. I. and Nikolaeva, E. M., *ibid.* **24**, 1674 (1969).
74. Marchenko, P. V., *Ukr. Khim. Zh.* **30**, 224 (1964).
75. Gurkina, T. V. and Igoshin, A. M., *Zh. Analit. Khim.* **20**, 778 (1965).
76. Srivastava, K. C. and Banerji, S. K., *Chim. Anal. (Paris)* **51**, 28 (1969).
77. Trinder, N., *Analyst* **91**, 587 (1966).
78. Umland, F. and Meckenstock, K., *Z. Anal. Chem.* **177**, 244 (1960).
79. Röbisch, G., *Anal. Chim. Acta* **47**, 539 (1969).
80. Uhlemann, E. and Pohl, V., *ibid.* **65**, 319 (1973).
81. Akki, S. B. and Khopkar, S. M., *Bull. Chem. Soc. Japan* **45**, 167 (1972).
82. Meyer, S. and Koch, O. G., *Arch. Eisenhüttenw.* **29**, 677 (1958).
83. Koch, O. G., *Anal. Chim. Acta* **62**, 462 (1972).
84. Koch, K. H., Ohls, K. and Riemer, G., *Z. Anal. Chem.* **237**, 167 (1968).

Chapter 29

MAGNESIUM

Magnesium (Mg, at.wt. 24·31) occurs in its compounds exclusively in the +II oxidation state. The hydroxide, $Mg(OH)_2$, begins to precipitate at pH 9·6 and shows no amphoteric properties: it is more soluble than the phosphate or oxinate. Magnesium forms rather weak tartrate, citrate, and EDTA complexes.

29.1 Methods of Separation

29.1.1 Precipitation

It is generally necessary to separate magnesium from the Analytical Group I–III metals before its determination. These metals are often removed by precipitation as hydroxides, sulphides, 8-hydroxyquinolinates, or dithiocarbamates (at pH 8–9), while magnesium (along with Ca, Sr, and Ba) remains in solution [1, 2].

For separation from amphoteric metals (e.g. Al, Sn, and Pb), magnesium is precipitated with an alkali metal hydroxide solution [2]. Iron may be used as a collector for traces of magnesium. When $Mg(OH)_2$ is precipitated with alkalis in the presence of tartrate or citrate, Fe, Cu, Ni, Mn, and certain other metals remain in solution. Titanium is masked as the peroxide complex in the alkaline solution.

Traces of magnesium are separated from metals forming soluble cyanide complexes (Ni, Zn, Mn, and others) by precipitation as the phosphate from alkaline cyanide medium containing lanthanum as collector [3].

Small quantities of magnesium have also been precipitated as the ammonium arsenate [4] or the 8-hydroxyquinolinate [5].

Activated carbon has been proposed as a collector for isolation of magnesium traces as $Mg(OH)_2$ [5a].

Magnesium (together with Al, Ti, V, Ca, and other metals) remains in the aqueous solution during electrolysis with a mercury cathode [6].

29.1.2 Extraction and Ion-Exchange

In contrast to the other Group I–III metals, magnesium remains in the aqueous phase during extraction with chloroform solutions of 8-hydroxyquinoline [6,7], cupferron [8], DDTC [9], or acetylacetone [10].

Ion-exchangers are important in the separation of magnesium from calcium. After both metals have been retained on a cation-exchanger, the magnesium is more readily eluted with dilute acid [11,12].

EDTA complexes of magnesium and calcium have been separated from the more stable EDTA complexes of Fe, Al, Zn, Cu, and Mn by elution of the latter complexes at pH 3·6 from a Dowex 50 cation-exchange column [13].

Separation of magnesium from calcium on ion-exchange resins is successful [14,15]. Phosphate, which precipitates magnesium, is readily removed on an anion-exchange column [16].

29.2 Methods of Determination

Spectrophotometric methods for determining magnesium are generally unselective, and thus necessitate preliminary separation of magnesium from other metals. Some methods (e.g. the Titan Yellow method) involve formation of coloured compounds by adsorption of organic dyes on magnesium hydroxide; others are based on soluble coloured magnesium complexes formed with some reagents (e.g. Eriochrome Black T) in ammoniacal media.

Tikhonov [17] has reviewed spectrophotometric methods for determining magnesium.

29.2.1 TITAN YELLOW METHOD

When precipitated from solutions containing certain coloured high molecular-weight organic reagents, magnesium hydroxide gives coloured adsorption compounds [18]. The compound formed with $Mg(OH)_2$ and Titan Yellow provides the basis of a commonly used spectrophotometric method for determining magnesium [10,19–24]. Commercial preparations of the sodium salt of Titan Yellow (**I**) (also known as Thiazole Yellow

or Clayton Yellow) from different sources display differences in their absorption spectra [22–24]. The absorption maximum of a solution of Titan Yellow at pH > 12 is at ~410 nm, whereas a colloidal solution of the adsorption compound formed with $Mg(OH)_2$ and Titan Yellow is pink and has its absorption maximum at 545 nm. At that wavelength, the absorbance of the free reagent is negligible. The absorption spectra of the reagent and of its magnesium compound are shown in Fig. 29.1.

The molar absorptivity of the pseudosolution of the magnesium compound with high-quality Titan Yellow is ~$3·6 \times 10^4$ at 545 ($a = 1·52$).

The intensity and reproducibility of the colour obtained are affected by a number of parameters; namely the method of pH adjustment, the excess of Titan Yellow, the protective colloid used, the temperature of the solution, and the time of standing.

Solutions containing more than 2 µg of Mg per ml are unstable and soon become turbid as the magnesium hydroxide coagulates. Protective colloids [e.g. poly(vinyl alcohol)] prevent coagulation. Although they

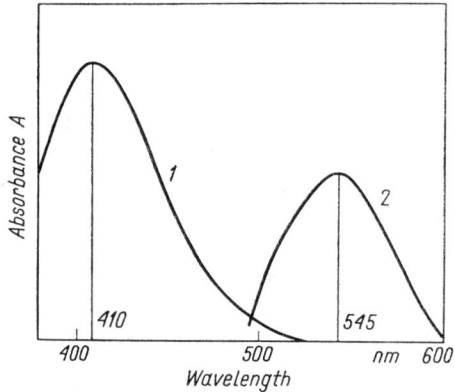

Fig. 29.1 Absorption spectra of Titan Yellow (1) and its magnesium compound (2) (reagent solution as reference)

are often used, natural substances such as gelatin, gum arabic, and starch are not suitable anticoagulants since they usually contain considerable amounts of magnesium.

The absorbance of the coloured solution increases with increasing NaOH concentration. The reproducibility of the results depends on careful addition of NaOH solution from a burette to the sample solution with vigorous stirring.

Immediately after the start of the colour reaction, an increase in absorbance is noticed, but between 10 and 30 minutes the colour of the solution remains almost constant. After this it weakens progressively. Hydroxylamine is reported to stabilize the colour [20].

Temperature greatly affects the reproducibility of the results. Heating a coloured solution from 20 to 30°C increases the absorbance by $\sim 40\%$ [19].

A synergistic effect on the determination of magnesium is exerted by calcium. Although this metal by itself gives no colour reaction with Titan Yellow, its presence with the magnesium causes increased absorbance. Since no further increase occurs above a certain concentration of calcium, the increased absorbance is exploited by adding excess of calcium to the sample and the standard solutions.

Species which decrease the quantity of $Mg(OH)_2$ precipitated, in other words ammonium salts and anions which precipitate magnesium (e.g. phosphate), interfere in the Titan Yellow method. A number of metal cations also interfere. Decreased absorbance in the presence of aluminium, zinc, or tin is believed to be due to adsorption of hydroxo aluminate, zincate, or stannate ions on the Mg(OH)$_2$ which reduces the amount of dye adsorbed [10]. The increase in colour caused by some metals [e.g. Fe, Cu, Mn(II), and Ni] results from the colour of their hydroxides and from their also forming coloured adsorption compounds with dyes [25]. Some interfering metals may be masked, e.g. copper with cyanide, iron(III) with triethanolamine [26], and aluminium with lactic or tartaric acid [27]. A high concentration of salts in the sample solution interferes by increasing the solubility of Mg(OH)$_2$ [19].

Hunter [28] determined magnesium indirectly by extracting into n-butanol the excess of Titan Yellow not adsorbed, and measuring the absorbance of the extract. Babko and Lutokhina [29] centrifuged the Mg(OH)$_2$–Titan Yellow compound, dissolved the precipitate in an EDTA solution, and measured the absorbance of the released dye.

Magnesium has been determined by the Titan Yellow method in cast iron [1,30], nickel and its alloys [4,9], nickel compounds [30a], titanium and its alloys [8], aluminium alloys [31], silicate minerals [23,32], carbonate rocks [32], rock salt [5,33], soil extracts [22,28], bones and teeth [34], plant materials [16,28,35], and blood [36].

Reagents

1. Titan Yellow: 0·01% solution. The solution is stable for at least one week.
2. Standard magnesium solution: 1 mg/ml. Dissolve 1·014 g of MgSO$_4$·7H$_2$O (or 0·495 g of magnesium sulphate ignited at 400–500°C) in water containing 1 ml of conc. H$_2$SO$_4$, and dilute the solution with water in a volumetric flask to 1 litre.
3. Calcium chloride: 2% CaCl$_2$.2H$_2$O solution, magnesium-free. Add sodium hydroxide to this solution till its concentration is $\sim 1M$. After 1 hour, filter off the precipitated Mg(OH)$_2$, and slightly acidify the solution with dilute hydrochloric acid.

Procedure

To the sample solution in a 50-ml volumetric flask and containing not more than 30 μg of Mg, add 2 ml of the calcium chloride solution, 5 ml of the Titan Yellow solution, 5 ml of 1% poly(vinyl alcohol) solution, and water to ~ 35 ml. While swirling the solution vigorously, add $1M$ NaOH dropwise from a burette till the solution changes colour. Then add 5 ml more NaOH solution with continued mixing, and dilute the solution with water to the mark. After 15 minutes, measure the absorbance of the coloured solution at 545 nm against a blank solution containing the same amount of the calcium solution.

29.2.2 ERIOCHROME BLACK T METHOD

The reaction of a blue solution of Eriochrome Black T (Erio T) (**II**) in an alkaline medium (pH 7·5–11·5) with magnesium ions to form a pink

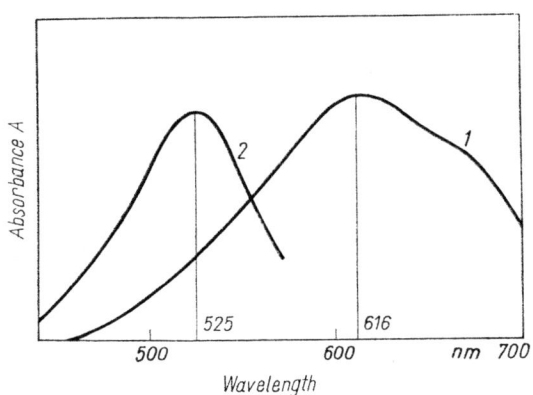

complex has been employed in the spectrophotometric determination of magnesium [37–40].

Depending on the pH of the medium, the ratio of magnesium to reagent in the complex varies from 1:1 to 1:3, causing changes in the absorbance and absorption maximum of the complex solution [41]. Since alteration of the pH also slightly changes the blue colour of the Eriochrome Black T ($\lambda_{max} \approx 615$ nm), the pH of the sample and the standard solutions must be carefully controlled.

The sensitivity of this method depends on the quality and excess of the Erio T used. Sensitivity decreases rapidly with increasing ionic strength.

Figure 29.2 shows the absorption spectra of Erio T and its magnesium complex at pH 9·6. The molar absorptivity of the complex at pH 9·6 and at $\lambda_{max} \approx 525$ nm is $1·8 \times 10^4$ (specific absorptivity 0·76).

Fig. 29.2 Absorption spectra of Eriochrome Black T (1) and its magnesium complex (2) (reagent solution as reference) at pH 9·6

Although the method has greatest sensitivity at pH 10·4 ($\varepsilon = \sim 2·3 \times 10^4$), a pH of 9·6 is more convenient since calcium (in fourfold amount relative to magnesium) does not interfere at the lower pH value. Larger quantities of calcium must be separated, e.g. by precipitation as the oxalate [38] or sulphate (in 90% methanol) [37], before the determination of magnesium.

The Analytical Group I–III metals must also be separated first. Small amounts of Fe, Cu, Zn, and Ni may be masked with cyanide. Alkali metals in moderate concentrations do not interfere in the method. Phosphate should be removed by ion-exchange separation, for example [3].

Differences in the stabilities of the Eriochrome Black T complexes of magnesium and calcium at various pH values facilitate simultaneous determination of both metals by measurement of the absorbances at two pH values [41,42].

The Eriochrome Black T method has been used to determine magnesium in aluminium and its alloys [2, 43], zinc and its alloys [44], nickel [45,46], uranium [47], indium [48], non-ferrous metallurgy products [49], nickel, zinc, and manganese salts [3], sodium chloride [5a], water [50], and biological samples [50a].

Variations of the method include extraction of the magnesium Eriochrome Black T complex into butanol [51], and the extractive spectrophotometric determination of magnesium with a related reagent (Erio T without the sulphonic acid group) [52].

Reagents

1. Eriochrome Black T: 0·02% solution in methanol; this solution is stable for about one week.
2. Standard magnesium solution: 1 mg/ml (p. 332).
3. Buffer solution: pH 9·6. Dissolve 60 g of NH_4Cl in water, add 120 ml of conc. ammonia, and dilute the solution with water to 1 litre.

Procedure

The sample solution must be slightly alkaline (pH 8–9), contain not more than 60 μg of Mg, and be free from the Group I–III metals. To this solution in a 50-ml volumetric flask, add ~ 20 mg of ascorbic acid and ~ 20 mg of potassium cyanide (to mask possible traces of Fe, Ni, Cu, Zn, and other metals). After 5 minutes, add 1 ml of buffer solution (pH 9·6) and 5 ml of the Eriochrome Black T reagent. Dilute the solution with water to 50 ml, and measure the absorbance at 520 nm against a reagent blank.

Note. Under these conditions a fourfold amount of calcium relative to magnesium does not interfere.

29.2.3 OTHER METHODS

Several other organic dyes which react with $Mg(OH)_2$ analogously to Titan Yellow have been recommended, e.g. Magneson I (**III**) [53,54],

$O_2N-\langle\rangle-N=N-\langle\rangle-OH$ with HO on the second ring   (**III**)

Magneson II [26,29,30], Magneson IREA [55], Phenazo [26,29,56], Picraminazo [57], Brilliant Yellow [58,59], several polymethine dyes [60–62a], and Azovan Blue [63].

The largest group comprises methods in which (as with Eriochrome Black T) magnesium ions form coloured chelates with chromogenic organic reagents (within a pH range from 9·5 to 12). Mann and Yoe [64,65] have suggested Xylidyl Blue I (Magon Sulphonate) and Xylidyl Blue II (Magon).

Xylidyl Blue I, X = SO₃Na
Xylidyl Blue II, X = H

These reagents and their applications have also been investigated by other authors [6, 66–68].

Flaschka and Sawyer [69], among other authors [70–72], have determined magnesium with Calmagite, an azo reagent resembling Eriochrome Black T (molar absorptivity 2.0×10^4 at 540 nm). Further spectrophotometric azo reagents for magnesium include Solochrome Black PV [73], o,o'-dihydroxyazobenzene [74], Chromotrope 2R [75], 5-(3-nitrophenylazo)salicylic acid [76], 2-(2-hydroxy-5'-sulphophenylazo)chromotropic acid [77], Pontacyl Violet 4BSN [78], Calcon [79], Calcichrome (formula, p. 187) [80], Arsenazo I [81], and Chlorophosphonazo III ($\varepsilon = 4.8 \times 10^4$ at 570 nm) [82]. Methylthymol Blue [83] and Thymolphthalexone [84] are triphenylmethane reagents which have been used similarly. 8-Hydroxyquinoline has been employed in extractive spectrophotometric methods for determining magnesium. Magnesium oxinate is not extracted by chloroform alone, but passes quantitatively into the organic phase in the presence of "butylcellosolve" (2-butoxyethanol) [85,86], butyl carbitol [87], or butylamine [7,86,88]. The molar absorptivity in the method using butylamine is 7.0×10^3 at 388 nm [87]. Magnesium oxinate is also extracted by trichloroethane [89]. Magnesium may be determined indirectly by precipitation with oxine and determination of the oxine in the precipitate (or the excess of oxine) by its colour reaction with iron(III) or with sulphanilic acid and nitrite [90].

Umland *et al.* [91,92] have developed a more selective method for magnesium, using CMAB-oxine, a derivative of 8-hydroxyquinoline.

References

1. Graue, G., Marotz, R. and Zöhler, A., *Angew. Chem.* **67**, 123 (1955).
2. Selzer, G. and Ariel, M., *Anal. Chim. Acta* **19**, 496 (1958).
3. Marczenko, Z. and Kasiura K., *Chem. Anal.* (*Warsaw*) **7**, 775 (1962).
4. Hegedüs, A. J. and Bali, M., *Mikrochim. Acta* **1961**, 721.
5. Marczenko, Z. and Stępień, A., *Chem. Anal.* (*Warsaw*) **5**, 247 (1960).
5a. Marczenko, Z., Mojski, M. and Balcerzak, M., *Mikrochim. Acta* **1975 I**, 539.
6. Maurice, M. J., *Anal. Chim. Acta* **20**, 181 (1959).
7. Newman, E. J. and Watson, C. A., *Analyst* **88**, 506 (1963).
8. Challis, H. J. and Wood, D. F., *ibid.* **79**, 762 (1954).
9. Noddack, W., Eckert, G. and Riedel, K., *Z. Anal. Chem.* **147**, 417 (1955).
10. Abrahamczik, E., *Angew. Chem.* **61**, 96 (1949).
11. Campbell, D. N. and Kenner, C. T., *Anal. Chem.* **26**, 560 (1954).
12. Mann, C. K., *ibid.* **32**, 67 (1960).
13. Chwastowska, J. and Szymczak, S., *Chem. Anal.* (*Warsaw*) **14**, 1161 (1969).
14. Povondra, P. and Přibil, R., *Talanta* **10**, 713 (1963).

15. Fritz, J. S. and Waki, H., *Anal. Chem.* **35**, 1079 (1963).
16. Bradfield, E. G., *Analyst* **85**, 666 (1960); **86**, 269 (1961).
17. Tikhonov, V. N., *Zh. Analit. Khim.* **26**, 1616 (1971).
18. Kuznetsov, V. I., *ibid.* **11**, 81 (1956).
19. Kenyon, O. A. and Oplinger, G., *Anal. Chem.* **27**, 1125 (1955).
20. Van Wesemael, J. C., *Anal. Chim. Acta* **25**, 238 (1961).
21. Bradfield, E. G., *ibid.* **27**, 262 (1962).
22. Hall, R. J., Gray, G. A. and Flynn, L. R., *Analyst* **91**, 102 (1966).
23. King, H. G., and Pruden, G., *ibid.* **92**, 83 (1967).
24. King, H. G., Pruden, G. and Janes, N. F., *ibid.* **92**, 695 (1967).
25. Bussmann, A., *Z. Anal. Chem.* **148**, 413 (1956).
26. Babko, A. K., and Lutokhina, N. V., *Ukr. Khim. Zh.* **28**, 389 (1962).
27. Babko, A. K. and Romanova, N. V., *Zavodsk. Lab.* **34**, 1435 (1968); *Zh. Analit. Khim.* **24**, 786 (1969).
28. Hunter, J. G., *Analyst* **75**, 91 (1950).
29. Babko, A. K. and Lutokhina, N. V., *Zh. Analit. Khim.* **17**, 416, 922 (1962).
30. Babko, A. K. and Lutokhina, N. V., *Ukr. Khim. Zh.* **25**, 226 (1959).
30a. Cyrankowska, M., *Chem. Anal. (Warsaw)* **19**, 309 (1974).
31. Jean, M., *Anal. Chim. Acta* **7**, 338 (1952).
32. Evans, W. H., *Analyst* **93**, 306 (1968).
33. Bahr, H., *Chem. Anal. (Warsaw)* **10**, 1045 (1965).
34. McCann, H. G., *Anal. Chem.* **31**, 2091 (1959).
35. Chenery, E. M., *Analyst* **89**, 365 (1964).
36. Butler, E. J., Forbes, D. H., Munro, C. S. and Russell, J. C., *Anal. Chim. Acta* **30**, 524 (1964).
37. Harvey, A. E., Jr., Komarmy, J. M. and Wyatt, G. M., *Anal. Chem.* **25**, 498 (1953).
38. Dirscherl, W. and Breuer, H., *Mikrochemie* **40**, 322 (1953).
39. Gasser, J. K., *Analyst* **80**, 482 (1955).
40. Pollard, F. H. and Martin, J. V., *ibid.* **81**, 348 (1956).
41. Young, A., Sweet, T. R. and Baker, B. B., *Anal. Chem.* **27**, 356, 418 (1955).
42. Menon, V. P. and Das, M. S., *Analyst* **83**, 434 (1958).
43. Pohl, H., *Aluminium* **33**, 260 (1957).
44. Pohl, H., *Metall* **10**, 709 (1956).
45. Pohl, H., *Z. Anal. Chem.* **155**, 263 (1957).
46. Koźlicka, M. and Kubica, M., *Chem. Anal. (Warsaw)* **12**, 841, 1315 (1967).
47. Krot, N. N., Smirnov-Averin, A. P. and Kozlov, A. G., *Zh. Analit. Khim.* **14**, 352 (1959).
48. Kasiura, K., *Chem. Anal. (Warsaw)* **11**, 141 (1966).
49. Shkrobot, E. P., Shebarshina, N. I. and Blyakhman, A. A., *Zavodsk. Lab.* **35**, 539 (1969).
50. Impedovo, S., Traini, A. and Papoff, P., *Talanta* **18**, 97 (1971).
50a. Hunter, G., *Analyst* **83**, 93 (1958).
51. Zolotov, Yu. A. and Bagreev, V. V., *Zh. Analit. Khim.* **22**, 1423 (1967).
52. Flaschka, H., *Mikrochim. Acta* **1956**, 784.
53. Rusconi, Y., Monnier, D. and Wenger, P. E., *Helv. Chim. Acta* **31**, 1549 (1948).
54. Budanova, L. M., Nenasheva, L. A. and Matrosova, T. V., *Zavodsk. Lab.* **22**, 1419 (1956).
55. Chwastowska, J., *Chem. Anal. (Warsaw)* **7**, 731 (1962).
56. Kuznetsov, V. I., Budanova, L. M. and Nenasheva, L. A., *Zavodsk. Lab.* **24**, 1053 (1958).
57. Gusev, S. I., Sokolova, E. V. and Bitovt, Z. A., *Zh. Analit. Khim.* **16**, 674 (1961).
58. Taras, M., *Anal. Chem.* **20**, 1156 (1948).
59. Cherkesov, A. I. and Lushinov, Yu. V., *Zavodsk. Lab.* **30**, 1053 (1964).
60. Asmus, E. and Schnabel, H., *Z. Anal. Chem.* **200**, 197 (1964).
61. Asmus, E. and Klank, W., *ibid.* **206**, 88 (1964).

62. Asmus, E. and Kuchenbecker, H., *ibid.* **248**, 291 (1969).
62a. Asmus, E. and Ortlepp, W., *ibid.* **266**, 343 (1973).
63. Chauhan, U. P. and Pande, P. C., *Talanta* **14**, 575 (1967).
64. Mann, C. K. and Yoe, J. H., *Anal. Chem.* **28**, 202 (1956).
65. Mann, C. K. and Yoe, J. H., *Anal. Chim. Acta* **16**, 155 (1957).
66. Apple, R. F. and White, J. C., *Talanta* **8**, 419 (1961).
67. Abbey, S. and Maxwell, J. A., *Anal. Chim. Acta* **27**, 233 (1962).
68. Svoboda, V. and Chromý, V., *ibid.* **54**, 121 (1971).
69. Flaschka, H. and Sawyer, P., *Talanta* **9**, 249 (1962).
70. Harrison, F. H., *Metallurgia*, **70**, 251 (1964).
71. Ingman, F. and Ringbom, A., *Microchem. J.* **10**, 545 (1966).
72. Hofer, A. and Heidinger, R., *Z. Anal. Chem.* **230**, 95 (1967).
73. Khalifa, H., and Bishara, S. W., *ibid.* **184**, 11 (1961).
74. Diehl, H., Olsen, R., Spielholtz, G. I. and Jensen, R., *Anal. Chem.* **35**, 1144 (1963).
75. Shibata, S., Uchiumi, A., Sasaki, S. and Goto, K., *Anal. Chim. Acta* **44**, 345 (1969).
76. Betteridge, D. and Yoe, J. H., *Talanta* **9**, 355 (1962).
77. Bezděková, A. and Buděšínský, B., *Collection Czech. Chem. Commun.* **31**, 199 (1966).
78. Uesugi, K., *Bull. Chem. Soc. Japan* **38**, 337 (1965).
79. Reilley, C. N. and Hildebrand, G. P., *Anal. Chem.* **31**, 1763 (1959).
80. Ishii, H. and Einaga, H., *Bull. Chem. Soc. Japan* **40**, 1531 (1967).
81. Hattori, T., Tsukahara, I. and Yamamoto, T. *Japan Analyst* **15**, 35 (1966).
82. Ferguson, J. W., Richard, J. J., O'Laughlin, J. W. and Banks, C. V., *Anal. Chem.* **36**, 796 (1964)
83. Metcalfe, J., *Analyst* **90**, 409 (1965).
84. Kish, P. P., Kotelyanskaya, L. I. and Kish, E. V., *Zh. Analit. Khim.* **26**, 487 (1971).
85. Luke, C. L. and Campbell, M. E., *Anal. Chem.* **26**, 1778 (1954); **28**, 1443 (1956).
86. Jankowski, S. J. and Freiser, H., *ibid.* **33**, 776 (1961).
87. Athavale, V. T., Bhasin, R. L. and Jangida, B. L., *Analyst* **87**, 217 (1962).
88. Umland, F. and Hoffmann, W., *Anal. Chim. Acta* **17**, 234 (1957).
89. Toribara, T. Y., Koval, L. and Olive, J. F., *Talanta* **10**, 1277 (1963).
90. Willson, A. E., *Anal. Chem.* **23**, 754 (1951).
91. Umland, F. and Meckenstock, K. U., *Angew. Chem.* **71**, 373 (1959); *Z. Anal. Chem.* **177**, 244 (1960).
92. Umland, F., Poddar, B. K. and Meckenstock, K. U., *ibid.* **185**, 362 (1962).

Chapter 30

MANGANESE

Manganese (Mn, at.wt. 54·94) occurs in its compounds in the oxidation states +II, III, IV, VI and VII. The hydroxide, $Mn(OH)_2$, precipitates at pH $\sim$ 8·5 and exhibits no amphoteric properties. On standing in air, the white $Mn(OH)_2$ precipitate slowly darkens as it is oxidized to hydrous manganese dioxide. Manganese(II) forms complexes of low stability with EDTA, cyanide, tartrate, and ammonia. Manganese(III) occurs in sulphate, cyanide, phosphate, and pyrophosphate complexes, for example. Manganese(IV) complexes (e.g. with formaldoxime) are also known. Brown hydrous MnO_2 is sparingly soluble in alkalis and in non-reducing acids. Manganese(VII) occurs in the violet permanganate ion, MnO_4^-, a powerful oxidant. When reduced with alcohol in an alkaline medium, MnO_4^- is converted into the green manganate ion, MnO_4^{2-}. On acidification, the manganate disproportionates into MnO_4^- and hydrous MnO_2.

30.1 Methods of Separation

30.1.1 PRECIPITATION

Sodium hydroxide quantitatively precipitates traces of manganese as the hydrous dioxide, MnO_2, with iron(III), lanthanum, or magnesium as carrier. Manganese can also be separated as hydrous MnO_2 by treatment with excess of ammonia in the presence of hydrogen peroxide. This method separates manganese from Ti, V, and other metals which form soluble peroxide complexes. Addition of H_2O_2 or bromine is not essential for the precipitation of manganese traces from excess of ammonia since dissolved oxygen will effect the oxidation.

From a neutral or slightly ammoniacal medium, manganese(II) is coprecipitated as the sulphide with iron(III) or lanthanum [1]. Traces of manganese have also been separated with 8-hydroxyquinoline as precipitant and iron(III) as carrier [2].

30.1.2 EXTRACTION

With sodium diethyldithiocarbamate, Mn(II) forms a pale yellow precipitate. Contact with air and excess of Na-DDTC converts this precipitate into brown-violet $Mn(DDTC)_3$, which is readily soluble in chloroform, ethyl acetate, or a solution of isoamyl alcohol in carbon tetrachloride [3–6]. The extraction of $Mn(DDTC)_3$ (optimum pH 6–8) is a convenient way of separating manganese before its determination by other methods. Diethylammonium diethyldithiocarbamate, which is soluble in chloroform, is sometimes used instead of Na-DDTC [7,8].

Traces of manganese have been extracted into chloroform as the 8-hydroxyquinolinate [2].

In their analysis of steel, Kinnunen and Merikanto [9] have extracted manganese as the thiocyanate complex. Iron(III) is masked with ammonium fluoride, NH_4SCN is added to the neutral solution, and manganese is extracted with a mixture of TBP and ether (3+2) and then stripped with hydrochloric acid. In the presence of pyridine, $Mn(Py)_4(SCN)_2$ is extracted into chloroform from a neutral solution containing NH_4SCN [10]. Manganese can also be extracted with TBP in xylene from $1M$ HCl containing $AlCl_3$ as salting-out agent [11].

Extraction is useful in the separation of MnO_4^- from MnO_4^{2-} ions. The MnO_4^- ions are extracted from $4M$ KOH with pyridine, while the manganate remains in the aqueous phase [12].

Manganese(VII) has been separated by chloroform extraction of the ion pair formed by MnO_4^- and the tetraphenylarsonium ion [6].

30.1.3 OTHER METHODS

Volatile permanganic acid, $HMnO_4$, can be distilled from $\sim 10M$ H_2SO_4 containing KIO_4 and HNO_3. CO_2 is used as carrier gas [13,14].

Manganese(II) forms a weaker chloride complex than bivalent Fe, Co, Cu, and Zn (the nickel complex is, however, even weaker), enabling Mn to be separated from these metals on anion- and cation-exchange columns [15, 16].

Manganese has been retained on a cation-exchanger while Ce(IV), V(V), and Cr(III) are masked as their anionic complexes with Tiron [17]. Separation of Mn(II) from many other metals [including Co(II) and Ni(II)] has been achieved by sequential elution from a cation-exchanger with a (1+1) mixture of $1M$ ammonium chloride and $1M$ ammonium thiocyanate [17a].

30.2 Methods of Determination

The well-known and often-used spectrophotometric method for determining manganese which is based on the coloured permanganate ion is highly selective but rather insensitive. Greater sensitivity is achieved in the method using formaldoxime, which is a cheap and readily available reagent. Even more sensitive is the extractive photometric PAN method.

30.2.1 PERMANGANATE METHOD

Oxidation of manganous ions by powerful oxidants in acidic solutions to yield violet MnO_4^- ions constitutes the basis of this spectrophotometric method [18–21]. The reactions with potassium periodate and ammonium persulphate, two popular reagents for the oxidation of Mn^{2+}, are:

$$2Mn^{2+} + 5IO_4^- + 3H_2O \rightarrow 2MnO_4^- + 5IO_3^- + 6H^+$$
$$2Mn^{2+} + 5S_2O_8^{2-} + 8H_2O \rightarrow 2MnO_4^- + 10HSO_4^- + 6H^+$$

With persulphate as oxidant, it is necessary to add a small amount of silver or cobalt ions to catalyse the oxidation of manganese(II) to MnO_4^-. Alternative oxidants include sodium bismuthate [22] and ceric ammonium nitrate (with silver as catalyst) [23]. Bane [24] oxidizes manganese with sodium perxenate (Na_4XeO_6), highly active in cold 0.1–$0.2M$ HNO_3. Strickland and Spicer [19] have elucidated the mechanism of the periodate oxidation of manganese(II). Gottschalk [21] has investigated the mechanism of, and optimum conditions for, the silver-catalysed persulphate oxidation of manganese(II).

The oxidation of manganese with periodate or persulphate is carried out in sulphuric acid or nitric acid, or in a mixture of the two acids. The concentration of the acid affects the rate of oxidation of Mn(II). With KIO_4, higher concentrations of H_2SO_4 and HNO_3 can be used than with $(NH_4)_2S_2O_8$. Excessive acid concentrations, especially with persulphate as the oxidant, slow the reaction and increase the risk of volatilizing $HMnO_4$. The optimum concentrations of the acids and other reagents are detailed in the procedure below.

Periodate oxidation yields a more stable solution of MnO_4^- than does persulphate oxidation. In the case of minute amounts of manganese, however, persulphate oxidation takes place more rapidly.

Reducing species, including chloride, present in the solution interfere in the determination of manganese. They are removed beforehand by evaporating the sample solution with sulphuric acid to white fumes. Large amounts of coloured metal ions (e.g. Ce^{4+}, Ni^{2+}, Co^{2+}, Cu^{2+}, $Cr_2O_7^{2-}$, and UO_2^{2+}) also interfere. Ferric ions are masked as a colourless complex by phosphoric acid. The presence of phosphoric acid also prevents the precipitation of hydrous MnO_2 or manganese and iron periodate (or iodate). Interference from coloured ions other than Ce(IV) is overcome by measuring the absorbance before and after reduction of the Mn(VII) by sodium azide (20–50 mg) [21], hydrogen peroxide [25], or sodium nitrite [26].

Turbidity (AgCl) due to traces of chloride present during silver-catalysed persulphate oxidation is prevented by the addition of a little mercuric sulphate [18].

The permanganate method is one of the less sensitive colorimetric methods. The molar absorptivity is 2.4×10^3 ($a = 0.044$) at 528 nm. The absorption spectrum of MnO_4^- is shown in Fig. 30.1.

Permanganate ions react with the tetraphenylarsonium cation to form an ion-pair, $[(C_6H_5)_4As]^+ [MnO_4]^-$, which is soluble in chloroform and certain other solvents. Extractive variations of the permanganate method have been developed based on this ion-pair [6, 27–30].

Shevchuk et al. [31] have extracted MnO_4^- ions from $0.1M$ H_2SO_4 into a benzene solution of di-n-decylamine or di-n-hexylamine.

The permanganate method has been used to determine manganese in cast iron and steel [21, 26, 30, 32–34], non-ferrous metals [30], aluminium ores and alloys [35], magnesium alloys [4], uranium [24], rhenium

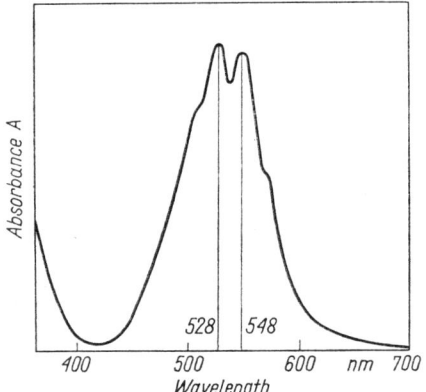

Fig. 30.1 Absorption spectrum of permanganate in $0.5M$ H_2SO_4

and its compounds [36], cement [37], glass [21,38], blast furnace ores and slags [39], calcium carbonate [29], petroleum [40], foodstuffs [41], and water [21].

Larger manganese contents have been determined as MnO_4^- by differential spectrophotometry [25, 42]. The permanganate method has also been applied in automatic procedures for determining manganese in steel and slags [43, 44].

Reagents

1. Potassium periodate, KIO_4.
2. Ammonium persulphate, $(NH_4)_2S_2O_8$.
3. Silver nitrate: about $0.001M$ solution. Dissolve 85 mg of $AgNO_3$ in 500 ml of water. Keep the solution in an amber glass bottle.
4a. Standard manganese solution: 1 mg/ml. Dissolve 2·873 g of $KMnO_4$ in water containing 2 ml of conc. H_2SO_4, and dilute the solution with water to 1 litre.
4b. Standard manganese solution: 1 mg/ml. Dissolve 2·749 g of anhydrous $MnSO_4$ in water containing 1 ml of conc. H_2SO_4, and dilute the solution with water to 1 litre. The anhydrous salt is obtained from hydrated manganese(II) sulphate by drying at 150°C and subsequent ignition at about 400°C. Alternatively, acidify precisely 91 ml of exactly $1N$ ($0.2M$) $KMnO_4$ with 5 ml of conc. H_2SO_4, and add perhydrol dropwise till the solution is decolorized. Heat the solution to decompose the excess of H_2O_2, cool, and dilute with water to 1 litre.

Procedure

Oxidation of Mn with periodate. Dilute the sample solution containing not more than 500 μg of Mn (free from chloride and other reducing agents) in a beaker with water to about 35 ml. Add 5 ml of conc. H_2SO_4, 1 ml of conc. HNO_3, 2 ml of conc. H_3PO_4, and 0·25 g of KIO_4, and stir. Heat the solution nearly to boiling, and keep at about 90°C for 10 minutes. Cool the solution, transfer it to a 50-ml volumetric flask, and dilute to the mark with water. Measure the absorbance of the solution at 528 nm against water or a reagent blank.

Oxidation of Mn with persulphate. To the sample solution as above, add 1 ml of conc. H_2SO_4, 1 ml of conc. HNO_3, 2 ml of conc. H_3PO_4, 2 ml of 0·001M $AgNO_3$, and 0·5 g of $(NH_4)_2S_2O_8$. Heat the solution nearly to boiling and keep at this temperature for about 5 minutes. Proceed further as described above.

30.2.2 Formaldoxime Method

If a solution containing manganese and formaldoxime (formula, p. 59) is made alkaline with ammonia or an alkali metal hydroxide, a colourless manganese(II) complex is formed. Oxygen instantaneously transforms this complex into a brown-red complex of manganese (IV).

If the initial solution contains either MnO_4^- ions or a suspension of hydrous MnO_2, formaldoxime first reduces the manganese to Mn(II) and then complexes it.

Irrespective of the reagent used for pH adjustment and of the excess of formaldoxime, only one coloured complex is formed in the solution. Its absorption spectrum is shown in Fig. 2.6, p. 59. The Mn:CH_2NOH ratio in the coloured complex is 1:6, corresponding to $[Mn(CH_2NO)_6]^{2-}$ [45,46]. The stability of the complex is shown by the fact that the colour reaction is not affected by tartrate, oxalate, phosphate, pyrophosphate, sulphide, cyanide, or EDTA. Reductants such as ascorbic acid, hydroxylamine or sulphite do not hinder the rapid formation of the coloured manganese complex.

The molar absorptivity of the complex is $1·12 \times 10^4$ ($a = 0·20$) at $\lambda_{max} = 455$ nm. Thus, the formaldoxime method is more than four times as sensitive as the permanganate method.

In 0·04–0·05M NaOH medium, the coloured complex can be heated for 15 minutes at 90°C without any change in colour. The heat resistance of the complex is less at other concentrations of NaOH and in ammoniacal solutions.

Formaldoxime also forms intensely coloured complexes with Ce, Cu, Fe, V, Ni, and Co. Cerium and copper formaldoximates decompose in less than a minute at 70°C. The violet iron complex decomposes at 70°C in 20 minutes. The vanadium complex decomposes at 90°C in 5 minutes of heating. Cobalt and nickel formaldoximates are more resistant to heat:

15 minutes heating at 90°C reduces their absorbances to 40% and 4% of the initial values, respectively.

Cyanide prevents the formation of formaldoximates of Ni, Co, Cu. and Fe(II). Larger quantities of iron(III) should be separated from manganese by precipitation with a $Zn(OH)_2$ suspension, or by extraction with phenylacetic acid in $CHCl_3$ (which also extracts interfering copper and uranium) [47].

Formaldoxime forms colourless or slightly coloured complexes with Al, Ti, U, Mo, Cr, and the platinum metals. Larger quantities of aluminium and titanium are masked by tartrate.

Use of cyanide as a masking agent and heating the coloured solution to 90°C make this method specific for manganese [45].

The reagent is added either as the acid solution obtained from formalin and hydroxylamine hydrochloride or as solid $(CH_2NOH)_3.HCl$. The solid form is more convenient for determining traces of manganese when it is important to limit the volume of the final coloured solution.

Formaldoxime has been applied in the spectrophotometric determination of manganese in natural waters [48], biological material [49,50], plant material [51], foodstuffs [41], alkalis [2], tin [1], nickel alloys [3], glass [52] silicate and carbonate minerals [53], ores and rocks [53a], and animal feeds [16].

The formaldoxime method has also been automated for the determination of manganese in water [54] and in silicate minerals [55].

Reagents

1. Formaldoxime: $1M$ solution ($\sim 5\%$). Mix 7·9 g of formalin (38% solution of CH_2O) with 7·0 g of $NH_2OH.HCl$ and dilute with water to 100 ml. The solution is acidic, $1M$ in HCl, and stable.
2. Formaldoxime hydrochloride: crystalline $(CH_2NOH)_3.HCl$. Dissolve 105 g of $NH_2OH.HCl$ in 110 ml of water. Add 45 g of paraformaldehyde, and stir. Heat the solution under reduced pressure (~ 25 mmHg) at about 40°C. When large quantities of crystals begin to precipitate, stop the heating, add 80 ml of anhydrous ethanol, and stir. Leave overnight, then filter off the crystals on a Büchner funnel, and wash them with anhydrous ethanol. Recrystallize the product from anhydrous ethanol and dry in a vacuum desiccator.
3. Standard manganese solution: 1 mg/ml (p. 341).
4. $Zn(OH)_2$ suspension. Preparation and purification from Fe, Mn, and other metals: dissolve 2 g of zinc nitrate in 50 ml of water, add 5 mg of lanthanum (in the form of the nitrate solution) as a collector, some H_2O_2 solution, and ammonia in excess. Filter off the coagulated precipitate of La, Fe, Mn, and other metal hydroxides. Evaporate off the excess of ammonia from the filtrate, and slightly acidify the solution with dilute HCl. Precipitate $Zn(OH)_2$ by adding dilute NaOH solution till the pH is 8–9. Filter off the precipitate and wash with water until Cl^-

ions are no longer present in the washings. Mix the precipitate with 100 ml of water.

Procedure

To a weakly acidic solution containing not more than 100 µg of Mn, add 20% potassium sodium tartrate solution (in the presence of Al, Ti, Cr, or U), about 20 mg of ascorbic acid [in the presence of traces of Fe(III)], and about 50 mg of KCN [in the presence of Ni, Co, Fe(II), or Cu]. Next add 2 ml of $1M$ formaldoxime followed immediately by 2 ml more $1M$ NaOH than the volume required for neutralization. Dilute the solution with water to the mark in a 50-ml volumetric flask. After 10 minutes, measure the absorbance of the coloured solution at 455 nm, using water or a blank solution as the reference.

Notes. (1) If there is more iron than manganese present, the iron is removed by extraction of its chloride or thiocyanate complex, or by precipitation (along with Al and Ti) with the $Zn(OH)_2$ suspension, or by extraction with phenylacetic acid.

(2) Before the addition of $Zn(OH)_2$, the solution is adjusted to pH 2–3 with dilute NaOH solution. After $Zn(OH)_2$ is added, the solution is heated for about 15 minutes at ~70°C, whereupon the coagulated precipitate is filtered off and washed with hot water.

(3) If the sample solution is likely to contain traces of vanadium or cerium, the coloured solution should be heated at 90°C for 15 minutes. Any precipitate present after cooling should be filtered off.

30.2.3 Pyridylazonaphthol (PAN) Method

1-(2-Pyridylazo)-2-naphthol (formula, p. 46) reacts with manganous ions in weakly alkaline solution (pH 8–10) to form a chelate which is sparingly soluble in water. When the suspension is shaken with chloroform, the red-violet complex and the excess of the orange reagent pass into the organic layer. The coloured extract constitutes the basis of this extractive spectrophotometric method for determining manganese [56–60].

The method is highly sensitive. The molar absorptivity of the chloroform solution of the manganese-PAN complex is 5.8×10^4 ($a = 1.05$) at $\lambda_{max} = 564$ nm. As can be seen in Fig. 30.2, the excess of free reagent interferes only slightly at this wavelength.

The optimum pH for the reaction and extraction is $9.2(\pm 0.4)$, which is obtained with an ammoniacal buffer [60]. The aqueous solution should contain hydroxylamine to prevent oxidation of the manganese(II).

Since the distribution coefficients of PAN and its manganese complex are high, it is sufficient to extract the aqueous solution with only one portion of chloroform. Carbon tetrachloride, benzene, diethyl ether, and isoamyl alcohol are alternative extractants. Only slight changes are observed in the value of λ_{max} for the complex in the different solvents.

Tartrate is used to mask hydrolysable metals (e.g. Al, Cr, Zr, In, Bi, and Fe) but does not interfere with the reaction between manganese and PAN. Citrate causes low results. A moderate amount of cyanide is added

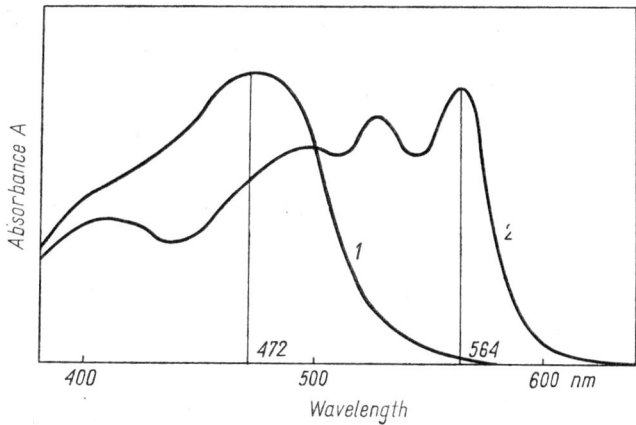

Fig. 30.2 Absorption spectra of 1-(2-pyridylazo)-2-naphthol (PAN) (1) and its manganese complex (2) in chloroform

to mask a number of metals such as Cu, Ni, Cd, Fe(II), Zn, and Co. Iron(III) is masked by fluoride.

According to Donaldson and Inman [61], in the presence of tartrate and cyanide (as in this procedure) 1-mg amounts of Ni, Cu, Cd, Cr(III), V(V), Al, Fe(III), In, Ti, Sn(IV), and Sb can be tolerated. Zinc, lead, and cobalt can be tolerated in quantities not larger than 50, 100, and 500 µg, respectively.

PAN has been used to determine manganese in beryllium [59], in platinum and gold [60], cadmium [60a], and in niobium, tantalum, molybdenum and tungsten [61]. In the last case, macro quantities of Nb, Ta, Mo, and W were first separated by extraction of their cupferronates into chloroform.

Reagents

1. 1-(2-Pyridylazo)-2-naphthol (PAN): 0·1% solution in ethanol.
2. Standard manganese(II) solution: 1 mg/ml (p. 341).
3. Hydroxylamine hydrochloride: 10% solution, freshly prepared. The solution (neutralized to pH ~ 7 with NaOH) must be purified by scrubbing with dithizone.
4. Ammoniacal cyanide buffer, pH 9·8. Dissolve 10·5 g of NH_4Cl in water, add 60 ml of conc. ammonia and 0·75 g of KCN, and dilute the solution with water to 500 ml. Store the solution in a polyethylene bottle.
5. Potassium sodium tartrate: 10% solution.

Procedure

Put a weakly acidic solution containing not more than 30 µg of Mn in a separating funnel. Add 1 ml of tartrate solution and 2 ml of hydroxylamine solution, and dilute with water to ~ 25 ml. Add 5 ml of buffer and 2 ml of PAN solution. After 2 minutes, shake the aqueous solution with chloroform for 1 minute. Transfer the chloroform extract to a 50-ml (or smaller, according to the amount of manganese) volumetric flask and dilute to the mark with the solvent. Measure the absorbance of the coloured solution at 564 nm against a reagent blank.

30.2.4 Other Methods

The brown chloroform-soluble manganese(III) diethyldithiocarbamate complex is suitable not only for the extractive separation of manganese but also for its spectrophotometric determination [62–67]. This relatively insensitive method (molar absorptivity 3.8×10^3 at 400 nm) has been used to determine manganese in steel [63,64], alkali metal halides [65], and organic matter [66,67]. Manganese has also been determined as the morpholinium 3-oxapentamethylenedithiocarbamate complex [68].

The coloured chloroform-soluble manganese(III) complexes with hydroxamic acids form the basis of extractive colorimetric methods for the determination of manganese [69–71]. Benzohydroxamic acid forms a red-brown complex ($\varepsilon = 3.6 \times 10^3$ at $\lambda_{max} = 500$ nm) [69].

Other extractive spectrophotometric methods exploit coloured manganese complexes with TTA [72–76], 8-hydroxyquinaldine [77], 8-mercaptoquinoline (thio-oxine) ($\varepsilon = 7.0 \times 10^3$ at $\lambda_{max} = 413$ nm) [88], and dithizone in the presence of pyridine ($\varepsilon = 5.7 \times 10^4$ at 510 nm) [79,80]. Minczewski et al. [81] have extracted the ion-pair formed between Rhodamine 6G and the tris-(5,7-dichloro-8-hydroxyquinoline)manganese(II) chelate ($\varepsilon = 7.0 \times 10^4$ at $\lambda_{max} = 540$ nm in benzene).

Calcichrome [82] and PAR [83–85] have been recommended as reagents for the spectrophotometric determination of manganese in aqueous media.

In cold dilute H_2SO_4 (HNO_3 or $HClO_4$) containing oxidants (BrO_3^-, $Cr_2O_7^{2-}$), Mn(II) reacts with pyrophosphate to give the violet manganese(III) pyrophosphate complex (molar absorptivity 7.5×10^2 at 520 nm) [86,87]. A similar complex is formed by heating manganese(II) with a mixture of 72% $HClO_4$ and 85% H_3PO_4 (1+1) [88,89]. The anionic manganese(III) pyrophosphate complex can be extracted into chloroform in the presence of n-dodecylamine [90]. The pyrophosphate complex has been used for the determination of manganese in cast iron, steel, iron minerals, and non-ferrous alloys [87–89].

Coloured manganese(III) complexes are also formed in the presence of oxidants with sulphate [91], dipicolinate [92], EDTA [93], and triethanolamine [94].

Chung and Meloan [95] have based their determination of manganese in alkaline medium on the green manganate ion, MnO_4^{2-}.

Oxidation products of Mn(II) react with o-tolidine to produce a yellow colour, which can be utilized in the determination of manganese [96].

There are very sensitive catalytic spectrophotometric methods for the determination of manganese in which traces of Mn(VII) catalyse the oxidation of various organic substances by another oxidant (e.g. KIO_4), with the formation of coloured reaction products. In these methods, the amount of colour produced depends on the reaction time.

References

1. Marczenko, Z. and Kasiura, K., *Chem. Anal. (Warsaw)* **10**, 449 (1965).
2. Marczenko, Z., *Mikrochim. Acta* **1965**, 281.
3. Eckert, G., *Z. Anal. Chem.* **148**, 14 (1955).
4. Tikhonov, V. N. and Nikitina, A. P., *Zavodsk. Lab.* **28**, 662 (1962).
5. Usatenko, Yu. I. and Fedash, N. P., *Tr. Komis. po Analit. Khim. Akad. Nauk SSSR* **14**, 183 (1963).
6. Dinstl, G. and Hecht, F., *Mikrochim. Acta* **1962**, 321.
7. Wyatt, P. F., *Analyst* **78**, 656 (1953).
8. Clinch, J. and Guy, M. J., *ibid.* **83**, 429 (1958).
9. Kinnunen, J. and Merikanto, B., *Chemist-Analyst* **43**, 93 (1954).
10. Rane, A. T., *J. Radioanal. Chem.* **8**, 117 (1971).
11. Yadav, A. A. and Khophar, S. M., *Sepn. Sci.* **4**, 349 (1969).
12. Hornig, H. C., Zimmerman, G. L. and Libby, W. F., *J. Am. Chem. Soc.* **72**, 3808 (1950).
13. Strickland, J. D. and Spicer, G., *Anal. Chim. Acta* **3**, 543 (1949).
14. Pijck, J. and Hoste, J., *ibid.* **26**, 501 (1962).
15. Kraus, K. A. and Moore, G. E., *J. Am. Chem. Soc.* **75**, 1460 (1953).
16. Tušl, J., *Chem. Listy* **61**, 1500 (1967).
17. Golovatyi, R. N. and Oshchapovskii, V. V., *Ukr. Khim. Zh.* **28**, 518 (1962).
17a. Matsui, H., *Anal. Chim. Acta* **69**, 216 (1974).
18. Nydahl, F., *ibid.* **3**, 144 (1949).
19. Strickland, J. D. and Spicer, G., *ibid.* **3**, 517 (1949).
20. Waterbury, G. R., Hayes, A. M. and Martin, D. S., Jr., *J. Am. Chem. Soc.* **74**, 15 (1952).
21. Gottschalk, G., *Z. Anal. Chem.* **212**, 303 (1965).
22. Přibil, R. and Hornychová, E., *Chem. Listy* **44**, 101 (1950).
23. Rao, G. G., Murty, K. S. and Rao, P. V., *Talanta* **11**, 955 (1964).
24. Bane, R. W., *Analyst* **90**, 756 (1965).
25. Young, I. G. and Hiskey, C. F., *Anal. Chem.* **23**, 506 (1951).
26. Cooper, M. D., *ibid.* **25**, 411 (1953).
27. Bock, R. and Beilstein, G. M., *Z. Anal. Chem.* **192**, 44 (1963).
28. Matuszek, J. M., Jr. and Sugihara, T. T., *Anal. Chem.* **33**, 35 (1961).
29. Richardson, M. L., *Analyst* **87**, 435 (1962).
30. Goto, H. and Kakita, Y., *Z. Anal. Chem.* **254**, 18 (1971).
31. Shevchuk, I. A., Nikol'skaya, N. N. and Simonova, T. N., *Ukr. Khim. Zh.* **32**, 635 (1966).
32. Meyer, S. and Koch, O. G., *Mikrochim. Acta* **1964**, 216.
33. Mal'tsev, V. F. and Luk'yanenko, L. P., *Zh. Analit. Khim.* **20**, 394 (1965).
34. Gaunt, J. A. and Diehl, H., *Talanta* **19**, 1 (1972).
35. Lacroix, S. and Labalade, M., *Anal. Chim. Acta* **3**, 262 (1949).
36. Ryabchikov, D. I., Lazarev, A. I. and Lazareva, V. I., *Zh. Analit. Khim.* **19**, 1110 (1964).
37. Diamond, J. J., *Anal. Chem.* **28**, 328 (1956).
38. Hecht, F. and Gottlieb, A., *Mikrochemie* **35**, 329 (1950).
39. Neuberger, A., Schöffmann, E. and Herkenhoff, K., *Arch. Eisenhüttenw.* **29**, 35, 547 (1958).
40. Steinke, E. D., Jones, R. A. and Brandt, M., *Anal. Chem.* **33**, 101 (1961).

41. Młodecki, H., Chmielnicka, J., Pawłowska, B. and Piotrowska, A. *Chem. Anal. (Warsaw)* **10**, 1267 (1965).
42. Barkovskii, V. F. and Vtorygina, I. N., *Zh. Analit. Khim.* **17**, 865 (1962).
43. Scholes, P. H. and Thulbourne, C., *Analyst* **88**, 702 (1963); **89**, 466 (1964).
44. Scholes, P. H., *Z. Anal. Chem.* **222**, 162 (1966).
45. Marczenko, Z., *Chem. Anal. (Warsaw)* **6**, 477 (1961); *Anal. Chim. Acta* **31**, 224 (1964).
46. Okač, A. and Bartušek, M., *Z. Anal. Chem.* **178**, 198 (1960).
47. Adam, J. and Přibil, R., *Talanta* **19**, 1105 (1972).
48. Goto, K., Komatsu, T. and Furukawa, T., *Anal. Chim. Acta* **27**, 331 (1962).
49. Bartley, W., Notton, B. M. and Werkheiser, W. C., *Biochem. J.* **67**, 291 (1957).
50. Holeyšovská-Kozáková, H., *Chem. Listy* **54**, 967 (1960).
51. Bradfield, E. G., *Analyst* **82**, 254 (1957).
52. Gottlieb, A. and Hecht, F., *Mikrochemie* **35**, 337 (1950).
53. Riley, J. P. and Williams, H. P., *Mikrochim. Acta* **1959**, 804.
53a. Adam, J., *Hutn. Listy*, **28**, 592 (1973).
54. Henriksen, A., *Analyst* **91**, 647 (1966).
55. Abdullah, M. I., *Anal. Chim. Acta* **40**, 526 (1968).
56. Berger, W. and Elvers, H., *Z. Anal. Chem.* **171**, 185 (1959).
57. Shibata, S., *Anal. Chim. Acta* **23**, 367 (1960); **25**, 348 (1961).
58. Betteridge, D., Fernando, Q. and Freiser, H., *Anal. Chem.* **35**, 294 (1963).
59. Pollock, E. N. and Zopatti, L. P., *Anal. Chim. Acta* **28**, 68 (1963).
60. Marczenko, Z., Kasiura, K. and Krasiejko, M., *Mikrochim. Acta* **1969**, 625; *Chem. Anal. (Warsaw)* **14**, 1277 (1969).
60a. Krasiejko, M. and Marczenko, Z., *Mikrochim. Acta* **1975 I**, 585.
61. Donaldson, E. M. and Inman, W. R., *Talanta* **13**, 489 (1966).
62. Specker, H., Hartkamp, H. and Kuchtner, M., *Z. Anal. Chem.* **143**, 425 (1954).
63. Specker, H. and Hartkamp, H., *ibid.* **145**, 260 (1955).
64. Meyer, S. and Koch, O.G., *Mikrochim. Acta* **1958**, 744.
65. Blank, A. B., Bulgakova, A. M. and Sizonenko, N. T., *Zh. Analit. Khim.* **16**, 715 (1961).
66. Healy, W. B., *Anal. Chim. Acta* **34**, 238 (1966).
67. Dittel, F., *Z. Anal. Chem.* **228**, 412 (1967).
68. Beyer, W., Wawschinek, O. and Ott, R. D., *Mikrochim. Acta* **1967**, 233.
69. Miller, D. O. and Yoe, J. H., *Talanta* **7**, 107 (1960); *Anal. Chim. Acta* **26**, 224 (1962).
70. Dutta, R. L., *J. Indian Chem. Soc.* **34**, 311 (1957); **37**, 167 (1960).
71. Ksandr, Z. and Neuwirt, J., *Collection Czech. Chem. Commun.* **27**, 1381 (1962).
72. De, A. K. and Rahaman, S., *Anal. Chem.* **35**, 159 (1963).
73. Onishi, H. and Toita, Y., *Talanta* **11**, 1357 (1964).
74. Yoshida, H., Nagai, H. and Onishi, H., *ibid.* **13**, 37 (1966).
75. Akaiwa, H. and Kawamoto, H., *Japan Analyst* **16**, 359 (1967).
76. Onishi, H. and Toita, Y., *ibid.* **21**, 756 (1972).
77. Motojima, K., Hashitani, H. and Imahashi, T., *Anal. Chem.* **34**, 571 (1962).
78. Bankovskii, Yu. A., Ievin'sh, A. F. and Luksha, E. A., *Zh. Analit. Khim.* **14**, 222 (1959).
79. Akaiwa, H. and Kawamoto, H., *Anal. Chim. Acta* **40**, 407 (1968).
80. Marczenko, Z. and Mojski, M., *ibid.*, **54**, 469 (1971); *Chem. Anal. (Warsaw)* **16**, 865 (1971).
81. Minczewski, J. Chwastowska, J. and Lachowicz, E., *Chem. Anal. (Warsaw)* **18**, 199 (1973).
82. Ishii, H. and Einaga, H., *Japan Analyst* **15**, 1124 (1966).
83. Yotsuyanagi, T., Goto, K., Nagayama, M. and Aomura, K., *ibid.* **18**, 477 (1969).
84. Tataev, O. A. and Anisimova, L. G., *Zh. Analit. Khim.* **26**, 184 (1971).
85. Nonova, D. and Evtimova, B., *Talanta* **20**, 1347 (1973).
86. Gottschalk, G., *Z. Anal. Chem.* **211**, 344 (1965).

87. Rao, P. K. and Chowdhury, G. S., *ibid.* **246**, 19 (1969).
88. Schröder, H., *Metall* **9**, 100 (1955).
89. Knoeck, J. and Diehl, H., *Talanta* **14**, 1083 (1967).
90. Shevchuk, I. A. and Simonova, T. N., *Zh. Analit. Khim.* **23**, 1386 (1968).
91. Purdy, W. C. and Hume, D. N., *Anal. Chem.* **27**, 256 (1955).
92. Hartkamp, H., *Z. Anal. Chem.* **199**, 183 (1964).
93. Přibil, R. and Hornychová, E., *Chem. Listy* **44**, 101 (1950); *Collection Czech. Chem. Commun.* **15**, 456 (1950).
94. Nightingale, E. R., Jr., *Anal. Chem.* **31**, 146 (1959).
95. Chung, O. K. and Meloan, C. E., *ibid.* **39**, 525 (1967).
96. Malý, J. and Fadrus, H., *Analyst* **99**, 128 (1974).

Chapter 31

MERCURY

Mercury (Hg, at.wt. 200·59) occurs in its compounds in the $+I$ and $+II$ oxidation states. Mercury(II) is similar in its chemical properties to copper(II) and lead (II), whereas mercury(I) resembles silver and gold(I). Mercury(II) forms stable, water-soluble, halide complexes.

31.1 Methods of Separation

31.1.1 Extraction

The extraction of mercury as its yellow dithizonate is a popular separation method and is discussed in detail below. Extraction of mercury dithizonate from a strong acid medium (sulphuric acid is the most suitable) separates mercury from Cu, Bi, Zn, Ni, Pb, and other metals (except Au, Pt, and Pd). In the presence of chloride, mercury can also be separated from silver since chloride and HCl at moderate concentrations do not interfere with the extraction of mercury dithizonate. Mercury is separated from larger quantities of copper by extracting $Hg(HDz)_2$ from the aqueous solution with small portions of dithizone (in CCl_4 or $CHCl_3$) until the violet colour of $Cu(HDz)_2$ appears in the extract.

Mercury can be separated by the extraction of HgI_2 and HgI_3^- with cyclohexanone or aliphatic ketones from an acid medium [1]. Owing to the high stability of HgI_2 and HgI_3^-, an iodide:mercury molar ratio of 2–3 is sufficient.

From an EDTA–cyanide medium buffered with carbonate, mercury can be extracted into CCl_4 or $CHCl_3$ as the diethyldithiocarbamate [2]. Only bismuth, copper, and thallium(III) interfere seriously.

The optimum conditions for the separation of mercury by extraction of its halide complexes with TBP and other solvents have been investigated [3–5].

31.1.2 Precipitation

In the absence of larger quantities of other Analytical Group II metals, mercury can be separated by precipitation as the sulphide by saturating an acidic aqueous solution with hydrogen sulphide [6]; cadmium, arsenic, and copper are suitable scavengers. Traces of mercury are also quantitatively precipitated as the sulphide from a neutral or weakly alkaline medium; a suitable metal sulphide or hydroxide may serve as the collector [7].

Jackwerth et al. [8] have enriched traces of mercury in silver by collecting the Hg-EDTA complex on silver iodide.

Mercury (together with noble metals such as Ag, Au, and Pt) can be removed from solution by reduction to the metal with hydrazine, hydroxylamine, or stannous chloride. Mercury is also deposited by cementation with less noble metals (Cu, Zn, Fe) or by electrolysis at the platinum cathode [6]. Similarly, traces of mercury have been deposited on copper wire [9,10].

31.1.3 Volatilization and Other Methods

Stannous chloride reduces mercury to the volatile elementary form, which can then be distilled with steam or air as carrier gas and trapped in a solution of $KMnO_4$ [11,12]. Brookes and Solomon [13] have distilled mercury by igniting the sample in a quartz still. Mercury can be volatilized from organic and inorganic samples by mixing them with Na_2CO_3 and Na_2O_2 and heating at 600–800°C [14]. Traces of mercury vapour have been trapped in metallic gold [15].

Mercury can also be separated as volatile mercuric chloride, which is distilled either from concentrated sulphuric acid in a stream of chlorine or from a mixture of sulphuric and hydrochloric acids [16].

De and Majumdar [17] have separated mercury(II) from silver and lead retained on a Dowex-50 cation-exchanger by elution first of lead with $0.25M$ ammonium acetate, then of silver with $0.5M$ NH_3, and finally of mercury with $4M$ NH_3 solution. Fritz and Garralda [18] have separated the Hg, Bi, Cd, and Pb retained on a cation-exchanger by elution with 0.1–$0.6M$ HBr. Most other metals remain in the column.

Separation methods have been rewieved [18a].

31.2 Methods of Determination

The extractive spectrophotometric dithizone method has excellent sensitivity and selectivity, and is particularly useful in determining traces of mercury in biological materials and foodstuffs. Extractive methods using ion-pairs of anionic mercury(II) halide complexes and basic dyes are worthy of mention by virtue of their sensitivity.

31.2.1 Dithizone Method

Mercury ions in an acid medium react with excess of dithizone (formula, p. 41) to form the orange-yellow dithizonate, $Hg(HDz)_2$, which is soluble in CCl_4 or $CHCl_3$, and which is the basis of this spectrophotometric method [19–21]. The molar absorptivity at $\lambda_{max} = 485$ nm is 7.1×10^4 ($a = 0.35$). With a deficiency of dithizone, the violet dithizonate, HgDz, is formed in a neutral or alkaline medium [22,23].

Mercury dithizonate is readily formed even when dithizone is shaken with $5M$ H_2SO_4 containing mercury. Concentrations of HCl greater than

$1M$ impede the formation of mercury dithizonate, owing to the formation of stable mercury chloride complexes. The $Hg(HDz)_2$ formed is resistant to the action of dilute alkalis (e.g. $2M$ NaOH) which are used to wash free dithizone from the extract. A carbon tetrachloride solution of $Hg(HDz)_2$ is sensitive to light, its colour changing through brown to greenish-blue. The orange-yellow colour is slowly restored when the extract is left in the dark, or when it is shaken with dilute sulphuric acid. When the organic phase is shaken with dilute acetic acid, which dissolves to some extent in CCl_4, the yellow colour of $Hg(HDz)_2$ is stabilized.

Briscoe and Cooksey [24] have pointed out that, in dilute hydrochloric acid containing a large excess of mercury(II) in relation to dithizone, the yellow mixed HgClHDz complex is formed. This complex reacts with dithizone to give $Hg(HDz)_2$.

Dithizone (in CCl_4) extracts, along with mercury, Pt(II), Au, Pd, Ag, Po and Cu from strongly acidic solutions (pH $\leqslant$ 0). Bismuth and other metals are not extracted at this acidity, even though they are present in large amounts relative to mercury.

Copper is removed by masking the mercury as strong bromide or iodide complexes in acid media. Only copper dithizonate is extracted from $1M$ HCl containing KI. The pH of the solution is then increased, weakening the mercury halide complexes and allowing the extraction of mercury with dithizone. Alternatively, the mercury and copper can be extracted from acid solution, and the mercury stripped with an acidic solution of KI or KBr. The mercury is re-extracted from the aqueous phase after the addition of ammonia.

A convenient way of determining mercury in the presence of copper by selective extraction takes advantage of the fact that the rate of formation of mercury dithizonate is considerably higher than that of copper dithizonate. The aqueous solution (pH $\sim$ 0) is shaken with small portions of dithizone in CCl_4 until the final portion is either green or has the violet colour of copper dithizonate.

From a slightly acidic medium (pH $\sim$ 4) containing EDTA, mercury can be extracted in the presence of Cu, Bi, Ni, Zn, and Pb. Silver is removed either by masking with chloride or thiocyanate, or by shaking the CCl_4 extract containing Hg and Ag dithizonates with $1M$ HCl for 20 seconds, the silver passing quantitatively into the aqueous phase while all the mercury remains in the extract [7].

Interference by palladium is prevented by prior separation with dimethylglyoxime either as extractant or as precipitant. In the selective extraction method, palladium does not interfere since its dithizonate is formed rather slowly [$Pd(HDz)_2$ is grey-green]. Before mercury is determined, gold and platinum should also be separated, e.g. by selective reduction to the elements.

Free dithizone is stripped from the extract with either dilute ammonia or $0.2M$ NaOH. Subsequently, the absorbance of the yellow-orange

Hg(HDz)$_2$ solution, or of the corresponding green dithizone solution after the Hg(HDz)$_2$ has been decomposed with aqueous KI, is measured. Sequential reversion can be used to determine mercury after bismuth and silver [25].

Organomercury(II) ions (e.g. phenyl and methyl mercury) also react with dithizone and have been determined by this method [26–29], which can also be used to determine mercury(II) in the presence of organomercury(II) compounds [28,30].

The dithizone method is widely used in the spectrophotometric determination of mercury in biological materials [31–34], especially in urine [32,35–37] and in foodstuffs [38–42a]. Mercury has also been determined with dithizone in coal and its products [43,44], sulphide minerals [45], ores [46], uranium compounds [47], tin [7], silver [8], selenium [48], cadmium [48a], Hg–Cd–Te thin films [49], alkalis [11], industrial wastes [50], and air [51].

Di-2-naphthylthiocarbazone [34,52,53], an analogue of dithizone, has also been applied in the determination of mercury.

Reagents

1. Dithizone: 0·001% solution in CCl$_4$ (p. 492).
2. Standard mercury solution: 1 mg/ml. Dissolve 1·713 g of Hg(NO$_3$)$_2$·H$_2$O in water containing 1 ml of conc. HNO$_3$, and dilute the solution with water in a volumetric flask to 1 litre. The stability of dilute solutions of mercury has been discussed by Toribara *et al.* [54], and Feldman [55].
3. Acetic acid: 2M ($\sim$ 10%) solution. Purify the solution by shaking with a solution of dithizone in CCl$_4$.
4. Potassium iodide solution: pH $\sim$ 4. Dissolve 15 g of KI and 5 g of potassium hydrogen phthalate, and dilute with water to 250 ml. Add 0·1M thiosulphate solution dropwise to remove free iodine. Remove traces of metals from the solution by shaking with dithizone in CCl$_4$.

Procedure

Add HCl to the solution, which should be free from Au, Pt, and Pd, and contain not more than 50 µg of Hg, to make the HCl concentration $\sim$ 1M, and extract mercury in a separating funnel with small portions of the dithizone solution (1 ml of 0·001% H$_2$Dz is equivalent to 3·9 µg of Hg). The last portion of dithizone should not change its green colour unless copper is present, in which case it may take on a violet colour as the copper dithizonate starts to form.

Remove free dithizone from the combined extracts by shaking with dilute NH$_3$ solution (2 drops of conc. NH$_3$ solution in 25 ml of water), then shake the extract with dilute CH$_3$COOH. Dilute the yellow-orange Hg(HDz)$_2$ solution with carbon tetrachloride in a 50-ml (or smaller) volumetric flask, and measure the absorbance of the solution at 485 nm against carbon tetrachloride.

Notes. (1) To avoid interference by possible copper contamination, measure the absorbance of the extract at 620 nm (λ_{max} for dithizone), shake the extract with KI solution buffered at pH~4 to decompose Hg(HDz)$_2$ and liberate the equivalent amount of dithizone, and remeasure the absorbance of the extract at 620 nm. The difference in the absorbance values correponds to the mercury content.

(2) Mercury is more rapidly extracted with dithizone from HNO$_3$ or H$_2$SO$_4$ media than from HCl media.

31.2.2 Other Methods

Mercury(II) forms anionic complexes with I$^-$, Br$^-$, and Cl$^-$ ions which react with basic dyes to give ion-association compounds which are extractable into organic solvents. Sensitive extractive photometric methods for determining mercury are based on such reactions with the following dyes: Rhodamine B [56], Crystal Violet [57,58], Methylene Blue [59], Victoria Blue B [60], Bindschedler's Green [61], Methyl Green [62], Brilliant Green [63], various antipyrine dyes [64], and Variamine Blue B [65]. Dichloroethane [61], benzene [58], toluene [57], chloroform [59], and nitrobenzene [65] are used as extractants. The cationic iron(II) complex with 2,2'-bipyridyl forms ion-association complexes with [HgI$_4$]$^{2-}$ or [HgBr$_4$]$^{2-}$ which have also been utilized in the determination of mercury [66,67].

A number of organic sulphur compounds are recommended as spectrophotometric reagents for mercury. These include *p*-dimethylaminobenzylidenerhodanine (rhodanine) [6], thio-Michler's ketone (the molar absorptivity of the mercury complex in isoamyl alcohol is 8.8×10^4 at 550 nm) [68], thiothenoyltrifluoroacetone [69,70], thiodibenzoylmethane [71], diethyldithiocarbamate of copper(II) [72,73], and Mercupral [copper(II) complex with tetraethylthiuramdisulphide] [74–76]. The last two reagents give indirect methods in which the coloured organic solution of the reagent is decolorized when shaken with an aqueous solution containing mercury(II), owing to displacement of copper from its complexes by mercury.

Azo dyes have also been suggested for the spectrophotometric determination of mercury, e.g. Cadion [77], Zincon [78], PAR [79], Sulpharsazen [80], Azoxin-H and Azoxin-C ($\varepsilon = 4.38 \times 10^4$ at 540 nm) [81], *o*-(2-thiazolylazo)-4-methoxyphenol (TAMP) [82], 5-nitro-2-furfurylidenesemicarbazone [83], and 2-antipyrylazo-5-diethylaminophenol [84].

Other organic reagents used in spectrophotometric methods for determining mercury are diphenylcarbazone [34,85–87], Methylthymol Blue [88], Xylenol Orange [89], and Metalphthalein (*o*-Cresolphthalein Complexone) [90].

Mercury and its compounds present in air have been determined by absorption by iodinated active carbon. Subsequently, the mercury vapour is desorbed by igniting the carbon, and is passed through a filter paper impregnated with selenium sulphide [91]. The mercury content is determined from the degree of darkening of the stain.

Mercury(II) displaces iron(II) from ferrocyanide ions, thus giving a colour reaction with either 2,2'-bipyridyl or 1,10-phenanthroline present in the solution. This underlies methods for the indirect determination of mercury [92, 93].

References

1. Jackwerth, E. and Specker, H., *Z. Anal. Chem.* **167**, 269 (1959).
2. Hakkila, E. A. and Waterbury, G. R., *Anal. Chem.* **32**, 1340 (1960).
3. Morris, D. F. and Williams, J. H., *Talanta* **9**, 623 (1962).
4. Specker, H., Jackwerth, E. and Kloppenburg, H. G., *Z. Anal. Chem.* **183**, 81 (1961).
5. Gupta, C. B. and Tandon, S. N., *J. Radioanal. Chem.* **11**, 59 (1972).
6. Jangg, G., *Z. Anal. Chem.* **183**, 255 (1961).
7. Marczenko, Z. and Kasiura, K., *Chem. Anal. (Warsaw)* **10**, 449 (1965).
8. Jackwerth, E., Döring, E. and Lohmar, J., *Z. Anal. Chem.* **253**, 195 (1971).
9. Brandenbenger, H. and Bader, H., *Helv. Chim. Acta* **50**, 1409 (1967).
10. Schaller, K. H., Strasser, P., Woitowitz, R. and Szadkowski, D., *Z. Anal. Chem.* **256**, 123 (1971).
11. Kemula, W., Brachaczek, W. and Hulanicki, A., *Chem. Anal. (Warsaw)* **5**, 215 (1960).
12. Kimura, Y. and Miller, V. L., *Anal. Chim. Acta* **27**, 325 (1962).
13. Brookes, H. E. and Solomon, L. E., *Analyst* **84**, 622 (1959).
14. Jerie, H., *Mikrochim. Acta* **1970**, 1089.
15. Anderson, D. H., Evans, J. H., Murphy, J. J. and White, W. W., *Anal. Chem.* **43**, 1511 (1971).
16. Ensslin, F., Dreyer, H. and Lessmann, O., *Z. Anal. Chem.* **149**, 25 (1956).
17. De, A. K. and Majumdar, S. K., *Talanta* **10**, 201 (1963).
18. Fritz, J. S. and Garralda, B. B., *Anal. Chem.* **34**, 102 (1962).
18a. Chilov, S., *Talanta* **22**, 205 (1975).
19. Melles, J. L. and de Bree, W., *Rec. Trav. Chim.* **72**, 576 (1953).
20. Friedeberg, H., *Anal. Chem.* **27**, 305 (1955).
21. Irving, H. M., Andrew, G. and Risdon, E. J., *J. Chem. Soc.* **1949**, 541.
22. Bréant, M., *Bull. Soc. Chim. France* **1956**, 948.
23. Nowicka-Jankowska, T. and Irving, H. M., *Anal. Chim. Acta* **54**, 489 (1971).
24. Briscoe, G. B. and Cooksey, B. G., *J. Chem. Soc. (A)* **1969**, 205.
25. Chalmers, R. A. and Dick, D. M., *Anal. Chim. Acta* **32**, 117 (1965).
26. Miller, V. L., Lillis, D. and Csonka, E., *Anal. Chem.* **30**, 1705 (1958).
27. Irving, H. M. and Cox, J. J., *J. Chem. Soc.* **1963**, 466.
28. Kiwan, A. M. and Fonda, M. F., *Anal. Chim. Acta* **40**, 517 (1968).
29. Ingman, F., *Talanta* **18**, 744 (1971).
30. Gorgia, A. and Monnier, D., *Anal. Chim. Acta* **55**, 247 (1971).
31. Barrett, F. R., *Analyst* **81**, 294 (1956).
32. Hintzsche, E., *Chem. Techn.* **8**, 670 (1956).
33. Wanntorp, H. and Dyfverman, A., *Arkiv Kemi* **9**, 7 (1956).
34. Vignoli, L., Badre, R., Morel, M. C. and Ardorino, J., *Chim. Anal. (Paris)* **45**, 53 (1963).
35. Gray, D. J., *Analyst* **77**, 436 (1952).
36. Rolfe, A. C., Russell, F. R. and Wilkinson, N. T., *ibid.*, **80**, 523 (1955).
37. Miller, V. L. and Swanberg, F., Jr., *Anal. Chem.* **29**, 391 (1957).
38. Abbott, D. C. and Johnson, E. I., *Analyst* **82**, 206 (1957).
39. Joint Mercury Residues Panel, *ibid.* **86**, 608 (1961).
40. Hordyńska, S., Legatowa, B. and Bernstein, I., *Chem. Anal. (Warsaw)* **7**, 567 (1962).
41. Analytical Methods Committee, *Analyst* **85**, 643 (1960); **90**, 515 (1965).

42. Fujita, M., Takeda, Y., Terao, T., Hoshino, O. and Ukita, T., *Anal. Chem.* **40**, 2042 (1968).
42a. Nabrzyski, M., *ibid.* **45**, 2438 (1973).
43. Vasilevskaya, A. E., Shcherbakov, V. P. and Karakozova, E. V., *Zh. Analit. Khim.* **19**, 1200 (1964).
44. Shcherbakov, W. P. and Vasilevskaya, A. E., *ibid.* **19**, 308 (1964).
45. Leong, P. C. and Ong, H. P., *Anal. Chem.* **43**, 940 (1971).
46. Kroužek, E. and Povondra, P., *Chem. Listy* **52**, 1825 (1958).
47. Mareček, J. and Singer, E., *Z. Anal. Chem.* **203**, 336 (1964).
48. Pollock, E. N., *Talanta* **11**, 1548 (1964).
48a. Krasiejko, M. and Marczenko, Z., *Mikrochim. Acta* **1975 I**, 585.
49. Marczenko, Z., Mojski, M. and Czarnecka, I., *Chem. Anal. (Warsaw)* **18**, 189 (1973).
50. Joint A.B.C.M.—S.A.C. Committee, *Analyst* **81**, 176 (1956).
51. Drew, R. G. and King, E., *ibid.* **82**, 461 (1957).
52. Truhaut, R. and Boudène, C., *Bull. Soc. Chim. France* **1959**, 1850.
53. Tiptsova, V. G., Andreichuk, A. M. and Bazhanova, L. A., *Zh. Analit. Khim.* **20**, 1200 (1965).
54. Toribara, T. Y., Shields, C. P. and Koval, L., *Talanta* **17**, 1025 (1970).
55. Feldman, C., *Anal. Chem.* **46**, 99 (1974).
56. Imai, H., *J. Chem. Soc. Japan, Pure Chem. Sect.* **90**, 275 (1969).
57. Kothny, E. L., *Analyst* **94**, 198 (1969).
58. Blyum, I. A., Brushtein, N. A. and Oparina, L. I., *Zh. Analit. Khim.* **26**, 48 (1971).
59. Ganchev, N. and Atanasova, B. V., *Compt. Rend. Acad. Bulg. Sci.* **21**, 359 (1968).
60. Pilipenko, A. T., Kish, P. P. and Vitenko, G. M., *Ukr. Khim. Zh.* **38**, 481 (1972).
61. Tsubouchi, M., *Anal. Chem.* **42**, 1087 (1970).
62. Tarayan, V. M., Ovsepyan, E. N. and Lebedeva, S. P., *Zh. Analit. Khim.* **26**, 1745 (1971).
63. Tarayan, V. M., Ovsepyan, E. N. and Karimyan, N. S., *Izv. Vyssh. Ucheb. Zaved., Khim. Khim. Tekhnol.* **16**, 358 (1973).
64. Busev, A. I. and Khintibidze, L. S., *Zh. Analit. Khim.* **22**, 694, 857 (1967).
65. Tsubouchi, M., *Bull. Chem. Soc. Japan* **43**, 2812 (1970).
66. Kotsuji, K., *ibid.* **38**, 402 (1965).
67. Yamamoto, Y., Kikuchi, S., Hayashi, Y. and Kumamaru, T., *Japan Analyst* **16**, 931 (1967).
68. Cheng, K. L. and Goydish, B. L., *Microchem. J.* **10**, 158 (1966).
69. Hashitani, H. and Katsuyama, K., *Japan Analyst* **19**, 355 (1970).
70. Solanke, K. R., and Khopkar, S. M., *Indian J. Chem.* **11**, 485 (1973).
71. Uhlemann, E. and Schuknecht, B., *Anal. Chim. Acta* **69**, 79 (1974).
72. Vašak, V. and Šedivec, V., *Chem. Listy* **45**, 437 (1951).
73. Tertoolen, J. W., Buijze, C. and van Kolmeschate, G. J., *Chemist-Analyst* **52**, 100 (1963).
74. Michal, J. and Zýka, J., *Collection Czech. Chem. Commun.* **22**, 1135 (1957).
75. Michal, J., Pavlíková, E. and Zýka, J., *Z. Anal. Chem.* **159**, 321 (1958).
76. Gershuns, A. L., Vail', E. I., Mirnaya, A. P., Rastrepina, I. A. and Sigalova, L. V., *Zavodsk. Lab.* **27**, 1465 (1961).
77. Chavanne, P. and Geronimi, C., *Anal. Chim. Acta* **19**, 442 (1958).
78. Morris, A. G., *Analyst* **82**, 34 (1957).
79. Ueda, J., *J. Chem. Soc. Japan, Pure Chem. Sect.* **92**, 418 (1971).
80. Lukin, A. M., Smirnova, K. A. and Petrova, G. S., *Tr. Vses. Nauchno-issled. Inst. Khim. Reakt.* **1966**, 290; *Anal. Abstr.* **15**, 3203 (1968).
81. Cherkesov, A. I., Tonkoshkurov, V. S., Postoronko, A. I. and Ryzhov, V. N., *Zh. Analit. Khim.* **25**, 466 (1970).
82. Kai, F., *Anal. Chim. Acta* **44**, 242 (1969).
83. Bagdasarov, K. N., Anisimova, L. G. and Tataev, O. A., *Zh. Analit. Khim.* **23**, 1002 (1968); *Zavodsk. Lab.* **34**, 390 (1968).

84. Kolosova, I. V., *Izv. Vyssh. Ucheb. Zaved, Khim. Khim. Tekhnol.* **12**, 1329 (1969).
85. Kemula, W. and Janowski, A., *Chem. Anal. (Warsaw)* **3**, 587 (1958).
86. Balt, S. and van Dalen, E., *Anal. Chim. Acta* **27**, 416 (1962).
87. Okutani, T., *Bull. Chem. Soc. Japan* **41**, 1728 (1968).
88. Iritani, N. and Miyahara, T., *Japan Analyst* **12**, 1183 (1963).
89. Cabrera-Martin, A., Peral-Fernández, J. L., Vicente-Pérez, S. and Burriel-Marti, *Talanta* **16**, 1023 (1969).
90. Komatsu, S. and Nomura, T., *J. Chem. Soc. Japan, Pure Chem. Sect.* **88**, 542 (1967).
91. Sergeant, G. A., Dixon, B. E. and Lidzey, R. G., *Analyst* **82**, 27 (1957).
92. Karas, V. and Pinter, T., *Croat. Chem. Acta* **30**, 141 (1958).
93. Prasał, Z., *Chem. Anal. (Warsaw)* **7**, 617 (1962).

Chapter 32

MOLYBDENUM

Molybdenum (Mo, at.wt. 95·94) is an amphoteric element with predominantly acidic properties. Molybdenum occurs principally in the +VI oxidation state as molybdate (MoO_4^{2-}) ions which form condensed species in acid media. In strongly acidic solutions, molybdenyl cations MoO_2^{2+} occur. Depending on the reducing agent used and on the reduction conditions, molybdenum(VI) is reduced to Mo(V) or Mo(III). Molybdenum(VI) gives fluoride, peroxide, citrate, oxalate, and chloride complexes, as well as forming heteropoly acids with Si, P, V, As(V), Ge, etc. White MoO_3 is volatile at above 550°C.

32.1 Methods of Separation

32.1.1 Precipitation

Hydrogen sulphide precipitates molybdenum from acid medium as MoS_3; antimony(V), arsenic, or copper (2–4 mg) are suitable collectors. Molybdenum is separated in this way from Cr, V, Ti, U, etc. In the presence of tartaric acid, it is also possible to separate molybdenum from tungsten and niobium [1, 2]. Thioacetamide may be used in place of hydrogen sulphide [3, 4].

In the presence of EDTA, 8-hydroxyquinoline precipitates molybdenum from an acetate-buffered medium while Fe(III), Al, Zn, Ni, Co, Mn, Pb, Cd, Bi, and Cu remain in solution [5].

Traces of molybdenum have been coprecipitated as the cupferronate in the presence of iron [6], with thorium hydroxide [7], and with lead phosphate or arsenate [8] as collector.

During the precipitation of hydrous MnO_2 from dilute HNO_3 (by treating Mn^{2+} with MnO_4^-), traces of molybdenum(VI) are coprecipitated together with Sb, Sn, Bi, Tl(III), and Au [9].

Traces of molybdenum have also been separated with organic scavengers such as Methylene Blue [10], and a mixture of Methyl Violet and tannin [11].

Molybdenum can be separated from Fe, Ti, and certain other metals by precipitation (double precipitation if necessary) with excess of NaOH. When an Na_2CO_3 melt is leached with water, molybdenum is one of the metals which pass into solution.

32.1.2 EXTRACTION

Extraction into chloroform of the α-benzoinoximate, $MoO_2(C_{14}H_{12}O_2N)_2$, from acid medium (0·01–2M HCl) is a selective method for separating molybdenum [12–17].

$$\langle\!\!\!\bigcirc\!\!\!\rangle\!-\!\underset{\underset{OH}{|}}{CH}\!-\!\underset{\underset{NOH}{||}}{C}\!-\!\langle\!\!\!\bigcirc\!\!\!\rangle \quad (I)$$

Under the conditions employed in the extraction of the molybdenum α-benzoinoxime (I) (cupron) complex, only W, V, Cr, and Pd are also extracted. However, in the presence of phosphoric acid and iron(II), tungsten and vanadium are not extracted [14].

From 6–7M HCl media, molybdenum is separated by extraction of the chloride complex into diethyl ether, MIBK, or isoamyl alcohol [18–22]. Fe(III), Au, As(III), Ga, Sb(V), Ge, Ti(III), and Sn(IV) are extensively or completely extracted, but tungsten is not extracted in the presence of phosphoric acid [21].

Yatirajam and Ram separate molybdenum from many elements by extraction of its thiosulphate complex [22a] or of phosphomolybdenum blue [22b].

From acid solutions, small amounts of molybdenum have been separated by extraction in the form of chelates with cupferron [23–25], BPHA [25], acetylacetone [26], TTA [27,28], DDTC [29], mono-2-ethylhexylphosphoric acid [30], di-n-butylphosphoric acid [31], and xanthate [31a].

The extraction of molybdenum as a dithiol or thiocyanate complex is discussed later in connection with its determination.

Busev and Rodionova [32] have made a comprehensive survey of extraction methods for separating molybdenum.

32.1.3 ION-EXCHANGE

Hall and Bryson [33] have separated Mo, Cr, and V from other components in steel by passing an acetate solution (pH 2·5–3) through a strongly basic anion-exchanger. The following elements are successively eluted from the column: vanadium with 0·6M NaOH, chromium with 8M HCl, and molybdenum with 1M HCl.

Molybdenum(VI), W(VI), Ti, Zr, Nb, and Ta are retained as anionic fluoride complexes in anion-exchange columns [34,35]. The individual metals are selectively eluted with media containing HCl, HF, NH_4Cl, or NH_4F.

After retention (as MoO_2^{2+}) with other cations on a cation-exchanger, molybdenum can be eluted with hydrogen peroxide solution [10,36], while most of the other metals remain in the column. Molybdenum has been separated from several other metals by elution of its anionic citrate complex from a cation-exchanger [37].

32.2 Methods of Determination

Many different organic compounds have been recommended as spectrophotometric reagents for the determination of molybdenum. The two methods which are discussed here in more detail, i.e. the classical thiocyanate method and the more sensitive and selective dithiol method, are usually used in conjunction with extraction.

32.2.1 Dithiol Method

Dithiol (toluene-3,4-dithiol) (II) reacts with molybdenum(VI) in strongly acidic media (4–12M HCl, 3–7M H$_2$SO$_4$) to form a green, sparingly-soluble complex, which can be extracted with both polar (esters) and non-polar (petroleum ether, benzene, carbon tetrachloride) organic solvents. According to Gilbert and Sandell [38], the composition of the molybdenum dithiol complex is III.

The presence of iron(II) in the solution promotes the reaction between dithiol and molybdenum and also increases the extraction of the dithiolate.

Solid dithiol (m.p. 31°C) and its solutions are unstable since they are readily oxidized by atmospheric oxygen [39]. Dithiol is stored in sealed glass ampoules. Its aqueous solutions are stabilized by adding a reducing agent, e.g. thioglycollic or ascorbic acid, and by storage at low temperatures (in a refrigerator). An aqueous suspension of stable zinc dithiolate can be used alternatively as the reagent [40,41].

Green solutions of molybdenum(VI) dithiolate in organic solvents (usually amyl, isoamyl, or butyl acetate) are the basis of this spectrophotometric determination of molybdenum [23,42,43]. The complex exhibits two absorption maxima in the visible spectrum (Fig. 32.1), the less intense maximum being at 440 nm and the more intense at 675 nm. The molar absorptivity at 675 nm is $2 \cdot 1 \times 10^4$ ($a = 0 \cdot 22$).

Granger [44] obviated the necessity for the extraction by using an aqueous butanol medium, in which both molybdenum dithiolate and dithiol are soluble. Oelschläger [45] stabilized the molybdenum dithiolate suspension in the aqueous phase with a protective colloid.

Tungsten reacts with dithiol in a similar manner to molybdenum. Hence, when molybdenum is determined in the presence of tungsten, the latter is masked by citric or tartaric acid [43]. In his analysis of tungsten ores, Jeffery [12] has determined molybdenum as the dithiolate, and simultaneously corrected for the extracted tungsten by measuring the absorbance of the extract at 680 nm and 630 nm.

In the determination of small amounts of molybdenum in tungsten and its compounds, Buss et al. [4,46] have precipitated MoS$_3$, under

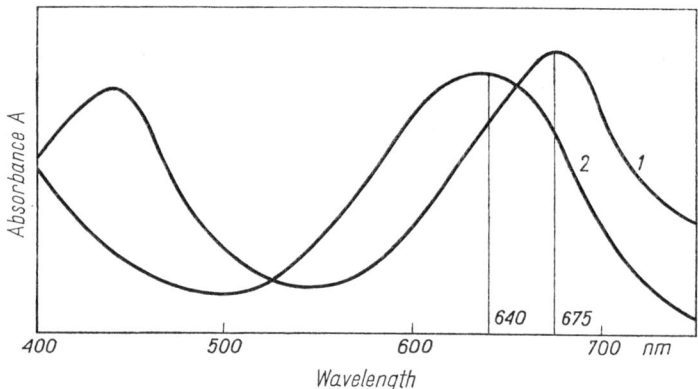

Fig. 32.1 Absorption spectra of dithiol complexes of molybdenum (1) and of tungsten (2) in amyl acetate

pressure and in the presence of tartaric acid, with thioacetamide as precipitant and antimony as collector.

Tin(II) dithiolate is also extracted with amyl acetate; however, this dithiolate does not interfere as it absorbs only slightly and is different in colour from molybdenum dithiolate.

Iron(III), which interferes in the determination of molybdenum, is reduced to Fe(II) with iodide, the iodine liberated being reduced with thiosulphate.

Most heavy metals form sparingly soluble dithiolates which are not, however, extracted with molybdenum dithiolate [46a].

Oxidants which decompose dithiol interfere in the determination of molybdenum.

The dithiol method has been used to determine molybdenum in tungsten and its compounds [4,35,46], vanadium and its compounds [47], copper [48], uranium [44], niobium [44,49,50], tantalum [50], titanium dioxide [51], silicate rocks [9,12,43,52], ores [12,40], soils [41,43,53], biological materials [9,45,54], milk [55], and water [9,56,57].

Reagents

1. Dithiol: 0·2% solution. Dissolve 0·2 g of dithiol and 0·2 g of ascorbic acid in 100 ml of 0·2M NaOH. The solution is unstable and must be replaced after 2 days.
2. Standard molybdenum solution: 1 mg/ml.
    a. Dissolve 1·500 g of the trioxide, MoO_3, in 25 ml of 2M NaOH, acidify the solution slightly with hydrochloric acid, and dilute with water in a volumetric flask to 1 litre.
    b. Dissolve 2·522 g of $Na_2MoO_4.2H_2O$ in water containing 1 ml of conc. HCl, and dilute the solution with water in a volumetric flask to 1 litre.

Procedure

To a weakly acidic sample solution (~20 ml) containing not more than 100 μg of Mo, add 10 ml of conc. HCl and 5 ml of 20% KI solution. Mix, and let the solution stand for 5 minutes, then add $0.1M$ potassium thiosulphate dropwise till the solution is decolorized. Add 5 ml of 20% citric acid solution and 5 ml of dithiol solution. After 10 minutes, shake the aqueous solution for 30 seconds with two 15-ml portions of amyl acetate. Dilute the green extract with the solvent to 50 ml in a volumetric flask, and measure the absorbance of the solution at 675 nm, using the solvent as reference.

Note. If the initial solution contains no iron, ~5 mg of iron should be added.

32.2.2 Thiocyanate Method

Molybdenum(VI) reacts in acid media (HCl, H_2SO_4, $HClO_4$) with thiocyanate ions in the presence of reducing agents to give an orange-red colour [58,59]. The system may contain several molybdenum(V) complexes [60]. Perrin [61] has determined the Mo:SCN ratio as 1:3 in the coloured complex (aqueous acetone medium), and assigned the complex the formula $MoO(SCN)_3$. The 1:2 complex is yellow, and a colourless 1:1 complex has been postulated. Since the mechanism of the colour reaction is complicated, it is essential to keep the reaction conditions constant if good reproducibility is to be obtained. The concentrations of thiocyanate, reducing agent, and acid are critical.

The most commonly used reducing agent is stannous chloride. With this reductant, the presence of iron(III) in the sample solution is indispensable. In the absence of iron the results obtained are low and of poor reproducibility. Ascorbic acid [17,62], hydrazine, thiourea [63], and potassium iodide in the presence of copper(I) [64] are also used for the reduction of molybdenum(VI). With these weaker reducing agents more intensely coloured solutions are obtained. Stannous chloride is a strong reducing agent and is believed to reduce some of the molybdenum(VI) to an oxidation state lower than Mo(V). This molybdenum does not participate in the colour reaction.

Thiocyanate itself reduces molybdenum(VI) under the catalytic influence of traces of copper: the colour is developed within 3–5 minutes in 2.5–$3M$ HCl [65]. Under these conditions, no colour reaction is given by tungsten. Photochemical reduction of molybdenum with thiocyanate in sunlight has also been exploited [66].

The absorbance of the coloured aqueous or aqueous–acetone solution (the acetone can also serve as a reducing agent) [58] may be measured directly, or alternatively, the absorbance of a suitable organic extract (usually isoamyl alcohol, MIBK [67], or isopropyl ether) is measured. Extraction can be carried out conveniently with a denser-than-water solution of isoamyl alcohol and carbon tetrachloride $(1+1)$ [68]. The molybdenum thiocyanate complex can also be extracted into chloroform in the presence of organic bases [69–71]. The anionic thiocyanate complex forms an ion-association complex with the basic dye Crystal Violet, providing a sensitive method for determining molybdenum ($\varepsilon = 2\cdot3 \times 10^5$) [71a].

The molar absorptivity of the thiocyanate complex in isoamyl alcohol (absorption spectrum shown in Fig. 57.1, p. 570) is $2\cdot0 \times 10^4$ (specific absorptivity 0·21) at 470 nm.

Iron(III) does not interfere in the determination of molybdenum since, under the reaction conditions, it is reduced to Fe(II). Tungsten is masked with citric or tartaric acid, and titanium is masked with fluoride. Larger quantities of Re, U, V, Co, Cu, and Bi interfere. Before its determination by the thiocyanate method, molybdenum is often separated from other elements by extraction or precipitation.

Molybdenum has been determined by the thiocyanate method in the following: iron and steels [16,27,59,72–73], tungsten and its compounds [16,62,64], niobium and tantalum [16], lead [8], copper ores [17], aluminium alloys [74], uranium compounds [75,75a], beryllium [76], rhenium compounds [62], minerals and ores [28, 65], rocks [76a], soils [6,77], plant material [68], and sea water [7].

The thiocyanate method has also been used in a differential procedure for determining larger quantities of molybdenum in tungsten alloys [78] and in niobium alloys [79], as well as in the automatic determination of molybdenum in steel [80].

Reagents

1. Potassium thiocyanate: 10% solution.
2. Standard molybdenum solution: 1 mg/ml (p. 361).
3. Stannous chloride: 10% $SnCl_2.2H_2O$ solution in HCl (1+9).
4. Ferric alum: 0·5% solution in $0\cdot5M$ H_2SO_4.
5. α-Benzoinoxime: 0·2% solution in chloroform.

Procedure

Separation of Mo with α-benzoinoxime. Shake the sample solution ($\sim 1M$ in HCl) in a separating funnel for 2 minutes with two portions of α-benzoinoxime solution (the organic phase should be 1/3–1/4 the volume of the aqueous phase). Wash the combined extracts with $\sim 1\cdot5M$ HCl. Add 0·5 ml of conc. H_2SO_4 to the organic extract in a beaker, and evaporate off the chloroform. Mineralize the organic matter by adding conc. HNO_3 dropwise and heating to fumes. Dilute the cooled residue with 10–20 ml of water, heat to boiling, and then cool the sample. Depending on the amount of molybdenum, either all or an aliquot of this solution is used in the subsequent determination.

Determination of Mo. Add 30 ml of weakly acidic sample solution (HCl, H_2SO_4), containing not more than 100 μg of Mo, to a separating funnel.

Add successively 5 ml of conc. HCl, 5 ml of 20% citric acid solution, 2 ml of iron(III) solution, 5 ml of thiocyanate solution, and, with swirling, 5 ml of stannous chloride solution. After 5 minutes, extract the solution with two portions of isoamyl alcohol. Make up the combined extracts to the mark with the solvent in a 50-ml (or smaller) volumetric flask, and measure the absorbance at 470 nm, using the solvent as reference.

32.2.3 Other Methods

Apart from dithiol and thiocyanate, several other compounds with sulphur as the ligand atom are used in spectrophotometric methods for determining molybdenum.

Thioglycollic acid (mercaptoacetic acid, thioacetic acid) (**IV**) reacts with molybdenum to form a yellow complex [81,82] which can be extracted

$$\begin{array}{c} H_2C-C-OH \\ | \quad \| \\ SH \quad O \end{array} \quad (IV) \qquad \begin{array}{c} CH_2-CH-CH_2 \\ | \quad \quad | \quad \quad | \\ SH \quad SH \quad SO_3H \end{array} \quad (V)$$

as an ion-association complex with tributylamine [83] or trioctylamine [84]. Busev et al. [85,86] have extracted the complex with various solvents in the presence of diphenylguanidine and its derivatives.

8-Mercaptoquinoline (thio-oxine) (formula, p. 58), an analogue of 8-hydroxyquinoline, forms the coloured $MoO_2(C_9H_6NS)_2$ complex which can be extracted from acid media into toluene or xylene [87–89].

In other methods, the following organic sulphur reagents have been used: various dithiocarbamates [90–92], mercaptopropionic acid and its derivatives [93,94], unithiol (**V**) [95,96], thiomalic acid [97,98], N,N'-diphenylthiocarbamohydroxamic acid [99], 2-amino-4-chlorobenzenethiol [100], and potassium ethyl xanthate [100a].

Spectrophotometric reagents for molybdenum which contain oxygen as a ligand atom (hydroxyl or carbonyl groups) include Tiron [101,102], catechol [103], gallic acid [104], pyrogallolsulphonic acid [105], chromotropic acid [106], chloranilic acid [20], pyrazine-2,3-dicarboxylic acid and pyridine derivatives [107], 2-methyl-3-hydroxy-γ-pyrone [108], and 6,7-dihydroxy-2,4-diphenylbenzopyrylium chloride ($\varepsilon = 5.04 \times 10^4$) [109].

Included in the same group of reagents are the following flavone derivatives: morin [110], quercetin [111], and dihydroquercetin [112], together with fluorone derivatives such as phenylfluorone [113], o-nitrophenylfluorone ($\varepsilon = 5.6 \times 10^4$ at 584 nm) [29,114], p-nitrophenylfluorone [115], methylfluorone [116], salicylfluorone [117], and other trihydroxyfluorones [118,119].

Various azo dyes have also been employed in the spectrophotometric determination of molybdenum, e.g. Stilbazogall I [120], Rezarson [121], Magneson IREA [122], Solochrome Violet R [123], Sulphonitrazo [124], Sulphochlorophenol S [125], Sulphonitrophenol K (formula, p. 595) ($\varepsilon = 5.0 \times 10^4$) [126], PAR ($\varepsilon = 2.74 \times 10^4$ at 530 nm) [127], and TAR [128].

Certain triphenylmethane dyes have been used similarly, e.g. Xylenol Orange [129], Chrome Azurol S [130], and Pyrocatechol Violet in the presence of quaternary ammonium salts [131,132].

The xanthene dyes Bromopyrogallol Red [133] and Gallein (in the presence of cetyltrimethylammonium bromide) [134] have also been recommended as reagents for molybdenum. Eberle and Lerner [135] have found that only molybdenum is extracted by a chloroform solution of oxine from sulphate solution (pH 0.85 ± 0.1) in the absence of halide ions. Molybdenum has also been determined with 8-hydroxyquinoline-5-sulphonic acid [136] and bromo-oxine [137].

Other organic reagents used for determining molybdenum include EDTA [138,139], phenylhydrazine [140], 3-hydroxy-1-*p*-sulphonatophenyl-3-phenyltriazine [141], and 1,10-phenanthroline [142,143]. Molybdenum can be reduced with hydrazine in boiling $5.5M$ HCl and extracted with isoamyl acetate from $7M$ HCl. The green colour is measured at 720 nm [144].

Lastly, molybdenum can be determined as the mixed heteropoly acids with tungsten [145], phosphorus and tungsten [146,147], and also as the coloured peroxymolybdate complex [148].

References

1. Henrickson, R. B. and Sandell, E. B., *Anal. Chim. Acta* **7**, 57 (1952).
2. Norwitz, G. and Codell, M., *Anal. Chem.* **25**, 1438 (1953).
3. McNerney, W. N. and Wagner, W. F., *ibid.* **29**, 1177 (1957).
4. Buss, H., Kohlschütter, H. W. and Miedtank, S., *Z. Anal. Chem.* **178**, 1 (1960).
5. Přibil, R. and Malát, M., *Collection Czech. Chem. Commun.* **15**, 120 (1950).
6. Sapek, B., *Chem. Anal. (Warsaw)* **15**, 651 (1970).
7. Kim, Y. S. and Zeitlin, H., *Anal. Chim. Acta* **46**, 1 (1969); **51**, 516 (1970).
8. Ustimov, A. M. and Gladyshev, V. P., *Tr. Komis. po Analit. Khim. Akad. Nauk SSSR* **15**, 275 (1965).
9. Chan, K. M. and Riley, J. P., *Anal. Chim. Acta* **36**, 220 (1966).
10. Marchenko, P. V. and Uzhvii, V. N., *Ukr. Khim. Zh.* **31**, 612 (1965).
11. Kuznetsov, V. I., Loginowa, L. G. and Myasoedova, G. V., *Zh. Analit. Khim.* **13**, 453 (1958).
12. Jeffery, P. G., *Analyst* **81**, 104 (1956); **82**, 558 (1957).
13. Jones, G. B., *Anal. Chim. Acta* **10**, 584 (1954).
14. Hoenes, H. J. and Stone, K. G., *Talanta* **4**, 250 (1960).
15. Peng, P. Y. and Sandell, E. B., *Anal. Chim. Acta* **29**, 325 (1963).
16. Luke, C. L., *ibid.* **34**, 302 (1966).
17. Adamiec, I., *Chem. Anal. (Warsaw)* **11**, 1175, 1183 (1966).
18. Alimarin, I. P. and Polyanskii, V. N., *Zh. Analit. Khim.* **8**, 266 (1953).
19. Codell, M., Mikula, J. J. and Norwitz, G., *Anal. Chem.* **25**, 1441 (1953).
20. Waterbury. G. R. and Bricker, C. E., *ibid.* **29**, 129 (1957).
21. Zharovskii, F. G., *Zh. Neorgan. Khim.* **2**, 623 (1957); *Ukr. Khim. Zh.* **23**, 767 (1957).
22. Greenland, L. P. and Lillie, E. G., *Anal. Chim. Acta* **69**, 335 (1974).
22a. Yatirajam, V. and Ram, J., *Mikrochim. Acta* **1973**, 77.
22b. Yatirajam, V. and Ram, J., *Talanta* **20**, 885 (1973).
23. Allen, S. H. and Hamilton, M. B., *Anal. Chim. Acta* **7**, 483 (1952).
24. Healy, W. B. and McCabe, W. J., *Anal. Chem.* **35**, 2117 (1963).
25. Pyatnitskii, I. V. and Kravtsova, L. F., *Ukr. Khim. Zh.* **35**, 77 (1969).
26. McKaveney, J. P. and Freiser, H., *Anal. Chem.* **29**, 290 (1957).

27. De, A. K. and Rahaman, M. S., *ibid.* **36**, 685 (1964).
28. Busev, A. I. and Rodionova, T. V., *Zh. Analit. Khim.* **23**, 877 (1968).
29. Shustova, M. B. and Shelikhina, E. I., *Zavodsk. Lab.* **29**, 810 (1967).
30. Kletenik, Yu. B., Bykhovskaya, I. A. and Sekretova, L. V., *Zh. Analit. Khim.* **24**, 707 (1969).
31. Kiss, A. B. and Hegedüs, A. J., *Mikrochim. Acta* **1966**, 771.
31a. Yatirajam, V. and Ram, T., *Talanta* **21**, 439 (1974).
32. Busev, A. I. and Rodionova, T. V., *Zh. Analit. Khim.* **26**, 578 (1971).
33. Hall, F. M. and Bryson, A., *Anal. Chim. Acta* **24**, 138 (1961).
34. Dixon, E. J. and Headridge, J. B., *Analyst* **89**, 185 (1964).
35. Sugawara, K. F., *Anal. Chem.* **36**, 1373 (1964).
36. Fritz, J. S. and Dahmer, L. H., *ibid.* **37**, 1272 (1965).
37. Klement, R., *Z. Anal. Chem.* **136**, 17 (1952).
38. Gilbert, T. W., Jr. and Sandell, E. B., *J. Am. Chem. Soc.* **82**, 1087 (1960).
39. Bode, H. and Schaaff, G., *Z. Anal. Chem.* **159**, 182 (1958).
40. Stepanova, N. A. and Yakunina, G. A., *Zh. Analit Khim.* **17**, 858 (1962).
41. Stanton, R. E. and Hardwick, A. J., *Analyst* **92**, 387 (1967).
42. Kawabuchi, K., *Japan Analyst* **14**, 52 (1965).
42a. Koyama, M., Emoto, K., Kawashima, M. and Fujinaga, T., *Chem. Anal. (Warsaw)* **17**, 679 (1972).
43. Clark, L. J. and Axley, J. H., *Anal. Chem.* **27**, 2000 (1955).
44. Granger, C. O., *Analyst* **83**, 609 (1958).
45. Oelschläger, W., *Z. Anal. Chem.* **188**, 190 (1962).
46. Buss, H., Kuhlschütter. H. W. and Walter, L., *ibid.* **191**, 273 (1962).
46a. Milham, P. J., Maksvytis, A. and Barkus, B., *Anal. Chem.* **44**, 2102 (1972).
47. Klug, O. N. and Metlenko, A. I., *Chem. Anal. (Warsaw)* **13**, 7 (1968).
48. Zopatti, L. P. and Pollock, E. N., *Anal. Chim. Acta* **32**, 178 (1965).
49. Hobart, E. W. and Hurley, E. P., *ibid.* **27**, 144 (1962).
50. Kallmann, S., Hobart, E. W. and Oberthin, H. K., *Talanta* **15**, 982 (1968).
51. Stonhill, L. G., *Chemist-Analyst* **47**, 68 (1958).
52. Kawabuchi, K. and Kuroda, R., *Talanta* **17**, 67 (1970).
53. North, A. A., *Analyst* **81**, 660 (1956).
54. Ssekaalo, H., *ibid.* **96**, 346 (1971).
55. Stanton, R. E. and Hardwick, A. J., *ibid.* **93**, 193 (1968).
56. Kawabuchi, K. and Kuroda, R., *Anal. Chim. Acta* **46**, 23 (1969).
57. Riley, J. P. and Taylor, D., *ibid.* **41**, 175 (1968).
58. Crouthamel, C. E. and Johnson, C. E., *Anal. Chem.* **26**, 1284 (1954).
59. Lounamaa, N., *Anal. Chim. Acta* **33**, 21 (1965).
60. Babko, A. K. and Drako, O. F., *Zh. Analit. Khim.* **12**, 342 (1957).
61. Perrin, D. D., *J. Am. Chem. Soc.* **80**, 3540 (1958).
62. Lazarev, A. I. and Lazareva, V. I., *Zavodsk. Lab.* **24**, 798 (1958).
63. Potrokhov, V. K. and Lebedeva, L. I., *Zh. Analit. Khim.* **21**, 182 (1966).
64. Hope, R. P., *Anal. Chem.* **29**, 1053 (1957).
65. Reznik, B. E., Ganzburg, G. M. and Sachko, V. V., *Zavodsk. Lab.* **28**, 277 (1962).
66. Prasad, J. and Suryanarayana, M., *Z. Anal. Chem.* **219**, 346 (1966).
67. Hibbits, J. O. and Williams, R. T., *Anal. Chim. Acta* **26**, 363 (1962).
68. Johnson, C. M. and Arkley, T. H., *Anal. Chem.* **26**, 572 (1954).
69. Wilson, A. M. and McFarland, O. K., *ibid.* **36**, 2488 (1964).
70. Kolling, O. W., *ibid.* **37**, 436 (1965).
70a. Yatirajam, V. and Ram, J., *Anal. Chim. Acta* **59**, 381 (1972).
71. Tananaiko, M. M. and Blukke, L. A., *Ukr. Khim. Zh.* **29**, 974 (1963).
71a. Ivanova, I. F., Bukhteeva, L. N., Tanago, L. I., Gribkovskaya, L. G. and Yurii, L. N., *Zavodsk. Lab.* **39**, 388 (1973).
72. Patil, S. P. and Shinde, V. M., *Anal. Chim. Acta* **67**, 473 (1973).

References

- 72a. Bermejo-Martinez, F. and Latas-Perez, P., *Z. Anal. Chem.* **264**, 139 (1973).
- 73. Boulin, R. and Jaudon, E., *Chim. Anal. (Paris)* **47**, 290 (1965).
- 74. Stross, W. and Clark, J., *Metallurgia* **67**, 47 (1963).
- 75. Kuehn, P. R., Howard, O. H. and Weber, C. W., *Anal. Chem.* **33**, 740 (1961).
- 75a. Korkisch, J. and Steffan, I., *Mikrochim. Acta* **1973**, 545.
- 76. Hibbits, J. O., Davis, W. F., Menke, M. R. and Kallmann, S., *Talanta* **4**, 104 (1960).
- 76a. Lillie, E. G. and Greenland, L. P., *Anal. Chim. Acta* **69**, 313 (1974).
- 77. Grigg, J. L., *Analyst* **78**, 470 (1953).
- 78. Malyutina, T. M., Dobkina, B. M. and Pisareva, V. A., *Zavodsk. Lab.* **31**, 648 (1965).
- 79. Privalova, M. M. and Tulina, M. D., *ibid.* **33**, 16 (1967).
- 80. Braithwaite, K. and Hobson, J. D., *Analyst* **93**, 633 (1968).
- 81. Will, F., III and Yoe, J. H., *Anal. Chem.* **25**, 1363 (1953).
- 82. Otterson, D. A. and Graab, J. W., *ibid.* **30**, 1282 (1958).
- 83. Ziegler, M. and Horn, H. G., *Z. Anal. Chem.* **166**, 362 (1959).
- 84. Přibil, R. and Adam, J., *Talanta* **18**, 349 (1971).
- 85. Busev, A. I., Rudzit, G. P., Chipen, G. I. and Grinshtein, V. Ya., *Zh. Analit. Khim.* **20**, 76 (1965).
- 86. Busev, A. I., Rudzit, G. P. and Dzintarnieks, M. Ya., *ibid.* **21**, 176 (1966).
- 87. Bankovskii, Yu. A., Shvarts, E. M. and Ievin'sh, A. F., *ibid.* **14**, 313 (1959).
- 88. Golubtsova, R. B., *ibid.* **14**, 493 (1959).
- 89. Magee, R. J. and Witwit, A. S., *Anal. Chim. Acta* **29**, 27 (1963).
- 90. Haas, W. and Schwarz, T., *Mikrochim. Acta* **1963**, 253.
- 91. Likussar, W., Beyer, W. and Wawschinek, O., *ibid.* **1968**, 735.
- 92. Sarma, V. B. and Suryanarayana, M., *Z. Anal. Chem.* **240**, 6 (1968).
- 93. Busev, A. I., Chzhan Fan' and Kuzyaeva, Z. P., *Zh. Analit. Khim.* **16**, 695 (1961).
- 94. Busev, A. I., Naku, A. and Rudzit, G. P., *ibid.* **19**, 337, 767 (1964).
- 95. Busev, A. I., Chang Fan and Kuzyaeva, Z. P., *Talanta* **9**, 113 (1962).
- 96. Novak, V. P., Bogovina, V. I., Bedovik, S. S. and Mal'tsev, V. F., *Zavodsk. Lab.* **37**, 1170 (1971).
- 97. Busev. A. I. and Chzhan Fan', *Zh. Analit. Khim.* **16**, 171 (1961).
- 98. Polyak, L. Ya. and Bashkirova, I. S., *ibid.* **22**, 200 (1967).
- 99. Maklakova, V. P. and Ryazanov, I. P., *Zavodsk. Lab.* **34**, 1049 (1968).
- 100. Kirkbright, G. F. and Yoe, J. H., *Talanta* **11**, 415 (1964).
- 100a. Arunachalam, M. K. and Kumaran, K., *ibid.* **21**, 355 (1974).
- 101. Yoe, J. H. and Will, F., III, *Anal. Chim. Acta* **6**, 450 (1952); **8**, 546 (1953).
- 102. Busev, A. I. and Rudzit, G. P., *Zh. Analit. Khim.* **19**, 569 (1964).
- 103. Haight, G. P., Jr., and Paragamian, V., *Anal. Chem.* **32**, 642 (1960).
- 104. Buchwald, H. and Richardson, E., *Talanta* **9**, 631 (1962).
- 105. Horák, J. and Okáč, A., *Collection Czech. Chem. Commun.* **29**, 188 (1964).
- 106. Tserkovnitskaya, I. A. and Kustova, N. A., *Zh. Analit. Khim.* **23**, 72 (1968).
- 107. Hartkamp, H., *Z. Anal. Chem.* **231**, 161 (1967); **241**, 66 (1968).
- 108. Jungnickel, H. E. and Klinger, W., *ibid.* **202**, 107 (1964).
- 109. Busev, A. I. and Chzhan Fan', *Zh. Analit. Khim.* **16**, 578 (1961).
- 110. Almássy, G. and Vigvári, M., *Acta Chim. Acad. Sci. Hung.* **20**, 243 (1959).
- 111. Goldstein, G., Manning, D. L. and Menis, O., *Anal. Chem.* **30**, 539 (1958).
- 112. Chan, F. L. and Moshier, R. W., *Talanta* **3**, 272 (1960).
- 113. Black, A. H. and Bonfiglio, J. D., *Anal. Chem.* **33**, 431 (1961).
- 114. Shustova, M. B. and Nazarenko, V. A., *Zh. Analit. Khim.* **18**, 964 (1963).
- 115. Bagdasarov, K. N., Shchemeleva, G. G. and Mikalauskas, T. V., *Zavodsk. Lab.* **29**, 1047 (1973).
- 116. Majumdar, A. K. and Savariar, C. P., *Anal. Chim. Acta* **22**, 158 (1960).
- 117. Nazarenko, V. A., Shustova, M. B. and Shelikhina, E. I., *Zh. Analit. Khim.* **25**, 2139 (1970).

118. Antonovich, V. P., Shelikhina, E. I., Zhadanov, B. V. and Nazarenko, V. A., *ibid.* **27**, 100 (1972).
119. Nazarenko, V. A., Shelikhina, E. I. and Antonovich, V. P., *ibid.* **27**, 307 (1972).
120. Pushinov, Yu. V. and Cherkesov, A. I., *Izv. Vyssh. Ucheb. Zaved., Khim. Khim. Tekhnol.* **8**, 559 (1965).
121. Lukin, A. M., Petrova, G. S. and Kaslina, N. A., *Zh. Analit. Khim.* **24**, 39 (1969).
122. D'yachenko, S. S., Agrinskaya, N. A. and Petrashen', V. I., *Zavodsk. Lab.* **36**, 23 (1970).
123. Korkisch, J. and Osman, M., *Z. Anal. Chem.* **171**, 349 (1959).
124. Dedkov, Yu. M., Rybina, T. F. and Yakovlev, P. Ya., *Zavodsk. Lab.* **38**, 787 (1972).
125. Elinson, S. V., Savvin, S. B. and Nezhnova, T. I., *Zh. Analit. Khim.* **22**, 531 (1967).
126. Savvin, S. B., Mineeva, V. A., Okhanova, L. A. and Pachadzhanov, D. N., *ibid.* **26**, 532 (1971).
127. Lassner, E., Püschel, R., Katzengruber, K. and Schedle, H., *Mikrochim. Acta* **1969**, 134, 527.
128. Szczygielska, M. and Kasiura, K., *Chem. Anal. (Warsaw)* **18**, 799 (1973).
129. Budeshinskii, B., *Zh. Analit. Khim.* **18**, 1071 (1963).
130. Horiuchi, Y. and Nishida, H., *Japan Analyst* **18**, 1092 (1969).
131. Kohara, H., Ishibashi, N. and Abe, K., *ibid.* **19**, 48 (1970).
132. Bailey, B. W., Chester, J. E., Dagnall, R. M. and West, T. S., *Talanta* **15**, 1359 (1968).
133. Honsa, I. and Suk, V., *Collection Czech. Chem. Commun.* **35**, 1283 (1970).
134. Leong, C. L., *Analyst* **95**, 1018 (1970).
135. Eberle, A. R. and Lerner, M. W., *Anal. Chem.* **34**, 627 (1962).
136. Busev, A. I. and Chang Fan, *Talanta* **9**, 107 (1962).
137. Elbeih, I. I. and Abou-Elnaga, M. A., *Can. J. Chem.* **46**, 1379 (1968).
138. Lassner, E. and Scharf, R., *Z. Anal. Chem.* **168**, 30 (1959).
139. Polyak, L. Ya. and Bashkirova, I. S., *Zh. Analit. Khim.* **21**, 682 (1966).
140. Bozsai, I., *Talanta* **10**, 543 (1963).
141. Sogani, N. C. and Bhattacharyya, S. C., *Anal. Chem.* **33**, 1273 (1961).
142. Havermans, E., Verbeek, F. and Hoste, J., *Anal. Chim. Acta* **26**, 326 (1962).
143. Bhadra, A. K. and Banerjee, S., *Talanta* **20**, 342 (1973).
144. Yatirajam, V. and Ram, J., *ibid.* **20**, 1207 (1973).
145. Reznik, B. E., Ganzburg, G. M. and Mal'tseva, G. V., *Zh. Analit. Khim.* **23**, 1848 (1968).
146. Reznik, B. E., Ganzburg, G. M. and Milovanova, V. F., *Zavodsk. Lab.* **33**, 18 (1967).
147. Mal'tseva, G. V. and Reznik, B. E. *Zh. Analit. Khim.* **28**, 1751 (1973).
148. Caiozzi, M., Zunino, H. and Sepúlveda, L. *Talanta* **16**, 1590 (1969).

Chapter 33
NICKEL

Nickel (Ni, at. wt. 58·71) usually occurs in the +II oxidation state, but some complexes contain nickel in higher oxidation states (+III and +IV). Nickel sulphide is precipitated at pH ~ 4. Nickel(II) hydroxide (precipitated at pH ~7) dissolves in ammonia owing to formation of ammine complexes, but is insoluble in excess of NaOH. Nickel (II) also forms stable cyanide, oxalate, and EDTA complexes.

33.1 Methods of Separation

33.1.1 Extraction

Nickel dimethylglyoximate, $Ni(HDm)_2$, is sparingly soluble in aqueous media but readily soluble in chloroform. This permits the specific separation of nickel by extraction, usually from weakly ammoniacal medium containing citrate or tartrate to prevent the precipitation of hydrolysable metals such as Fe(III) and Al [1-4]. Larger quantities of manganese interfere with the extraction of $Ni(HDm)_2$ since manganese(II) is readily oxidized, and can subsequently oxidize the nickel in the dimethylglyoxime complex, thus preventing its extraction into chloroform. This interference is prevented by adding hydroxylamine. Copper and cobalt are extracted on a limited scale, but are removed when the extract is shaken with dilute ammonia. Nickel can also be extracted with various other dioximes [5,6].

Extraction with dithizone in CCl_4 or $CHCl_3$ results in the separation of nickel along with traces of a number of other heavy metals [7-9]. By taking advantage of the differences in resistance of dithizonates to dilute hydrochloric acid, it is possible to separate nickel from zinc and cadmium [10].

33.1.2 Precipitation

The separation of nickel by precipitation as the dimethylglyoximate is almost specific [11,12]. For greater precision in isolating nickel from Co and Cu, the precipitation of $Ni(HDm)_2$ is carried out twice by introducing dimethylglyoxime into an acidic solution, and subsequently adding ammonia. In the case of trace amounts of nickel, palladium can be used as collector.

Hibbits and Kallmann [13] have precipitated nickel with benzotriazole, using cadmium as collector. Kraft [14] has isolated small amounts of nickel from nickel oxide by selective dissolution of metallic nickel in a solution of the silver thiocyanate complex (pH ~ 5).

When the sample is fused with sodium carbonate and then leached with water, nickel is one of the metals which remain quantitatively in the undissolved residue [15].

In the presence of KCN and an excess of NaOH (or KOH), nickel is precipitated as Ni(OH)$_2$, whereas cobalt remains in the solution (pH 13–14) as the cyanide complex. Suitable collectors for nickel are Mg(OH)$_2$ or La(OH)$_3$ [6].

33.1.3 Ion-Exchange

The fact that the chloride complexes of nickel are much weaker than those of Co, Cu, Zn, Fe(III), and certain other metals allows the separation of nickel from those metals on ion-exchanger columns [13,16,17]. When a solution in 8-9M HCl is passed through a strongly basic anion-exchanger (e.g. Dowex 1), nickel alone is eluted.

Kemula and Brajter [18] have retained the anionic nickel picolinic acid complex on an anion-exchanger. Copper and iron(III), which form neutral complexes, pass to the eluate.

33.2 Methods of Determination

Nickel ions have chromophoric properties and give colour reactions with many reagents. The most important photometric reagents for determining nickel are dioximes, which give specific and fairly sensitive methods. Although there exist more sensitive methods, e.g. those employing 1-(2-pyridylazo)-2-naphthol (PAN) and pyridine-2-aldehyde-2-quinolyl hydrazone (PAQH), the oxime methods are the most selective.

The two methods discussed below in detail are the dimethylglyoxime (+ oxidant) method for determining nickel in an aqueous medium and the extractive method using α-furildioxime.

33.2.1 Dimethylglyoxime Method

Dimethylglyoxime (H$_2$Dm, diacetyldioxime, Chugaev's reagent) reacts with nickel ions in a neutral or ammoniacal medium to form a pink,

$$2 \begin{array}{c} H_3C-C=NOH \\ | \\ H_3C-C=NOH \end{array} + Ni^{2+} \rightarrow \begin{array}{c} H \\ \cdot\cdot O \diagdown \diagup O \\ \uparrow \quad | \\ H_3C-C=N \quad N=C-CH_3 \\ | \quad \diagdown Ni \diagup \quad | \\ H_3C-C=N \quad N=C-CH_3 \\ | \quad \downarrow \\ O \quad O \\ \diagdown H \diagup \end{array} + 2H^+$$

flocculent precipitate which has for a long time been the basis of an excellent gravimetric method for determining nickel. The nickel dimethylglyoximate chelate is soluble in chloroform and other non-polar organic solvents. The extraction of Ni(HDm)$_2$ is primarily important in the separation of nickel, but relatively large quantities of nickel have been determined by means of the pale yellow chloroform solution of Ni(HDm)$_2$ [19, 20].

Curve 1 in Fig. 33.1 represents the absorption spectrum of nickel dimethylglyoximate in chloroform. The molar absorptivity at λ_{max} = 360 nm is 3.4×10^3, and at 400 nm is 1.8×10^3.

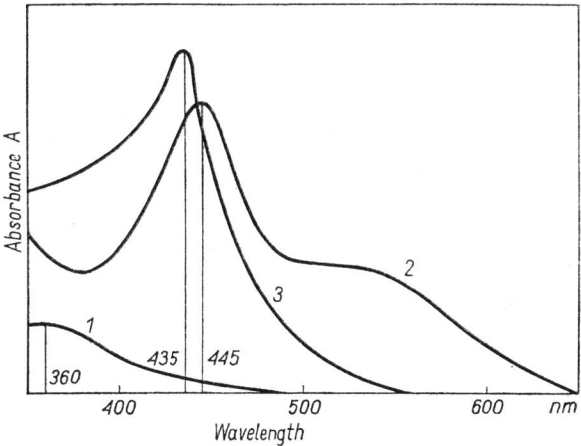

Fig. 33.1 Absorption spectra of nickel(II) dimethylglyoxime complex in CHCl$_3$ (1), nickel(IV) dimethylglyoxime complex in alkaline aqueous solution (2) and nickel(II) α-furildioxime complex in chloroform (3)

This method has been used to determine nickel in iron ore [21], copper and its alloys [19,22], niobium, tantalum, molybdenum, and tungsten [23], and cobalt [19]. In the last case, cobalt is masked as the stable cyanide complex before the extraction of Ni(HDm)$_2$.

In an alkaline medium and in the presence of oxidants, nickel forms a brown-red, water-soluble dimethylglyoxime complex which is the basis of the most popular spectrophotometric method for determining nickel [2,3,12]. It seems generally agreed that in this complex nickel is in the +IV oxidation state, but there are doubts about this and the composition has not yet been conclusively elucidated [24–28]. Two species are said to be formed [13,24]. The usual oxidants are bromine, iodine, or persulphate, but the oxidant used has no effect on the colour obtained. Since the formation of the coloured complex is rather slow, it is prudent to wait several minutes before measuring the absorbance, but it must be remembered that the coloured solutions are not stable. It is most important to add the reagents in the following sequence: dimethylglyoxime, oxidant, ammonia.

Curve 2 in Fig. 33.1 represents the absorption spectrum of the oxidized complex. The molar absorptivity is $1 \cdot 5 \times 10^4$ ($a = 0 \cdot 26$) at $\lambda_{max} = 445$ nm.

Coloured ions and metals (Fe^{2+}, Co^{2+}, Cu^{2+}) which form coloured, water-soluble complexes with dimethylglyoxime interfere in the nickel determination. However, the complexes of Fe(II), Co(II), and Cu(II) are decomposed by EDTA, and preliminary extraction as $Ni(HDm)_2$ enables nickel to be isolated from larger amounts of Cu, Co, Fe, Cr, Al, and Mn. The presence of hydroxylamine ensures the quantitative extraction of $Ni(HDm)_2$ and prevents interference from copper and manganese.

When the chloroform extract is shaken with dilute ($0 \cdot 5M$) hydrochloric acid, $Ni(HDm)_2$ is decomposed and the nickel is quantitatively stripped into the aqueous phase. After the chloroform has been removed, this aqueous extract is treated with dimethylglyoxime, oxidant, and ammonia (or NaOH solution) to form coloured $[Ni(Dm)_3]^{2-}$. The dimethylglyoxime is added as an alcoholic solution or a solution in dilute (e.g. $0 \cdot 2M$) NaOH.

Chowdhury and Sarma [29] have determined nickel by extracting nickel dimethylglyoximate with a mixture of benzene and amyl alcohol, and then shaking the extract with an alkaline dimethylglyoxime solution, aerial oxidation causing the nickel to pass into the aqueous phase as red $[Ni(Dm)_3]^{2-}$.

The anionic nickel dimethylglyoxime complex formed in an NaOH medium in the presence of an oxidant can be extracted into organic solvents in the presence of diphenylguanidinium cation. This extraction-photometric method is more sensitive than the determination in aqueous solution [29a].

Nickel has been determined with dimethylglyoxime in the following: steel [12,30], zinc and cadmium [31], tin [32], lead and antimony [33], zirconium [34], aluminium alloys [30], uranium and its compounds [35,36], tungsten and its alloys [37,38], beryllium [13], rocks [39], copper ores and concentrates [40], petroleum cracking catalysts [41], soils [29], sea water [42,43], foodstuffs [8,11,44,44a], and air [45].

Dimethylglyoxime (+oxidant) has been used to determine larger quantities of nickel in steel by differential spectrophotomery [46], and has also been applied in automated procedures [47].

Reagents

1. Dimethylglyoxime: 1% solution in ethanol.
2. Standard nickel solution: 1 mg/ml. Dissolve 6·730 g of $(NH_4)_2Ni(SO_4)_2 \cdot 6H_2O$ in water containing 2 ml of conc. H_2SO_4, and dilute the solution with water to 1 litre. It is also possible to prepare a more concentrated solution of a nickel salt (sulphate, nitrate, or chloride), determine the nickel concentration gravimetrically by precipitating the dimethylglyoximate, and then dilute the solution with water till it contains exactly 1 mg of Ni per ml.
3. Bromine water: saturated aqueous solution.
4. Potassium persulphate: 4% solution, freshly prepared.

Procedure

Extractive separation of Ni. To the solution containing not more than 100 μg of Ni, add 20% potassium sodium tartrate solution (1–5 ml depending on the quantities of hydrolysable metals in the solution), 2 ml of 10% $NH_2OH \cdot HCl$ solution, 2 ml of the dimethylglyoxime solution, and ammonia to pH 9–10. Shake the solution for about half a minute in a separating funnel with two portions of chloroform. Wash the combined extracts by shaking with dilute ammonia solution (1+50), then strip the nickel from the organic phase by shaking for 1 minute with $0.5M$ HCl. Discard the chloroform layer.

Determination of Ni. Quantitatively transfer the solution obtained to a 50-ml volumetric flask and add 1 ml of the dimethylglyoxime solution, 2 ml of bromine water (or persulphate solution), and 5 ml of conc. NH_3 solution in that order. Dilute the solution with water to the mark, and mix well. After 10 minutes, measure the absorbance at 445 nm against water.

33.2.2 α-FURILDIOXIME METHOD

α-Furildioxime (I) (sometimes called Neonickelone) reacts with nickel ions similarly to dimethylglyoxime and other α-dioximes, forming a chelate

which is sparingly water-soluble, but is extractable into chloroform and similar solvents. The yellow colour of the organic extract provides the basis of a specific spectrophotometric method for determining nickel [48–50]. The sensitivities of the dimethylglyoxime (+oxidant) and the α-furildioxime methods are similar, but the latter method is simpler since it obviates the necessity to preconcentrate the nickel.

The nickel α-furildioxime complex is formed and extracted quantitatively in the rather narrow pH range from 7·5 to 9·0. The pH can be adjusted most conveniently by adding small portions of sodium bicarbonate to the slightly acidic sample solution. Alternatively, a drop of a 1% solution of phenolphthalein can be added to the solution, and ammonia added carefully until the indicator just turns pink.

Chloroform, carbon tetrachloride, *o*-dichlorobenzene and ethyl acetate have been used as solvents. Solubility of the nickel α-furildioxime complex in carbon tetrachloride is rather low and hence this solvent can be used only for extracting limited amounts of nickel (< 1 μg Ni/ml CCl_4).

The absorbance maximum of the nickel α-furildioxime chelate in chloroform is at 435 nm (the molar absorptivity is 2.0×10^4; $a = 0.34$) (see Fig. 33.1).

Low and erratic results obtained after the extraction of nickel α-furildioximate with certain chloroform samples are probably due to the presence of trace oxidants from the decomposition of $CHCl_3$. Preliminary shaking of the solvent with thiosulphate solution before the extraction prevents such interferences. Copper and cobalt, which interfere by forming coloured, extractable complexes with α-furildioxime, are effectively masked with thiourea or thiosulphate and with ethylenediamine, respectively [51]. Tartrate is used to mask hydrolysable metals.

Fryer et al. [51a] have pointed out that some α-furildioxime (anti-) samples also contain the other two stereoisomers, γ- (amphi-) and β- (syn-), which react with nickel ions in a 1:1 ratio. These nickel complexes are also of a different colour and are soluble in ethanol. The melting points of α-, β-, and γ-furildioximes are 193°C, 152°C, and 183°C, respectively. The absorbance, therefore, depends also on the quality of the α-furildioxime sample used.

The α-furildioxime method has been used to determine nickel in steel [51], silicate and sulphide minerals [51], indium and aluminium [52], silver [53], tin [54], cadmium [54a], alkalis [9,55], beryllium [56], rhenium [57], petroleum oils [58], boiler feed water [59], and various metals [60].

Reagents

1. α-Furildioxime: 0·5% solution in alcohol. If the solution is coloured, shake it with activated carbon.
2. Standard nickel solution: 1 mg/ml (p. 372).
3. Sodium bicarbonate: saturated solution (~ 10%).
4. Chloroform. Shake the solvent with dilute (e.g. $0·1M$) $Na_2S_2O_3$, and then with water.

Procedure

Place a weakly acid (pH ~ 1) sample solution, containing not more than 100 μg Ni, in a separating funnel, and add 20% potassium sodium tartrate solution (0·2–1 ml, depending on the amount of hydrolysable metals present) and 1 ml of the α-furildioxime solution. Add the $NaHCO_3$ solution in small portions with stirring until carbon dioxide is no longer evolved. Allow to stand for 10 minutes, and then extract the nickel complex with two portions of chloroform. Dilute the extract with the solvent in a 50-ml volumetric flask (or smaller, depending on the amount of Ni) and measure the absorbance at 435 nm, using the solvent as a reference.

Note. Small amounts of copper in the sample should be masked with 1–2 ml of 5% thiourea solution before addition of the α-furildioxime.

33.2.3 OTHER METHODS

The most important group of reagents for nickel is composed of dioximes which contain the nickel-specific group

$$\begin{array}{c} -C=NOH \\ | \\ -C=NOH \end{array}$$

Apart from dimethylglyoxime and α-furildioxime, the following reagents have been recommended for the spectrophotometric determination of nickel: nioxime (II) (1,2-cyclohexanedionedioxime) [61,62], and its

$$
\begin{array}{cc}
\text{H}_2\text{C} \diagup \overset{\text{H}_2}{\underset{\text{C}}{\text{C}}} \diagdown \text{C=NOH} \\
| \quad\quad\quad | \\
\text{H}_2\text{C} \diagdown \underset{\text{C}}{\underset{\text{H}_2}{\text{C}}} \diagup \text{C=NOH}
\end{array} \text{(II)} \qquad
\begin{array}{c}
\bigcirc\!\!-\text{C=NOH} \\
| \\
\bigcirc\!\!-\text{C=NOH}
\end{array} \text{(III)}
$$

derivatives (4-methylnioxime [63], 4-isopropylnioxime [64], 4-butylnioxime [65], and 4-carboxynioxime [66]), heptoxime (1,2-cycloheptanedionedioxime) [61,67–69], 1,2,3-cyclohexanetrionetrioxime [70], and α-benzildioxime (III) (diphenylglyoxime, nickelone) [71–74]. The metal complexes of dioximes have been reviewed [75].

Other oxime reagents include nicotinamidoxime [76], pyridine-2,6-diacetoxime and *syn*-phenyl-2-pyridylketoxime [77]. Formaldoxime, which is commonly used to determine manganese, can also serve as a reagent for nickel since it forms a brown nickel complex in alkaline media ($\varepsilon = 1.84 \times 10^4$ at 473 nm) [78,79].

Several spectrophotometric methods for determining nickel are based on organic reagents containing sulphur as a ligand atom. Quinoxaline-2,3-dithiol reacts with nickel ions to give a red complex in ammoniacal media or a blue complex in acid media (molar absorptivity is 1.8×10^4 at 660 nm) [80–85]. Dithio-oxalic acid forms the nickel bis(dithio-oxalate) chelate which can be extracted as an ion-pair with DAM or with the triphenylmethylarsonium ion [86–88]. Further methods employ dithio-oxamide (rubeanic acid) and its derivatives [89,90], Na-DDTC [91], 2,2′-dimercaptodiethylsulphide [92], 2-mercaptobenzothiazole [93], thiothenoyltrifluoroacetone [93a], β-dithionaphthoic acid [94], 2-amino-1-cyclopentene-1-dithiocarboxylic acid [95], p-hydroxydithiobenzoic acid [96], and β-mercaptocinnamic acid and its derivatives [97].

High sensitivity is characteristic of spectrophotometric methods for nickel using azo reagents such as PAN ($\varepsilon = 5.3 \times 10^4$ at 565 nm) [15,98,99], PAR [100], 2-2-(thiazolylazo)-5-dimethylaminophenol (TAM) ($\varepsilon = 4.6 \times 10^4$ at 575 nm) [101], 1-(2-thiazolylazo)-2-naphthol (TAN) ($\varepsilon = 3.8 \times 10^4$ at 590 nm) [102,103], Solochrome Red ERS [104], pyridyl-2-azochromotropic acid [105], and Calcichrome [106].

Although nickel can be determined with dithizone as the Ni(HDz)$_2$ complex [7], the method based on the mixed dithizone–phenanthroline complex which is extracted into chloroform is much superior ($\varepsilon = 4.9 \times 10^4$ at 520 nm) [107,108]. Ducret and Pateau [109] have determined nickel in the presence of Fe(III) and Co by extracting the ion-association compound formed by the nickel 1,10-phenanthroline complex and trichloroacetate into 1,2-dichloroethane.

A sensitive spectrophotometric method is based on the extraction of the nickel complex with pyridine-2-aldehyde-2-quinolylhydrazone (IV) (PAQH) into benzene ($\varepsilon = 6.7 \times 10^4$ at 515 nm) [110–112].

<chemical structure of (IV): pyridine ring connected via CH=N-NH to a quinoline ring>

(IV)

The presence of thioglycollic acid prevents PAQH from reacting with cobalt.

Triphenylmethane dyes which have been recommended as reagents for nickel include Chrome Azurol S [113], and Xylenol Orange ($\varepsilon = 3.7 \times 10^4$ at 584 nm) [114–116].

Umland and Thierig [117] have developed a highly sensitive indirect method ($\varepsilon > 3.0 \times 10^5$) in which nickel is separated as the precipitate formed with dimethylglyoxime and diphenylboric acid, $Ni(HDm)_2 \cdot 2(C_6H_5)_2BOH$. After the mineralization of the organic moiety, boron is determined by the curcumin method (see p. 161).

The following organic reagents have also been used to determine nickel: Murexide [118], 8-hydroxyquinoline [119], TTA [120], isonitrosoacetophenone [121], salicylidene-o-aminophenol [122], N,N'-bis(o-aminoacetophenone)-ethylenedi-imine [123,124], 2-pyrrolaldehyde-ethylenedi-imine [125], diethylenetriamine [126], and EDTA [127,128].

Heller and Guyon [129] have determined nickel as a mixed heteropoly molybdophosphoric acid. Komar' and Palagina [130] have estimated higher concentrations of nickel from the colour of nickel(II) ions in a sulphate medium (molar absorptivity 4·8 at 390 nm).

References

1. Christopherson, H. and Sandell, E. B., *Anal. Chim. Acta* **10**, 1 (1954).
2. Claassen, A. and Bastings, L., *Rec. Trav. Chim.* **73**, 783 (1954).
3. Oelschläger, W., *Z. Anal. Chem.* **146**, 339, 346 (1955).
4. Oki, S., *Talanta* **18**, 1233 (1971).
5. Peshkova, V. M., Savostina, V. M., Astakhova, E. K. and Minaeva, N. A, *Tr. Komis. po Analit. Khim. Akad. Nauk SSSR* **15**, 104 (1965).
6. Peshkova, V. M., Belousova, M. Ya. and Novikova, I. S., *Vestn. Moskov. Univ. Khim.* **1968**, No. 5, 57.
7. Sherwood, R. M. and Chapman, F. W., Jr., *Anal. Chem.* **27**, 88 (1955).
8. Kenigsberg, M. and Stone, I., *ibid.* **27**, 1339 (1955).
9. Marczenko, Z., *Mikrochim. Acta* **1965**, 281.
10. Marczenko, Z., Mojski, M. and Kasiura, K., *Zh. Analit. Khim.* **22**, 1805 (1967).
11. Hoffman, I., *Analyst* **87**, 650 (1962).
12. Claassen, A. and Bastings, L., *ibid.* **91**, 725 (1966).
13. Hibbits, J. and Kallmann, S., *Talanta* **10**, 181 (1963).
14. Kraft, G., *Z. Anal. Chem.* **209**, 150 (1965).
15. Püschel, R. and Lassner, E., *Mikrochim. Acta* **1965**, 17, 751.
16. Liberman, A., *Analyst* **80**, 595 (1955).

References

17. Hazan, I. and Korkisch, J., *Anal. Chim. Acta* **32**, 46 (1965).
18. Kemula, W. and Brajter, K., *Chem. Anal. (Warsaw)* **11**, 373 (1966).
19. Nielsch, W., *Z. Anal. Chem.* **140**, 267 (1953); **143**, 272 (1954); **150**, 114 (1956); *Z. Metallk.* **50**, 234 (1959).
20. Nielsch, W. and Giefer, L., *Mikrochim. Acta* **1956**, 522.
21. Shcherbov, D. P. and Perminova, D. N., *Zavodsk. Lab.* **33**, 921 (1967).
22. Dozinel, C. M., *Chim. Anal. (Paris)* **44**, 436 (1962).
23. Penner, E. M. and Inman, W. R., *Talanta* **10**, 997 (1963).
24. Hooreman, M., *Anal. Chim. Acta* **3**, 635 (1949).
25. Babko, A. K., *Zh. Analit. Khim.* **3**, 284 (1948); *Zh. Neorgan. Khim.* **1**, 485 (1956).
26. Yatsimirskii, K. B. and Grafova, Z. M., *Zh. Obshch. Khim.* **23**, 935 (1953).
27. Okáč, A. and Šimek, M., *Chem. Anal. (Warsaw)* **3**, 253 (1958); *Chem. Listy* **52**, 1903, 2285 (1958).
28. Boreiko, M. K. and Kalinichenko, I. I., *Zh. Analit. Khim.* **23**, 1359 (1968).
29. Chowdhury, A. N. and Sarma, B. D., *Anal. Chem.* **32**, 820 (1960).
29a. Zolotov, Yu. A. and Vlasova, G. E., *Zh. Analit. Khim.* **28**, 1540 (1973).
30. Cooper, M. D., *Anal. Chem.* **23**, 875, 880 (1951).
31. Tiptsova, V. G. and Kopnina, O. I., *Zh. Analit. Khim.* **22**, 1108 (1967).
32. Coopins, W. C. and Price, J. W., *Metallurgia* **48**, 149 (1953).
33. Cyrankowska, M., *Chem. Anal. (Warsaw)* **15**, 209 (1970).
34. Chernikhov, Yu. A. and Dobkina, B. M., *Zavodsk. Lab.* **22**, 1019 (1956).
35. Haslam, J., Russell, F. R. and Wilkinson, N. T., *Analyst* **77**, 464 (1952).
36. Nowicka-Jankowska, T., Gołkowska, A., Pietrzak, I. and Żmijewska, W., *Chem. Anal. (Warsaw)* **3**, 977 (1958).
37. Norwitz, G. and Gordon, H., *Anal. Chem.* **37**, 417 (1965).
38. Green, T. E., *ibid.* **37**, 1595 (1965).
39. Sandell, E. B. and Perlich, R. W., *Ind. Eng. Chem., Anal. Ed.* **11**, 309 (1939).
40. Kunz, K., Duczymińska, E. and Ostachowska, J., *Rudy Metale* **9**, 35 (1964).
41. Blackwell,. A T., Daniel, A. M. and Miller, J. D., *Anal. Chem.* **28**, 1209 (1956).
42. Forster, W. and Zeitlin, H., *Anal. Chim. Acta* **35**, 42 (1966).
43. Kentner, E., Armitage, D. B. and Zeitlin, H., *ibid.* **45**, 343 (1969).
44. Rudnicki, A. and Niewiadomski, H., *Chem. Anal. (Warsaw)* **13**, 755 (1968).
44a. Mędrzycka, K. and Niewiadomski, H., *ibid.* **14**, 771 (1969).
45. Belyakov, A. A., *Zavodsk. Lab.* **26**, 158 (1960).
46. Barkovskii, V. F. and Vtorygina, I. N., *ibid.* **28**, 275 (1962).
47. Braithwaite, K. and Hobson, J. D., *Metallurgia* **81**, 205 (1970).
48. Gahler, A. R., Mitchell, A. M. and Mellon, M. G., *Anal. Chem.* **23**, 500 (1951).
49. Peshkova, V. M., Goncharova, G. A., Gribova, E. A. and Puzdrenkova, I. V., *Zh. Analit. Khim.* **8**, 114 (1953).
50. Taylor, C. G., *Analyst* **81**, 369 (1956).
51. Bodart, D. E., *Z. Anal. Chem.* **247**, 32 (1969).
51a. Fryer, F. A., Galliford, D. J. and Yardley, J. T., *Analyst* **88**, 188 (1963).
52. Peshkova, V. M., Bochkova, V. M. and Lazareva, L. I., *Zh. Analit. Khim.* **15**, 610 (1960).
53. Marczenko, Z. and Kasiura, K., *Chem. Anal. (Warsaw)* **9**, 87 (1964).
54. Marczenko, Z. and Kasiura, K., *ibid.* **10**, 449 (1965).
54a. Krasiejko, M. and Marczenko, Z., *Mikrochim. Acta*, **1975 I**, 585.
55. Mains, F. and Raggett, R. E., *Chemist-Analyst* **50**, 4 (1961).
56. Pollock, E. N. and Zopatti, L. P., *Anal. Chim. Acta* **28**, 68 (1963).
57. Ryabchikov, D. I., Lazarev, A. I. and Lazareva, V. I., *Zh. Analit. Khim.* **19**, 1110 (1964).
58. Forrester, J. S. and Jones, J. L., *Anal. Chem.* **32**, 1443 (1960).
59. Wilson, A. L., *Analyst* **93**, 83 (1968).
60. Peshkova, V. M., Astakhova, V. K., Dolmanova, I. F. and Savostina, V. M., *Acta Chim. Acad. Sci. Hung.* **53**, 121 (1967).

61. Ferguson, R. C. and Banks, C. V., *Anal. Chem.* **23**, 448, 1486 (1951).
62. Monnier, D. and Haerdi, W., *Anal. Chim. Acta* **20**, 444 (1959).
63. Blundy, P. D. and Simpson, M. P., *Analyst* **83**, 558 (1958).
64. McDowell, B. L., Meyer, A. S., Jr., Feathers, R. E., Jr. and White, J. C., *Anal. Chem.* **31**, 931 (1959).
65. Barling, M. M. and Banks, C. V., *ibid.* **36**, 2359 (1964).
66. Banks, C. V. and Laplante, J. P., *Anal. Chim. Acta* **27**, 101 (1962).
67. Peshkova, V. M. and Ignat'eva, N. G., *Zh. Analit. Khim.* **17**, 1086 (1962).
68. Gillis, J., Hoste, J. and van Moffaert, Y., *Chim. Anal. (Paris)* **36**, 43 (1954).
69. Savostina, V. M., Kobyakova, S. O. and Peshkova, V. M., *Zh. Analit. Khim.* **23**, 938 (1968).
70. Frierson, W. J. and Marable, N., *Anal. Chem.* **34**, 210 (1962).
71. Uzumasa, Y. and Washizuka, S., *Bull. Chem. Soc. Japan* **29**, 403 (1956).
72. Gregorowicz, Z., *Acta Chim. Acad. Sci. Hung.* **18**, 79 (1959).
73. Peshkova, V. M., Bochkova, V. M. and Astakhova, E. K., *Zh. Analit. Khim.* **16**, 596 (1961).
74. Magee, R. J. and Nasouri, F., *Microchem. J.* **9**, 324 (1965).
75. Egneus, B., *Talanta* **19**, 1387 (1972).
76. Tripathi, K. K. and Banerjea, D., *Z. Anal. Chem.* **176**, 91 (1960).
77. Hartkamp, H., *ibid.* **178**, 19 (1960).
78. Okàč, A. and Bartušek, M., *ibid.* **178**, 198 (1960).
79. Marczenko, Z. and Kasiura, K., *Chem. Anal. (Warsaw)* **6**, 353 (1961); *Anal. Chim. Acta* **31**, 224 (1964).
80. Skoog, D. A., Lai, M. and Furst, A., *Anal. Chem.* **30**, 365 (1958).
81. Burke, R. W. and Yoe, J. H., *ibid.* **34**, 1378 (1962).
82. Ayres, G. H. and Annand, R. R., *ibid.* **35**, 33 (1963).
83. Forster, W. and Zeitlin, H., *ibid.* **38**, 649 (1966).
84. Dalziel, J. A. and Slawinski, A. K., *Talanta* **15**, 367 (1968).
85. Burke, R. W. and Deardorff, E. R., *ibid.* **17**, 255 (1970).
86. Cameron, A. J. and Gibson, N. A., *Anal. Chim. Acta* **24**, 360 (1961).
87. Pilipenko, A. T., Maslei, N. N. and Skorokhod, E. G., *Zh. Analit. Khim.* **23**, 227 (1968).
88. Pilipenko, A. T. and Maslei, N. N., *Ukr. Khim. Zh.* **34**, 174 (1968).
89. Jacobs, W. D. and Yoe, J. H., *Anal. Chim. Acta* **20**, 332, 435 (1959).
90. Janssens, A. A., van de Cappelle, G. L. and Herman, M. A., *ibid.* **31**, 325 (1964).
91. Cluett, M. L. and Yoe, J. H., *Anal. Chem.* **29**, 1265 (1957).
92. Corsini, A. and Nieboer, E., *Talanta* **20**, 291 (1973).
93. Walliczek, E. G., *ibid.* **11**, 573 (1964).
93a. Mulye, R. R. and Khopkar, S. M., *Sepn. Sci.* **7**, 605 (1972).
94. Janik, B. and Gawron, H., *Mikrochim. Acta* **1967**, 843.
95. Yokoyama, M. and Takeshima, T., *Anal. Chem.* **40**, 1344 (1968).
96. Rudzit, G. P., Pastare, S. Ya. and Yanson, E. Yu., *Zh. Analit. Khim.* **25**, 2407 (1970).
97. Busev, A. I. and Vin', D. K., *ibid.* **21**, 1082, 1311 (1966).
98. Nakagawa, G. and Wada, H., *J. Chem. Soc. Japan, Pure Chem. Sect.* **84**, 636 (1963).
99. Püschel, R., Lassner, E. and Katzengruber, K., *Z. Anal. Chem.* **223**, 414 (1966).
100. Nonova, D. and Evtimova, B., *Anal. Chim. Acta* **49**, 103 (1970); **62**, 456 (1972).
101. Kasiura, K. and Sytniewska, Z., *Chem. Anal. (Warsaw)* **13**, 177 (1968).
102. Wada, H. and Nakagawa, G., *Anal. Lett.* **1**, 687 (1968).
103. Kasiura, K., *Chem. Anal. (Warsaw)* **14**, 375 (1969).
104. Janauer, G. E. and Korkisch, J., *Z. Anal. Chem.* **177**, 407 (1960); **179**, 241 (1961).
105. Majumdar, A. K. and Chatterjee, A. B., *Talanta* **13**, 821 (1966).
106. Ishii, H. and Einaga, H., *Japan Analyst* **16**, 322 (1967).
107. Math, K. S., Bhatki, K. S. and Freiser, H., *Talanta* **16**, 412 (1969).
108. Freiser, B. S. and Freiser, H., *ibid.* **17**, 540 (1970).

109. Ducret, L. and Pateau, L., *Anal. Chim. Acta* **20**, 568 (1959).
110. Singhal, S. P. and Ryan, D. E., *ibid.* **37**, 91 (1967).
111. Afghan, B. K. and Ryan, D. E., *ibid.* **41**, 167 (1968).
112. Frei, R. W., Jamro, G. H. and Navratil, O., *ibid.* **55**, 125 (1971).
113. Horiuchi, Y. and Nishida, H., *Japan Analyst* **16**, 576 (1967).
114. Tataev, O. A., Akhmedov, S. A. and Magomedova, B. A., *Zh. Analit. Khim.* **25**, 1229 (1970).
115. Bulatov, M. I., *ibid.* **24**, 1053 (1969).
116. Wet, de, W. J. and Behrens, G. B., *Anal. Chem.* **40**, 200 (1968).
117. Umland, F. and Thierig, D., *Z. Anal. Chem.* **197**, 151 (1963).
118. Shapiro, M. Ya., *Zh. Analit. Khim.* **14**, 365 (1959).
119. Mukhedkar, A. J. and Deshpande, N. V., *Anal. Chem.* **35**, 47 (1963).
120. De, A. K. and Rahaman, M. S., *Anal. Chim. Acta* **27**, 591 (1962).
121. Talwar, U. B. and Haldar, B. C., *ibid.* **51**, 53 (1970).
122. Ishii, H. and Einaga, H., *Bull. Chem. Soc. Japan* **42**, 1558 (1969).
123. Berge, H. and Mennenga, H., *Z. Anal. Chem.* **213**, 346 (1965).
124. Uhlemann, E. and Wischnewski, W., *Anal. Chim. Acta* **42**, 247 (1968).
125. Wawschinek, O. and Weiss, E., *Mikrochim. Acta* **1964**, 690.
126. Whealy, R. D. and Colgate, S. O., *Anal. Chem.* **28**, 1897 (1956).
127. Nielsch, W. and Böltz, G., *Anal. Chim. Acta* **11**, 367 (1954).
128. Brake, L. D., McNabb, W. M. and Hazel, J. F., *ibid.* **19**, 39 (1958).
129. Heller, R. L. and Guyon, J. C., *Talanta* **17**, 865 (1970).
130. Komar', N. P. and Palagina, L. S., *Zh. Analit. Khim.* **23**, 75 (1968).

Chapter 34

NIOBIUM

Niobium (Nb, at.wt. 92·91) hydrolyses (in the absence of complexing anions) over the pH range 0–14. Polymerized forms of niobium(V) give pseudo-solutions or separate as a white precipitate. When fused with KOH, niobium pentoxide (Nb_2O_5) forms the niobate, which is soluble in strongly alkaline media. Niobium(V) forms stable, soluble fluoride, tartrate, oxalate, and peroxide complexes, as well as weak sulphate complexes. The niobium complexes are more stable than the corresponding tantalum complexes. A niobium chloride complex is formed in HCl ($> 5M$). Niobium(V) can be reduced with difficulty to coloured ter- and quadrivalent species. In an acid medium, zinc metal reduces niobium(V), but not tantalum(V).

34.1 Separation of Niobium and Tantalum

34.1.1 PRECIPITATION

On being heated in acid solutions (or pseudosolutions), niobium and tantalum hydrolyse and coagulate to form hydrous oxides [1–4]. The following compounds may be used as collectors for traces of niobium or tantalum: $Zr(OH)_4$, hydrous MnO_2 (in acid solutions), and $Fe(OH)_3$ and $Mg(OH)_2$ (in alkaline solutions). When an alkaline melt (Na_2CO_3, NaOH) is leached, niobium and tantalum remain in the solid phase, while W, Mo, V, and Re pass into the aqueous solution [5].

From solutions containing not too much oxalate, tartrate, or EDTA, niobium and tantalum may be precipitated by cupferron [6], BPHA [7], various hydroxamic acids [8], and phenylarsonic acid [9,10]; zirconium is often used as collector.

When boric acid is added to a solution of Nb, Ta, W, Mo, and Zr fluoride complexes and the solution heated, the boron displaces the fluoride (formation of BF_4^-) from the Nb, Ta, and W complexes, causing the precipitation of these metals, while the zirconium and molybdenum remain in solution since their fluoride complexes are more stable [11].

Malissa [12] has separated niobium from tantalum by selective precipitation with dithiocarbamates.

34.1.2 EXTRACTION

Separation from many metals (e.g. Ti, Zr, Sn, Mo, U, W, and Fe) is achieved by extraction of the stable niobium and tantalum fluoride complexes with oxygenated solvents such as MIBK, di-isopropyl ketone, cyclohexanone, and TBP [13–17]. By suitable choice of acid concentrations (HF, HCl, H_2SO_4), tantalum can be separated from niobium and vice versa [17]. Niobium and tantalum can be stripped from the organic phase with a hydrogen peroxide solution.

Alimarin et al. [18,19] have separated niobium from tantalum by extracting into $CHCl_3$ or CCl_4 the ion-pairs formed with the tetraphenylarsonium or trioctylammonium ion in fluoride or chloride media. Niobium and tantalum have also been separated from each other and from other metals by extraction as complexes with cupferron [20,21], BPHA [22,23], 8-hydroxyquinoline [24,25], TTA [26], pyrrolidinedithiocarbamate [27], and TBP [28,28a].

Ion pairs of niobium with catechol and quaternary ammonium ions have been used for separating niobium from other refractory metals (Ti, W, Mo, V) [28b].

34.1.3 Ion-Exchange

Many methods for isolating and separating Nb and Ta are based on retention of their anionic fluoride complexes in an anion-exchange column (plastic apparatus) [29–33]. Dixon and Headridge [33], for instance, recommend running a sample solution in HF through a strongly basic anion-exchange column, thereby sorbing Ti, Zr, Nb, Ta, Mo(VI), and W(VI), and eluting Al, V(IV), Cr(III), Mn(II), Fe(III), Co(II), Ni, and Cu. The retained metals are successively washed out as follows: Ti and Zr with $0.01M$ HF and $9M$ HCl, W with $3M$ HF and $10M$ HCl, Nb with $0.2M$ HF and $7M$ HCl, Mo with $3M$ HF and $3M$ HCl, and Ta with $1M$ NH_4F and $4M$ NH_4Cl. From a medium $6M$ in HCl and $1M$ in HF, niobium is retained on an anion-exchanger while zirconium is eluted [34].

Speecke and Hoste [35] have separated niobium and tantalum on Dowex 1 and 2 with mixed oxalic and hydrochloric acid media. Either tantalum is eluted first with $0.01M$ $H_2C_2O_4$ + $2M$ HCl, or niobium is eluted first with $0.5M$ $H_2C_2O_4$ and $1M$ HCl. This procedure is suitable for determining niobium in tantalum and vice versa.

Fritz and Dahmer [36] have absorbed Mo, W, Nb, and Ta on a Dowex-50 cation-exchanger from a solution of 0.5–$1M$ H_2SO_4 and 1.5–2% H_2O_2, and eluted these metals with a dilute solution of HNO_3 and H_2O_2.

Savvin et al. [37] have separated tantalum from niobium on chelating ion-exchangers.

34.2 Methods of Determination

The increased interest in niobium has produced many papers describing various spectrophotometric methods. A description is given below of the extractive thiocyanate method and of the more recent methods based on Bromopyrogallol Red and PAR. These methods have high selectivity and sensitivity. Pashchenko et al. [37a] have compared the thiocyanate, PAR, Pyrocatechol Violet and Sulphochlorophenol S methods for niobium.

34.2.1 Thiocyanate Method

Niobium(V) reacts with thiocyanate in hydrochloric acid to form a yellow complex. The niobium is determined spectrophotometrically either after extraction of the complex [38–43] or in an aqueous acetone (20–60%) medium [44–47]. The sensitivities in both cases are similar, but the extraction method is less subject to interference by other metals. Diethyl ether is commonly used as the solvent ($\sim 98\%$ extraction) but other solvents, such as ketones, esters, higher alcohols, and other ethers, are also suitable.

The absorption maximum of an ethereal solution of the niobium thiocyanate complex is at 385 nm. The molar absorptivity of the complex is $3 \cdot 5 \times 10^4$ ($a = 0 \cdot 38$).

Niobium is extracted in the presence of stannous chloride which reduces iron(III) and interfering oxidants, and which also enhances the colour of the organic extract. The concentrations of $SnCl_2$, hydrochloric acid, and thiocyanate greatly affect the intensity of the colour and the reproducibility of the results. These concentrations should not be lower than 4% $SnCl_2$, $2M$ HCl, and 10% KSCN.

A mixture of diethyl ether and CCl_4 (1+1) makes a useful extractant since it is denser than water and gives 92% extraction [42].

Because some thiocyanic acid is also extracted by ether, its concentration in the aqueous phase is lower for the second extraction, leading to increased dissociation and hence decreased extraction of the niobium complex. This can be prevented by using ether saturated with thiocyanic acid.

Niobium is kept in solution by the addition of tartaric acid, the presence of which also prevents the hydrolysis of tantalum and the consequent occlusion of niobium in the precipitate [48]. The concentration of tartaric acid should be identical in both sample and standard solutions.

Tungsten, molybdenum, and vanadium interfere in the determination of niobium by the thiocyanate method. In contrast to the corresponding tungsten complex, the niobium thiocyanate complex is decomposed by oxalic acid [49]. Interference by iron, uranium, titanium and tantalum can be prevented if these metals are present in not greater than hundredfold amount relative to niobium [50]. Phosphate and fluoride interfere, but the latter may be masked with aluminium ions [51].

Niobium may be determined as the extractable mixed thiocyanate complexes with tributylamine [52], tetraphenylarsonate [53], BPHA [54], N-acetylsalicyloyl-N-phenylhydroxylamine [55], 2-benzylaminopyridine [56], and 1,10-phenanthroline [56a].

Various versions of the thiocyanate method have been used to determine niobium in steel [42,49,57,57a], tantalum and its compounds [17,58,59], cobalt alloys [49], uranium [60], ores [4,46], rocks and minerals [5,61-63], molybdenite [64], and thin films of Nb-Ti [64a].

Niobium in industrial niobium salts has been determined by differential spectrophotometry of the niobium thiocyanate complex in aqueous acetone [65].

Reagents

1. Potassium thiocyanate: 30% solution.
2. Standard niobium solution: 1 mg/ml. Fuse 0·1430 g of niobium pentoxide, Nb_2O_5, with 4 g of $K_2S_2O_7$ in a quartz or platinum crucible. Dissolve the melt in a hot 5% tartaric acid solution, allow to cool, and dilute with the tartaric acid solution to 100 ml in a volumetric flask. Working solutions are obtained by suitable dilution of the stock solution with 2% tartaric acid.
3. Stannous chloride: 20% solution in $2M$ HCl.

Procedure

To the sample solution containing not more than 100 μg of Nb (complexed with tartrate), add concentrated hydrochloric acid and the $SnCl_2$ and thiocyanate solutions until the solution is $\sim 3M$ in HCl, $\sim 5\%$ in $SnCl_2$, and 12% in KSCN. After 5 minutes, extract the niobium thiocyanate complex with two portions of diethyl ether. Transfer the extracts to a 50-ml (or smaller) volumetric flask (according to the niobium content), make up to the mark with ether, and measure the absorbance at 385 nm against a reagent blank solution.

Note. The ratio of the volumes of extractant to the volume of the aqueous phase must be identical for both the sample and the standard solutions.

34.2.2 BROMOPYROGALLOL RED METHOD

In a mixed EDTA-tartrate medium at pH ~ 6, Bromopyrogallol Red (BPR) (formula, p. 52) reacts with niobium to form a blue complex with the constituent ratio Nb:BPR = 1:2.

The sensitivity of the Bromopyrogallol Red method developed by Belcher et al. [66,67] is higher than that of other conventional spectrophotometric methods for determining niobium. At 610 nm, the molar absorptivity is 4.75×10^4 (specific absorptivity 0·51). The absorption maximum of the reagent is at 560 nm.

Since the niobium BPR complex is insoluble in water, gelatin is added to give a stable colloidal dispersion. The colour reaction proceeds so slowly that maximum absorbance is attained only after 90 minutes, after which it remains constant. Greatest sensitivity is obtained in solutions buffered between pH 5·8 and 6·6 with ammonium acetate.

Most interfering cations are masked with EDTA. Tartrate masks milligram quantities of Ta, Ti, W, Mo, Sb(V), and Sn(IV); phosphate masks U(VI) and Zr; fluoride masks Al and Th; and cyanide masks any silver present. Interference from cerium(IV) and vanadium(V) may be eliminated by adding ascorbic acid to reduce these ions to Ce(III) and V(IV). Thousandfold amounts of oxalate, fluoride, and phosphate do not interfere.

Thus, when suitable masking agents are employed, the BPR method becomes specific for niobium.

If the niobium has been separated from other metals, it can be determined without EDTA or tartrate as masking agent. In such circumstances, a 1:3 Nb:BPR complex is formed and the sensitivity is increased ($\varepsilon \sim 6.0 \times 10^4$) [66].

The niobium–BPR chelate can be extracted into isopentyl acetate containing di-n-octylmethylamine ($\varepsilon = 2.5 \times 10^4$) [68].

The Bromopyrogallol Red method has been used to determine niobium in steels [68], and various metals and alloys [69].

Reagents

1. Bromopyrogallol Red: 0.02% solution. Dissolve 20 mg of the reagent in 50 ml of ethanol, and dilute the solution with water to 100 ml in a volumetric flask. The solution should not be used if over a week old.
2. Standard niobium solution: 1 mg/ml (p. 383).
3. Acetate buffer: pH 6.0. Dissolve 80 g of ammonium acetate in water, add 6 ml of glacial acetic acid, and dilute the solution with water to 1 litre.
4. Gelatin: 1% solution. Dissolve 1 g of gelatin in hot water, allow to cool, and dilute to 100 ml.

Procedure

Add the sample solution (~ 5 ml) containing not more than 60 µg of Nb (complexed with oxalate or tartrate) to a 50-ml volumetric flask. Neutralize to pH ~ 6, and add 5 ml of 20% potassium sodium tartrate solution, 5 ml of 5% EDTA solution, 4 ml of the Bromopyrogallol Red solution, 5 ml of the acetate buffer, and 0.5 ml of the gelatin solution. Mix the solution and set aside for 90 minutes. Then dilute with water to the mark, and measure the absorbance at 610 nm against a reagent blank solution.

Note. If aluminium or thorium is present, add NaF; if uranium or zirconium is present, add Na_3PO_4; and if silver is present, add KCN.

34.2.3 Pyridylazoresorcinol (PAR) Method

4-(2-Pyridylazo)resorcinol (PAR) (formula, p. 46) reacts with the niobium tartrate complex in nearly neutral medium to form a red, water-soluble 1:1 anionic complex, which has been utilized in the spectrophotometric determination of niobium [70–74].

The absorbance of the reagent ($\lambda_{max} = 410$ nm) is negligible at the absorption maximum (550 nm) of the niobium complex. The molar absorptivity of the complex is 3.6×10^4 ($a = 0.38$). The complex is formed in an ammonium acetate medium (pH 5.8–6.4), one hour being allowed for the colour to develop.

In the presence of EDTA and KCN as masking agents, only U(VI), V(V), and Ta interfere, giving high results. Interference by vanadium is readily overcome by adding zinc ions, and uranium is masked with oxalate. The interference by tantalum is prevented by adding excess of tartrate. When readily hydrolysable ions (e.g. Bi, Ti, and Zr) are present, the order of addition of the reagents should be EDTA, PAR, and buffer.

Fluoride, oxalate, and citrate cause a very slight reduction in the absorbance. Fluoride can be masked with borate. Phosphate is the only anion which interferes seriously.

PAR has been used to determine niobium in tantalum [75], steels [75–78], magnet alloys [79], zirconium and its alloys [80–82], titanium alloys [80,81], copper alloys [83], and rocks [84,85], as well as to determine larger amounts of niobium by a differential spectrophotometric method [86,87].

Reagents

1. 4-(2-Pyridylazo)resorcinol (PAR): 0·02% aqueous solution.
2. Standard niobium solution: 1 mg/ml (p. 383).
3. Buffer solution: pH 5·8. Dissolve 80 g of ammonium acetate in water, add 6·5 ml of glacial acetic acid, and dilute the solution to 1 litre.

Procedure

To the acidic sample solution containing not more than 100 μg of Nb (complexed with tartrate) in a 50-ml volumetric flask, add 5 ml of 1% EDTA solution and 5 ml of the PAR solution. Neutralize the solution with ammonia to pH ~ 6, and add 5 ml of the buffer solution. Dilute the solution with water to the mark, and allow to stand for 1 hour. Measure the absorbance of the solution at 550 nm against a reagent blank solution.

Note. If vanadium is present, add 1 ml of $0·01M$ ZnSO$_4$ before the addition of EDTA (this is sufficient to cause masking of at least 1 mg of VO$_3^-$). If uranium (VI) is present, add ammonium oxalate.

34.2.4 OTHER METHODS

A large number of other azo compounds besides PAR have been employed as spectrophotometric reagents for niobium. As with PAR, the reactions are carried out in the presence of complexants (tartrate, oxalate, hydrogen peroxide) to keep the niobium in solution. The coloured species produced are generally ternary niobium complexes.

Alimarin, Savvin *et al.* [88–91] have compared many mono- and bisazo-dyes as potential reagents for determining niobium. Sulphochlorophenol S (formula, p. 50), which reacts with niobium in 1–3M HCl, proved to be one of the best ($\varepsilon = 3·3 \times 10^4$) [92–96]. Niobium has been determined with this reagent in steels [94] and in various metals and alloys [96].

High sensitivity is a characteristic of the method for determining niobium with Sulphonitrophenol M ($\varepsilon = 5·3 \times 10^4$) [90,91,97,97a].

Further azo reagents used for determining niobium include Sulphonitrophenol S [90,98], Lumogallion (formula, p. 270) [89,99,100], Picramine R [88,101], Picramine-epsilon (formula, p. 50) [102], Magneson IREA [103], Arsenazo I [104], Rezarson [105], Acid Chrome Violet K [106], PAN [107], TAR [84,108], 5-(2-thiazolylazo)-2-monoethylamino-*p*-cresol (TAAK) [109], and bromo derivatives of pyridylazo compounds [110].

The following triphenylmethane dyes have been suggested as colorimetric reagents for niobium: Xylenol Orange (in the presence of oxalate or tartrate) ($\varepsilon = 1 \cdot 4 \times 10^4$ at 570 nm) [111–113], Pyrocatechol Violet [114,115], and Methylthymol Blue [116].

Polyphenol reagents characteristic for niobium include pyrocatechol [117], pyrogallol [118–122], tribromopyrogallol [123], dibromogallic acid [124], and Tiron [125]. The niobium polyphenol complexes can be extracted as ion-association compounds by organic bases [126–128].

Other diverse organic spectrophotometric reagents for niobium include oxine [129,130], haematoxylin [131], benzohydroxamic acid [132], salicylohydroxamic acid [133], *o*-nitrophenylfluorone and other trihydroxyfluorones ($\varepsilon = 1 \cdot 3 - 1 \cdot 7 \times 10^5$) [134,135], Tichromin ($\varepsilon = 1 \cdot 7 \times 10^4$) [136,137], Tipyrogin [138], TTA [139], and various dithiocarbamates [140–142].

The yellow niobium peroxide complex formed in a conc. H_2SO_4 medium gives a less sensitive ($\varepsilon \sim 1 \cdot 0 \times 10^3$) method suitable for determining larger quantities of niobium [143–145].

With phosphoric and molybdic acids, niobium forms yellow heteropoly acids which can be reduced with stannous chloride to the corresponding heteropoly blues. These colour reactions have been applied to the determination of niobium [146–149].

References

1. Schäfer, H., *Angew. Chem.* **71**, 153 (1959).
2. Ikenberry, L., Martin, J. L. and Boyer, W. J., *Anal. Chem.* **25**, 1340 (1953).
3. Kassner, J. L., Garcia-Porrata, A. and Grove, E. L., *ibid.* **27**, 492 (1955).
4. Dorosh, V. M., *Zh. Analit. Khim.* **16**, 250 (1961).
5. Grimaldi, F. S., *Anal. Chem.* **32**, 119 (1960).
6. Hibbits, J. O., Oberthin, H., Liu, R. and Kallmann, S., *Talanta* **8**, 209 (1961).
7. Majumdar, A. K., and Mukherjee, A. K., *Anal. Chim. Acta* **19**, 23 (1958).
8. Majumdar, A. K. and Pal, B. K., *ibid.* **27**, 356 (1962).
9. Majumdar, A. K. and Mukherjee, A. K., *ibid.* **21**, 330 (1959).
10. Patrovský, V., *Collection Czech. Chem. Commun.* **30**, 1727 (1965).
11. Ponomarev, A. I. and Bykovskaya, Yu. I., *Zh. Analit. Khim.* **21**, 1427 (1966).
12. Malissa, H., *Mikrochim. Acta* **1958**, 726.
13. Stevenson, P. C. and Hicks, H. G., *Anal. Chem.* **25**, 1517 (1953).
14. Waterbury, G. R. and Bricker, C. E., *ibid.* **29**, 1474 (1957); **30**, 1007 (1958).

References

15. Chernikhov, Yu. A., Tramm, R. S. and Pevzner, K. S., *Zavodsk. Lab.* **22**, 637 (1956).
16. Kaplan, G. E. and Baram, I. I., *Zh. Neorgan. Khim.* **10**, 703 (1965).
17. Luke, C. L., *Anal. Chim. Acta* **34**, 165 (1966).
18. Alimarin, I. P. and Makarova, S. V., *Zh. Analit. Khim.* **17**, 1072 (1962).
19. Alimarin, I. P., Ivanov, N. A. and Gibalo, I. M., *ibid.* **24**, 1521 (1969).
20. Alimarin, I. P. and Gibalo, I. M., *Dokl. Akad. Nauk SSSR* **109**, 1137 (1956).
21. Reed, J. F., *Talanta* **10**, 347 (1963).
22. Alimarin, I. P., Petrukhin, O. M. and Tsê, Y. H., *Dokl. Akad. Nauk SSSR* **136**, 1073 (1961); *Zh. Neorgan. Khim.* **7**, 1191 (1962).
23. Erskine, J. S., Sink, M. L. and Varga, L. P., *Anal. Chem.* **41**, 70 (1969).
24. Alimarin, I. P., Bilimovich, G. N. and Tsui, H. H., *Zh. Neorgan. Khim.* **7**, 2725 (1962).
25. Pyatnitskii, I. V. and Sereda, E. S., *Zh. Analit. Khim.* **25**, 1552 (1970).
26. Jurriaanse, A. and Moore, F. L., *Anal. Chem.* **39**, 494 (1967).
27. Gibalo, I. M., Alimarin, I. P. and Davaadorzh, P., *Zh. Analit. Khim.* **19**, 467 (1964); *Vestn. Moskov. Univ. Khim.* **1965**, No 2, 73.
28. Ryabchikov, D. I. and Volynets, M. P., *Zh. Analit. Khim.* **14**, 700 (1959).
28a. De, A. K. and Sen, A. K., *Talanta* **13**, 853 (1966).
28b. Nardillo, A. M. and Catoggio, J. A., *Anal. Chim. Acta* **66**, 359 (1973).
29. Cabell, M. J. and Milner, I., *ibid.* **13**, 258 (1955).
30. Hague, J. L. and Machlan, L. A., *J. Res. Natl. Bur. Stds.* **62**, 11 (1959).
31. Kallmann, S., Oberthin, H. and Liu, R., *Anal. Chem.* **34**, 609 (1962).
32. Headridge, J. B. and Dixon, E. J., *Analyst* **87**, 32 (1962).
33. Dixon, E. J. and Headridge, J. B., *ibid.* **89**, 185 (1964).
34. Holloway, J. H. and Nelson, F., *J. Chromatog.* **14**, 255 (1964).
35. Speecke, A. and Hoste, J., *Talanta* **2**, 332 (1959).
36. Fritz, J. S. and Dahmer, L. H., *Anal. Chem.* **37**, 1272 (1965).
37. Savvin, S. B., Myasoedov, B. F. and Eliseeva, O. P., *Zh. Analit. Khim.* **24**, 1023 (1969).
37a. Pashchenko, E. N., Vasilieva, L. A., Mal'tsev, V. F. and Volkova, N. P., *Zavodsk. Lab.* **39**, 1297 (1973).
38. Alimarin, I. P. and Podval'naya, R. L., *Zh. Analit. Khim.* **1**, 30 (1946).
39. Lauw-Zecha, A. B., Lord, S. S., Jr. and Hume, D. N., *Anal. Chem.* **24**, 1169 (1952).
40. Troitskii, K. V., *Zh. Analit. Khim.* **12**, 349 (1957).
41. Mari, E. A., *Anal. Chim. Acta* **29**, 303, 312 (1963).
42. Minczewski, J. and Różycki C., *Chem. Anal. (Warsaw)* **9**, 601 (1964); **10**, 463, 701 (1965).
43. Djordjević, C. and Tamhina, B., *Anal. Chem.* **40**, 1512 (1968).
44. Freund, H. and Levitt, A. E., *ibid.* **23**, 1813 (1951).
45. Crouthamel, C. E., Hjelte, B. E. and Johnson, C. E., *ibid.* **27**, 507 (1955).
46. Marzys, A. E., *Analyst* **79**, 327 (1954); **80**, 194 (1955).
47. Bacon, A. and Milner, G. W., *Anal. Chim. Acta* **15**, 129 (1956).
48. Bukhsh, M. N. and Hume, D. N., *Anal. Chem.* **27**, 116 (1955).
49. McDuffie, B., Bandi, W. R. and Melnick, L. M., *ibid.* **31**, 1311 (1959).
50. Mundy, R. J., *ibid.* **27**, 1408 (1955).
51. Canada, D. C., *ibid.* **39**, 381 (1967).
52. Ziegler, M., Glemser, O. and Baeckmann, A., *Z. Anal. Chem.* **172**, 105 (1960).
53. Affsprung, H. E. and Robinson, J. L., *Anal. Chim. Acta* **37**, 81 (1967).
54. Villarreal, R. and Barker, S. A., *Anal. Chem.* **41**, 611 (1969).
55. Savariar, C. P. and Joseph, J., *Talanta* **17**, 45 (1970).
56. Pilipenko, A. T. and Eremenko, O. M., *Zh. Analit. Khim.* **25**, 1330 (1970).
56a. Lobanov, F. I., Zatonskaya, V. M. and Gibalo, I. M., *ibid.* **29**, 826 (1974).
57. White, G. and Scholes, P. H., *Metallurgia* **70**, 197 (1964).
57a. Westland, A. D. and Bezaire, J., *Anal. Chim. Acta* **66**, 187 (1973).

58. Hastings, J. and McClarity, T. A., *Anal. Chem.* **26**, 683 (1954).
59. Bergstresser, K. S., *ibid.* **31**, 1812 (1959).
60. Shrimal, S. K. and Varde, M. S., *Anal. Chim. Acta* **33**, 683 (1965).
61. Milner, G. W. and Smales, A. A., *Analyst* **79**, 315 (1954).
62. Ward, F. N. and Marranzino, A. P., *Anal. Chem.* **27**, 1325 (1955).
63. Esson, J., *Analyst* **90**, 488 (1965).
64. Minczewski, J. and Różycki, C., *Chem. Anal. (Warsaw)* **10**, 965 (1965).
64a. Godovskaya, K. I., Babenko, A. S. and Alferov, E. A., *Zavodsk. Lab.* **39**, 1169 (1973).
65. Malyutina, T. M., Futoryanskaya, E. L. and Vinokurova, F. A., *ibid* **28**, 540 (1962).
66. Belcher, R., Ramakrishna, T. V. and West, T. S., *Talanta* **12**, 681 (1965).
67. West, T. S., *Zh. Analit. Khim.* **21**, 913 (1966).
68. Ramakrishna, T. V., Rahim, S. A. and West, T. S., *Talanta* **16**, 847 (1969).
69. Williams, A. I., *Analyst* **92**, 43 (1967).
70. Belcher, R., Ramakrishna, T. V. and West, T. S., *Talanta* **9**, 943 (1962); **10**, 1013 (1963).
71. Alimarin, I. P. and Han, H. I., *Zh. Analit. Khim.* **18**, 182 (1963).
72. Elinson, S. V., Pobedina, L. I. and Rezova, A. T., *ibid.* **20**, 676 (1965).
73. Kuchmistaya, G. I., Dobkina, B. M. and Elinson, S. V., *ibid.* **25**, 742 (1970).
74. Široki, M. and Djordjevic, C., *Anal. Chem.* **43**, 1375 (1971).
75. Elinson, S. V., Pobedina, L. I. and Rezova, A. T., *Zavodsk. Lab.* **31**, 1434 (1965); **32**, 1314 (1966).
76. Staats, G. and Brück, H., *Z. Anal. Chem.* **230**, 271 (1967).
77. Pakalns, P., *Anal. Chim. Acta* **41**, 283 (1968).
78. Gagliardi, E. and Höllinger, W., *Mikrochim. Acta* **1972**, 136.
79. Bagdasarov, K. N. and Osmanov, Kh. A., *Zavodsk. Lab.* **34**, 1044 (1968).
80. Elinson, S. V. and Pobedina, L. I., *Zh. Analit. Khim.* **18**, 189 (1963).
81. Wood, D. F. and Jones, J. T., *Analyst* **93**, 131 (1968).
82. Pakalns, P. and Ivanfy, A. B., *Anal. Chim. Acta* **41**, 139 (1968).
83. Elinson, S. V., Pyatiletova, N. M. and Novikova, I. S., *Zavodsk. Lab.* **36**, 659 (1970).
84. Patrovský, V., *Chem. Listy* **59**, 1464 (1965).
85. Greenland, L. P. and Campbell, E. Y., *Anal. Chim. Acta* **49**, 109 (1970).
86. Elinson, S. V., Pobedina, L. I. and Rezova, A. T., *Zavodsk. Lab.* **37**, 391 (1971).
87. Dobkina, B. M., Kuchmistaya, G. I. and Nadezhdina, G. B., *Zh. Analit. Khim.* **27**, 194 (1972).
88. Alimarin, I. P., Savvin, S. B. and Dedkov, Yu. M., *ibid.* **19**, 328 (1964).
89. Alimarin, I. P. and Savvin, S. B., *Talanta* **13**, 689 (1966).
90. Alimarin, I. P., Savvin, S. B. and Okhanova, L. A., *ibid.* **15**, 601 (1968).
91. Savvin, S. B., Alimarin, I. P., Okhanova, L. A., and Belova, T. Ya., *Tr. Komis. po Analit. Khim. Akad. Nauk SSSR* **17**, 163 (1969).
92. Buděšínský, B. and Savvin, S. B., *Z. Anal. Chem.* **214**, 189 (1965).
93. Savvin, S. B., Bortsova, V. A. and Malkina, E. N., *Zh. Analit. Khim.* **20**, 947 (1965).
94. Savvin, S. B., Romanov, P. N. and Eremin, Yu. G., *ibid.* **21**, 1423 (1966).
95. Savvin, S. B., Alimarin, I. P., Belova, T. Ya. and Okhanova, L. A., *ibid.* **23**, 1117 (1968).
96. Elinson, S. V., Savvin, S. B. and Mirzoyan, N. A., *Zavodsk. Lab.* **34**, 136 (1968).
97. Savvin, S. B., Propistsova, R. F. and Okhanova, L. A., *Zh. Analit. Khim.* **24**, 1634 (1969).
97a. Gerkhardt, L. I., Okhanova, L. A., Savvin, S. B. and Vagan, V. F., *Zavodsk. Lab.* **39**, 769 (1973).
98. Yurchenko, E. I., Savvin, S. B., Zubasheva, L. V., Garan', V. F. and Mishinskaya, I. S., *ibid.* **32**, 12 (1966).
99. Alimarin, I. P. and Han, H. I., *Zh. Analit. Khim.* **18**, 82 (1963); *Vestn. Mosk. Univ. Khim.* **1964**. No. 1, 65; No. 2, 41.

100. Konusova, V. V., Opochanskaya, L. D. and Tsykhanskii, V. D., *Zh. Analit. Khim.* **24**, 400 (1969).
101. Elinson, S. V., Savvin, S. B., Dedkov, Yu. M. and Tsvetkova, V. T., *Zavodsk. Lab.* **32**, 654 (1966).
102. Dedkov, Yu. M., Dymova, M. S. and Yakovlev, P. Ya., *ibid.* **37**, 753 (1971).
103. Elinson, S. V. and Nezhnova, T. I., *Zh. Analit. Khim.* **26**, 1535 (1971).
104. Nikitina, E. I., *Zavodsk. Lab.* **27**, 663 (1961).
105. Lukin, A. M., Kaslina, N. A. and Petrova, G. S., *Zh. Analit. Khim.* **26**, 1782 (1971).
106. Tramm, R. S. and Pevzner, K. S., *Zavodsk. Lab.* **30**, 20 (1964).
107. Gagliardi, E. and Wolf, E., *Mikrochim. Acta* **1967**, 104.
108. Patrovský, V., *Talanta* **12**, 971 (1965).
109. Mal'tseva, L. S. and Elinson, S. V., *Zavodsk. Lab.* **39**, 385 (1973).
110. Elinson, S. V. and Mal'tseva, L. S., *Zh. Analit. Khim.* **22**, 79 (1967); *Tr. Komis. po Analit. Khim. Akad. Nauk SSSR* **17**, 175 (1969).
111. Cheng, K. L. and Goydish, B. L., *Talanta* **9**, 987 (1962).
112. Babko, A. K. and Shtokalo, M. I., *Zh. Analit. Khim.* **17**, 1068 (1962).
113. Elinson, S. V. and Pobedina, L. I., *ibid.* **18**, 734 (1963).
114. Mal'tsev, V. F., Pashchenko, E. N. and Volkova, N. P., *ibid.* **21**, 1205 (1966).
115. Tananaiko, M. M., Vdovenko, O. P. and Shevchuk, V. V., *Ukr. Khim. Zh.* **39**, 939 (1973).
116. Elinson, S. V. and Mirzoyan, N. A., *ibid.* **21**, 1436 (1966).
117. Patrovský, V., *Collection Czech. Chem. Commun.* **23**, 1774 (1958).
118. Hunt, E. C. and Wells, R. A., *Analyst* **79**, 345 (1954).
119. Wood, D. F. and Scholes, I. R., *Anal. Chim. Acta* **21**, 121 (1959).
120. Webb, H. W., Ashworth, V. and Hills, J. M., *Analyst* **88**, 142 (1963).
121. Geissler, M. and Lorenz, G., *Z. Anal. Chem.* **244**, 235 (1969).
122. Ali-Zade, I. D., Gami-Zade, G. A. and Akhmetova, G. A., *Zh. Analit. Khim.* **29**, 739 (1974).
123. Ackermann, G. and Koch, S., *Talanta* **9**, 1015 (1962).
124. Ackermann, G. and Koch, S., *ibid.* **16**, 95, 284, 288 (1969); **17**, 757 (1970).
125. Flaschka, H. and Lassner, E., *Mikrochim. Acta* **1956**, 778.
126. Catoggio, J. A. and Rogers, L. B., *Talanta* **9**, 377 (1962).
127. Yagnyatinskaya, G. Ya. and Nazarenko, V. A., *Zavodsk. Lab.* **34**, 1047 (1968); **36**, 158 (1970).
128. Gibalo, I. M. and Eremina, G. V., *Zh. Analit. Khim.* **26**, 1531 (1971).
129. Motojima, K. and Hashitani, H., *Anal. Chem.* **33**, 48 (1961).
130. Keil, R., *Z. Anal. Chem.* **229**, 267 (1967).
131. Kornilova, V. I. and Nazarchuk, T. N., *Ukr. Khim. Zh.* **29**, 1205 (1963).
132. Gibalo, I. M., Voskov, V. S. and Lobanov, F. I., *Zh. Analit. Khim.* **25**, 1918 (1970).
133. Alimarin, I. P. and Borzenkova, N. P., *Vestn. Mosk. Univ. Khim.* **1963**, No. 6, 65.
134. Yagnyatinskaya, G. Ya. and Nazarenko, V. A., *Zavodsk. Lab.* **32**, 510 (1966).
135. Nazarenko, V. A. and Yagnyatinskaya, G. Ya., *ibid.* **38**, 1427 (1972); **40**, 21 (1974).
136. Yakovlev, P. Ya., Basargin, N. N. and Panarina, N. A., *Zh. Analit. Khim.* **25**, 505 (1970).
137. Basargin, N. N., Yakovlev, P. Ya., Panarina, N. A. and Onuchina, G. V., *Zavodsk. Lab.* **37**, 143 (1971).
138. Basargin, N. N., Yakovlev, P. Ya. and Panarina, N. A., *Zh. Analit. Khim.* **25**, 746 (1970).
139. Savrova, O. D., Gibalo, I. M. and Lobanov, F. I., *Anal. Lett.* **5**, 669 (1972).
140. Gibalo, I. M., Alimarin, I. P. and Davaadorzh, P., *ibid.* **18**, 835 (1963).
141. Uvarova, K. A., Usatenko, Yu. I. and Klopova, Zh. G., *Zavodsk. Lab.* **36**, 909 (1970).
142. Uvarova, K. A., Usatenko, Yu. I. and Klopova, Zh. G., *ibid.* **38**, 1431 (1972).
143. Charlot, G. and Saulnier, J., *Chim. Anal. (Paris)* **35**, 51 (1953).
144. Schäfer, M. and Schutle, F., *Z. Anal. Chem.* **149**, 73 (1956).

145. Gorlach, V. F., Pyatnitskii, I. V. and Kostyuchenko, L. P., *Ukr. Khim. Zh.* **36**, 1260 (1970).
146. Norwitz, G. and Codell, M., *Anal. Chem.* **26**, 1230 (1954).
147. Guyon, J. C., Wallace, G. W., Jr. and Mellon, M. G., *ibid.* **34**, 640 (1962).
148. Shkaravskii, Yu. F., *Zh. Analit. Khim.* **18**, 196 (1963).
149. Zaboeva, M. I., Surin, I. G. and Serkova, A. V., *ibid.* **28**, 1736 (1973).

Chapter 35

NITROGEN

Nitrogen (N, at.wt. 14·01) is a chemically inert gas (N_2) at room temperature. It occurs in its compounds in the following oxidation states: $-III$ in ammonia, $-II$ in hydrazine, $-I$ in hydroxylamine, $+III$ in nitrite, and $+V$ in nitrate. Spectrophotometrically, it is usually determined as ammonia.

35.1 Methods for Separating Nitrogen as Ammonia

The distillation of ammonia (NH_3) from an alkaline medium is a commonly used method for separating small amounts of nitrogen [1–3]. Ammonia is a volatile weak base which escapes quantitatively when a strong base is added in excess and the solution is subsequently heated. The ammonia and steam driven off are condensed and collected in dilute acid.

Thaler and Sturm [4] have investigated the separation of ammonia by steam distillation, vacuum distillation, and isothermal diffusion methods.

Nitrate and nitrite nitrogen are reduced to ammonia by powdered Devarda's alloy (50% Cu, 45% Al, 5%Zn) in a cold alkaline medium [5–7]. The reaction is carried out in the same flask from which ammonia is afterwards distilled.

By Kjeldahl's method (1883), organic nitrogen is converted into ammonia which is separated by distillation. The method entails boiling a sample with concentrated sulphuric acid in the presence of K_2SO_4 (to raise the b.p.) and a catalyst (mercury, copper, or selenium) [1–3,6,8]. If the nitrogen is present in a group also containing oxygen, a reduction step is also necessary. The carbon in the organic compound is oxidized to CO_2 while the sulphuric acid is reduced to SO_2. Organic nitrogen is converted into ammonia, which remains in solution as the ammonium ion. After the mineralization, the acidic solution is diluted with water and made alkaline with sodium hydroxide before the ammonia is distilled.

35.2 Methods for Ammonia Determination

Two methods of similar sensitivity are commonly used for determining ammonia. These are the classical Nessler method and the modern indophenol method, both of which are generally preceded by the distillative separation of ammonia. The indophenol method is more convenient since the blue reaction product is soluble in water.

35.2.1 INDOPHENOL METHOD

The reaction of ammonia with hypochlorite and phenol in an alkaline medium (Berthollet's reaction) yields a blue product which is the basis of a sensitive and specific spectrophotometric method for determining nitrogen as ammonia [9–13]. The probable mechanism of the reaction is:

$$\text{C}_6\text{H}_5\text{–OH} \xrightarrow{\text{ClO}^-+\text{NH}_3} \text{O=C}_6\text{H}_4\text{=N–Cl} + \text{C}_6\text{H}_5\text{–OH} \rightarrow$$

$$\text{O=C}_6\text{H}_4\text{=N–C}_6\text{H}_4\text{–OH} \xrightarrow[\text{H}^+]{\text{OH}^-} \text{O=C}_6\text{H}_4\text{=N–C}_6\text{H}_4\text{–O}^-$$

The blue colour is due to the indophenol anion, formed in alkaline medium. Undissociated indophenol, present in acid solution, is yellow.

The intensity of the blue colour is greatly increased by adding a little acetone ($\sim$ 0·3 ml of acetone per 50 ml of solution). Curve 2 in Fig. 35.1 represents the absorption spectrum of the coloured solution in the presence of acetone. The molar absorptivity at λ_{max} = 625 nm is $4 \cdot 5 \times 10^3$ (a = 0·32).

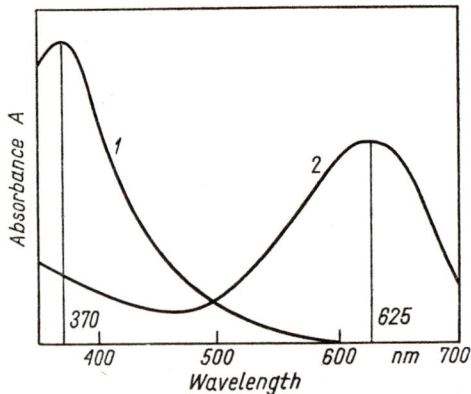

Fig. 35.1 Absorption spectra of the coloured products in the Nessler method (1) and the indophenol method (2) for determining ammonia

Namiki et al. [14] have extracted the indophenol with isobutyl or isoamyl alcohol after adding a considerable amount of sodium chloride to the aqueous solution as a salting-out agent. The organic extract is, however, less intensely coloured than a corresponding solution in aqueous acetone.

The ratio of sodium phenolate to hypochlorite is important in achieving maximum absorbance. Chloramine-T or hypobromite have been substituted for hypochlorite [14,15]. An analogous reaction occurs when phenol is replaced by thymol, the blue product being extractable into isoamyl alcohol, butanol, or diethyl ether [15–17]. It is sometimes possible to carry out the indophenol reaction without separating the ammonia (e.g. in natural waters). In the presence of EDTA, moderate quantities

(0·1–0·5 mg) of Ca, Mg, and Al do not interfere. The addition of tartrate or citrate prevents the precipitation of hydrolysable metals. Phosphate interferes in the colour reaction.

Since the sensitivity of the method is often limited by the high blank values caused by the presence of traces of ammonia in the reagents, purification of the reagents by distilling ammonia from their alkaline solutions may be necessary.

The indophenol method has been used to determine nitrogen (as ammonia) in tantalum alloys [18], refractory alloys [19], uranium-plutonium dioxides [20], titanium carbide [15], soil extracts [21,22], petroleum [23], plant materials [24], organic substances [10,11,24a], biological materials [25–27], sea water [9,16], boiler feed water [28], natural water [29], and air [30].

Many authors have used the indophenol method for the automated determination of ammonia [21–24,31].

Reagents

1. Phenol–acetone solution. Dissolve 70 g of phenol in 15 ml of ethanol, add 20 ml of acetone, and dilute the solution with ethanol to 100 ml.
2. Sodium phenolate solution. Immediately before use, mix 10 ml of solution 1 with 10 ml of 30% aqueous NaOH, and dilute with water to 50 ml.
3. Sodium hypochlorite: 2% solution. (Content of hypochlorite in the solution is checked iodometrically.)
4. Standard ammonia solution: 1 mg NH_3/ml. Dissolve 3·141 g of ammonium chloride (previously dried at $\sim 100°C$) in water, and dilute the solution with water to 1 litre.
   Standard ammonia solution: 1 mg N/ml. Prepare as above by dissolving 3·819 g of dried NH_4Cl.
5. Sodium hydroxide: 30% aqueous solution. Boil the solution for 10–15 minutes in an open vessel to remove traces of ammonia.

Procedure

Distillation of NH_3. Place the solution (≤ 50 ml) containing ammonia in a 70–150-ml still. Immerse the condenser outlet in a receiver containing 5 ml of water and 5 drops of $0·1M$ H_2SO_4. Pour 10–20 ml of 30% NaOH into the still, and dilute with water to 40–80 ml. Add a few fragments of porous porcelain to promote regular ebullition, and distil a quarter of the liquid volume from the still into the receiver.

Determination of NH_3. Place all or part of the distillate containing not more than 80 μg of NH_3 in a 50-ml volumetric flask. Add 5 ml of the sodium phenolate solution and 2 ml of the sodium hypochlorite solution. Dilute the solution to the mark with water, and mix thoroughly. After 30 minutes, measure the absorbance at 625 nm against water (or a reagent blank solution when traces of ammonia are being determined).

Note. For determination of traces of ammonia, the sample solution should be made slightly acidic with dilute H_2SO_4 (e.g. 0·5 ml of 0·1M H_2SO_4), and concentrated by evaporation, before the distillation.

35.2.2 Nessler's Method

In 1856, Nessler recommended an alkaline solution of mercuric iodide and potassium iodide as a reagent for the colorimetric determination of ammonia [32–35]. Nessler's reagent reacts with ammonia in an alkaline medium according to the following equation [34]:

$$2HgI_4^{2-} + NH_3 + 3OH^- \rightarrow NH_2Hg_2IO \downarrow + 7I^- + 2H_2O$$

Since the brown-orange sparingly-soluble product forms a stable dispersion only at very low concentrations, protective colloids such as gum arabic, gelatin, and poly(vinyl alcohol) are added.

The absorption spectrum of the compound formed by ammonia with Nessler's reagent is shown in Fig. 35.1. The molar absorptivity at λ_{max} = 370 nm is $6·8 \times 10^3$; at 400 nm, it is $\sim 5·0 \times 10^3$ (specific absorptivity 0·36).

As in the indophenol method, the determination of ammonia is usually preceded by a distillation from strongly alkaline solution. Nessler's method is, however, commonly used for determining ammonia directly in natural waters. Since calcium and magnesium present in water interfere by forming precipitates with the reagent, they are masked with tartrate.

Nessler's method has been used to determine nitrogen (as ammonia) in biological materials [34,36], plant materials [6,37], air [38], minerals [39], zirconium [40], chromic acid [7], tungsten [41], and water [42].

Reagents

1. Nessler's reagent. Dissolve 5 g of potassium iodide in 10 ml of water. Add saturated $HgCl_2$ solution until a permanent precipitate forms. Add 50 ml of 30% NaOH, dilute the solution to 200 ml with water, and mix well. Decant the clear solution from the precipitate, and store the solution in an amber-glass bottle.
2. Standard ammonia solution (p. 393).
3. Potassium sodium tartrate (Seignette salt): 20% solution. Remove traces of ammonia by making the solution alkaline with sodium hydroxide, and boiling for 10–15 minutes.

Procedure

Place the clear, colourless, neutral solution containing not more than 100 μg of NH_3 in a 50-ml volumetric flask, add 1 ml of the tartrate solution, 1 ml of 1% gum arabic solution, and 1 ml of Nessler's reagent, and dilute the solution to the mark with water. After 10 minutes, measure the absorbance of the pseudosolution at $\sim$ 400 nm, using a reagent blank solution as reference.

35.2.3 Other Methods for NH_3 Determination

Ammonia reacts with pyrazolone in pyridine to form a purple compound which is determined spectrophotometrically after extraction into carbon tetrachloride [43]. In the reaction of ammonia with bis-pyrazolone and chloramine-T, a yellow compound is formed which is extractable by trichloroethylene [44].

Zitomer and Lambert [45] have formed trichloramine from ammonia and hypochlorite, the excess of ClO^- being decomposed with nitrite, and the trichloramine reacted with a mixture of cadmium iodide and starch to give a blue complex ($\lambda_{max} = \sim 610$ nm). This method has been used to determine traces of ammonia in blood [46].

Ammonia has also been determined on the basis of its colour reaction with indanetrione [47], and with ruthenium(III) chloride and triphenylphosphine [48].

Finally, ammonia can be oxidized to nitrite, the latter then being determined by any suitable method [49].

35.3 Methods for Nitrite Determination

The classical Griess method for nitrite determination is very well known and very widely used. However, some of the more recent methods which are also based upon the formation of azo dyes may be better.

35.3.1 Modified Griess Method

In an acid medium, nitrite reacts with primary aromatic amines to form a diazonium salt. The salt is then coupled with a suitable aromatic compound to yield an azo dye which is the basis of the spectrophotometric method [50–52].

In the Griess method (1879), nitrite, sulphanilic acid (4-aminobenzenesulphonic acid), and 1-naphthylamine are reacted as follows:

$$HO_3S-\langle\bigcirc\rangle-NH_2 + NO_2^- + 2H^+ \rightarrow HO_3S-\langle\bigcirc\rangle-\overset{+}{N}\equiv N + 2H_2O$$

$$HO_3S-\langle\bigcirc\rangle-\overset{+}{N}\equiv N + \langle\bigcirc\bigcirc\rangle-NH_2 \rightarrow HO_3S-\langle\bigcirc\rangle-N=N-\langle\bigcirc\bigcirc\rangle-NH_2 + H^+$$

Optimum conditions for achieving a quantitative reaction and a stable coloured product are fulfilled when diazotization occurs in a strongly acidic medium and coupling in a weakly acidic medium obtained by adding sodium acetate [50,51].

The original Griess method is highly sensitive (the molar absorptivity at $\lambda_{max} = 520$ nm is 4.0×10^4, specific absorptivity 2·85), and is specific, but of only moderate precision. It is not applicable if too much nitrite is present, however, as side-reactions occur.

The solution in which nitrite is determined must not contain oxidants, reductants, or coloured substances. Neither should urea or aliphatic amines be present since they may react with nitrite to liberate free nitrogen. Copper ions catalyse the decomposition of the diazonium salt, thereby causing low results.

Besides its wide application in the analysis of food and water, the Griess method is suitable for the automated determination of nitrite and nitrate (after reduction to nitrite) in water [53], blood [54], and soil extracts [55].

1-Naphthylamine is a well-known carcinogen and its use is not recommended [55a]. The 7-sulphonic acid seems acceptable [56].

Reagents
1. Sulphanilic acid solution. Dissolve 0·50 g of sulphanilic acid in 120 ml of water and 30 ml of glacial acetic acid. Store in a brown bottle.
2. 1-Naphthylamine-7-sulphonic acid solution. Dissolve 0·50 g of the compound in 120 ml of water, warming on a water-bath. Filter, cool, and add 30 ml of glacial acetic acid. Store in a brown bottle.
3. Standard nitrite solution: 1 mg NO_2^-/ml. Dissolve 1·500 g of anhydrous $NaNO_2$ in water, add 1 ml of chloroform and 0·2 g of NaOH to stabilize the solution, and dilute with water to 1 litre.

Procedure
To the clear, colourless, neutral (pH 6·5–7·5) solution containing not more than 40 μg of nitrite (NO_2^-), and diluted to $\sim$ 40 ml in a 50-ml volumetric flask, add 2·0 ml of the sulphanilic acid solution, mix, and allow to stand for 10 minutes. Add 2·0 ml of the 1-naphthylamine-7-sulphonic acid solution, dilute to the mark and mix. After 20 minutes, measure the absorbance at 520 nm, using a reagent blank solution as reference.

35.3.2 Other Methods for NO_2^- Determination

Besides the classical reagents (sulphanilic acid and 1-naphthylamine) used in the Griess method, a number of other organic compounds are suitable for the diazotization and coupling reactions [57–66]. Many of these give more sensitive methods. For example, the reaction with 4-aminoazobenzene and 1-naphthylamine yields a blue dye [60], and when 8-aminoquinoline is used for both the diazotization and the subsequent coupling, an extractable dye is formed [62]. Sawicki *et al.* [57] have compared 52 methods based on the formation of azo dyes.

When determining traces of nitrite in sea water, Wada and Hattori [67] formed an azo dye in a large volume of sample, adsorbed the dye on an anion-exchange column, and eluted it with a small volume of acetic acid before measuring the absorbance.

Other worthwhile methods for determining nitrite include those based on colour reactions with thioglycollic acid [68], thiourea [+Fe(III)] [69], brucine [70], and *o*-tolidine [71].

35.4 Methods for Nitrate Determination

Most spectrophotometric methods for determining nitrate are based either on (1) nitration or oxidation of appropriate organic reagents to form coloured compounds, or on (2) determination as NO_2^- or NH_3 after reduction of the NO_3^- ion. Prevalent among methods belonging to the first group are those using phenoldisulphonic acid, brucine, and xylenols.

35.4.1 PHENOLDISULPHONIC ACID METHOD

Reaction between 1-phenol-2,4-disulphonic acid and nitric acid occurs when a dry sample (or dry residue from the evaporation of the sample solution) containing nitrate is mixed with a solution of the reagent in concentrated sulphuric acid.

$$\underset{\underset{SO_3H}{|}}{\overset{\overset{OH}{|}}{C_6H_3}}-SO_3H + HNO_3 \rightarrow \underset{\underset{SO_3H}{|}}{\overset{\overset{OH}{|}}{C_6H_2}}(O_2N)-SO_3H$$

The reaction product, nitrophenoldisulphonic acid, is pale yellow, but, when the solution is made alkaline, the intensely coloured anion which is the basis of this spectrophotometric method is formed [72–74].

The reaction is specific for nitrate. The absorption maximum of the nitrophenoldisulphonate anion is at 410 nm. The molar absorptivity is 9.4×10^3 ($a = 0.67$).

Neither the nature nor the excess of the reagent used to raise the pH (NH_3, NaOH, KOH) is important. Chloride causes low results owing to the reaction between HCl and HNO_3 when the phenoldisuphonic acid in concentrated sulphuric acid is added to the sample ($3Cl^- + NO_3^- + 4H^+ \rightarrow Cl_2 + NOCl + 2H_2O$). This effect is, however, negligible if the amount of chloride present is less than that of nitrate. Larger amounts of chloride should be separated beforehand by precipitation of AgCl with an Ag_2SO_4 solution or Ag_2O suspension. Since the presence of silver ions in the solution after removal of AgCl is detrimental, the excess is precipitated with sodium phosphate. If the concentration of chloride in the solution is known exactly, it is better to use a stoichiometric (or slightly lesser) quantity of silver sulphate as precipitant [73,74].

Nitrite interferes since it may be partially converted into nitrate under the conditions of the determination. When the concentration of nitrite is not higher than that of nitrate, the effect is negligible. Larger quantities of nitrite must be removed, e.g. by reduction with sodium azide, sulphamic acid, urea, or hydrazine [74].

If nitrite is determined in one part of the sample (e.g. by the modified Griess method), and if in the other the sum of nitrite plus nitrate is determined by the phenoldisulphonic acid method after oxidation of the nitrite, the amount of nitrate is obtained by difference.

Hora and Webber [75] have found that ammonium ions cause low results when nitrate is determined with phenoldisulphonic acid. They recommend preliminary expulsion of ammonia by heating the solution after it has been made alkaline with sodium hydroxide.

EDTA is added before the evaporation when larger quantities of Ca and Mg are present, thus preventing precipitation when the solution is finally made alkaline. To prevent loss of HNO_3 during evaporation to dryness, the solution is neutralized with NaOH or $CaCO_3$.

The phenoldisulphonic acid method has been used to determine nitrate in water [73], plant material [72], soil extracts [76], chemical reagents [74], and cellulose nitrates [77].

Reagents

1. Phenoldisulphonic acid: solution in conc. H_2SO_4. Dissolve 12·5 g of phenol in 75 ml of conc. H_2SO_4, add 37·5 ml of 13% oleum, and stir well. Heat the solution in a 250-ml conical flask for 2 hours on a boiling water-bath with occasional stirring.
2. Standard nitrate solution: 1 mg NO_3^-/ml. Dissolve in water 1·631 g of KNO_3^- (previously dried at 110°C) and dilute the solution with water to 1 litre.
3. Silver sulphate solution. Dissolve in water 1·10 g of Ag_2SO_4 and dilute the solution with water to 250 ml. One ml of solution is equivalent to 1 mg of Cl^-.
4. Calcium carbonate suspension. Mix 1 g of $CaCO_3$ with 100 ml of water.

Procedure

Neutralize the solution (containing not more than 200 µg of nitrate) in a small evaporating dish, add 1 ml of the $CaCO_3$ suspension, and evaporate to dryness on a water-bath. Treat the cooled residue with 2 ml of phenoldisulphonic acid reagent and stir well. After 5 minutes, dilute with 20 ml of water, and transfer the solution quantitatively to a 50-ml volumetric flask. Add conc. ammonia solution until the solution becomes intensely yellow, then add 5 ml more, and dilute the solution with water to the mark. Measure the absorbance of the yellow solution at 410 nm against water or a reagent blank solution (at low nitrate contents).

Notes. (1) If, after being made alkaline, the solution becomes turbid or a precipitate is formed, the solution should be filtered before the absorbance is measured.

(2) If the amount of chloride present exceeds that of nitrate, the former should be separated by adding a slightly less than stoichiometric quantity of silver sulphate to the acid solution. After 30 minutes, the precipitated AgCl is filtered off, and the filter paper is washed with a small volume of dilute Na_2SO_4. The combined filtrate and washings are neutralized, and analysed for nitrate as described above.

35.4.2 Other Methods for NO_3^- Determination

The following xylenols (dimethylphenols) yield coloured nitration products when heated with nitrate in sulphuric acid media: 2,4-xylenol [78], 2,6-xylenol [79–81], and 3,4-xylenol [82]. The nitrated derivatives of the first and last of these xylenols can be steam-distilled.

Brucine reacts with nitrate to form a yellow oxidation product [69,83–86]. Chloride interferes less in this reaction than in the phenoldisulphonic acid or xylenol reactions.

Chromotropic acid [87–90], bianthronyl [90a], salicylic acid [91], 1-aminopyrene [92], and p-diaminodiphenylsulphone + diphenylamine [93] are other reagents for nitrate. Norwitz [94] exploits the colour reaction between NO_3^- and ferrous sulphate in concentrated sulphuric acid.

Bloomfield et al. [95] have proposed an indirect method for the determination of nitrate based on the interference of NO_3^- in the formation of the rhenium α-furildioxime complex.

Nitrate has been determined after extraction into MIBK of the ion-pair formed by NO_3^- and the cationic copper(I) neocuproine complex [96]. Goffart and Duyckaerts [97] have extracted nitric acid with phosphine oxides in benzene, toluene, or n-octane.

Nitrate is often determined after quantitative reduction to nitrite or ammonia. Nitrate is reduced to nitrite by zinc [98, 99], cadmium [100–102], and hydrazine [53,103], and to ammonia with Devarda's alloy [5–7].

35.3 Determination of Other Nitrogen Compounds

Hydrazine ($NH_2.NH_2$) can be determined spectrophotometrically with p-dimethylaminobenzaldehyde [104–106], picryl chloride [107], and pyridylpyridinium dichloride [108]. Ashworth [109] has determined hydrazine turbidimetrically by its reaction with selenium(IV):

$$NH_2.NH_2 + SeO_2 \rightarrow Se + N_2 + 2H_2O$$

Hydroxylamine (NH_2OH) can be determined by its colour reaction with Nessler's reagent [110], or by oxidation to nitrite and formation of an azo dye [111]. It can also be determined by its reaction with diacetyl to give dimethylglyoxime, which then reacts with nickel [112].

Nitrogen dioxide absorbed in alkali can react to form azo dyes [113–121]. These diazotization reactions are suitable for the continuous automatic monitoring of atmospheric NO_2 [114,115,117]. Oxides of nitrogen in gases and cigarette smoke have been determined with brucine [122] and with $FeSO_4$ in conc. H_2SO_4 [123,124]. The amount of NO in liquid N_2O_4 (0°C) can be determined from the green colour of the N_2O_3 produced, but a special cell is necessary [125]. Comer and Jensen [126] have determined NO_2 and N_2O_4 in the atmosphere by measuring the absorbance of the brown NO_2 gas, and applying a correction for the colourless N_2O_4.

Azide (N_3^-) forms a red complex with iron(III) which provides the basis for its spectrophotometric determination [127].

Hyponitrite ($N_2O_2^{2-}$) has been determined by Holzapfel and Gürtler [128] by oxidation to nitrite and formation of an azo dye.

References

1. Kirk, P. L., *Anal. Chem.* **22**, 354, 611 (1950).
2. Kuck, J. A., Kingsley, A., Kinsey, D., Sheehan, F. and Swigert, G. F., *ibid.* **22**, 604 (1950).
3. Committee for the Stardardisation of Microchemical Apparatus, *ibid.* **23**, 523 (1951).
4. Thaler, H. and Sturm, W., *Z. Anal. Chem.* **244**, 379; **246**, 315 (1969); **250**, 120; **251**, 30 (1970).
5. Bremner, J. M. and Keeney, D. R., *Anal. Chim. Acta* **32**, 485 (1965).
6. Frankenburg, W. G., Gottscho, A. M., Kissinger, S., Bender, D. and Ehrlich, M., *Anal. Chem.* **25**, 1784 (1953).
7. McDonald, I. G. and Lench, A., *Analyst* **85**, 564 (1960).
8. Baker, P. R., *Talanta* **8**, 57 (1961).
9. Riley, J. P., *Anal. Chim. Acta* **9**, 575 (1953).
10. Scheurer, P. G. and Smith, F., *Anal. Chem.* **27**, 1616 (1955).
11. Bohnstedt, U., *Z. Anal. Chem.* **163**, 415 (1958).
12. Bolleter, W. T., Bushman, C. J. and Tidwell, P. W., *Anal. Chem.* **33**, 592 (1961).
13. Rommers, P. J. and Visser, J., *Analyst* **94**, 653 (1969).
14. Namiki, M., Kakita, Y. and Gotô, H., *Talanta* **11**, 813 (1964).
15. Klibus, A. Kh. and Nazarchuk, T. N., *Zh. Analit. Khim.* **16**, 79 (1961).
16. Roskam, R. T. and de Langen, D., *Anal. Chim. Acta* **30**, 56 (1964).
17. Gotô, H., Kakita, Y. and Atsuya, I., *Japan Analyst* **12**, 727 (1963).
18. Davis, W. F., Graab, J. W. and Merkle, E. J., *Talanta* **18**, 263 (1971).
19. Kallmann, S., Hobart, E. W., Oberthin, H. K. and Brienza, W. C., Jr., *Anal. Chem.* **40**, 332 (1968).
20. Sinclair, V. M., Davies, W. and Melhuish, K. R., *Talanta* **12**, 841 (1965).
21. Keay, J. and Menagé, P. M., *Analyst* **94**, 895 (1969); **95**, 379 (1970).
22. Selmer-Olsen, A. R., *ibid.* **96**, 565 (1971).
23. Heistand, R. N., *Anal. Chem.* **42**, 903 (1970).
24. Varley, J. A., *Analyst* **91**, 119 (1966).
24a. Strukova, M. P. and Veslova, G. I., *Zh. Analit. Khim.* **28**, 1025 (1973).
25. Müller-Beissenhirtz, W., *Z. Anal. Chem.* **212**, 145 (1965).
26. Wearne, T. J., *Anal. Chem.* **35**, 327 (1963).
27. Mann, L. T., Jr., *ibid.* **35**, 2179 (1963).
28. Tetlow, J. A. and Wilson, A. L., *Analyst* **89**, 453 (1964).
29. Emmet, R. T., *Anal. Chem.* **41**, 1648 (1969).
30. Leithe, W. and Petschl, G., *Z. Anal. Chem.* **230**, 344 (1967).
31. Davidson, J., Mathieson, J. and Boyne, A. W., *Analyst* **95**, 181 (1970).
32. Thompson, J. F. and Morrison, G. R., *Anal. Chem.* **23**, 1153 (1951).
33. Sarkar, P. B. and Ghosh, N. N., *Anal. Chim. Acta* **13**, 195 (1955); **14**, 209 (1956).
34. Massmann, W., *Z. Anal. Chem.* **193**, 332 (1963).
35. Möller, G., *ibid.* **245**, 155 (1969).
36. Williams, P. C., *Analyst* **89**, 276 (1964).
37. Middleton, K. R., *J. Appl. Chem.* **10**, 281 (1960).
38. Buck, M. and Stratmann, H., *Z. Anal. Chem.* **213**, 241 (1965).
39. Stevenson, F. J., *Anal. Chem.* **32**, 1704 (1960).
40. Rodgers, J. F. and Harter, G. J., *ibid.* **26**, 395 (1954).

References

41. Awasthi, S. P., Sahasranaman, S. and Sundaresan, M., *Analyst* **92**, 650 (1967).
42. Crosby, N. T., *ibid.* **93**, 406 (1968).
43. Kruse, J. M. and Mellon, M. G., *Anal. Chem.* **25**, 1188 (1953).
44. Procházková, L., *ibid.* **36**, 865 (1964).
45. Zitomer, F. and Lambert, J. L., *ibid.* **34**, 1738 (1962).
46. Seta, F. R. and Tamagno, B. E., *ibid.* **42**, 1443 (1970).
47. Jacobs, S., *Nature* **183**, 262 (1959); *Analyst* **85**, 257 (1960).
48. Hashmi, M. H., Ajmal, A. I. and Rashid, A., *Mikrochim. Acta* **1968**, 860.
49. Truesdale, V. W., *Analyst* **96**, 584 (1971).
50. Rider, B. F. and Mellon, M. G., *Ind. Eng. Chem., Anal. Ed.* **18**, 96 (1946).
51. Barnes, H. and Folkard, A. R., *Analyst* **76**, 599 (1951).
52. Nelson, J. L., Kurtz, L. T. and Bray, R. H., *Anal. Chem.* **26**, 1081 (1954).
53. Henriksen, A., *Analyst* **90**, 83 (1965).
54. Litchfield, M. H., *ibid.* **92**, 132 (1967).
55. Henriksen, A. and Selmer-Olsen, A. R., *ibid.* **95**, 514 (1970).
55a. Scott, T. S., *Carcinogen and Chronic Toxic Hazards of Aromatic Amines*, Elsevier, Amsterdam (1962).
56. Bunton, N. G., Crosby, N. T. and Patterson, S. J., *Analyst* **94**, 585 (1969).
57. Sawicki, E., Stanley, T. W., Pfaff, J. and D'Amico, A., *Talanta* **10**, 641 (1963).
58. Lambert, J. L. and Zitomer, F., *Anal. Chem.* **32**, 1684 (1960).
59. Bark, L. S. and Catterall, R., *Mikrochim. Acta* **1960**, 553.
60. Sawicki, E. and Noe, J. L., *Anal. Chim. Acta* **25**, 166 (1961).
61. Combs, H. F. and Grove, E. L., *Talanta* **9**, 452 (1962).
62. Foris, A. and Sweet, T. R., *Anal. Chem.* **37**, 701 (1965).
63. Kieruczenkowa, A., *Chem. Anal. (Warsaw)* **12**, 1031 (1967).
64. Garcia, E. E., *Anal. Chem.* **39**, 1605 (1967).
64a. Celardin, F., Marcantonatos, M. and Monnier, D., *Anal. Chim. Acta* **68**, 61 (1974).
65. Macchi, G. R. and Cescon, B. S., *Anal. Chem.* **42**, 1809 (1970).
66. Szekely, E., *Talanta* **15**, 795 (1968).
67. Wada, E. and Hattori, A., *Anal. Chim. Acta* **56**, 233 (1971).
68. Ziegler, M. and Glemser, O., *Z. Anal. Chem.* **144**, 187 (1955).
69. Hutchinson, K. and Boltz, D. F., *Anal. Chem.* **30**, 54 (1958).
70. Fadrus, H. and Malý, J., *Z. Anal. Chem.* **246**, 239 (1969).
71. Ghimicescu, C. and Dorneanu, V., *Talanta* **19**, 1474 (1972).
72. Johnson, C. M. and Ulrich, A., *Anal. Chem.* **22**, 1526 (1950).
73. Taras, M. J., *ibid.* **22**, 1020 (1950).
74. Marczenko, Z. and Nowicka-Jankowska, T., *Przemysł Chem.* **33**, 421 (1954).
75. Hora, F. B. and Webber, P. J., *Analyst* **85**, 567 (1960).
76. Hahn, F. L., *Anal. Chim. Acta* **7**, 68 (1952).
77. Gardon, J. L. and Leopold, B., *Anal. Chem.* **30**, 2057 (1958).
78. Buckett, J., Duffield, W. D. and Milton, R. F., *Analyst* **80**, 141 (1955).
79. Andrews, D. W., *ibid.* **89**, 730 (1964).
80. Hartley, A. M. and Asai, R. I., *Anal. Chem.* **35**, 1207 (1963).
81. Montgomery, H. A. and Dymock, J. F., *Analyst* **87**, 374 (1962).
82. Holler, A. C. and Huch, R. V., *Anal. Chem.* **21**, 1385 (1949).
83. Fisher, F. L., Ibert, E. R. and Beckman, H. F., *ibid.* **30**, 1972 (1958).
84. Robinson, J. B., Allen, M. V. and Gacoka, P., *Analyst* **84**, 635 (1959).
85. Jenkins, D. and Medsker, L. L., *Anal. Chem.* **36**, 610 (1964).
86. Fadrus, H. and Malý, J., *Z. Anal. Chem.* **202**, 164 (1964).
87. West, P. W. and Lyles, G. L., *Anal. Chim. Acta* **23**, 227 (1960).
88. Batten, J. J., *Anal. Chem.* **36**, 939 (1964).
89. West, P. W. and Ramachandran, T. P., *Anal. Chim. Acta* **35**, 317 (1966).
90. Robinson, J. W. and Hsu, C. J., *ibid.* **44**, 51 (1969).
90a. Nawratil, B., Marcantonatos, M. and Monnier, D., *ibid.* **68**, 217 (1974).
91. Řezáč, Z. and Kadič, K., *Z. Anal. Chem.* **190**, 305 (1962).

92. Sawicki, E., Johnson, H. and Stanley, T. W., *Anal. Chem.* **35**, 1934 (1963).
93. Szekely, E., *Talanta* **14**, 941 (1967).
94. Norwitz, G., *Anal. Chem.* **34**, 227 (1962).
95. Bloomfield, R. A., Guyon, J. C. and Murmann, R. K., *ibid.* **37**, 248 (1965).
96. Yamamoto, Y., Okamoto, N. and Tao, E., *J. Chem. Soc. Japan, Pure Chem. Sect.* **89**, 399 (1968).
97. Goffart, J. and Duyckaerts, G., *Anal. Chim. Acta* **36**, 499 (1966); **38**, 529; **39**, 57 (1967).
98. Chow, T. J. and Johnstone, M. S., *ibid.* **27**, 441 (1962).
99. Matsunaga, K. and Nishimura, M., *ibid.* **45**, 350 (1969).
100. Morris, A. W. and Riley, J. P., *ibid.* **29**, 272 (1963).
101. Lambert, R. S. and DuBois, R. J., *Anal. Chem.* **43**, 955 (1971).
102. Elliott, R. J. and Porter, A. G., *Analyst* **96**, 522 (1971).
103. Terrey, D. R., *Anal. Chim. Acta* **34**, 41 (1966).
104. Watt, G. W. and Chrisp, J. D., *Anal. Chem.* **24**, 2006 (1952).
105. Freier, R. and Resch, G., *Z. Anal. Chem.* **149**, 177 (1956).
106. Pilz, W. and Stelzl, E., *ibid.* **219**, 416 (1966).
107. Riley, J. P., *Analyst* **79**, 76 (1954).
108. Asmus, E., Ganzke, J. and Schwarz, W., *Z. Anal. Chem.* **253**, 102 (1971).
109. Ashworth, M. R., *Mikrochim. Acta* **1961**, 5.
110. Fishbein, W. N., *Anal. Chim. Acta* **37**, 484 (1967).
111. Lee, D. F. and Roughan, J. A., *Analyst* **96**, 798 (1971).
112. Pittwell, L., *Mikrochim. Acta* **1974**, in press.
113. Saltzman, B. E., *Anal. Chem.* **26**, 1949 (1954).
114. Jacobs, M. B. and Hochheiser, S., *ibid.* **30**, 426 (1958).
115. Yanagisawa, S., Yamate, N., Mitsuzawa, S. and Mori, M., *Bull. Chem. Soc. Japan* **39**, 2173 (1966).
116. Fauth, M. I. and Richardson, A. C., *Microchem. J.* **12**, 534 (1967).
117. Häntzsch S., Nietruch F. and Prescher, K. E., *Mikrochim. Acta* **1969**, 550.
118. Huygen I. C., *Anal. Chem.* **42**, 407 (1970).
119. Fisher, G. E. and Becknell, D. E., *ibid.* **44**, 863 (1972).
120. Halstead, C. J., Nation, G. H. and Turner, L., *Analyst* **97**, 55 (1972).
121. Bultez, A., *Analusis*, **2**, 190 (1973).
122. Smith, G. A., Sullivan, P. J. and Irvine, W. J., *Analyst* **92**, 456 (1967).
123. Norwitz, G., *ibid.* **91**, 553 (1966).
124. Scherbak, M. P. and Smith, T. A., *ibid.* **95**, 964 (1970).
125. Wright, C. M., Orr, A. A. and Balling, W. J., *Anal. Chem.* **40**, 29 (1968).
126. Comer, S. W. and Jensen, A. V., *ibid.* **36**, 799 (1964).
127. Roberson, C. E. and Austin, C. M., *ibid.* **29**, 854 (1957).
128. Holzapfel, H. and Gürtler, O., *J. Prakt. Chem.* **35**, 59, 68, 70 (1967).

Chapter 36
OSMIUM

Osmium (Os, at.wt. 190·2), the heaviest of the platinum metals, occurs in the +VIII, VI, IV and III oxidation states. Powerful oxidants liberate volatile, poisonous OsO_4. When fused with alkalis, the metal forms the osmate, OsO_4^{2-}, which disproportionates in acid into OsO_4 and Os(IV). Osmium (IV) forms halide complexes. Tin(II) reduces osmium compounds to metallic osmium.

36.1 Methods of Separation

The separation of osmium and other platinum metals is discussed in the chapter on platinum (p. 431). Methods for separating osmium are similar to those for separating ruthenium (p. 462). Allan and Beamish [1] have discussed fire assay for osmium.

36.1.1 DISTILLATION OF OsO_4

Osmium is often separated from other metals by distillation of volatile osmium tetroxide (m.p. 39·5°C, b.p. 130°C) from a sulphuric or nitric acid medium [1–6]. Hydrochloric acid inhibits the distillation by yielding less volatile osmium chloride complexes. Osmium can be separated from ruthenium since the former is more readily oxidized to OsO_4 than is the latter to RuO_4 (which is also volatile). Nitric acid oxidizes osmium to osmium tetroxide, but fails to oxidize ruthenium if the HNO_3 concentration does not exceed 40%. Ruthenium is oxidized at higher concentrations of nitric acid and in the presence of a powerful oxidant (e.g. bromine, $KMnO_4$, $NaBiO_3$, or conc. $HClO_4$). OsO_4, but not RuO_4, distils from solutions containing hydrogen peroxide [2,5].

The distilled OsO_4 may be absorbed in an alkaline solution, in $6M$ HCl saturated with SO_2, or in hydrochloric acid containing thiourea [1]. For the quantitative separation of osmium, it is sufficient to distil 1/5 of the liquid from the still.

36.1.2 EXTRACTION OF OsO_4

Osmium tetroxide can be extracted from acid solution with chloroform (distribution coefficient 24) or carbon tetrachloride (distribution coefficient 13) [7,8]. Since RuO_4 is also extracted, it should be reduced with iron(II) before the oxidation of osmium with nitric acid [7]. The low distribution coefficients necessitate two extractions.

Osmium can be extracted from HBr media with MIBK [9]. Meier et al. [10] have thoroughly investigated the extraction of osmium from HCl or HBr solutions with various oxygenated and inert solvents containing amines; distribution coefficients of the order of 500 have been obtained.

36.1.3 Ion-Exchange

Osmium can be separated by ion-exchange on cation-exchangers [11] and on anion-exchange paper [12].

36.2 Methods of Determination

The simple but rather insensitive thiourea method for the spectrophotometric determination of osmium is popular. A sensitive method using diphenylcarbazide is presented in more detail below. Spectrophotometric methods for determining osmium have been reviewed [13,14].

36.2.1 Diphenylcarbazide Method

Goldstein et al. [8] have recommended diphenylcarbazide (formula, p. 215), a well-known reagent for chromium(VI), for the spectrophotometric determination of osmium. Diphenylcarbazide (in ethanol) reacts with osmium tetroxide (in chloroform) to form a blue-green complex having an absorption maximum at 560 nm. A freshly prepared solution of diphenylcarbazide does not absorb between 400 and 700 nm.

The molar absorptivity of the osmium complex in chloroform is 3.1×10^4 ($a = 0.16$).

A considerable excess of the reagent is essential if maximum sensitivity and reproducibility are to be obtained. Acetone and ethanol are equally good solvents for the reagent: with methanol or acetic acid, the absorbances obtained are higher, but the precision is lower.

Osmium reacts rather slowly with diphenylcarbazide, but moderate heating ($\sim 50°C$) accelerates the reaction [8,15]. The chloroform solution of OsO_4 must be added to the solution of diphenylcarbazide, and not vice versa.

Of the platinum metals, only ruthenium interferes in this method. Separation of osmium from ruthenium by selective distillation or extraction as the tetroxide also permits its isolation from other metals.

Goldstein et al. [16] have modified their original method to achieve an almost fivefold increase in sensitivity. This modified method enables very minute traces (< 5 μg) [15] of osmium to be determined. The osmium is oxidized to OsO_4 and added to an aqueous solution of diphenylcarbazide. After the solution has been heated to 65°C, the complex is extracted into chloroform. The molar absorptivity of this chloroform solution is 1.5×10^5.

Iron(III), Cu, Ru, and Au interfere, as do larger quantities of Ni, Cr(VI), Mo(VI), Ir, and chloride.

Reagents

1. Diphenylcarbazide. Prepare a fresh 0·2% solution in ethanol.
2. Standard osmium solution: 1 mg/ml. Carefully break an accurately weighed glass ampoule containing $\sim$ 0·5 g of OsO_4 in a beaker containing $\sim$ 100 ml of water acidified with 3 ml of H_2SO_4 (1+1). Wash, dry, and weigh the glass fragments of the ampoule, and calculate the weight of OsO_4 by difference. Dilute the osmium solution with water till 1 ml contains precisely 1 mg of Os. Perform all these operations in a fume cupboard, and keep the osmium solution in a bottle with a precision-ground stopper on account of the toxic properties and offensive odour of the tetroxide.

Procedure

Extractive separation of Os. Adjust the sample solution containing not more than 200 µg of Os (in an oxidation state lower than +VIII) with sulphuric acid so that the volume is $\sim$ 5 ml and the concentration of H_2SO_4 is $\sim$ 1M. Oxidize the osmium by adding 5% $KMnO_4$ solution dropwise until the sample solution has a stable pink colour. Decolorize the solution by dropwise addition of 2% ammonium ferrous sulphate solution. Immediately add 3 ml of conc. HNO_3 and 2 ml of water, and extract OsO_4 with two 20-ml portions of chloroform. Wash the combined extracts with 10 ml of 0·1M H_2SO_4.

Determination of Os. Transfer the washed chloroform solution to a 50-ml flask containing 10 ml of diphenylcarbazide solution. Heat the flask at a temperature of 50°C for 15 minutes. Make the cooled solution up to the mark with chloroform. Measure the absorbance of the coloured solution at 560 nm against a reagent blank solution.

36.2.2 OTHER METHODS

A well-known less-sensitive method for the determination of osmium is based on the reaction with thiourea in a sulphuric or hydrochloric acid medium to give a red complex $[Os(NH_2.CS.NH_2)_6]^{3+}$, having an absorption maximum at 480 nm, in which osmium is tervalent [1,7,17]. In the initial solution, osmium may be in any oxidation state. Since the other platinum metals interfere, it is advantageous to separate osmium by distillation of OsO_4 directly into a solution of thiourea in 6M HCl [1].

The thiourea method has been used to determine osmium in meteorites [18], organic compounds [19], and osmium hexafluoride [20]. Some related reagents which have also been utilized in the spectrophotometric determination of osmium are: ditolylthiourea [21], 2-mercaptobenzimidazole (phenylenethiourea) [22], o-(2-benzoylthiourido)benzoic acid [23], and selenourea [24].

Other sulphur reagents include 2-mercaptobenzothiazole [25], 2-thione-5-mercapto-1,3,4-thiadiazolidine [26], triazine derivatives [27,28], Bismuthiol II [29], thiosalicylamide [30], and rubeanic acid [31].

Steele and Yoe [32] have determined osmium with 1-naphthylamine-4,6,8-trisulphonic acid. In an acid medium (pH ~ 1), a violet water-soluble complex is formed ($\varepsilon = 2.98 \times 10^4$ at $\lambda_{max} = 575$ nm). The reagent itself does not absorb at 575 nm. The isomeric 1-naphthylamine-3,5,7-trisulphonic acid is a less suitable reagent for osmium [33].

The following are various organic reagents suggested for the spectrophotometric determination of osmium: m-aminobenzoic acid [34], o-aminophenol-p-sulphonic acid [35], 2-amino-8-naphthol-3,6-disulphonic acid [36], nitroso-R salt [37], 3-nitroso-2,6-pyridinediol ($\varepsilon = 2.4 \times 10^4$ at 550 nm) [38], p-(morpholino)-N-(4'-hydroxy-3'-methoxy)benzylidineaniline [39], quinisatinoxime [40], acenaphthenequinone monoxime [41], anthranilic acid [42], TAR [43], Tiron [44], TTA [45], pyrocatechol [46], pyrogallol [47], and 3-nitroso-4-hydroxycoumarin [48].

Less sensitive extraction spectrophotometric methods are based on ion-association complexes of hexachloro-osmate(IV) with the tetraphenylarsonium ion [49], 3,4-dichlorobenzyltriphenylphosphonium ion [50], DAM or DAPM [51].

Lastly, osmium can be determined spectrophotometrically as the thiocyanate [52–54] and bromide [55] complexes.

References

1. Allan, W. J. and Beamish, F. E., *Anal. Chem.* **24**, 1569, 1608 (1952).
2. Westland, A. D. and Beamish, F. E., *ibid.* **26**, 739 (1954).
3. Geilmann, W. and Neeb, R., *Z. Anal. Chem.* **156**, 411, 420 (1957).
4. Van Loon, J. C. and Beamish, F. E., *Anal. Chem.* **36**, 1771 (1964).
5. Chung, K. S. and Beamish, F. E., *Talanta* **15**, 823 (1968).
6. Alimarin, I. P., Khvostova, V. P. and Shlenskaya, V. I., *Zh. Analit. Khim.* **25**, 2167 (1970).
7. Sauerbrunn, R. D. and Sandell, E. B., *Anal. Chim. Acta* **9**, 86 (1953); *J. Am. Chem. Soc.* **75**, 4170 (1953).
8. Goldstein, G., Manning, D. L., Menis, O. and Dean, J. A., *Talanta* **7**, 296, 301 (1961).
9. Berg, E. W. and Moseley, H. E., *Anal. Chim. Acta* **47**, 360 (1969).
10. Meier, H., Zimmerhackl, E., Albrecht, W., Bösche, D., Hecker, W., Menge, P., Ruckdeschel, A., Unger, E. and Zeitler, G., *Mikrochim. Acta* **1969**, 557, 573, 826, 839.
11. Beamish, F. E., *Talanta* **14**, 991 (1967).
12. Taylor, H. and Beamish, F. E., *ibid.* **15**, 497 (1968).
13. Beamish, F. E. and McBryde, W. A., *Anal. Chim. Acta* **9**, 349 (1953); **18**, 551 (1958).
14. Beamish, F. E., *Talanta* **12**, 789 (1965).
15. Marczenko, Z. and Balcerzak, M., unpublished work.
16. Goldstein, G., Manning, D. L., Menis, O. and Dean, J. A., *Talanta* **7**, 307 (1961).
17. Ayres, G. H. and Wells, W. N., *Anal. Chem.* **22**, 317 (1950).
18. Sen Gupta, J. G., *Anal. Chim. Acta* **42**, 481 (1968).
19. Dwyer, F. P. and Gibson, N. A., *Analyst* **76**, 104 (1951).
20. Jensen, K. J., *Anal. Chem.* **37**, 1430 (1965).

21. Geilmann, W. and Neeb, R., *Z. Anal. Chem.* **152**, 96 (1956).
22. Bera, B. C. and Chakrabartty, M. M., *Anal. Chem.* **38**, 1419 (1966).
23. Majumdar, A. K. and Bhowal, S. K., *Anal. Chim. Acta* **35**, 479 (1966).
24. Pilipenko, A. T. and Sereda, I. P., *Zh. Analit. Khim.* **16**, 73 (1961); *Zh. Neorgan. Khim.* **6**, 413 (1961).
25. Bera, B. C. and Chakrabartty, M. M., *Microchem. J.* **11**, 420 (1966).
26. Gregorowicz, Z. and Klima, Z., *Z. Anal. Chem.* **239**, 87 (1968).
27. Lazăr, C., Popa, G. and Cristescu, C., *Anal. Chim. Acta* **47**, 166 (1969).
28. Popa, G., Lazăr, C. and Cristescu, C., *Talanta* **17**, 635 (1970).
29. Majumdar, A. K. and Bhowal, S. K., *Anal. Chim. Acta* **62**, 223 (1972).
30. Sur, K., Mazumdar, M. and Shome, S. C., *ibid.* **59**, 306 (1972).
31. Bhowal, S. K., *ibid.* **69**, 465 (1974).
32. Steele, E. L. and Yoe, J. H., *Anal. Chem.* **29**, 1622 (1957); *Anal. Chim. Acta* **20**, 205 (1959).
33. Wingfield, H. C. and Yoe, J. H., *ibid.* **14**, 446 (1956).
34. Majumdar, A. K. and Sen Gupta, J. G., *Z. Anal. Chem.* **179**, 13 (1961).
35. Majumdar, A. K. and Sen Gupta, J. G., *Anal. Chim. Acta* **22**, 306 (1960).
36. Agarwala, B. V. and Ghose, A. K., *Talanta* **20**, 129 (1973).
37. Nath, S. K. and Agarwal, R. P., *Chim. Anal. (Paris)* **47**, 257 (1965); **48**, 439 (1966).
38. McDonald, C. W. and Carter, R., Jr., *Anal. Chem.* **41**, 1478 (1969).
39. Ayres, G. H. and McDonald, C. W., *Anal. Chim. Acta* **30**, 40 (1964).
40. Ayres, G. H. and Briggs, T. C., *ibid.* **26**, 340 (1962).
41. Sindhwani, S. K. and Singh, R. P., *Microchem. J.* **18**, 627 (1973).
42. Majumdar, A. K. and Sen Gupta, J. G., *Anal. Chim. Acta* **20**, 532 (1959).
43. Ivanov, V. M., Busev, A. I., Popova, L. V. and Bogdanovich, L. I., *Zh. Analit. Khim.* **24**, 1064 (1959).
44. Majumdar, A. K. and Savariar, C. P., *Anal. Chim. Acta* **21**, 146 (1959).
45. Rangnekar, A. V. and Khopkar, S. M., *Bull. Chem. Soc. Japan* **41**, 600 (1968).
46. Jasim, F., Magee, R. J. and Wilson, C. L., *Mikrochim. Acta* **1962**, 160.
47. Faye, G. H., *Anal. Chem.* **37**, 259, 696 (1965).
48. Manku, G. S., Bhat, A. N. and Jain, B. D., *Talanta* **16**, 1421 (1969).
49. Neeb, R., *Z. Anal. Chem.* **154**, 23 (1957).
50. Neeb, R. and Khan-Boluki, K., *ibid.* **215**, 392 (1966).
51. Busev, A. I. and Akimov, V. K., *Zh. Analit. Khim.* **17**, 979 (1962); *Talanta* **11**, 1657 (1964).
52. Wiersma, J. H. and Lott, P. F., *Anal. Chem.* **39**, 674 (1967).
53. Qureshi, M. and Mathur, K. N., *Z. Anal. Chem.* **242**, 159 (1968).
54. Shlenskaya, V. I. and Khvostova, V. P., *Zh. Analit. Khim.* **23**, 237 (1968).
55. Preetz, W. and Pfeifer, H. L., *Z. Anal. Chem.* **247**, 37 (1969).

Chapter 37

OXYGEN

Oxygen (O, at.wt. 16·00) is a gas, O_2 (in ozone, O_3). It occurs in most compounds in the −II oxidation state and in the −I state in peroxides. With other elements, it forms numerous oxide complexes such as CrO_4^{2-}, MoO_4^{2-}, WO_4^{2-}, MnO_4^-, VO^{2+}, UO_2^{2+}, SO_4^{2-}, and NO_3^-. Volatile oxygen compounds include OsO_4 and CO_2. One oxygen compound of great importance in analysis is hydrogen peroxide (perhydrol is a 30% H_2O_2 solution in water), the peroxide complexes of a number of metals (e.g. Ti, V, Nb, U, and Zr) being familiar.

37.1 Determination of Oxygen

Methods for the determination of oxygen exploit its oxidizing properties.

A colorimetric modification of the well-known Winkler titrimetric method for determining oxygen in water is based on oxidizing the manganese in $Mn(OH)_2$ with oxygen. After the addition of potassium iodide and acidification of the solution with sulphuric acid, an equivalent amount of iodine is liberated. The iodine is determined spectrophotometrically either as a blue complex with starch [1] (cf. p. 296), or after the iodine has been extracted into toluene or chloroform [2].

Oxygen impurities (0·1–0·0001%) in various gases can be determined by the colour reaction with anthraquinone-2-sulphonate in alkaline solution [3]. The red solution of the reagent (reduced with zinc amalgam) is decolorized when oxidized by oxygen. Various modifications of this reaction are suitable for the continuous spectrophotometric determination of oxygen in gases [4–8].

The colour reaction with Indigo Carmine is useful for determining oxygen dissolved in water [9–12]. The yellow reduced form of the reagent (leucobase) turns red ($\lambda_{max} = 555$ nm) under the influence of oxygen.

Other spectrophotometric methods for the determination of oxygen use the following reagents: 3,3′-dimethylnaphthidine [13,14], Methylene Blue [15], Safranine T [16], Methyl Viologen [17], leuco-Berbelin Blue [18], DCTA [+Mn(II)] [19], and pyrogallol [20].

The reaction between copper or copper(I) in ammoniacal solution and oxygen can also be used [21–23]. Dissolved oxygen can be determined by its oxidizing action on the tris(4,7-dihydroxy-1,10-phenanthroline) iron(II) complex [24].

37.2 Determination of Ozone

Since ozone is a stronger oxidant than oxygen, it is determined on the basis of colour redox reactions with reagents which are not oxidized by oxygen.

Ozone reacts specifically with eugenol (4-allyl-2-methoxyphenol) to release formaldehyde, which can be determined with pararosaniline [25]. The iodine liberated from slightly acidic iodide solutions by ozone can be determined colorimetrically [26–28].

Reagents which have been used for the spectrophotometric determination of ozone include diphenylaminesulphonate [29], tetramethyl-p-phenylenediamine [30], leuco-Malachite Green [31], Methyl Red [32], and 4,4'-dimethoxystilbene [33].

Hauser and Bradley [34] have absorbed ozone in 1,2-di(4-pyridyl)-ethylene dissolved in glacial acetic acid. The pyridine-4-aldehyde produced gives a pink colour with 3-methyl-2-benzothiazolinone hydrazone (MBTH).

Hofmann and Stern [35] have suggested a less sensitive method in which ozone oxidizes manganese(II) in phosphoric acid medium to the violet manganese(III) diphosphate complex. With smaller amounts of ozone, this complex may be reacted further with o-tolidine.

37.3 Determination of Hydrogen Peroxide

Spectrophotometric methods for determining hydrogen peroxide are based on its capacity to form stable peroxide complexes as well as on its oxidizing and reducing properties.

The widely known, but relatively insensitive, titanium method [36–39], based on the orange-yellow titanium peroxide complex formed in acid (H_2SO_4) medium, has been used to determine hydrogen peroxide in hydrocarbons [37] and in diethyl ether [38]. The titanium 8-hydroxyquinolinate method is more sensitive ($\varepsilon = 3{\cdot}06 \times 10^3$ at 450 nm) [40]. In an alternative method having high sensitivity, H_2O_2 reacts with 1,2-di(4-pyridyl)ethylene to form pyridine-4-aldehyde which, in turn, gives a colour reaction with 3-methyl-2-benzothiazolinone hydrazone (MBTH) ($\varepsilon = 3{\cdot}65 \times 10^4$ at 442 nm) [41].

Hydrogen peroxide decolorizes solutions of vanadium or uranium benzohydroxamate by ligand substitution reactions [42].

The oxidizing properties of hydrogen peroxide are the basis of several methods. In a solution containing Fe^{2+} and SCN^-, hydrogen peroxide produces the equivalent amount of the iron(III) thiocyanate complex [36]. By oxidizing iron(II), hydrogen peroxide decolorizes solutions of the iron(II) 1,10-phenanthroline and iron(II) bathophenanthroline complexes [43]. From an iodide solution containing molybdate as catalyst, hydrogen peroxide liberates an equivalent amount of iodine, which can be determined spectrophotometrically [44].

A sensitive method for determining hydrogen peroxide depends on the oxidation of the colourless leuco base of phenolphthalein in alkaline medium containing copper(II), to form the familiar red colour [45]. The alkaline phenolphthalein solution is converted into the leuco base by heating with zinc dust.

In an indirect method taking advantage of the reducing properties of hydrogen peroxide, the sample solution is reacted with a known amount of dichromate, and the excess of Cr(VI) is determined spectrophotometrically with o-dianisidine [46]. Hydrogen peroxide can also be determined [47] from the decrease in absorbance (at 418 nm) of an alkaline ferricyanide solution as a result of the reaction

$$2\,Fe(CN)_6^{3-} + H_2O_2 + 2\,OH^- \rightarrow 2\,Fe(CN)_6^{4-} + 2\,H_2O + O_2.$$

In a similar method, H_2O_2 decolorizes (reduces) a green alkaline solution of manganate (MnO_4^{2-}) [48].

References

1. Pieters, H. A. and Hanssen, W. J., *Anal. Chim. Acta* **2**, 712 (1948).
2. Pepkowitz, L. P. and Shirley, E. L., *Anal. Chem.* **25**, 1718 (1953).
3. Brady, L. J., *ibid.* **20**, 1033 (1948).
4. Struszyński, M., Minczewski, J., Waszak, S. and Wacławik, J., *Przemysł Chem.* **32**, 449 (1953).
5. Stafford, C., Puckett, J. E., Grimes, M. D. and Heinrich, B. J., *Anal. Chem.* **27**, 2012 (1955).
6. Karasek, F. W., Loyd, R. J., Lupfer, D. E. and Houser, E. A., *ibid.* **28**, 233 (1956).
7. Silverman, L. and Bradshaw, W., *Anal. Chim. Acta* **14**, 514 (1956).
8. Wacławik, J. and Waszak, S., *Chem. Anal. (Warsaw)* **2**, 376 (1957); **4**, 343 (1959); **8**, 633 (1963).
9. Loomis, W. F., *Anal. Chem.* **26**, 402 (1954); **28**, 1347 (1956).
10. Buchoff, L. S., Ingber, N. M. and Brady, J. H., *ibid.* **27**, 1401 (1955).
11. Meyling, A. H. and Frank, G. H., *Analyst* **87**, 60 (1962).
12. St. John, P. A., Winefordner, J. D. and Silver, W. S., *Anal. Chim. Acta* **30**, 49 (1964).
13. Banks, J., *Analyst* **79**, 170 (1954); **84**, 700 (1959).
14. Fadrus, H. and Malý, J., *ibid.* **96**, 591 (1971).
15. Hamlin, P. A. and Lambert, J. L., *Anal. Chem.* **43**, 618 (1971).
16. Aleskovskii, V. B., Koval'tsov, V. A., Fedorov, I. N. and Tsyplyatnikov, G. P., *Zavodsk. Lab.* **28**, 1440 (1962).
17. Sweetser, P. B., *Anal. Chem.* **39**, 979 (1967).
18. Altmann, H. J., *Z. Anal. Chem.* **262**, 97 (1972).
19. Sastry, G. S., Hamm, R. E. and Pool, K. H., *Anal. Chem.* **41**, 857 (1969).
20. Williams, D. D., Blachly, C. H. and Miller, R. R., *ibid.* **24**, 1819 (1952).
21. Brooks, F. R., Dimbat, M., Treseder, R. S. and Lykken, L., *ibid.* **24**, 520 (1952).
22. Rezaeva, L. T., *Zh. Analit. Khim.* **17**, 874 (1962).
23. Pilarczyk, H., Miller, G. and Paterok, N., *Chem. Anal. (Warsaw)* **32**, 899 (1967).
24. Poe, D. P. and Diehl, H., *Talanta* **21**, 1065 (1974).
25. Sachdev, S. L., Lodge, J. P. and West, P. W., *Anal. Chim. Acta* **58**, 141 (1972).
26. Hunold, G. A. and Pietrulla, W., *Z. Anal. Chem.* **165**, 20 (1959).
27. Cohen, I. C., Smith, A. F. and Wood, R., *Analyst* **93**, 507 (1968).
28. Wierzbicki, T. and Pieprzyk, H., *Chem. Anal. (Warsaw)* **15**, 1041 (1970).
29. Bovee, H. H. and Robinson, R. J., *Anal. Chem.* **33**, 1115 (1961).
30. Galster, H., *Z. Anal. Chem.* **186**, 359 (1962).
31. Koppe, P. and Muhle, A., *ibid.* **210**, 241 (1965).

32. Czerniec, J., Gregorowicz, Z., Fliegier, J. and Czichoń, P., *Chem. Anal. (Warsaw)* **16**, 1125 (1971).
33. Bravo, H. A. and Lodge, J. P., Jr., *Anal. Chem.* **36**, 671 (1964).
34. Hauser, T. R. and Bradley, D. W., *ibid.* **38**, 1529 (1966); **39**, 1184 (1967).
35. Hofmann, P. and Stern, P., *Anal. Chim. Acta* **45**, 149 (1969); **47**, 113 (1969).
36. Egerton, A. C., Everett, A. J., Minkoff, C. J., Rudrakanchana, S. and Salooja, K. C., *ibid.* **10**, 422 (1954).
37. Pobiner, H., *Anal. Chem.* **33**, 1423 (1961).
38. Wolfe, W. C., *ibid.* **34**, 1328 (1962).
39. Csányi, L. J., *ibid.* **42**, 680 (1970).
40. Cohen, I. R. and Purcell, T. C., *ibid.* **39**, 131 (1967).
41. Hauser, T. R. and Kolar, M. A., *ibid.* **40**, 231 (1968).
42. Meloan, C. E., Mauck, M. and Huffman, C., *ibid.* **33**, 104 (1961).
43. Bailey, R. and Boltz, D. F., *ibid.* **31**, 117 (1959).
44. Ovenston, T. C. and Rees, W. T., *Analyst* **75**, 204 (1950).
45. Dukes, E. K. and Hyder, M. L., *Anal. Chem.* **36**, 1689 (1964).
46. Buscaróns, F., Artigas, J. and Rodriguez-Roda, C., *Anal. Chim. Acta* **23**, 214 (1960).
47. Aziz, F. and Mirza, G. A., *Talanta* **11**, 889 (1964).
48. Bubyreva, N. S., Bukhareva, V. I., Kirakosyan, A. K., Pantyukhova, T. A. and Rozenblyum, N. D., *Zavodsk. Lab.* **35**, 1044 (1969).

Chapter 38

PALLADIUM

Palladium (Pd, at.wt. 106·4) is a platinum metal; it occurs in the +II and +IV oxidation states. Palladium(II) compounds are the more stable. Unlike the other platinum metals, palladium is soluble in conc. HNO_3. Brown-red $Pd(OH)_2$ precipitates at pH ~ 4, but dissolves in excess of an alkali metal hydroxide. Palladium(II) gives stable ammine, nitrite, cyanide, chloride, bromide, and iodide complexes. Palladium(II and IV) is reduced to the metal by sulphur dioxide, iron(II), and ethanol.

38.1 Methods of Separation

Methods for separating the platinum metals (including palladium) are discussed in the chapter on platinum (p. 431).

38.1.1 Extraction

Palladium dimethylglyoximate is specifically extracted from dilute acid with chloroform [1]. Dithiocarbamate complexes of palladium are also extractable into chloroform [2–4].

Extraction of palladium halide complexes (Cl^-, Br^-, I^-, SCN^-) affords convenient separations [5–9]. The iodide is extracted from ~ $3M$ H_2SO_4 with MIBK [5]. Higher alcohols [8], cyclohexanone [8], and mesityl oxide [9] are other suitable extractants. Forsythe et al. [6] have extracted the palladium thiocyanate complex in the presence of pyridine. Egli [7] has extracted the palladium chloride complex with 2-chloropyridine from hydrochloric acid medium.

Mizuike et al. [10] have extracted microgram quantities of palladium from acid solutions by amalgamation with mercury.

38.1.2 Precipitation and Other Methods

The precipitation of palladium dimethylglyoximate from an acid medium is an excellent separation method [11]. Nickel has been used as a collector for microgram quantities of palladium [12]. The optimum pH for the precipitation is 6·5 (acetate medium). If gold is separated beforehand by reduction with oxalic acid, and if copper is masked with EDTA, the separation of traces of palladium with nickel as the collector is specific. 3-Nitroso-4-hydroxycoumarin has been used for precipitation of palladium [12a].

Palladium metal can be coprecipitated with tellurium [13]. Traces of palladium can also be precipitated as the sulphide with lead as a collector, or as the hydroxide with iron as a collector. The enrichment of palladium on AgCN is highly selective [14,15].

Ion-exchange separation methods are based on the palladium chloride complex which is retained on strongly basic anion-exchangers, and on the cationic ammine complex which is retained on cation-exchangers [16].

Like other platinum metals, palladium is often separated from ores and concentrates by fire assay and cupellation methods with lead [17], tin [18], copper [19], or an alloy of iron, nickel, and copper [20] as the collector. Palladium is then separated from the collector in ion-exchange columns [18–20].

38.2 Methods of Determination

Recently, several new methods for the spectrophotometric determination of palladium have been reported. A sensitive dithizone method and a less sensitive iodide method are discussed below in more detail. A simple modification of the iodide method, however, gives an exceptionally sensitive method involving the formation of the blue starch-iodine complex. Beamish [21] has critically evaluated the methods for determining palladium.

38.2.1. IODIDE METHODS

In an acid medium (HCl, H_2SO_4) containing excess of iodide, palladium forms a brown-red complex, $[PdI_4]^{2-}$, which provides the basis for a moderately sensitive colorimetric method [1,22,23]. The concentration of hydrochloric or sulphuric acid (up to $10N$) does not affect the colour. A reductant (e.g. ascorbic acid) is added to reduce the iodine liberated by atmospheric oxygen.

The molar absorptivity of the complex is $1·02 \times 10^4$ at $\lambda_{max} = 410$ nm ($a = 0·096$).

The PdI_4^{2-} complex may be extracted as ion-association complexes with DAM [24,25], tetraethyleneglycol dimethyl ether [26], and triphenylarsine (in cyclohexane) [27], but no increase in sensitivity is achieved.

The presence of only a small excess of iodide results in the formation of PdI_2, which is sparingly soluble in acidic aqueous solutions. The extractability of PdI_2 into benzene or di-isopropyl ether has been made the basis of a very sensitive method for determining palladium [28]. The palladium iodide is back-extracted with dilute ammonia into the aqueous phase in which, after acidification, the iodide ions are oxidized with bromine to iodate ions. These, in turn, react with added potassium iodide to form iodine, which is determined by its colour reaction with starch (see determination of iodide, p. 296). One palladium atom in PdI_2 is thus equivalent to 12 iodine atoms.

The apparent molar absorptivity in this method is 2.2×10^5 ($a = 2.1$) at 590 nm (the absorption maximum of the blue starch-iodine complex).

Combining either of the iodide methods with the preliminary extraction of palladium dimethylglyoximate makes the method specific for palladium.

Reagents

1. Potassium iodide: 20% and 0.01% solution (iodine-free), freshly prepared.
2. Standard palladium solution: 1 mg/ml. Dissolve 0.1000 g of metallic palladium in aqua regia (3 ml of conc. HCl + 1 ml of conc. HNO_3). Evaporate the solution nearly to dryness, add 3 ml of conc. HCl, and evaporate to half volume. Dilute the solution with water to 100 ml.
3. Dimethylglyoxime: 1% solution in ethanol.
4. Phenol: 20% solution in glacial acetic acid.
5. Starch: 1% solution (p. 298).

Procedure

Extractive separation of Pd. Acidify the sample solution with hydrochloric acid (to $\sim 0.2M$ HCl), add 2 ml of the dimethylglyoxime solution and 5 ml of $0.1M$ EDTA, mix well, and allow to stand for 10 minutes. Extract the palladium dimethylglyoximate with two portions of $CHCl_3$, shaking for 1 minute. Wash the combined chloroform extracts with two portions of $0.2M$ HCl, and evaporate the organic phase to dryness on a water-bath. Mineralize the residue by heating with a few drops of conc. H_2SO_4 and conc. HNO_3. Expel the nitric acid, allow the residual solution to cool, dilute it with water, and heat it until it clears.

Determination of Pd (*as* $[PdI_4]^{2-}$). To the sample solution containing not more than 0.5 mg of Pd, add 5 ml of HCl (1+1), 10 ml of 20% KI solution, and 2 ml of 1% ascorbic acid solution. Dilute the solution to the mark with water in a 50-ml volumetric flask, and measure the absorbance at 410 nm against water.

Determination of Pd (*by the indirect starch–iodine method*). To the sample solution containing not more than 20 µg of Pd, add 2.5 ml of 0.01% KI solution and 5 ml of H_2SO_4 (1+3), and dilute with water to ~ 30 ml. Shake the solution with two portions of benzene for 1 minute. Wash the combined benzene extracts with two portions of $1M$ H_2SO_4. Strip the palladium complex from the benzene solution by shaking for 15 seconds with 10 ml of ammonia solution (1+9). Place the ammoniacal solution in a 50-ml volumetric flask, acidify slightly with sulphuric acid, add 3 drops of bromine water, and allow to stand for 1 minute. Add 3 drops of phenol solution, and after 1 minute add 2 ml of 0.5% KI solution (freshly prepared), and 2 ml of the starch solution. Dilute the solution with water to the mark, and measure the absorbance at 590 nm against a reagent blank solution or water.

38.2.2 DITHIZONE METHOD

When an acidic solution of palladium(II) is shaken with an excess of dithizone (H_2Dz) (formula, p. 41) in CCl_4, the grey-green palladium dithizonate, $Pd(HDz)_2$, soluble in CCl_4 and $CHCl_3$, is formed. With a deficiency of dithizone, the red dithizonate, PdDz, which is soluble only in $CHCl_3$, is formed. The grey-green $Pd(HDz)_2$ is useful in analysis [12,29,30]. It is resistant both to acids (e.g. $3M$ H_2SO_4 and $6M$ HCl) and to ammoniacal solutions (up to $3M$ NH_3). This enables free dithizone to be stripped from the organic phase with dilute ammonia.

Figure 38.1 shows the absorption spectrum of $Pd(HDz)_2$ in CCl_4. The compound shows two absorption maxima in the visible spectrum. The molar absorptivity is $4 \cdot 25 \times 10^4$ at $\lambda_{max} = 450$ nm (specific absorptivity 0·40) [29].

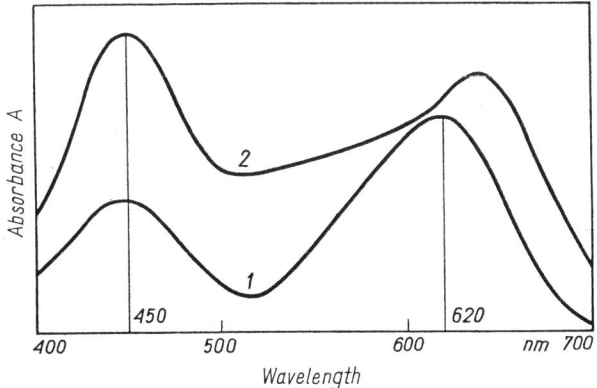

Fig. 38.1 Absorption spectra of dithizone (1) and palladium dithizonate (2) in CCl_4

$Pd(HDz)_2$ is formed rather slowly. Since complete extraction of the metal complex from the aqueous solution at $\sim 50\%$ excess of H_2Dz is achieved only after shaking has been continued for 4 minutes, a mechanical shaker is recommended. The extraction rate increases with increasing excess of dithizone.

Unlike other dithizone systems, with palladium there is only a slight difference between the colours of the complex and the free reagent. Hence, the single-colour method is used for determining palladium.

In an acid medium suitable for the formation of $Pd(HDz)_2$, other noble metals [Au, Pt(II), Hg, and Ag] and copper also react with dithizone. Traces of palladium may be conveniently isolated, before the spectrophotometric determination with dithizone, by coprecipitation with nickel dimethylglyoximate from an acetate medium at pH $\sim 6 \cdot 5$ [12].

Only copper(II) and gold(III) interfere by forming precipitates with dimethylglyoxime under the conditions for precipitating Ni and Pd [platinum(IV) interferes only in hot solution]. Copper can be masked with

EDTA and gold removed after reduction to the element with oxalic acid, which does not attack the palladium.

The separated palladium dimethylglyoximate precipitate dissolves sparingly even in $6M$ HCl. However, when a suspension of Pd(HDm)$_2$ in $2M$ HCl is shaken with a solution of dithizone in carbon tetrachloride, Pd(HDz)$_2$ is formed. The rates of formation of Pd(HDz)$_2$ during the shaking of dithizone with the palladium solution in $2M$ HCl and the shaking of dithizone with the Pd(HDm)$_2$ suspension in $2M$ HCl are similar.

Reagents

1. Dithizone: $\sim 0.01\%$ solution in CCl$_4$ (p. 492).
2. Standard palladium solution: 1 mg/ml (p. 414).
3. Dimethylglyoxime: 1% solution in ethanol.
4. Nickel solution: $\sim$ 1 mg/ml. Dissolve 0·67 g of (NH$_4$)$_2$Ni(SO$_4$)$_2$·6H$_2$O in 100 ml of water.

Procedure

Separation of Pd with a collector. To the sample solution (100 ml maximum) in dilute hydrochloric acid (e.g. $0.1M$) containing not more than 80 μg of Pd and heated to $\sim 80°$C, add a macerated filter paper and 1 ml of 5% oxalic acid solution. Keep the solution at $\sim 80°$C for 1 hour, then allow to cool. Filter off the precipitate (if any) of elemental gold and silver chloride together with the paper. To the filtrate add successively 2 mg of nickel (as sulphate solution), 2 ml of 20% potassium sodium tartrate solution (to mask Fe, Al, Ti, etc.), 2 g of sodium acetate, 1 ml of $0.1M$ EDTA, and 2 ml of the dimethylglyoxime solution (pH $\sim$ 6·5). After 30 minutes, filter off the precipitate of nickel and palladium dimethylglyoximates. Wash the precipitate from the filter paper into a beaker, add 1 ml of conc. HCl, and evaporate to 5–10 ml, depending on the quantity of palladium in the solution.

Determination of Pd. Shake the solution containing the Pd(HDm)$_2$ suspension for 5 minutes with 10 ml of 0·01% dithizone solution (1 ml of this solution corresponds to 21 μg of Pd). Wash the organic extract by shaking with two portions of $2M$ HCl, and strip the residual dithizone with dilute ammonia solution (1+50). Transfer the organic phase to a volumetric flask of suitable capacity, dilute to the mark with CCl$_4$, and measure the absorbance at 450 nm, using CCl$_4$ as the reference.

38.2.3 Other Methods

Many organic spectrophotometric reagents for palladium incorporate sulphur as a ligand atom. Apart from dithizone, which has been discussed above, examples are: thiourea [31], thioglycollic acid [32,32a], mercaptoquinoline (thio-oxine) [33,34], thionalide [35], thiomalic acid [36], Bismuthiol I and Bismuthiol II [37], 2,3-quinoxalinedithiol [38,38a], p-dimethylaminobenzylidenerhodanine (molar absorptivity 4.3×10^4 at 515 nm) [39], thio-Michler's ketone ($\varepsilon = 1\cdot 6 \times 10^5$ at 520 nm in isoamyl alcohol) [40], diphenyldithiophosphoric acid [41], derivatives of dithio-oxamide (rubeanic acid) [42–45], derivatives of mercaptopropionic acid [46], phenyl-substituted dithiolthiones [47], pyrazolinedithiocarbamate [47a], 1-phenyl-tetrazoline-5-thione [48], 2-mercaptobenzoxazole [49], 5-amino-2-benzimidazolethiol [50], o-mercaptobenzoic acid [50a], and β-mercaptohydrocinnamic acid and its esters [51].

Oxime reagents give highly selective extraction methods for determining palladium. Pre-eminent is α-furildioxime ($\varepsilon = 2\cdot 25 \times 10^4$ at 380 nm in $CHCl_3$) [52–54], but the following have also been used: phenyl-α-pyridylketoxime [55], 2,2'-dipyridylketoxime [56], 2,2'-diquinolylketoxime [57], alkyl ketoximes [58], glyoxime [59], dimethylglyoxime ($\varepsilon = 1\cdot 7 \times 10^3$ at 380 nm) [60], α-benzildioxime and nioxime [61], 2-pyridinealdoxime [62], di(2-pyridyl)glyoxime [63], o-hydroxyacetophenone oxime [64], and 4-heptanone oxime [64a].

Certain nitroso compounds are sensitive spectrophotometric reagents for palladium. Of particular value is p-nitrosodimethylaniline [65,66], which reacts with palladium in the cold at pH 2–2·5 to form a red complex (molar absorptivity $8\cdot 6 \times 10^4$ at 525 nm). The corresponding platinum complex is not formed unless the solution is heated. Further reagents of this group are p-nitrosodiphenylamine [13], 2-nitroso-1-naphthol [10,67], nitroso-R salt [68,69], 3-nitroso-2,6-pyridinediol [70], isonitrosoacetylacetone [71], and isonitrosoacetophenone [72].

Recently, many azo compounds have been suggested as sensitive reagents for palladium, e.g. PAR ($\varepsilon = 1\cdot 8 \times 10^4$ at 510 nm) [73–75], PAN [76,76a], 2-pyridylazo compounds [77,78], TAN [79], and other 2-thiazolylazo compounds [80,81], Sulphonitrophenol M [82], Tropaeolins [83], o-aminoazo compounds [84], and several 2,7-bisazo-derivatives of chromotropic acid [85,86].

Arsonic acid azo compounds worthy of particular notice include Thoron I [87], Arsazen [88], Arsenazo III [89–91], and Palladiazo. The last reagent shows a high degree of specificity for palladium ($\varepsilon = 5\cdot 7 \times 10^4$ at 640 nm) [91–94].

Other spectrophotometric methods for the determination of palladium are based on Chrome Azurol S [95,96], Xylenol Orange [97], Aluminon [98], Pontachrome Azure Blue B [99], pyridine-2-aldehyde-2-pyridylhydrazone [100,101], pyridine-2-aldehyde-2-quinolylhydrazone [102], picolinealdehyde-2-quinolylhydrazone [103], 3-hydroxy-1-p-sulphonatophenyl-3-phenyl-triazine [104], benzoselenadiazole [105], naphthoselenadiazole [106], 8-aminoquinoline [107], Ferron [108], Alizarin S [109], TTA [110,111], NTA (Complexone I) [112], and EDTA [113].

Dagnall et al. [114] have developed a very sensitive method ($\varepsilon = 1\cdot 25 \times 10^5$) based upon the ternary complex formed by palladium with pyridine and Rose Bengal Extra (anionic dye) which is extractable into chloroform.

A group of methods exploits the coloured palladium complexes with bromide [115–119], chloride [119–121], thiocyanate [122], or azide [123], usually in the presence of tin(II). These complexes are extractable as ion-association complexes with DAM [117], tri-n-octylamine [120], tetraphenylarsonium ion [121], and Methylene Blue [123].

References

1. Fraser, J. G., Beamish, F. E. and McBryde, W. A., *Anal. Chem.* **26**, 495 (1954).
2. Bode, H., *Z. Anal. Chem.* **143**, 182; **144**, 165 (1955).
3. Coburn, H. G., Beamish, F. E. and Lewis, C. L., *Anal. Chem.* **28**, 1297 (1956).
4. Pyle, J. T. and Jacobs, W. D., *ibid.* **36**, 1796 (1964).
5. Duke, J. F. and Stawpert, W., *Analyst* **85**, 671 (1960).
6. Forsythe, J. H., Magee, R. J. and Wilson, C. L., *Talanta* **3**, 324, 330 (1960).
7. Egli, R. A., *Z. Anal. Chem.* **194**, 401 (1963).
8. Golub, A. M. and Pomerants, G. V., *Ukr. Khim. Zh.* **31**, 104 (1965).
9. Khopkar, S. M., *Anal. Chem.* **38**, 360 (1966).
10. Mizuike, A., Sakamoto, T. and Onishi, N., *Mikrochim. Acta* **1971**, 783.
11. Ayres, G. H. and Berg, E. W., *Anal. Chem.* **25**, 980 (1953).
12. Marczenko, Z. and Krasiejko, M., *Chem. Anal.* (*Warsaw*) **9**, 291 (1964).
12a. Manku, G. S., Bhat. A. N. and Jain, B. D., *Talanta* **16**, 1421 (1969).
13. Marhenke, E. R. and Sandell, E. B., *Anal. Chim. Acta* **28**, 259 (1963).
14. Jackwerth, E., Graffmann, G. and Lohmar, J., *Z. Anal. Chem.* **247**, 149 (1969).
15. Jackwerth, E., *ibid.* **251**, 353 (1970).
16. MacNevin, W. M. and Crummett, W. B., *Anal. Chem.* **25**, 1628 (1953); *Anal. Chim. Acta* **10**, 323 (1954).
17. Fraser, J. G. and Beamish, F. E., *Anal. Chem.* **26**, 1474 (1954).
18. Faye, G. H. and Inman, W. R., *ibid.* **33**, 278 (1961).
19. Banbury, L. M. and Beamish, F. E., *Z. Anal. Chem.* **211**, 178 (1965).
20. Plummer, M. E., Lewis, C. L. and Beamish, F. E., *Anal. Chem.* **31**, 254, 1141 (1959).
21. Beamish, F. E., *Talanta* **12**, 743 (1965).
22. Frantsevich-Zabludovskaya, T. F., Kalimanova, L. P. and Sharafan, G. I., *Zh. Analit. Khim.* **18**, 1083 (1963).
23. Morrow, J. J. and Markham, J. J., *Anal. Chem.* **36**, 1159 (1964).
24. Kreimer, S. E., Butylkin, L. P. and Stogova, A. V., *Zh. Analit. Khim.* **15**, 467 (1960).
25. Pilipenko, A. T. and Ol'khovich, P. F., *Ukr. Khim. Zh.* **34**, 83, 286 (1968).
26. Ziegler, M. and Pape, G., *Z. Anal. Chem.* **197**, 354 (1963).
27. Senise, P. and Levi, F., *Anal. Chim. Acta* **30**, 509 (1964).
28. Marczenko, Z. and Limbach, A., unpublished work.
29. Minczewski, J., Krasiejko, M. and Marczenko, Z., *Chem. Anal.* (*Warsaw*) **15**, 43 (1970).
30. Young, R. S., *Analyst* **76**, 49 (1951).
31. Nielsch, W., *Mikrochim. Acta* **1954**, 532.
32. Pilipenko, A. T. and Maslei, N. N., *Ukr. Khim. Zh.* **33**, 730 (1967).
32a. Diamantatos, A., *Anal. Chim. Acta* **61**, 233 (1972).
33. Bankovskii, Yu. A. and Ievin'sh, A. F., *Zh. Analit. Khim.* **13**, 507 (1958).
34. Lystsova, G. G., *Zavodsk. Lab.* **28**, 543 (1962).
35. Busev, A. I. and Naku, A., *Zh. Analit. Khim.* **18**, 1479 (1963).
36. Wagner, V. L., Jr. and Yoe, J. H., *Talanta* **2**, 223 (1959).
37. Majumdar, A. K. and Chakrabartty, M. M., *Anal. Chim. Acta* **19**, 372, 482 (1958).
38. Ayres, G. H. and Janota, H. F., *Anal. Chem.* **31**, 1985 (1959); **36**, 138 (1964).

References

38a. Dalziel, J. A. and Slawinski, A. K., *Talanta* **19**, 1190 (1972).
39. Ayres, G. H. and Narang, B. D., *Anal. Chim. Acta* **24**, 241 (1961).
40. Cheng, K. L. and Goydish, B. L., *Microchem. J.* **10**, 158 (1966).
41. Busev, A. I. and Shishkov, A. N., *Zh. Analit. Khim.* **23**, 1675 (1968).
42. Jacobs, W. D., *Anal. Chem.* **32**, 512 (1960); **33**, 1279 (1961).
43. Jacobs, W. D., Wheeler, C. M. and Waggoner, W. H., *Talanta* **9**, 243 (1962).
44. Pyle, J. T. and Jacobs, W. D., *ibid.* **9**, 761 (1962).
45. Goeminne, A., Herman, M. and Eeckhaut, Z., *Anal. Chim. Acta* **28**, 512 (1963).
46. Busev, A. I. and Naku, A., *Zh. Analit. Khim.* **18**, 500, 1233 (1963).
47. Busev, A. I. and Evsikov, V. V., *ibid.* **25**, 953 (1970).
47a. Busev, A. I., Byrko, V. M., Kvesitadze, A. G. and Simonova, V. N., *ibid.* **27**, 1802 (1972).
48. Chechneva, A. N. and Radushev, A. V., *ibid.* **23**, 1059 (1968).
49. Arita, T. and Yoe, J. H., *Anal. Chim. Acta* **29**, 500 (1963).
50. Sen Gupta, J. G., *Talanta* **8**, 729 (1961).
50a. Khosla, M. M. and Rao, S. P., *Microchem. J.* **18**, 640 (1973).
51. Busev, A. I. and Vin', D. Kh., *Zh. Analit. Khim.* **20**, 976 (1965).
52. Menis, O. and Rains, T. C., *Anal. Chem.* **27**, 1932 (1955).
53. Peshkova, V. M., Shlenskaya, V. I. and Sokolov, S. S., *Tr. Komis. po Analit. Khim., Akad. Nauk SSSR* **11**, 328 (1960).
54. Marczenko, Z. and Krasiejko, M., *Chem. Anal. (Warsaw)* **15**, 1233 (1970).
55. Sen, B., *Anal. Chem.* **31**, 881 (1959).
56. Holland, W. J. and Bozic, J., *ibid.* **40**, 433 (1968).
57. Stupavsky, S. and Holland, W. J., *Mikrochim. Acta* **1972**, 122.
58. Holland, W. J., Dimenna, R. A. and Walker, R. J., *ibid.* **1972**, 183; **1973**, 591.
59. Ayres, G. H. and Martin, J. B., *Anal. Chim. Acta* **35**, 181 (1966).
60. Davis, W. F., *Talanta* **16**, 1330 (1969).
61. Pshenitsyn, N. K. and Ivonina, O. M., *Zavodsk. Lab.* **24**, 1185 (1958).
62. Pflaum, R. T., Wehking, M. W. and Jensen, R. E., *Talanta* **11**, 1193 (1964).
63. Holland, W. J. and Soules, D., *Anal. Lett.* **2**, 167 (1969).
64. Poddar, S. N., *Anal. Chim. Acta* **28**, 586 (1963).
64a. Holland, W. J. and Walker, R. J., *Mikrochim. Acta* **1973**, 193.
65. Yoe, J. H. and Kirkland, J. J., *Anal. Chem.* **26**, 1335 (1954).
66. Johnson, R. W., *Analyst* **86**, 185 (1961).
67. Cheng, K. L., *Anal. Chem.* **26**, 1894 (1954).
68. Shamir, J. and Schwartz, A., *Talanta* **8**, 330 (1961).
69. Rollins, O. W. and Oldham, M. M., *Anal. Chem.* **43**, 262 (1971).
70. McDonald, C. W. and Bedenbaugh, J. H., *Mikrochim. Acta* **1970**, 474.
71. Talwar, U. B. and Haldar, B. C., *Anal. Chem.* **38**, 1929 (1966).
72. Talwar, U. B. and Haldar, B. C., *Anal. Chim. Acta* **39**, 264 (1967).
73. Busev, A. I. and Ivanov, V. M., *Zh. Analit. Khim.* **19**, 232 (1964).
74. Flaschka, H. and Hicks, J., *Microchem. J.* **11**, 517 (1966).
75. Ivanov, V. M., *Zh. Analit. Khim.* **22**, 763 (1967).
76. Busev, A. I. and Kiseleva, L. V., *Vestn. Mosk. Univ. Khim.* **1958**, No. 4, 179.
76a. Ivanov, V. M., Figurovskaya, V. N. and Busev, A. I., *Zavodsk. Lab.* **38**, 1311 (1972).
77. Gusev, S. I. and Vin'kova, V. A., *Zh. Analit. Khim.* **22**, 376, 552, 1039 (1967).
78. Shibata, S., Ishiguro, Y. and Nakashima, R., *Anal. Chim. Acta* **64**, 305 (1973).
79. Busev, A. I., Ivanov, V. M. and Krysina, L. S., *Vestn. Mosk. Univ. Khim.* **1968**, No. 1, 80.
80. Busev, A. I., Ivanov, V. M., Krysina, L. S., Zholondkovskaya, T. N. and Abramova, T. I., *Tr. Komis. po Analit. Khim. Akad. Nauk. SSSR* **17**, 368 (1969).
81. Busev, A. I., Krysina, L. S., Zholondkovskaya, T. N., Pribylova, G. A. and Krysin, E. P., *Zh. Analit. Khim.* **25**, 1575 (1970).
82. Savvin, S. B., Propistsova, R. F. and Okhanova, L. A., *ibid.* **24**, 1634 (1969).

83. Saxena, K. K. and Dey, A. K., *Anal. Chem.* **40**, 1280 (1968).
84. Dedkov, Yu. M. and Levina, G. P., *Zh. Analit. Khim.* **26**, 558 (1971).
85. Savvin, S. B., Sokolovskaya, L. A., Okhanova, L. A. and Propistsova, R. F., *Tr. Komis. po Analit. Khim. Akad. Nauk SSSR* **17**, 187 (1969).
86. Baiulescu, Gh., Greff, C. and Dănet, F., *Analyst* **94**, 354 (1969).
87. Sangal, S. P. and Dey, A. K., *Microchem. J.* **7**, 257 (1963).
88. Pilipenko, A. T., Maslei, N. N. and Filimonova, V. V., *Ukr. Khim. Zh.* **38**, 268 (1972).
89. Sen Gupta, J. G., *Anal. Chem.* **39**, 18 (1967); *Anal. Chim. Acta* **42**, 481 (1968).
90. Savvin, S. B., Propistsova, R. F. and Okhanova, L. A., *Talanta* **16**, 423 (1969).
91. Pérez-Bustamante, J. A., Morell García, C. and Burriel-Martí, F., *Anal. Chim. Acta* **44**, 95 (1969).
92. Pérez-Bustamante, J. A. and Burriel-Martí, F., *ibid.* **37**, 49, 62 (1967); **51**, 277 (1970).
93. Pérez-Bustamante, J. A. and Burriel-Martí, F., *Talanta* **18**, 183, 717 (1971).
94. Bocanegra Sierra, L., Pérez-Bustamante, J. A. and Burriel-Martí, F., *Anal. Chim. Acta* **59**, 231, 249 (1972).
95. Sangal, S. P. and Dey, A. K., *Mikrochim. Acta* **1963**, 993.
96. Ishida, R., *Bull. Chem. Soc. Japan* **42**, 1011 (1969).
97. Otomo, M., *ibid.* **36**, 889 (1963).
98. Munshi, K. N. and Dey, A. K., *Talanta* **11**, 1265 (1964).
99. Uesugi, K., Shigematsu, T. and Tabushi, M., *Anal. Chim. Acta* **60**, 79 (1972).
100. Bell, C. F. and Rose, D. R., *Talanta* **12**, 696 (1965).
101. Cameron, A. J. and Gibson, N. A., *Anal. Chim. Acta* **40**, 413 (1968).
102. Heit, M. L. and Ryan, D. E., *ibid.* **34**, 407 (1966).
103. Jensen, R. E. and Pflaum, R. T., *ibid.* **37**, 397 (1967).
104. Sogani, N. C. and Bhattacharyya, S. C., *Anal. Chem.* **29**, 397 (1957).
105. Bunting, T. G. and Meloan, C. E., *ibid.*, **40**, 435 (1968).
106. Lau, H. K. and Lott, P. E., *Talanta* **17**, 717 (1970).
107. Gustin, V. K. and Sweet, T. R., *Anal. Chem.* **35**, 44 (1963).
108. Singh, T. and Dey, A. K., *Talanta* **18**, 225 (1971).
109. Sangal, S. P., *Chim. Anal. (Paris)* **50**, 131 (1968).
110. De, A. K. and Rahaman, M. S., *Analyst* **89**, 795 (1964).
111. Rangnekar, A. V. and Khopkar, S, M., *Bull. Chem. Soc. Japan* **38**, 1696 (1965).
112. Desideri, P. G. and Pantani, F., *Talanta* **8**, 235 (1961).
113. MacNevin, W. M. and Kriege, O. H., *Anal. Chem.* **26**, 1768 (1954).
114. Dagnall, R. M., El-Ghamry, M. T. and West, T. S., *Talanta* **15**, 1353 (1968).
115. Ayres, G. H. and Tuffly, B. L., *Anal. Chem.* **24**, 949 (1952).
116. Pantani, F. and Piccardi, G., *Anal. Chim. Acta* **22**, 231 (1960).
117. Danilova, V. N. and Lisichenok, S. L., *Zh. Analit. Khim.* **26**, 1157 (1971).
118. Diamantatos, A., *Anal. Chim. Acta* **63**, 220 (1973).
119. Dalziel, J. A., Donaldson, J. D. and Woodget, B. W., *Talanta* **16**, 1477 (1969).
120. Khattak, M. A. and Magee, R. J., *Anal. Chim. Acta* **35**, 17 (1966).
121. Nasouri, F. G. and Witwit, A. S., *Talanta* **16**, 1492 (1969).
122. Pilipenko, A. T., Ol'khovich, P. F. and Bondarenko, V. Yu., *Zh. Analit. Khim.* **39**, 480 (1973).
123. Kuroda, R., Yoshikuni, N. and Kamimura, Y., *Anal. Chim. Acta* **60**, 71 (1972).

Chapter 39
PHOSPHORUS

Phosphorus (P, at.wt. 30·97) is a non-metal. The most important compounds are those of phosphorus(V), such as phosphate (derived from orthophosphoric acid, H_3PO_4) and the rather less common condensed forms, pyro-, meta-, and polyphosphate. Phosphorus(V) gives stable heteropoly acids with Mo(VI), W(VI), V(V) etc. Phosphorus occurs in the +III, +I and −III oxidation states in phosphite, hypophosphite and phosphine, respectively.

39.1 Methods of Separation

When a sample is dissolved, the phosphorus usually passes into solution as orthophosphate. Rather than isolate the phosphate, it may be better to isolate the interfering elements, leaving the phosphate to be determined in the mother liquor. Examples of such separations include distillation of Si, As, and Ge as volatile halides [1–3] or of boron as trimethyl borate [4], precipitation of heavy metals as sulphides from an acid medium, retention of cations on a strongly acidic cation-exchanger (phosphate is eluted) [5–6], and electrolytic separation of metals at a mercury cathode [7].

39.1.1 Extraction

The separation of phosphate from other elements, in particular from silicon, is often achieved by extracting phosphorus as a heteropoly acid from a slightly acidic solution (pH ~ 1·4). Higher alcohols, esters, and ethers are suitable extractants [8–12]. Extraction with a mixture of butanol and chloroform enables molybdophosphoric acid to be separated from molybdoarsenic acid [13,14]. Isobutyl acetate extracts molybdophosphoric acid, but not molybdosilicic acid, from a solution at pH 0·3–1·0 [11].

When determining traces of phosphorus in silicon tetrachloride, Lancaster and Everingham [15] shook the sample with concentrated sulphuric acid, causing the phosphorus to pass into the acid layer.

39.1.2 Precipitation

Fedorov and Linkova [16] have separated traces of phosphorus in metallic chromium by adding $Ca(NO_3)_2$ to the alkaline solution of chromium(VI), thereby precipitating calcium phosphate with $Ca(OH)_2$ as collector.

Traces of phosphate can also be precipitated with aluminium [2] or iron [17] hydroxide as collector.

39.2 Methods of Determining Orthophosphate

Spectrophotometric methods for determining orthophosphate are not numerous. Microgram quantities of phosphorus are conventionally determined by the phosphomolybdenum blue method, either with or without an extraction. The molybdovanadophosphoric acid method is suitable for determining relatively large quantities of phosphorus.

39.2.1 PHOSPHOMOLYBDENUM BLUE METHOD

In an acid medium containing excess of molybdate, orthophosphate forms pale yellow molybdophosphoric acid, which absorbs intensely in the ultraviolet. This reaction is useful for determining phosphate at fairly high concentrations [13,18,18a], but the sensitivity can be increased considerably by the use of aqueous acetone [18,19].

A sensitive, spectrophotometric method for determining phosphorus has been based on the reduction of molybdophosphoric acid to phosphomolybdenum blue under mild conditions to prevent reduction of the free molybdic acid [9,20–26]. Reductants employed include hydrazine [3,27], stannous chloride [2,20], ascorbic acid [28], metol [29], and various other reagents [8,30–33]. In the preparation of phosphomolybdenum blue, a single reagent consisting of ammonium molybdate, hydrazine sulphate, and sulphuric acid is convenient (see procedure below).

Molybdophosphoric acid is reduced either in aqueous medium ($\sim 0.5M$ H_2SO_4) [1,2] or in the organic phase (usually butanol) after molybdophosphoric acid has been extracted [8,20]. Alternatively, the phosphomolybdenum blue may be formed in the aqueous phase, then extracted into butanol [34].

The absorbance of the phosphomolybdenum blue is very dependent on the medium (water, butanol, or other oxygenated organic solvent), the reductant, and the acidity of the aqueous phase. The molar absorptivity of the blue solution in butanol after hydrazine reduction is 2.5×10^4 ($a = 0.81$) at $\lambda_{max} = 780$ nm (see Fig. 7.1, p. 133).

Extraction of phosphomolybdenum blue slightly displaces the absorption maximum towards shorter wavelengths. The reduced species is more readily extractable than molybdophosphoric acid itself [34].

Interference in the determination of phosphorus by the phosphomolybdenum blue method comes primarily from arsenic(V), silicon, and germanium, which also react with molybdate to form the corresponding acids which are reduced to the respective heteropoly blues [35]. Arsenic does not interfere when reduced to As(III) with sulphite or thiourea. The precipitation of readily hydrolysable metals [Nb, Ta, Ti, Zr, Sn(IV), W, Bi] occludes phosphate. Titanium and zirconium catalyse reduction of molybdic acid during the formation of phosphomolybdenum blue [28]. In the presence of vanadium(V), molybdovanadophosphoric acid is produced. Large amounts of vanadium(V) are reduced with Mohr's salt (ammonium ferrous sulphate) to the +IV oxidation state before the molybdate is added; the molybdophosphoric acid is extracted and reduced in the organic phase [36].

Shen and Dyroff [37] have taken advantage of the difference in the rates of formation of the phosphomolybdenum and silicomolybdenum blues to determine phosphorus in the presence of silicon. The interference of silicon can be prevented by use of sufficiently acid medium ($> 0.7N$ H_2SO_4) [38].

Oxalic, tartaric, and citric acids, and EDTA affect the completeness of reduction of molybdophosphoric acid. Phosphate can be determined in presence of silicate, and vice versa, depending on the order of addition of the reagents, when these complexing acids or mannitol are used to mask the molybdate, the effect being determined by the kinetics and thermodynamics of the system [18]. Before the determination of phosphate, any nitrate must first be reduced to ammonia which is then distilled from alkaline medium [39].

Phosphomolybdenum blue may be extracted with chloroform in the presence of dioctylamine, trioctylamine [40,41], or propylene carbonate [42].

The phosphomolybdenum blue method may be employed for determining orthophosphate in the presence of phosphate esters [43] and polyphosphate [44]. Condensed phosphates are converted into orthophosphate by boiling for 15 minutes in a $2.5M$ H_2SO_4 medium [45].

The phosphomolybdenum blue method has been used to determine phosphorus in iron and steels [7,20,28,46–48], nickel and its alloys [13], silicates [49], copper [13], aluminium alloys [50,51], titanium alloys [52], chromium [16], silicon [1,3], germanium [1], boron [4], arsenic [53], bauxite [54], niobium and tantalum oxides [55], neodymium and yttrium oxides [56], coke [57], petroleum products [58], textiles [59], organic compounds [60–62a], soil and plant materials [63], biological materials [29,30, 64–66], trade effluents [45,67], and water [2,67,68].

This method is also suitable for the automated spectrophotometric determination of phosphorus in steel and slag [69,70], detergents [71], and water [72].

Analogous methods for the determination of phosphate in water have been based on the mixed phosphorus, antimony, and molybdenum heteropoly blue [73–73b].

Reagents

1. Molybdenum reagent. (a) Dissolve 1·0 g of ammonium molybdate in 100 ml of $2M$ H_2SO_4 (1+8); (b) dissolve 0·10 g of hydrazine sulphate in 100 ml of water. Immediately before use, mix 10 ml of solution (a)

with 10 ml of solution (b), and dilute to 100 ml with water. Solutions (a) and (b) should not be stored longer than 3–4 days.
2. Standard phosphorus(V) solution: 1 mg/ml. Dissolve in water 4·390 g of potassium dihydrogen phosphate (KH_2PO_4) dried at 110°C, add 1 ml of chloroform (to prevent the formation of mould), and dilute with water in a volumetric flask to 1 litre.
3. Ammonium molybdate: 10% solution adjusted with ammonia to pH 7·4 ($\pm 0·2$).

Procedure

Extractive separation of P. Evaporate the sample solution freed from As (e.g. by extraction as $AsCl_3$, cf. p. 134) nearly to dryness, dilute with 20 ml of water, add 2 ml of the ammonium molybdate solution and adjust the pH to 1·4 ($\pm 0·1$) with $0·5M$ H_2SO_4. After 5 minutes, transfer the solution to a separating funnel and extract the molybdophosphoric acid with two 10-ml portions of butanol. Wash the alcoholic extract with $0·05$ M H_2SO_4.

Determination of P. Evaporate an aliquot of the extract (or an aqueous sample solution freed from As, Ge, and Si), containing not more than 70 μg of P, to dryness in a beaker with nitric acid. Add 25 ml of the molybdenum reagent, and place the beaker on a boiling water-bath for 10 minutes. Transfer the cooled solution to a separating funnel, and extract the phosphomolybdenum blue with two portions of butanol. Dilute the extract to the mark with the solvent in a 50-ml (or smaller) volumetric flask, and measure the absorbance at 780 nm, using the solvent as reference.

Notes: (1) The absorbance of the phosphomolybdenum blue may be measured in the aqueous solution. In this case, the coloured aqueous solution is diluted to the mark in the volumetric flask with the molybdenum reagent.
(2) Molybdophosphoric acid may be extracted, then stripped with dilute ammonia solution (1+50), the solution acidified with nitric acid and evaporated, and the complex reduced to the heteropoly blue.

39.2.2 Molybdovanadophosphoric Acid Method

Addition of molybdate to an acidic solution containing orthophosphate and vanadate results in the formation of yellow-orange molybdovanadophosphoric acid having the P:V:Mo ratio of 1:1:11 [74–77].

The absorption maximum of the compound is in the ultraviolet at 315 nm ($\varepsilon = 2·0 \times 10^4$). At 400 nm, $\varepsilon = 2·5 \times 10^3$ (specific absorptivity 0·08). In the molybdovanadophosphoric acid method, the absorbance is measured either at 315 nm (sensitivity as high as that in the phosphomolybdenum blue method), or between 400 and 470 nm (much lower sensitivity).

The colour depends on the acidity of the solution and on the concentrations of the reagents used. The most suitable acid concentration is from 0·5 to $1·0M$ (HNO_3, H_2SO_4, $HClO_4$, or HCl). In insufficiently acid solutions, the yellow colour is produced even in the absence of orthophosphate;

in excessively acid solutions, the formation of molybdovanadophosphoric acid proceeds very slowly. The concentrations of vanadate and molybdate in the final solution should be $\sim 0.002M$ and $0.01M$ respectively. Since the reagents also produce a slight colour in the absence of phosphate, the absorbance must be measured against a reagent blank solution. The coloured solutions obtained are stable.

In $0.8M$ HNO_3 silicon does not interfere provided it is not present in greater amount than phosphorus. In more acid media, even more silicon can be tolerated [78]. At higher concentrations, silicic acid can be converted into the inert polymeric form by heating the sample solution to fumes with conc. perchloric acid.

Pyrophosphoric acid does not interfere. Arsenic(V) produces a colour only 1/100 as intense as that from phosphoric acid.

Large amounts of iron(III) interfere, but may be masked with fluoride, the excess of which is complexed with boric acid. Reductants and certain coloured metal ions [e.g.Cr(VI), Ni, Co, Cu, and U(VI)] also interfere. Molybdovanadophosphoric acid may be separated from many coloured ions by extraction with oxygenated organic solvents [79]. Reductants must be absent.

The molybdovanadophosphoric acid method has been used to determine phosphorus in iron and steels [80–85], copper alloys [80, 81, 86–88], aluminium and nickel [88], white metals [88], niobium, zirconium, titanium, and tungsten [89], phosphate rocks [6], phosphorous acid [90], silicate rocks [91], fertilizers [92,93], silicon tetrachloride [15], plant tissue [94–96], organic compounds [61,77,97,98], biological samples [66,99], and air [100].

Gee and Deitz [101] have determined large quantities of phosphate by differential spectrophotometry.

Reagents

1. Ammonium metavanadate: 0.25% solution. Dissolve 1.25 g of NH_4VO_3 in 250 ml of hot water. Cool the solution and add 10 ml of conc. HNO_3. Allow the solution to stand overnight, filter (if necessary), and dilute with water to 500 ml. Store the solution in a polyethylene container.
2. Ammonium molybdate: 5% solution. Dissolve 25 g of $(NH_4)_6Mo_7O_{24} \cdot 4H_2O$ in 250 ml of water (at $\sim 50°C$). Allow the solution to stand overnight, filter (if necessary), dilute with water to 500 ml, and store in a polyethylene container.
3. Standard phosphorus(V) solution: 1 mg/ml (p. 424).

Procedure

To the slightly acidic sample solution containing not more than 0.70 mg of P, add successively 5 ml of nitric acid (1+1), 5 ml of the vanadate solution, and 5 ml of the molybdate solution, mixing the solution after the addition of each reagent. Dilute the solution to volume with water in a 50-ml volumetric flask. After 30 minutes, measure the absorbance at 400 nm against a reagent blank solution.

39.2.3 Other Methods of Determining Orthophosphate

With basic dyes, molybdophosphoric acid forms ion-association complexes which are the basis of sensitive extraction methods for determining phosphorus spectrophotometrically [102–108]. The following basic dyes have been used: Safranine T [102,103], Crystal Violet and Methyl Violet [104,106–109], Brilliant Green and Malachite Green [104–106], and Rhodamine 6G [104]. The molar absorptivity with Crystal Violet is 2.7×10^5 at 582 nm [106]. The coloured suspension of the compound formed by molybdophosphoric acid and Malachite Green has been exploited similarly [110]. Numerous sensitive indirect methods have been devised for the determination of orthophosphate. The molybdenum in an extract of molybdophosphoric acid (P:Mo = 1:12) has been determined with thiocyanate [111,112], phenylfluorone [113], 2-amino-4-chlorobenzenethiol [114], and Sulphonitrophenol S or dithiol [115]. Babko et al. [116] have extracted the compound formed between molybdophosphoric acid and 8-hydroxyquinoline, coupled the oxine with diazosulphanilic acid, and measured the absorbance of the resulting dye. Phosphate has also been determined by precipitation of uranium(VI) magnesium phosphate and subsequent reaction of the UO_2^{2+} ions to form the dark red ferrocyanide [117].

The indirect lanthanum chloranilate method, in which phosphate displaces the coloured chloranilate anion, is less sensitive [118,119].

Lastly, turbidimetric methods for the determination of phosphate have been based on sparingly soluble magnesium, lead, and bismuth phosphates [120].

39.3 Determination of Other Phosphorus Compounds

Traces of phosphine in the atmosphere can be determined by passing a known volume of air through a filter paper (or tube packed with silica gel) impregnated with silver nitrate, the amount of PH_3 being estimated from the intensity (or length) of the resulting black stain by comparison with standards (as in the Gutzeit method for determining As) [121].

Hypophosphite can be determined in the presence of phosphate from the colour the former gives with ammonium molybdate under controlled conditions [122,123]. Hypophosphite is determined by the reduction in colour of the iron(III)-thiocyanate complex [124].

Condensed phosphates can be separated on anion-exchangers and converted into orthophosphate, which is then determined by a suitable method [125–128].

Colorimetric methods for the determination of polymetaphosphate and pyrophosphate have been based on the effect of these anions on the reactions of iron with thiocyanate [129,130] and with 1,10-phenanthroline [131].

Shevchuk et al. [132] have extracted the iron(III) pyrophosphate complex with a chloroform solution of n-dodecylamine, shaken the extract with a sulphosalicylic acid solution, and measured the absorbance of the resulting iron(III) sulphosalicylate complex.

Salts of chloranilic acid are useful for estimating polymetaphosphate and pyrophosphate in the presence of orthophosphate [133].

Yoza and Ohashi [134,135] have determined various phosphorus oxyacids, using the molybdate reagent containing both Mo(V) and Mo(VI) described by Lucena-Conde and Prat [136].

Pollard et al. have separated phosphite, hypophosphite, and phosphate in an anion-exchange column [137].

References

1. Luke, C. L. and Campbell, M. E., *Anal. Chem.* **25**, 1588 (1953).
2. Levine, H., Rowe, J. J. and Grimaldi, F. S., *ibid.* **27**, 258 (1955).
3. Pohl, F. A. and Bonsels, W., *Mikrochim. Acta* **1962**, 97.
4. Marczenko, Z., *Chem. Anal. (Warsaw)* **9**, 1093 (1964).
5. Fischer, W., Paul, R. and Abendroth, H. J., *Anal. Chim. Acta* **13**, 38 (1955).
6. Schafer, H. N., *Anal. Chem.* **35**, 53 (1963).
7. Gates, O. R., *ibid.* **26**, 730 (1954).
8. Ging, N. S., *ibid.* **28**, 1330 (1956).
9. Ruf, E., *Z. Anal. Chem.* **151**, 169 (1956); **161**, 1 (1958).
10. Andersson, L. H., *Acta Chem. Scand.* **13**, 1743 (1959).
11. Paul, J., *Anal. Chim. Acta* **23**, 178 (1960); *Mikrochim. Acta* **1965**, 830, 836.
12. Umland, F. and Wünsch, G., *Z. Anal. Chem.* **225**, 362 (1967).
13. Marczenko, Z. and Lenarczyk, Ł., *Chem. Anal. (Warsaw)* **19**, 679 (1974).
14. Ross, H. H. and Hahn, R. B., *Talanta* **7**, 276 (1961).
15. Lancaster, W. A. and Everingham, M. R., *Anal. Chem.* **36**, 246 (1964).
16. Fedorov, A. A. and Linkova, F. V., *Zavodsk. Lab.* **26**, 535 (1960).
17. Novikov, A. I. and Shchekoturova, E. K., *Radiokhimiya* **10**, 366 (1968).
18. Chalmers, R. A. and Sinclair, A. G., *Anal. Chim. Acta* **34**, 412 (1966).
18a. Halász, A., Pungor, E. and Polyák, K., *Talanta* **18**, 577 (1971).
19. Bernhart, D. N. and Wreath, A. R., *Anal. Chem.* **27**, 440 (1955).
20. Lueck, C. H. and Boltz, D. F., *ibid.* **28**, 1168 (1956).
21. Babko, A. K. and Evtushenko, L. M., *Zavodsk. Lab.* **23**, 423 (1957).
22. Sims, R. P., *Analyst* **86**, 584 (1961).
23. Namiki, H., *Bull. Chem. Soc. Japan* **37**, 484 (1964).
24. Crouch, S. R. and Malmstadt, H. V., *Anal. Chem.* **39**, 1084 (1967).
25. Pakalns, P., *Anal. Chim. Acta* **40**, 1 (1968).
26. Goryushina, V. G., Esenina, N. V. and Snesarev, K. A., *Zh. Analit. Khim.* **24**, 1699 (1969); **25**, 1610 (1970).
27. Hahn, F. L. and Luckhaus, R., *Z. Anal. Chem.* **149**, 172 (1956).
28. Jean, M., *Anal. Chim. Acta* **14**, 172 (1956).
29. Pilz, W., *Mikrochim. Acta* **1965**, 34.
30. Shin, Y. S., *Anal. Chem.* **34**, 1164 (1962).
31. Vinogradova, N. B., Dubovskaya, L. V. and Zhukovskii, Yu. G., *Zh. Analit. Khim.* **19**, 997 (1964).
32. Sudakov, F. P., Galankina, N. F. and Khamrakulova, M. I., *ibid.* **25**, 548 (1970).
33. Sudakov, F. P. and Tverdova, N. A., *ibid.* **26**, 573 (1971).

34. Klitina, V. I., Sudakov, F. P. and Alimarin, I. P., *ibid.* **20**, 1145 (1965).
35. Duval, L., *Chim. Anal. (Paris)* **48**, 290 (1966).
36. Goryushina, V. G. and Biryukova-Gailis, E. Ya., *Zavodsk. Lab.* **24**, 402 (1958).
37. Shen, C. Y. and Dyroff, D. R., *Anal. Chem.* **34**, 1367 (1962).
38. Rockstein, M. and Herron, P. W., *ibid.*, **23**, 1500 (1951).
39. Duff, E. J. and Stuart, J. L., *Analyst* **96**, 802 (1971).
40. Klitina, V. I., Sudakov, F. P. and Alimarin, I. P., *Zh. Analit. Khim.* **21**, 338 (1966).
41. Sudakov, F. P., Klitina, V. I. and Maslova, N. T., *ibid.* **21**, 1089 (1966); *Vestn. Mosk. Univ. Khim.* **1966**, No. 1, 98.
42. Jakubiec, R. J. and Boltz, D. F., *Mikrochim. Acta* **1970**, 1199.
43. Golterman, H. L. and Wurtz, I. M., *Anal. Chim. Acta* **25**, 295 (1961).
44. Tewari, K. K. and Krishnan, P. S., *ibid.* **22**, 111 (1960).
45. Analytical Methods Committee for Trade Effluents, *Analyst* **83**, 50 (1958).
46. Lounamaa, N. and Fugmann, W., *Z. Anal. Chem.* **199**, 352 (1964).
47. Theakston, H. M. and Bandi, W. R., *Anal. Chem.* **38**, 1764 (1966).
48. Heslop, R. B. and Pearson, E. F., *Anal. Chim. Acta* **39**, 209 (1967).
49. Chalmers, R. A., *Analyst* **78**, 32 (1953).
50. Davey, M. L., *Metallurgia* **65**, 151 (1962).
51. Mukai, K., *Talanta* **19**, 489 (1972).
52. Codell, M. and Mikula, J. J., *Anal. Chem.* **25**, 1444 (1953).
53. Goryushina, V. G. and Esenina, N. V., *Zh. Analit. Khim.* **21**, 239 (1966).
54. Erdey, L. and Fleps, V., *Acta Chim. Acad. Sci. Hung.* **11**, 195 (1957).
55. Zaboeva, M. I. and Spitsyn, P. K., *Zavodsk. Lab.* **33**, 554 (1967).
56. Alikina, N. A., Barkovskii, V. F. and Shvarev, V. S., *Zh. Analit. Khim.* **24**, 1848 (1969).
57. Kirk, B. P. and Wilkinson, H. C., *Talanta* **19**, 80 (1972).
58. Gedansky, S. J., Bowen, J. E. and Milner, O. I., *Anal. Chem.* **32**, 1447 (1960).
59. McAloren, J. T. and Reynolds, G. F., *Talanta* **10**, 145 (1963).
60. Kirsten, W. J. and Carlsson, M. E., *Microchem. J.* **4**, 3 (1960).
61. Debal, E., *Chim. Anal. (Paris)* **45**, 66 (1963).
62. Tölg, G., *Z. Anal. Chem.* **194**, 20 (1963).
62a. Chalmers, R. A. and Thomson, D. A., *Anal. Chim. Acta* **18**, 575 (1958).
63. Van Schouwenburg, J. C. and Walinga, I., *ibid.* **37**, 271 (1967).
64. Chen, P. S., Jr., Toribara, T. Y. and Warner, H., *Anal. Chem.* **28**, 1756 (1956).
65. Bauminger, B. B. and Walters, G., *Analyst* **91**, 205 (1966).
66. Tušl, J., *ibid.* **97**, 111 (1972).
67. Fogg, D. N. and Wilkinson, N. T., *ibid.* **83**, 406 (1958).
68. Johnson, D. L. and Pilson, M. E., *Anal. Chim. Acta* **58**, 289 (1972).
69. Scholes, P. H. and Thulbourne, C., *Analyst* **89**, 466 (1964).
70. Scholes, P. H., *Z. Anal. Chem.* **222**, 162 (1966).
71. Lundgren, D. P., *Anal. Chem.* **32**, 824 (1960).
72. Henriksen, A., *Analyst* **90**, 29 (1965).
73. Murphy, J. and Riley, J. P., *Anal. Chim. Acta* **27**, 31 (1962).
73a. Isaeva, A. B., *Zh. Analit. Khim.* **24**, 1854 (1969).
73b. Nnadi, L. A. and Tabatabai, M. A., *Anal. Lett.* **6**, 555 (1973).
74. Kitson, R. E. and Mellon, M. G., *Ind. Eng. Chem., Anal. Ed.* **16**, 379 (1944).
75. Quinlan, K. P. and DeSesa, M. A., *Anal. Chem.* **27**, 1626 (1955).
76. Michelson, O. B., *ibid.* **29**, 60 (1957).
77. Salvage, T. and Dixon, J. P., *Analyst* **90**, 24 (1965).
78. Lew, R. B. and Jakob, F., *Talanta* **10**, 322 (1963).
79. Heslop, R. B. and Ramsey, A. C., *Anal. Chim. Acta* **47**, 305 (1969).
80. Baghurst, H. C. and Norman, V. J., *Anal. Chem.* **27**, 1070 (1955); **29**, 778 (1957).
81. Elwell, W. T. and Wilson, H. N., *Analyst* **81**, 136 (1956).
82. Lindley, G., *Anal. Chim. Acta* **25**, 334 (1961).
83. Schwarz, H., *Mikrochim. Acta* **1969**, 677.

References

84. Pakalns, P., *Anal. Chim. Acta* **49**, 511 (1970).
85. Blazejak-Ditges, D., *Mikrochim. Acta* **1972**, 65.
86. Lutwak, H. K., *Analyst* **78**, 661 (1953).
87. Bilińska, U. and Terpiłowski, J., *Chem. Anal. (Warsaw)* **5**, 17 (1960).
88. Pakalns, P., *Anal. Chim. Acta* **51**, 497 (1970).
89. Pakalns, P., *ibid.* **50**, 103 (1970).
90. Demarcq, M. C. and Portas, M., *Chim. Anal. (Paris)* **48**, 654 (1966).
91. Baadsgaard, H. and Sandell, E. B., *Anal. Chim. Acta* **11**, 183 (1954).
92. Schüller, H. and Russ, H., *Z. Anal. Chem.* **186**, 410 (1962).
93. Docherty, A. C., Farrow, S. G. and Skinner, J. M., *Analyst* **97**, 36 (1972).
94. Koter, M. and Panak, H., *Chem. Anal. (Warsaw)* **5**, 317 (1960).
95. Varley, J. A., *Analyst* **91**, 119 (1966).
96. Basson, W. D., Stanton, D. A. and Böhmer, R. G., *ibid.* **93**, 166 (1968).
97. Christopher, A. J., Fennell, T. R. and Webb, J. R., *Talanta* **11**, 1323 (1964).
98. Christopher, A. J. and Fennell, T. R., *Microchem. J.* **12**, 593 (1967).
99. Pulss, G., *Z. Anal. Chem.* **176**, 412 (1960).
100. Talvitie, N. A., Perez, E. and Illustre, D. P., *Anal. Chem.* **34**, 866 (1962).
101. Gee, A. and Deitz, V. R., *ibid.* **25**, 1320 (1953).
102. Ducret, L. and Drouillas, M., *Anal. Chim. Acta* **21**, 86 (1959).
103. Sudakov, F. P., Kletina, V. I. and Dan'shova, T. Ya., *Zh. Analit. Khim* **21**, 1333 (1966).
104. Babko, A. K., Shkaravskii, Yu. F. and Kulik, V. I., *ibid.* **21**, 196 (1966).
105. Chalaya, Z. I. and Yakumova, M. N., *Zavodsk. Lab.* **32**, 792 (1966).
106. Babko, A. K., Shkaravskii, Yu. F. and Ivashkovich, E. M., *Ukr. Khim. Zh.* **33**, 951 (1967).
107. Shkaravskii, Yu. F., Lynchak, K. A., and Chernogorenko, V. B., *Zavodsk. Lab.* **36**, 524 (1970).
108. Babko, A. K., Shkaravskii, Yu. F. and Ivashkovich, E. M., *Zh. Analit. Khim.* **26**, 854 (1971).
109. Pilipenko, A. T. and Shkaravskii, Yu. F., *ibid.* **29**, 716 (1974).
110. Altmann, H. J., Fürstenau, E., Gielewski, A. and Scholtz, L., *Z. Anal. Chem.* **256**, 274 (1971).
111. Sugawara, K. and Kanamori, S., *Bull. Chem. Soc. Japan* **34**, 258 (1961).
112. Umland, F. and Wünsch, G., *Z. Anal. Chem.* **213**, 186 (1965).
113. Halász, A., Polyák, K. and Pungor, E., *Talanta* **18**, 691 (1971).
114. Djurkin, V., Kirkbright, G. F. and West, T. S., *Analyst* **91**, 89 (1966).
115. Malyutina, T. M., Savvin, S. B., Orlova, V. A., Mineeva, B. A. and Kirillova, T. I. *ibid.* **29**, 925 (1974).
116. Babko, A. K., Shkaravskii, Yu. F. and Ivashkovich, E. M., *Ukr. Khim. Zh.* **33**, 397 (1967).
117. Ghimicescu, Gh. and Dorneanu, V., *Talanta* **19**, 263 (1972).
118. Hayashi, K., Danzuka, T. and Ueno, K., *ibid.* **4**, 244 (1960).
119. Wynne, E. A., Burdick, R. D. and Fine, L. H., *Microchem. J.* **5**, 185 (1961).
120. Podobed, N. D. and Tyupina, M. F., *Zh. Analit. Khim.* **23**, 1573 (1968).
121. Nelson, J. P. and Milun, A. J., *Anal. Chem.* **29**, 1665 (1957).
122. Scanzillo, A. P., *ibid.* **26**, 411 (1954).
123. Anton, A., *ibid.* **37**, 1422 (1965).
124. Volokhova, V. I., Vakhidov, R. S. and Luk'yanitsa, A. I., *Zavodsk. Lab.* **39**, 932 (1973).
125. Lindenbaum, S., Peters, T. V., Jr. and Rieman, W., III, *Anal. Chim. Acta* **11**, 530 (1954).
126. Peters, T. V., Jr. and Rieman, W., III, *ibid.* **14**, 131 (1956).
127. Ohashi, S. and Takada, S., *Bull. Chem. Soc. Japan* **34**, 1516 (1961).
128. Wernet, J., Ebert, J. and Adrian, R., *Z. Anal. Chem.* **212**, 155 (1965).
129. Maurice, J., *Bull. Soc. Chim. France* **1959**, 819.

130. Kolloff, R. H., Ward, H. K. and Ziemba, V. F., *Anal. Chem.* **32**, 1687 (1960).
131. Chess, W. B. and Bernhart, D. N., *ibid.* **30**, 111 (1958).
132. Shevchuk, I. A., Skripnik, N. A. and Enal'eva, L. Ya., *Zavodsk. Lab.* **33**, 288 (1967).
133. Hoffmann, E. and Saracz, A., *Z. Anal. Chem.* **190**, 326 (1962).
134. Ohashi, S. and Yoza, N., *Bull. Chem. Soc. Japan* **36**, 707 (1963).
135. Yoza, N. and Ohashi, S., *ibid.* **37**, 33, 37 (1964).
136. Lucena-Conde, F. and Prat, L., *Anal. Chim. Acta* **16**, 473 (1957).
137. Pollard, F. H., Rogers, D. E., Rothwell, M. T. and Nickless, G., *J. Chromatog.* **9**, 227 (1962).

Chapter 40

PLATINUM

Platinum (Pt, at.wt. 195·09) occurs in its compounds in the +II and +IV oxidation states, compounds of platinum(IV) being the more stable. The hydroxide Pt(OH)$_4$ dissolves in excess of NaOH. Platinum(IV) forms chloride, iodide, cyanide, nitrite, and certain other complexes. Platinum(II and IV) is more difficult to reduce to the metal than is gold(III), for kinetic reasons. Zinc and aluminium in acid solution, and formaldehyde in an alkaline medium, are suitable reductants. Of the other platinum metals, palladium resembles platinum most closely, and osmium and ruthenium resemble it least.

40.1 Methods of Isolation and Separation of Platinum Metals

40.1.1 FIRE ASSAY AND CUPELLATION

These methods are used to isolate small amounts of the noble metals (platinum metals, gold, silver) from ores and concentrates [1-7]. The finely powdered sample is fused with a mixture of the oxide of a metal collector, fluxes (soda, borax, silica), and an organic reductant (flour, graphite). Lead in the form of massicot (PbO) is the classical collector. During fusion, reduction of the massicot yields minute drops of lead. Noble metals (in particular, Pt, Pd, Au, and Ag) dissolve in the lead, which falls to the bottom as a separate, heavy phase. The resulting alloy is mechanically separated from the lighter slag, and placed in a porous crucible (cupel) of bone ash, magnesite, or other refractory material. During subsequent heating, the lead is oxidized to molten lead monoxide, which is absorbed by the porous crucible walls (the process is known as cupellation) until a button of alloy consisting of only noble metals remains in the cupel.

Silver is often employed in addition to lead as the collector [8, 9]. Iron, nickel, copper, tin, or their alloys, are suitable alternative collectors [10-13]. The noble metals are isolated from these collectors by wet chemical methods after dissolution of the alloy in acid.

Tertipis and Beamish [14] have compared the separation of traces of Pt, Pd, Rh, and Ir from ores and concentrates by fire assay and cupellation with lead, by fire assay with a Cu-Ni-Fe alloy collector, and by wet chemical methods in which the sample is dissolved in aqua regia and then noble metals isolated from other metals on ion-exchangers. All three methods give satisfactory separations of platinum and palladium but the fire assay methods (especially those using lead) give rather poor results for iridium and rhodium.

40.1.2 Precipitation

In weak acid, less noble metals such as zinc, magnesium, iron, and copper reduce Pt, Pd, Rh, Ir, and Au to the elementary form (this process is known as cementation).

Platinum, as well as palladium, gold, and rhodium, can be precipitated from solution by the following reducing agents: formic acid, hydrazine, stannous chloride, and calomel. Tellurium and selenium are suitable collectors for traces [6,15]. When an acid solution containing powdered tellurium is boiled, Pt and Pd are precipitated quantitatively, whereas Rh and Ir remain in the solution [16].

Platinum metal sulphides can be coprecipitated from an acid medium with copper, lead, or rhenium [15,17,18].

The other platinum metals may be separated from platinum by distillation of the volatile tetroxides of Ru and Os, followed by double precipitation of Pd, Rh, and Ir hydroxides (with iron as collector) at pH 6 [16,19].

After volatile OsO_4 and RuO_4 have been separated, the remaining platinum metals can be masked as the soluble nitrite complexes, while a number of base metals are removed as sparingly soluble hydroxides [20].

Separation of platinum from the other noble metals as sparingly soluble ammonium chloroplatinate is prone to error [19].

40.1.3 Ion-Exchange

Platinum metals (Pt, Pd, Rh, Ir, Ru) form stable anionic chloride complexes which allow their separation from a number of common metals on both anion- and cation-exchangers [6,10,12,21–25]. Platinum has been separated from Rh or Ir by anion-exchange chromatography in the bromide system [25a].

Blasius and Rexin [26] have separated Pt, Pd, Rh, and Ir on anion- and cation-exchangers by taking advantage of differences in the stabilities of their chloride, nitrite and hydroxy complexes. Koster and Schmuckler [27] have separated noble metals from base metals on a chelating resin.

40.1.4 Extraction

Platinum metal complexes with thiocyanate [28–30], bromide [31], iodide [32], chloride [33], and nitrite [34] can be selectively extracted.

Platinum and palladium can be separated from the other platinum metals by extraction as dithizonates [35] and dithiocarbamates [36,37].

Diamantatos [37a] has developed a solvent extraction scheme for separating Pt, Pd, Rh, and Ir.

40.2 Methods of Determination

Beamish and McBryde have reviewed spectrophotometric methods for determining platinum [38,39]. The methods are generally rather unselective. The most popular method employs $SnCl_2$, but it is not very sensitive. More sensitive methods are based on dithizone and *p*-nitrosodimethylaniline.

40.2.1 STANNOUS CHLORIDE METHOD

Stannous chloride reacts with platinum(IV) in dilute hydrochloric acid to yield a yellow-orange complex which is the basis of this spectrophotometric method for determining platinum [40–45]. The absorption maximum of the complex is at 403 nm. The molar absorptivity is 1.3×10^4 ($a = 0.067$).

The most suitable concentration of hydrochloric acid is between 0.8 and $1.5M$; the concentration of $SnCl_2$ is not critical.

The coloured complex is extractable into higher alcohols, ethers, and esters, amyl acetate containing resorcinol as stabilizer being particularly recommended [40]. The molar absorptivity in amyl acetate is similar to that in aqueous medium, but the absorption maximum is shifted slightly towards shorter wavelengths. The high distribution coefficient facilitates concentration of platinum in a small volume of the organic solvent.

Palladium(II) and platinum(IV) react similarly with stannous chloride. If, however, the sample solution containing palladium and platinum is first made alkaline with ammonia, then acidified with hydrochloric acid till $\sim 1M$ in HCl, and finally treated with stannous chloride, only the platinum complex is extracted. Ruthenium, rhodium, and iridium react with $SnCl_2$ to form much less intensely coloured complexes.

The difference in the positions of the absorption maxima of the complexes formed in the presence of stannous chloride permits the simultaneous determination of platinum and palladium [46], as well as platinum and rhodium [47].

Khattak and Magee [48] have increased the sensitivity considerably by extracting the Pt–Sn–Cl complex with chloroform or benzene in the presence of tri-n-octylamine or other long-chain high molecular-weight amines.

The $SnCl_2$ method has been used to determine platinum in catalysts [49,50], ferronickel alloys [10], plutonium [51], and cathode slurries in the manufacture of hydrogen peroxide [52].

Reagents

1. Stannous chloride: 25% solution in HCl (1+3). When covered with xylene, the solution is protected from atmospheric oxygen and may be stored for a long time.
2. Standard platinum(IV) solution: 1 mg/ml. Dissolve 0·1000 g of suitably pure platinum in 4 ml of aqua regia (3 ml of conc. HCl + 1 ml of conc.

HNO$_3$) and evaporate the solution nearly to dryness. Add 5 ml of conc. HCl and 0·1 g of sodium chloride, and evaporate the solution to dryness. Dissolve the solid in 20 ml of HCl (1+1), and dilute the solution to volume with water in a 100-ml volumetric flask. Working solutions are obtained by appropriate dilutions of the stock solution with dilute hydrochloric acid (e.g. 0·2M).

3. Amyl acetate containing 1% resorcinol.

Procedure

Place the sample solution containing not more than 0·7 mg of Pt in a 50-ml volumetric flask. Add 5 ml of conc. HCl and 10 ml of the SnCl$_2$ solution, dilute the solution to the mark with water, and measure the absorbance at 403 nm against water.

In the extraction modification of the method, add to the acidified sample solution (see above) 10 ml of 25% ammonium chloride solution, 10 ml of the SnCl$_2$ solution, and water to ~ 40 ml. Shake the solution with two portions of amyl acetate. Dilute the combined extracts with the solvent to the mark in a volumetric flask of suitable capacity, and measure the absorbance at 400 nm against the solvent.

40.2.2 OTHER METHODS

A sensitive but unselective method for determining platinum is based on the reaction with *p*-nitrosodimethylaniline [53] or p-nitrosodiethylaniline

$$(CH_3)_2N-\!\!\left\langle\;\;\right\rangle\!\!-NO$$

[54]. When a platinum(IV) solution (pH 2-3) is heated with excess of the reagent, an orange-red complex forms slowly ($\varepsilon = 5\cdot66 \times 10^4$ at $\lambda_{max} = 525$ nm). Because palladium reacts rapidly with the reagent even in the cold, it should be separated before the determination of platinum. Alternatively, platinum can be determined from the difference between the absorbances before and after the solution is heated [53]. The other platinum metals must be separated beforehand. These reagents have been used to determine platinum in gold [54], reforming catalysts [55], and glass [56].

Dithizone reacts with platinum(II) in an acid medium (1–4M HCl, 1–5M H$_2$SO$_4$) containing tin(II), to form a brown-yellow dithizonate, Pt(HDz)$_2$, which is soluble in chloroform, carbon tetrachloride, and benzene [35,57–60]. The molar absorptivity of the complex is $3\cdot2 \times 10^4$ at 490 nm and has a similar value at 710 nm. Under the conditions for the determination of platinum, the coloured palladium, gold, and mercury dithizonates may also be formed. Gold can be separated from platinum by reduction with sulphur dioxide or extraction with ethyl acetate. Silver does not react with dithizone when hydrochloric acid is present.

Besides dithizone, other organosulphur reagents have been used for the spectrophotometric determination of platinum. Ayres *et al.* [61,62] have proposed 2,3-quinoxalinedithiol which reacts in aqueous dimethylformamide containing $SnCl_2$ as a reducing agent to give a blue complex ($\varepsilon = 2\cdot75 \times 10^4$ at 624 nm). Other reagents of the same group used similarly include 1,4-diphenylthiosemicarbazide [63,64], dibenzyldithio-oxamide [65], thiosalicylamide [66], N,N'-bis(3-dimethylaminopropyl)dithio-oxamide [67], 1,2,4-triazoline-3-thione [68], o-hydroxythiobenzhydrazide [69], 1-phenyltetrazoline-5-thione [70], 5-mercapto-1,3,4-thiadiazolidine-2-thione [71], p-dimethylaminobenzylidenerhodanine (well-known reagent for silver) [72], Sulphochlorophenolazorhodanine ($\varepsilon = 3\cdot2 \times 10^4$) [73,74], and other azo derivatives of rhodanine and thiorhodanine [75].

3,4-Diaminobenzoic acid [76,77], anthranilic acid [78], PAN [79,80], TTA [81], o-phenylenediamine [82,83], acenaphthenequinone monoxime [84], tetrabromofluorescein ethyl ester ($\varepsilon = 8\cdot0 \times 10^4$ at 555 nm) [85] and 3-nitroso-4-hydroxycoumarin [86] are diverse organic spectrophotometric reagents for platinum.

Extractive photometric methods have been based on the ion-association complexes formed by $PtCl_6^{2-}$ with the tris-1,10-phenanthroline iron(II) ion [87], by $PtBr_6^{2-}$ with DAM [88] or with DAPM [89], and by $Pt(SCN)_6^{2-}$ with Crystal Violet [90].

The coloured bromide [91–93] and iodide [39] complexes of platinum formed in aqueous media are also employed in spectrophotometric methods.

References

1. Beamish, F. E. and Van Loon, J. C., *Recent Advances in the Analytical Chemistry of the Noble Metals*, Pergamon, Oxford (1972).
2. Hoffman, I. and Beamish, F. E., *Anal. Chem.* **28**, 1188 (1956).
3. Plummer, M. E., Lewis, C. L. and Beamish, F. E., *ibid.* **31**, 254 (1959).
4. Plummer, M. E. and Beamish, F. E., *ibid.* **31**, 1141 (1959).
5. Agrawal, K. C. and Beamish, F. E., *Talanta* **11**, 1449 (1964).
6. Beamish, F. E., *ibid.* **5**, 1 (1960); **14**, 991, 1133 (1967).
7. Georgiev, G. T. and Apostolov, D., *Zh. Analit. Khim.* **27**, 506 (1972).
8. Barefoot, R. R. and Beamish, F. E., *Anal. Chim. Acta* **9**, 49 (1953).
9. Hoffman, I., Westland, A. D., Lewis, C. L. and Beamish, F. E., *Anal. Chem.* **28**, 1174 (1956).
10. Coburn, H. G., Beamish, F. E. and Lewis, C. L., *ibid.* **28**, 1297 (1956).
11. Faye, G. H. and Inman, W. R., *ibid.* **33**, 278 (1961).
12. Banbury, L. M. and Beamish, F. E., *Z. Anal. Chem.* **211**, 178 (1965).
13. Faye, G. H. and Moloughney, P. E., *Talanta* **19**, 269 (1972).
14. Tertipis, G. G. and Beamish, F. E., *ibid.* **10**, 1139 (1963).
15. Beyermann, K., *Z. Anal. Chem.* **200**, 183 (1964).
16. Westland, A. D. and Beamish, F. E., *Mikrochim. Acta* **1957**, 625.
17. Pshenitsyn, N. K. and Prokof'eva, I. V., *Zh. Neorgan. Khim.* **3**, 996 (1958).
18. Ujihara, Y. and Hirano, S., *Bull. Chem. Soc. Japan* **37**, 66 (1964).
19. Gilchrist, R., *Anal. Chem.* **25**, 1617 (1953).
20. Payne, S. T., *Analyst* **85**, 698 (1960).
21. MacNevin, W. M. and Crummett, W. B., *Anal. Chem.* **25**, 1628 (1953).
22. Blasius, E. and Wachtel, U., *Z. Anal. Chem.* **142**, 341 (1954).
23. Zachariasen, H. and Beamish, F. E., *Talanta* **4**, 44 (1960).
24. Korkisch, J. and Klakl, H., *ibid.* **15**, 339 (1968).

25. Strel'nikova, N. P., Kashlinskaya, S. É., Litvinskaya, I. I. and Rylova, N. A., *Zavodsk. Lab.* **34**, 926 (1968).
25a. Chwastowska, J., Dybczyński, R. and Maleszewska, H., *Chem. Anal. (Warsaw)* **16**, 891 (1971).
26. Blasius, E. and Rexin, D., *Z. Anal. Chem.* **179**, 105 (1961).
27. Koster, G. and Schmuckler, G., *Anal. Chim. Acta* **38**, 179 (1967).
28. Forsythe, J. H., Magee, R. J. and Wilson, C. L., *Talanta* **3**, 330 (1960).
29. Berg, E. W. and Lau, E. Y., *Anal. Chim. Acta* **27**, 248 (1962).
30. Senise, P. and Pitombo, L. R., *Talanta* **11**, 1185 (1964).
31. Berg, E. W. and Sanders, J. R., *Anal. Chim. Acta* **38**, 377 (1967).
32. Faye, G. H. and Inman, W. R., *Anal. Chem.* **35**, 985 (1963).
33. Shendrikar, A. D. and Berg, E. W., *Anal. Chim. Acta* **47**, 299 (1969).
34. Khan, M. A. and Morris, D. F., *Chem. Ind. (London)* **1968**, 1802.
35. Young, R. S., *Analyst* **76**, 49 (1951).
36. Pyle, J. T. and Jacobs, W. D., *Anal. Chem.* **36**, 1796 (1964).
37. Lee, A. S., Beamish, F. E. and Bapat, M. G., *Mikrochim. Acta* **1969**, 329.
37a. Diamantatos, A., *Anal. Chim. Acta* **67**, 317 (1973).
38. Beamish, F. E. and McBryde, W. A. E., *ibid.* **9**, 349 (1953); **18**, 551 (1958).
39. Beamish, F. E., *Talanta* **12**, 743 (1965).
40. Ayres, G. H. and Meyer, A. S., Jr., *Anal. Chem.* **23**, 299 (1951); **25**, 1622 (1953).
41. Meyer, A. S., Jr. and Ayres, G. H., *J. Am. Chem. Soc.* **77**, 2671 (1955).
42. Elizarova, G. L. and Matvienko, L. G., *Zh. Analit. Khim.* **25**, 301 (1970).
43. Milner, O. I. and Shipman, G. F., *Anal. Chem.* **27**, 1476 (1955).
44. Egli, R. A., *Z. Anal. Chem.* **194**, 401 (1963).
45. Khopkar, S. M., *Anal. Chem.* **38**, 360 (1966).
46. Pilipenko, A. T. and Sereda, I. P., *Ukr. Khim. Zh.* **27**, 524 (1961).
47. Ayres, G. H., Tuffly, B. L. and Forrester, J. S., *Anal. Chem.* **27**, 1742 (1955).
48. Khattak, M. A. and Magee, R. J., *Talanta* **12**, 733 (1965).
49. Maziekien, I., Ermanis, L. and Walsh, T. J., *Anal. Chem.* **32**, 645 (1960).
50. Rees, T. D. and Hill, S. R., *Talanta* **15**, 1312 (1968).
51. Smith, M. E., *Anal. Chem.* **30**, 912 (1958).
52. Struszyński, M. and Chwastowska, J., *Chem. Anal. (Warsaw)* **3**, 949 (1958).
53. Yoe, J. H. and Kirkland, J. J., *Anal. Chem.* **26**, 1335, 1340 (1954).
54. Marczenko, Z. and Krasiejko M., *Chem. Anal. (Warsaw)* **15**, 1233 (1970).
55. Conrad, A. L. and Evans, J. K., *Anal. Chem.* **32**, 46 (1960).
56. Fuller, C. W., Himsworth, G. and Whitehead, J., *Analyst* **96**, 177 (1971).
57. Miyamoto, M., *Japan Analyst* **9**, 925 (1960).
58. Goryushina, V. G. and Gailis, E. Ya., *Zavodsk. Lab.* **20**, 161 (1954).
59. Kemula, W., Brachaczek, W. and Hulanicki, A., *Chem. Anal. (Warsaw)* **3**, 913 (1958).
60. Kawahata, M., Mochizuki, H. and Misaki, T., *Japan Analyst* **11**, 1020 (1962).
61. Ayres, G. H. and McCrory, R. W., *Anal. Chem.* **36**, 133 (1964).
62. Janota, H. F. and Ayres, G. H., *ibid.* **36**, 138 (1964).
63. Tolubara, A. I. and Usatenko, Yu. I., *Zavodsk. Lab.* **32**, 807 (1966).
64. Radushev, A. V. and Statina, L. A., *Zh. Analit. Khim.* **28**, 2360 (1973).
65. Pyle, J. T. and Jacobs, W. D., *Talanta* **9**, 761 (1962).
66. Sur, K. and Shome, S. C., *Anal. Chim. Acta* **57**, 201 (1971).
67. Jacobs, W. D., *Anal. Chem.* **33**, 1279 (1961).
68. Radushev, A. V., Chechneva, A. N. and Kovalev, E. G., *Zh. Analit. Khim.* **23**, 1410 (1968).
69. Gangopadhyay, P. K., Das, H. R. and Shome, S. C., *Anal. Chim. Acta* **66**, 460 (1973).
70. Chechneva, A. N. and Radushev, A. V., *Zh. Analit. Khim.* **23**, 1059 (1968).
71. Gregorowicz, Z. and Klima, Z., *Z. Anal. Chem.* **239**, 87 (1968).
72. Piercy, F. E. and Ryan, D. E., *Can. J. Chem.* **41**, 667 (1963).

References

73. Propistsova, R. F., Savvin, S. B. and Rozovskii, Yu. G., *Zh. Analit. Khim.* **26**, 2424 (1971).
74. Savvin, S. B., Propistsova, R. F. and Rozovskii, Yu. G., *ibid.* **27**, 1554 (1972).
75. Basargin, N. N., Rozovskii, Yu. G. and Merzlyakova, A. N., *Zavodsk. Lab.* **38**, 260 (1972).
76. Johnson, L. D. and Ayres, G. H., *Anal. Chem.* **38**, 1218 (1966).
77. Keil, R., *Z. Anal. Chem.* **254**, 191 (1971).
78. Majumdar, A. K. and Sen Gupta, J. G., *ibid.* **177**, 265 (1960).
79. Ivanov, V. M., Figurovskaya, V. N. and Busev, A. I., *Zavodsk. Lab.* **38**, 1311 (1972).
80. Ivanov, V. M., Figurovskaya, V. N. and Busev, A. I., *ibid.* **39**, 270 (1973).
81. De, A. K. and Rahaman, M. S., *Analyst* **89**, 795 (1964).
82. Sen Gupta, J. G., *Anal. Chim. Acta* **23**, 462 (1960).
83. Golla, E. D. and Ayres, G. H., *Talanta* **20**, 199 (1973).
84. Sindhwani, S. K. and Singh, R. P., *ibid.* **20**, 248 (1973).
85. El-Ghamry, M. T. and Frei, R. W., *ibid.* **16**, 235 (1969).
86. Manku, G. S., Bhat, A. N. and Jain, B. D., *ibid.* **16**, 1421 (1969).
87. Yamamoto, Y., Tsubouchi, M., Okimura, I. and Takagi, T., *J. Chem. Soc. Japan, Pure Chem. Sect.* **88**, 745 (1968).
88. Danilova, V. N. and Lisichenok, S. L., *Zh. Analit. Khim.* **24**, 1061 (1969).
89. Akimov, V. K., Emel'yanova, I. A. and Busev, A. I., *ibid.* **26**, 2416 (1971).
90. Pilipenko, A. T. and Ol'khovich, P. F., *Ukr. Khim. Zh.* **37**, 1146 (1971).
91. Pantani, F. and Piccardi, G., *Anal. Chim. Acta* **22**, 231 (1960).
92. Ginzburg, S. I. and Sal'skaya, L. G., *Zh. Analit. Khim.* **17**, 492 (1962).
93. Vorlíček, J. and Doležal, J., *Z. Anal. Chem.* **262**, 365 (1972).

Chapter 41

RARE EARTH ELEMENTS (LANTHANIDES AND YTTRIUM)

The lanthanides are a group of 15 elements (with atomic numbers 57–71) which exhibit very similar chemical properties. They occur naturally in two main groups—the cerium group, elements 57–62, associated with ceria, and the yttrium group, elements 63–71, which are associated with yttria. Because yttrium has very similar chemical properties to the lanthanides in the second group, although it is not itself a lanthanide (having no f electrons), it is included with the lanthanides in the group of elements known as the rare earth elements (REE). Scandium is in the same family in the periodic table, but much more closely resembles aluminium in its properties, and it will be dealt with in a separate chapter (45). The two groups of REE are the cerium group which includes cerium (Ce), lanthanum (La), praseodymium (Pr), neodymium (Nd), promethium (Pm), samarium (Sm), and europium (Eu), and the yttrium group which consists of yttrium (Y), gadolinium (Gd), terbium (Tb), dysprosium (Dy), holmium (Ho), erbium (Er), thulium (Tm), ytterbium (Yb), and lutetium (Lu). The two groups have slightly different properties.

All the rare earth elements occur in the $+III$ oxidation state in compounds, and can be separated and determined in this form to provide what is known as the total REE. Samarium, europium, and ytterbium also occur in the unstable $+II$ oxidation state, whereas cerium, praseodymium, and terbium can be found in the $+IV$ oxidation state. Owing to the considerable stability of cerium(IV), which facilitates its separation from other REE and its spectrophotometric determination in a mixture of REE, cerium is also discussed in a separate chapter (15).

Rare earth element hydroxides, $M(OH)_3$, precipitate from nitrate solution at pH values above 6·3–7·8 and reveal no amphoteric properties. Like thorium, the rare earth elements yield acid-insoluble fluorides and oxalates, and soluble EDTA, tartrate, and citrate complexes.

41.1 Methods of Separation

41.1.1 PRECIPITATION

Rare earth elements may be isolated by precipitation as oxalates, fluorides, or hydroxides. When the oxalates [1–3] are precipitated from a weakly acidic medium (pH 1–4), calcium is used as a collector. Thorium is also precipitated with the REE, but it can be masked with EDTA at pH 3·2 [1].

When the medium is more acidic than that for the precipitation of oxalates, REE are separated as fluorides [5–7]. Again, thorium is coprecipitated, and calcium is a suitable collector. Both methods give good separation of REE from Fe, Al, Ti, Zr, U(VI), Nb, Ta, and certain other metals. However, the calcium oxalate precipitate retains some manganese, and the precipitation of rare earth elements as fluorides in the presence of alkali metal ions is not quantitative [6]. Unlike the oxalate, the fluoride precipitate is slimy and difficult to collect by filtration.

In $0.2M$ Na_2CO_3, the light lanthanides (cerium group) are precipitated quantitatively, while the remaining lanthanides and scandium are only partly precipitated [8].

41.1.2 Extraction

The yttrium group elements can be separated from the cerium group elements by extraction methods based on the thiocyanate [9,10], cupferron [11], di-n-butyl phosphate [12], and TBP [13] complexes. The cerium group elements have been extracted with di-isoamyl methylphosphonate [14].

REE have been isolated from other elements by extraction with TBP from a nitrate medium [15,16]. Complexes of REE with TTA have been extracted with a solution of TOPO in toluene [17].

41.1.3 Ion-Exchange

The most important methods for separating the rare earth elements are based on ion-exchange, this separation being one of the major triumphs of the technique. The first methods were developed by Spedding *et al.* in the work on fission products [18].

Dybczyński and Minczewski [19,20] have separated REE on the strongly basic anion-exchanger Amberlite IRA-400 by taking advantage of the small differences in the stabilities of the EDTA complexes of the various elements.

Anion-exchange separation of the REE in mixed nitric acid and water-miscible alcohols is also feasible [21–25]. Mixed nitric acid–acetone [26] and 2-hydroxyisobutyric acid–alcohol [27] eluents have been used similarly. Korkisch and Arrhenius [28], using $5M$ HNO_3 and conc. CH_3COOH (1+9) as a medium, have retained REE, U, and Th on an anion-exchanger while most other elements were eluted.

EDTA complexes have also found application in cation-exchange separations. From an EDTA solution at pH 2·1, Dowex 50 cation-exchanger retains rare earth elements while thorium is eluted as the more stable EDTA complex [1]. Since the EDTA complex of uranium(VI) is very weak, traces of rare earths in uranium can be eluted from a cation-exchanger (Amberlite IRC-50) by a solution of EDTA while uranium is sorbed on the column [29]. Topp and Young [30] have separated the lighter rare earth elements by eluting them from a cation-exchanger with EDTA or NTA (nitrilotriacetic acid) solution.

Several authors [31–33] have separated rare earth elements by eluting them from cation-exchange columns with α-hydroxyisobutyrate. Complexes with citrate [34,34a], thiocyanate [35], oxalate [36] and hydroxyethylenediaminetriacetic acid [37] have been used similarly. The original work [18] used citrate.

Fe, Al, Ti, U(VI), Ca, Mg, and certain other metals can be eluted from a cation-exchanger with $1\cdot0$–$1\cdot75M$ HCl, the REE being subsequently eluted with $3M$ HCl [38–40].

Extensive reviews of the methods for isolating and separating the rare earth elements have been published [41,42].

41.2 Methods of Determination

The methods presented below can be used to determine either the total REE or the individual lanthanides [including cerium(III)] and yttrium. When the total REE are determined, the result is expressed in terms of the element (e.g. lanthanum, cerium, yttrium) upon which the calibration curve was based.

Since the REE in the +III oxidation state have only very slight chromophoric properties, more sensitive spectrophotometric methods utilize coloured organic reagents. The Arsenazo I and Arsenazo III methods are the most commonly used, but the Xylenol Orange method is also important.

After comparing 16 organic reagents giving colour reactions with ytterbium (as a typical member of the yttrium group), Babko *et al.* [43] have concluded that the Arsenazo III and the Xylenol Orange methods are among the best.

41.2.1 Arsenazo I Method

Arsenazo I (formula, p. 48) reacts in a neutral medium with tervalent REE ions to form a water-soluble pink-violet 1:1 complex, which is stable between pH 4 and 9. A solution of the reagent alone is red. With Arsenazo I, the REE are usually determined in solutions buffered at pH 7–8 with triethanolamine [7,44–47].

Figure 41.1 shows the absorption spectra for Arsenazo I and its lanthanum complex at pH 8. The molar absorptivity of the lanthanum complex is $2\cdot7 \times 10^4$ at 570 nm ($a = 0\cdot20$). The molar absorptivities of the complexes of the other lanthanides range from $2\cdot5 \times 10^4$ to $2\cdot7 \times 10^4$. Since the reagent absorbs slightly at 570–580 nm, where the absorbance of the complex is measured, a reagent blank solution should be used as reference.

Many metals react with Arsenazo I in neutral solution [Fe(III), Al, Ti, U, Cu, Zr, Th, Mo, Sn, Bi, and others]. Interfering metals may first be separated by various techniques such as mercury-cathode electrolysis or extraction. However, separation of the REE as oxalates or fluorides is more dependable. Interference from calcium and magnesium may be overcome by determining the REE at pH 5, although at this pH the sensitivity of the reaction between Arsenazo I and the rare earth elements is almost halved [44]. Of the common anions, fluoride, phosphate, and oxalate interfere.

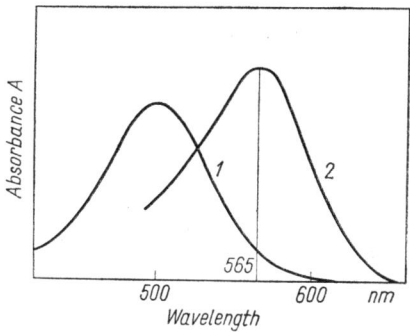

Fig. 41.1 Absorption spectra of Arsenazo I (1) and its lanthanum complex (2) (reagent solution as reference) at pH 8

Thorium, which interferes, can be determined separately at pH 1·7 with Arsenazo I [48,49]. Kuznetsov and Petrova [50] have simultaneously determined the total concentration of REE and the concentration of thorium by measuring the absorbances of the Arsenazo I complexes at two temperatures. As the temperature is raised, the absorbance of the REE complexes (but not of the Th complex) increases.

Arsenazo I has been used to determine rare earth elements in uranium [51], steels [52], and ores [2].

Reagents
1. Arsenazo I: 0·05% solution.
2a. Standard lanthanum solution: 1 mg/ml. Dissolve 0·1173 g of lanthanum oxide, La_2O_3 (dehydrated and freed from CO_2 by ignition), in 5 ml of hot HCl (1+1). Dilute the solution to volume with water in a 100-ml volumetric flask.
  b. Standard cerium solution: 1 mg/ml. Dissolve 0·1228 g of ceric oxide, CeO_2 (dehydrated by ignition), in 5 ml of hot HCl (1+1), and add ~ 0·1 g of $NH_2OH \cdot HCl$ to the solution. Dilute the colourless solution to volume with water in a 100-ml volumetric flask. For the preparation of a standard cerium solution from cerium nitrate, see p. 200.
  c. Standard yttrium solution: 1 mg/ml. Dissolve 0·1270 g of yttrium oxide, Y_2O_3 (dehydrated by ignition), in 5 ml of hot HCl (1+1). Dilute the solution to volume with water in a 100-ml volumetric flask.
3. Triethanolamine buffer. Mix 200 ml of 15% triethanolamine, 160 ml of $1M$ HNO_3, and 40 ml of water. Adjust the solution to pH 7·2($\pm$0·1) with dilute ammonia or dilute nitric acid.

Procedure

Separation of REE as oxalates. To the acid solution (50–150 ml) add 5–20 ml (depending on the concentration of other metals in the solution) of 8% $H_2C_2O_4$ solution. Adjust the pH of the solution to 2·0–2·5, heat to ~ 80°C, and add 5 mg of calcium (as a chloride solution) dropwise with stirring. Heat the solution for 1 hr, but do not boil. After two hours (or on the next day), filter off the precipitate and wash it thoroughly with 1% $H_2C_2O_4$ solution and water. Ignite the precipitate to the oxides, and dissolve it in a small amount of hot $4M$ HCl.

To separate the REE from calcium, dilute this solution with water to 10–15 ml, add 3 mg of aluminium (as a salt solution), and coprecipitate REE hydroxides and aluminium hydroxide with ammonia (pH ~ 9). Dissolve the precipitate in a small amount of hot $2M$ HCl.

Determination of REE. To the acid solution, obtained as detailed above and containing not more than 150 μg of REE, add 5 ml of 3% sulphosalicylic acid solution (to mask aluminium). After 2 minutes, add 5 ml of the Arsenazo I solution, 10 ml of the triethanolamine buffer, water to 40 ml, and ammonia till the pH is 7·2 ($\pm 0·1$). Transfer the solution to a 50-ml volumetric flask, make up to the mark with water, and measure the absorbance at 580 nm, using a reagent blank solution as the reference.

41.2.2 Arsenazo III Method

In weakly acidic media the lanthanides and yttrium react with Arsenazo III (formula, p. 48) to form coloured complexes which are the basis of this sensitive method [53–56]. In weakly acidic solution, the reagent is violet, whereas its complexes with the rare earth elements are green. The maximum absorbance for cerium(III) is obtained at pH 2·3–2·7. The optimum pH values for the various REE differ slightly.

Figure 41.2 shows the absorption spectra for Arsenazo III and its cerium(III) complex. The method has the advantage that the reagent does not absorb at the wavelength of the absorption maximum of the complex (650 nm). The molar absorptivity of the Arsenazo III complex with cerium is $5·6 \times 10^4$ (specific absorptivity 0·40). The other REE complexes have similar molar absorptivities.

Chloride, sulphate, and phosphate do not interfere in the determination of rare earth elements with Arsenazo III. Neither do small amounts (less than 1 mg per 50 ml of solution) of titanium, aluminium, calcium, and iron [reduced to Fe(II) with ascorbic acid], but larger quantities of these metals and certain other metals (Th, Zr, U, Bi, Cu) should be removed.

The REE are separated from the calcium used as the collector in the precipitation of the oxalates or fluorides by reprecipitation as the hydroxides with ammonia. In this instance, iron(III) serves as the collector.

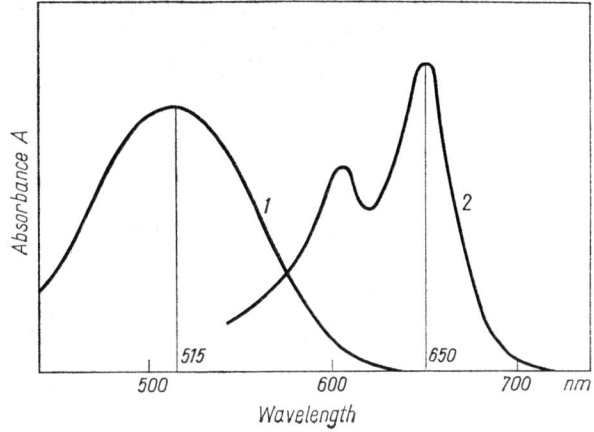

Fig. 41.2 Absorption spectra of Arsenazo III (1) and its cerium complex (2) (reagent solution as reference) at pH 2·6

When EDTA is used to mask interfering metals, the properties of the Arsenazo III complexes of REE are significantly different from those in the absence of EDTA [57,58].

Total REE or individual elements have been determined by the Arsenazo III method in ores [59], rocks and minerals [60], chromium and its alloys [61], and natural water [62].

Reagents

1. Arsenazo III: 0·05% aqueous solution.
2. Standard lanthanum, cerium(III), and yttrium solutions: 1 mg/ml (p. 441).
3. Formate buffer: pH 3·5. Dissolve 60 ml of formic acid and 28 g of NaOH in water, and dilute the solution with water to 1 litre.

Procedure

To the acid sample solution (pH ~ 1) containing not more than 80 μg of REE, add 2 ml of 1% ascorbic acid solution. After a few minutes, add 1 ml of the formate buffer and 4 ml of the Arsenazo III solution, and dilute with water to ~ 40 ml. Adjust the coloured solution to pH 2·6 (±0·1) with ~ 0·1M NaOH. Transfer the solution to a 50-ml volumetric flask, make up to the mark with water, and measure the absorbance at 650 nm against a reagent blank solution.

41.2.3 OTHER METHODS

The following other azo dyes containing the arsonic acid group have been recommended as spectrophotometric reagents for either the individual or the total REE: Thoron I [63], Thoron II [64], Arsenazo M (the molar absorptivity of the La-complex is 8.6×10^4 at 640 nm) [65], Dicarboxyarsenazo III [66], Carboxynitrazo [67], and Arsenazo-p-NO$_2$ [68], together with the related reagents Chlorophosphonazo III [69, 69a] and Chlorophosphonazo DAL [70]. Further azo reagents include PAN [71,72], PAR [73,74], SPADNS [75], Chromotrope 2B [76], Chromotrope 2R [77], Eriochrome Black T [78], 1-[(5-methyl-2-pyridyl)azo]-2-naphthol [79], Stilbazo [80], and Diantipyrylazo [81].

Many authors have recommended Xylenol Orange for the determination of REE [82–88]. Higher sensitivity is obtained in the presence of cetylpyridinium bromide [87,88]: the ternary lanthanum complex has a molar absorptivity of 9.2×10^4 at 625 nm [87]. Other triphenylmethane dyes employed similarly are Pyrocatechol Violet [89], Aluminon [90,91], Methylthymol Blue [92–95], Chrome Azurol S [96,97], and Eriochrome Cyanine R [98].

Many other organic spectrophotometric reagents have been proposed for the determination of the rare earth metals, e.g. Alizarin Red S [99–101], quinalizarin [102], 8-hydroxyquinoline [103–105], Bromopyrogallol Red [106–108], salicylfluorone [109], carmine [110], and Thymolphthalexone [111]. Methods based on REE complexes with Tiron [112–114], TTA [115], dibenzoylmethane [116], chromotropic acid [117], and EDTA [118] have low sensitivity.

There exists a group of methods exploiting ion-association complexes formed between the REE anionic complexes (with TTA, dibromo-oxine, salicylic acid, tetrabromofluorescein, and salicylfluorone) and organic bases such as Rhodamine B [119–121], 1,10-phenanthroline [122–124], o-dihydroxychromenols [125], and pyridine [126]. Some of these methods are characterized by remarkably high sensitivity.

The spectrophotometric determination of yttrium with molybdophosphoric acid [127], and the indirect determination of europium with 2,2'-bipyridyl [reduction of iron(III) with Eu(II)] [128] also deserve mention.

Tervalent REE ions (in chloride, nitrate, and perchlorate solutions) absorb ultraviolet, visible, and infrared radiation. The absorption bands are narrow with sharp maxima, and in most cases they do not overlap. The absorption values are, however, very low. The molar absorptivities, being between 1 and 10, preclude any possibility of determining low concentrations of REE. The simultaneous spectrophotometric determination of various REE is, however, possible in the case of more concentrated (mg/ml) REE salt solutions [129–135].

References

1. Gordon, L., Firsching, F. H. and Shaver, K. J., *Anal. Chem.* **28**, 1476 (1956).
2. Zaikovskii, F. V. and Bashmakova, V. S., *Zh. Analit. Khim.* **14**, 50 (1959).
3. Zaikovskii, F. V., Furtova, E. V. and Sadova, G. F., *ibid.* **17**, 202 (1962).
4. Matsui, M., *Bull. Chem. Soc. Japan* **39**, 1114 (1966).
5. Kallmann, S., Oberthin, H. K. and Hibbits, J. O., *Anal. Chem.* **32**, 1278 (1960).
6. Butler, J. R. and Hall, R. A., *Analyst* **85**, 149 (1960).

References

7. Onishi, H. and Banks, C. V., *Talanta* **10**, 399 (1963).
8. Upor, E. and Nagy, G., *Acta Chim. Acad. Sci. Hung.* **68**, 313 (1971).
9. Yoshida, H., *J. Inorg. Nucl. Chem.* **24**, 1257 (1962).
10. Fischer, W., Bramekamp, K. J. Klinge, N. and Pohlmann, H. P., *Z. Anorg. Chem.* **329**, 44 (1964).
11. Tishchenko, N. A., Lauer, R. S. and Poluektov, N. S., *Ukr. Khim. Zh.* **30**, 390 (1964).
12. Scadden, E. M. and Ballou, N. E., *Anal. Chem.* **25**, 1602 (1953).
13. Slyusareva, R. L., Kondrat'eva, L. I., Peizulaev, Sh. I. and Nikolaeva, O. N., *Zh. Analit. Khim.* **26**, 894 (1971).
14. Mikhlin, E. B. and Korpusov, G. V., *Zh. Neorgan. Khim.* **10**, 2787 (1965).
15. Edge, R. A., *Anal. Chim. Acta* **27**, 396 (1962); **28**, 278 (1963).
16. Robinson, F. V. and Topp, N. E., *J. Inorg. Nucl. Chem.* **26**, 473 (1964).
17. Taketatsu, T. and Banks, C. V., *Anal. Chem.* **38**, 1524 (1966).
18. Spedding, F. H. et al., *J. Am. Chem. Soc.* **69**, 2777, 2786, 2812 (1947).
19. Dybczyński, R., *Chem. Anal. (Warsaw)* **4**, 531 (1959); *J. Chromatog.* **14**, 79 (1964).
20. Minczewski, J. and Dybczyński, R., *Chem. Anal. (Warsaw)* **6**, 275 (1961); *J. Chromatog.* **7**, 98, 568 (1962).
21. Edge, R. A., *J. Chromatog.* **5**, 526 (1961); **8**, 419 (1962); *Anal. Chim. Acta* **29**, 321 (1963).
22. Korkisch, J., Hazan, I. and Arrhenius, G., *Talanta* **10**, 865 (1963).
23. Fritz, J. S. and Greene, R. G., *Anal. Chem.* **36**, 1095 (1964).
24. Hazan, I., Ahluwalia, S. S. and Korkisch, J., *Z. Anal. Chem.* **206**, 324 (1964).
25. Molnár, F., Horváth, A. and Khalkin, V. A., *J. Chromatog.* **26**, 215, 225 (1967).
26. Alstad, J. and Brunfelt, A. O., *Anal. Chim. Acta* **38**, 185 (1967).
27. Fairs, J. P., *J. Chromatog.* **26**, 232 (1967).
28. Korkisch, J. and Arrhenius, G., *Anal. Chem.* **36**, 850 (1964).
29. Krawczyk, I., *Nukleonika* **5**, 649 (1960); *Acta Chim. Acad. Sci. Hung.* **27**, 269 (1961).
30. Topp, N. E. and Young, D. D., *J. Chromatog.* **14**, 469 (1960).
31. Wolfsberg, K., *Anal. Chem.* **34**, 518 (1962).
32. Massart, D. L. and Hoste, J., *Anal. Chim. Acta* **28**, 378 (1963).
33. Deelstra, H. and Verbeek, F., *J. Chromatog.* **17**, 558 (1965).
34. Radhakrishna, P., *Anal. Chim. Acta* **8**, 140 (1953).
34a. Nevoral, V., *Z. Anal. Chem.* **268**, 189 (1974).
35. Hamaguchi, H., Kuroda, R. and Onuma, N., *Talanta* **10**, 120 (1963).
36. Krawczyk-Obojska, I., *Chem. Anal. (Warsaw)* **12**, 13, 225 (1967).
37. Merciny, E. and Duyckaerts, G., *J. Chromatog.* **22**, 164 (1966); **26**, 471 (1967).
38. Chung, K. S. and Riley, J. P., *Anal. Chim. Acta* **28**, 1 (1963).
39. Strelow, F. W. E., *Anal. Chem.* **38**, 127 (1966); *Anal. Chim. Acta* **34**, 387 (1966).
40. Strelow, F. W. E., and Baxter, C., *Talanta* **16**, 1145 (1969).
41. Ryabchikov, D. I. and Terent'eva, E. A., *Usp. Khim.* **16**, 461 (1947) **24**, 260 (1955); **29**, 1285 (1960).
42. Nowicka-Jankowska, T. and Radwan, Z., *Chem. Anal. (Warsaw)* **8**, 127, 307, 481 (1963).
43. Babko, A. K., Akhmedli, M. K. and Granovskaya, P. B., *Ukr. Khim. Zh.* **32**, 879 (1966).
44. Fritz, J. S., Richard, M. J. and Lane, W. J., *Anal. Chem.* **30**, 1776 (1958).
45. Kuteinikov, A. F., *Zavodsk. Lab.* **24**, 1050 (1958).
46. Shibata, S., Takeuchi, F. and Matsumae, T., *Anal. Chim. Acta* **21**, 177 (1959).
47. Hiiro, K., Russell, D. S. and Berman, S. S., *ibid.* **37**, 209 (1967).
48. Kuteinikov, A. F. and Lanskoi, G. A., *Zh. Analit. Khim.* **14**, 686 (1959).
49. Onishi, H., Nagai, H. and Toita, Y., *Anal. Chim. Acta* **26**, 528 (1962).
50. Kuznetsov, V. I. and Petrova, T. V., *Zh. Analit. Khim.* **14**, 404 (1959).

51. Banks, C. V., Thompson, J. A. and O'Laughlin, J. W., *Anal. Chem.* **30**, 1792 (1958).
52. Bornong, B. J. and Moriarty, J. L., *ibid.* **34**, 871 (1962).
53. Savvin, S. B., *Talanta* **8**, 673 (1961); *Zavodsk. Lab.* **29**, 131 (1963).
54. Budanova, L. M. and Pinaeva, S. N., *Zh. Analit. Khim.* **20**, 320 (1965).
55. Cherkesov, A. I. and Alykov, N. M., *ibid.* **20**, 1312 (1965).
56. Spitsyn, P. K. and Shvarev, V. S., *ibid.* **25**, 1503 (1970); **26**, 1313 (1971).
57. Vdovenko, M. E. and Lisichenok, S. L., *Zavodsk. Lab.* **33**, 1327 (1967).
58. Spitsyn, P. K., Shvarev, V. S. and Popyvanova, T. P., *Zh. Analit. Khim.* **26**, 86 (1971).
59. Goryushina, V. G., Savvin, S, B. and Romanova, E. V., *ibid.* **18**, 1340 (1963).
60. Kirillov, A. I., Makarenko, O. P. and Vlasov, N. A., *Izv. Vyssh. Ucheb. Zaved. Khim. Khim. Tekhnol.* **14**, 1479 (1971).
61. Wood, D. F. and Adams, M. R., *Analyst* **95**, 556 (1970).
62. Poluektov, N. S., Kirillov, A. I., Makarenko, O. P. and Vlasov, N. A., *Zavodsk. Lab.* **37**, 536 (1971).
63. Sangal, S. P., *Microchem. J.* **8**, 304 (1964); **9**, 9 (1965).
64. Sentyurina, N. N., *Zh. Analit. Khim.* **17**, 442 (1962).
65. Savvin, S. B., Propistsova, R. F. and Strel'nikova, R. V., *ibid.* **24**, 31 (1969).
66. Buděšínský, B. and Haas, K., *Z. Anal. Chem.* **210**, 263 (1965).
67. Savvin, S. B., Petrova, T. V. and Romanov, P. N., *Talanta* **19**, 1437 (1972); *Zh. Analit. Khim.* **28**, 272 (1973).
68. Perišić, N. U., Muk, A. A. and Canić, V. D., *Anal. Chem.* **45**, 798 (1973).
69. O'Laughlin, J. W. and Jensen, D. F. *Talanta* **17**, 329 (1970).
69a. Taketatsu, T., Kaneko, M. and Kono, N., *ibid.* **21**, 87 (1974).
70. Buděšínský, B. and Menclová, B., *ibid.* **14**, 688 (1967).
71. Shibata, S., *Anal. Chim. Acta* **28**, 388 (1963).
72. Inczédy, J., Nemeshegyi, G. and Erdey, L., *Acta Chim. Acad. Sci. Hung.* **43**, 1 (1965).
73. Sommer, L. and Novotná, H., *Talanta* **14**, 457 (1967).
74. Munshi, K. N. and Dey, A. K., *Mikrochim. Acta* **1971**, 751.
75. Munshi, K. N. and Dey, A. K., *Microchem. J.* **8**, 152 (1964).
76. Dey, A. K., Sangal, S. P., Sinha, S. N. and Munshi, K. N., *ibid.* **9**, 282 (1965).
77. Shah, V. L. and Sangal, S. P., *Chim. Anal. (Paris)* **53**, 47 (1971).
78. Akhmedli, M. K., Granovskaya, P. B. and Neimatova, R. A., *Zh. Analit. Khim.* **28**, 278 (1973).
79. Shibata, S., Sasaki, S. and Ishiguro, Y., *Mikrochim. Acta* **1973**, 325.
80. Serdyuk, L. S. and Fedorova, G. P., *Zh. Analit. Khim.* **15**, 287 (1960).
81. Buděšínský, B. and Vrzalová, D., *Anal. Chim. Acta* **36**, 246 (1966).
82. Tonosaki, K. and Otomo, M., *Bull. Chem. Soc. Japan* **35**, 1683 (1962).
83. Prajsnar, D., *Chem. Anal. (Warsaw)* **8**, 71 (1963).
84. Študlar, K., *Collection Czech. Chem. Commun.* **29**, 1499 (1964).
85. Tikhonova, A. A. and Timofeeva, N. I., *Zh. Analit. Khim.* **21**, 289 (1966).
86. Vdovenko, M. E., *Zavodsk. Lab.* **32**, 785 (1966).
87. Svoboda, V. and Chromý, V., *Talanta* **13**, 237 (1966).
88. Wet, de, W. J. and Behrens, G. B., *Anal. Chem.* **40**, 200 (1968).
89. Young, J. P., White, J. C. and Ball, R. G., *ibid.* **32**, 928 (1960).
90. Holleck, L., Eckardt, D. and Hartinger, L., *Z. Anal. Chem.* **146**, 103 (1955).
91. Serdyuk, L. S. and Fedorova, G. P., *Zh. Analit. Khim.* **25**, 172 (1970).
92. Tereshin, G. S., Rubinshtein, A. R. and Tananaev, I. V., *ibid.* **20**, 1082 (1965).
93. Poluektov, N. S. and Efryushina, N. P., *ibid.* **23**, 1802 (1968).
94. Beschetnova, E. T., Golovina, A. P. and Zorov, N. B., *ibid.* **28**, 1943 (1973).
95. Poluektov, N. S., Ovchar, L. A. and Lauer, R. S., *ibid.* **28**, 1958 (1973).
96. Ganago, L. I. and Alinovskaya, L. A., *ibid.* **25**, 904 (1970).
97. Cattrall, R. W. and Slater, S. J., *Microchem. J.* **16**, 602 (1971).

98. Poluektov, N. S., Melent'eva, E. V., Sandu, M. A. and Lauer, R. S., *Zh. Analit. Khim.* **26**, 2354 (1971).
99. Rinehart, R. W., *Anal. Chem.* **26**, 1820 (1954).
100. Kawashima, T., Ogawa, H. and Hamaguchi, H., *Talanta* **8**, 552 (1961).
101. Sangal, S. P., *Microchem. J.* **12**, 321 (1967).
102. Poluektov, N. S., Makarenko, O. P., Kirillov, A. I. and Lauer, R. S., *Zh. Analit. Khim.* **28**, 285 (1973).
103. Serdyuk, L. S. and Lazorina, S. M., *ibid.* **21**, 561 (1966).
104. Keil, R., *Z. Anal. Chem.* **245**, 362 (1969).
105. Pyatnitskii, I. V. and Gawrilova, E. F., *Zh. Analit. Khim.* **25**, 445 (1970).
106. Herrington, J. and Steed, K. C., *Anal. Chim. Acta* **22**, 180 (1960).
107. Suk, V., *Collection Czech. Chem. Commun.* **31**, 367 (1966).
108. Sandu, M. A., Poluektov, N. S. and Lauer, R. S., *Ukr. Khim. Zh.* **37**, 820 (1971).
109. Zaikovskii, F. V. and Sadova, G. F., *Zh. Analit. Khim.* **16**, 29 (1961).
110. Poluektov, N. S., Lauer, R. S. and Sandu, M, A., *ibid.* **25**, 2118 (1970).
111. Prajsnar, D., *Chem. Anal. (Warsaw)* **11**, 1111 (1966).
112. Tserkasevich, K. V. and Poluektov, N. S., *Zh. Analit. Khim.* **19**, 1309 (1964).
113. Taketatsu, T. and Toriumi, N., *Talanta* **17**, 465 (1970).
114. Taketatsu, T. and Yamauchi, T., *ibid.* **18**, 647 (1971).
115. Mishchenko, V. T., Lauer, R. S., Efryushina, N. P. and Poluektov, N. S., *Zh. Analit. Khim.* **20**, 1073 (1965).
116. Kononenko, L. I., Tishchenko, M. A. and Drobyazko, V. N., *ibid.* **26**, 729 (1971).
117. Zelinski, S. and Puach, V., *ibid.* **25**, 2342 (1970).
118. Mishchenko, V. T. and Poluektov, N. S., *ibid.* **17**, 825 (1962).
119. Poluektov, N. S. and Sandu, M. A., *ibid.* **24**, 1828 (1969).
120. Poluektov, N. S. and Mishchenko, V. T., *ibid.* **24**, 1434 (1969).
121. Poluektov, N. S., Bel'tyukova, S. V. and Meshkova, S. B., *ibid.* **27**, 266 (1972).
122. Poluektov, N. S. and Sandu, M. A., *ibid.* **25**, 1510 (1970).
123. Alinovskaya, L. A. and Ganago, L. I., *ibid.* **25**, 2347 (1970).
124. Poluektov, N. S., Sandu, M. A. and Lauer, R. S., *ibid.* **26**, 499 (1971).
125. Poluektov, N. S., Sandu, M. A. and Lauer, R. S., *ibid.* **25**, 899 (1970).
126. Kononenko, L. I. and Piontkovskaya, T. P., *ibid.* **24**, 379 (1969).
127. Madison, B. L. and Guyon, J. C., *Anal. Chim. Acta* **42**, 415 (1968).
128. Nowicka-Jankowska, T. and Szyszko, H., *Chem. Anal. (Warsaw)* **16**, 3 (1971).
129. Moeller, T. and Brantley, J. C., *Anal. Chem.* **22**, 433 (1950).
130. Holleck, L. and Hartinger, L., *Angew. Chem.* **67**, 648 (1955).
131. Banks, C. V. and Klingman, D. W., *Anal. Chim. Acta* **15**, 356 (1956).
132. Stewart, D. C. and Kato, D., *Anal. Chem.* **30**, 164 (1958).
133. Maeck, W. J., Kussy, M. E. and Rein, J. E., *ibid.* **37**, 103 (1965).
134. Auer-Welsbach, H., *Monatsh. Chem.* **96**, 1611 (1965).
135. Krotova, L. V., *Zh. Analit. Khim.* **21**, 789 (1966).

Chapter 42

RHENIUM

Rhenium (Re, at.wt. 186·2) occurs in various oxidation states but the +VII, +V, and +IV states are the most important. Rhenium(VII) compounds are the most stable. The colourless perrhenate ReO_4^- ion exhibits weak oxidizing properties. The chemical properties of rhenium resemble those of molybdenum and manganese. Rhenium(V and IV) forms halide, oxalate, and tartrate complexes.

42.1 Methods of Separation

The main requirement in connection with the determination of small quantities of rhenium is separation from molybdenum, which is usually the major interfering element. Small amounts of rhenium are normally found along with molybdenum in natural materials, such as molybdenite.

42.1.1 Extraction

Traces of rhenium can be separated from larger amounts of molybdenum, by shaking a chloroform solution of tetraphenylarsonium chloride with an aqueous solution at pH 8–9 containing perrhenate ions, thereby forming the chloroform-soluble ion-association complex $[(C_6H_5)_4As^+][ReO_4^-]$ [1–4]. Molybdate ions remain in the aqueous solution. The distribution coefficient for rhenium is $\sim 10^3$. At concentrations greater than 0·1M, chloride impairs the extraction of rhenium. Tetraphenylarsonium perrhenate is extracted over a wide pH range, but a medium at pH 8–9 is most suitable for the separation from molybdenum. EDTA is added to prevent hydrolysis of other metals.

$AuCl_4^-$, MnO_4^-, ClO_4^-, and SCN^- are extracted together with ReO_4^-. Tetraphenylphosphonium chloride may be substituted for tetraphenylarsonium chloride [1].

From acid medium, ReO_4^- can be extracted as its ion-pair with the tributylammonium ion [6], tribenzylammonium ion [7], or nitron [8]. From alkaline media, rhenium(VII) is extracted with pyridine or methyl-substituted pyridine derivatives [9].

Perrhenate and chloro-complexes of rhenium(IV) have been separated by solvent extraction using organophosphorus compounds (TBP or TOPO in hexane) [9a].

The separation of rhenium and molybdenum has also been achieved by extraction with TTA [10], acetylacetone [11], and cupferron [12]. Yatirajam and Kakkar [13–15] have discussed the separation of rhenium from molybdenum in a number of papers.

42.1.2 Precipitation

Rhenium can be separated from many metals by precipitation with hydrogen sulphide from 5–6M HCl (or H_2SO_4) [16]. Arsenic(III) is a suitable collector since it does not interfere in the subsequent determination of rhenium by the thiocyanate method.

Molybdenum can be coprecipitated with iron at pH < 7·5 while ReO_4^- ions remain in the solution [17].

In Ranskii's method [18], the sample containing rhenium is sintered at 600–700°C for 2–3 hours with CaO and $Ca(NO_3)_2$ or $KMnO_4$ [19,20], after which the sinter is leached with hot dilute bromine water. Most metals (including molybdenum which is then present as $CaMoO_4$) remain in the solid residue while perrhenate ions pass into solution.

Perrhenate is also leached with water from a sodium carbonate or alkali metal hydroxide melt.

42.1.3 Distillation

Rhenium heptoxide, Re_2O_7, is sufficiently volatile to be separated by distillation at 260–280°C from a concentrated sulphuric acid medium [3,21–24]. It may be accompanied by As(III), Hg, Se and to a lesser degree by Sb, Te, and Mo. Technetium is masked by reduction with hydroxylamine [22]. In the presence of hydrochloric acid, rhenium distils as the oxychloride at a lower temperature, but As, Ge, Hg, Sn, Se, Mo, Te, and Tl are wholly or partly codistilled. In view of the partial codistillation of molybdenum, rhenium cannot be directly determined in the distillate. In these distillation methods, steam, carbon dioxide, nitrogen, or air is used as the carrier gas [24a].

It follows from this discussion that some rhenium(VII) may be lost during the evaporation of acidic solutions [3].

42.1.4 Ion-Exchange

When a slightly acidic solution is run onto a cation-exchange column, most metals (Fe, Cu, Ni, Mn, Al, etc.) are retained as cations, while rhenium is eluted as the ReO_4^- anion [25].

Rhenium can be separated from manganese, technetium, molybdenum and tungsten by selective elution of the rhenium from an anion-exchange column with ammonium thiocyanate in dilute hydrochloric acid [26–28], or with mixed aqueous–organic solvent eluents [29].

42.2 Methods of Determination

A common spectrophotometric method for the determination of rhenium involves extraction of the thiocyanate complex into higher alcohols or ethers. The α-furildioxime method has similar sensitivity, the rhenium complex being measured in either aqueous acetone medium or in a chloroform extract.

Recently, many new methods for determining rhenium have been published. Pollock and Zopatti [30] have compared several suitable spectrophotometric reagents.

42.2.1 THIOCYANATE METHOD

The red-orange colour formed when $SnCl_2$ is added to a hydrochloric acid solution containing rhenium(VII) and thiocyanate is the basis for this spectrophotometric method for rhenium [1,31–33]. The reaction is complex, and the colour obtained depends on the concentrations of the reducing agent and of thiocyanate, and on the time.

Ryabchikov *et al.* [34,35] have shown that two complexes can be formed: a greenish-yellow and a red-orange complex with absorption maxima at 350 nm and 430 nm, respectively. These authors ascribe the formula $[ReO_2(SCN)_4]^{3-}$ to the red-orange complex. According to Iordanov and Pavlova [36], further complexes can be formed.

Hence, reproducible results in the determination of rhenium are only obtained when the reaction conditions are kept constant. Maximum colour intensity requires the presence of a small excess of $SnCl_2$, but higher concentrations of $SnCl_2$ result in considerably reduced absorbance; $2M$ HCl in the aqueous solution is the most suitable acid concentration. The thiocyanate concentration should not be lower than 1% KSCN.

Extraction of the thiocyanate complex with isoamyl alcohol, diethyl ether, or di-isopropyl ether enhances the sensitivity of the method because the high distribution coefficients facilitate concentration of the complex in a small volume of organic extractant.

The molar absorptivity of the rhenium thiocyanate complex in isoamyl alcohol obtained as described in the procedure below is 3.8×10^4 ($a = 0.21$) at $\lambda_{max} = 430$ nm. The absorption spectrum is shown in Fig. 42.1.

When rhenium is present in chloroform solution as the tetraphenylarsonium perrhenate, the thiocyanate complex can be developed directly by adding hydrochloric acid, thiocyanate, $SnCl_2$, and isoamyl alcohol [2]. The tributylammonium perrhenate extract in dichloroethane gives an analogous colour reaction [6].

Akimov *et al.* [36a] have extracted the thiocyanate complex of rhenium with DAPM in chloroform.

The following species interfere in the determination of rhenium by the thiocyanate method: Mo, W, V, and Cu, and oxidizing and reducing agents.

The thiocyanate method has been used to determine rhenium in molybdenite and other molybdenum ores [1,18,20], minerals [37], copper ores and concentrates [38], iron, nickel, and manganese alloys [25], and coal [19].

Reagents

1. Potassium thiocyanate: 20% solution.
2. Standard rhenium solution: 1 mg/ml. Dissolve 0·1554 g of potassium perrhenate, $KReO_4$, in water and dilute the solution to the mark with water in a 100-ml volumetric flask.
3. Stannous chloride: fresh 0·1% solution prepared by dissolving 0·1 g of $SnCl_2.2H_2O$ in 100 ml of hydrochloric acid (1+3).
4. Tetraphenylarsonium chloride: 1% solution.

Procedure

Extractive separation of Re. Adjust $\sim$ 40 ml of the perchlorate-free sample solution to pH 8–8·5 with $NaHCO_3$. If hydrolysable metals are present, add EDTA first. Transfer the solution to a separating funnel, add 1 ml of the tetraphenylarsonium chloride solution, and extract rhenium by shaking for 2 minutes with two 5-ml portions of $CHCl_3$. Before draining the extract, allow the phases to separate completely by standing for 30 minutes. Wash the combined extracts with 1 ml of 1% $NaHCO_3$ solution, and again allow the phases to stand for 30 minutes before separating them. Carefully separate the chloroform phase and evaporate it to dryness in a platinum crucible containing a little Na_2CO_3 dissolved in a few drops of water. Fuse the residue with 0·5 g of Na_2CO_3, dissolve the cooled melt in water, and neutralize the solution with 1 ml of H_2SO_4 (1+3).

Determination of Re. Acidify $\sim$ 30 ml of solution containing not more than 200 μg of Re with 8 ml of conc. HCl, and add 5 ml of the thiocyanate solution. Heat the solution to 50°C in a beaker on a water-bath, add 1 ml of the $SnCl_2$ solution with vigorous stirring, and keep the solution at 50°C for 20 minutes on the water-bath. After cooling to room temperature, transfer the solution to a separating funnel and extract with two portions of isoamyl alcohol. Dilute the extracts to the mark with the solvent in a 50-ml or smaller (depending on the amount of rhenium) volumetric flask, and measure the absorbance at 430 nm, using isoamyl alcohol or water as the reference.

42.2.2 α-Furildioxime Method

In an acid medium (HCl, H_2SO_4) containing stannous chloride as a reducing agent, α-furildioxime (formula, p. 373) reacts with rhenium to form a red complex [39–42].

The coloured complex is sparingly soluble in aqueous medium, but dissolves in solutions containing 24–26% of acetone [40]. Provided its concentration is not greater than 26%, acetone accelerates the reaction.

Heating the solution (to not higher than 60°C since volatile acetone is present) also increases the reaction rate. According to Meloche et al. [40], maximum absorbance is attained without heating only after 45 minutes. The optimum acidity in the medium corresponds to $0\cdot8$–$1M$ HCl. The reaction with rhenium requires a considerable excess of α-furildioxime, the quality of which affects the absorbance [43]. The minimum amount of stannous chloride necessary is 5 ml of 10% $SnCl_2$ solution per 50 ml of the final solution. The procedure given below employs these optimum conditions.

The molar absorptivity of the complex in aqueous acetone at λ_{max} = 530 nm is $4\cdot1\times10^4$ ($a = 0\cdot22$) (see Fig. 42.1).

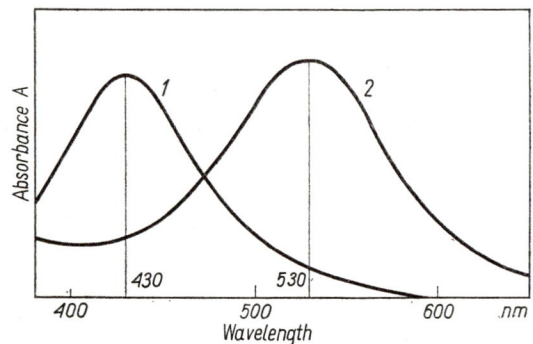

Fig. 42.1 Absorption spectra of the rhenium thiocyanate complex in isoamyl alcohol (1) and the rhenium α-furildioxime complex in aqueous acetone (2)

The rhenium α-furildioxime complex can be extracted into chloroform ($\varepsilon = 3\cdot7\times10^4$) [39]. The distribution coefficient is 150. Addition of isoamyl alcohol clears the chloroform extract.

Thiocyanate, nitrate, fluoride, and oxidants interfere in this method for determining rhenium. Palladium and copper form sparingly soluble α-furildioximates. Normally, rhenium should be separated from molybdenum before the determination, but Peshkova et al. [44] have described a modified method which enables rhenium to be determined in the presence of small amounts of molybdenum.

α-Furildioxime has been used to determine rhenium in ores and concentrates [8,44], and tungsten and molybdenum alloys [45].

Reagents

1. α-Furildioxime: 0·5% solution in acetone. If coloured, purify the solution by shaking with activated carbon.
2. Standard rhenium solution: 1 mg/ml (p. 451).
3. Stannous chloride: a freshly prepared 10% solution in HCl (1+9).

Procedure

Place ~ 25 ml of the slightly acidic sample solution, containing not more than 200 μg of Re, in a 50-ml volumetric flask. Add 4 ml of conc. HCl and 12 ml of the α-furildioxime solution, and mix well. Add 5 ml of the $SnCl_2$ solution and dilute to the mark. Heat the flask and its contents in a water-bath at 60°C for 20 minutes with occasional shaking. Cool the solution in a cold water-bath to room temperature, and measure the absorbance at 530 nm against a reagent blank solution or water.

Extraction into chloroform. After heating at 60°C as above and cooling to room temperature, transfer the coloured solution to a separating funnel and extract the rhenium complex by shaking for 1 minute with two portions of chloroform. Dilute the combined extracts in a suitable volumetric flask with chloroform and isoamyl alcohol (4+1). Measure the absorbance of the extract at 530 nm, using the solvent as reference.

42.2.3 OTHER METHODS

Further oximes used as spectrophotometric reagents for rhenium are 4-methylnioxime [46] (I), α-pyridildioxime [47], dimethylglyoxime [48–51], methyl-2-pyridylketoxime [52], phenyl-2-pyridylketoxime [53] (II), and biacetyl monoxime [54]. In each case the colour reaction occurs in the presence of $SnCl_2$. With the exception of 4-methylnioxime, these reagents are less sensitive than α-furildioxime.

In 3–5M HCl containing $SnCl_2$, thiourea forms a yellow cationic rhenium complex which is soluble in water, thus affording the basis of a less sensitive method for determining rhenium ($\varepsilon \sim 6.0 \times 10^3$) [25,55–57]. Since the reaction is slow, heating to 70°C is advantageous. Above 75°C, browning of the solution can be observed as the thiourea complex decomposes to rhenium sulphide. Several complexes can be formed in the rhenium thiourea system. Which species predominates depends on the excess of reagent and the acidity. The thiourea method is suitable for the differential spectrophotometric determination of rhenium as a major constituent [58]. Pollock [59] has used 1-phenyl-2-thiourea for determining rhenium ($\varepsilon = 9.5 \times 10^3$).

Geilmann and Neeb [60] recommend 2,4-diphenylthiosemicarbazide(III) which forms a red rhenium complex ($\lambda_{max} = 510$ nm) which can be extracted with chloroform. Small quantities of molybdenum do not interfere seriously. The sensitivity of the method is similar to that of the thiocyanate method.

Borisova et al. [61] have examined 14 thiourea derivatives for spectrophotometric determination of rhenium. 1,4-Diphenylthiosemicarbazide appeared to be the most sensitive reagent.

Organosulphur reagents for rhenium include p-thiocresol [62], dithiol [63], pyrazolinedithiocarbamates [64], morpholine-N-dithiocarbamate [65], diethyldithiophosphoric acid [66]. Bankovskii et al. [67] have determined rhenium with thio-oxine. The method is remarkable in that a 5000-fold amount of molybdenum can be tolerated. The colour reaction is carried out in $9M$ HCl, the acidity being so great that only platinum metals can interfere. The molar absorptivity of the chloroform extract of the complex is 8.5×10^3 at 438 nm. Thio-oxine has been used to determine rhenium in titanium alloys [68].

Rhenium(VII) reacts with diphenylcarbazide in $8M$ HCl to form a violet complex which can be extracted with chloroform ($\varepsilon = 1.9 \times 10^4$ at 540 nm) [55,69]. The Re(VII) is reduced by diphenylcarbazide to Re(V). The 1:1 complex is the product of the reaction between rhenium(V) and the diphenylcarbazone produced. The reaction is similar to that of chromium(VI) with diphenylcarbazide (see p. 215).

The most sensitive spectrophotometric methods for the determination of rhenium involve the extraction of ion-association complexes formed by the perrhenate ion with basic organic dyes, e.g. fuchsin, the ReO_4^- complex of which is extracted at pH 4.5–7.5 by amyl acetate or chloroform [3].

Other dyes used similarly include Methyl Violet [70,71], Crystal Violet and Malachite Green [71], Brilliant Green [71,72], Methyl Green [73], Victoria Blue 4R [74], Rhodamine B [70], Butylrhodamine B [75], Safranine T [76], Nile Blue A [77], and various antipyrine dyes [78–80]. The association compound formed by ReO_4^- and the 2,2'-bipyridyl complex of iron(II) is also extractable [81,82].

Akimov et al. [83] have determined rhenium as the extractable $ReBr_6^{2-}$ ion-association complex with DAPM.

Less sensitive methods for determining rhenium using sulphite [55,84] and sulphate [85,86] are also worthy of mention.

References

1. Tribalat, S., *Anal. Chim. Acta* **3**, 113 (1949); **4**, 228 (1950).
2. Beeston, J. M. and Lewis, J. R., *Anal. Chem.* **25**, 651 (1953).
3. Beyermann, K., *Z. Anal. Chem.* **183**, 91 (1961).
4. Thierig, D. and Umland, F., *ibid.* **240**, 19 (1968).
5. Bock, R. and Jainz, J. Z., *ibid.* **198**, 315 (1963).
6. Ziegler, M. and Schroeder, H., *ibid.* **212**, 395 (1965).
7. Yatirajam, V. and Kakkar, L. R., *Anal. Chim. Acta* **52**, 555 (1970).

8. Lazarev, A. I., Lazareva, V. I., Zak, S. Sh. and Usatenko, T. M., *Zavodsk. Lab.* **28**, 1316 (1962).
9. Rimshaw, S. J. and Malling, G. F., *Anal. Chem.* **33**, 751 (1961).
9a. Jordanov, N. and Pavlova, M., *Chem. Anal. (Warsaw)* **17**, 819 (1972).
10. De, A. K. and Rahaman, M. S., *Talanta* **12**, 343 (1965).
11. Yatirajam, V. and Kakkar, L. R., *Anal. Chim. Acta* **44**, 468 (1969).
12. Meyer, R. J. and Rulfs, C. L., *Anal. Chem.* **27**, 1387 (1955).
13. Yatirajam, V. and Kakkar, L. R., *Anal. Chim. Acta* **47**, 568 (1969); **54**, 152 (1971).
14. Yatirajam, V. and Kakkar, L. R., *Talanta* **17**, 759 (1970).
15. Yatirajam, V. and Kakkar, L. R., *Mikrochim. Acta* **1970**, 708; **1971**, 479.
16. Geilmann, W. and Bode, H., *Z. Anal. Chem.* **130**, 222 (1950); **132**, 250; **133**, 177 (1951).
17. Novikov, A. I., *Zh. Analit. Khim.* **16**, 588 (1961).
18. Ranskii, B. N., *Zavodsk. Lab.* **24**, 803 (1958).
19. Kuznetsova, V. V., *Zh. Analit. Khim.* **16**, 736 (1961).
20. Tsyvina, B. S. and Davidovich, N. K., *Zavodsk. Lab.* **26**, 930 (1960).
21. Geilmann, W. and Bode, H., *Z. Anal. Chem.* **130**, 323 (1950).
22. Koyama, M., *Bull. Chem. Soc. Japan* **34**, 1766 (1961).
23. Jordanov, N. and Pavlova, M., *Mikrochim. Acta* **1963**, 477.
24. Basińska, M. and Rutkowski, W., *Chem. Anal. (Warsaw)* **13**, 799 (1968).
24a. Marczenko, Z. and Ramsza, A., *unpublished work*.
25. Ryabchikov, D. I. and Lazarev, A. I., *Zh. Analit. Khim.* **10**, 228 (1955); **16**, 366 (1961).
26. Pirs, M. and Magee, R. J., *Talanta* **8**, 395 (1961).
27. Hamaguchi, H., Kawabuchi, K. and Kuroda, R., *Anal. Chem.* **36**, 1654 (1964); *J. Chromatog.* **17**, 567 (1965).
28. Kawabuchi, K., *J. Chem. Soc. Japan, Pure Chem. Sect.* **85**, 787 (1964).
29. Korkisch, J. and Feik, F., *Anal. Chim. Acta* **37**, 364 (1967).
30. Pollock, E. N. and Zopatti, L. P., *ibid.* **32**, 418 (1965).
31. Geilmann, W. and Bode, H., *Z. Anal. Chem.* **128**, 489 (1948).
32. Ryabchikov, D. I. and Borisova, L. V., *Talanta* **10**, 13 (1963).
33. Koźlicka, M., Wójtowicz, M. and Adamiec, I., *Chem. Anal. (Warsaw)* **15**, 247 (1970).
34. Ryabchikov, D. I., Zarinskii, V. A. and Nazarenko, I. I., *Zh. Neorgan. Khim.* **6**, 641 (1961); **7**, 931 (1962).
35. Ryabchikov, D. I. and Nazarenko, I. I., *Zh. Analit. Khim.* **19**, 229 (1964).
36. Iordanov, N. and Pavlova, M., *ibid.* **19**, 221 (1964); **20**, 591 (1965); **22**, 212 (1967).
36a. Akimov, V. K., Kliot, L. Ya. and Busev, A. I., *ibid.* **28**, 118 (1973).
37. Tribalat, S., Pamm, I. and Jungfleisch, M. L., *Anal. Chim. Acta* **6**, 142 (1952).
38. Skiba, H. and Wójtowicz, M., *Chem. Anal. (Warsaw)* **10**, 183 (1965).
39. Peshkova, V. M. and Gromova, M. I., *Vestn. Mosk. Univ., Ser. Fiz.-Mat. i Estestven. Nauk* **1952**, No. 10, 85.
40. Meloche, V. W., Martin, R. L. and Webb, W. H., *Anal. Chem.* **29**, 527 (1957).
41. Peskhova, V. M. and Chon Un Am, *Vestn. Mosk. Univ. Khim.* **1960**, No. 4, 59.
42. Peshkova, V. M. and Ignat'eva, N. G., *Zh. Analit. Khim.* **22**, 757 (1967).
43. Fryer, F. A., Galliford, D. J. and Yardley, J. T., *Analyst* **88**, 188, 191 (1963).
44. Peshkova, V. M., Ignat'eva, N. G. and Ozerova, G. P., *Zh. Analit. Khim.* **18**, 496 (1963).
45. Cotton, T. M. and Woolf, A. A., *Anal. Chem.* **36**, 248 (1964).
46. Kassner, J. L., Ting, S.-F, and Grove, E. L., *Talanta* **7**, 269 (1961).
47. Trusell, F. and Thompson, R. J., *Anal. Chem.* **36**, 1870 (1964).
48. Kenna, B. T., *ibid.* **33**, 1130 (1961).
49. Döge, H. G. and Grosse-Ruyken, H., *Mikrochim. Acta* **1967**, 98.
50. Koźlicka, M., Wójtowicz, M. and Adamiec, I., *Chem. Anal. (Warsaw)* **15**, 701 (1970).
51. Narayanan, A. and Umland, F., *Mikrochim. Acta* **1972**, 451.

52. Thompson, R. J., Gore, R. H. and Trusell, F., *Anal. Chim. Acta* **31**, 590 (1964).
53. Guyon, J. and Murmann, R. K., *Anal. Chem.* **36**, 1058 (1964).
54. Narayanan, A. and Subbaraman, P. R., *Indian J. Chem.* **5**, 436 (1967).
55. Ryabchikov, D. I. and Borisova, L. V., *Talanta* **10**, 13 (1963).
56. Borisova, L. V., *Zh. Analit. Khim.* **24**, 1361 (1969).
57. Nemodruk A. A. and Bezrogova, E. V., *ibid.* **24**, 1534 (1969).
58. Malyutina, T. M., Dobkina, B. M. and Chernikhov, Yu. A., *Zavodsk. Lab.* **26**, 259 (1960).
59. Pollock, E. N., *Anal. Chim. Acta* **47**, 367 (1969).
60. Geilmann, W. and Neeb, R., *Z. Anal. Chem.* **151**, 401 (1956).
61. Borisova, L. V., Plastinina, E. I. and Ermakov, A. N., *Zh. Analit. Khim.* **29**, 743 (1974).
62. Al-Kayssi, M. and Magee, R. J., *Talanta* **10**, 1047 (1963).
63. Koyama, M., Emoto, K., Kawashima, M. and Fujinaga, T., *Chem. Anal. (Warsaw)* **17**, 679 (1972).
64. Busev, A. I., Byr'ko, V. M. and Kondakova, G. K., *Zh. Analit. Khim.* **22**, 1028 (1967).
65. Likussar, W. and Beyer, W., *Mikrochim. Acta* **1973**, 211.
66. Lazarev, A. I. and Rodzaevskii, V. V., *Zh. Analit. Khim.* **16**, 243 (1961).
67. Bankovskii, Yu. A., Ievin'sh, A. F. and Luksha, E. A., *ibid.* **14**, 714 (1959).
68. Egorova, K. I. and Gurevich, A. N., *Zavodsk. Lab.* **29**, 789 (1963).
69. Ryabchikov, D. I. and Borisova, L. V., *Zh. Analit. Khim.* **18**, 851 (1963); *Zavodsk. Lab.* **29**, 785 (1963).
70. Poluektov, N. S., Kononenko, L. I. and Lauer, R. S., *Zh. Analit. Khim.* **13**, 396 (1958).
71. Pilipenko, A. T. and Obolonchik, V. A., *Ukr. Khim. Zh.* **24**, 506 (1958); **25**, 359 (1959); **26**, 99 (1960).
72. Fogg, A. G., Burgess, C. and Burns, D. T., *Analyst* **95**, 1012 (1970).
73. Tarayan, V. M., Bartanyan, S. V. and Eliazyan, L. A., *Zh. Analit. Khim.* **24**, 1040 (1969).
74. Pilipenko, A. T., Kish, P. P. and Zheltvai, I. I., *Ukr. Khim. Zh.* **37**, 477 (1971).
75. Blyum, I. A. and Dushina, T. K., *Zavodsk. Lab.* **28**, 903 (1962).
76. Pilipenko, A. T. and Nguen Mong Shin', *Ukr. Khim. Zh.* **32**, 1211 (1966).
77. Gagliardi, E. and Füsselberger, E., *Mikrochim. Acta* **1972**, 385.
78. Busev, A. I. and Ogareva, M. B., *Zh. Analit. Khim.* **21**, 574 (1966).
79. Busev, A. I., Ogareva, M. B. and Dzintarnieks, M. E., *ibid.* **22**, 205 (1967).
80. Vasil'ev, V. P., Kashirina, F. D. and Khranina, E. N., *Izv. Vyssh. Zaved., Khim. Khim. Tekhnol.* **14**, 503 (1971).
81. Kotsuji, K., Sakurai, T. and Yamamoto, Y., *J. Chem. Soc. Japan, Pure Chem. Sect.* **86**, 741 (1965).
82. Ackermann, G. and Pitzler, G., *Z. Anal. Chem.* **248**, 298 (1969).
83. Akimov, V. K., Busev, A. I. and Emel'yanova, I. A., *Zh. Analit. Khim.* **25**, 1938 (1970).
84. Lazarev, A. I., *ibid.* **14**, 362 (1959).
85. Iordanov, N. and Pavlova, M., *ibid.* **24**, 865 (1969).
86. Borisova, L. V. and Ermakov, A. N., *ibid.* **25**, 1128 (1970).

Chapter 43
RHODIUM

Rhodium (Rh, at.wt 102·91) is one of the platinum metals. It may occur in the +III, +IV and +VI oxidation states, the compounds of rhodium(III) being the most stable. It forms halide, cyanide, and ammine complexes. Powerful reducing agents (e.g. zinc, magnesium, and formic acid) reduce rhodium(III) to the metal. Strong oxidants [e.g. $(NH_4)_2S_2O_8$, and $NaBiO_3$] oxidize Rh(III) to Rh(IV). Rhodium(IV) disproportionates into Rh(III) and Rh(VI) (rhodate, RhO_4^{2-}).

43.1 Methods of Separation

Methods for the separation and isolation of the platinum metals have been discussed earlier (Platinum, p. 431).

43.1.1 Ion-Exchange

Unlike iridium(IV), platinum(IV), and palladium(II), rhodium(III) can exist as a cationic complex in hydrochloric acid medium. This rhodium complex passes through an anion-exchange column, while the other platinum metals are retained in the column [1–5]. In a cation-exchange column, the reverse happens [6,7]. In dilute hydrochloric acid ($0·3M$) containing thiourea, another cationic rhodium complex, which can be retained on Dowex 50, is formed. The anionic iridium complex formed under the same conditions passes to the eluate. Rhodium is subsequently eluted with hot $6M$ HCl [8].

43.1.2 Extraction

Rhodium(III) is separated from iridium(IV) by extraction with isoamyl alcohol from an acid medium containing hydrobromic acid and tin(II) [9–11]. The extraction of rhodium from a hydrochloric acid medium by TBP [12,13] ($SnCl_2$ may be added first [14]) also enables these two similar metals to be separated. Rhodium has also been separated from Ir, Pt, and Pd by extraction of the chloride complexes of the latter group of metals with dichloroethane in the presence of DAPM [15].

Rhodium can be separated from iridium by solvent extraction with tri-n-octylamine in benzene [16].

The rhodium complex with piperidinedithiocarbamate can be extracted from a nitrite medium [17].

43.1.3 Precipitation and Other Methods

Rhodium(III) may be separated from iridium(IV) by reduction to the metal [iridium being reduced only to Ir(III)] with antimony powder in hot, dilute H_2SO_4 [18], copper powder in $1M$ HCl [19], or sodium borohydride ($NaBH_4$) [20].

Rhodium has also been separated from iridium by precipitation with thioacetanilide [21], and formamidinesulphinic acid [22]. Rhodium coprecipitates with $Fe(OH)_3$ [23] and with organic collectors [24].

Fire assay methods for rhodium in ores and concentrates have been thoroughly investigated [25–27].

43.2 Methods of Determination

A critical evaluation of spectrophotometric methods for determining rhodium is contained in review papers by Beamish and McBryde [28,29].

One of the more popular methods is based on the reaction between rhodium(III) and $SnCl_2$ in HCl, but replacement of this reagent by $SnBr_2$ or SnI_2 results in greatly increased sensitivity.

43.2.1 Tin(II) Chloride, Bromide, or Iodide Method

When $SnCl_2$ is added to a solution of rhodium(III) in hydrochloric acid and the system heated, the solution turns red. At acidities lower than $2M$ HCl, the red complex is reversibly converted into a yellow complex. The spectrophotometric determination of rhodium is based on the more stable red complex, despite its lower absorptivity [30–33].

The absorption maximum of the red rhodium complex with $SnCl_2$ is at 470 nm (molar absorptivity 4.2×10^3; $a = 0.041$).

If rhodium is present in solution as the sulphate complex, the solution should be heated for some time after the addition of HCl to allow the rhodium sulphate complex to be converted into the chloride complex.

The red complex changes to yellow on extraction with esters, higher alcohols, and tri-n-octylamine in benzene ($\lambda_{max} = 415$ nm) [34].

Since the remaining platinum metals and other transition metals interfere, they must be separated before rhodium is determined by this method.

The $SnCl_2$ method has been used to determine rhodium in uranium alloys [5,7,35], plutonium [36], and platinum concentrate [37].

A considerable increase in sensitivity is obtained by replacing stannous chloride with stannous bromide [9,11,38,39] or iodide [40]. Potassium bromide or iodide is added to the rhodium solution in hydrochloric acid, and tin(II) is added after the solution has been heated.

In the bromide method, a yellow-orange complex is obtained ($\varepsilon = 2.9 \times 10^4$ at $\lambda_{max} = 427$ nm).

The rhodium–tin(II) iodide complex is red and has an absorption maximum at 460 nm ($\varepsilon = 3.9 \times 10^4$; $a = 0.38$) [40].

In the iodide method, the optimum concentration of HCl is $1M$, and the colour intensity decreases with increasing acidity. The concentration of potassium iodide should not be lower than 4% in the final coloured solution. The quantity of stannous chloride only slightly affects the absorbance. The iodine liberated by air in the initial stage of the procedure is reduced when the $SnCl_2$ is added.

As in the tin(II) chloride method, the remaining platinum metals and other transition metals interfere.

The rhodium–tin(II) bromide complex can be extracted into chloroform as an ion-pair with DAM, diphenylguanidine, or tribenzylamine [41]. The corresponding iodide complex can be extracted with dichloroethane in the presence of DAM [42].

Reagents

1. Stannous chloride: 10% solution in $2M$ HCl.
2. Potassium iodide: 20% solution.
3. Standard rhodium solution: 1 mg/ml. Fuse 0·1000 g of metallic rhodium powder with 2 g of $K_2S_2O_7$ in a quartz crucible. Dissolve the melt in hot $1M$ HCl and make the solution up to 100 ml with this acid in a volumetric flask.

Procedure

Tin(II) chloride method. To the sample solution containing not more than 1 mg of Rh, add 10 ml of the stannous chloride reagent. Heat the solution for 1 hour, nearly to boiling. Cool the solution, dilute to the mark with $2M$ HCl in a 50-ml volumetric flask, and measure the absorbance at 470 nm against water as the reference.

Tin(II) iodide method. To the sample solution containing not more than 100 μg of Rh, add 10 ml of the KI solution. Mix well, and heat for 15 minutes in a boiling water-bath. To the cooled solution, add 10 ml of the $SnCl_2$ solution. Dilute the solution in a 50-ml volumetric flask with dilute hydrochloric acid so that the final HCl concentration is $1M$. Place the unstoppered flask in the boiling water-bath for 2 minutes. Cool the solution rapidly and measure its absorbance at 460 nm against a reagent blank solution.

43.2.2 OTHER METHODS

Many reagents with sulphur as the ligand atom have been proposed for the spectrophotometric determination of rhodium, e.g. thiomalic acid [43], 2-mercaptobenzoxazole [44], 2-mercapto-4,5-dimethylthiazole [18,45], 5-amino-2-benzimidazolethiol [46], 2-diethylaminoethanethiol [47,48], thiosalicylamide [49], N,N'-bis(3-dimethylaminopropyl)dithio-oxamide [50], the allyl ether of thio-oxine (Allthiox) (molar absorptivity $7·0 \times 10^3$ at 400 nm) [51,52], and Sulphoallthiox [53].

PAN was the first of the azo reagents to be applied to the determination of rhodium ($\varepsilon = 1\cdot15 \times 10^4$ at 598 nm, in chloroform) [54–56]. Rhodium has been determined with this reagent in the presence of the other platinum metals, gold, and silver [55–57]. Further azo reagents recommended are PAR [58], TAR [59], TAN [60], Thoron I [61], and Sulphochlorophenolazorhodanine [62].

Diverse organic reagents for rhodium include *p*-nitrosodimethylaniline [63], *p*-nitrosodiphenylamine [64], diphenylcarbazone (in methanolic dimethylformamide) [65], 1-nitroso-2-naphthol [66], nitroso-R salt [67], TTA [68], Xylenol Orange [69], Eriochrome Cyanine R [70], tropolone [71], oximidobenzotetronic acid [72], and diphenylselenium oxide [73].

Lastly, there are the well-known methods based on the following inorganic reagents: thiocyanate [74], azide [75], hypochlorite [76], and hypobromite [77].

References

1. MacNevin, W. M. and Crummett, W. B., *Anal. Chem.* **25**, 1628 (1953).
2. Berman, S. S. and McBryde, W. A. E., *Can. J. Chem.* **36**, 845 (1958).
3. Marks, A. G. and Beamish, F. E., *Anal. Chem.* **30**, 1464 (1958).
4. Coufalík, F. and Svach, M., *Z. Anal. Chem.* **173**, 113 (1960).
5. Evans, H. B., Bloomquist, C. A. and Hughes, J. P., *Anal. Chem.* **34**, 1692 (1962).
6. MacNevin, W. M. and McKay, E. S., *ibid.* **29**, 1220 (1957).
7. Karttunen, J. O. and Evans, H. B., *ibid.* **32**, 917 (1960).
8. Berg, E. W. and Senn, W. L., Jr., *ibid.* **27**, 1255 (1955).
9. Kashlinskaya, S. E., Litvinskaya, I. I. and Strel'nikova, N. P., *Zavodsk. Lab.* **33**, 925 (1967).
10. Lee, A. S., Beamish, F. E. and Bapat, M. G., *Mikrochim. Acta* **1969**, 329.
11. Tertipis, G. G. and Beamish, F. E., *Anal. Chem.* **34**, 623 (1962).
12. Berg, E. W. and Senn, W. I., Jr., *Anal. Chim. Acta* **19**, 109 (1958).
13. Wilson, R. B. and Jacobs, W. D., *Anal. Chem.* **33**, 1650 (1961).
14. Kalinin, S. K., Katykhin, G. S., Nikitin, M. K. and Yakovleva, G. A., *Zh. Analit Khim.* **25**, 535 (1970).
15. Busev, A. I. and Akimov, V. A., *ibid.* **18**, 610 (1963).
16. Kanert, G. A. and Chow, A., *Anal. Chim. Acta* **69**, 355 (1974).
17. Fedorenko, N. V. and Filimonova, V. N., *Zavodsk. Lab.* **30**, 402 (1964).
18. Westland, A. D. and Beamish, F. E., *Mikrochim. Acta* **1956**, 1474.
19. Tertipis, G. G. and Beamish, F. E., *Anal. Chem.* **32**, 486 (1960).
20. McKay, E. S. and Cordell, R. W., *Talanta* **18**, 841 (1971).
21. Jackson, E., *Analyst* **84**, 106 (1959).
22. Prokof'eva, I. V. and Bukanova, A. E., *Zh. Analit. Khim.* **20**, 598 (1965).
23. Novikov, A. I. and Rustamov, S., *Radiokhimiya* **13**, 134 (1971).
24. Kuznetsov, V.I. and Marugin, V. A., *Zavodsk. Lab.* **36**, 1320 (1970).
25. Sant, B. R. and Beamish, F. E., *Anal. Chem.* **33**, 304 (1961).
26. Faye, G. H and Inman, W. R , *ibid.* **34**, 972 (1962).
27. Banbury, L. M. and Beamish, F. E., *Z. Anal. Chem.* **218**, 263 (1966).
28. Beamish, F. E. and McBryde, W. A. E., *Anal. Chim. Acta* **9**, 349 (1953); **18**, 551 (1958).
29. Beamish, F. E., *Talanta* **12**, 789 (1965).
30. Maynes, A. D. and McBryde, W. A. E., *Analyst* **79**, 230 (1954).
31. Ayres, G. H., Tuffly, B. L. and Forrester, J. S., *Anal. Chem.* **27**, 1742 (1955).

32. Kalinin, S. K., Stolyarov, K. P. and Yakovleva, G. A., *Zh. Analit. Khim.* **25**, 133 (1970).
33. Kalinin, S. K. and Yakovleva, G. A., *ibid.* **25**, 312 (1970).
34. Khattak, M. A. and Magee, R. J., *Anal. Chim. Acta* **45**, 297 (1969).
35. Gardner, R. D. and Hues, A. D., *Anal. Chem.* **31**, 1488 (1959).
36. Smith, M. E., *ibid.* **30**, 912 (1958).
37. Hofer, A., *Z. Anal. Chem.* **238**, 183 (1968).
38. Berman, S. S. and Ironside, R., *Can. J. Chem.* **36**, 1151 (1958).
39. Pantani, F. and Piccardi, G., *Anal. Chim. Acta* **22**, 231 (1960).
40. Berg, E. W. and Youmans, H. L., *ibid.* **25**, 366 (1961).
41. Pilipenko, A. T., Danilova, V. N. and Lisichenok, S. L., *Zh. Analit. Khim.* **25**, 1154 (1970).
42. Pilipenko, A. T. and Ol'khovich, P. F., *Ukr. Khim. Zh.* **34**, 397 (1968).
43. Wagner, V. L., Jr. and Yoe, J. H., *Talanta* **2**, 239 (1959).
44. Ryan, D. E., *Anal. Chem.* **22**, 599 (1950).
45. Ryan, D. E., *Analyst* **75**, 557 (1950); **76**, 731 (1951).
46. Sen Gupta, J. G., *Talanta* **8**, 785 (1961).
47. Srivastava, S. C., *Anal. Chem.* **35**, 1165 (1963).
48. Srivastava, S. C. and Good, M. L., *Anal. Chim. Acta* **32**, 309 (1965).
49. Sur, K. and Shome, S. C., *ibid.* **48**, 145 (1969).
50. Jacobs, W. D., *Anal. Chem.* **32**, 514 (1960).
51. Dedkov, Yu. M., Lozovskaya, L. V. and Slotintseva, M. G., *Zh. Analit. Khim.* **27**, 512 (1972).
52. Dedkov, Yu. M., Eliseeva, O. P., Ermakov, A. N., Savvin, S. B. and Slotintseva, M. G., *ibid.* **27**, 730 (1972).
53. Dedkov, Yu. M. and Slotintseva, M. G., *ibid.* **28**, 2367 (1973).
54. Stokeley, J. R. and Jacobs, W. D., *Anal. Chem.* **35**, 149 (1963).
55. Busev, A. I., Grössl, V. G. and Ivanov, V. M., *Anal. Lett.* **1**, 267 (1968).
56. Busev, A. I., Gresl', V. G. and Ivanov, V. M., *Zavodsk. Lab.* **34**, 388 (1968).
57. Figurovskaya, V. N., Busev, A. I. and Ivanov, V. M., *ibid.* **39**, 132 (1973).
58. Busev, A. I., Ivanov, V. M. and Grössl, V. G., *Anal. Lett.* **1**, 595 (1968).
59. Busev, A. I., Ivanov, V. M. and Gresl', V. G., *Zh. Analit. Khim.* **23**, 1570 (1968).
60. Ivanov, V. M., Busev, A. I., Gresl', V. G. and Zagruzina, A. N., *ibid.* **26**, 1553 (1971).
61. Shrivastava, S. C., Munshi, K. N. and Dey, A. K., *Microchem. J.* **14**, 37 (1969).
62. Propistsova, R. F. and Savvin, S. B., *Zh. Analit. Khim.* **28**, 1768 (1973).
63. Wilson, R. B. and Jacobs, W. D., *Anal. Chem.* **33**, 1652 (1961).
64. Stokeley, J. R. and Jacobs, W. D., *Talanta* **10**, 43 (1963).
65. Ayres, G. H. and Johnson, F. L., Jr., *Anal. Chim. Acta* **23**, 448 (1960).
66. Watanabe, K., *J. Chem. Soc. Japan, Pure Chem. Sect.* **77**, 1008 (1956).
67. Rollins, O. W. and Oldham, M. M., *Anal. Chem.* **43**, 146 (1971).
68. Rangnekar, A. V. and Khopkar, S. M., *Bull. Chem. Soc. Japan* **39**, 2169 (1966)
69. Otomo, M., *Japan Analyst* **17**, 125 (1968).
70. Shrivastava, S. C., Garg, V. C. and Dey, A. K., *Z. Anal. Chem.* **248**, 305 (1969).
71. Rizvi, G. H., Gupta, B. P. and Singh, R. P., *Mikrochim. Acta* **1972**, 459.
72. Manku, G. S., Bhat, A. N. and Jain, B. D., *Talanta* **14**, 1229 (1967).
73. Ziegler, M. and Schroeder, H., *Mikrochim. Acta* **1967**, 782.
74. Forsythe, J. H., Magee, R. J. and Wilson, C. L., *Talanta* **3**, 330 (1960).
75. Majumdar, A. K. and Mitra, B. K., *Anal. Chim. Acta* **33**, 670 (1965).
76. Ayres, G. H. and Young, F., *Anal. Chem.* **24**, 165 (1952).
77. Pantani, F., *Talanta* **9**, 15 (1962).

Chapter 44

RUTHENIUM

Ruthenium (Ru, at.wt. 101·07), one of the platinum metals, occurs in various oxidation states, the +VIII, +VI, +IV and +III states being the most common. Powerful oxidants, such as Cl_2 in an alkaline medium or $KMnO_4$, convert ruthenium compounds into the tetroxide, RuO_4, which is volatile and poisonous. Ruthenium(VI) is stable only in alkaline solutions (as orange ruthenate, RuO_4^{2-}). Ruthenium(IV and III) chloride complexes occur in hydrochloric acid solutions. Iron(II) and iodide reduce Ru(IV) to Ru(III). Zinc reduces ruthenium in weak acid media to the metal.

44.1 Methods of Separation

Methods for the isolation and separation of the platinum metals have been discussed earlier (Platinum, p. 431).

44.1.1 DISTILLATION OF RuO_4

The foremost method for the separation of ruthenium is the highly selective distillation of volatile ruthenium tetroxide [1–6a]. Only OsO_4, and to some extent Re_2O_7, can distil with ruthenium. $KMnO_4$, $NaBiO_3$, $HClO_4$, and $K_2Cr_2O_7$ are used to oxidize ruthenium(III and IV) in sulphuric or condensed phosphoric acid media. The distilled RuO_4 is absorbed either in acid solutions containing reductants or in alkaline solutions where the ruthenate is formed. Menis and Powell [4] have separated RuO_4 by heating a dry sample mixed with $NaBiO_3$ in a quartz boat placed in a tube flushed with moist oxygen.

Osmium is separated from ruthenium by a preliminary distillation from boiling sulphuric acid containing hydrogen peroxide. Under these conditions OsO_4 escapes while ruthenium remains in a lower oxidation state [1]. Alternatively, ruthenium and osmium may be oxidized to the tetroxides with $KMnO_4$, then iron(II) added to reduce the Ru(VIII) and the excess of MnO_4^-. After nitric acid (to 5–6M) has been added, the OsO_4 is distilled. Once the OsO_4 is removed, ruthenium is oxidized to RuO_4 which is then distilled [7].

44.1.2 Extraction and Other Methods

Ruthenium tetroxide is extractable into inert solvents (CCl_4, C_6H_6, $CHCl_3$) from dilute H_2SO_4 or HNO_3 [7–10]. Besides the oxidants listed above, silver(II) oxide, AgO, oxidizes ruthenium in HNO_3 or H_2SO_4 media. Ruthenium is stripped from CCl_4 with either $1M$ H_2SO_4 containing sulphite, or a KOH solution. In a quick method, it may be re-extracted with aqueous NaSCN, and the absorbance of the resulting blue ruthenium thiocyanate complex measured directly [11].

The chloride and bromide complexes of ruthenium(III) can be extracted from HCl and HBr solutions by amines, organophosphorus compounds, alcohols, and ketones [12,13].

The isolation of ruthenium and osmium by fire assay and cupellation methods with an iron-nickel-copper alloy or tin as collector has been investigated [14,15].

Ruthenium may be separated from uranium by precipitating the sulphide from a solution saturated with hydrogen sulphide under pressure [16].

44.2 Methods of Determination

There are several well-known spectrophotometric methods for determining ruthenium. In view of the specific isolation of ruthenium as RuO_4, the generally poor selectivity of these methods is of little consequence. Beamish and McBryde [17,18] have critically reviewed spectrophotometric methods for determining ruthenium.

44.2.1 Diphenylthiosemicarbazide Method

1,4-Diphenylthiosemicarbazide, a derivative of thiourea, has been used by Hara and Sandell [3] for the spectrophotometric determination of ruthe-

$$S=C\underset{NH-NH-C_6H_5}{\overset{NH-C_6H_5}{\diagup}}$$

nium. Heating a solution of the reagent and ruthenium in the presence of stannous chloride as reductant results in the formation of a red-violet chloroform-soluble complex in which the ruthenium is probably tervalent.

The molar absorptivity of the complex in chloroform is $1 \cdot 01 \times 10^4$ ($a = 0 \cdot 10$) at $\lambda_{max} = 560$ nm. Under the conditions of the ruthenium determination the absorbance of the reagent is negligible.

The higher the hydrochloric acid concentration, the faster the colour develops. The optimum acidity is between $5 \cdot 5$ and $6 \cdot 5M$ HCl. The maximum absorbance is obtained after heating the solution for 10–15 minutes at 100°C. The optimum quantity of stannous chloride is $0 \cdot 5$–$0 \cdot 7$ ml of 5% $SnCl_2 \cdot 2H_2O$ solution per 50 ml of the coloured solution.

A tenfold amount of osmium relative to ruthenium does not interfere in this method. The presence of rhenium causes slightly high results.

Before the determination with 1,4-diphenylthiosemicarbazide, ruthenium should be separated as RuO_4 by distillation or extraction. The solution from which RuO_4 is distilled must contain neither HNO_3 nor HCl. These acids are expelled from the sample solution by evaporation with sulphuric acid to white fumes.

1,4-Diphenylthiosemicarbazide has been used to determine ruthenium in meteorites [3,19].

2,4-Diphenylthiosemicarbazide has been employed similarly [20,21], but this reagent is somewhat inferior to 1,4-diphenylthiosemicarbazide because the colour of its ruthenium complex is less stable.

Reagents
1. 1,4-Diphenylthiosemicarbazide: saturated solution in methanol.
2. Standard ruthenium solution: 1 mg/ml. Fuse 0·1000 g of suitably pure, powdered ruthenium with 2 g of Na_2O_2 in a silver crucible. Dissolve the melt in water. Slightly acidify the solution with hydrochloric acid (1+1), and filter off any AgCl precipitate. Dilute the solution to the mark with water in a 100-ml volumetric flask.

   Alternatively, the standard solution may be prepared from ruthenium(III) or (IV) chloride or sulphate. In these cases, standardization (e.g. by gravimetric analysis) of the solution is indispensable.
3. Stannous chloride, $SnCl_2.2H_2O$: a freshly prepared 5% solution in HCl (1+1).

Procedure

Distillative separation of Ru. Place ~ 25 ml of sample solution (~ 0·5M in H_2SO_4 and containing not more than 200 μg of Ru) in a 50-ml distillation flask. Add 1 ml of 1% $KMnO_4$ solution (the solution should not be decolorized) and 0·5 g of $NaBiO_3$. Connect the still with a simple condenser and a receiver containing 0·6 ml of the $SnCl_2$ solution in 20 ml of HCl (1+1). Heat the still, while bubbling air through the liquid contents at a rate of 2 bubbles/second, and boil the solution for 5 minutes.

Determination of Ru. Dilute the distillate to ~ 50 ml with 6·5M HCl, add 5 ml of the 1,4-diphenylthiosemicarbazide solution, mix well, and heat in a boiling water-bath for 12 minutes. Transfer the cooled solution to a separating funnel, and shake with two portions of chloroform for 1 minute. Place the combined extracts in a 50-ml or smaller volumetric flask (depending on the amount of ruthenium), dilute to the mark with the solvent, and measure the absorbance at 560 nm against a reagent blank solution.

44.2.2 OTHER METHODS

Thiourea, $S=C(NH_2)_2$, reacts with ruthenium(III) in a hot acidic medium to form a water-soluble blue complex. The method based on this reaction

[22–25] is simple but insensitive ($\varepsilon \sim 2\cdot 5 \times 10^3$ at 620 nm). Related reagents which have been proposed for the determination of ruthenium include diphenylthiourea [26], 2-mercaptobenzimidazole (phenylenethiourea) [27], and selenourea [28].

The method for determining ruthenium with rubeanic acid (dithiooxamide) [29–31] is more than twice as sensitive as the thiourea method. Jacobs and Yoe [32] have proposed a related reagent, N,N'-bis(3-dimethylaminopropyl)dithio-oxamide.

A group of spectrophotometric methods for determining ruthenium is based on nitroso-compounds such as 1-nitroso-2-naphthol and 2-nitroso-1-naphthol [4,33–35], nitroso-R salt [36,37], p-nitrosodimethylaniline [15,38], and 3-nitroso-2,6-pyridinediol [39].

Banks and O'Laughlin [40] have developed a method with 1,10-phenanthroline, a well-known reagent for iron(II). In hydrochloric acid medium containing hydroxylamine as a reducing agent, the yellow ruthenium(II) complex [Ru(phen)$_3$]$^{2+}$, is formed on prolonged heating (the molar absorptivity is $1\cdot 85 \times 10^4$ at 448 nm). The tris(1,10-phenanthroline) ruthenium(II) complex can be extracted into various organic solvents with several anions, such as perchlorate, iodide, or thiocyanate [41].

8-Hydroxyquinoline forms a yellow complex with ruthenium which can be extracted at pH 4–6·5 into CHCl$_3$ ($\varepsilon = 1\cdot 2 \times 10^4$ at 430 nm) [42,43]. Gupta *et al.* [44] have recommended 8-hydroxyquinoline-N-oxide and its chloro-, bromo-, and nitro-derivatives.

Many varied organic reagents have been used in spectrophotometric methods for determining ruthenium: e.g. 2,4,6-tri(2'-pyridyl)-s-triazine (TPTZ) ($\varepsilon = 1\cdot 81 \times 10^4$ at 510 nm) [45], 2,3-diaminopyridine [46], 3,4-diaminobenzoic acid [47], 2-amino-8-naphthol-6-sulphonic acid [48], 1-naphthylamine-3,5,7-trisulphonic acid [2], acenaphthenequinone monoxime [49], thiosalicylamide [50], thiovioluric acid [51], diphenylthiovioluric acid [52], tropolone [53], o-hydroxythiobenzhydrazide [54], oximidobenzotetronic acid [55], 3-nitroso-4-hydroxy-5,6-benzocoumarin [56], diphenylcarbazone [57], anthranilic acid [58], acetylacetone [59], TTA [60], Xylenol Orange [61], TAR ($\varepsilon = 1\cdot 55 \times 10^4$) [62], Thoron I [63], and EDTA + H$_2$O$_2$ [64].

The ruthenium(III) thiocyanate complex is the basis of a less sensitive method ($\varepsilon \sim 4 \times 10^3$ at 590 nm) [11]. In the presence of pyridine, the complex can be extracted with MIBK [65]. Pilipenko *et al.* [66] have determined ruthenium as an aqueous suspension of the compound formed by the Ru(SCN)$_4^-$ complex and Crystal Violet. Ruthenium has also been determined as the bromide complex [67].

Suitable methods for determining larger amounts of ruthenium are based on the orange colour ($\lambda_{max} = 465$ nm) of the ruthenate ion (RuO$_4^{2-}$) [9,68,69], and on the greenish-yellow colour ($\lambda_{max} = 380$ nm) of the perruthenate ion (RuO$_4^-$) [70–72]. Ruthenate is formed when RuO$_4$ is absorbed in solutions of alkali hydroxides, or when RuO$_4$ is stripped from an organic phase with alkali. Perruthenate is formed in an alkaline medium containing hypochlorite or periodate.

El Guebely [73] oxidizes ruthenium(III) with chlorine to ruthenium (IV) and boils the solution to drive off the excess of chlorine. The ruthenium(IV) is then made to oxidize iodide to iodine, the absorbance of which is measured at 410 nm.

References

1. Westland, A. D. and Beamish, F. E., *Anal. Chem.* **26**, 739 (1954).
2. Steele, E. L. and Yoe, J. H., *Anal. Chim. Acta* **20**, 211 (1959).
3. Hara, T. and Sandell, E. B., *ibid.* **23**, 65 (1960).
4. Menis, O. and Powell, R. H., *Anal. Chem.* **34**, 166 (1962).
5. Chung, K. S. and Beamish, F. E., *Talanta* **15**, 823 (1968).
6. Alimarin, I. P., Khvostova, V. P. and Shlenskaya, V. I., *Zh. Analit. Khim.* **25**, 2167 (1970).
6a. Kiba, T., Terada, K., Kiba, T. and Suzuki, K., *Talanta* **19**, 451 (1972).
7. Surasiti, C. and Sandell, E. B., *Anal. Chim. Acta* **22**, 261 (1960).
8. Martin, F. S., *J. Chem. Soc.* **1954**, 2564.
9. Anderson, C. J., Del Grosso, R. and Ortner, M. H., *Anal. Chem.* **33**, 646 (1961).
10. Bezděk, M. and Mencl, J., *Collection Czech. Chem. Commun.* **30**, 711 (1965).
11. Belew, W. L., Wilson, G. R. and Corbin, L. T., *Anal. Chem.* **33**, 886 (1961).
12. Meier, H., Bösche, D., Zimmerhackl, E., Albrecht, W., Hecker, W., Menge, P., Ruckdeschel, A., Unger, E. and Zeitler, G., *Mikrochim. Acta* **1969**, 1083, 1107.
13. Berg, E. W. and Moseley, H. E., *Anal. Chim. Acta* **47**, 360 (1969).
14. Kavanagh, J. M. and Beamish, F. E., *Anal. Chem.* **32**, 490 (1960).
15. Faye, G. H., *ibid.* **37**, 696 (1965).
16. Rao, M. N., *Z. Anal. Chem.* **227**, 326 (1967).
17. Beamish, F. E. and McBryde, W. A. E., *Anal. Chim. Acta* **9**, 349 (1953); **18**, 551 (1958).
18. Beamish, F. E., *Talanta* **12**, 789 (1965).
19. Sen Gupta, J. G., *Anal. Chim. Acta* **42**, 481 (1968).
20. Geilmann, W. and Neeb, R., *Z. Anal. Chem.* **152**, 96 (1956).
21. Scharner, P. and Baresel, D., *Mikrochim. Acta* **1969**, 304.
22. Ayres, G. H. and Young, F., *Anal. Chem.* **22**, 1277 (1950).
23. Yaffe, R. P. and Voigt, A. F., *J. Am. Chem. Soc.* **74**, 2500, 2503 (1952).
24. Musil, A. and Pietsch, R., *Z. Anal. Chem.* **137**, 259 (1953).
25. Berg, E. W. and Moseley, H. E., *Anal. Lett.* **2**, 259 (1969).
26. Knight, S. B., Parks, R. L., Leidt, S. C. and Parks, K. L., *Anal. Chem.* **29**, 571 (1957).
27. Pilipenko, A. T., Sereda, I. P. and Semenyuk, E. P., *Zh. Analit. Khim.* **25**, 1958 (1970).
28. Pilipenko, A. T. and Sereda, I. P., *ibid.* **16**, 73 (1961).
29. Ayres, G. H. and Young, F., *Anal. Chem.* **22**, 1281 (1950).
30. Yaffe, R. P. and Voigt, A. F., *J. Am. Chem. Soc.* **74**, 3163 (1952).
31. Lingane, P. J., *Anal. Chim. Acta* **47**, 529 (1969).
32. Jacobs, W. D. and Yoe, J. H., *Talanta* **2**, 270 (1959).
33. Manning, D. L. and Menis, O., *Anal. Chem.* **34**, 94 (1962).
34. Konečný, C., *Anal. Chim. Acta* **29**, 423 (1963).
35. Kesser, G., Meyer, R. J. and Larsen, R. P., *Anal. Chem.* **38**, 221 (1966).
36. Miller, D. J., Srivastava, S. C. and Good, M. L., *ibid.* **37**, 739 (1965).
37. Nath, S. and Agarwal, R. P., *Chim. Anal. (Paris)* **47**, 257 (1965).
38. Currah, J. E., Fischel, A., McBryde, W. A. E. and Beamish, F. E., *Anal. Chem.* **24**, 1980 (1952).

39. McDonald, C. W. and Bedenbaugh, J. H., *Mikrochim. Acta* **1970**, 612.
40. Banks, C. V. and O'Laughlin, J. W., *Anal. Chem.* **29**, 1412 (1957).
41. Takamatsu, T., *Bull. Chem. Soc. Japan* **47**, 118 (1974).
42. Jasim, F., Magee, R. J. and Wilson, C. L., *Rec. Trav. Chim.* **79**, 541 (1960).
43. Hashitani, H., Katsuyama, K. and Motojima, K., *Talanta* **16**, 1553 (1969).
44. Gupta, R. D., Manku, G. S., Bhat, A. N. and Jain, B. D., *ibid.* **17**, 772 (1970); *Anal. Chim. Acta* **50**, 109 (1970).
45. Embry, W. A. and Ayres, G. H., *Anal. Chem.* **40**, 1499 (1968).
46. Ayres, G. H. and Eastes, D. T., *Anal. Chim. Acta* **44**, 67 (1969).
47. Ayres, G. H. and Arno, J. A., *Talanta* **18**, 411 (1971).
48. Popa, G. and Lazăr, C., *Anal. Chim. Acta* **33**, 676 (1965).
49. Sindhwani, S. K. and Singh, R. P., *ibid.* **55**, 409 (1971).
50. Sur, K. and Shome, S. C., *ibid.* **48**, 145 (1969).
51. Chawla, R. S. and Singh, R. P., *Mikrochim. Acta* **1972**, 496.
52. Chawla, R. S., Singh, R. P. and Trikha, K. C., *Talanta* **18**, 1245 (1971).
53. Rizvi, G. H., Gupta, B. P. and Singh, R. P., *Anal. Chim. Acta* **54**, 295 (1971).
54. Shome, S. C. and Gangopadhyay, P. K., *ibid.* **65**, 216 (1973).
55. Manku, G. S., Bhat, A. N. and Jain, B. D., *Talanta* **14**, 1229 (1967); **16**, 1421 (1969).
56. Kohli, N. and Singh, R. P., *ibid.* **21**, 638 (1974).
57. Bhatia, P. G., Manku, G. S., Bhat, A. N. and Jain, B. D., *Mikrochim. Acta* **1971**, 788.
58. Majumdar, A. K., and Sen Gupta, J. G., *Z. Anal. Chem.* **178**, 401 (1961).
59. Brandštetr, J. and Vřeštal, J., *Collection Czech. Chem. Commun.* **26**, 392 (1961).
60. Rangnekar, A. V. and Khopkar, S. M., *Mikrochim. Acta* **1968**, 272.
61. Shrivastava, S. C. and Dey, A. K., *Chim. Anal. (Paris)* **51**, 131 (1969).
62. Ivanov, V. M., Busev, A. I., Popova, L. V. and Bogdanovich, L. I., *Zh. Analit. Khim.* **24**, 1064 (1969).
63. Shrivastava, S. C., Munshi, K. N. and Dey, A. K., *Microchem. J.* **14**, 37 (1969).
64. Ezerskaya, N. A. and Solovykh, T. P., *Zh. Analit. Khim.* **24**, 422 (1969).
65. Forsythe, J. H., Magee, R. J. and Wilson, C. L., *Talanta* **3**, 324 (1960).
66. Pilipenko, A. T., Ol'khovich, P. F. and Bondarenko, V. Yu., *Ukr. Khim. Zh.* **37**, 1269 (1971).
67. Sereda, I. P. and Maslei, N. N., *ibid.* **32**, 755 (1966).
68. Marshall, E. D. and Rickard, R. R., *Anal. Chem.* **22**, 795 (1950).
69. Norkus, P. and Jankauskas, J., *Zh. Analit. Khim.* **27**, 2014 (1972).
70. Stoner, G. A., *Anal. Chem.* **27**, 1186 (1955).
71. Larsen, R. P. and Ross, L. E., *ibid.* **31**, 176 (1959).
72. Dinstl, G. and Hecht, F., *Mikrochim. Acta* **1963**, 895.
73. El Guebely, M. A., *Anal. Chim. Acta* **15**, 580 (1956).

Chapter 45

SCANDIUM

Scandium (Sc, at.wt. 44·96), resembles aluminium more than yttrium and the lanthanides, and is discussed separately here.

Scandium occurs in its compounds exclusively in the +III oxidation state. Some of its chemical properties resemble those of the lanthanides and yttrium (cf. p. 438). Scandium hydroxide, $Sc(OH)_3$, precipitates at a pH as low as 4·8 and dissolves in alkaline medium: in this respect scandium resembles aluminium.

45.1 Methods of Separation

45.1.1 PRECIPITATION

Along with rare earth elements, scandium may be separated as the sparingly soluble oxalate or fluoride. Suitable collectors in the oxalate and the fluoride methods are lanthanum and calcium, respectively.

Nazarenko and Biryuk [1] have isolated >99·5% of the scandium from a number of elements (Al, Sn, W, Mo, Nb, Ta, V, Ti) by precipitating the hydroxide with KOH in the presence of H_2O_2. Iron, which is used as collector, is subsequently removed by extraction as the chloride complex into ether.

Precipitation of scandium with ammonium tartrate in dilute ammonia permits the separation of this element from the lanthanides [1–4]. Yttrium is employed as a collector for microgram quantities of scandium. Larger amounts of non-lanthanide elements (>20 mg of Al, 20 mg of Fe, 2 mg of Zr, or 2 mg of Th) prevent the quantitative separation of scandium as the basic tartrate. Before being determined, scandium is separated from yttrium by extraction as the thiocyanate complex.

45.1.2 EXTRACTION

Three portions of diethyl ether extract practically all the scandium from an acid solution (0·1–0·2M HCl) containing ~ 50% of NH_4SCN [2,5]. Yttrium and the lanthanides remain quantitatively in the aqueous phase, but Fe(III), Co, Ga, In, Mo, Re, and certain other ions should be separated beforehand since they are also extracted.

TBP extracts scandium from 9–11M HCl, while yttrium, the lanthanides, aluminium, beryllium, uranium, and chromium remain in the aqueous phase [6–8].

Scandium can also be separated from the rare earths and other elements by extraction with TTA [9–11] or BPHA [12].

45.1.3 Ion-Exchange

The anionic scandium sulphate complex is absorbed by strongly basic anion-exchangers (e.g. Dowex 1) [13,14]. Under the conditions employed, yttrium, the lanthanides, aluminium, and beryllium are eluted, but thorium, zirconium, and uranium(VI) are retained along with scandium. Scandium can also be adsorbed on anion-exchangers as the thiocyanate complex [15].

Kuroda and Hikawa [16] have retained scandium, yttrium, and the lanthanides on a strongly basic anion-exchanger from a medium comprising glacial acetic acid and $3M$ HCl (9:1 v/v). First yttrium and the lanthanides are run off the column, then scandium is eluted with $1M$ hydrochloric acid.

Scandium is selectively eluted from a strongly acidic cation-exchange column with dilute sulphuric acid or acidic ammonium sulphate solution [17,18]. Lanthanides, yttrium, and other metals (e.g. Al, Ca, Cd, Co, Cu, In, Mg, Ni, and Zn) which do not form anionic sulphate complexes remain in the column. A thiocyanate solution acidified with hydrochloric acid also elutes scandium from cation-exchangers [19].

45.2 Methods of Determination

Several authors [3,20,21] have compared spectrophotometric methods for determining scandium. The methods based on Arsenazo III, Xylenol Orange, and Sulphonazo are generally the best. Some methods show sufficiently high selectivity to enable scandium to be determined in the presence of yttrium and the lanthanides.

45.2.1 Xylenol Orange Method

Xylenol Orange (XO) (formula, p. 51) reacts in a slightly acidic medium with scandium to form a red-violet complex which is the basis of a sensitive spectrophotometric method for determining scandium [22–26]. The reagent is yellow-orange in acidic solution (pH 1–5), but it turns red-violet above pH ~ 6.

Maximum absorbance of the scandium–Xylenol Orange complex is obtained between pH 2·5 and 2·7. As the acidity rises, the absorbance rapidly drops; with increase in pH, the absorbance slowly decreases. The absorption maximum of the complex occurs at 555–560 nm ($\varepsilon = 2·9 \times 10^4$; $a = 0·65$). At this wavelength the reagent absorbs imperceptibly. The absorption spectra of Xylenol Orange and its complex with scandium at pH 2·6, are shown in Fig. 45.1. Solutions of the complex are stable with respect to time. With a moderate excess of reagent, a 1:1 complex is formed between scandium and Xylenol Orange. When the excess of the reagent is very large, a less intensely coloured complex, the absorption maximum of which occurs at 525 nm, is formed.

Higher selectivity is gained by determining scandium with Xylenol Orange at pH 2·0 rather than at pH 2·6, where the reaction is most sensitive.

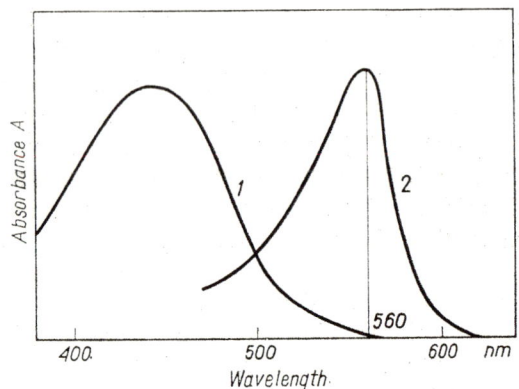

Fig. 45.1 Absorption spectra of Xylenol Orange (1) and its scandium complex (2) (reagent solution as reference) at pH 2·6

At pH 2·6, Th, Zr, Ti, Fe(III), Bi, In, Al, and Y (but not the lanthanides) interfere. Reducing the pH to 2·0 minimizes these interferences and even masks yttrium completely. (According to Belopol'skii and Popov [4], the presence of yttrium slightly increases the absorbance of the scandium complex; therefore, yttrium should be added to the standard solutions in quantities approximately the same as those in the sample solutions).

Iron(III) and cerium(IV) are masked by reduction with ascorbic acid [22]. Of the common anions, oxalate, sulphate, fluoride, and phosphate interfere in the determination of scandium.

Xylenol Orange has been used to determine scandium in minerals and coal ash [4], magnesium and its alloys [22], copper [23], and copper and nickel alloys [27].

Reagents

1. Xylenol Orange: 0·05% aqueous solution.
2. Standard scandium solution: 1 mg/ml. Dissolve 0·1533 g of scandium oxide Sc_2O_3 in 10 ml of hot $2M$ HCl, and dilute the solution to volume with water in a 100-ml volumetric flask. Working solutions are obtained by suitable dilutions of the stock solution with $\sim 0·01M$ HCl.
3. Chloroacetate buffer: pH ~ 3. Dissolve in water 3 g of NaOH, add 15 ml of chloroacetic acid, and dilute the solution with water to 250 ml.

Procedure

To the slightly acidic (pH ~ 1) sulphate-free sample solution containing not more than 60 μg of Sc, add 2 ml of 1% ascorbic acid solution, 2 ml of the chloroacetate buffer and 5 ml of the Xylenol Orange solution. Dilute the solution with water to ~ 40 ml, and adjust its pH to 2·0 (±0.1). Transfer the solution to a 50-ml volumetric flask, and make up to the mark with water. After 10 minutes, measure the absorbance of the solution at 560 nm against a reagent blank solution.

Notes. (1) If yttrium is known to be present in the sample solution [introduced, for example, as a collector (2–10 mg) during the precipitation of basic scandium tartrate], the same amount of yttrium should be added to the standard solutions used to prepare the calibration curve.

(2) The addition of the chloroacetate buffer is intended to facilitate the adjustment of the pH of the solution.

45.2.2 OTHER METHODS

Many azo dyes have been recommended as spectrophotometric reagents for scandium.

Arsenazo III (formula, p. 48) [20,28–30] gives similar sensitivity to Xylenol Orange: the optimum pH is between 1·5 and 3·0 (the molar absorptivity is $2·9 \times 10^4$ at 640 nm). Arsenazo I [9,20,31,32] is less sensitive and less selective (pH 6–8, $\varepsilon = 1·6 \times 10^4$ at 570 nm).

Other notable azo reagents used for determining scandium include Chlorophosphonazo III [33,34], Lumogallion [35–37], TAR [38], PAR [39], Sulphochlorophenol S [21,30], Sulphochlorophenol R [21,40], Azonol A [41], Stilbazochrome ($\varepsilon = 2·3 \times 10^4$) [20,42], Magneson IREA [36], *p*-nitrophenylazochromotropic acid [43], α-(2,4-dihydroxyphenylazo)-pyridine [44], and Nitrobromoarsenazo [45].

Sulphonazo (I) has been suggested as a very selective reagent for scandium [3,20]. In an acetate medium (pH 5–6), Sulphonazo does not react

(I)

with yttrium, and may, therefore, be applied in the determination of scandium after its coprecipitation as the tartrate with yttrium. The molar absorptivity is $1·0 \times 10^4$ at 575 nm.

The following triphenylmethane dyes have been employed similarly to Xylenol Orange: Chrome Azurol S [46], Pyrocatechol Violet [47], Methylthymol Blue [48,49], Eriochrome Azurol G [50], Chromal Blue G [51], and Eriochrome Brilliant Violet B [52].

Further organic reagents for scandium are Bromopyrogallol Red [53], propylfluorone ($\varepsilon = 3·2 \times 10^4$) [1,54], quercetin [55], morin [56], Murexide (ammonium purpurate) [57], glyoxal bis(2-hydroxyanil) (GBHA) [58], indoferron [59], anthrarufin-2,6-disulphonic acid [8], 8-hydroxyquinoline [60] and its dihalo-derivatives [61], Alizarin Red S [7], 1,5-di-(2'-hydroxy-3',5',6'-trichlorophenyl)-3-acetylformazan ($\varepsilon = 2·7 \times 10^4$ at 675 nm) [62], and 1,5-diantipyrinyl-3-cyanoformazan [63].

References

1. Nazarenko, V. A. and Biryuk, E. A., *Zavodsk. Lab.* **28**, 401 (1962).
2. Fischer, W., Steinhauser, O., Hohmann, E., Bock, E. and Borchers, P., *Z. Anal. Chem.* **133**, 57 (1951).
3. Brudz', V. G., Titov, V. I., Osiko, E. P., Drapkina, D. A. and Smirnova, K. A., *Zh. Analit. Khim.* **17**, 568 (1962).
4. Belopol'skii, M. P. and Popov, N. P., *Zavodsk. Lab.* **30**, 1441 (1964).
5. Vickery, R. C., *J. Chem. Soc.* **1956**, 3113.
6. Peppard, D. F., Faris, J. P., Gray, P. R., and Mason, G. W., *J. Phys. Chem.* **57**, 294 (1953).
7. Eberle, A. R. and Lerner, M. W., *Anal. Chem.* **27**, 1551 (1955).
8. MacDonald, J. C. and Yoe, J. H., *Anal. Chim. Acta* **28**, 264 (1963).
9. Onishi, H. and Banks, C. V., *ibid.* **29**, 240 (1963).
10. Ashbrook, A. W., *Analyst* **88**, 113 (1963).
11. Zolotov, Yu. A., Shakhova, N. V. and Alimarin, I. P., *Zh. Analit. Khim.* **23**, 1321 (1968).
12. Alimarin, I. P., and Tze Yung-Schaing, *Talanta*, **8**, 317 (1961).
13. Hamaguchi, H., Ohuchi, A., Onuma, N. and Kuroda, R., *J. Chromatog.* **16**, 396 (1964).
14. Hamaguchi, H., Ohuchi, A., Shimizu, T., Onuma, N. and Kuroda, R., *Anal. Chem.* **36**, 2304 (1964).
15. Hamaguchi, H., Onuma, N., Kishi, M. and Kuroda, R., *Talanta* **11**, 495 (1964).
16. Kuroda, R. and Hikawa, I., *J. Chromatog.* **25**, 408 (1966).
17. Strelow, F. W. and Bothma, C. J., *Anal. Chem.* **36**, 1217 (1964).
18. Kuroda, R., Nakagomi, Y. and Ishida, K., *J. Chromatog.* **22**, 143 (1966).
19. Hamaguchi, H., Kuroda, R., Aoki, K., Sugisita, R. and Onuma, N., *Talanta* **10**, 153 (1963).
20. Cherkesov, A. I. and Alykov, N. M., *Zh. Analit. Khim.* **19**, 943, 1067 (1964).
21. Ryabchikov, D. I., Savvin, S. B. and Dedkov, Yu. M., *ibid.* **19**, 1210 (1964).
22. Volodarskaya, R. S. and Derevyanko, G. N., *Zavodsk. Lab.* **29**, 148 (1963).
23. Berman, S. S., Duval, G. R. and Russell D. S., *Anal. Chem.* **35**, 1392 (1963).
24. Kon'kova, O. V., *Zh. Analit. Khim.* **19**, 73 (1964).
25. Antonovich, V. P. and Nazarenko, V. A., *ibid.* **23**, 1143 (1968).
26. Kranz, M., Duczmal, W. and Łangowska, K., *Chem. Anal. (Warsaw)* **16**, 399 (1971).
27. Postnikova, I. S., *Zavodsk. Lab.* **36**, 542 (1970).
28. Kuznetsov, V. I., Ni Chzhe-Min, Myasoedova, G. V. and Okhanova, L. A., *Acta Chim. Sinica* **27**, 74 (1961).
29. Savvin, S. B., *Zavodsk. Lab.* **29**, 131 (1963).
30. Nazarenko, V. A. and Antonovich, V. P., *Zh. Analit. Khim.* **24**, 1008 (1969).
31. Nazarenko, V. A. and Biryuk, E. A., *Ukr. Khim. Zh.* **29**, 198 (1963).
32. Shimizu, T., *Anal. Chim. Acta* **37**, 75 (1967).
33. Fadeeva, V. I. and Alimarin, I. P., *Zh. Analit. Khim.* **17**, 1020 (1962).
34. Fadeeva, V. I. and Kuchinskaya, O. I., *Vestn. Mosk. Univ. Khim.* **1967**, No. 1, 67.
35. Akhmedli, M. K. and Gambarov, D. G., *Zh. Analit. Khim.* **22**, 276 (1967).
36. Alimarin, I. P., Pigaga, A. K. and Gibalo, I. M., *Vestn. Mosk. Univ. Khim.* **1969**, No. 5, 94.
37. Alimarin, I. P., Gibalo, I. M. and Pigaga, A. K., *Zh. Analit. Khim.* **25**, 2336 (1970).
38. Shimizu, T. and Momo, E., *Anal. Chim. Acta* **52**, 146 (1970).
39. Sommer, L. and Hniličková, M., *ibid.* **27**, 241 (1962).
40. Ryabchikov, D. I., Savvin, S. B. and Dedkov, I. M., *Zavodsk. Lab.* **31**, 154 (1965).
41. Buděšínský, B. and Svecová, J., *Anal. Chim. Acta* **49**, 231 (1970).
42. Alykov, N. M. and Cherkesov, A. I., *Zh. Analit. Khim.* **20**, 870 (1965).
43. Sangal, S. P., *Microchem. J.* **8**, 313 (1964).
44. Busev, A. I. and Chang Fan, *Talanta* **9**, 101 (1962).

45. Busev, A. I., Lunina, G. E. and Basargin, N. N., *Zh. Analit. Khim.* **21**, 1414 (1966).
46. Ishida, R. and Hasegawa, N., *Bull. Chem. Soc. Japan* **40**, 1153 (1967).
47. Onosova, S. P. and Kuntsevich, G. K., *Zh. Analit. Khim.* **20**, 802 (1965).
48. Akhmedli, M. K. and Gambarov, D. G., *ibid.* **22**, 1183 (1967).
49. Antonovich, V. P. and Nazarenko, V. A., *ibid.* **23**, 1460 (1968).
50. Uesugi, K., *Bull. Chem. Soc. Japan* **42**, 2398 (1969).
51. Uesugi, K., *ibid.* **42**, 2051 (1969).
52. Uesugi, K., *Anal. Chim. Acta* **49**, 597 (1970).
53. Shimizu, T., *Talanta* **14**, 473 (1967); *Bull. Chem. Soc. Japan* **42**, 1561 (1969).
54. Biryuk, E. A. and Nazarenko, V. A., *Zh. Analit. Khim.* **14**, 298 (1959).
55. Hamaguchi, H., Kiroda, R., Sugisita, R., Onuma, N. and Shimizu, T., *Anal. Chim. Acta* **28**, 61 (1963).
56. Nazarenko, V. A. and Antonovich, V. P., *Zh. Analit. Khim.* **24**, 358 (1969).
57. Sangal, S. P., *Microchem. J.* **11**, 508 (1966).
58. Okáč, A. and Vrchlabský, M., *Z. Anal. Chem.* **195**, 338 (1963).
59. Shimizu, T. and Ogami, K., *Talanta* **16**, 1527 (1969).
60. Umland, F. and Puchelt, H., *Anal. Chim. Acta* **16**, 334 (1957).
61. Nishikawa, Y., Hiraki, K., Goda, S. and Shigematsu, T., *J. Chem. Soc. Japan, Pure Chem. Sect.* **83**, 1264 (1962).
62. Dziomko, V. M., Ostrovskaya, V. M. and Kon'kova, O. V., *Zh. Analit. Khim.* **25**, 267 (1970).
63. Buděšínský, B. W. and Svec, J., *Microchem. J.* **16**, 253 (1971).

Chapter 46

SELENIUM

Selenium (Se, at.wt. 78·96) forms selenide($-$II), selenite(IV) and selenate(VI) ions. Selenium(IV) compounds are the most stable. The dioxide, SeO_2, dissolves in water as selenious acid. Selenium dioxide is volatile and sublimes readily (unlike TeO_2). On dissolution in nitric acid, selenium is oxidized to Se(IV). Powerful oxidizing agents (e.g. aqua regia) oxidize selenium to Se(VI). Moderate reducing agents reduce Se(IV) and Se(VI) to the element. Selenium compounds are, in general, more easily reduced and less easily oxidized than the corresponding tellurium compounds. Selenium(IV) forms volatile chloride and bromide compounds.

46.1 Methods of Separation

46.1.1 Distillation

Selenium is usually separated by distillation as volatile selenium bromide ($SeBr_4$) [1–5] or chloride ($SeCl_4$) [1,5,6], which are distilled from concentrated hydrobromic acid medium [in the presence of bromine to prevent the reduction of Se(IV)] and from concentrated hydrochloric acid medium (see p. 476), respectively. Perchloric or sulphuric acid is added to the still, and distillation is continued until all the HBr or HCl is driven off, i.e. till white fumes of H_2SO_4 or $HClO_4$ appear. Passage of nitrogen through the liquid promotes the distillation. During the distillation, tellurium remains quantitatively in the still, but arsenic, germanium, and antimony are distilled with the selenium.

Selenium may be separated from various non-volatile materials as the volatile oxide, SeO_2, which forms when a stream of oxygen is passed over the sample in a quartz tube heated to 1000°C. The SeO_2 sublimed onto the cold part of the tube is dissolved and determined by any convenient method [7].

46.1.2 Precipitation

Selenium is easily separated by reduction to the element with stannous chloride, sulphur dioxide, hypophosphite, or hydrazine [1]. Arsenic [8–10] or tellurium [1,10–11] are suitable collectors for traces of selenium. When selenium is precipitated from 1–8M HCl with SO_2, the following are wholly or partly reduced to the element: Te, Au, platinum metals, Hg, Bi, Sb, Sn, and Cu [12].

It is possible to use selective reduction to separate selenium(IV) from selenium(VI). In 0·5M HCl in a closed vessel at 100°C, Se(IV) is reduced to the element by sulphur dioxide, whereas Se(VI) is not reduced until the acidity is increased to $4M$ HCl [13]. Hydrazine reduces Se(IV) in an almost neutral medium: after acidification of the solution to 2·1–2·5M HCl, Se(VI) is reduced to the element along with some tellurium(IV) [14]. Thiourea reduces selenium whereas with tellurium it yields a soluble complex [15,15a].

Traces of selenite can be coprecipitated with iron(III), manganese, or aluminium hydroxide [16–18]. Jackwerth [19] has coprecipitated selenite with $PbSO_4$.

When samples mixed with Na_2CO_3 and MgO are ignited at 800°C, a sinter is formed, from which water leaches selenate while sparingly soluble magnesium tellurate remains in the solid residue [20].

46.1.3 Extraction and Other Methods

Jordanov and Futekov [21] found that the selenium(IV) chloride complex in 6–7M HCl reacts with monoketones to form compounds which can be extracted with chloroform. $SeCl_2(C_8H_7O)_2$ is formed with acetophenone, for example. Such extractions separate selenium from tellurium.

Selenium(IV) can be extracted from hydrochloric acid medium with TBP [22,23]. Extraction of selenium dithiolate from 7M HCl yields a separation from many elements [24].

Although ion-exchange resins have found little application in methods for separating selenium, some such methods are mentioned in the chapter on tellurium.

46.2 Methods of Determination

The sensitive method based on 3,3'-diaminobenzidine is the most widely used. Selenium is determined either in the aqueous medium or after extraction of the coloured complex into toluene. A less sensitive method involving the coloured sol of elementary selenium is suitable for determining larger quantities of this element.

Murashova and Sushkova [25] have reviewed spectrophotometric methods for determining selenium.

46.2.1 Selenium Sol Method

Reduction to a brown-yellow sol of elementary selenium in acid medium containing a protective colloid has been made the basis of a simple but rather insensitive method for determining selenium [26–29].

Suitable protective colloids for preventing coagulation of the selenium sol are gum arabic, gelatine, and poly(vinyl alcohol).

Stannous chloride [6], ascorbic acid [2,27], thiourea [15,28], or hydrazine [29], are used to reduce selenium(IV). In 3–4M HCl, stannous chloride rapidly reduces selenium in the cold. Depending on the reducing agent and acid strength, pseudosolutions of different colours are obtained.

The molar absorptivity of the selenium sol obtained with $SnCl_2$ in 3M HCl containing poly(vinyl alcohol) is 1.7×10^3 at 400 nm ($a = 0.022$). Towards longer wavelengths, the absorptivity of the sol decreases; in the ultraviolet it increases. At 325 nm the absorptivity is twice as great, and at 450 nm half as great, as that at 400 nm.

Interference in this method for determining selenium comes from Te, Hg, Au, and platinum metals, all of which are easily reduced to the element.

As the coloured sol, selenium has been determined in steel [9,29], lead [28], sulphuric acid [6], wastes [2,27], and air [26].

Reagents

1. Stannous chloride, $SnCl_2.2H_2O$: 20% solution in 3M HCl.
2. Standard selenium solution: 1 mg/ml. Dissolve in water 1·405 g of selenium dioxide, SeO_2 (resublimed and stored over phosphorus pentoxide); dilute the solution with water in a volumetric flask to 1 litre.

Procedure

Distillation separation of Se. To the sample solution of selenium(IV) in a 50–100-ml still, add just enough conc. HCl to bring its concentration to 7M. Add 10 ml of H_2SO_4 (1+1) and a few fragments of porous porcelain, and connect the still to a condenser, the end of which is immersed in a small amount of 2M HCl in the receiver. Distil until white fumes of sulphuric acid appear in the still. To the cooled still, add 5–10 ml of conc. HCl and distil again till fumes appear. The receiver should be cooled with ice-water.

Determination of Se. To the sample solution (in ~ 3M HCl) containing not more than 1·0 mg of Se in a volume of ~ 35 ml, add 5 ml of 2% poly(vinyl alcohol) solution, and mix well. Add 2·5 ml of the $SnCl_2$ solution with stirring. Dilute to 50 ml in a volumetric flask. Measure the absorbance of the coloured pseudosolution at 400 nm, using water as the reference.

46.2.2 3,3′-Diaminobenzidine Method

Selenium(IV) reacts with 3,3′-diaminobenzidine (DAB) in acid medium to form yellow piazselenol, which is sparingly soluble in water and which is utilized for the spectrophotometric determination of selenium [30,31].

$$H_2N-C_6H_3(NH_2)-C_6H_3(NH_2)-NH_2 + H_2SeO_3 \rightarrow H_2N-C_6H_3-C_6H_3(N=Se)-N + 3H_2O$$

At suitable reagent ratios, symmetrical dipiazselenol may also be formed [32].

According to Hoste and Gillis [30], the colour reaction should be carried out in $0 \cdot 1 M$ HCl, 50 minutes being allowed for colour development before the absorbance of the aqueous pseudosolution is measured. In Cheng's extractive spectrophotometric method [31], the time for reaction in a formate medium at pH 2–3 is 30 minutes, after which the solution is neutralized to pH 6–7, and the piazselenol is extracted into toluene. The colour reaction may be accelerated by heating the solution, but this also decomposes the reagent [32,33]. Formation of the complex requires a considerable excess of 3,3′-diaminobenzidine.

Within the pH range 5–10, the distribution coefficient of piazselenol between toluene and water is high, and one portion of toluene extracts practically all the selenium complex into the organic phase. The free reagent (DAB) is also extracted. Related solvents such as benzene and xylene may be substituted for toluene.

The two absorption maxima of piazselenol occur at 340 and 420 nm. Since DAB absorbs strongly at 340 nm but negligibly at 420 nm, absorbances are measured at 420 nm. Figure 46.1 shows the absorption spectrum of piazselenol in toluene.

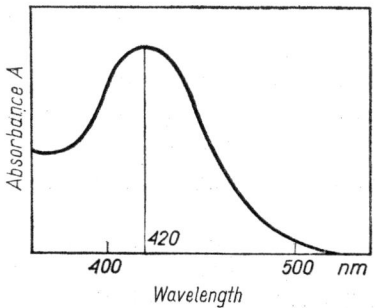

Fig. 46.1 Absorption spectrum of selenium 3,3′-diaminobenzidine complex in toluene (reagent solution as reference)

The molar absorptivity of the toluene solution of piazselenol at 420 nm is $1 \cdot 02 \times 10^4$ (specific absorptivity $0 \cdot 13$).

This method is almost specific for selenium. Tellurium does not react with DAB, but vanadium(V) and iron(III) oxidize DAB, giving coloured oxidation products. Iron(III) can be masked with fluoride or phosphate. EDTA is used as masking agent to prevent hydrolysis of metals in the neutral medium. Substances capable of reducing selenium to the element interfere in the determination of selenium by the 3,3′-diaminobenzidine method.

Selenium has been determined by this method in copper [7,8,34], steel [34,35], lead [8], antimony and bismuth tellurides [36], tellurium and its compounds [14,15], thin films [37], silicates [17], soils [38], sulphide ores [39], sulphuric acid [40], plant materials [41], biological materials [24,33,42,43], organic compounds [44], water [17], and air [45].

Other aromatic *o*-diamines react similarly to DAB with selenium(IV) in acid medium. The *o*-phenylenediamine method [46-50] is more sensitive than the DAB method (λ_{max} = 330 nm). This reagent has been used to determine selenium in steroids [47], sulphuric acid [48], non-ferrous alloys [49], and tellurium oxide [50]. The following are analogous reagents for selenium: 2,3-diaminonaphthalene [51], 4-dimethylamino-1,2-phenylenediamine [52], 4,5-diamino-6-thiopyrimidine [53], and 4-methyl-1,2-phenylenediamine [53a].

Reagents

1. 3,3'-Diaminobenzidine tetrahydrochloride: 0·5% solution in boiled and cooled water. The solution is stable for a few hours, then turns brown under the influence of atmospheric oxygen.
2. Standard selenium solution: 1 mg/ml (p. 476).

Procedure

To the sample solution containing not more than 200 μg of Se in a volume of ∼ 30 ml, add 2 ml of 10% formic acid solution and 5 ml of diaminobenzidine solution, and adjust the pH to 2·0–2·5. Let the solution stand for 30 minutes, then neutralize with ammonia to pH 6–7. Extract the piazselenol by shaking for 1 minute with two portions of toluene (avoid vigorous shaking, to prevent the formation of a stable emulsion). Dilute the extract to the mark in a 50-ml or smaller volumetric flask (depending on the amount of selenium). Mix the solution well and measure its absorbance at 420 nm against water or a reagent blank solution (in the case of small amounts of selenium).

Note. If metal ions are present (Al, Bi, Cu, Ni, etc.), add at the start of the procedure 2–10 ml of 5% EDTA solution as masking agent. To mask Fe(III), add also sodium fluoride (0·05–0·5 g depending on the amount of iron).

46.2.3 OTHER METHODS

Busev *et al.* [54–56] have proposed several organosulphur compounds as reagents for the spectrophotometric determination of selenium, e.g. 2-mercaptobenzimidazole (phenylenethiourea), and *N*-mercaptoacetyl-*p*-anisidine. To the same group of reagents for selenium belong Bismuthiol II (an excellent reagent for tellurium, see p. 524) [57,58], 1,4-diphenylthiosemicarbazide [59,60], 2-mercaptobenzothiazole [61], and thioglycollic acid [62]. Mabuchi and Nakahara [63] have proposed dithizone for the spectrophotometric determination of selenium (molar absorptivity ∼ 7·0 × 10^4). With this sensitive reagent, selenium has been determined in ores [64] and tellurium [65].

Sensitive indirect methods for the determination of selenium take advantage of the oxidation by selenium(IV) of phenylhydrazine-*p*-sulphonic acid [66,67] and 1,1'-diphenylhydrazine [68]. Osburn *et al.* [69] have oxidized NH_2OH with selenium(IV) to HNO_2, which yielded an azo dye

($\varepsilon = 19 \cdot 3 \times 10^4$ at 544 nm) after diazotization and coupling reactions. Selenious acid oxidizes iodide to iodine, which can be determined as the I_3^- ion, as the compound with starch, or by extraction into benzene [70,71].

Langmyhr *et al.* have determined selenium(IV) as the complexes with 1,1'-dianthrimide [72,73] or 2,2'-dianthrimide [74] in conc. H_2SO_4 medium. Elemental selenium forms a green addition compound in hot conc. H_2SO_4 [75].

Other methods for determining selenium are based on the cyclohexanone–chloro complex of selenium(IV) [76], a yellow selenium(IV)–molybdate complex [77], and an indirect method involving the determination of copper in precipitated Cu_2Se with DDTC [78].

References

1. Bock, R. and Jacob, D., *Z. Anal. Chem.* **200**, 81 (1964).
2. Fogg, D. N. and Wilkinson, N. T., *Analyst* **81**, 525 (1956).
3. Handley, R. and Johnson, C. M., *Anal. Chem.* **31**, 2105 (1959).
4. Barcza, L., *Z. Anal. Chem.* **199**, 10 (1964).
5. Dolique, R. and Pérahia, S., *Bull. Soc. Chim. France* **13**, 44 (1946).
6. Kotarski, A. and Marczenko, Z., *Chem. Anal. (Warsaw)* **5**, 235 (1960).
7. Gebauhr, W. and Spang, A., *Z. Anal. Chem.* **175**, 175 (1960).
8. Luke, C. L., *Anal. Chem.* **31**, 572 (1959).
9. Léontovitch, N., *Chim. Anal. (Paris)* **42**, 329 (1960).
10. Ivankova, A. I. and Blyum, I. A., *Zavodsk. Lab.* **27**, 371 (1961).
11. Voronkova, M. A. and Sidorenko, G. A., *Zh. Analit. Khim.* **22**, 1085 (1967).
12. Bode, H., *Z. Anal. Chem.* **153**, 335 (1956).
13. Bode, H. and Stemmer, H. D., *ibid.* **155**, 96 (1957).
14. Veale, C. R., *Analyst* **85**, 130, 133 (1960).
15. Jílek, A., Vřeštal, J. and Havíř, J., *Chem. Zvesti* **10**, 110 (1956).
15a. Murty, A. S., *Indian J. Chem.* **3**, 298 (1965).
16. Sakharov, A. A., *Zh. Analit. Khim.* **15**, 614 (1960).
17. Chau, Y. K. and Riley, J. P., *Anal. Chim. Acta* **33**, 36 (1965).
18. Russell, B. G., Lubbe, W. V., Wilson, A., Jones, E., Taylor, J. D. and Steele, T. W., *Talanta* **14**, 957 (1967).
19. Jackwerth, E., *Z. Anal. Chem.* **235**, 235 (1968).
20. Knyazeva, R. N. and Kleiman, V. Ya., *Zavodsk. Lab.* **31**, 410 (1965).
21. Jordanov, N. and Futekov, L., *Talanta* **12**, 371 (1965); **13**, 163 (1966); **15**, 850 (1968).
22. Inarida, M., *Bull. Chem. Soc. Japan* **39**, 403 (1966).
23. Yadav, A. A. and Khopkar, S. M., *Chem. Anal. (Warsaw)* **16**, 299 (1971).
24. Koval'skii, V. V. and Ermakov, V. V., *Zh. Analit. Khim.* **21**, 447 (1966).
25. Murashova, V. I. and Sushkova, S. G., *ibid.* **24**, 729 (1969).
26. Berton, A., *Chim. Anal. (Paris)* **35**, 91 (1953).
27. Committee on Methods for Analysis of Trade Effluents, *Analyst* **81**, 607 (1956).
28. Gladyshev, V. P., *Zavodsk. Lab.* **24**, 275 (1958).
29. Léontovitch, N., *Chim. Anal. (Paris)* **41**, 56 (1959).
30. Hoste, J. and Gillis, J., *Anal. Chim. Acta* **12**, 158 (1955).
31. Cheng, K. L., *Anal. Chem.* **28**, 1738 (1956).
32. Barcza, L., *Mikrochim. Acta* **1964**, 967.
33. Cummins, L. M., Martin, J. L. and Maag, D. D., *Anal Chem.* **37**, 430 (1965).
34. Cheng, K. L., *Chemist-Analyst* **45**, 67 (1956).
35. Nivière, P., *Chim. Anal. (Paris)* **47**, 125 (1965).
36. Cheng, K. L. and Goydish, B. L., *Anal. Chem.* **35**, 1965 (1963).

37. Marczenko, Z., Mojski, M. and Czarnecka, I., *Chem. Anal. (Warsaw)* **18**, 189 (1973).
38. Stanton, R. E. and McDonald, A. J., *Analyst* **90**, 497 (1965).
39. Barcza, L. and Zsindely, S., *Z. Anal. Chem.* **199**, 117 (1964).
40. Danzuka, T. and Ueno, K., *Anal. Chem.* **30**, 1370 (1958).
41. Służewska, L., *Roczniki Państw. Zakł. Hig.* **15**, 303 (1964).
42. Cummins, L. M., Martin, J. L., Maag, G. W. and Maag, D. D., *Anal. Chem.* **36**, 382 (1964).
43. Kelleher, W. J., and Johnson, M. J., *ibid.* **33**, 1429 (1961).
44. Domenech, R., *Chim. Anal. (Paris)* **51**, 440 (1969).
45. West, P. W. and Cimerman, Ch., *Anal. Chem.* **36**, 2013 (1964).
46. Ariyoshi, H., Kiniwa, M. and Toei, K., *Talanta* **5**, 112 (1960).
47. Throop, L. J., *Anal. Chem.* **32**, 1807 (1960).
48. Toei, K. and Ito, K., *Talanta* **12**, 773 (1965).
49. Shkrobot, E. P. and Shebarshina, N. I., *Zavodsk. Lab.* **35**, 417 (1969).
50. Goszczyńska, H. and Kowalczyk, M., *Chem. Anal. (Warsaw)* **15**, 561 (1970).
51. Lott, P. F., Cukor, P., Moriber, G. and Solga, J., *Anal. Chem.* **35**, 1159 (1963).
52. Demeyere, D. and Hoste, J., *Anal. Chim. Acta* **27**, 288 (1962).
53. Chan, F. L., *Talanta* **11**, 1019 (1964).
53a. Kawashima, T. and Ueno, A., *Anal. Chim. Acta* **58**, 219 (1972).
54. Busev, A. I. and Khoang Min' Tyau, *Zh. Analit. Khim.* **17**, 1091 (1962); **18**, 360, 1370 (1963).
55. Busev, A. I., Babenko, N. L. and Khoang Min' Tyau, *ibid.* **18**, 1094 (1963).
56. Busev, A. I., *Talanta* **11**, 485 (1964).
57. Stantscheff, P., *Z. Anal. Chem.* **220**, 33 (1966).
58. Navrátil, O. and Šorfa, J., *Collection Czech. Chem. Commun.* **34**, 975 (1969).
59. Sushkova, S. G. and Murashova, V. I., *Zh. Analit. Khim.* **21**, 1475 (1966).
60. Murashova, V. I., Sushkova, S. G. and Bakunina, L. I., *Zavodsk. Lab.* **33**, 280 (1967).
61. Bera, B. C. and Chakrabartty, M. M., *Analyst* **93**, 50 (1968).
62. Kirkbright, G. F. and Ng, W. K., *Anal. Chim. Acta* **35**, 116 (1966).
63. Mabuchi, H. and Nakahara, H., *Bull. Chem. Soc. Japan* **36**, 151 (1963).
64. Shcherbov, D. P., Ivankova, A. I. and Gladysheva, G. P., *Zavodsk. Lab.* **33**, 683 (1967).
65. Kasterka, B. and Dobrowolski, J., *Chem. Anal. (Warsaw)* **15**, 303 (1970).
66. Kirkbright, G. F. and Yoe, J. H., *Anal. Chem.* **35**, 808 (1963).
67. Rigin, V. I., Mel'nichenko, N. N. and Yanitskii, V. K., *Zavodsk. Lab.* **33**, 1370 (1967).
68. Murashova, V. I. and Sushkova, S. G., *Zh. Analit. Khim.* **19**, 1503 (1964).
69. Osburn, R. L., Shendrikar, A. D. and West, P. W., *Anal. Chem.* **43**, 594 (1971).
70. Lambert, J. L., Arthur, P. and Moore, T. E., *ibid.* **23**, 1101 (1951).
71. Tumanov, A. A. and Shakhverdi, N. M., *Zavodsk. Lab.* **37**, 147 (1971).
72. Langmyhr, F. J. and Omang, S. H., *Anal. Chim. Acta* **23**, 565 (1960).
73. Langmyhr, F. J. and Myhrstad, J. A., *ibid.* **35**, 212 (1966).
74. Langmyhr, F. J. and Dahl, I., *ibid.* **29**, 377 (1963); **35**, 24 (1966).
75. Wiberley, S. E., Bassett, L. G., Burrill, A. M. and Lyng, H., *Anal. Chem.* **25**, 1586 (1953).
76. Cresser, M. S. and West, T. S., *Talanta* **16**, 416 (1969).
77. Shakhova, Z. F., Gavrilova, S. A. and Zakharova, V. F., *Vestn. Mosk. Univ. Khim.* **1966**, No. 6, 64.
78. Bode, H. and Mösenthin, H., *Z. Anal. Chem.* **164**, 232 (1958).

Chapter 47
SILICON

Silicon (Si, at.wt. 28.09) occurs in its compounds in the +IV oxidation state, e.g. in silica (SiO_2) or in silicic acids. The formation of complexes (heteropoly acids) with molybdic acid and of volatile SiF_4 is important analytically.

47.1 Methods of Separation

47.1.1 Distillation

Volatile silicon tetrafluoride, SiF_4, distils from a hot $HClO_4$ or H_2SO_4 medium containing excess of HF in a closed apparatus made of platinum, silver, or Teflon [1–3]. The expelled SiF_4 is absorbed in NaOH or H_3BO_3 solution. As explained in the chapter on fluorine (p. 254) the silicon tetrafluoride is distilled only under certain circumstances.

Separation of SiF_4 by the Conway microdiffusion method is successful, but slow [4,5]. Housenholder and Russell [5] have recommended decomposing samples (e.g. copper, aluminium, iron alloys, dolomite, or titanium dioxide) with acids in a polystyrene Petri dish to which hydrofluoric acid is subsequently added. The SiF_4 evolved is trapped in a film of dried NaOH on the lid of the Petri dish. The closed Petri dish is maintained at 70°C for 18 hours to achieve quantitative separation of silicon. Szabó et al. [5a] have designed a distillation outfit for preconcentration of silicon traces.

47.1.2 Extraction

Molybdosilicic acid can be selectively extracted from a strongly acidic medium (e.g. $3M$ H_2SO_4) with oxygenated organic solvents [6,7], silicon being determined directly as yellow molybdosilicic heteropoly acid or as silicomolybdenum blue. The molybdosilicic acid must be formed at lower acidity ($[H^+] < 0.7M$) and the acidity raised just before extraction. Molybdosilicic acid will not form if the initial acidity is too high [7a,7b]. In the presence of high molecular-weight amines, molybdosilicic acid or its reduced form (heteropoly blue) may be extracted with toluene, chloroform, or a mixture of $CHCl_3$ and isoamyl alcohol [8–10].

Silicon can be quantitatively extracted as hexafluorosilicic acid with trioctylamine in xylene [11].

47.1.3 Ion-Exchange

Anion- and cation-exchangers are applicable both for separating small amounts of silicon and for removing interfering anions and cations from silicon [12–17]. Wickbold [12] has enriched traces of silicon by retaining SiF_6^{2-} on a strongly basic anion-exchanger. The silicon is then displaced from the fluoride complex (and the column) by elution with boric acid solution. Silicon has been separated from phosphate, arsenate, and other anions [13], and from tungsten and silver [14], on strongly basic anion-exchangers [13,14]. Weakly basic anion-exchangers have been used to separate silicon from interfering anions as soluble silicic acid [13,15].

47.1.4 Precipitation and Other Methods

Marczenko and Kasiura [18] have coprecipitated microgram amounts of silica with niobium, which precipitates as niobic acid. This method enables silicon to be quantitatively separated from major quantities of phosphorus(V), arsenic(V), iron(III), and aluminium.

Milligram amounts of silica are quantitatively precipitated without any collector by evaporation with perchloric acid to fumes.

When samples are fused with alkalis (Na_2CO_3, NaOH) and the melt is leached with water, silicon passes into solution as sodium silicates, while many elements (e.g. Fe, Ti, Cu, Ni, and Zr) remain in the solid residue, which retains only negligible amounts of silicate.

Silica can be separated from elementary silicon by selective dissolution of the former in 40% HBF_4 at $< 70°C$ [19]. Traces of silicon have been coprecipitated from $8M$ HNO_3 with ammonium molybdophosphate [20].

47.2 Methods of Determination

Silicon is determined spectrophotometrically almost exclusively as yellow molybdosilicic heteropoly acid (less sensitive method) or, after reduction, as silicomolybdenum blue (sensitive method).

47.2.1 Molybdosilicic Acid Method

In alkaline solutions, silica exists in the form of silicate ions (e.g. SiO_3^{2-}). In dilute solutions (up to 0·1 mg of Si/ml) between pH 1 and 8, water-soluble monomeric silicic acid is the stable form. In more concentrated solutions of the same acidity, monosilicic acid condenses to disilicic acid and polysilicic acids which can be transformed into colloidal species.

Soluble monosilicic acid reacts with molybdic acid in a medium at pH 1–2, ratio of $[H^+]:[MoO_4^{2-}] > 3$, at least $0·05M$ excess of molybdenum, and ionic strength $< 0·5$, to form the yellow soluble β-molybdosilicic acid complex, $H_4(SiMo_{12}O_{40})$ [21]. The yellow colour is the basis of a rather insensitive spectrophotometric method for silicon [21–25a]. The β-form spontaneously changes into the α-form, slowly at room temperature, faster on heating [21]. The presence of trichloroacetic acid [26], acetone [23], or sulpholane [26a] enhances the colour of molybdosilicic acid. The latter two reagents also stabilize the β-heteropoly acid [23,26a]. The absorption maximum of the complex is in the ultraviolet, at ~ 300 nm. At 400 nm, the molar absorptivity is $2 \cdot 2 \times 10^3$ ($a = 0 \cdot 08$) (in presence of acetone).

At lower acidity and lower $[H^+]:[MoO_4^{2-}]$ ratios ($< 1 \cdot 5$), the reaction product is α-molybdosilicic acid, which is more stable than the β-form but which absorbs only half as intensely at 400 nm. The α- and β-forms have the same molar absorptivity at 329 nm [27]. A third modification of molybdosilicic acid (γ-form) has also been shown to exist [28]. Silicon is often determined from the colour of the α-form of molybdosilicic acid [29-32]. Besides the acidity, the temperature, the Si:Mo concentration ratio, and the degree of condensation of the molybdate ions determine the predominant heteropoly acid modification [33-35]. In a medium of pH > 7, MoO_4^{2-} ions are stable but undergo condensation when acidified, the degree of condensation being a function of the $[H^+]:[MoO_4^{2-}]$ ratio.

Precise results are obtained in this method only if the conditions are rigorously kept the same for both samples and standards.

Silica is converted into monosilicic acid by fusing with sodium carbonate or hydroxide and acidifying the alkaline solution produced when the melt is dissolved in water. Alternatively, an acidic sample solution may be made alkaline with sodium hydroxide and heated to convert colloidal silica into silicate. Soluble monosilicic acid is formed after appropriate dilution and acidification (see above).

On heating with dilute hydrofluoric acid silica is transformed into soluble hexafluorosilicic acid, H_2SiF_6. Aluminium chloride or boric acid introduced subsequently masks the excess of HF and decomposes H_2SiF_6 to monosilicic acid (the more stable AlF_6^{3-} or BF_4^- complexes being produced).

Phosphorus(V) and germanium(IV), which give yellow heteropoly acids, interfere (molybdoarsenic acid is practically colourless). Before silicon is determined, germanium and arsenic may be separated by volatilization or extraction of $GeCl_4$ and $AsCl_3$ (see p. 274 and p. 131). Molybdophosphoric and molybdoarsenic acids are separated from molybdosilicic acid by extraction with isobutyl acetate, optimally at pH 0·3-1·0. Ethyl acetate [36] and butanol [37] have been used similarly.

Molybdophosphoric acid is rapidly decomposed by mannitol, tartrate, citrate, or oxalate, whereas the corresponding decomposition of molybdosilicic acid is much slower. Since molybdosilicic acid forms only slowly in the presence of these carboxylic acids, phosphorus(V) is masked by adding tartrate after molybdosilicic acid has been formed [23,38].

Ferric ions also interfere in the determination of silicon. Large quantities are separated beforehand by extraction and smaller ones are dealt with by using a higher pH than usual in the α-molybdosilicic acid method [29].

The molybdosilicic acid method has been used to determine silicon in iron and steel [33,39,40], aluminium [41], various other metals [42], rocks and refractory materials [23,31,31a], anhydrite [9], bauxite [43], semiconductors [44], silicates [45], hydrofluoric acid and fluorides [46,47], organic compounds [48], and water [7].

Reagents

1. Ammonium molybdate: 8% solution.
2. Standard silicon solution: 0·1 mg/ml. Fuse 0·2140 g of ignited and comminuted silica, SiO_2, with 2 g of sodium carbonate in a platinum crucible. Dissolve the melt in water, dilute the solution with water to $\sim$ 900 ml, acidify with $1M$ H_2SO_4 to pH $\sim$ 1·5, and make up the solution to 1 litre in a volumetric flask with water. The solution obtained is stable.
   Store the reagents in polyethylene bottles.

Procedure

In a 50-ml volumetric flask place 10 ml of a 1:1 mixture of ammonium molybdate solution and $1M$ sulphuric acid, and add 5 ml of acetone and an aliquot of sample solution (e.g. neutralized solution from sodium carbonate fusion) containing not more than 0·5 mg of Si. Dilute to the mark with water, mix well, let stand for 15 minutes, and measure the absorbance at 400 nm against a reagent blank.

47.2.2 Silicomolybdenum Blue Method

Molybdosilicic heteropoly acid, formed by treating monosilicic acid with molybdic acid, reacts with suitable reducing agents to yield intensely coloured silicomolybdenum blue, upon which a sensitive method for determining silicon is based [24,49–53]. The reaction conditions are adjusted so that only molybdosilicic acid, and not unreacted molybdic acid, is reduced.

Stannous chloride or oxalate [51,52,54], iron(II) (Mohr's salt) [55,56], ascorbic acid [56], sodium sulphite [57], metol [58], and various other reagents have been used as reductants. Photochemical reduction of molybdosilicic acid has also been reported [59]. To prevent partial reduction of molybdic acid, molybdosilicic acid is reduced in sufficiently acidic medium. Molybdosilicic acid is produced in a slightly acidic medium but, once formed, it is relatively slowly decomposed if the acidity is strongly increased (up to $1\cdot 5M$ H_2SO_4). The most suitable acidity for the reduction depends on the reducing agent used. The various forms of molybdosilicic acid (α, β) and the various reductants yield products which differ in absorption spectra, absorption maxima, and stability [60,61].

Molybdosilicic acid can also be reduced after extraction into an oxygenated organic solvent (e.g. isoamyl alcohol) (see procedure below) [62,63].

Alternatively, silicomolybdenum blue may be formed in the aqueous phase and then extracted [64]. The heteropoly blue exhibits similar molar absorptivities in both the organic phase and in the aqueous solution, but the absorption maximum is shifted slightly towards shorter wavelengths in organic solvents.

Silicomolybdenum blue produced by extraction of molybdosilicic acid into isoamyl alcohol and reduction with $SnCl_2$ exhibits a broad absorption maximum (cf. Curve 3 in Fig. 7.1, p. 133) at 750 nm (the molar absorptivity is $1 \cdot 7 \times 10^4$; specific absorptivity 0·60).

Phosphorus(V), germanium, and arsenic(V), which form corresponding heteropoly acids and heteropoly blues, must be separated or masked (see above, p. 483) before the determination of silicon. It is possible to separate silicomolybdenum and phosphomolybdenum blues by extraction [57].

Before determining silicon, Wilson [65] separated Group III metals by double precipitation as benzoates at pH 3·5. Iron does not interfere in the silicomolybdenum blue method if it is reduced to Fe(II) [58].

The silicon traces present in the reagents and water used interfere in the determination of microgram amounts of silicon. "Analytically pure" hydrochloric, sulphuric, and hydrofluoric acids, and distilled water contain $2 \times 10^{-5}\%$, $7 \times 10^{-5}\%$, $4 \times 10^{-2}\%$, and $2 \times 10^{-6}\%$ of Si, respectively [18]. These reagents may be considerably purified by distillation in quartz (but not the HF!) or platinum apparatus. Platinum, Teflon, and polyethylene vessels should be used and the silicon in a reagent blank solution should be taken into account when traces of silicon are determined.

Reduced molybdosilicic acid may be extracted with chloroform in the presence of amines [8], or propylene carbonate [66]. Also tri-n-octylamine solution in toluene has been used as extractant [67].

Silicon has been determined by the silicomolybdenum blue method in iron and steel [55,56,68–70], nickel and its alloys [71–74], platinum and gold [75], copper and its alloys [5,76–78], alumina [79], aluminium [5,74,80], uranium and its compounds [2,73,81–83], titanium and its alloys [84,85], molybdenum and its compounds [86], chromium [87], zirconium [74], refractory metals [3], semiconductor metals [64], boron [88], tellurium [89], gallium phosphide [15], phosphate minerals [90], phosphoric acid [2,5], various chemical reagents [18,91], alkalis [92,93], organic compounds [94,95], biological materials [96,97], plants [98], and water [8,16,58,60,99–102].

Jeffery and Wilson [103] have determined spectrophotometrically the fraction of silica which remains in solution during the gravimetric determination of silicon in minerals.

The silicomolybdenum blue method is suitable for the automatic determination of silicon in steel [104,105] and in water [106,107], as well as for the determination of silicon by differential spectrophotometry [108].

Reagents

1. Ammonium molybdate: 10% solution adjusted with ammonia to pH 7·4 ($\pm 0·2$).
2. Standard silicon solution: 0·1 mg/ml (p. 484).
3. Stannous chloride, $SnCl_2 \cdot 2H_2O$: 50% solution in HCl (1+1).
4. Niobium solution: ~1 mg of Nb in 1 ml. Heat 0·145 g of Nb_2O_5 in a platinum crucible with 5 ml of conc. HF until the oxide dissolves. Evaporate the solution to ~1 ml, add 2 ml of H_2SO_4 (1+1), and heat to white fumes. Let cool, rinse the walls of the crucible with water and heat to fumes again. Repeat the operation once more to remove hydrogen fluoride completely. Pour the niobium solution in conc. H_2SO_4 into 30 ml of 5% aqueous ammonium oxalate solution, and dilute the clear solution of niobium oxalate complex to ~100 ml with water.

Store the reagent solutions in polyethylene bottles.

Procedure

Separation of Si with a niobium collector. To the sample solution, containing not more than 50 μg of Si, in a Teflon or platinum vessel, add 2 ml of the niobium solution and 5 ml of conc. $HClO_4$ and carefully evaporate the solution to fumes, expelling most of the perchloric acid. Dilute the residue with 10–20 ml of $HClO_4$ (1+50), stir till the salts dissolve, add some macerated filter paper, and filter off the precipitate and wash it with very dilute perchloric acid. Ignite the filter paper and precipitate in a platinum crucible. Add 1 ml of 5% HF to the cooled crucible and heat in a sealed vessel in a water-bath at ~70°C for 30 minutes. Quantitatively transfer the solution from the crucible to a polyethylene beaker, dilute with water to ~10 ml, and add 5 ml of 3% boric acid solution.

Determination of Si. To the solution obtained as described above, add 1 ml of the molybdate solution and adjust the pH to 1·4 ($\pm 0·1$) with $0·5M\ H_2SO_4$. After 5 minutes, transfer the solution to a separating funnel, add 10 ml of H_2SO_4 (1+4), and extract molybdosilicic acid with two 15-ml portions of isoamyl alcohol. Wash the combined extracts by shaking with 10 ml of $0·5M\ H_2SO_4$ for 15 seconds. Transfer the organic phase to a 50-ml volumetric flask, and add two drops of the $SnCl_2$ solution, ~2 ml of diethyl ether (to clarify the solution), and isoamyl alcohol to the mark. After 5 minutes, measure the absorbance of the blue solution at 750 nm, using a reagent blank as reference.

47.2.3 OTHER METHODS

Sensitive methods for determining silicon are based on the formation of molybdosilicate ion-association complexes with certain basic dyes [109–113]. The complex with Crystal Violet can be extracted with a mixture of cyclohexanol and isoamyl alcohol [109]. This complex is also soluble in acetone (molar absorptivity $1·4 \times 10^5$) [112]. The complex with Rhodamine B can be separated by flotation with di-isopropyl ether and dissolved in ethanol [110, 111]. A basic antipyrine dye, Chrompyrazole, has also been recommended [113].

There exist several indirect methods for the spectrophotometric determination of silicon. After extraction as molybdosilicic acid, the molybdenum has been determined with phenylfluorone [114] or with 2-amino-4-chlorobenzenethiol [115]. When silicic acid is added to a solution which contains hexafluorotitanic acid and hydrogen peroxide, a yellow titanium peroxide complex is formed [116].

Lastly, there are methods for the determination of silicon which are based on the mixed molybdovanadosilicic [117] and molybdostannosilicic acids [118].

References

1. Ehrlich, P. and Keil, T., *Z. Anal. Chem.* **166**, 254 (1959).
2. Holt, B. D., *Anal. Chem.* **32**, 124 (1960).
3. Stobart, J. A., *Analyst* **94**, 1142 (1969).
4. Alon, A., Bernas, B. and Frenkel, M., *Anal. Chim. Acta* **31**, 279 (1964).
5. Housenholder, R. and Russell, R. G., *Anal. Chem.* **36**, 2279 (1964).
5a. Szabó, Z. G., Zapp, E.É. and Perczel, S., *Microchim. Acta* **1974**, 167.
6. Ruf, E., *Z. Anal. Chem.* **151**, 169 (1956); **161**, 1 (1958).
7. Schink, D. R., *Anal. Chem.* **37**, 764 (1965).
7a. Rockstein, M. and Herron, P. W., *ibid.* **23**, 1500 (1951).
7b. Sinclair, A. G., *Ph.D. Thesis*, University of Aberdeen (1960).
8. Sonnenschein, W., *Z. Anal. Chem.* **168**, 18 (1959).
9. Wilson, H. N. and Skinner, J. M., *Rec. Trav. Chim.* **79**, 574 (1960).
10. Sudakov, F. P., Klitina, V. I. and Maslova, N. T., *Vestn. Mosk. Univ. Khim.* **1966**, No. 1, 98; *Zh. Analit. Khim.* **21**, 1089 (1966).
11. Pal'shin, E. S., Palei, P. N., Davydov, A. V. and Ivanova, L. A., *Zh. Analit. Khim.* **24**, 797 (1969).
12. Wickbold, R., *Z. Anal. Chem.* **171**, 81 (1959).
13. Andersson, L. H., *Arkiv Kemi* **19**, 243 (1963).
14. Toy, C. H. and Van Santen, R. T., *Anal. Chem.* **36**, 151 (1964).
15. Luke, C. L., *ibid.* **36**, 2036 (1964).
16. Duce, F. A. and Yamamura, S. S., *Talanta* **17**, 143 (1970).
17. Nemodruk, A. A., Palei, P. N. and Bezrogova, E. V., *Zh. Analit. Khim.* **25**, 319 (1970).
18. Marczenko, Z. and Kasiura, K., *Chem. Anal. (Warsaw)* **9**, 321 (1964).
19. Houda, M. and Jošt, F., *Collection Czech. Chem. Commun.* **31**, 776 (1966).
20. Barkovskii, V. F., Radovskaya, T. L. and Zaporozhets, A. S., *Zh. Analit. Khim.* **23**, 1853 (1968).
21. Strickland, J. D., *J. Am. Chem. Soc.* **74**, 862, 868, 872 (1952).
22. Lacroix, S. and Labalade, M., *Anal. Chim. Acta* **3**, 383 (1949).
23. Chalmers, R. A. and Sinclair, A. G., *ibid.* **33**, 384 (1965); **34**, 412 (1966).
24. Kratochvil, V., *Chem. Listy* **59**, 672 (1965); **60**, 1238 (1966).
25. Halász, A. and Pungor, E., *Talanta* **18**, 557, 569 (1971).
25a. Halász, A., Pungor, E. and Polyák, K., *ibid.* **18**, 577 (1971).
26. Dellamonica, E. S., Bingham, E. W. and Zittle, C. A., *Anal. Chem.* **30**, 1986 (1958).
26a. Flaschka, H. and Tice, J. J., IV, *Talanta* **20**, 423 (1973).
27. Garrett, H. E. and Walker, A. J., *Analyst* **89**, 642 (1964).
28. Kemula, W. and Rosołowski, S., *Roczniki Chem.* **34**, 3 (1960); *Chem. Anal. (Warsaw)* **5**, 419 (1960).

29. Andersson, L. H., *Acta Chem. Scand.* **12**, 495 (1958); **14**, 1571 (1960).
30. Ringbom, A., Ahlers, P. E. and Siitonen, S., *Anal. Chim. Acta* **20**, 78 (1959).
31. Bloxam, T. W., *Analyst* **86**, 420 (1961).
31a. Privalova, M. M., Makhova, G. P. and Tulina, M. D., *Zh. Analit. Khim.* **29**, 279 (1974).
32. Andersson, L. H., *Arkiv Kemi* **19**, 257 (1963).
33. Jean, M., *Chim. Anal. (Paris)* **38**, 37 (1956).
34. Kemula, W., Rosołowski, S. and Wolfram, W., *Chem. Anal. (Warsaw)* **3**, 593 (1958).
35. Govett, G. J., *Anal. Chim. Acta* **25**, 69 (1961).
36. Huré, J. and Ortis, T., *Bull. Soc. Chim. France* **1949**, 834.
37. Andersson, L. H., *Acta Chem. Scand.* **13**, 1743 (1959).
38. DeSesa, M. A. and Rogers, L. B., *Anal. Chem.* **26**, 1278 (1954).
39. Meyer, S. and Koch, O. G., *Mikrochim. Acta* **1961**, 82; **1963**, 929.
40. Macher, F. and Glász, M., *ibid.* **1964**, 104.
41. Armand, M. and Berthoux, J., *Anal. Chim. Acta* **5**, 380 (1951).
42. Bril, J., *Mikrochim. Acta* **1966**, 1047.
43. Fresenius, W. and Schneider, W., *Z. Anal. Chem.* **214**, 341 (1965).
44. Revenko, V. G., Bagreev, V. V., Zolotov, Yu. A., Kopanskaya, L. S. and Pal'shin, E. S., *Zh. Analit. Khim.* **26**, 2235 (1971).
45. Langer, K., *Z. Anal. Chem.* **245**, 139 (1969).
46. Graff, P. R. and Langmyhr, F. J., *Anal. Chim. Acta* **21**, 429 (1959).
47. Augustyn, W. and Sosin, Z., *Chem. Anal. (Warsaw)* **2**, 305 (1957).
48. Debal, E., *Talanta* **19**, 15 (1972).
49. Čelechovský, J., *Chem. Listy* **48**, 391 (1954).
50. Blasius, E. and Czekay, A., *Z. Anal. Chem.* **147**, 1 (1955).
51. Eckert, G., *ibid.* **161**, 421 (1958).
52. Shakhova, Z. F., Dorokhova, E. N. and Chuyan, N. K., *Zh. Analit. Khim.* **21**, 707 (1966).
53. Hargis, L. G., *Anal. Chem.* **42**, 1494, 1497 (1970).
54. Babko, A. K. and Evtushenko, L. M., *Zavodsk. Lab.* **23**, 423 (1957).
55. Andrew, T. R. and Gentry, C. H., *Analyst* **81**, 339 (1956).
56. Keller, H. and Sauer, K. H., *Arch. Eisenhüttenw.* **36**, 533 (1965).
57. Paul, J. and Pover, W. F., *Anal. Chim. Acta* **22**, 185 (1960); **23**, 178 (1960).
58. Mullin, J. B. and Riley, J. P., *ibid.* **12**, 162 (1955).
59. Nemodruk, A. A. and Bezrogova, E. V., *Zh. Analit. Khim.* **24**, 1704 (1969).
60. Morrison, I. R. and Wilson, A. L., *Analyst* **88**, 88, 100 (1963).
61. Andersson, L. H., *Arkiv Kemi* **19**, 223 (1963).
62. Pakalns, P. and Flynn, W. W., *Anal. Chim. Acta* **38**, 403 (1967).
63. Kakita, Y. and Goto, H., *Talanta* **14**, 543 (1967).
64. Nazarenko, V. A. and Flyantikova, G. V., *Zavodsk. Lab.* **24**, 663 (1958).
65. Wilson, A. D., *Analyst* **88**, 18 (1963).
66. Trudell, L. A. and Boltz, D. F., *Anal. Chim. Acta* **52**, 343 (1970).
67. Dorokhova, E. N., Zhukova, L. B., Tereshchenko, A. P. and Krasnoshchekov, V. V., *Vestn. Mosk. Univ. Khim.* **1973**, 604.
68. Sanders, W. F. and Cramer, C. H., *Anal. Chem.* **29**, 1139 (1957).
69. Meyer, S. and Koch, O. G., *Mikrochim. Acta* **1961**, 134.
70. Sauer, K. H. and Keller, H., *Arch. Eisenhüttenw.* **41**, 961 (1970).
71. Gann, W., *Z. Anal. Chem.* **150**, 254 (1956).
72. Eckert, G., *ibid.* **162**, 408 (1958).
73. Harrison, F. H., *Metallurgia* **66**, 300 (1962).
74. Pakalns, P., *Anal. Chim. Acta* **54**, 281 (1971).
75. Marczenko, Z. and Lenarczyk, Ł., *Chem. Anal. (Warsaw)* **19**, 679 (1974).
76. Luke, C. L., *Anal. Chem.* **25**, 148 (1953).
77. Sturton, J. M., *Anal. Chim. Acta* **32**, 394 (1965).

References

78. Pakalns, P., *ibid.* **40**, 328 (1968).
79. Szabó, Z. G., Zapp, E. É. and Perczel, S., *Mikrochim. Acta* **1974**, 167.
80. Gołkowska, A., *Chem. Anal. (Warsaw)* **10**, 749 (1965).
81. Nowicka-Jankowska, T., Gołkowska, A., Pietrzak, I. and Żmijewska, W., *ibid.* **3**, 983 (1958).
82. Boirie, C. and Platzer, R., *Acta Chim. Acad. Sci. Hung.* **33**, 267 (1962).
83. Rajković, D., *Z. Anal. Chem.* **255**, 190 (1971).
84. Codell, M., Clemency, C. and Norwitz, G., *Anal. Chem.* **25**, 1432 (1953).
85. Barkovskii, V. F. and Radovskaya, T. L., *Zavodsk. Lab.* **35**, 160 (1969).
86. Fukker, K. and Hegedüs, A. J., *Mikrochim. Acta* **1961**, 227.
87. Minczewski, J. and Chwastowska, J., *Chem. Anal. (Warsaw)* **6**, 715 (1961).
88. Marczenko, Z., *ibid.* **9**, 1093 (1964).
89. Dobrowolski, J. and Kwiatkowska-Sienkiewicz, K., *ibid.* **15**, 647 (1970).
90. Greenfield, S., *Analyst* **84**, 380 (1959).
91. Kemula, W. and Wolfram, W., *Chem. Anal. (Warsaw)* **3**, 897 (1958).
92. Kenyon, O. A. and Bewick, H. A., *Anal. Chem.* **25**, 145 (1953).
93. Marczenko, Z., *Mikrochim. Acta* **1965**, 281.
94. Jenik, J. and Jureček, M., *Collection Czech. Chem. Commun.* **26**, 967 (1961).
95. Christopher, A. J., Fennel, T. R. and Webb, J. R., *Talanta* **11**, 1323 (1964).
96. King, E. J., Stacy, B. D., Holt, P. F., Yates, D. M. and Pickles, D., *Analyst* **80**, 441 (1955).
97. Tůma, J., *Mikrochim. Acta* **1962**, 513.
98. Volk, R. J. and Weintraub, R. L., *Anal. Chem.* **30**, 1011 (1958).
99. Webber, H. M. and Wilson, A. L., *Analyst* **89**, 632 (1964); **94**, 110 (1969).
100. Baker, P. M. and Farrant, B. R., *ibid.* **93**, 732 (1968).
101. Morrison, I. R. and Wilson, A. L., *ibid.* **94**, 54 (1969).
102. Fanning, K. A. and Pilson, M. E., *Anal. Chem.* **45**, 136 (1973).
103. Jeffery, P. G. and Wilson, A. D., *ibid.* **85**, 478 (1960).
104. Scholes, P. H. and Thulbourne, C., *ibid.* **89**, 466 (1964).
105. Scholes, P. H., *Z. Anal. Chem.* **222**, 162 (1966).
106. Wilson, A. L., *Analyst* **90**, 270 (1965).
107. Brewer, P. G. and Riley, J. P., *Anal. Chim. Acta* **35**, 514 (1966).
108. Tikhonov, V. N. and Chernysheva, A. N., *Zh. Analit. Khim.* **20**, 487 (1965).
109. Babko, A. K., Shkaravskii, Yu. F. and Gołkowska, A., *Chem. Anal. (Warsaw)* **11**, 1091 (1966).
110. Gołkowska, A., *ibid.* **14**, 803 (1969); **15**, 59 (1970).
111. Gołkowska, A. and Pszonicki, L., *Talanta* **20**, 749 (1973).
112. Babko, A. K. and Ivashkovich, E. M., *Zh. Analit. Khim.* **27**, 120 (1972).
113. Zhivopistsev, V. P. and Bondareva, E. G., *Zavodsk. Lab.* **37**, 1409 (1971).
114. Halász, A., Polyák, K. and Pungor, E., *Talanta* **18**, 691 (1971).
115. Trudell, L. A. and Boltz, D. F., *ibid.* **19**, 37 (1972).
116. Fukamauchi, H., *Z. Anal. Chem.* **229**, 413 (1967).
117. Lew, R. B. and Oyung, W., *Anal. Chem.* **36**, 1857 (1964).
118. Pilipenko, A. T., Nabivanets, B. I. and Shainskaya, L. G., *Zavodsk. Lab.* **37**, 664 (1971).

Chapter 48

SILVER

Silver (Ag, at.wt. 107·87) occurs in its compounds in the +I oxidation state. So far, silver(II) is only of limited value in spectrophotometry. Silver(I) sulphide and halides are sparingly soluble. Ammine, cyanide, and thiosulphate complexes of silver are formed. In the presence of excess of Cl⁻ or SCN⁻, traces of silver form soluble complexes.

48.1 Methods of Separation

48.1.1 Extraction

Silver is often separated from other metals, in particular from mercury and copper, by extraction with dithizone. The details of this separation are discussed below in the dithizone method for determining silver.

Other important methods for separating silver involve extraction of the diethyldithiocarbamate into chloroform (pH 4–11, EDTA as masking agent) [1], extraction into MIBK of the silver ion-association complex with n-butylamine and salicylic acid [2], extraction into methylene chloride of the silver complex with saccharin and tri-n-butylamine [3], and extraction of silver with 40% TBP in butanol from an aqueous medium of $1M$ $HNO_3 + 2M$ $LiNO_3$ [4].

48.1.2 Precipitation

Sandell and Neumayer [5] recommend separating traces of silver by coprecipitation with tellurium. Tellurium tetrachloride and $SnCl_2$ as a reductant are added to a solution of the sample in $2M$ HCl. Good separation of silver is also obtained by precipitation of the sulphide with mercury, copper, or lead as scavenger [6,7].

Zagórski and Kempiński [8] have separated silver from aqueous media at pH 4–5 by reduction to the element (cementation) with zinc dust and as the sparingly soluble silver salt of *p*-dimethylaminobenzylidenerhodanine, the excess of reagent serving as the collector. Mizuike and Fukuda [9,10] have also coprecipitated silver on rhodanine and on dithizone.

When a silver solution containing iron(II) sulphate is adjusted to pH ~ 8 with ammonia, the silver is quantitatively coprecipitated with the ferrous hydroxide [11].

Silver can also be precipitated as the bromide or chloride: thallium(I) is a suitable collector [12,13].

48.1.3 Ion-Exchange

Silver ions can be retained on a strongly basic anion-exchange column in the chloride form, and subsequently eluted with ammonia [7,14] or an acetone–nitric acid–water mixture [15].

Silver may be separated from Ce, Zr, Th, Bi, and Fe(III) on Dowex-50 cation-exchanger by converting these metals into anionic complexes, or separated from Cu, U(VI), Al, and Zn by selective elution with nitric acid [16]. After retention of lead, silver, and mercury(II) on Dowex 50, first lead is eluted with $0.25M$ CH_3COONH_4 and then silver with $0.5M$ CH_3COONH_4 [17].

Vydra [18] has selectively retained silver [as $Ag(NH_3)_2^+$] from solutions at pH $>$ 8.5 (in the presence of EDTA) on a silica column.

48.2 Methods of Determination

Of all the spectrophotometric methods for determining silver, the following two are particularly important: the extractive dithizone method, and the non-extractive rhodanine method. Both are of similar selectivity, but the dithizone method is more sensitive and more accurate. Lisitsyna and Shcherbov [18a] have reviewed photometric methods for determining silver.

48.2.1 Dithizone Method

Dithizone (formula, p. 41) reacts with silver ions in acid medium (H_2SO_4, HNO_3, $HClO_4$) to form the orange-yellow dithizonate, AgHDz, which is soluble in carbon tetrachloride and other inert solvents. A solution of dithizone in carbon tetrachloride extracts silver rapidly even from a $4M$ H_2SO_4 medium.

This dithizonate (AgHDz), upon which is based the spectrophotometric determination of silver [19–22], exhibits a molar absorptivity of 3.05×10^4 ($a = 0.28$), at $\lambda_{max} = 462$ nm (the absorption spectrum of AgHDz is shown in Fig. 2.2 on p. 43)

At higher pH values and in the presence of excess of silver, the purple secondary dithizonate, Ag_2Dz, is formed. This complex is readily converted into yellow AgHDz by adding excess of dithizone and acidifying the aqueous solution. Once formed, the yellow dithizonate is resistant to even 5% NaOH solution.

From a strongly acidic chloride-free medium, noble metals [Au, Pt(II), Pd, Hg] and copper are extracted together with the silver [23]. The presence of more than traces of chloride in the acid solution prevents the formation of silver dithizonate [24]. From a medium comprising $0.12M$ H_2SO_4 + $0.25M$ HNO_3, 25 µg of silver in the presence of 25 µg of Cl^- were quantitatively extracted with dithizone. However, in the presence of 250 µg of Cl^-, only 60% of the silver in the aqueous solution was extracted.

The AgHDz can be decomposed by shaking the carbon tetrachloride extract with 1M HCl, thereby separating the silver from the co-extracted metals [24]. Similarly, an acidified thiocyanate solution strips silver from the extract containing Ag, Hg, and Cu dithizonates [25].

Dithizone extracts silver from solutions containing chloride but at higher pH values (2–5). In the presence of EDTA (pH 4·7), silver, and also Hg and Au, can be extracted from a solution containing considerable quantities of Cu, Bi, Zn, Cd, Ni, and Pb; boiling the solution first for 2 minutes causes the gold to be reduced by EDTA to the element [26]. Mercury can be volatilized by igniting the sample before the determination of silver.

Silver is determined with dithizone by the monocolour, mixed-colour, and extractive titration methods. In the extractive titration technique, the end-point colour change is from yellow to green. Since silver dithizonate is more stable than copper dithizonate, a solution of violet $Cu(HDz)_2$ may be used in the extractive titration technique instead of dithizone. The colour change is then more easily observed—from yellow to violet. When this variation is employed, the presence of copper ions in the solution is of no consequence.

When determining silver traces in metallic bismuth, Sinyakova and Tsvetkova [25] have found that the extraction of minute quantities of silver (e.g. 1 µg) is promoted by the presence of relatively larger quantities of mercury in the solution.

The dithizone method has been used to determine silver in waste water [7], uranium compounds [27], tin [24,28], bismuth [25,28], gold [29], lead [28,30], copper [12,28], copper concentrates [31], and zinc, cadmium, thallium, nickel, cobalt, antimony, arsenic, manganese, and iron [28]. Silver has also been determined in an aqueous acetone medium with dithizone [32].

For the determination of silver in selenium and tellurium, a related reagent, di-2-naphthylthiocarbazone, has been recommended ($\varepsilon = 4\cdot 73 \times 10^4$ at 505 nm) [33].

Reagents

1. Dithizone (H_2Dz): 0·01% solution in CCl_4. Dissolve enough reagent to provide 50 mg of active dithizone in 100 ml of CCl_4. Filter the solution through a filter paper into a 500-ml separating funnel. Shake the green solution with 100 ml of aqueous ammonia (1+50). Discard the brown CCl_4 layer containing oxidation products of dithizone, acidify the orange ammoniacal dithizone solution with 1M HCl, and shake it with 200 ml of CCl_4 till decolorization of the aqueous phase. Dilute the green CCl_4 solution of dithizone to 500 ml with the solvent, and keep in an amber-glass bottle under a layer of 1M H_2SO_4 (the solution of dithizone is not very stable). Prepare working solutions (e.g. 0·001% H_2Dz) by suitable dilution of the stock solution (0·01%). Determine the concentration of dithizone in the CCl_4 solution either by

measuring the absorbance of the green solution or by extractive titration with a standard silver solution as follows. Place in a separating funnel 100 µg of Ag (10 ml of a 0·01 mg/ml solution), add ∼ 0·001% dithizone solution in portions from an amber-glass burette, and shake. Drain the resulting orange silver dithizonate solution from the separating funnel, and continue shaking with successive dithizone portions. Toward the end-point, add 0·5–0·2 ml portions of the dithizone solution. The first extractive titration is only approximate. The titration of 100 µg of silver requires 23·75 ml of exactly 0·001% dithizone solution.

2. Standard silver solution: 1 mg/ml. Dissolve 1·575 g of dried (110°C) silver nitrate in water containing 1 ml of conc. HNO_3, and dilute the solution with water in a volumetric flask to 1 litre.

3. Carbon tetrachloride. Regeneration of the solvent used or purification of the commercial product is carried out as follows. Shake a portion of the solvent (about 400 ml) in a 500-ml separating funnel successively with H_2SO_4 (1+2), 10% NaOH solution, water, 1% $KMnO_4$ solution, H_2SO_4 (1+9), water, and 5% Na_2SO_3 solution. Dehydrate by mixing the clear CCl_4 layer in a conical flask with 50 g of anhydrous K_2SO_4. Pour the solvent into a still, add 1 g of $Na_2S_2O_3.5H_2O$, and distil the CCl_4, collecting the fraction boiling at 76–78°C.

It is often sufficient to purify commercial grade (or old) carbon tetrachloride (or chloroform) by shaking with dilute (e.g. 0·1N) $Na_2S_2O_3$ and then with water.

Procedure

Adjust the acidity of the chloride-free solution containing not more than 80 µg of Ag with H_2SO_4 or HNO_3 till the concentration is 1–2M, in relation to acid, and extract silver (along with other noble metals) with small portions of dithizone in CCl_4 (1 ml of 0·001% H_2Dz solution corresponds to 4·2 µg of Ag). The last portion of dithizone added should not change from green to yellow, although it may turn violet [due to Cu(HDz)$_2$].

Shake the combined orange-yellow carbon tetrachloride extracts for 20 seconds with 1M HCl. Separate the aqueous layer (containing Ag), add (optionally) a little EDTA, adjust to pH 4–5 with ammonia, and extract silver with portions of 0·001% dithizone solution in CCl_4. Remove free dithizone from the extract with dilute ammonia solution (2 drops of conc. ammonia solution in 25 ml of water). Dilute the AgHDz solution with carbon tetrachloride in a 50-ml (or smaller) volumetric flask, and measure the absorbance at 462 nm, using CCl_4 as reference.

48.2.2 RHODANINE METHOD

p-Dimethylaminobenzylidenerhodanine (*p*-dimethylaminobenzalrhodanine, rhodanine) reacts with silver ions in an acid medium to form a red compound which is sparingly soluble in water. The nature of the product has

$$\begin{array}{c} HN\text{——}C=O \\ | \quad\quad | \\ S=C\diagdown_S\diagup C=CH-\!\!\!\bigcirc\!\!\!-N(CH_3)_2 \end{array}$$

not yet been conclusively elucidated. It was formerly thought that the silver atom substituted for the hydrogen atom in the cyclo-amino group of the reagent. Now, some authors [34–36] believe that a chelate is formed, the silver being co-ordinated by the sulphur atom of the thiocarbonyl group and the adjacent nitrogen atom; others [37], however, believe that the species is a linear polymerized complex having the structure

$$\left[\begin{array}{c} \text{structure} \end{array}\right]_n$$

The method for determining silver is based on the absorption of the red pseudosolution of the rhodanine complex with silver in the presence of an excess of yellow rhodanine [38–40]. The diethyl analogue of dimethylrhodanine is also used since both reagents display the same analytical properties [5]. Figure 48.1 shows the absorption spectra of the reagent and its complex with silver. The absorption maximum is at about 450 nm (the molar absorptivity is $2 \cdot 0 \times 10^4$; $a = 0 \cdot 18$).

The absorbance depends on the acidity of the solution, the concentration of rhodanine, and time. The optimum acidity corresponds to $0 \cdot 05 M$

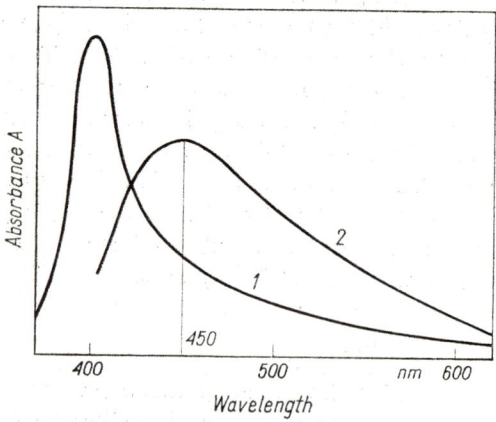

Fig. 48.1 Absorption spectra of p-dimethylaminobenzylidenerhodanine (1) and its silver compound (2) at pH ~1·8

HNO$_3$ (pH ~ 1·8) in the final solution. At the optimum rhodanine concentration (0·001–0·002%), maximum absorbance is obtained in 5 minutes; the colour fades after this. The temperature of the solution should be kept constant since large variations affect the colour intensity.

The acetone (optimum concentration is 5%) added as the solvent for the rhodanine helps to stabilize both the complex and the free reagent sols, but too high a concentration of acetone causes coagulation and precipitation of the complex, while too low a concentration results in coagulation and precipitation of the reagent. Since high concentrations of salts accelerate coagulation, it is advisable to separate silver preliminarily, e.g. by reduction with stannous chloride in the presence of tellurium as collector [5]. Protective colloids (gum arabic or gelatine) are added as stabilizing agents.

Gold, platinum, palladium, and mercury react similarly to silver in acid medium, but copper(II) does not interfere in the reaction.

Anions which form sparingly soluble silver salts interfere. Chloride and thiocyanate should be absent. To overcome the interference from chloride, Castagna and Chauveau [35] recommend using an ammoniacal medium.

Ringbom and Linko [41] have determined silver indirectly by precipitation with rhodanine while other metals are masked with EDTA. Free rhodanine is washed from the precipitate with ethanol. The silver-rhodanine complex is dissolved in KCN solution, and the absorbance of the yellow rhodanine released is measured.

The rhodanine method has been used to determine silver in ores [40–42], lead oxide [43], photographs [44], plant materials [5], soils [42], and biological materials [35].

Reagents

1. Rhodanine: 0·03% solution in acetone.
2. Standard silver solution: 1 mg/ml (p. 493).

Procedure

To an acid solution (containing not more than 120 μg of Ag) add just enough HNO$_3$ to bring its concentration to 0·05M after dilution to 50 ml. Add 1 ml of 1% gum arabic solution and dilute with water to 30–40 ml. Add 2·5 ml of rhodanine solution, make up the solution with water to 50 ml in a volumetric flask, and mix thoroughly. After 5 minutes, measure the absorbance at 450 nm, using a reagent blank solution as reference.

48.2.3 OTHER METHODS

Numerous organic reagents for the spectrophotometric determination of silver contain sulphur (like dithizone and rhodanine).

Copper diethyldithiocarbamate, Cu(DDTC)$_2$ [45,46] (see p. 241), and mercupral (cupric complex of Dicupral, i.e. tetraethylthiuram disulphide)

[47–50] are employed in indirect determination of silver. A chloroform solution of Cu(DDTC)$_2$ or a benzene solution of mercupral becomes deco-

$$(C_2H_5)_2N-\underset{\underset{S}{\|}}{C}-S-S-\underset{\underset{S}{\|}}{C}-N(C_2H_5)_2$$

lorized when shaken with an aqueous solution which contains silver, owing to the formation of the more stable, colourless, silver complexes. The reaction with Cu(DDTC)$_2$ is

$$Cu(DDTC)_2 + 2Ag^+ \rightarrow 2AgDDTC + Cu^{2+}$$

Other organosulphur reagents used for the direct determination of silver include dithiol [5], thio-Michler's ketone [52], diallyldithiocarbamidohydrazine (Dalzin) [53], and 2-amino-6-methylthio-4-pyrimidine carboxylic acid [54].

Dagnall and West [55,56] have recommended Pyrogallol Red and Bromopyrogallol Red. The cationic complex of silver and 1,10-phenanthroline reacts with the latter dye to yield an ion-association complex which can be extracted into nitrobenzene ($\varepsilon = 3\cdot2 \times 10^4$ at 590 nm). These reagents have been used to determine silver in tellurium [57] and non-ferrous metals [58].

Ion-association compounds of the 1,10-phenanthroline–silver complex with halogenated derivatives of fluorescein [59,60] and with Sulpharsazen [61] are also highly coloured.

Markham [62] has determined silver by extracting into benzene the ion-pair formed between Ag(CN)$_2^-$ and the basic dye, Crystal Violet.

Other colorimetric reagents include 2,3-naphthotriazole [63], Nile Blue A [64], and phenylenediamine [65].

Larger amounts of silver can be determined turbidimetrically after reduction to yield a silver sol [66], as the silver sulphide sol [3], or as the silver chloride sol [14].

Methods based on coloured compounds of silver(II) are also worthy of notice. In the presence of persulphate, an olive-green silver(II) anionic complex with pyridine-2,6-dicarboxylic acid [67] and coloured silver(II) complexes with 2,2'-bipyridyl and terpyridyl [68] are formed.

References

1. Bode, H., *Z. Anal. Chem.* **144**, 165 (1955).
2. Betteridge, D. and West, T. S., *Anal. Chim. Acta* **26**, 101 (1962).
3. Ziegler, M., Sbrzesny, H. and Glemser, O., *Z. Anal. Chem.* **173**, 411; **175**, 321 (1960).
4. Yadav, A. A. and Khopkar, S. M., *Mikrochim. Acta* **1971**, 464.
5. Sandell, E. B. and Neumayer, J. J., *Anal. Chem.* **23**, 1863 (1951); *Anal. Chim. Acta* **5**, 445 (1951).
6. Süpfle, K. and Werner, R., *Mikrochemie* **36/37**, 866 (1951).
7. Pierce, T. B., *Analyst* **85**, 166 (1960).
8. Zagórski, Z. and Kempiński, O., *Chem. Anal. (Warsaw)* **4**, 422 (1959).

References

9. Mizuike, A. and Fukuda, K., *Anal. Chim. Acta* **44**, 193 (1969).
10. Fukuda, K. and Mizuike, A., *ibid.* **51**, 77 (1970).
11. Triché, H. and Rocques, G., *Bull. Soc. Chim. France* **1955**, 1507.
12. Angermann, W. and Bastius, H., *Neue Hütte* **9**, 36 (1964).
13. Tiptsova-Yakovleva, V. G. and Dvortsan, A. G., *Zh. Analit. Khim.* **24**, 1141 (1969).
14. Kemula, W., Brajter, K., Cieślik, S. and Lipińska-Kostrowicka, H., *Chem. Anal. (Warsaw)* **5**, 225 (1960).
15. Chao, T. T., Fishman, M. J. and Ball, J. W., *Anal. Chim. Acta* **47**, 189 (1969).
16. Rangnekar, A. V. and Khopkar, S. M., *Mikrochim. Acta* **1965**, 642.
17. De, A. K. and Majumdar, S. K., *Talanta* **10**, 201 (1963).
18. Vydra, F., *ibid.* **10**, 753 (1963).
18a. Lisitsyna, D. N. and Shcherbov, D. P., *Zh. Analit. Khim.* **28**, 1174 (1973).
19. Erdey, L., Rády, G. and Fleps, V., *Acta Chim. Acad. Sci. Hung.* **5**, 133 (1954).
20. Trémillon, B., *Bull. Soc. Chim. France* **1954**, 1156, 1160.
21. Schweitzer, G. K. and Dyer, F. F., *Anal. Chim. Acta* **22**, 172 (1960).
22. Dyer, F. F. and Schweitzer, G. K., *ibid.* **23**, 1 (1960).
23. Friedeberg, H., *Anal. Chem.* **27**, 305 (1955).
24. Marczenko, Z. and Kasiura, K., *Chem. Anal. (Warsaw)* **10**, 449 (1965).
25. Sinyakova, S. I. and Tsvetkova, L. A., *Tr. Komis. po Analit. Khim. Akad. Nauk SSSR* **12**, 191 (1960).
26. Goryushina, V. G. and Gailis, E. Ya., *Zavodsk. Lab.* **22**, 905 (1956).
27. Mareček, J. and Singer, E., *Z. Anal. Chem.* **203**, 336 (1964).
28. Cyrankowska, M., *Chem. Anal. (Warsaw)* **6**, 649 (1961).
29. Marczenko, Z., Kasiura, K. and Krasiejko, M., *ibid.* **14**, 1277 (1969).
30. Jones, P. D. and Newman, E. J., *Analyst* **87**, 66 (1962).
31. Ostachowska, J. and Kunz, K., *Rudy Metale Nieżelazne* **11**, 426 (1966).
32. Makovskii, M. E., *Zh. Analit. Khim.* **25**, 1226 (1970).
33. Tiptsova, V. G., Andreichuk, A. M. and Bazhanova, L. A., *ibid.* **21**, 1179 (1966).
34. Bagdasarov, K. N., Shelepin, O. E. and Gorbachevskaya, T. M., *Izv. Vyssh. Ucheb. Zaved., Khim. Khim. Tekhnol.* **14**, 995 (1971).
35. Castagna, M. and Chauveau, J., *Bull. Soc. Chim. France* **1961**, 1165.
36. Navrátil, O. and Kotas, J., *Collection Czech. Chem. Commun.* **30**, 2736 (1965).
37. Stephen, W. I. and Townshend, A., *J. Chem. Soc.* **1965**, 3738.
38. Allen, J. A. and Holloway, D. G., *Nature* **166**, 274 (1950).
39. Cave, G. C. and Hume, D. N., *Anal. Chem.* **24**, 1503 (1952).
40. Struszyński, M., Nowicka, T. and Marczenko, Z., *Przemysl Chem.* **32**, 574 (1953).
41. Ringbom, A. and Linko, E., *Anal. Chim. Acta* **9**, 80 (1953).
42. Bhattathiripad, K. M. and Joshi, R. G., *Z. Anal. Chem.* **242**, 247 (1968).
43. Dicker, E. S. and Johnson, E. A., *Analyst* **82**, 285 (1957).
44. Shishkina, N. N., *Zh. Analit. Khim.* **15**, 431 (1960).
45. Kreimer, S. E., Lomekhov, A. S. and Stogova, A. V., *ibid.* **17**, 674 (1962).
46. Hattori, T. and Kuroha, T., *Japan Analyst* **11**, 723 (1962).
47. Michal, J. and Zýka, J., *Collection Czech. Chem. Commun.* **22**, 1135 (1957).
48. Michal, J., Pavlíková, E. and Zýka, J., *Z. Anal. Chem.* **160**, 277 (1958).
49. Patrovský, V., *Chem. Listy* **57**, 268 (1963).
50. Hirano, S., Mizuike, A. and Ujihira, Y., *Japan Analyst* **12**, 160 (1963).
51. Dux, J. P. and Feairheller, W. R., *Anal. Chem.* **33**, 445 (1961).
52. Cheng, K. L., *Mikrochim. Acta* **1967**, 820.
53. Dutt, N. K. and Sen Sarma, K. P., *Z. Anal. Chem.* **177**, 7 (1960).
54. Chung, O. K. and Meloan, C. E., *Anal. Chem.* **39**, 383 (1967).
55. Dagnall, R. M. and West, T. S., *Talanta* **8**, 711 (1961).
56. Dagnall, R. M. and West, T. S., *ibid.* **11**, 1533, 1627 (1964).
57. Dobrowolski, J. and Szwabski, S., *Chem. Anal. (Warsaw)* **15**, 1033 (1970).
58. Lukashenkova, N. V., Tolmacheva, N. S., and Shkrobot, E. P., *Zavodsk. Lab.* **39**, 541 (1973).

59. El-Ghamry, M. T. and Frei, R. W., *Anal. Chem.* **40**, 1986 (1968).
60. Stolyarov, K. P. and Firyulina, V. V, *Zh. Analit. Khim.* **26**, 1731 (1971).
61. Stolyarov, K. P. and Firyulina, V. V., *ibid.* **24**, 1494 (1969).
62. Markham, J. J., *Anal. Chem.* **39**, 241 (1967).
63. Wheeler, G. L., Andrejack, J., Wiersma, J. H. and Lott, P. F., *Anal. Chim. Acta* **46**, 239 (1969).
64. Likussar, W. and Raber, H., *ibid.* **50**, 173 (1970).
65. Rebertus, R. L. and Levin, V., *Anal. Chem.* **40**, 2053 (1968).
66. Hepenstrick, H., *Helv. Chim. Acta* **32**, 364 (1949).
67. Hartkamp, H., *Z. Anal. Chem.* **184**, 98 (1961).
68. Gagliardi, E. and Presinger, P., *Mikrochim. Acta* **1964**, 1175.

Chapter 49

STRONTIUM AND BARIUM

Strontium (Sr, at.wt. 87·62) and barium (Ba, at.wt. 137·34) occur in solutions exclusively in the +II oxidation state, as does calcium, their congener. The basicity and solubility in water increase from $Ca(OH)_2$ to $Ba(OH)_2$. Barium chromate and sulphate are less soluble than the corresponding strontium compounds, The stability of relatively weak complexes (with EDTA or tartrate, for example) diminishes in the sequence Ca, Sr, Ba.

49.1 Methods of Separation

Before spectrophotometric determination, it is usually necessary to separate strontium or barium from the Analytical Group I–III metals. Suitable methods are discussed in the chapter on calcium (p. 182).

49.1.1 Ion-Exchange

Most ion-exchange methods for separating the alkaline earth metals from one another involve retention on strongly acidic cation-exchangers and selective elution with appropriate complexants, exploiting the differences in stability of alkaline earth metal complexes. Complexones such as EDTA [1–5], EGTA [1,6], and DCTA [1,7–9] are suitable eluents, but other complexing agents are also employed, e.g. lactate [1,10–12], citrate [1,11,13], and sulphate [14]. Strelow [15] has separated barium from strontium and other elements by cation-exchange chromatography in mixed HCl–organic solvent eluents. The enrichment and determination of strontium in sea water [8,9], drinking water [12], and milk [6,11] have been extensively discussed.

Nelson and Kraus [16] have separated Ca, Sr, and Ba by elution from a strongly basic anion-exchanger with citrate. Calcium and strontium have been separated by using a mixed medium comprising $0·25M$ HNO_3 and 95% (v/v) methanol [17]. Barium and strontium, or calcium and magnesium, can be separated with pyridine-2,6-dicarboxylate as eluent [18]. Winowski [19] has quantitatively retained barium on the chromate form of an anion-exchanger, while calcium passes through.

49.1.2 Extraction

Gorbenko et al. [20,21] have developed a method for separating calcium, strontium, and barium by extraction with Azo-azoxy BN (see p. 183) and

TBP in carbon tetrachloride. First calcium is extracted from $0.05M$ NaOH, then strontium from $0.8M$ NaOH, while barium remains in the aqueous phase. Strontium is 95% extracted from $0.8-2M$ NaOH, in the presence of a tenfold excess of Azo-azoxy BN [21]. When the related reagent Azo-azoxy PMP is used, the extraction of strontium is quantitative [22].

Strontium may be separated from other metals by extraction with TTA in MIBK [23,24] or with TTA and TBP in carbon tetrachloride [25].

In the leaching of $Ca(NO_3)_2$ from $Sr(NO_3)_2$, acetone is a better solvent than ethanol-ether $(1+1)$ [26].

49.1.3 Precipitation

Traces of strontium have been coprecipitated as the chromate with barium [27] and as the oxalate with calcium [28]. Enrichment of strontium on hydrous MnO_2 has also been applied [29,30].

Potassium rhodizonate selectively precipitates strontium in the presence of calcium [31,32]. Barium has been separated from strontium by precipitating $BaSO_4$ in the presence of EDTA (pH $\sim$ 8) [33], or $BaCrO_4$ in the presence of MEDTA (pH 6-7) [34], and strontium has been separated from calcium by precipitating $SrSO_4$ from an EDTA solution (pH $\sim$ 5) [33].

Strontium impurities have been isolated from calcium salts by precipitating calcium as $Ca(OH)_2$ with dilute NaOH. After double precipitation of $Ca(OH)_2$, the strontium coprecipitated has dropped to 3 ppm [35].

49.2 Determination of Strontium

Savvin et al. [36] have discussed organic spectrophotometric reagents for strontium and barium.

Of the bisazo derivatives of chromotropic acid, Nitro-orthanilic S is particularly suitable for strontium [37-39].

$$O_2N-\underset{HO_3S}{\overset{SO_3H}{\bigcirc}}-N=N-\underset{HO_3S}{\overset{HO\ \ OH}{\bigcirc\bigcirc}}-N=N-\underset{}{\overset{HO_3S}{\bigcirc}}-NO_2$$

The colour reaction ($\lambda_{opt} = 650$ nm) underlying the determination of strontium is carried out in 60% acetone at pH 2·8. Strontium has been determined in barium salts by this method [38,39].

The related reagents Chlorophosphonazo III [40], Sulphonazo III [41], Dimethylsulphonazo III [42], and Dimethylsulphonazo DAL [43] have also been employed for the determination of strontium in slightly acidic media (pH 2-5). Arsenazo III can also be used [44].

In alkaline solutions, strontium is determined spectrophotometrically with Murexide (see p. 186) [45,46] and Metalphthalein (o-Cresolphthalein Complexone, see p. 501) [46].

In indirect methods, strontium is precipitated with a known amount of dilituric acid [47,48] or chloranilic acid [49]. The strontium content is determined from the absorbance of the excess of coloured reagent. In another indirect method, $SrCrO_4$ is precipitated, the precipitate dissolved in acid, and the dichromate ions determined either directly or from the colour produced with diphenylcarbazide [50].

49.3 Determination of Barium

One of the foremost spectrophotometric reagents for barium is Sulphonazo III (Orthanilic S) [41,42,51–54], a bisazoderivative of chromotropic

acid. The colour reaction is carried out in a weakly acidic or neutral medium (pH 2–8), in 60% acetone (or ethanol) or in aqueous solution [52]. The molar absorptivity depends on the pH, the medium, and the excess of reagent ($\varepsilon = 4 \cdot 0 \times 10^4$ at 640 nm, pH 8, 60% acetone; $\varepsilon = 1 \cdot 1 \times 10^4$, pH 2·4, water). In aqueous solution, 100 μg of strontium and 300 μg of calcium can be tolerated.

The following bisazo derivatives of chromotropic acid have been similarly proposed as reagents for barium: Nitro-orthanilic S (see p. 500) [39,54], Dimethylsulphonazo DAL [43], Chlorophosphonazo III (see p. 50) [55], Carboxyarsenazo (50% ethanol, pH 5·6, $\varepsilon = 4 \cdot 1 \times 10^4$ at 640 nm) [56], and Arsenazo III [44].

Barium can also be determined spectrophotometrically as its complex with Metalphthalein (o-Cresolphthalein Complexone) [46,57].

The colour reaction is carried out in a solution of pH 11·3 ($\lambda_{max} = 575$ nm).

Barium can be determined indirectly by precipitation as $BaCrO_4$, dissolution of the precipitate in acid, and measurement of the absorbance of the dichromate ions or of the chromium–diphenylcarbazide complex [58,59].

References

1. Strelow, F. W. E. and Weinert, C. H., *Talanta* **17**, 1 (1970).
2. Bovy, R. and Duyckaerts, G., *Anal. Chim. Acta* **11**, 134 (1954).
3. Duyckaerts, G. and Lejeune, R., *J. Chromatog.* **3**, 58 (1960).
4. Bouquiaux, J. J. and Gillard, J. H., *Anal. Chim. Acta* **30**, 273 (1964).
5. Sednev, M. P., Starobinets, G. L. and Akulovich, A. M., *Zh. Analit. Khim.* **21**, 23 (1966).
6. Brandt, P. J. and van't Riet, B., *Anal. Chem.* **38**, 1790 (1966).
7. Šulcek, Z., Povondra, P. and Štangl, R., *Talanta* **9**, 647 (1962).
8. Noshkin, V. E. and Mott, N. S., *ibid.* **14**, 45 (1967).
9. Andersen, N. R. and Hume, D. N., *Anal. Chim. Acta* **40**, 207 (1968).
10. Lerner, M. and Rieman, W., III, *Anal. Chem.* **26**, 610 (1954).
11. Milton, G. M. and Grummitt, W. E., *Can. J. Chem.* **35**, 541 (1957).
12. Knapstein, H., *Z. Anal. Chem.* **175**, 255 (1960).
13. Ibbett, R. D., *Analyst* **92**, 417 (1967).
14. Christova, R. and Kruschevska, A., *Anal. Chim. Acta* **36**, 392 (1966).
15. Strelow, F. W. E., *Anal. Chem.* **40**, 928 (1968).
16. Nelson, F. and Kraus, K. A., *J. Am. Chem. Soc.* **77**, 801 (1955).
17. Fritz, J. S., Waki, H. and Garralda, B. B., *Anal. Chem.* **36**, 900 (1964).
18. Bennett, W. E. and Skovlin, D. O., *ibid.* **38**, 518 (1966).
19. Winowski, Z., *Chem. Anal. (Warsaw)* **13**, 583 (1968).
20. Gorbenko, F. P. and Lapitskaya, E. V., *Zh. Analit. Khim.* **23**, 1139 (1968); *Zavodsk. Lab.* **34**, 1051 (1968).
21. Gorbenko, F. P. and Nadezhda, A. A., *Ukr. Khim. Zh.* **34**, 625 (1968).
22. Gorbenko, F. P. and Nadezhda, A. A., *Zh. Analit. Khim.* **24**, 671 (1969).
23. Johnson, W. C., Jr., *Anal. Chem.* **38**, 954 (1966).
24. Akaza, I., *Bull. Chem. Soc. Japan* **39**, 971 (1966).
25. Sekine, T. and Dyrssen, D., *Anal. Chim. Acta* **37**, 217 (1967).
26. Baranov, V. I. and Vilenskiĭ, V. D., *Zh. Analit. Khim.* **17**, 295 (1962).
27. Beneš, J. and Kyrš, M., *Collection Czech. Chem. Commun.* **33**, 2822 (1968).
28. Beneš, J., *ibid.* **35**, 591 (1970).
29. Shipman, W. H., *Anal. Chem.* **38**, 1175 (1966).
30. Rudnev, N. A., Pustovalova, M. N. and Malofeeva, G. I., *Zh. Analit. Khim.* **25**, 1085 (1970).
31. Weiss, H. V. and Shipman, W. H., *Anal. Chem.* **29**, 1764 (1957).
32. Boni, A. L., *ibid.* **35**, 744 (1963).
33. Afanas'eva, L. I., *Zh. Analit. Khim.* **14**, 294 (1959).
34. Firsching, F. H. and Werner, P. H., *Talanta* **19**, 790 (1972).
35. Patti, F. and Hernandez, J. A., *Anal. Chim. Acta* **55**, 325 (1971).
36. Savvin, S. B., Akimova, T. G. and Dedkova, V. P., *Organicheskie reagenty dlya opredeleniya Ba^{2+} i SO_4^{2-} (Organic Reagents for Ba^{2+} and SO_4^{2-} Determination)*, Izdat. Nauka, Moscow (1971).
37. Dedkov, Yu. M., Makarova, V. P., Vinokurova, F. A., Chashchikhina, M. V. and Savvin, S. B., *Zh. Analit. Khim.* **20**, 440 (1965).
38. Kreshkov, A. P. and Kuznetsov, V. V., *Zavodsk. Lab.* **34**, 134 (1968).
39. Kreshkov, A. P. and Kuznetsov, V. V., *Izv. Vyssh. Zaved., Khim. Khim. Tekhnol.* **12**, 1186 (1969); *Zh. Analit. Khim.* **25**, 49, 874 (1970).
40. Lukin, A. M., Zelichenok, S. L. and Chernysheva, T. V., *ibid.* **19**, 1513 (1964).
41. Kemp, P. J. and Williams, M. B., *Anal. Chem.* **45**, 124 (1973).
42. Buděšínský, B., Vrzalová, D. and Bezdeková, A., *Acta Chim. Acad. Sci. Hung.* **52**, 37 (1967).
43. Buděšínský, B. and Vrzalová, D., *Talanta* **13**, 1217 (1966).
44. Michaylova, V. and Kouleva, N., *ibid.* **21**, 523 (1974).

45. Russell, D. S., Campbell, J. B. and Berman, S. S., *Anal. Chim. Acta* **25**, 81 (1961).
46. Pollard, F. H. and Martin, J. V., *Analyst* **81**, 348 (1956).
47. Karnaukhov, A. S., Mizera, M. and Paloush, R., *Zh. Analit. Khim.* **15**, 502 (1960).
48. Palouš, R., Kharnauchoff, A. and Mizera, M., *Anal. Chim. Acta* **24**, 96 (1961).
49. Lucchesi, P. J., Lewin, S. Z. and Vance, J. E., *Anal. Chem.* **26**, 521 (1954).
50. Nozaki, T., *J. Chem. Soc. Japan, Pure Chem. Sect.* **80**, 1278 (1959).
51. Savvin, S. B., Dedkov, Yu. M. and Makarova, V. P., *Zh. Analit. Khim.* **17**, 43 (1962).
52. Buděšínský, B. and Vrzalová, D., *Z. Anal. Chem.* **210**, 161 (1965).
53. Slovák, Z., Fischer, J. and Borák, J., *Talanta* **15**, 831 (1968).
54. Savvin, S. B., Dedkova, V. P. and Akimova, T. G., *Tr. Komis. po Analit. Khim. Akad. Nauk. SSSR* **17**, 322 (1969).
55. Lukin, A. M., Smirnova, K. A. and Chernysheva, T. V., *Zh. Analit. Khim.* **21**, 1300 (1966).
56. Basargin, N. N. and Nogina, A. A., *ibid.* **25**, 2320 (1970).
57. Cohen, A. I. and Gordon, L., *Anal. Chem.* **28**, 1445 (1956).
58. Agterdenbos, J., *Z. Anal. Chem.* **159**, 202 (1958).
59. Anand, K. S., Dayal, P. and Anand, O. N., *ibid.* **247**, 310 (1969).

Chapter 50

SULPHUR

Sulphur (S, at.w. 32·06) occurs in the oxidation states −II (in sulphide), +IV (in SO_2 and sulphite), and +VI (in H_2SO_4 and sulphate). Sulphur dissolves in certain organic solvents (e.g. CS_2 and C_6H_6); when dissolved in an alkali metal sulphide solution, it yields yellow polysulphides (S^{2-} + $nS \to S_{n+1}^{2-}$); when heated with an alkaline solution of sulphite, it forms thiosulphate; with cyanide it gives thiocyanate. Sulphide, sulphite, and thiosulphate are reducing agents. Persulphate (peroxodisulphate) has strongly oxidizing properties. Complexes are formed by sulphide (e.g. with As, Sb, Mo, and Pt) and also by thiosulphate [e.g. with Ag, Cu, and Fe(III)]. The most stable sulphur species is sulphate, which forms sparingly soluble compounds with Ba and Pb, for example, and stable complexes with Zr and Th.

50.1 Methods of Separation

Many methods for separating sulphur (in various forms) from other elements are associated with methods for its determination and are discussed with them below.

Of the methods for separating sulphur, those based on distillation are the most important. Oxy-compounds of sulphur are reduced to hydrogen sulphide, which is carried in an inert gas stream (e.g. nitrogen) to a receiver containing zinc or cadmium ions [1,2] (cf. Methylene Blue method).

On ignition of the sample in air or oxygen, sulphur is converted into volatile sulphur dioxide, which is subsequently absorbed and determined spectrophotometrically [3,4] (cf. pararosaniline method).

A very selective method for separating sulphur is to precipitate sulphate as $BaSO_4$ from dilute hydrochloric acid medium. Chromate is a suitable collector for traces.

Sulphate is separated from heavy metals when the solution is passed through a strongly acidic cation-exchanger [5,6]. Hydrogen sulphide in air has been concentrated by using an anion-exchanger [7].

Elementary sulphur can be separated from other elements by leaching with acetone [8], pyridine, or carbon disulphide.

50.2 Methods of Determination

Sulphur is one of the elements which exist as numerous species. Accordingly, a number of spectrophotometric methods for determining sulphur in its various forms have been developed. Two particularly sensitive methods are discussed below, namely the Methylene Blue method in which sulphur is first converted into hydrogen sulphide, and the pararosaniline method in which sulphur dioxide is the reacting species. The classical turbidimetric ($BaSO_4$) method is also discussed. All three methods are highly selective.

Szekeres [8a] has reviewed the analytical chemistry of the sulphur acids.

50.2.1 Turbidimetric ($BaSO_4$) Method

When the concentration of sulphate in the solution is sufficiently low, the barium sulphate formed after the addition of barium chloride does not coagulate to form a precipitate, but remains as a fine suspension producing a turbidity. The determination of small quantities of sulphate (or any other sulphur compound after oxidation to SO_4^{2-}) is based either on comparison of the turbidity with a set of standards or on its photoelectric estimation [9–13]. This turbidimetric method for determining sulphate is simple and fast, but of rather low precision and sensitivity.

The $BaSO_4$ suspension is formed in dilute hydrochloric acid medium. The turbidity varies with time, but is almost stable for a certain period after 10–15 minutes. Since the crystals of barium sulphate are liable to age on prolonged standing, the standards should be prepared simultaneously or the absorbance should be measured at a definite time after the addition of barium chloride, e.g. after 15 minutes. The determination is carried out at room temperature; rise in temperature increases the rate of aging of the precipitate.

The addition of 20–30% of ethanol decreases the solubility of barium sulphate and enhances the degree of dispersion [14]. Some authors [15,16] recommend the addition of protective colloids to stabilize the suspension, but Wimberley [12] refutes the necessity. Stabilization is also obtained by adding glycerol (with or without ethanol) [10,11]. Phosphoric acid also helps to stabilize the suspension in the presence of glycerol [13].

Sulphate (sulphur) has been determined by this turbidimetric method in plant material [11,15–17], soils [15,17,18], organic compounds [19], reagent-grade salts [14], petrol [20], foods [21], perhydrol [22], and water [23]. The method has also been applied to the automatic determination of sulphate during the production of phosphoric acid [24].

Reagents

1. Barium chloride: 2% solution.
2. Standard sulphur solution: 1 mg/ml. Dissolve in water 5·437 g of K_2SO_4 (previously ignited at 400–500°C), and dilute the solution with water to 1 litre.

Procedure

Place the sample solution containing not more than 120 μg of S (present as sulphate) in a measuring cylinder. Acidify with 2 ml of $2M$ HCl, add 10 ml of ethanol and water to ~ 40 ml, pour in quickly 5 ml of $BaCl_2$ solution, and stir well. After 15 minutes, compare the turbidity with a series of standards prepared simultaneously. The cylinders should be observed from above, against black paper, in a uniformly brightly illuminated place.

The absorption of the turbid solutions may also be measured photoelectrically.

50.2.2 METHYLENE BLUE METHOD

Sulphur in the form of sulphate (or in any other oxy-compound), is reduced by a mixture of hydriodic acid and hypophosphite to hydrogen sulphide, which is carried in a nitrogen stream to a zinc acetate solution where it is trapped as ZnS. Then, in an acid medium, the sulphide reacts with *p*-aminodimethylaniline and iron(III) to yield Methylene Blue [1,25–31]:

$$(CH_3)_2\overset{+}{N}=\!\!\underset{S}{\overset{N}{\bigcirc\!\!\bigcirc}}\!\!-N(CH_3)_2$$

Tin(II) [26] or chromium(II) [27] are alternative reagents for the reduction of sulphur compounds to sulphide. *p*-Aminodiethylaniline is an alternative chromogenic reagent [31a].

To form Methylene Blue, an acid solution of the amine is added with stirring to the sulphide solution, then an iron(III) solution is immediately added, and the solution thoroughly mixed. A concentration of $0.6M$ HCl or $0.3M$ H_2SO_4 gives the most suitable acidity for the reaction [1,28,29]. The colour reaction converts only 65–70% of the hydrogen sulphide into Methylene Blue [1,32].

Methylene Blue is a cationic dye (MB^+) which can be extracted into chloroform as the $[MB^+][ClO_4^-]$ ion-pair in the presence of an excess of perchlorate [28]. Chloride and sulphate do not form any extractable ion-association complexes with Methylene Blue.

The quantitative extraction of Methylene Blue from the aqueous phase requires a 10^3–10^4 fold excess of perchlorate and necessitates shaking with 2 or 3 portions of $CHCl_3$. Equilibrium is attained after shaking for 20–30 seconds. The percentage of extraction with chloroform depends on whether perchlorate ions are present in the solution when the Methylene Blue is formed, or whether they are added afterwards. In the first case, the extraction is 95–97% complete; in the second, about 80%. The optimum acidity for the Methylene Blue formation is also suitable for the chloroform extraction.

The absorption spectra of Methylene Blue in chloroform (curve 1) and in aqueous solution (curve 2) are shown in Fig. 50.1. The absorption maximum in CHCl$_3$ (λ_{max} = 650 nm) is the sharper and the more intense ($\varepsilon = 3.5 \times 10^4$; $a = 1.1$). The absorption maximum in aqueous solution is at a slightly longer wavelength (662 nm).

The Methylene Blue method has been used to determine sulphur in various reagents (including barium chloride) [28], plant materials [25,33],

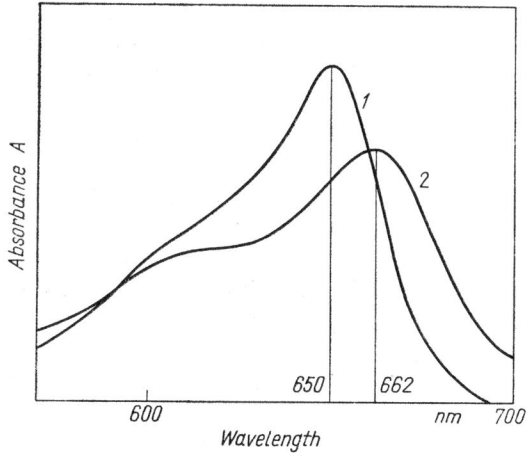

Fig. 50.1 Absorption spectra of ion-association complex of Methylene Blue with ClO$_4^-$ in chloroform (1) and Methylene Blue in aqueous solution (2)

soils [25,32], water [7,25], alkalis [26], selenium [34,35], iron alloys [36–38], cobalt [37], chromium [38], titanium tetrachloride [39], hydrocarbons [40–42], organic compounds [43], and beer [44].

If the determination of only sulphide sulphur is required, the weighed sample is treated in a distillation flask with sulphuric acid (1+2) instead of the reducing agent, and the H$_2$S is distilled off in the nitrogen stream. The rest of the procedure is as given below.

Sulphide sulphur has been determined in air [7,30,45], biological materials [46], and natural water [47].

When determining sulphur in organic compounds, Takeuchi et al. [43] used a silver gauze to collect the sulphur oxides obtained on combustion of the sample in oxygen. Heating in a stream of hydrogen reduces the sulphur oxides to H$_2$S, which is absorbed in an alkaline zinc acetate solution and determined by the Methylene Blue method.

Ducret and Ratouis [48] have run traces of SO$_4^{2-}$ through an anion-exchanger in the thiocyanate form, thereby displacing thiocyanate ions into the eluate. Addition of Methylene Blue yields the [MB$^+$][SCN$^-$] ion-pair which can be extracted into 1,2-dichloroethane from a medium at pH 1.

Reagents

1. *p*-Aminodimethylaniline: ~$0.005M$ solution. Dissolve 0.50 g of $(CH_3)_2NC_6H_4NH_2 \cdot 0.5H_2SO_4$ in HCl (1+1), and dilute the solution with the same acid to 500 ml.
2. Standard sulphur solution: 1 mg/ml (p. 505).
3. Reducing reagent. Dissolve 4.0 g of $NaH_2PO_2 \cdot H_2O$ (sodium hypophosphite) in 25 ml of glacial acetic acid and 100 ml of conc. HI. Reflux the solution in a 250-ml round-bottomed flask for 1 hour with nitrogen passed through at a rate of 3 or 4 bubbles per second. Allow the solution to cool (still with nitrogen passing), transfer the reagent to a bottle with a ground-glass stopper, and store in the dark.
4. Zinc acetate: $0.25M$ solution. Dissolve 27.5 g of zinc acetate and 7.0 g of sodium acetate in 500 ml of water.
5. Iron(III): ~$0.25M$ solution. Dissolve 30 g of $Fe(NH_4)(SO_4)_2 \cdot 12H_2O$ (ferric alum) in $1M$ HCl, and make up the solution with the same acid to 250 ml.
6. Nitrogen-wash solution. Dissolve 2 g of $KMnO_4$ and 4 g of $Hg(NO_3)_2$ in 100 ml of dilute HNO_3 (1+99); boil the solution in an open vessel for 20 minutes, then cool.
7. Sodium perchlorate: $1M$ solution. Dissolve 31 g of $NaClO_4$ in 250 ml of water.

Procedure

Place the sample containing not more than 25 μg of S in a flask [(1) in the apparatus shown in Fig. 50.2]. If the sample is liquid, make it alkaline with sodium hydroxide, and evaporate to dryness in the flask while bubbling nitrogen through. Change the water in the washer (3) every few determinations. Add to the receiver (4) (a measuring cylinder, test-tube, or conical flask) 2.5 ml of the zinc acetate solution and 15 ml of water, and connect to the apparatus as shown in Fig. 50.2.

Pipette 5 ml of the reducing reagent into the flask (1) containing the sample. Quickly connect the flask to the rest of the apparatus, and simultaneously connect the nitrogen supply (purified by passage through the wash solution at a rate of 1 or 2 bubbles per second) to the side-tube (2). Bring the solution in the flask (1) to the boil within 1–2 minutes, and continue the boiling for 15–20 minutes.

Disconnect the receiver (4) together with the delivery tube (5), and add 2.5 ml of the $NaClO_4$ solution. Stir well, then add through the delivery tube (5) to the bottom of the receiver 2.5 ml of the *p*-aminodimethylaniline solution and 0.5 ml of the iron(III) solution. Stopper the receiver and mix the solution. After 15 minutes, transfer the coloured solution to a separating funnel and extract with 2 or 3 portions of $CHCl_3$. Place the combined extracts in a volumetric flask (the capacity depending on the sulphur content), dilute to the mark with chloroform, and measure the absorbance at 650 nm against a reagent blank solution.

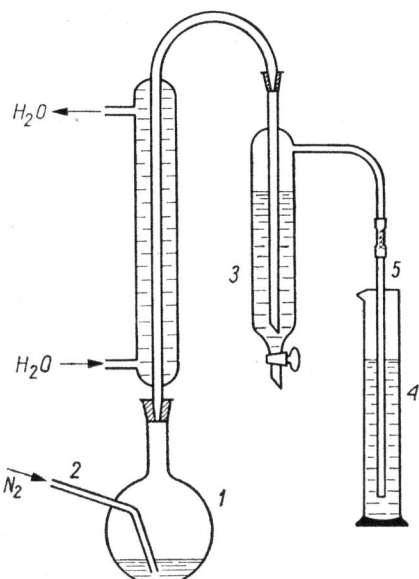

Fig. 50.2 Apparatus for distillation and absorption of hydrogen sulphide

Note. If high sensitivity is not required, the final extraction is unnecessary. In this case no $NaClO_4$ solution is added. After the appearance of colour in the receiver, the solution is transferred to a volumetric flask, diluted to the mark with water, and the absorbance is measured at 662 nm.

50.2.3 PARAROSANILINE METHOD

Sulphur dioxide reacts with sodium tetrachloromercurate to form a stable dichlorosulphitomercurate complex:

$$[HgCl_4]^{2-} + SO_2 + H_2O \rightarrow [HgCl_2SO_3]^{2-} + 2Cl^- + 2H^+$$

Combined in this complex, sulphur dioxide displays high stability and resistance to oxidation. The complex reacts with excess of formaldehyde to give hydroxymethylsulphonic acid:

$$[HgCl_2SO_3]^{2-} + HCHO + 2H^+ \rightarrow HO.CH_2.SO_3H + HgCl_2$$

In strongly acidic media, pararosaniline exists as a colourless species:

The reaction of colourless pararosaniline with hydroxymethylsulphonic acid yields a purple compound which is the basis of this sensitive spectrophotometric method for determining sulphur dioxide, sulphite, or any other form of sulphur converted into SO_2 [49–57]. The coloured reaction product is essentially

$$H_2N-C_6H_4C=C_6H_4=\overset{+}{N}H.CH_2.SO_3H\overset{-}{Cl}$$
$$H_2N-C_6H_4$$

with di- and tri-substituted species also formed.

The molar absorptivity of the product at $\lambda_{max} = 560$ nm is $\sim 3.0 \times 10^4$ ($a = 0.47$).

The sensitivity of the reaction is increased threefold in the presence of dimethylformamide. It is probable that in the presence of this solvent the reaction of pararosaniline with formaldehyde and sulphur dioxide affords a trisubstituted product [53].

The pararosaniline method has been developed by West and Gaeke [49]. Because of its sensitivity and specificity, it has found widespread application, e.g. in the determination of sulphur (or SO_2) in air [55,58], organic and inorganic compounds [59,60], polymers [61], soils [62], rocks [63], selenium [64], uranium and zirconium compounds [65], nickel [66], and copper [4,67].

The pararosaniline method has been employed in the continuous determination of atmospheric sulphur dioxide [68,69].

To determine SO_2 in gases, a sample is bubbled through a solution of tetrachloromercurate. The interference of nitrogen dioxide in the determination of SO_2 is eliminated by adding sulphamic acid to the trapping solution [58,70].

Total sulphur is determined in metals and other solids by combustion in a stream of oxygen to SO_2, which is subsequently absorbed in tetrachloromercurate solution [4,60,66,67].

Methods analogous to the pararosaniline method involve the determination of sulphur dioxide with *p*-aminoazobenzene [71], and *p*-nitroaniline [72].

Reagents

1. Pararosaniline hydrochloride: 1% solution.
2. Pararosaniline (bleached): 0.04% solution. To 4 ml of the 1% pararosaniline solution add 6 ml of conc. HCl, and dilute the solution with water to 100 ml while stirring.
3. Formaldehyde: 0.2% solution, freshly prepared. Dilute 5 ml of 40% CH_2O solution with water to 1 litre.
4. Sodium tetrachloromercurate Na_2HgCl_4, $0.1M$ solution. Dissolve 27.2 g

of $HgCl_2$ and 11·7 g of NaCl in water and dilute the solution with water to 1 litre.
5. Standard SO_2 solution: 0·1 mg/ml. Dissolve 0·1625 g of $NaHSO_3$ in 0·1M tetrachloromercurate solution, and dilute with the same reagent to 1 litre. The solution is unstable.

Procedure

Place 20–30 ml of the sodium tetrachloromercurate solution containing not more than 50 µg of SO_2 (after absorption of SO_2 from a gas mixture or separation by distillation) in a 50-ml volumetric flask. Add 5 ml of the bleached pararosaniline solution and 5 ml of formaldehyde solution. Dilute to the mark with the Na_2HgCl_4 solution, mix, and allow to stand for 30 minutes. Measure the absorbance of the coloured solution at 560 nm, using a reagent blank solution as reference.

50.2.4 OTHER METHODS

Sulphate can be determined indirectly by using a suspension of barium chloranilate. In the formation of $BaSO_4$ at pH $\sim$ 4, sulphate ions displace an equivalent amount of coloured chloranilate ions [73–77]:

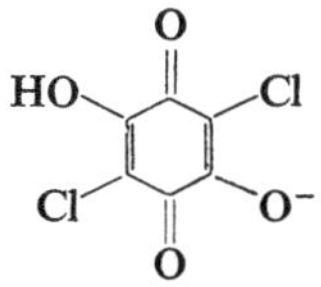

The absorbance of the solution is measured at 530 nm. The reaction is usually carried out in aqueous alcohol since alcohol lowers the solubility of barium chloranilate. This method has been used to determine sulphate in biological materials [3], organic compounds [6], water [74], naphthas [78], and coal ash [77]. Barium chloranilate has also been used in the automatic spectrophotometric determination of sulphate in water [79].

Barium rhodizonate [80] and the barium complex with Nitro-orthanilic S [81] have been used similarly as reagents for sulphate. Sulphate ions will displace an equivalent amount of chromate ions from sparingly soluble barium chromate. The chromium may then be determined either with diphenylcarbazide [82,83] (see p. 215) or directly from the colour of the chromate ions [84,85]. The iodate displaced from barium iodate has been determined colorimetrically as the blue starch-iodine complex [85a].

In another indirect method, sulphate is precipitated as $PbSO_4$ from aqueous acetone or aqueous dioxan, the precipitate dissolved, and the lead determined with dithizone [86].

Sulphate displaces alizarin [87], Xylenol Orange [88], and Amaranth (an azo dye) [89] from their complexes with thorium or zirconium.

When the sulphuric acid formed by removal of the cations from solution on an ion-exchange resin is heated with sucrose, a green product is formed. This reaction is suitable for determining sulphate [5].

Norwitz [90] has developed a turbidimetric method in which sulphate is reduced to hydrogen sulphide, which reacts with an ammoniacal solution of lead citrate to form a yellow-brown suspension of PbS. Other suggested turbidimetric methods involve reaction of sulphate ions with 4-amino-4'-chlorobiphenyl [91,92] or 2-aminoperimidine [93].

The concentration of sulphuric acid in the range 85–99% or 80–99.5% can be determined on the basis of the colour reactions with quinalizarin [94] or 1,1'-dianthrimide [95].

Elemental sulphur reacts quantitatively with cyanide in aqueous acetone to form thiocyanate, which can then be determined as the $FeSCN^{2+}$ complex [96,97].

Hart [8] has determined elemental sulphur turbidimetrically by adding a solution of sulphur in acetone to water.

Absorptiometric methods for determining sulphide are based on the colloidal suspensions formed with lead [98], bismuth [99], and iron(III)+NTA [100].

Wroński [101] has determined sulphide from the blue colour obtained when hydrogen sulphide reacts with the colourless thiofluorescein-silver complex.

Indirect methods for determining sulphide are based on the absorbance (at 598 nm) of the dithizone released when a chloroform solution of silver dithizonate is shaken with the aqueous solution of sulphide [102], or the absorbance (at 525 nm) of the chloranilate released by sulphide from mercuric chloranilate [103,104].

Hydrogen sulphide reduces molybdate in $0.2M$ H_2SO_4 to molybdenum blue. This reaction is suitable for the colorimetric determination of various sulphur compounds after their reduction [105].

Sulphite and sulphur dioxide can be determined on the basis of a change in colour produced by their reducing effect on iodine and starch [106], dichromate [107], and iron(III) in the presence of 1,10-phenanthroline [108,109] or 2,4,6-tri(2'-pyridyl)-s-triazine (TPTZ) [110]. Okutani and Utsumi [111] have added a known amount of mercury(II) to a solution containing sulphite, and extracted into benzene the excess of mercury as the coloured complex with diphenylcarbazone. Sulphite may also be determined indirectly with mercuric chloranilate [104]. Other reagents for sulphite include 2,6-dichlorophenolindophenol [112], 5,5'-dithiobis(2-nitrobenzoic acid) [113], and trinitrobenzoic acid [114].

The bleaching of Methylene Blue by thiosulphate affords a sensitive spectrophotometric method [115].

Urban [116] has developed a method for determining thiosulphate in the presence of polythionates. In a fast reaction with cyanide, catalysed by $CuCl_2$, thiosulphate forms thiocyanate ions which are determined via the colour reaction with iron(III). Cyanolysis (thiocyanate formation) with trithionate and tetrathionate takes 5–15 minutes and 1.5 minutes, respectively. Various methods for determining polythionates by decomposition with cyanide or sulphite have been developed [116–119].

Using anion-exchangers, Iguchi [120] has separated various polythionates; Pollard et al. [121] have similarly separated thiosulphate, sulphite, and polythionates.

A spectrophotometric method for determining persulphate takes advantage of its oxidation of Alcian Blue (a derivative of Phthalocyanine) [122].

References

1. Gustafsson, L., *Talanta* **4**, 227, 237 (1960).
2. Pomeroy, R., *Anal. Chem.* **26**, 571 (1954).
3. Buck, M., *Z. Anal. Chem.* **184**, 427 (1961).
4. Pugh, H. and Waterman, W. R., *Anal. Chim. Acta* **55**, 97 (1971).
5. Ohlweiler, O. A. and Meditsch, J. O., *ibid.* **25**, 233 (1961).
6. Stoffyn, P. and Keane, W., *Anal. Chem.* **36**, 397 (1964).
7. Pacz, D., M. and Guagnini, O. A., *Mikrochim. Acta* **1971**, 220.
8. Hart, M. G., *Analyst* **86**, 472 (1961).
8a. Szekeres, L., *Talanta* **21**, 1 (1974).
9. Tananaev, I. V. and Rudnev, N. A., *Zh. Analit. Khim.* **5**, 82, 281 (1950).
10. Toennies, G. and Bakay, B., *Anal. Chem.* **25**, 160 (1953).
11. Steinbergs, A., *Analyst* **78**, 47 (1953).
12. Wimberley, J. W., *Anal. Chim. Acta* **42**, 327 (1968).
13. Pantaleeva, E. P. and Krupina, I. N., *Zh. Analit. Khim.* **25**, 1989 (1970).
14. Keily, H. J. and Rogers, L. B., *Anal. Chem.* **27**, 759 (1955).
15. Butters, B. and Chenery, E. M., *Analyst* **84**, 239 (1959).
16. Garrido, M. L., *ibid.* **89**, 61 (1964).
17. Chaudhry, I. A. and Cornfield, A. H., *ibid.* **91**, 528 (1966).
18. Massoumi, A. and Cornfield, A. H., *ibid.* **88**, 321 (1963).
19. Zdybek, G., McCann, D. S. and Boyle, A. J., *Anal. Chem.* **32**, 558 (1960).
20. Klipp, R. W., *ibid.* **33**, 1912 (1961).
21. Beswick, G. and Johnson, R. M., *Talanta* **17**, 709 (1970).
22. Kemula, B., Brachaczek, W. and Kornacki, J., *Chem. Anal. (Warsaw)* **3**, 939 (1958).
23. Blanc, P., Bertrand, P. and Liandier, L., *Chim. Anal. (Paris)* **37**, 305 (1955).
24. Claudy, H. N., Karasek, F. W., Ayers, B. O. and Skinner, J. G., *Anal. Chem.* **31**, 1255 (1959).
25. Johnson, C. M. and Nishita, H., *ibid.* **24**, 736 (1952).
26. Budd, M. S. and Bewick, H. A., *ibid.* **24**, 1536 (1952).
27. Stone, H. W. and Forstner, J. L., *J. Am. Chem. Soc.* **79**, 1840 (1957).
28. Marczenko, Z. and Chołuj-Lenarczyk, Ł., *Chem. Anal. (Warsaw)* **10**, 729 (1965).
29. Hofmann, K. and Hamm, R., *Z. Anal. Chem.* **232**, 167 (1967).
30. Zutshi, P. K. and Mahadevan, T. N., *Talanta* **17**, 1014 (1970).
31. Mecklenburg, W. and Rosenkränzer, F., *Z. Anorg. Allgem. Chem.* **86**, 143 (1914).
31a. Rees, T. D., Gyllenspetz, A. B. and Docherty, A. C., *Analyst* **96**, 201 (1971).
32. Skerrett, E. J. and Dickes, G. J., *ibid.* **86**, 69 (1961).
33. Steinbergs, A., Iismaa, O., Freney, J. R. and Barrow, N. J., *Anal. Chim. Acta* **27**, 158 (1962).
34. Muschaweck, J. and Siebke, H., *Z. Anal. Chem.* **209**, 325 (1965).
35. Sjöborg, B. L., *Talanta* **14**, 693 (1967).
36. Kriege, O. H. and Wolfe, A. L., *ibid.* **9**, 673 (1962).
37. Tyou, P. and Humblet, L., *ibid.* **3**, 232 (1960).

38. Fedorov, A. A., Krichevskaya, A. M. and Linkova, F. V., *Zavodsk. Lab.* **27**, 1460 (1961); **36**, 1433 (1970).
39. Zinchenko, V. A. and Gertseva, N. M., *Zh. Analit. Khim.* **22**, 1080 (1967).
40. Farley, L. L. and Winkler, R. A., *Anal. Chem.* **40**, 962 (1968).
41. Kovalenko, N. P. and Martynenko, T. V., *Zavodsk. Lab.* **35**, 1051 (1969).
42. Barcicki, J., Kogutowski, W. and Zajdel, J., *Chem. Anal. (Warsaw)* **15**, 987 (1970).
43. Takeuchi, T., Fujishima, I. and Wakayama, Y., *Mikrochim. Acta* **1965**, 635.
44. Sinclair, A., Hall, R. D., Burns, D. T. and Hayes, W. P., *Talanta* **18**, 972 (1971).
45. Jacobs, M. B., Braverman, M. M. and Hochheiser, S., *Anal. Chem.* **29**, 1349 (1957).
46. Patel, S. S. and Spencer, C. P., *Anal. Chim. Acta* **27**, 278 (1962).
47. Grasshoff, K. M. and Chan, K. M., *ibid.* **53**, 442 (1971).
48. Ducret, L. and Ratouis, M., *ibid.* **21**, 91 (1959).
49. West, P. W. and Gaeke, G. C., *Anal. Chem.* **28**, 1816 (1956).
50. Nauman, R. V., West, P. W., Tron, F. and Gaeke, G. C., *ibid.* **32**, 1307 (1960).
51. Terraglio, F. P. and Manganelli, R. M., *ibid.* **34**, 675 (1962).
52. Pate, J. B., Lodge, J. P. and Wartburg, A. F., *ibid.* **34**, 1660 (1962).
53. Huitt, H. A. and Lodge, J. P., Jr., *ibid.* **36**, 1305 (1964).
54. Nietruch, F. and Prescher, K. E., *Z. Anal. Chem.* **226**, 259; **231**, 28 (1967).
55. Scaringelli, F. P., Saltzman, B. E. and Frey, S. A., *Anal. Chem.* **39**, 1709 (1967).
56. Arikawa, Y., Ozawa, T. and Iwasaki, I., *Bull. Chem. Soc. Japan* **41**, 1454 (1968).
57. King, H. G. and Pruden, G., *Analyst* **94**, 43 (1969).
58. Pate, J. B., Ammons, B. E., Swanson, G. A. and Lodge, J. P., Jr., *Anal. Chem.* **37**, 942 (1965).
59. Seefield, E. W. and Robinson, J. W., *Anal. Chim. Acta* **22**, 61 (1960).
60. Dokládalová, J., *Mikrochim. Acta* **1965**, 344; *Z. Anal. Chem.* **208**, 92 (1965).
61. Majewska, J., *Chem. Anal. (Warsaw)* **12**, 183 (1967); **13**, 29 (1968).
62. Bloomfield, C., *Analyst* **87**, 586 (1962).
63. Sen Gupta, J. G., *Anal. Chem.* **35**, 1971 (1963).
64. Acs, L. and Barabas, S., *ibid.* **36**, 1825 (1964).
65. Larsen, R. P., Ross, L. E. and Ingber, N. M., *ibid.* **31**, 1596 (1959).
66. Burke, K. E. and Davis, C. M., *ibid.* **34**, 1747 (1962).
67. Barabas, S. and Kaminski, J., *ibid.* **35**, 1702 (1963).
68. Helwig, H. L. and Gordon, C. L., *ibid.* **30**, 1810 (1958).
69. Yanagisawa, S., Mitsuzawa, S., and Mori, M., *Japan Analyst* **17**, 580 (1968).
70. West, P. W. and Ordoveza, F., *Anal. Chem.* **34**, 1324 (1962).
71. Kniseley, S. J. and Throop, L. J., *ibid.* **38**, 1270 (1966).
72. Bethge, P. O. and Carlson, M., *Talanta* **16**, 144 (1969).
73. Bertolacini, R. J. and Barney, J. E., II, *Anal. Chem.* **29**, 281 (1957).
74. Procházková, L., *Z. Anal. Chem.* **182**, 103 (1961).
75. Agterdenbos, J. and Martinius, N., *Talanta* **11**, 875 (1964).
76. Carlson, R. M., Rosell, R. A. and Vallejos, E., *Anal. Chem.* **39**, 688 (1967).
77. Schafer, H. N., *ibid.* **39**, 1719 (1967).
78. Klipp, R. W. and Barney, J. E., II, *ibid.* **31**, 596 (1959).
79. Gales, M. E., Jr., Kaylor, W. H. and Longbottom, J. E., *Analyst* **93**, 97 (1968).
80. Babko, A. K. and Litvinenko, V. A., *Zh. Analit. Khim.* **18**, 237 (1963).
81. Basargin, N. N., Men'shikova, V. L., Belova, Z. S. and Myasishcheva, L. G., *Zh. Analit. Khim.* **23**, 732 (1968).
82. Alimarin, I. P. and Sheskol'skaya, A. Ya., *ibid.* **19**, 166 (1964).
83. Iwasaki, I., Utsumi, S., Hagino, K., Tarutani, T. and Ozawa, T., *Bull. Chem. Soc. Japan* **30**, 847 (1957); *J. Chem. Soc. Japan, Pure Chem. Sect.* **79**, 32 (1958).
84. Broekhuysen, J. and Béchet, J., *Anal. Chim. Acta* **13**, 277 (1955).
85. Middleton, K. R., *Analyst* **87**, 444 (1962).
85a. Hinze, W. L. and Humphrey, R. E., *Anal. Chem.* **45**, 814 (1973).
86. Baranowski, R., Ciba, J., Czerniec, J. and Gregorowicz, Z., *Chem. Anal. (Warsaw)* **10**, 499 (1965).

References

87. Babko, A. K. and Markova, L. V., *Zavodsk. Lab.* **24**, 524 (1958).
88. Palatý, V., *Talanta* **10**, 307 (1963).
89. Lambert, J. L., Yasuda, S. K. and Grotheer, M. P., *Anal. Chem.* **27**, 800 (1955).
90. Norwitz, G., *Analyst* **96**, 494 (1971).
91. Martin, J. M. and Stephen, W. I., *Anal. Chim. Acta* **39**, 175, 525 (1967).
92. Mendes-Bezerra, A. E. and Uden, P. C., *Analyst* **94**, 308 (1969).
93. Stephen, W. I., *Anal. Chim. Acta* **50**, 413 (1970).
94. Zimmerman, E. and Brandt, W. W., *Talanta* **1**, 374 (1958).
95. Langmyhr, F. J. and Skaar, O. B., *Anal. Chim. Acta* **23**, 28 (1960).
96. Bartlett, J. K. and Skoog, D. A., *Anal. Chem.* **26**, 1008 (1954).
97. Erdenbaeva, M. I. and Usenova, Z. M., *Zh. Analit. Khim.* **21**, 378 (1966).
98. Goryushina, V. G. and Biryukova, E. Ya., *Zavodsk. Lab.* **31**, 1303 (1965).
99. Dean, G. A., *Analyst* **91**, 530 (1966).
100. Rahim, A. S. and West, T. S., *Talanta* **17**, 851 (1970).
101. Wroński, M., *Chem. Anal. (Warsaw)* **5**, 457 (1960).
102. Kirsten, W. J., *Mikrochim. Acta* **1955**, 1086.
103. Hoffmann, E., *Z. Anal. Chem.* **185**, 372 (1962).
104. Humphrey, R. E. and Hinze, W., *Anal. Chem.* **43**, 1100 (1971).
105. Stratmann, H., *Mikrochim. Acta* **1954**, 668; **1956**, 1031.
106. Katz, M., *Anal. Chem.* **22**, 1040 (1950).
107. Sussman, S. and Portnoy, I. L., *ibid.* **24**, 1652 (1952).
108. Stephens, B. G. and Lindstrom, F., *ibid.* **36**, 1308 (1964).
109. Attari, A., Igielski, T. P. and Jaselskis, B., *ibid.* **42**, 1282 (1970).
110. Stephens, B. G. and Suddeth, H. A., *Analyst* **95**, 70 (1970).
111. Okutani, T. and Utsumi, S., *Bull. Chem. Soc. Japan* **40**, 1386 (1967).
112. Pfleiderer, G., Stock, A., Ötting, F. and Diemair, W., *Z. Anal. Chem.* **239**, 225 (1968).
113. Humphrey, R. E., Ward, M. H. and Hinze, W., *Anal. Chem.* **42**, 698 (1970).
114. Blasius, E. and Ziegler, K., *Z. Anal. Chem.* **269**, 15 (1974).
115. Quentin, K. E. and Pachmayr, F., *ibid.* **200**, 250 (1964).
116. Urban, P. J., *ibid.* **179**, 415, 422; **180**, 110, 116 (1961).
117. Koh, T. and Iwasaki, I., *Biull. Chem. Soc. Japan* **39**, 352 (1966).
118. Iwasaki, I. and Suzuki, S., *ibid.* **39**, 576 (1966).
119. Kelly, D. P., Chambers, L. A. and Trudinger, P. A., *Anal. Chem.* **41**, 898 (1969).
120. Iguchi, A., *Bull. Chem. Soc. Japan* **31**, 597 (1958).
121. Pollard, F. H., Nickless, G. and Glover, R B., *J. Chromatog.* **15**, 533 (1964).
122. Villegas, E., Pomeranz, Y. and Shellenberger, J. A., *Anal. Chim. Acta* **29**, 145 (1963).

Chapter 51

TANTALUM

Tantalum (Ta, at.wt. 180·95) is very similar to niobium in chemical properties (cf. p. 380). It occurs in the +V oxidation state, and forms fluoride, tartrate, oxalate, and peroxide complexes which are, however, not as strong as the corresponding niobium complexes. Tantalum(V) is harder to reduce than niobium(V).

51.1 Methods of Separation

Methods for separating niobium and tantalum from other elements and from each other are discussed in the chapter on niobium (p. 380).

51.2 Methods of Determination

The well known pyrogallol method for determining tantalum is quite selective but rather insensitive. For traces, the extractive spectrophotometric methods using Methyl Violet or other basic dyes are suitable. Babko and Shtokalo [1] have reviewed spectrophotometric methods for determining tantalum.

51.2.1 PYROGALLOL METHOD

In an acid medium (HCl, H_2SO_4), pyrogallol

[benzene ring with three OH groups]

reacts with tantalum(V) to form a yellow complex, the absorption maximum of which occurs in the near ultraviolet. This reaction has long been the basis of the simple and selective, though not very sensitive, colorimetric method for determining tantalum [2–6].

The absorption spectrum of the tantalum complex and the interferences due to other elements (particularly niobium and titanium) vary depending on the reaction conditions. In a $4M$ HCl and $0·02M$ $(NH_4)_2C_2O_4$ medium, the colour due to pyrogallol complexes with niobium and titanium is insignificant. In this medium, a ternary tantalum complex with pyrogallol and oxalic acid and a colourless niobium oxalate complex are formed [5].

The molar absorptivity of the tantalum–pyrogallol complex in $4M$ HCl and $0.02M$ $(NH_4)_2C_2O_4$ solution is 2.4×10^3 ($a = 0.013$) at $\lambda_{max} = 335$ nm. It is advisable, however, to measure the absorbance at longer wavelengths to avoid interference by the excess of pyrogallol, the absorption maximum of which occurs at 315 nm.

When tantalum is determined with pyrogallol, Mo, W, Sb, U, and fluoride interfere seriously. Fluoride can be masked with boric acid [7]. Selectivity can be improved by extracting the tantalum pyrogallol complex into ethyl acetate in the presence of long-chain quaternary ammonium salts [7,8].

To reduce the interference by niobium, tartrate is sometimes added [9].

The colour reaction is carried out in a reducing medium because pyrogallol is readily oxidized to dark coloured products by atmospheric oxygen.

Horák and Okáč [10] have used pyrogallolsulphonic acid instead of pyrogallol, the former being less liable to oxidation.

Pyrogallol has been used to determine tantalum in ores and minerals [2,4,11,12], niobium [13,14], steels [15–17], magnetic alloys [17], zirconium alloys [18], titanium alloys [19], and beryllium and its oxide [20].

Reagents

1. Pyrogallol: 20% solution. Dissolve 20 g of resublimed pyrogallol in water, add 10 ml of conc. HCl and 2 g of $SnCl_2.2H_2O$ (dissolved in 5 ml of conc. HCl), and dilute the solution with water to 100 ml.
2. Standard tantalum solution: 1 mg/ml. Fuse 0.1221 g of tantalum oxide Ta_2O_5 with 4 g of $K_2S_2O_7$ in a quartz or platinum crucible. Dissolve the melt in 4% $(NH_4)_2C_2O_4$ solution and dilute to the mark with the same reagent in a 100-ml volumetric flask. Working solutions are obtained by suitable dilution of the stock solution with 2% $(NH_4)_2C_2O_4$ solution.
3. Ammonium oxalate and hydrochloric acid solution containing 15 g of $(NH_4)_2C_2O_4$ and 760 ml of conc. HCl per litre.

Procedure

Place the sample solution (~ 10 ml) containing not more than 3 mg of Ta in a 50-ml volumetric flask. Add 25 ml of the $(NH_4)_2C_2O_4$ and HCl solution, and 10 ml of the pyrogallol solution. Dilute the solution with water to the mark. After 30 minutes, measure the absorbance of the solution at 350 nm against a reagent blank solution.

51.2.2 METHYL VIOLET METHOD

In dilute hydrofluoric acid medium, tantalum forms an anionic complex, $[TaF_6]^-$, which combines with the basic dye Methyl Violet (MV), (formula, p. 53), to form an ion-pair which can be extracted with benzene. Poluektov *et al.* [21] have developed a sensitive spectrophotometric method for determining tantalum, based on this coloured extract. Rutkowski and Wąsowicz [22] have investigated the optimum reaction conditions.

Maximum absorbance of the coloured extract is obtained when the reaction medium is 0·2–0·3M HF. When the volumes of the aqueous phase and benzene are 30 ml and 10 ml, respectively, 80% of the tantalum is extracted. Extraction of free Methyl Violet is only slight. A mixture of toluene and acetone (8:1 or 9:1) has been recommended as a less toxic extractant [22a].

The molar absorptivity of the complex in the benzene extract obtained under the conditions specified in the procedure below is $\sim 7\cdot5 \times 10^4$ (specific absorptivity 0·42) at $\lambda_{max} = 605$ nm.

Low concentrations of hydrochloric and sulphuric acids do not interfere with the extraction. In the presence of nitric acid, however, more free Methyl Violet is extracted.

Concentrations of niobium up to 0·2 mg/ml can be tolerated in this method. Rhenium at concentrations >5 μg/ml causes high results for tantalum. High concentrations of molybdenum and aluminium cause low results since they mask the hydrofluoric acid as stable fluoride complexes. Moderate amounts of Zr, Ti, W, Fe, Cu, Ca, and Mg do not interfere.

Tantalum has been determined by the Methyl Violet method in zirconium, hafnium, and niobium [23], and in various concentrates [24]. Larger amounts of tantalum have been determined by differential spectrophotometry [24,25].

Reagents

1. Methyl Violet: 0·1% solution in 0·2M HF. Store the solution in a polyethylene bottle.
2. Standard tantalum solution: 1 mg/ml. Dissolve 0·1000 g of suitably pure tantalum in 5 ml of conc. HF and a few drops of conc. HNO_3. Evaporate the solution (in a platinum or Teflon vessel) to dryness, add a few drops of conc. HCl and 2 ml of conc. HF, and evaporate to dryness again. Dissolve the residue in 1 ml of conc. HF and dilute the solution accurately with water (with vigorous stirring) to 100 ml. Store the solution in a polyethylene bottle. Working solutions are obtained by suitable dilution of this stock solution with 0·2M HF.

Procedure

To the sample solution (in a polyethylene separating funnel) containing 1–20 μg of Ta, add 2 ml of the Methyl Violet solution and sufficient hydrofluoric acid to make its concentration 0·2–0·3M in a volume of 30 ml (pH 2·1–2·3). Add 10 ml of benzene and shake for 1 minute. After separation of the phases measure the absorbance of the benzene extract at 605 nm against benzene, or (in the case of small amounts of tantalum) a reagent blank solution.

51.2.3 OTHER METHODS

Many other basic dyes besides Methyl Violet have been employed in extraction–spectrophotometric methods for the determination of tantalum as TaF_6^- [26]. Mention may be made of the triphenylmethane dyes Crystal Violet (molar absorptivity $6 \cdot 6 \times 10^4$) [26,27], Malachite Green [28–30], and Brilliant Green ($\varepsilon = 6 \cdot 8 \times 10^4$) [31,32]; the xanthene dyes Butylrhodamine B ($\varepsilon = 5 \cdot 7 \times 10^4$), Rhodamine 6G [26,33–35], and Rhodamine 3B [26]; and various other basic dyes such as Methylene Blue ($\varepsilon = 6 \cdot 0 \times 10^4$) [36], Methyl Green [36], Acridine Orange [36], Victoria Blue B [37], Nile Blue A [38], and Meldola's Blue [39]. Ion-association complexes with these dyestuffs are extractable into benzene [29,31,35], xylene [28], chloroform [27,32], chlorobenzene [27,38,39], or dichloroethane [27,36].

Some fluorones are sensitive spectrophotometric reagents for tantalum; e.g. phenylfluorone [40–42] and p-dimethylaminophenylfluorone [43,44].

PAR, the well-known reagent for niobium, is also applicable to the determination of tantalum ($\varepsilon \sim 1 \cdot 7 \times 10^4$ at 535 nm) [45–47].

Further diverse organic reagents for the spectrophotometric determination of tantalum include pyrocatechol [48,49], dibromogallic acid [50], haematoxylin [1], quercetin [51], morin [1], and Arsenazo I [1,52].

Guyon [53] has determined tantalum as tantalomolybdenum blue.

References

1. Babko, A. K. and Shtokalo, M. I., *Ukr. Khim. Zh.* **29**, 963 (1963); **30**, 220 (1964).
2. Dinnin, J. I., *Anal. Chem.* **25**, 1803 (1953).
3. Hunt, E. C. and Wells, R. A., *Analyst* **79**, 345 (1954).
4. Marzys, A. E., *ibid.* **80**, 194 (1955).
5. Babko, A. K. and Lukachina, V. V., *Ukr. Khim. Zh.* **28**, 371, 779 (1962).
6. Sarry, B. and Lange, A., *Z. Anal. Chem.* **241**, 186 (1968).
7. Catoggio, J. A. and Rogers, L. B., *Talanta* **9**, 387 (1962).
8. Scott, B. B., *Analyst* **91**, 506 (1966).
9. Dobkina, B. M. and Petrova, E. I., *Zavodsk. Lab.* **25**, 1064 (1959).
10. Horák, J. and Okáč, A., *Collection Czech. Chem. Commun.* **28**, 2563 (1963).
11. Bykova, V. S. and Skrizhinskaya, V. I., *Zavodsk. Lab.* **26**, 523 (1960).
12. Webb, H. W., Ashworth, V. and Hills, J. M., *Analyst* **88**, 142 (1963).
13. Theodore, M. L., *Anal. Chem.* **30**, 465 (1958).
14. Chernikhov, Yu. A., Tramm, R. S. and Pevzner, K. S., *Zavodsk. Lab.* **25**, 398 (1959).
15. Tietze, B., *Mikrochim. Acta* **1972**, 658.
16. Kidman, L., Darwent, C. L. and White, G., *Metallurgia* **62**, 125 (1960).
17. Kidman, L. and White, G., *ibid.* **64**, 153 (1961).
18. Wood, D. F. and Scholes, I. R., *Anal. Chim. Acta* **21**, 121 (1959).
19. Norwitz, G., Codell, M. and Mikula, J. J., *ibid.* **11**, 173 (1954).
20. Hibbits, J. O., Oberthin, H., Liu, R. and Kallmann, S., *Talanta* **8**, 209 (1961).
21. Poluektov, N. S., Kononenko, L. I. and Lauer, R. S., *Zh. Analit. Khim.* **13**, 396 (1958).
22. Rutkowski, W. and Wąsowicz, S., *Chem. Anal.* (*Warsaw*) **11**, 971 (1966).
22a. Dobkina, B. M., Kuchmistaya, G. I., Nadezhdina, G. B. and Davydova, N. M., *Zavodsk. Lab.* **39**, 671 (1973).
23. Lauer, R. S. and Poluektov, N. S., *ibid.* **25**, 903 (1959).

24. Kuchmistaya, G. I., Nadezhdina, G. B. and Dobkina, B. M., *ibid.* **36**, 275 (1970).
25. Dobkina, B. M., Zubynina, K. B., Malyutina, T. M. and Sazikova, G. B., *Zh. Analit. Khim.* **22**, 1510 (1967).
26. Makarova, S. V. and Alimarin, I. P., *ibid.* **19**, 564, 847 (1964).
27. Alimarin, I. P. and Makarova, S. V., *ibid.* **19**, 90 (1964).
28. Kakita, Y. and Gotô, H., *Anal. Chem.* **34**, 618 (1962).
29. Eberle, A. R. and Lerner, M. W., *ibid.* **39**, 662 (1967).
30. Grossmann, O., *Z. Anal. Chem.* **245**, 135 (1969).
31. Tsukahara, I., *Japan Analyst* **16**, 583 (1967).
32. Nevzorov, A. N., Onoprienko, N. S. and Mordvinova, S. N., *Zavodsk. Lab.* **36**, 1176 (1970).
33. Pavlova, N. N. and Blyum, I. A., *ibid.* **28**, 1305 (1962).
34. Dorosh, V. M., *Zh. Analit. Khim.* **18**, 961 (1963).
35. Pavlova, N. N. and Blyum, I. A., *Zavodsk. Lab.* **32**, 1196 (1966).
36. Tarayan, V. M., Ovsepyan, E. N. and Barkhudaryan, S. R., *Izv. Vyssh. Ucheb. Zaved., Khim. Khim. Tekhnol.* **13**, 1573 (1970); *Zh. Analit. Khim.* **27**, 19 (1972).
37. Kirkbright, G. F., Mayhew, M. D. and West, T. S., *Anal. Chem.* **40**, 2210 (1968).
38. Gagliardi, E. and Wolf, E., *Mikrochim. Acta* **1969**, 888.
39. Pilipenko, A. T. and Nguen Dyk Tu, *Ukr. Khim. Zh.* **34**, 1291 (1968).
40. Luke, C. L., *Anal. Chem.* **31**, 904 (1959).
41. Hill, J. H., *Analyst* **91**, 659 (1966).
42. Bingham, C. D., Maseda, M. S. and Johnson, B. G., *Anal. Chem.* **41**, 1144 (1969).
43. Nazarenko, V. A., and Shustova, M. B., *Zavodsk. Lab.* **23**, 1283 (1957).
44. Patrovský, V., *Chem. Listy* **59**, 1464 (1965).
45. Alimarin, I. P. and Khan' Si-I, *Zh. Analit. Khim.* **18**, 182 (1963).
46. Elinson, S. V. and Rezova, A. T., *ibid.* **19**, 1078 (1964).
47. Elinson, S. V., Pobedina, L. I. and Rezova, A. T., *Zavodsk. Lab.* **37**, 521 (1971).
48. Rosotte, R. and Jaudon, E., *Chim. Anal. (Paris)* **41**, 229 (1959).
49. Babko, A. K. and Lukachina, V. V., *Ukr. Khim. Zh.* **27**, 682 (1961).
50. Ackermann, G. and Koch, S., *Talanta* **16**, 95, 284, 288 (1969); **17**, 757 (1970).
51. Popa, G., Negoiu, D. and Baiulescu, G., *Z. Anal. Chem.* **165**, 16 (1959).
52. Nikitina, E. I., *Zh. Analit. Khim.* **13**, 72 (1958).
53. Guyon, J. C., *Anal. Chim. Acta* **30**, 395 (1964).

Chapter 52
TELLURIUM

Tellurium (Te, at.wt. 127·60) occurs in the $-II$, $+IV$, and $+VI$ oxidation states in telluride, tellurite, and tellurate, respectively. In acidic non-complexing medium (e.g. H_2SO_4), Te^{4+} cations exist. The oxide, TeO_2, is only slightly volatile (it escapes when ignited in a crucible at 800–900°C) and is not readily soluble in water. Tellurium dissolves in nitric acid to give tellurite. Reducing agents, such as $SnCl_2$, SO_2, NH_2OH, and hypophosphite, reduce tellurium(IV) to the element. Tellurium can be oxidized to Te(VI) only by powerful oxidants. Tellurium(IV) forms chloride and bromide compounds which are not as volatile as the corresponding selenium compounds.

52.1 Methods of Separation

52.1.1 PRECIPITATION

A frequently used method for separating tellurium from most elements consists in its reduction in acid medium to the element [1–3], usually by stannous chloride, sulphur dioxide, or hypophosphorous acid. Bode and Hettwer [2] have investigated more closely the precipitation of tellurium from 0·1–0·7M HCl with SO_2: Se, Au, Ag, Pd, and Pt are quantitatively precipitated, while Hg, Ru, Rh, Os, Ir, and Pb are only partially precipitated. Selenium and arsenic are suitable collectors for traces of tellurium [1].

Selenium is quantitatively removed from tellurium by evaporating to dryness a solution of both elements in hydrochloric or hydrobromic acid (in the presence of a small amount of free bromine). Selenium is thereby volatilized while tellurium remains in the vessel.

Knyazeva and Kleiman [4] have separated tellurium from selenium after sintering the sample with Eschka mixture (1 part by weight of Na_2CO_3 + 2 parts of MgO) for 40 minutes at 800°C. When the sinter is leached with water, readily-soluble magnesium selenate passes into solution while sparingly-soluble magnesium tellurate remains in the precipitate.

Bock and Tschöpel [1] recommend coprecipitating traces of tellurium with $Fe(OH)_3$.

52.1.2 EXTRACTION

Tellurium(IV) can be extracted from 4–8M HCl with the following solvents: methyl ethyl ketone [5], MIBK [6], TBP [7], and octanol [8]. Extraction with TBP from 2–10M HCl separates tellurium(IV) from tellurium(VI) [9]. The extraction of tellurium as the bromide [8] and iodide [1,8,10] complexes is also possible.

Ion-association complexes of the tellurium(IV) chloride complex and DAPM [11], 4,4′-methylenediaminediantipyrine [11], DAM [12], or the tetraphenylphosphonium ion [1] are extractable into non-polar solvents from 5–7M HCl.

Separation of tellurium from selenium is achieved by extraction of tellurium into CCl_4 with DDTC (pH 8·5) [13,14], or from 2M H_2SO_4 with a solution of thionaphthenic acid in CCl_4 [15].

52.1.3 Ion-Exchange

Busev et al. [16] have separated tellurium from many metals on an AV-17 anion-exchanger by using an aqueous LiCl medium.

Te(IV), Sb(V), and Sn(IV) can be separated on Dowex-1 anion-exchanger by first eluting tellurium and then antimony with 0·1M $H_2C_2O_4$ (pH 4·8), and finally eluting tin with 1M H_2SO_4 [17].

Šimek [18] has separated tellurium(IV) from much larger quantities of selenium(IV) by sorption of the anionic tellurium chloride complex (formed in 6M HCl) on a strongly basic anion–exchanger.

From <0·05M HCl, Amberlite IR-120 cation-exchanger retains tellurium but not selenium [19]. Similarly, in 0·01–0·1M HCl tellurium is retained while platinum metals are eluted [20]. Pt, Pd, Cu, and Ni (but not Te) are retained in the cation-exchange column from ammoniacal solutions [20].

52.2 Methods of Determination

The extractive spectrophotometric method with Bismuthiol II is sensitive and highly selective. A method based on the coloured sol of elementary tellurium is simple and less sensitive, but is very well suited to the determination of larger quantities of tellurium.

Murashova and Sushkova [21] have reviewed spectrophotometric methods for determining tellurium.

52.2.1 Tellurium Sol Method

When small amounts of tellurium(IV) are reduced to the element in acid medium, the absorbance of the resulting coloured sol is a suitable basis for the determination of tellurium [22–24]. To stabilize the pseudosolutions, the reduction is carried out in the presence of protective colloids, e.g. gum arabic, gelatine, or poly(vinyl alcohol).

When stannous chloride is used as reducing agent [14,18,25], 1·5–3M HCl is the most suitable medium. With hypophosphite as reducing agent [23], the acidity of the solution should be kept within the limits 0·1–0·4M HCl or H_2SO_4. Since the colour (brown, blue, red) of the sol depends on the acidity and the quality and concentration of the reducing agent, the reaction conditions must be precisely the same for the sample solutions and for the standard solutions.

The molar absorptivity of the tellurium sol formed by reduction with stannous chloride under the conditions described in the procedure below is 5.4×10^3 ($a = 0.043$) at 400 nm. The pseudosolution does not have an absorption maximum in the near ultraviolet or in the visible spectrum. At 340 nm the absorbance is 15% higher, and at 500 nm it is 20% lower, than the absorbance at 400 nm.

Interference in the determination of tellurium with $SnCl_2$ as reductant comes from other elements also precipitated by the reducing agent, namely Se, Hg, Au, and the platinum metals. Selenium is readily removed from tellurium by double evaporation to dryness in hydrochloric or hydrobromic acid.

Tellurium has been determined as a coloured sol in (*inter alia*) bismuth [25], selenium [25a], steel and cast iron [26,26a], and urine [10].

Reagents

1. Stannous chloride, $SnCl_2 \cdot 2H_2O$: 20% solution in $2M$ HCl.
2. Standard tellurium solution: 1 mg/ml. Place in a beaker 1·000 g of suitably pure powdered tellurium. Add 50 ml of conc. HCl, and introduce conc. HNO_3 in small portions and with heating until the tellurium dissolves. Then add 100 ml of water, and boil the solution for 5 minutes (to expel oxides of nitrogen). Add 20 ml of conc. HCl, and dilute the solution with water in a volumetric flask to 1 litre.
3. Selenium(IV) solution: 1 mg/ml (p. 476).

Procedure

Precipitation separation of Te. Heat a clear sample solution (~ 100 ml, $\sim 1M$ HCl) in a beaker to boiling. Add 2 mg of selenium [as the selenium(IV) solution], then the $SnCl_2$ solution dropwise until the yellow colour disappears [if the sample solution contains iron(III)], and 2 more ml of $SnCl_2$. Stand the beaker for ~ 3 hours in a hot water-bath, and allow to cool overnight. Filter off the precipitate of selenium and tellurium on a sintered-glass crucible, and wash with dilute $SnCl_2$ solution and water. Dissolve the precipitate in 5 ml of conc. HCl containing 3 or 4 drops of perhydrol. Evaporate the solution to dryness in the beaker. Add 2 ml of conc. HCl and evaporate to dryness again (to eliminate volatile selenium). Dissolve the residue in the beaker in dilute HCl.

Determination of Te. Acidify ~ 30 ml of sample solution containing not more than 0·7 mg of Te till $\sim 2M$ in HCl. Add 5 ml of 2% poly(vinyl alcohol) and stir well. Add 5 ml of $SnCl_2$ solution while stirring, and make up the solution with water in a volumetric flask to 50 ml. Mix thoroughly and measure the absorbance at 400 nm, using water as reference.

52.2.2 Bismuthiol II Method

Bismuthiol II (5-mercapto-3-phenyl-1,3,4-thiadiazole-2-thione, potassium salt), a reagent used for the gravimetric determination of bismuth and certain other metals, was independently adopted in 1960-1961 by Jankovský and Kšir [27] and by Cheng [28,29] as a spectrophotometric reagent for tellurium. The extractive spectrophotometric method for determining tellurium outlined below is a modification of the method of Yoshida et al. [30].

Bismuthiol II reacts in acid medium with tellurium(IV) to form a neutral 1:4 complex which is extractable into chloroform. The probable course of the reaction is as follows:

$$4 \text{(Bismuthiol II)} + Te^{4+} \rightarrow \text{(complex)} + 4H^+$$

Free Bismuthiol II, which absorbs in the same spectral region as the tellurium(IV) complex, is extracted into chloroform together with the complex, but is then stripped with an aqueous buffer solution at pH $\sim$ 8.

The absorption maximum of the tellurium–Bismuthiol II complex occurs at 330 nm ($\varepsilon = 3.6 \times 10^4$; $a = 0.31$). At 400 nm, the molar absorptivity is $\sim 8.0 \times 10^3$. The absorption spectrum of the complex is shown in Fig. 52.1.

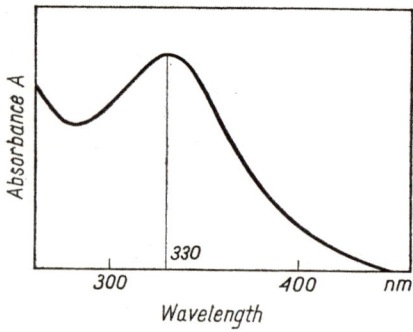

Fig. 52.1 Absorption spectrum of tellurium–Bismuthiol II complex in chloroform

More than 99% of the complex is extracted into chloroform from $3M$ HCl. Extraction from a weakly acid medium (pH 3·5) is less efficient (95%).

In the determination of tellurium by extracting the Bismuthiol II complex from $3M$ HCl into chloroform, As(V), Se(IV), Cu, Hg, and Pd interfere. The effect of selenium is considerable: with equal quantities (by weight) of selenium and tellurium present, the absorbance is more than doubled. In the case of copper, the increase in absorbance is 70%, in the case of mercury, it is ~ 15%.

Oxidizing substances which decompose Bismuthiol II also interfere. Nitric acid must not be present even in trace amounts. Aqueous solutions of the reagent cannot be stored because they are oxidized by atmospheric oxygen and become turbid as a precipitate forms.

In Jankovský and Kšir's method [27], the tellurium–Bismuthiol II complex is extracted into benzene from an acetate medium at pH ~ 4. In this case, selenium interference is less. Many metals which interfere at this pH can be masked with EDTA.

In Cheng's method [28], the colour reaction is carried out in a weakly acid medium (pH 2), and the pH is then raised to 6·5 before extraction of the complex into chloroform. At this pH, extraction of free Bismuthiol II is negligible. EDTA is again used as masking agent. In a modified method, Cheng and Goydish [29] achieved a 20% increase in sensitivity by adding ammonium sulphate to the aqueous solution.

The Bismuthiol II method has also been investigated by others [12, 31,32]. The method has been used for determining tellurium in ores [27] and thin films [33].

5-Mercapto-3-(2-naphthyl)-1,3,4-thiadiazole-2-thione, an analogue of Bismuthiol II, has also been proposed as a reagent [34,35].

Reagents

1. Bismuthiol II: 0·25% solution in cold water which has been previously boiled. The solution is stable for one day.
2. Standard tellurium solution: 1 mg/ml (p. 523).
3. Buffer solution (pH 8). Dissolve 7 g of sodium dihydrogen orthophosphate, 7 g of borax, and 5 g of EDTA in 500 ml of water, and, using a pH-meter, adjust the solution to pH 8·0 ($\pm 0\cdot 1$) by adding NaOH solution.

Procedure

Add $6M$ HCl (previously boiled and cooled) to the sample solution containing not more than 120 μg of Te till the concentration of HCl is $3M$. Add 2 ml of Bismuthiol II solution, and stir thoroughly. After 1 minute, extract tellurium by shaking with two portions of chloroform for 1 minute. Wash the extract with the pH 8 buffer solution. Dilute the extract to the mark with chloroform in a 50-ml or smaller volumetric flask (depending on the amount of tellurium). Measure the absorbance of the yellow solution at 330 nm, using a reagent blank solution as reference.

52.2.3 Other Methods

Tellurium(IV) reacts with thiourea, $S=C(NH_2)_2$, (the concentration of which in the final solution should be $\sim 10\%$) in $\sim 1M$ H_2SO_4, HNO_3, or H_3PO_4 to form a yellow complex suitable for spectrophotometry [36–39]. The absorption maximum of the complex occurs in the ultraviolet at 310–320 nm. Thiourea also gives colour reactions with Bi, Sb, Sn, Pd, Hg, Os, and Pt. Hikime [40] has extracted with TBP the ion-association complex formed by the cationic tellurium–thiourea complex and thiocyanate ions.

The following thiourea derivatives have been proposed for the spectrophotometric determination of tellurium: *sym*-diphenylthiourea [41,42], 1,4-diphenylthiosemicarbazide [43,44], dinaphthylthiourea [44], and diphenylthiocarbazide [44].

Other organosulphur reagents used similarly include DDTC [13,45,46], thioglycollic acid [47], mercaptobenzothiazole [48], and tetramethylthiuram disulphide [49].

Methods based on the yellow colour of tellurium(IV) complexes with iodide [50–53], bromide [54–57], and chloride [58] in strongly acidic solutions are less sensitive. The absorption maxima of these compounds occur in the ultraviolet. The molar absorptivity of the tellurium bromide complex (in $7M$ H_2SO_4 and $0.05M$ NaBr) is $\sim 3.0 \times 10^3$ at 440 nm [57].

The extractive spectrophotometric methods based on ion-association complexes composed of the tellurium bromide complex and Butylrhodamine B [3,7,59] or Victoria Blue 4R [60] (ε is $\sim 8.0 \times 10^4$ at 602 nm) exhibit high sensitivity. In the case of the analogous complex formed with DAPM [61,62], the sensitivity is much lower.

Ingamels and Sandell [63] have based a method for determining tellurium on the reaction between tellurium(IV) and the Fe(III)–1,10-phenanthroline complex in the presence of dichromate as catalyst:

$$Te(IV) + 2Fe(phen)_3^{3+} \rightarrow Te(VI) + 2Fe(phen)_3^{2+}$$

The equivalent amount of red ferroin is formed. The method has been applied to the determination of small amounts of tellurium in selenium [64,65].

Tellurium(IV) may also be determined from the change in the colour caused by the redox reaction with dichromate [66].

Furthermore, the colour reactions of tellurium(IV) with 1,1'-dianthrimide [67], and of elemental tellurium with hot concentrated sulphuric acid [68], have been exploited.

Lastly, there are possibilities for determining tellurium spectrophotometrically as the heteropoly complexes formed with such elements as Mo, W, V, and Si [69–71].

References

1. Bock, R. and Tschöpel, P., *Z. Anal. Chem.* **246**, 81 (1969).
2. Bode, H. and Hettwer, E., *ibid.* **173**, 285 (1960).
3. Ivankova, A. I. and Blyum, I. A., *Zavodsk. Lab.* **27**, 371 (1961).
4. Knyazeva, R. N. and Kleĭman, V. Ya., *ibid.* **31**, 410 (1965).
5. Atanassowa, B. W., *Mikrochim. Acta* **1966**, 970.
6. Jordanov, N. and Havesov, I., *Z. Anal. Chem.* **248**, 296 (1969).
7. Alekseeva, L. S., Murashova, V. I., Bakunina, L. I. and Skripchuk, V. G., *Zavodsk. Lab.* **37**, 1299 (1971).
8. Iofa, B. Z. and Ridvan, M., *Radiokhimiya*, **13**, 534 (1971).
9. Inarida, M., *Japan Analyst* **7**, 449 (1958).
10. Hanson, C. K., *Anal. Chem.* **29**, 1204 (1957).
11. Busev, A. I., Babenko, N. L. and Khoang Min' Tyau, *Zh. Analit. Khim.* **18**, 1094 (1963).
12. Pollock, E. N., *Anal. Chim. Acta* **40**, 285 (1968).
13. Bode, H., *Z. Anal. Chem.* **144**, 90,165 (1955).
14. Pavlova, V. N., Vasil'eva, N. G., and Kashlinskaya, S. E., *Zavodsk. Lab.* **27**, 965 (1961).
15. Zolotov, Yu. A. and Alekperova, A. A., *Zh. Analit. Khim.* **26**, 131 (1971).
16. Busev, A. I., Bagbanly, I. L., Bagbanly, S. I., Guseinov, I. K. and Rustamov, M. Kh., *ibid.* **25**, 1374 (1970).
17. Smith, G. W. and Reynolds, S. A., *Anal. Chim. Acta* **12**, 151 (1955).
18. Šimek, M., *Chem. Listy* **60**, 817 (1966).
19. Aoki, F., *Bull. Chem. Soc. Japan.* **26**, 480 (1953).
20. Strel'nikova, N. P. and Lystsova, G. G., *Zavodsk. Lab.* **26**, 142 (1960).
21. Murashova, V. I. and Sushkova, S. G., *Zh. Analit. Khim.* **24**, 729 (1969).
22. Johnson, R. A., *Anal. Chem.* **25**, 1013 (1953).
23. Johnson, R. A., Kwan, F. P. and Westlake, D., *ibid.* **25**, 1017 (1953).
24. Johnson, R. A. and Andersen, B. R., *ibid.* **27**, 120 (1955).
25. Sinyakova, S. I. and Krol', Ch. Ya., *Tr. Komis. po Analit. Khim. Akad. Nauk SSSR* **12**, 206 (1960).
25a. Futekov, L. and Atanasova, B., *Talanta* **19**, 817 (1972).
26. Maneschi, S. and Gallazzi, C., *Anal. Chim. Acta* **54**, 461 (1971).
26a. Koch, O. G., *ibid.* **64**, 156 (1973).
27. Jankovský, J. and Kšir, O., *Talanta* **5**, 238 (1960).
28. Cheng, K. L., *ibid.* **8**, 301 (1961).
29. Cheng, K. L. and Goydish, B. L., *ibid.* **13**, 1210 (1966).
30. Yoshida, H., Taga, M. and Hikime, S., *ibid.* **13**, 185 (1966).
31. Navrátil, O. and Šorfa, J., *Collection Czech. Chem. Commun* **34**, 975 (1969).
32. Bock, R. and Spalek, U., *Z. Anal. Chem.* **254**, 183 (1971).
33. Marczenko, Z., Mojski, M. and Czarnecka, I., *Chem. Anal. (Warsaw)* **18**, 189 (1973).
34. Busev, A. I. and Simonova, L. N., *Zh. Analit. Khim.* **22**, 1850 (1967).
35. Dang Chung Tuan, Simonova, L. N. and Busev, A. I., *Zavodsk, Lab.* **37**, 408 (1971).
36. Nielsch, W. and Böltz, G., *Z. Metallkunde* **45**, 380 (1954).
37. Nielsch, W. and Giefer, L., *Z. Anal. Chem.* **144**, 191; **145**, 347 (1955); **155**, 401 (1957).
38. Jílek, A., Vřeštál, J. and Havíř, J., *Chem. Zvesti* **10**, 110 (1956).
39. Gladyshev, V. P., *Zavodsk. Lab.* **24**, 275 (1958).
40. Hikime, S., *Bull. Chem. Soc. Japan.* **33**, 761 (1960).
41. Yoshida, H. and Hikime, S., *Talanta* **11**, 1349 (1964).
42. Russell, B. G., Lubbe, W. V., Wilson, A., Jones, E., Taylor, J. D. and Steele, T. W., *ibid.* **14**, 957 (1967).
43. Mel'chekova, Z. E. and Murashova, V. I., *Zh. Analit. Khim.* **25**, 556 (1970).
44. Murashova, V. I., Mel'chekova, Z. E., Novikov, E. G. and Podgornaya, I. V., *ibid.* **24**, 1205 (1969).

45. Inarida, M., *J. Chem. Soc. Japan, Pure Chem. Sect.* **79**, 968 (1958).
46. Luke, C. L., *Anal. Chem.* **31**, 572 (1959).
47. Kirkbright, G. F. and Ng, W. K., *Anal. Chim. Acta* **35**, 116 (1966).
48. Tutkuvene, V. E. and Ramanauskas, E. I., *Zavodsk. Lab.* **35**, 1045 (1969).
49. Tutkuvene, V. E. and Ramanauskas, E. I., *Zh. Analit. Khim.* **21**, 564 (1966).
50. Johnson, R. A. and Kwan, F. P., *Anal. Chem.* **23**, 651 (1951).
51. Geiersberger, K. and Durst, A., *Z. Anal. Chem.* **135**, 11 (1952).
52. Brown, E. G., *Analyst* **79**, 50 (1954).
53. Murashova, V. I., *Zh. Analit. Khim.* **17**, 80 (1962); **21**, 345 (1966).
54. Fletcher, N. W. and Wardle, R., *Analyst* **82**, 743 (1957).
55. Shitareva, G. G., *Zavodsk. Lab.* **27**, 1196 (1961).
56. Murashova, V. I., Sushkova, S. G. and Bakunina, L. I., *ibid.* **33**, 280 (1967).
57. Bakunina, L. I. and Murashova, V. I., *Zh. Analit. Khim.* **25**, 142 (1970).
58. Hanson, M. W., Bradbury, W. C. and Carlton, J. K., *Anal. Chem.* **29**, 490 (1957).
59. Murashova, V. I. and Skripchuk, V. G., *Zh. Analit. Khim.* **27**, 340 (1972).
60. Kish, P. P. and Kremeneva, S. G., *ibid.* **25**, 2200 (1970).
61. Busev, A. I. and Babenko, N. L., *ibid.* **18**, 972 (1963).
62. Busev, A. I., Babenko, N. L. and Chepik, M. N., *ibid.* **19**, 871, 926, 1057 (1964).
63. Ingamels, C. O. and Sandell, E. B., *Microchem. J.* **3**, 3 (1959).
64. Etten, N. and Muschaweck, J., *Z. Anal. Chem.* **206**, 17 (1964).
65. Baresel, D., Corinth, U. and Stähr, A., *ibid.* **252**, 14 (1970).
66. Ma, T. S. and Zoellner, W. G., *Mikrochim. Acta* **1971**, 329.
67. Skaar, O. B. and Langmyhr, F. J., *Anal. Chim. Acta* **23**, 175 (1960).
68. Wiberley, S. E., Bassett, L. G., Burrill, A. M. and Lyng, H., *Anal. Chem.* **25**, 1586 (1953).
69. Ganelina, E. Sh., *Zh. Analit. Khim.* **18**, 551 (1963).
70. Ku Tkhan' Long, Sudakov, F. P. and Shakhova, Z. F., *ibid.* **19**, 734, 968 (1964).
71. Polotebnova, N. A. and Derkach, L. V., *Tr. Komis. po Analit. Khim. Akad. Nauk SSSR* **16**, 31 (1968).

Chapter 53
THALLIUM

Thallium (Tl, at.wt. 204·37) occurs in its compounds in the +I and +III oxidation states. Thallous compounds are the more stable. Thallium(I) forms thiosulphate and ammine complexes of low stability, and several sparingly soluble compounds, e.g. white TlCl, yellow TlI, and black Tl_2S.

Thallium(III) ions, which are colourless, can exist only in strongly acidic media since the brown hydroxide, $Tl(OH)_3$, which has no amphoteric properties, is precipitated at pH values as low as 0·3. Thallium(III) yields halide, oxalate, and tartrate complexes. Tl(I) is oxidized to Tl(III) only by powerful oxidants, e.g. MnO_4^-, aqueous Cl_2, and aqueous Br_2.

53.1 Methods of Separation

53.1.1 Extraction

Of the thallium halide complexes, the bromide complex is the most important in extraction separations [1–4]. Thallium(III) and gold(III) are extracted quantitatively from 1–3M HBr with diethyl or di-isopropyl ether, or with butyl acetate. At that concentration of hydrobromic acid, Fe(III), Ga, In, Hg, and Te are extracted in negligible quantities. After extraction with MIBK, Ga, In, Fe (III), and Sn are stripped with a 1·5M HBr–formamide solution, leaving only Tl(III) in the ketone phase [5].

Extraction of thallium(III) from hydrochloric acid is less selective [6–11]. For diethyl ether and di-isopropyl ether, the most suitable media are 4M HCl and 6M HCl, respectively. Other oxygen-containing solvents, such as amyl acetate and TBP, are also suitable extractants. Au(III), Fe(III) Sb(V), Sn, Hg, Mo, As, Ge, and Ga are partly extracted along with the thallium from hydrochloric acid medium. N-Benzylaniline has also been used for extraction of the chloro-complex from 1M HCl, to separate Tl(III) from Ga and In [12]. Extraction from 2M HBr was also proposed [13].

Extraction of thallium (III and I) iodide complexes is of minor analytical importance [6].

Thallium(I) can be extracted into a chloroform solution of dithizone from an ammoniacal citrate–cyanide medium (pH ~ 11) [14–16]. Lead, Bi, and Sn(II) are also extracted [16].

Schweitzer *et al*. [17–19] have carried out detailed investigations of a large number of extractable thallium(I) chelates. DDTC [20,21], oxine [21], thio-oxine [21,22], and TTA [21] are particularly applicable. Thallium has also been extracted with alkylphosphoric acids [23].

53.1.2 Precipitation and Other Methods

Traces of thallium(III) can be precipitated from alkaline solution as $Tl(OH)_3$ with iron(III), manganese [3], or lanthanum hydroxide [24] as a collector. Ferric hydroxide also adsorbs thallium(I) [25]. Thallium is separated quantitatively from a dilute HNO_3 medium with hydrous MnO_2 produced by the reaction between Mn^{2+} and MnO_4^- [26].

From an acidic sulphide medium, traces of thallium(I) are precipitated as Tl_2S with bismuth as collector [27]. Minute amounts of thallium(I) may also be coprecipitated with lead iodide [28].

Thallium (as $TlCl_4^-$) is coprecipitated from $0.2M$ HCl with the precipitate formed by p-dimethylaminoazobenzene and Methyl Orange [29].

Geilmann and Neeb [3] have isolated traces of thallium by sublimation. The volatility of thallium causes losses during ignition or fusion with sodium carbonate, borax, or sodium hydroxide [3,29]. No thallium is lost, however, when samples are decomposed with acids (HCl, H_2SO_4, $HClO_4$) or during concentration of the acid solutions by evaporation.

A strongly acidic cation-exchanger retains Tl(I), but not Hg, Bi, Cu, Fe, Pb, or Zn, from an EDTA solution at pH 4: the thallium may be eluted with $2M$ HCl [30]. Thallium(I) is also retained from tartrate, citrate, or pyrophosphate solutions (pH 3–5), while Fe, Cu, Zn, Cd, Pb, and Sb pass to the eluate. In this case, thallium has been eluted with $6M$ HCl [31]. Mixed HCl–acetone eluents have been used to separate Tl(III) from Al, Ga, and In [32].

From $0.1M$ HCl medium, thallium(III) is adsorbed as an anionic chloride complex on a strongly basic anion-exchanger [33].

53.2 Methods of Determination

The spectrophotometric methods for determining thallium which are discussed in detail are the sensitive Rhodamine B and Brilliant Green extraction methods, and an indirect starch–iodine method. Methods involving other basic dyes are also noteworthy.

53.2.1 Rhodamine B Method

Thallium(III) in the form of the $[TlCl_4]^-$ or $[TlBr_4]^-$ complex ions reacts in an acid medium (1–2M HCl) with Rhodamine B (p. 53) to form a red-violet fluorescent complex, which is soluble in benzene, di-isopropyl ether, and isoamyl alcohol. Onishi [15,34] has exploited the coloured extracts in the spectrophotometric determination of thallium. A mixture of benzene and carbon tetrachloride $(2+1)$ is a convenient extractant since it is denser than water and less prone to emulsification.

The molar absorptivity of a (C_6H_6 + CCl_4) solution of the ion-pair [Rhod. B]$^+$ [$TlCl_4$]$^-$ is 9.7×10^4 (specific absorptivity 0·40) at λ_{max} = 560 nm.

Bromine water or ceric sulphate is used to oxidize thallium(I) to thallium(III). Excess of bromine is distilled off or reacted with phenol. Excess of cerium(IV) is reduced with hydroxylamine [35].

Under the reaction conditions employed for thallium(III), Rhodamine B also reacts with gold(III), iron(III), antimony(V), mercury(II), and gallium. Hence, thallium is usually separated from interfering elements beforehand, e.g. by extraction with dithizone in chloroform from an alkaline cyanide medium. In this separation, the metals which are extracted with thallium do not interfere in the Rhodamine B reaction [15,36]. Thallium dithizonate is stripped from the chloroform with dilute nitric acid or, alternatively, the solvent is evaporated and the dithizone mineralized with a mixture of H_2SO_4 and HNO_3. The preliminary extraction of thallium into diethyl ether as the bromide complex has also been applied [37,38].

When gold and mercury are removed from aqueous solution by cementation on metallic copper, antimony is reduced partly to the metal and partly to Sb(III), while thallium is reduced to Tl(I). Iron(III) may be masked with phosphoric acid.

The Rhodamine B method has been used to determine thallium in tin-cadmium alloys [37], zinc and cadmium [38], lead [26], selenium [39], and silicate minerals [36].

Reagents

1. Rhodamine B: 0·1% solution in $2M$ HCl.
2. Standard thallium(I) solution: 1 mg/ml.
   (a) Dissolve 1·303 g of thallous nitrate, $TlNO_3$, (dried at 110°C) in water containing 2 ml of conc. HNO_3, and dilute the solution with water in a volumetric flask to 1 litre.
   (b) Dissolve 1·036 g of thallous oxide, Tl_2O, in 15 ml of hot HNO_3 (1 + 1). Dilute the solution with water in a volumetric flask to 1 litre.
3. Ascorbic acid: 2% solution, freshly prepared.
4. Dithizone: 0·01% solution in chloroform (p. 492).

Procedure

Separation of Tl with dithizone. To the acidic sample solution containing not more than 100 μg Tl, add 1 ml of ascorbic acid solution. After 3 minutes, add 5 ml of 10% sodium citrate solution, ammonia to make the pH 9–10, 2·5 ml of 10% potassium cyanide solution, and water to 20–30 ml. Extract thallium with three portions of the dithizone solution, shaking for 2 minutes with each portion. Add 5 or 6 drops of conc. H_2SO_4 to the combined chloroform extracts in a beaker, and evaporate off the chloroform. Then heat more intensely and add conc. HNO_3 dropwise to mineralize the organic residue. Evaporate some of the sulphuric acid. Cool the residue and dissolve it in 10–20 ml of $2M$ HCl.

Determination of Tl. To the solution in $2M$ HCl, add 1 ml of saturated bromine water and heat the solution (without boiling) until the yellow colour of the free bromine in the solution disappears. Cool the solution, transfer it to a separating funnel, add 2 ml of Rhodamine B solution, and shake the solution for 1 minute with two portions of a mixture of benzene and carbon tetrachloride (2+1). Make up the combined extracts to the mark with the solvent in a 50-ml volumetric flask (or smaller depending on the amount of thallium). Measure the absorbance of the coloured solution at 560 nm, using the solvent or a reagent blank solution as reference.

53.2.2 Brilliant Green Method

Brilliant Green is a basic triphenylmethane dye (p. 52). Its ion-association complexes with $TlCl_4^-$ or $TlBr_4^-$ are soluble in amyl acetate, di-isopropyl ether, benzene, and toluene. The intensely coloured extract constitutes the basis of a sensitive spectrophotometric method for determining thallium [11,40–43]. The method has the advantage that free Brilliant Green is soluble in water, but is not extracted.

Thallium is extracted from hydrochloric acid after oxidation to Tl(III) with bromine, the excess of which is removed either by boiling or by reaction with phenol or sulphosalicylic acid. The thallium–Brilliant Green ion-association complex is then formed by shaking the organic extract with a solution of Brilliant Green in 0.1–$0.2M$ HCl. Alternatively, the association complex may be formed in the aqueous phase and then extracted.

The molar absorptivity of the Brilliant Green–$TlCl_4^-$ complex in di-isopropyl ether is 1.06×10^5 at $\lambda_{max} = 630$ nm.

Complex anions such as $SbCl_6^-$, $GaCl_4^-$, $FeCl_4^-$, $HgBr_4^{2-}$, WO_4^{2-}, ClO_4^-, $Cr_2O_7^{2-}$, and SCN^- also form extractable association complexes with Brilliant Green. However, the method detailed below incorporates the preseparation of thallium(III) by extraction with di-isopropyl ether from $6M$ HCl. Under such conditions, only gold and antimony interfere.

Traces of thallium in minerals and ores [41], urine [42], cadmium [44], and indium [11] have been determined by the Brilliant Green method.

Reagents

1. Brilliant Green: 0.02% solution in $0.15M$ HCl.
2. Standard thallium(I) solution: 1 mg/ml (p. 531).
3. Phenol: 10% solution in glacial acetic acid.

Procedure

Extractive separation of Tl. To the sample solution containing not more than 100 µg of Tl, add 5 drops of saturated bromine water and conc. HCl till the HCl concentration is $6M$. Transfer the solution quantitatively to a separating funnel and add 5 drops of the phenol solution. After 5 minutes, extract thallium by shaking for 1 minute with two portions of di-isopropyl ether. Wash the combined ethereal extracts by shaking with 3 ml of 6M HCl.

Determination of Tl. Shake the clear ethereal extract (obtained as above) for ~ 30 seconds with 5 ml of Brilliant Green solution. Wash the aqueous solution by shaking with 5 ml of di-isopropyl ether. Place the combined ethereal solutions in a 50-ml volumetric flask (or smaller depending on the amount of thallium), make up to the mark with ether, and measure the absorbance at 630 nm against the solvent or a reagent blank solution.

53.2.3 STARCH–IODINE METHOD

In a hydrochloric acid medium (optimum concentration ~ 0·5M HCl), thallium(I) is oxidized with bromine to thallium(III). This spectrophotometric method for determining thallium is based on the oxidation by the thallium(III) of iodide to iodine, and the subsequent formation of the blue complex with starch [14,45].

The molar absorptivity of the starch–iodine complex is $3·9 \times 10^4$ ($a = 0·19$) at $\lambda_{max} = 590$ nm.

The oxidation of thallium(I) with bromine occurs quantitatively in the cold. The excess of bromine is reacted with phenol. Too great an excess of bromine interferes since a suspension of bromophenol is then produced. The excess of starch does not interfere. The amount of potassium iodide added should, however, always be the same. Atmospheric oxygen slowly liberates iodine from the iodide, but the resulting increase in absorbance amounts to scarcely 7% in one hour. Other oxidants which oxidize iodide to iodine interfere. Small quantities of Fe(III) may be masked with phosphoric acid.

The iodine liberated by thallium(III) may be extracted into chloroform or carbon tetrachloride and the absorbance of the violet extract measured. This method is, however, much less sensitive than the starch-iodine method.

Reagents

1. Potassium iodide: 0·5% solution, freshly prepared.
2. Starch: 1% solution (p. 298).
3. Standard thallium(I) solution: 1 mg/ml (p. 531).
4. Phenol: 10% solution in glacial acetic acid.

Procedure

To 20–40 ml of acidic (pH ~ 1) sample solution containing not more than 300 µg of Tl, add 1–2 ml of conc. HCl and 5 drops of saturated bromine water, and stir. After 1 minute, add 5 drops of phenol solution and mix well. One minute later, add 2 ml of KI solution and 2 ml of starch solution, and dilute the solution to the mark with water in a 50-ml volumetric flask. Mix the solution thoroughly, and measure its absorbance at 590 nm against a reagent blank solution.

53.2.4 Other Methods

Apart from the methods already mentioned, using Rhodamine B and Brilliant Green, there are many spectrophotometric methods based on extraction of the ion-association complexes formed by basic dyes and $TlCl_4^-$ or $TlBr_4^-$. Of particular importance are Methyl Violet [27,46–49] and Crystal Violet [50–55]; others include Victoria Blue 4R [56], Methylene Blue [57], Safranine T [58], Meldola's Blue [59], various thionine dyes [60], and 2-(4-antipyrylazo)-5-diethyl-m-aminophenol [61]. These reagents have been used to determine thallium in minerals and ores [27,46], tungsten [49], lead [47], antimony and cadmium [51], zinc and its alloys [54], and biological materials [52]. Fogg, Burgess, and Burns [62] have reviewed the applications of basic dyes to the determination of thallium. The association complex formed between $TlBr_4^-$ and the 2,2'-bipyridyl–iron(II) chelate has also been extracted into dichloroethane [63].

Dithizone has been utilized not only for the separation of thallium but also for its determination [16,24,64]. The molar absorptivity of TlHDz in chloroform is 3.7×10^4 at $\lambda_{max} = 505$ nm. Interfering elements include Pb, Bi, and Sn(II).

The determination of thallium with dithiocarbamates is very selective but insensitive [65,66]. By displacement of the thallium in the complex (in CCl_4) with copper, the sensitivity is enhanced about 20-fold [66].

Various other organic spectrophotometric reagents for determining thallium include Xylenol Orange [67,68], Eriochrome Cyanine R [69], PAR and TAR [70], PAN [71], 5-(2-pyridylazo)-5-diethyl-m-aminophenol (PAAP) and 3,5-dibromo-PAAP [72], 3,6-disulpho-TAN [73], 6-(2-quinolylazo)-3,4-dimethylphenol [74], quercetin [75], thiosalicylamide [76], thiodibenzoylmethane [77], and 3-hydroxy-1,3-diphenyltriazine [78].

Thallium may be determined from thallium(III) redox colour reactions with p-aminophenol [79], aminodiphenylamine (+ α-naphthol) [80], p-toluidine [67], and p-anisidine [81].

Betteridge and Yoe [82] have determined thallium as the yellow thallium(III) iodide complex, which may also be extracted into benzene in the presence of DAM or DAPM (the molar absorptivity is 1.2×10^4 at 400 nm) [83].

There exist several indirect spectrophotometric methods for determining thallium. Thallium(III) can be precipitated as $[Co(NH_3)_6]TlCl_6$, and the cobalt determined with nitroso-R salt [30]. It is possible to precipitate thallium(I) as the dipicrylaminate, to dissolve the precipitate in acetone, and to measure the orange colour of the dipicrylaminate ion in alkaline solution [84]. Thallium(I) can also be precipitated as Tl_2CrO_4, the precipitate dissolved, and chromium determined with diphenylcarbazide [85]. Lastly, Hargis and Boltz [86] have precipitated thallium(I) molybdophosphate, dissolved the precipitate, and measured the absorbance of the phosphomolybdenum blue formed by reduction.

References

1. Bock, R., Kusche, H. and Bock, E., *Z. Anal. Chem.* **138**, 167 (1953).
2. Pohl, F. A. and Kokes, K., *Mikrochim. Acta* **1957**, 318.
3. Geilmann, W. and Neeb, K. H., *Z. Anal. Chem.* **165**, 251 (1959).
4. Gurkina, T. V. and Litvinova, E. Ya., *Zh. Analit. Khim.* **24**, 374 (1969).
5. Dean, J. A. and Eskew, J. B., *Anal. Lett.* **4**, 737 (1971).
6. Irving, H. M. and Rossotti, F. J., *Analyst* **77**, 801 (1952).
7. Horrocks, D. L. and Voigt, A. F., *J. Am. Chem. Soc.* **79**, 2440 (1957).
8. Srivastava, T. N. and Rupainwar, D. C., *Bull. Chem. Soc. Japan* **38**, 1792 (1965).
9. Zolotov, Yu. A., Alimarin, I. P. and Sukhanovskaya, A. I., *Zh. Analit. Khim.* **20**, 165 (1965).
10. De, A. K. and Sen, A. K., *Talanta* **14**, 629 (1967).
11. Marczenko, Z., Kałowska, H. and Mojski, M., *ibid.* **21**, 93 (1974).
12. Khosla, M. M., Singh, S. R. and Rao, S. P., *ibid.* **21**, 411 (1974).
13. Khosla, M. M. and Rao, S. P., *Anal. Chim. Acta* **68**, 470 (1974).
14. Sill, C. W. and Peterson, H. E., *Anal. Chem.* **21**, 1268 (1949).
15. Onishi, H., *Bull. Chem. Soc. Japan* **30**, 567 (1957).
16. Clarke, R. S., Jr. and Cuttitta, F., *Anal. Chim. Acta* **19**, 555 (1958).
17. Schweitzer, G. K. and Norton, A. D., *ibid.* **30**, 119 (1964).
18. Schweitzer, G. K. and Cochran, G. T., *ibid.* **30**, 413 (1964).
19. Schweitzer, G. K. and Davidson, J. E., *ibid.* **35**, 467 (1966).
20. Bode, H., *Z. Anal. Chem.* **144**, 165 (1955).
21. Bagreev, V. V. and Zolotov, Yu. A., *Zh. Analit. Khim.* **17**, 852 (1962).
22. Kuznetsov, V. I., Bankovskii, Yu. A. and Ievin'sh, A. F., *ibid.* **13**, 267 (1958).
23. Levin, I. S. and Rodina, T. F., *ibid.* **23**, 673, 1315 (1968).
24. Marczenko, Z. and Kasiura, K., *Chem. Anal. (Warsaw)* **10**, 449 (1965).
25. Bochvarova, M., Do Kim Chung and Khalkin, V. A., *Zh. Analit. Khim.* **26**, 890 (1971).
26. Luke, C. L., *Anal. Chem.* **31**, 1680 (1959).
27. Rudnev, N. A., Malofeeva, G. I. and Rasskazova, V. S., *Zavodsk. Lab.* **27**, 20 (1961).
28. Truhaut, R. and Boudène, C., *Bull. Soc. Chim. France* **1957**, 1504.
29. Kuznetsov, V. I. and Myasoedova, G. V., *Zh. Analit. Khim.* **10**, 211 (1955); *Zh. Prikl. Khim.* **29**, 1875 (1956).
30. Nozaki, T., *J. Chem. Soc. Japan, Pure Chem. Sect.* **77**, 493 (1956).
31. Ginzburg, L. B. and Shkrobot, E. P., *Zavodsk. Lab.* **21**, 1289 (1955).
32. Strelow, F. W. and Victor, A. H., *Talanta* **19**, 1019 (1972).
33. Matthews, A. D. and Riley, J. P., *Anal. Chim. Acta* **48**, 25 (1969).
34. Onishi, H., *Bull. Chem. Soc. Japan* **29**, 945 (1956).
35. Miketuková, V. and Kohliček, J., *Z. Anal. Chem.* **208**, 7 (1965).
36. Minczewski, J., Wieteska, E. and Marczenko, Z., *Chem. Anal. (Warsaw)* **6**, 515 (1961).
37. Woolley, J. F., *Analyst* **83**, 477 (1958).
38. Van Aman, R. E. and Kanzelmeyer, J. H., *Anal. Chem.* **33**, 1128 (1961).
39. Miyamoto, M., *Japan Analyst* **10**, 102 (1961).
40. Voskresenskaya, N. T., *Zh. Analit. Khim.* **11**, 585 (1956).
41. Voskresenskaya, N. T., *Zavodsk. Lab.* **24**, 395 (1958).
42. Ariel, M. and Bach, D., *Analyst* **88**, 30 (1963).
43. Fogg, A. G., Burgess, C. and Burns, D. T., *ibid.* **98**, 347 (1973).
44. Krasiejko, M. and Marczenko, Z., *Mikrochim. Acta*, **1975 I**, 585
45. Haddock, L. A., *Analyst* **60**, 394 (1935).
46. Blyum, I. A. and Ul'yanova, I. A., *Zavodsk. Lab.* **23**, 283 (1957).
47. Cyrankowska, M. and Downarowicz, J., *Chem. Anal. (Warsaw)* **12**, 137 (1967).
48. Bashilova, N. I. and Khomutova, T. V., *Zh. Analit. Khim.* **24**, 999 (1969).

49. Vadasdi, K., Buxbaum, P. and Salamon, A., *Anal. Chem.* **43**, 318 (1971).
50. Blyum, I. A., Solov'yan, I. T. and Shebalkova, G. N., *Zavodsk. Lab.* **27**, 950 (1961).
51. Lomonosov, S. A. and Mil'shtein, F. Ya., *ibid.* **33**, 14 (1967).
52. Wawschinek, O., Beyer, W. and Paletta, B., *Mikrochim. Acta* **1968**, 201.
53. Kothny, E. L., *Analyst* **94**, 198 (1969).
54. Chainani, P. A., Murugaiyan, P. and Venkateswarlu, C., *Anal. Chim. Acta* **57**, 67 (1971).
55. Alexandrow, A. and Dimitrow, A., *Mikrochim. Acta* **1972**, 681.
56. Kish, P. P. and Monich, E. E., *Zh. Analit. Khim.* **25**, 272 (1970).
57. Tarayan, V. M., Ovsepyan, E. N. and Artsruni, V. Zh., *Zavodsk. Lab.* **35**, 1435 (1969).
58. Pilipenko, A. T. and Nguyen Mong Shin, *Zh. Analit. Khim.* **23**, 934 (1968).
59. Pilipenko, A. T. and Nguyen Dyk Tu, *Ukr. Khim. Zh.* **34**, 703 (1968); **35**, 303 (1969).
60. Tarayan, V. M., Ovsepyan, E. N. and Artsruni, V. Zh., *Zh. Analit. Khim.* **25**, 691 (1970).
61. Gusev, A. I. and Kurepa, G. A., *ibid.* **27**, 294 (1972).
62. Fogg, A. G., Burgess, C. and Burns, D. T., *Talanta* **18**, 1175 (1971).
63. Kotsuji, K., Yoshimura, Y. and Ueda, S., *Anal. Chim. Acta* **42**, 225 (1968).
64. Dyfverman, A., *ibid.* **21**, 357 (1959).
65. Foley, W. T. and Pottie, R. F., *Anal. Chem.* **28**, 1101 (1956).
66. Keil, R., *Z. Anal. Chem.* **258**, 97 (1972).
67. Otomo, M., *Bull. Chem. Soc. Japan* **38**, 1044 (1965).
68. Dwivedi, C. D. and Dey, A. K., *Mikrochim. Acta* **1968**, 708.
69. Joshi, A. P. and Munshi, K. N., *Microchem. J.* **12**, 447 (1967).
70. Hniličková, M. and Sommer, L., *Talanta* **16**, 83 (1969).
71. Rodina, T. F., Kolomiichuk, V. S. and Levin, I. S., *Zh. Analit. Khim.* **28**, 1090 (1973).
72. Gusev, S. I. and Kurepa, G. A., *ibid.* **24**, 1148 (1969).
73. Busev, A. I., Zholondkovskaya, T. N., Krysina, L. S. and Golubkova, N. A., *ibid.* **27**, 2165 (1972).
74. Rakhmatullaev, K., Tashmamatov, Kh. and Rakhmatullaeva, M. A., *ibid.* **29**, 1020 (1974).
75. Golovina, A. P. and Tiptsova, V. G., *ibid.* **17**, 524 (1962).
76. Shome, S. C. and Mazumdar, M., *Anal. Chim. Acta* **46**, 155 (1969).
77. Uhlemann, E. and Schuknecht, B., *ibid.* **69**, 79 (1974).
78. Shome, S. C., Das, H. R. and Das, B., *Anal. Chem.* **38**, 1522 (1966).
79. Gladyshev, V. P. and Tolstikov, G. A., *Zavodsk. Lab.* **22**, 1166 (1956).
80. Kreshkov, A. P., Senetskaya, L. P. and Karagodina, A. M., *Zh. Analit. Khim.* **21**, 415 (1966).
81. Pfrepper, G., *Z. Anal. Chem.* **193**, 179 (1963).
82. Betteridge, D. and Yoe, J. H., *Anal. Chim. Acta* **27**, 1 (1962).
83. Busev, A. I. and Tiptsova, V. G., *Zh. Analit. Khim.* **14**, 550 (1959).
84. Gorbenko-Germanov, D. S. and Zenkova, R. A., *ibid.* **20**, 1020 (1965).
85. Zimmer, H., *Z. Anal. Chem.* **165**, 268 (1959).
86. Hargis, L. G. and Boltz, D. F., *Anal. Chem.* **37**, 240 (1965).

Chapter 54

THORIUM

Thorium (Th, at.wt. 232·04) occurs in solutions exclusively in the +IV oxidation state. In analytical properties it resembles zirconium and titanium, as well as the rare earth elements (REE). In non-complexing media at pH < 1, it exists in solution as colourless Th^{4+} ions. It is less readily hydrolysed than titanium or zirconium. The hydroxide $Th(OH)_4$ (precipitating at pH 3·5–4) shows no amphoteric properties. Thorium forms stable complexes with oxalate [in a weakly acidic medium $Th(C_2O_4)_2$ is precipitated], tartrate, citrate, and EDTA and less stable complexes with sulphate, nitrate and carbonate.

54.1 Methods of Separation

54.1.1 Precipitation

Traces of thorium can be precipitated as thorium hydroxide with ammonia (pH > ~ 4), iron(III) being a suitable collector [1]. Thorium can be separated from REE by double precipitation of the hydroxide.

The precipitation of thorium oxalate from a weakly acid medium (pH 1–4) with calcium as collector is a much favoured separation method [2–4]. The REE and uranium(IV) are also precipitated quantitatively, but many metals (e.g. Fe, Zr, Ti, Al, Nb, and Mo) remain in solution as soluble oxalate complexes.

The separation of thorium as sparingly soluble thorium fluoride [5–8] is equally selective, the solubility of the fluoride being lower than that of the oxalate. However, the solubility of thorium fluoride is enhanced in the presence of alkali metals owing to the formation of double fluorides [7]. Usually, the sample solution in hydrofluoric acid is evaporated to small volume and diluted with water to precipitate thorium, REE, and uranium(IV) fluoride. Lanthanum, cerium, or calcium are used as collectors for traces of thorium. Since the fluoride precipitate is difficult to filter off, centrifugation is advisable.

Thorium can be separated from rare earth elements and other metals by precipitation of the iodate, $Th(IO_3)_4$, from ~ $1M$ HNO_3 in the presence of tartaric acid and hydrogen peroxide [9]. The iodate precipitate also contains zirconium and cerium(IV). Mercury(II) [6,8] and cerium(IV) have been used as collectors.

Sill and Willis [10] have coprecipitated traces of thorium from a strongly acidic medium with barium sulphate. Barium sulphate precipitated at pH ⩾ 3 does not retain thorium.

Other reagents for the precipitative separation and preconcentration of thorium include sebacic acid [3], anthranilic acid [11], and Methyl Violet + SPADNS [12].

54.1.2 Extraction

Thorium is one of the few multivalent metals [others are U(VI), Ce(IV), and Au(III)] which are extractable as nitrate complexes from nitric acid solutions by oxygen-containing solvents [13–15]. Zirconium is not readily extracted. The presence of lithium and aluminium nitrates improves the extraction of thorium. Sulphate, phosphate, and tartrate do not interfere, but fluoride must be masked, e.g. by aluminium ($\rightarrow$ AlF_6^{3-}). Ross and White [16] have extracted thorium from a nitrate medium with a solution of TOPO in cyclohexane.

The following solutions also selectively extract thorium: TTA in CCl_4 or MIBK [17], butyric acid in $CHCl_3$ [18], N-benzylaniline in $CHCl_3$ [19], and HDEHP in CCl_4 [20].

54.1.3 Ion-Exchange

Hydrochloric acid (3–6M) elutes REE and most other elements except thorium from strongly acidic cation-exchangers [21–26]. Thorium is eluted with 3M H_2SO_4 [21,24,27], ammonium carbonate [1], or ammonium sulphate [28] solution. Cation-exchange chromatography in the following media has also been used to separate thorium: hydrobromic acid [29], dimethyl sulphoxide [30], and nitric acid + TOPO + methanol [31].

Korkisch et al. [32–36] and other authors [37–39] have adsorbed thorium on strongly basic anion-exchangers as the nitrate complex, while most other metals are not retained. Other complexants such as chloride [36,40–42], thiocyanate [43], sulphate [44], and carbonate [45] have also been employed in the anion-exchange separation of thorium. Mixed aqueous–organic solvent media are used in numerous cases [32–36,41–43].

Separation of thorium and uranium with liquid ion-exchangers is also feasible [46,46a].

54.2 Methods of Determination

Among the many spectrophotometric methods for determining thorium, those based on azo compounds containing arsonic acid groups are the most important. Although the Thoron I method is the one most commonly employed, it is much less sensitive and less selective than the newer Arsenazo III method.

54.2.1 THORON I METHOD

Thoron I (Thorin, Thoronol, APANS) (formula, p. 48) reacts with thorium(IV) in acid medium to yield a red, water-soluble complex, which forms the basis for the spectrophotometric determination [47–49]. Thoron I in acid solution is orange. The absorption spectra of the reagent and its thorium complex are shown in Fig. 54.1. Thorium reacts with excess of Thoron I in the molar ratio 1:2 [50].

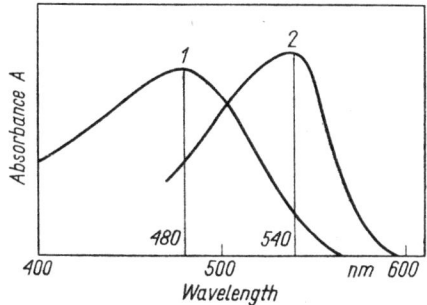

Fig. 54.1 Absorption spectra of Thoron I (1) and its thorium complex (2) (reagent solution as reference) in $0.25M$ HCl

Between pH 0·4 and 1·0, the absorbance of the complex is constant. A 0.2–$0.3M$ HCl medium is optimum. The absorbance decreases with increasing acidity. The molar absorptivity of the complex in $0.25M$ HCl is 1.7×10^4 ($a = 0.07$) at $\lambda = 540$ nm.

Fluoride, oxalate, phosphate, and (to a lesser degree) sulphate interfere. In a $0.01M$ sulphate solution, the absorbance of the Th–Thoron I complex decreases to 60%, falling to 15% in $0.05M$ sulphate.

Zirconium and hafnium also form coloured Thoron I complexes, but these metals can be masked with tartaric acid (especially *meso*-tartaric acid) [51–53]; 2 ml of 5% tartaric acid solution in 50 ml of solution effectively masks 1 mg of zirconium, while not affecting the colour of the thorium–Thoron I complex.

Uranium(IV) interferes in the determination of thorium, but does not interfere significantly when oxidized to U(VI). Up to 5 mg of aluminium, 5 mg of iron(II), 5 mg of cerium, and 2 mg of titanium can be tolerated. Ascorbic acid is used to reduce iron(III) [54].

The Thoron I method has been used to determine thorium in monazite sands and concentrates [5,16,24,51], silicate rocks [9,27,32,55], ores [15,53,56], ilmenite sands [6], lanthanum oxide [57], zirconium minerals [52], bismuth alloys [40], tungsten [58], magnesium alloys [59], and water [2]. Stuart [60] has employed Thoron I in the automatic spectrophotometric determination of thoria in uranium.

Reagents

1. Thoron I: 0·1% aqueous solution.

2. *Standard thorium solution*: 1 mg/ml. Dissolve 2·5 g of $Th(NO_3)_4 \cdot 4H_2O$ in water containing 5 ml of conc. HNO_3, and dilute the solution with water to 1 litre in a volumetric flask. Determine the concentration of thorium in the solution by evaporating an aliquot to dryness (completing the evaporation in a platinum crucible), and igniting at 1000°C to ThO_2. Dilute the solution with water till the thorium concentration is exactly 1 mg/ml. Working solutions are obtained by appropriate dilution of the standard solution with $\sim 0\cdot 1M$ HNO_3.

Procedure

Place the sulphate-free sample solution (containing not more than 300 μg of Th) in dilute HCl in a 50-ml volumetric flask. Add 2 ml of 1% ascorbic acid solution, 2 ml of 5% tartaric acid solution, 4 ml of the Thoron I solution, and sufficient hydrochloric acid to make its concentration $0\cdot 25M$ after dilution to the mark with water. Measure the absorbance of the coloured solution at 540 nm, using a reagent blank solution as reference.

54.2.2 Arsenazo III Method

Arsenazo III (formula, p. 48) reacts with thorium in strongly acidic solution to form a grey-green water-soluble complex. This reaction was adapted by Savvin *et al.* for the spectrophotometric determination of thorium [61–64]. With excess of Arsenazo III, a 2:1 complex with thorium is formed [65]. According to Nemodruk and Kochetkova [66], a complex with a higher ratio of Arsenazo III to thorium can also be formed.

The Arsenazo III method is very sensitive, and the absorbance varies only slightly with change in HCl concentration between 1 and $10M$. The maximum absorbance is obtained in $8M$ HCl, but the reaction in $3M$ HCl containing oxalate is more selective.

The molar absorptivity of the complex in $3M$ HCl is $1\cdot 15 \times 10^5$ (specific absorptivity 0·50) at $\lambda_{max} = 655$ nm. Figure 54.2 shows the absorption spectra of Arsenazo III and its thorium complex in $3M$ HCl.

The Arsenazo III method for determining thorium has high selectivity. In the presence of oxalic acid as a masking agent, thorium can be determined in $2\cdot 5$–$3\cdot 5M$ HCl in the presence of zirconium, hafnium, and niobium [61].

Uranium(IV) is oxidized to U(VI) by adding $KMnO_4$ (followed by ascorbic acid to decolorize the solution). At high concentrations of chloride (HCl + LiCl), thorium can be determined in the presence of a 100-fold amount of uranium(VI). Titanium can cause interference [67,68].

Under the conditions outlined in the procedure below, 5 mg of zirconium, 5 mg of titanium, and 5 mg of iron can be tolerated. Aluminium and REE do not interfere. Hence, in many materials, thorium can be determined without preliminary separation.

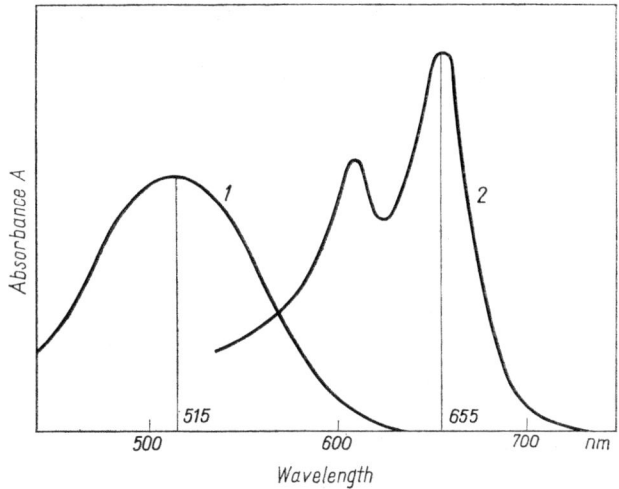

Fig. 54.2 Absorption spectra of Arsenazo III (1) and its thorium complex (2) (reagent solution as reference) in $3M$ HCl

Fluoride, phosphate, and (to a lesser degree) sulphate interfere in the reaction with Arsenazo III. The presence of $0.02–0.05M$ H_2SO_4 in the $3M$ HCl medium reduces the absorbance of the thorium complex by 15–20%.

Thorium can be determined with Arsenazo III in the organic phase after extraction with HDEHP in cyclohexane. The colour is developed by adding aqueous Arsenazo III and isopropanol [69].

Arsenazo III has been used to determine thorium in titanium minerals [70], zirconium minerals [62,70,71], uranium materials [71–73], silicate rocks and ores [39,74–76], geological samples [77], niobium products [78], REE compounds [20,79], biological materials [80,81] and natural waters [82,83].

Reagents

1. Arsenazo III: 0.05% solution.
2. Standard thorium solution: 1 mg/ml (p. 540)

Procedure

Separation of Th as oxalate. To 50–150 ml of acid sample solution, add 5–20 ml (depending on the amounts of other metals present) of 8% oxalic acid solution. Adjust the pH to ~3, heat the solution to ~80°C, and add dropwise, while stirring, 5 mg of calcium (as calcium chloride solution). Keep the solution at 80–90°C for 1 hour. After 3–4 hours, filter off the precipitate, and wash it thoroughly with 1% oxalic acid solution and water. Dissolve the precipitate in a small quantity of hot $2M$ HCl.

Determination of Th. Place the sulphate-free acid sample solution containing not more than 70 µg of Th in a 50-ml volumetric flask. Add 12 ml of conc. HCl and 5 ml of 8% oxalic acid solution, and mix well. Add 4 ml of the Arsenazo III solution, dilute to the mark with water, and measure the absorbance at 655 nm against a reagent blank solution.

54.2.3 Other Methods

Arsenazo I (Neothoron), another azo reagent containing the arsonic acid group, is fairly often used for determining thorium. This reagent is more sensitive but less selective than Thoron I [4,84–86]. The colour reaction ($\lambda_{max} \sim 575$ nm) with thorium is carried out in dilute HCl (pH 1–2). Interference from Zr, Ti, and REE is eliminated by adding tartaric acid [4]. Arsenazo I has been used to determine thorium in minerals and ores [87–89].

Several other azo dyes with arsonic acid groups are also recommended as spectrophotometric reagents for thorium, viz. Thoron II [90], Arsenazo II [91,92], Dibromoarsenazo II [93], Arsenazo DAL [94], and Arsenazo IV [95]. The related reagent, Chlorophosphonazo III, is also suitable for determining thorium [96–98].

Other azo reagents for thorium include PAR [99], TAR [100], SPADNS [101,102], Chromotrope 2B [103], Chromotrope 2C [104], SNADNS derivatives [105,106], Calmagite [107], Eriochrome Black T (molar absorptivity 3.5×10^4 at 700 nm) [108,109], Solochrome Fast Red [110], Solochrome Fast Grey [101], Solochrome Black WDFA [111], and Acid Alizarin Black SN [112].

Thorium has also been determined colorimetrically with triphenylmethane dyes such as Xylenol Orange [113–115], Methylthymol Blue ($\varepsilon = 5.0 \times 10^4$ at 580 nm) [116,117], and Chrome Azurol S [118–120]. The xanthene dye Bromopyrogallol Red has been used similarly [121,122].

Thorium reacts in weakly acid media with quercetin ($\varepsilon = 3.3 \times 10^4$ at 422 nm) [123,124] and with morin [124–126].

Other organic spectrophotometric reagents for determining thorium are Alizarin S [127], Naphthazarin [128], oxine [129], and indo-oxine [130]. Thorium forms a ternary complex with 5,7-dibromo-8-hydroxyquinoline and Rhodamine B which can be extracted into benzene [131].

Methods for determining thorium based on the mixed thorium–molybdophosphoric heteropoly acid also deserve mention [132,133].

References

1. Toribara, T. Y. and Koval, L., *Talanta* **14**, 403 (1967).
2. Taylor, A. E. and Dillon, R. T., *Anal. Chem.* **24**, 1624 (1952).
3. Carron, M. K., Skinner, D. L. and Stevens, R. E., *ibid.* **27**, 1058 (1955).
4. Zaíkovskií, F. V. and Gerkhardt, L. I., *Zh. Analit. Khim.* **13**, 274, 513 (1958).
5. Banks, C. V., Byrd, C. H. and Klingman, D. W., *Anal. Chem.* **25**, 416, 992 (1953).
6. Athavale, V. T., Oke, K. P. and Tillu, M. M., *Anal. Chim. Acta* **21**, 528 (1959).
7. Butler, J. R. and Hall, R. A., *Analyst* **85**, 149 (1960).
8. Almodóvar, I., *Anal. Chim. Acta* **33**, 426 (1965).

References

9. Grimaldi, F. S., Jenkins, L. B. and Fletcher, M. H., *Anal. Chem.* **29**, 848 (1957).
10. Sill, C. W. and Willis, C. P., *ibid.* **36**, 622 (1964).
11. Morachevskiĭ, Yu. V. and Tserkovnitskaya, I. A., *Zh. Analit. Khim.* **14**, 55 (1959).
12. Sudhalatha, K., *Talanta* **10**, 934 (1963).
13. Banks, C. V. and Edwards, R. E., *Anal. Chem.* **27**, 947 (1955).
14. Hesford, E., McKay, H. A. and Scargill, D., *J. Inorg. Nucl. Chem.* **4**, 321 (1957).
15. Everest, D. A. and Martin, J. V., *Analyst* **84**, 312 (1959).
16. Ross, W. J. and White, J. C., *Anal. Chem.* **31**, 1847 (1959).
17. Goldstein, G., Menis, O. and Manning, D. L., *ibid.* **32**, 400 (1960).
18. Pietsch, R. and Sinic, H., *Mikrochim. Acta* **1968**, 1287.
19. Khosla, M. M. and Rao, S. P., *Anal. Chim. Acta* **54**, 315 (1971).
20. Antonov, A. V., Shtenke, A. A. and Shvarev, V. S., *Zh. Analit. Khim.* **24**, 848 (1969).
21. Nietzel, O. A., Wessling, B. W. and DeSesa, M. A., *Anal. Chem.* **30**, 1182 (1958).
22. Strelow, F. W. E., *ibid.* **31**, 1201 (1959).
23. Vetejška, K., *Collection Czech. Chem. Commun.* **25**, 1895 (1960).
24. Chung, K. S. and Riley, J. P., *Anal. Chim. Acta* **28**, 1 (1963).
25. Alimarin, I. P., Medvedeva, A. M. and Burlova, M. A., *Zh. Analit. Khim.* **18**, 468 (1963); **19**, 1332 (1964).
26. Alimarin, I. P. and Medvedeva, A. M., *ibid.* **22**, 436 (1967).
27. Korkisch, J. and Antal, P., *Z. Anal. Chem.* **173**, 126 (1960).
28. Nabivanets, B. I. and Kudritskaya, L. N., *Zh. Analit. Khim.* **21**, 40 (1966).
29. Strelow, F. W. and Boshoff, M. D., *Anal. Chim. Acta* **62**, 351 (1972).
30. Qureshi, M. and Husain, K., *ibid.* **57**, 387 (1971).
31. Korkisch, J. and Orlandini, K. A., *Anal. Chem.* **40**, 1952 (1968).
32. Korkisch, J. and Antal, P., *Z. Anal. Chem.* **171**, 22 (1959).
33. Korkisch, J. and Janauer, G. E., *Talanta* **9**, 957 (1962).
34. Korkisch, J. and Tera, F., *Anal. Chem.* **33**, 1264 (1961); *Z. Anal. Chem.* **186**, 290 (1962).
35. Korkisch, J. and Arrhenius, G., *Anal. Chem.* **36**, 850 (1964).
36. Hazan, I., Ahluwalia, S. S. and Korkisch, J., *Z. Anal. Chem.* **206**, 324 (1964).
37. Fritz, J. S. and Garralda, B. B., *Anal. Chem.* **34**, 1387 (1962).
38. Archimbaud, M. and Noel, A., *Chim. Anal. (Paris)* **53**, 161 (1971).
39. Gumbar, K. K. and Bozhichko, M. A., *Radiokhimiya* **13**, 155 (1971).
40. Milner, G. W. and Edwards, J. W., *Anal. Chim. Acta* **17**, 259 (1957).
41. Urubay, S., Janauer, G. E. and Korkisch, J., *Z. Anal. Chem.* **193**, 165 (1963).
42. Cummings, T. F. and Korkisch, J., *Anal. Chim. Acta* **40**, 520 (1968).
43. Pietrzyk, D. J. and Kiser, D. L., *Anal. Chem.* **37**, 1578 (1965).
44. Nagle, R. A. and Murthy, T. K., *Analyst* **84**, 37 (1959).
45. Taketatsu, T., *Talanta* **10**, 1077 (1963).
46. Cospito, M. and Rigali, L., *Anal. Chim. Acta* **57**, 107 (1971).
46a. De, A. K. and Ray, U. S., *Sepn. Sci.* **7**, 419 (1972).
47. Thomason, P. F., Perry, M. A. and Byerly, W. M., *Anal. Chem.* **21**, 1239 (1949).
48. Margerum, D. W., Byrd, C. H., Reed, S. A. and Banks, C. V., *ibid.* **25**, 1219 (1953).
49. Clinch, J., *Anal. Chim. Acta* **14**, 162 (1956).
50. Palmer, A. R., *ibid.* **19**, 458 (1958).
51. Grimaldi, F. S. and Fletcher, M. H., *Anal. Chem.* **28**, 812 (1956).
52. Fletcher, M. H., Grimaldi, F. S. and Jenkins, L. B., *ibid.* **29**, 963 (1957).
53. Everest, D. A. and Martin, J. V., *Analyst* **82**, 807 (1957).
54. Hadobás, B., *Acta Chim. Acad. Sci. Hung.* **28**, 207 (1961).
55. Culkin, F. and Riley, J. P., *Anal. Chim. Acta* **32**, 197 (1965).
56. Arnfelt, A. L. and Edmundsson, I., *Talanta* **8**, 473 (1961).
57. Rajković, D., *Analyst* **97**, 114 (1972).
58. Norwitz, G., *Metallurgia* **66**, 297 (1962).

59. Mayer, A. and Bradshaw, G., *Analyst* **77**, 154 (1952).
60. Stuart, W. A., *ibid.* **91**, 208 (1966).
61. Savvin, S. B., *Dokl. Akad. Nauk SSSR* **127**, 1231 (1959); *Talanta* **8**, 673 (1961).
62. Luk'yanov, V. F., Savvin, S. B. and Nikol'skaya, I. V., *Zavodsk. Lab.* **25**, 1155 (1959).
63. Kuznetsov, V. I. and Savvin, S. B., *Radiokhimiya* **3**, 79 (1961).
64. Savvin, S. B., *Zavodsk. Lab.* **29**, 131 (1963); *Talanta* **11**, 1, 7 (1964).
65. Muk, A. and Savvin, S. B., *Zh. Analit. Khim.* **26**, 98 (1971).
66. Nemodruk, A. A. and Kochetkova, N. E., *ibid.* **17**, 330 (1962).
67. Nikol'skaya, I. V., Luk'yanov, V. F. and Maksimov, A. V., *ibid.* **27**, 1212 (1972).
68. Pakalns, P., *Anal. Chim. Acta* **58**, 463 (1972).
69. Cerrai, E. and Ghersini, G., *ibid.* **37**, 295 (1967).
70. Pakalns, P., *ibid.* **65**, 223 (1973).
71. Karalova, Z. K., Shibaeva, N. P. and Pyzhova, Z. I., *Zh. Analit. Khim.* **21**, 1133 (1966).
72. Palei, P. N., Karalova, Z. K., Shibaeva, N. P. and Pyzhova, Z. I., *ibid.* **21**, 126 (1966).
73. Korkisch, J. and Dimitriadis, D., *Mikrochim. Acta* **1974**, 25.
74. Savvin, S. B. and Bagreev, V. V., *Zavodsk. Lab.* **26**, 412 (1960).
75. Furtova, E. V., Sadova, G. F., Ivanova, V. N. and Zaikovskii, F. V., *Zh. Analit. Khim.* **19**, 94 (1964).
76. Abbey, S., *Anal. Chim. Acta* **30**, 176 (1964).
77. Korkisch, J. and Dimitriadis, D., *Talanta* **20**, 1199 (1973).
78. Vladimirova, V. M. and Davidovich, N. K., *Zavodsk. Lab.* **26**, 1210 (1960).
79. Luk'yanov, V. F. and Nikol'skaya, I. V., *Zh. Analit. Khim.* **23**, 1567 (1968).
80. Petrow, H. G. and Strehlow, C. D., *Anal. Chem.* **39**, 265 (1967).
81. Bazzano, E. and Ghersini, G., *Anal. Chim. Acta* **38**, 457 (1967).
82. Aksel'rod, F. M. and Lyalikov, Yu. S., *Zh. Analit. Khim.* **20**, 514 (1965).
83. Korkisch, J. and Dimitriadis, D., *Talanta* **20**, 1303 (1973).
84. Takahashi, T. and Miyake, S., *ibid.* **3**, 155 (1959).
85. Kuznetsov, V. I. and Nikol'skaya, I. V., *Zh. Analit. Khim.* **15**, 299 (1960).
86. Onishi, H., Nagai, H. and Toita, Y., *Anal. Chim. Acta* **26**, 528 (1962).
87. Gerkhardt, L. I., *Zh. Analit. Khim.* **14**, 434 (1959).
88. Zaikovskii, F. V., *ibid.* **14**, 440 (1959).
89. Holcomb, H. P., and Yoe, J. H., *Microchem. J.* **4**, 463 (1960).
90. Kuznetsov, V. I. and Savvin, S. B., *Radiokhimiya* **1**, 583 (1959).
91. Kuznetsov, V. I. and Savvin, S. B., *Zh. Analit. Khim.* **15**, 175 (1960).
92. Savvin, S. B., Volynets, M. P., Balashov, Yu. A. and Bagreev, V. V. *ibid.* **15**, 446 (1960).
93. Savvin, S. B., Basargin, N. N. and Makarova, V. P., *ibid.* **18**, 61 (1963).
94. Buděšínský, B. and Menclová, B., *Talanta* **14**, 523 (1967).
95. Vasilenko, V. D. and Shanya, M. V., *Izv. Vyssh. Ucheb. Zaved., Khim. Khim. Tekhnol.* **11**, 138 (1968).
96. Fadeeva, V. I. and Alimarin, I. P., *Zh. Analit. Khim.* **17**, 1020 (1962).
97. Yamamoto, T., *Anal. Chim. Acta* **63**, 65 (1973).
98. Lukin, A. M., Titov, V. I., Chernyshova, T. V. and Evdokimova, N. N., *Zavodsk. Lab.* **39**, 1174 (1973).
99. Busev, A. I. and Ivanov, V. M., *Izv. Vyssh. Ucheb. Zaved., Khim. Khim. Tekhnol.* **4**, 914 (1961).
100. Sakai, T. and Tonosaki, K., *Bull. Chem. Soc. Japan* **42**, 2718 (1969).
101. Banerjee, G., *Anal. Chim. Acta* **16**, 56 (1957).
102. Cooper, J. A. and Vernon, M. J., *ibid.* **23**, 351 (1960).
103. Sangal, S. P., *Bull. Chem. Soc. Japan* **38**, 141 (1965).
104. Majumdar, A. K. and Savariar, C. P., *Z. Anal. Chem.* **174**, 269 (1960).
105. Datta, S. K., *ibid.* **150**, 347 (1956); **168**, 347 (1959).

106. Datta, S. K., *ibid.* **173**, 369, 377 (1960).
107. Curcio, P. J. and Lott, P. F., *Anal. Chim. Acta* **26**, 487 (1962).
108. Lott, P. F., Cheng, K. L. and Kwan, B. C., *Anal. Chem.* **32**, 1702 (1960).
109. Korkisch, J. and Janauer, G. E., *Mikrochim. Acta* **1961**, 880.
110. Korkisch, J. and Janauer, G. E., *Anal. Chem.* **33**, 1930 (1961); *Z. Anal. Chem.* **182**, 26 (1961).
111. Janauer, G. E. and Korkisch, J., *Talanta* **9**, 427 (1962).
112. Kusakul, P. and West, T. S., *Anal. Chim. Acta* **32**, 301 (1965).
113. Buděšínský, B., *Collection Czech. Chem. Commun.* **27**, 226 (1962).
114. Mukherji, A. K., *Microchem. J.* **11**, 243 (1966).
115. Buděšínský, B. W. and Svec, J., *Anal. Chim. Acta* **61**, 465 (1972).
116. Vasilenko, V. D. and Shanya, M. V., *Zh. Analit. Khim.* **20**, 636 (1965).
117. Adam, J. and Přibil, R., *Talanta* **16**, 1596 (1969).
118. Ishida, R., *J. Chem. Soc. Japan, Pure Chem. Sect.* **86**, 1169 (1965).
119. Shijo, Y. and Takeuchi, T., *Japan Analyst* **18**, 469 (1969).
120. Evtimova, B., *Anal. Chim. Acta* **68**, 222 (1974).
121. Vasilenko, V. D., Shanya, M. V. and Bolbas, V. I., *Zh. Analit. Khim.* **22**, 1818 (1967).
122. Tonosaki, K. and Sakai, T., *Bull. Chem. Soc. Japan* **42**, 456 (1969).
123. Menis, C., Manning, D. L. and Goldstein, G., *Anal. Chem.* **29**, 1426 (1957).
124. Babko, A. K., Chan Ty Kh'eu, Volkova, A. I. and Get'man, T. E., *Ukr. Khim. Zh.* **35**, 292, 642 (1969).
125. Fletcher, M. H. and Milkey, R. G., *Anal. Chem.* **28**, 1402 (1956).
126. Perkins, R. W. and Kalkwarf, D. R., *ibid.* **28**, 1989 (1956).
127. Sarma, D. V. and Rao, B. S. R., *Anal. Chim. Acta* **13**, 142 (1955).
128. Moeller, T. and Tecotzky, T., *Anal. Chem.* **27**, 1056 (1955).
129. Goto, K., Russell, D. S. and Berman, S. S., *ibid.* **38**, 493 (1966).
130. Tomic, E. and Khalifa, H., *Z. Anal. Chem.* **156**, 326 (1957).
131. Mishchenko, V. T. and Zavarina, T. V., *Zh. Analit. Khim.* **25**, 1533 (1970).
132. Madison, B. L. and Guyon, J. C., *Anal. Chem.* **39**, 1706 (1967).
133. Murata, K., Yokoyama, Y. and Ikeda, S., *Anal. Chim. Acta* **48**, 349 (1969).

Chapter 55

TIN

Tin (Sn, at.wt. 118·69) occurs in its compounds in the $+II$ and $+I^V$ oxidation states. Tin(II) is unstable. The hydroxide, $Sn(OH)_2$, precipitates at pH ~ 2 but is amphoteric, redissolving in NaOH (pH 13) to form stannite. Tin(II) forms oxalate and chloride complexes. Tin(IV) hydrous oxide (metastannic acid) precipitates even at pH 0·5 and redissolves at pH ~ 9 as stannate. Tin(IV) gives stable halide, oxalate, and tartrate complexes.

55.1 Methods of Separation

55.1.1 DISTILLATION

Distillation of $SnBr_4$ is a convenient method for separating small amounts of tin [1–4]. In a procedure (described on p. 551) developed by Onishi and Sandell [1], first arsenic(III), germanium(IV), and antimony(III) are distilled off as chlorides from a medium comprising H_2SO_4, HCl, and H_3PO_4. Tin does not distil in the presence of phosphoric acid. Hydrobromic acid is then added to the still, and $SnBr_4$ is distilled off.

De Bruyne and Hoste's [3] procedure starts with the distillation of $AsCl_3$ at 109°C from $11M$ HCl. Phosphoric acid is then added to mask tin, and $SbCl_3$ is distilled at ~ 160°C. Finally, tin is demasked by adding HBr, and $SnBr_4$ is distilled at ~ 140°C.

Martinet [4] has separated tin from a silicate mineral by mixing the sample with ammonium iodide and subliming off SnI_4 at dull red heat.

55.1.2 PRECIPITATION

Microgram quantities of tin are very often coprecipitated as hydrous stannic oxide with hydrous MnO_2, which is formed by reacting MnO_4^- and Mn^{2+} in dilute nitric acid (see procedure, p. 549). Antimony is also quantitatively precipitated, and Au, Tl, Bi, and W are wholly or partly precipitated [5–8]. Traces of tin may also be precipitated from dilute nitric acid with antimony as collector [9]. Separation from molybdenum is achieved by coprecipitating Sn(IV) with ferric hydroxide by adding ammonia to make the pH 6–8 [10].

Baumgärtel and Gärtner [11] have isolated small amounts of tin in ores by coprecipitation with beryllium (pH 8). EDTA is used to mask iron, aluminium, and other elements.

Furthermore, tin can be precipitated from acid medium (HCl, H_2SO_4) as the sulphide. Copper, molybdenum, or cadmium are suitable scavengers [12,13].

55.1.3 Extraction

Tin(IV) iodide is extracted from acid medium ($HClO_4$, H_2SO_4) with benzene, toluene, or di-isopropyl ether; antimony and indium remain in the aqueous phase [14–17].

Tin(II) can be extracted into chloroform as complexes with dithiocarbamates, but Sb(III) and As(III) are also extracted [5,10,12,13,18].

Tin(IV) is separated from lead, zinc, and antimony(V) by extraction of the cupferronate from $0.5M$ HNO_3 into chloroform [19-21]. Extraction of the tin–BPHA complex facilitates separation from indium [21–23].

Because tin(IV) oxinate is extracted into chloroform at pH 0·85 only in the presence of halide, tin can be specifically separated by first extracting other metal oxinates at pH 0·85, then adding NH_4Cl and extracting tin oxinate [24].

Ross and White [25] have extracted tin(IV) from hydrochloric acid into cyclohexane containing TOPO or TEHPO. Levin and Tarasova [26] have extracted tin(II) with di-(2-ethylhexyl) dithiophosphoric acid. Hofer [27] has extracted tin from $5-9M$ HCl with a xylene solution of Amberlite LA-2 liquid anion-exchanger.

55.1.4 Ion-Exchange

Strongly basic anion-exchangers retain tin from $2-7M$ hydrochloric acid. Depending on the accompanying metals, tin is eluted either with $0.5-1M$ H_2SO_4 [28,29] or with $6M$ NaOH [30]. A malonate eluent has been used to separate tin(IV) from antimony(V) and lead(II) [31].

55.2 Methods of Determination

Below are described the sensitive phenylfluorone and Pyrocatechol Violet methods, together with the much less sensitive dithiol method. Babko and Karnaukhova [32] have compared several spectrophotometric methods for determining tin.

55.2.1 Phenylfluorone Method

Tin(IV) reacts with phenylfluorone (formula, p. 275) in a not too acidic medium to form a sparingly soluble complex. At low concentrations of tin, this complex occurs in solution as a sol suitable for the spectrophotometric determination of tin [5,13,29,33–35]. Protective colloids (e.g. gum arabic) [5,13] or ethanolic media [29] have been used to stabilize the system. The reagent solution is yellow, whereas the pseudosolution of the tin phenylfluorone complex is orange-red.

A pH of 1·0–1·2 is the most suitable for the reaction. When the solution is too acid, the reaction between tin(IV) and phenylfluorone does not proceed to completion, and maximum absorbance is not attained. Conversely, when the solution is insufficiently acid, the free reagent precipitates, causing a pink turbidity.

The molar absorptivity of the pseudosolution is $7·7 \times 10^4$ at $\lambda = 510$ nm (specific absorptivity 0·65). The absorption spectra of phenylfluorone and its complex with tin are shown in Fig. 55.1.

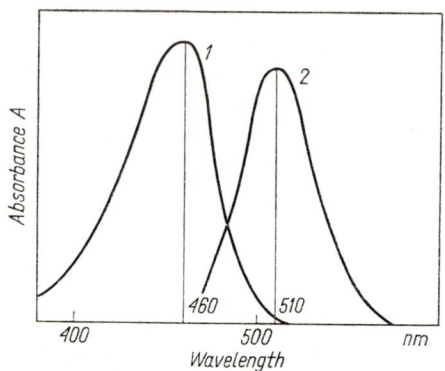

Fig. 55.1 Absorption spectra of phenylfluorone (1) and its tin complex (2) (reagent solution as reference) at pH 1·1

The phenylfluorone method is of low selectivity. Numerous multivalent metals (e.g. Sb, Ge, Zr, Ga, Fe, Ta, Nb, Mo, and Ti) interfere. Small amounts of Ti, Mo, Nb, and Ta can be masked with hydrogen peroxide. Antimony, which often accompanies tin, can be masked with citric acid.

When the amounts of tin in the solution are very small, the introduction of a stabilizing colloid is not necessary. At lower concentrations of tin, full colour development may take up to 2 hours [29].

The phenylfluorone method has been used to determine tin in iron and steels [13,36,37], ferromolybdenum [9], nickel [27], lead [5,19,38], copper and lead alloys [5], zinc and zinc concentrates [20,39], silver alloys [40], zirconium alloys [41], rhenium [42], ores [2,11], minerals [28], silicates [43], rocks, sediments, and soils [29], and biological materials [34].

Asmus *et al.* [44–48] have recommended a related reagent, 3′-pyridylfluorone, for determining tin. The method has remarkably high sensitivity (ε is $\sim 1·1 \times 10^5$ at 545 nm), and has been applied in the analysis of steels [45,46], hydrogen peroxide solutions [47], and organotin compounds [48]. Another fluorone derivative which has found similar application is *p*-nitrophenylfluorone [10].

Reagents

1. Phenylfluorone: 0·01% solution. Dissolve 10 mg of reagent in methanol, add 1 ml of 2*M* HCl, and dilute the solution with methanol to 100 ml.

2. *Standard tin(IV) solution*: 1 mg/ml. Dissolve 1·000 g of suitably pure tin in 50 ml of conc. H_2SO_4. After dissolution of the metal heat the solution to fuming until it clarifies. Carefully dilute the cooled solution with $\sim 0.5M$ H_2SO_4 to 1 litre in a volumetric flask. Working solutions are obtained by suitable dilution of the stock solution with $0.25M$ H_2SO_4.
3. *Citrate solution*. Dissolve 147 g of trisodium citrate and 105 g of citric acid in water, and dilute the solution with water to 1 litre.
4. *Gum arabic*: 1% solution.

Procedure

Separation of Sn with hydrous MnO_2. Heat the chloride-free sample solution containing 3 ml of conc. HNO_3 per 100 ml to almost boiling. Add 1 ml of 1% $KMnO_4$ solution and 2 ml of 1% $Mn(NO_3)_2$ solution, and heat the solution for 30 minutes without boiling. Filter off the precipitate on a filter paper and wash with hot HNO_3 (1+100). Dry and burn off the filter paper, and fuse the precipitate in a nickel crucible with 0·3 g of Na_2O_2 and a pellet of NaOH. Heat the melt till dark red. Leach the cooled fusion cake with hot water, transfer the whole contents of the crucible to a beaker, and add 5 ml of H_2SO_4 (1+4) and 2 or 3 drops of 3% H_2O_2 solution. Evaporate the solution to reduce its volume.

Determination of Sn. To the solution (obtained as above, for instance) containing not more than 50 μg of Sn, add 1 ml of 3% H_2O_2 solution, 2 ml of gum arabic solution, 5 ml of the phenylfluorone solution, 4 ml of the citrate solution, and then water to ~ 40 ml. After stirring, adjust the pH of the solution to 1·1 (± 0.1) (with dilute NaOH solution). Make up the solution with water in a volumetric flask to 50 ml. After 30 minutes, measure the absorbance at 510 nm, using a reagent blank as reference.

Note. The amount of citrate added masks a 25-fold amount of antimony relative to tin.

55.2.2 Pyrocatechol Violet Method

Pyrocatechol Violet (PV) reacts with tin(IV) in weakly acidic medium (pH 2–4·5) to form a red water-soluble complex (Sn:PV = 1:2). This reaction is the basis of a spectrophotometric method for determining tin [49–54].

Within the pH range suitable for tin determination, many other multivalent metals react with PV [e.g. Al, Ga, In, Fe(III), Sb, Bi, Th, Zr, Mo, and W]. However, separation of tin by extraction of SnI_4 from sulphuric acid medium makes the method almost specific for tin.

The method is sensitive: the molar absorptivity at 552 nm is 6.85×10^4 [53]. The absorbance of free reagent is insignificant at pH 3·8. The formation of the complex is slow, requiring 15–30 minutes.

In the presence of gelatine, the absorption maximum of the complex is shifted from 552 to 640 nm [50]. Dagnall *et al.* [55] have determined tin with Pyrocatechol Violet in the presence of cetyltrimethylammonium bromide (CTAB) at pH 2·2, thereby increasing the sensitivity ($\varepsilon = 9.56 \times 10^4$ at 662 nm), and shifting the absorption maximum of the complex. Ashton et al. [56] have also investigated addition of CTAB.

After extraction of tin with TEHPO in cyclohexane, the colour reaction can be carried out in the organic phase by adding a solution of Pyrocatechol Violet in ethanol [25].

Tin in iron and steels [51,56], uranium alloys [57], food [58], and organic matter [59,60] has been determined by the Pyrocatechol Violet method.

Reagents

1. Pyrocatechol Violet: 0·05% aqueous solution. This solution is stable for a week.
2. Standard tin(IV) solution: 1 mg/ml (p. 549).

Procedure

Separation of Sn by extraction [53]. To the solution containing tin, add sulphuric acid until its concentration reaches 4·5M. Transfer the cooled solution to a separating funnel, add ~2 g of KI, mix well, and shake for ~2 minutes with 10–15 ml of toluene. Discard the aqueous phase and wash the toluene extract (gently) with 5 ml of 4·5M H_2SO_4 containing 0·5 g of KI. (The toluene layer will be pink with extracted iodine). Strip the tin by shaking the toluene extract for ~30 seconds with two portions (10 and 5 ml) of 0·2M NaOH.

Determination of Sn. Acidify the alkaline tin solution (or an aliquot containing not more than 60 µg of Sn) obtained as above, with 3 ml of HCl (1+1) and decolorize (reduce iodine) by adding 2% ascorbic acid solution dropwise. Add 5 ml of the Pyrocatechol Violet solution and 10 ml of 20% sodium acetate solution. Adjust the pH to 3·8±0·1 with ammonia. Transfer the solution to a 50-ml volumetric flask, dilute to the mark with water, and mix thoroughly. After 30 minutes, measure the absorbance of the solution at 552 nm against a reagent blank.

55.2.3 Dithiol Method

Dithiol (4-methyl-1,2-dimercaptobenzene) reacts with an acid solution (dilute HCl or H_2SO_4) of tin(II) to form a pink chelate

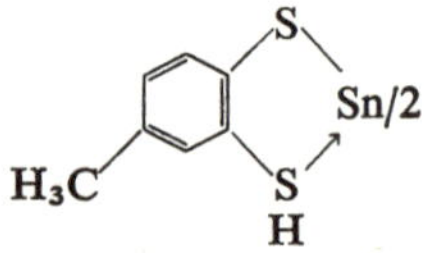

which is sparingly soluble in water, but which remains as a coloured suspension at suitably low concentrations [1,61,62]. It is rather difficult to obtain a clear pseudosolution of tin dithiolate, despite introduction of protective colloids (e.g. gum arabic) into the solution before the dithiol is added [62,63]. Thioglycollic acid is used to reduce tin(IV).

The dithiol method is rather insensitive: $\varepsilon \sim 5.8 \times 10^3$ ($a \sim 0.049$) at $\lambda_{max} = 535$ nm.

Dithiol dissolves in dilute NaOH solution to form a colourless solution which is susceptible to atmospheric oxidation. Hence, thioglycollic acid is added to stabilize the dithiol solution. In acid solutions, the reagent forms a fine suspension.

Tin dithiolate may be extracted with organic solvents (e.g. ethers); however, these extractions are not usually exploited in the colorimetric determination of tin because the absorbances of the pale yellow extracts are much less intense than those of the pink aqueous pseudosolutions.

A number of heavy metals (e.g. Bi, Ag, Hg, Cd, Cu, Mo, and W) also react with dithiol in acid solution to give coloured products. Fluoride and phosphate interfere.

The dithiol method has been used to determine tin in steel [1], silicate minerals [1,4], organic compounds [64], organic matter [65], and foods [66,67].

Reagents

1. Dithiol: freshly prepared 0·15% solution in 2% NaOH containing 5 drops of thioglycollic acid per 100 ml of solution. When stored, the solution turns brown under the influence of atmospheric oxygen.
2. Standard tin solution: 1 mg/ml (p. 549).

Procedure

Distillative separation of Sn [1]. To a 100-ml still containing the sample solution, add 25 ml of H_2SO_4 (1+1), 3 ml of conc. H_3PO_4, 20 ml of HCl (1+1), and 1 g of hydrazine sulphate. Dip the end of the condenser into a beaker of water. Bubble a slow stream of CO_2 through the solution in the still, and heat to boiling. When the temperature in the still has reached 160°C, add dropwise 20 ml of $6M$ HCl at a rate of 1 drop every 4 seconds while maintaining the temperature between 155 and 165°C. Then stop the heating, remove the receiver and rinse the condenser with water. Discard the distillate which contains As(III), Ge(IV), and Sb(III).

Dip the condenser of the distillation apparatus into 10 ml of water in the beaker. Add dropwise to the still a mixture of 15 ml of HCl (1+1) and 7 ml of 48% HBr. Bubble CO_2 through slowly, heating the still to maintain the temperature between 145 and 160°C. The distillation should take 15–20 minutes. Stop heating and rinse the condenser with a small amount of water, collecting the washings in the receiver. Add to the distillate 2 ml of H_2SO_4 (1+1) and 5 ml of conc. HNO_3, and cover the beaker. When the vigorous reaction (decomposition of HBr) has ceased, remove the cover glass and evaporate the solution to white fumes.

Determination of Sn. Place the acid sample solution (0·5–1M H_2SO_4), containing not more than 0·5 mg of Sn in a 50-ml volumetric flask. Add 5 drops of thioglycollic acid and 2 ml of 5% gum arabic solution, and mix well. Add 1 ml of dithiol solution, mix, dilute with water to the mark, and mix again. After 10 minutes, measure the absorbance of the pseudosolution at 535 nm against a reagent blank solution.

Note. In the case of smaller amounts of tin it is advisable to reduce the quantity of dithiol.

55.2.4 OTHER METHODS

Haematoxylin (strictly speaking its oxidized form, haematein) is considered a good reagent for determining tin [32,68–70]. The molar absorptivity of the coloured pseudosolution is $4·3 \times 10^4$ at 580 nm [32]. This reagent has been used to determine tin in, *inter alia*, iron and steel [15], lead alloys [17], and propellants [69].

Ducret and Maurel [71] have determined tin by extracting the coloured association complex formed by $SnCl_4^{2-}$ and Crystal Violet at pH 1 into 4-heptanone ($\varepsilon = 8·5 \times 10^4$). Rhodamine B [72], Butyl Rhodamine B [73], Brilliant Green [74], Malachite Green [75], and antipyrine dyes [76] have been used in similar extractive methods.

Other organic spectrophotometric reagents for tin include quercetin [32,77–80], gallein [81], 3,5-dinitrocatechol + basic dyes [82], Xylenol Orange [83], Bromopyrogallol Red [84], salicylideneamino-2-thiophenol [85], dimercaptothiopyrone derivatives [86], PAN [87], PAR ($\varepsilon = 4·85 \times 10^4$ at 515 nm) [88], Lumogallion [89], oxine [31,90], bromo-oxine [91], dithizone [92,93], and 2-(2-hydroxyphenyl)benzothiazoline [94].

Tin has been determined as the ternary dimethylglyoxime complex with iron(II) and tin [95,96].

Low sensitivity methods for determining larger amounts of tin are based on the yellow iodide [97,98] and bromide complexes [99].

Tin may also be determined by taking advantage of the reduction of molybdosilicic acid to silicomolybdenum blue by tin(II). Tin(IV) is first reduced to Sn(II) with granular zinc in $\sim 6M$ HCl [100,101].

References

1. Onishi, H. and Sandell, E. B., *Anal. Chim. Acta* **14**, 153 (1956).
2. Ginzburg, L. B. and Shkrobot, É. P., *Zavodsk. Lab.* **23**, 527 (1957).
3. De Bruyne, P. and Hoste, J., *Bull. Soc. Chim. Belg.* **70**, 221 (1961).
4. Martinet, B., *Chim. Anal. (Paris)* **43**, 483 (1961).
5. Luke, C. L., *Anal. Chem.* **28**, 1276 (1956); **31**, 1803 (1959).
6. Ogden, D. and Reynolds, G. F., *Analyst* **89**, 538 (1964).
7. Reynolds, G. F. and Tyler, F. S., *ibid.* **89**, 579 (1964).
8. Pyburn, C. M. and Reynolds, G. F., *ibid.* **93**, 375 (1968).
9. Silaeva, E. V. and Kurbatova, V. I., *Zavodsk. Lab.* **27**, 1462 (1961).
10. Nazarenko, V. A. and Lebedeva, N. V., *Zh. Analit. Khim.* **10**, 289 (1955); *Zavodsk. Lab.* **28**, 268 (1962).

References

11. Baumgärtel, E. and Gärtner, P., *Z. Anal. Chem.* **208**, 416 (1965).
12. Kovács, E. and Guyer, H., *ibid.* **186**, 267 (1962); **208**, 255 (1965).
13. Luke, C. L., *Anal. Chim. Acta* **37**, 97 (1967).
14. Gilbert, D. D. and Sandell, E. B., *Microchem. J.* **4**, 491 (1960).
15. Specker, H. and Graffmann, G., *Z. Anal. Chem.* **228**, 401 (1967).
16. Kosaric, N. and Leliaert, G., *Mikrochim. Acta* **1961**, 806.
17. Shirodker, R. and Schibilla, E., *Z. Anal. Chem.* **248**, 173 (1969).
18. Wyatt, P. F., *Analyst* **80**, 368 (1955).
19. Gur'ev, S. D. and Tsaraeva, N. F., *Zavodsk. Lab.* **24**, 1195 (1958).
20. Rączka, E. and Suchy, H., *Rudy Metale Nieżelazne* **6**, 274 (1961).
21. Lyle, S. J. and Shendrikar, A. D., *Anal. Chim. Acta* **36**, 286 (1966).
22. Rakovskii, E. E. and Petrukhin, O. M., *Zh. Analit. Khim.* **18**, 539 (1963).
23. Jordanov, N., Mareva, S. and Koeva, M., *Anal. Chim. Acta* **59**, 75 (1972).
24. Pollock, E. N. and Zopatti, L. P., *Anal. Chem.* **37**, 290 (1965).
25. Ross, W. J. and White, J. C., *ibid.* **33**, 424 (1961).
26. Levin, I. S. and Tarasova, V. A., *Zh. Analit. Khim.* **28**, 1341 (1973).
27. Hofer, A., *Z. Anal. Chem.* **240**, 229 (1968).
28. Khalizova, V. A., Alekseeva, A. Ya., Smirnova, E. P. and Krasyukova, N. G. *Zh. Analit. Khim.* **25**, 1525 (1970).
29. Smith, J. D., *Analyst* **95**, 347 (1970); *Anal. Chim. Acta* **57**, 371 (1971).
30. Ariel, M. and Kirova, E., *Talanta* **8**, 214 (1961).
31. Dawson, J. and Magee, R. J., *Mikrochim. Acta* **1958**, 325, 330.
32. Babko, A. K. and Karnaukhova, N. N., *Zh. Analit. Khim.* **22**, 868 (1967).
33. Bennett, R. L. and Smith, H. A., *Anal. Chem.* **31**, 1441 (1959).
34. Oelschläger, W., *Z. Anal. Chem.* **174**, 241 (1960).
35. Dymov, A. M., Ivanov, I. G. and Romantseva, T. I., *Zh. Analit. Khim.* **26**, 2360 (1971).
36. Amsheev, A. A., *Zavodsk. Lab.* **34**, 789 (1968).
37. Leblond, A. M. and Boulin, R., *Chim. Anal.* (*Paris*) **50**, 171 (1968).
38. Hofer, A. and Landl, B., *Z. Anal. Chem.* **244**, 103 (1969).
39. Górski, L. and Hołyńska, B., *Chem. Anal.* (*Warsaw*) **11**, 395 (1966).
40. Chwastowska, J. and Skorko-Trybuła, Z., *ibid.* **9**, 123 (1964).
41. Rajković, D., *Z. Anal. Chem.* **263**, 334 (1973).
42. Ryabchikov, D. I., Lazarev, A. I. and Lazareva, V. I., *Zh. Analit. Khim.* **19**, 1110 (1964).
43. Agterdenbos, J. and Vlogtman, J., *Talanta* **19**, 1295 (1972).
44. Asmus, E. and Kraetsch, J., *Z. Anal. Chem.* **223**, 401 (1966).
45. Asmus, E. and Kossmann, U., *ibid.* **245**, 137 (1969).
46. Asmus, E. and Weinert, H., *ibid.* **249**, 179 (1970).
47. Asmus, E. and Jahny, J., *ibid.* **255**, 186 (1971).
48. Asmus, E., Kropp, B. and Moczko, F. M., *ibid.* **256**, 276 (1971).
49. Ross, W. J. and White, J. C., *Anal. Chem.* **33**, 421 (1961).
50. Malát, M., *Z. Anal. Chem.* **187**, 404 (1962).
51. Tanaka, K., *Japan Analyst* **11**, 332 (1962); **13**, 725 (1964).
52. Yakovlev, P. Ya. and Razumova, G. P., *Zavodsk. Lab.* **31**, 1307 (1965).
53. Newman, E. J. and Jones, P. D., *Analyst* **91**, 406 (1966).
54. Wakley, W. D. and Varga, L. P., *Anal. Chem.* **44**, 169 (1972).
55. Dagnall, R. M., West, T. S. and Young, P., *Analyst* **92**, 27 (1967).
56. Ashton, A., Fogg, A. G. and Burns, D. T., *ibid.* **98**, 202 (1973); *Z. Anal. Chem.* **264**, 133 (1973).
57. Ferlin, C. and Lelièvre, B., *Bull. Soc. Chim. France* **1968**, 3415.
58. Adcock, L. H. and Hope, W. G., *Analyst* **95**, 868 (1970).
59. Analytical Methods Committee, *ibid.* **92**, 320 (1967).
60. Corbin, H. G., *Anal. Chem.* **45**, 534 (1973).
61. Farnsworth, M. and Pekola, J., *ibid.* **26**, 735 (1954).

62. Ovenston, T. C. and Kenyon, C., *Analyst* **80**, 566 (1955).
63. Board, P. W. and Elbourne, R. G., *ibid.* **89**, 555 (1964).
64. Farnsworth, M. and Pekola, J., *Anal. Chem.* **31**, 410 (1959).
65. Analytical Methods Committee, *Analyst* **93**, 414 (1968).
66. Dickinson, D. and Holt, R., *ibid.* **79**, 104 (1954).
67. Bernstein, I. and Gilewska, C., *Roczniki Państw. Zakł. Hig.* **5**, 245, 312 (1954); **6**, 243 (1955).
68. Teicher, H. and Gordon, L., *Anal. Chem.* **25**, 1182 (1953).
69. Norwitz, G., *Analyst* **86**, 835 (1961).
70. Asmus, E., Altmann, H. J. and Thomasz, E., *Z. Anal. Chem.* **216**, 3 (1966).
71. Ducret, L. and Maurel, H., *Anal. Chim. Acta* **21**, 79 (1959).
72. Arnesen, R. T. and Selmer-Olsen, A. R., *ibid.* **33**, 335 (1965).
73. Shumova, T. I. and Blyum, I. A., *Zavodsk. Lab.* **34**, 659 (1968).
74. Busev, A. I., Shestidesyatnaya, N. L. and Zimomrya, G. G., *Zh. Analit. Khim.* **26**, 1517 (1971).
75. Ackermann, G. and Köthe, J., *Chem. Anal. (Warsaw)* **17**, 445 (1972).
76. Zhivopistsev, V. P., Selezneva, E. A., Minina, V. S. and Bragina, Z. I., *ibid.* **26**, 761 (1971).
77. Liška, K., *Collection Czech. Chem. Commun.* **21**, 1439 (1956).
78. Kirk, R. S. and Pocklington, W. D., *Analyst* **94**, 71 (1969).
79. Wunderlich, E. and Bosse, G., *Erzmetall.* **24**, 537 (1971).
80. Engberg, A., *Analyst* **98**, 137 (1973).
81. Jones, J. C., *ibid.* **93**, 214 (1968).
82. Nazarenko, V. A., Lebedeva, N. V. and Vinarova, L. I., *Zh. Analit. Khim.* **28**, 1100 (1973).
83. Danilova, V. N., *Zavodsk. Lab.* **29**, 407 (1963).
84. Thierig, D. and Umland, F., *Z. Anal. Chem.* **221**, 229 (1966).
85. Gregory, G. R. and Jeffery, P. G., *Analyst* **92**, 293 (1967).
86. Usatenko, Yu. I., Arishkevich, A. M. and Moroz, A. A., *Zh. Analit. Khim.* **22**, 1823 (1967).
87. Pilloni, G., *Anal. Chim. Acta* **37**, 497 (1967).
88. Kasiura, K. and Olesiak, K., *Chem. Anal. (Warsaw)* **14**, 139 (1969).
89. Marchenko, P. V. and Obolonchik, N. V., *Zh. Analit. Khim.* **22**, 725 (1967).
90. Eberle, A. R. and Lerner, M. W., *Anal. Chem.* **34**, 627 (1962).
91. Ruf, E., *Z. Anal. Chem.* **162**, 9 (1958).
92. Aldridge, W. N and Cremer, J. E., *Analyst* **82**, 37 (1957).
93. Vancea, M. and Volusniuc, M., *Stud. Cercet. Chim. Cluj* **13**, 203 (1962).
94. Uhlemann, E. and Pohl, V., *Anal. Chim. Acta* **65**, 319 (1973).
95. Babko, A. K., Mikhel'son, P. B. and Kirpa, I. M., *Ukr. Khim. Zh.* **28**, 963 (1962).
96. Elinson, S. V. and Tsvetkova, V. T., *Zavodsk. Lab.* **37**, 662 (1971).
97. Paul, A. D. and Gibson, J. A., Jr., *Anal. Chem.* **36**, 2321 (1964).
98. Tanaka, K. and Takagi, N., *Anal. Chim. Acta* **48**, 357 (1969).
99. Nielsch, W. and Böltz, G., *Z. Anal. Chem.* **142**, 109; **143**, 161 (1954).
100. Baker, I., Miller, M. and Gibbs, R. S., *Ind. Eng. Chem., Anal. Ed.* **16**, 269 (1944).
101. Marczenko, Z., *Chem. Anal. (Warsaw)* **2**, 160 (1957).

Chapter 56
TITANIUM

Titanium (Ti, at.wt. 47·90) occurs in non-complexing acid solutions as the titanic ion Ti^{4+} or the titanyl ion TiO^{2+} (in a less acid medium). At pH $\geqslant 1$ basic salts precipitate, followed by the hydroxide, which displays very weak amphoteric properties. Titanium(IV) yields stable fluoride, citrate, tartrate, oxalate, EDTA and peroxide complexes, and weak sulphate and chloride complexes. Unstable titanous compounds [violet Ti^{3+} ions obtained by reduction of titanium(IV) with zinc] are of limited value for spectrophotometric determination.

56.1 Methods of Separation

56.1.1 Precipitation

Precipitation of titanium(IV) as the hydroxide $Ti(OH)_4$ with an excess of sodium hydroxide solution separates this species from V(V), Cr(VI), Mo(VI), W(VI), PO_4^{3-} and Al [1–4]. Double precipitation is best. Traces of titanium are separated with sodium hydroxide in the presence of iron [2,3], manganese, magnesium [3] or lanthanum [4] as collector. Titanium has been separated from iron and aluminium with sodium hydroxide in the presence of triethanolamine [5].

When a sodium carbonate fusion cake is leached with water, titanium hydrolyses and remains quantitatively in the undissolved residue.

Precipitation of titanium with ammonia in the presence of EDTA separates this metal from Fe, Mn, Al and other metals [6,6a].

Mention should also be made of some other precipitation methods for separating titanium, namely precipitation of titanium arsenate together with zirconium arsenate as collector [7,8], of titanium as cupferronate with an iron carrier [9] or as oxinate with an aluminium carrier [10], and lastly separation of titanium with tannin from an ammoniacal medium in the presence of EDTA [11].

56.1.2 Extraction

Titanium is extracted in the form of cupferronate [12–15] and separated from other metals, in particular from niobium and tantalum [13] and also from vanadium [14].

Traces of titanium (along with Fe, Al and Mn) can be extracted as the oxinate into chloroform [10]. DDTC has also been used for the extractive separation of titanium from certain metals [16]. Titanium (together with zirconium) is extracted from 13M H_2SO_4 medium with TBP in CCl_4 [17], and dialkylphosphoric acids have also been used [18,19].

It is possible to separate titanium(III) from titanium(IV) (masked with oxalic acid) by extracting the 1,10-phenanthroline complex of titanium(III) with di-(2-ethylhexyl)phosphoric acid in kerosene [20].

In a number of methods discussed below, the extractive separation of titanium is used in the spectrophotometric determination.

56.1.3 Ion-Exchange

Ion-exchangers play a very important part in the separation of titanium from other metals. Anionic titanium fluoride [21,22], sulphosalicylate [23] and ascorbate [24,25] complexes are retained on a column of strongly basic anion-exchanger. The titanium can be eluted with 3M HCl [21], 1M H_2SO_4 [23], or a hydrogen peroxide solution in 0·05M H_2SO_4 [25]. Ti, Zr, Nb, Ta, Mo, and W can be separated by ion-exchange and successive elution with suitable eluents [26,27]. Titanium, zirconium, and molybdenum have been separated by virtue of the differences in stability of their oxalate complexes as a function of acidity (H_2SO_4) [28].

Cation-exchangers are used for the separation of titanium from other metals [29-31]. The titanium may be eluted with dilute NH_4F-H_2SO_4 solution [30]. Dosch and Conrad [31] have separated titanium and zirconium by sorbing both of them on a Dowex-50 column and then eluting first titanium with 25% $HClO_4$ and subsequently zirconium with a mixture of $HClO_4$ and NH_4F.

56.2 Methods of Determination

At higher concentrations titanium is determined by the hydrogen peroxide method. The methods using chromotropic acid, Tiron and diantipyrylmethane are about ten times as sensitive. The thiocyanate (with extraction) and fluorone methods are the most sensitive.

56.2.1 Hydrogen Peroxide Method

In acid media titanium ions give with hydrogen peroxide a yellow-orange complex $[TiO.H_2O_2]^{2+}$ which forms the basis of the well-known spectrophotometric method for determining titanium. The method is simple and fairly selective, but not very sensitive. The molar absorptivity at $\lambda_{max} = 410$ nm (see Fig. 56.1) is $\sim 7 \times 10^2$ ($a = 0.015$). The complex has been thoroughly studied by Babko et al. [32,33] and Mori et al. [34].

Provided the H_2O_2 concentration is $>0.09M$, the nature and concentration of the acid used (HNO_3, H_2SO_4, $HClO_4$, HCl) are not important [35].

Sulphuric acid (0·7–1·8M) is the most suitable medium for the reaction. Hydrochloric acid is less suitable owing to the yellow colour of the chloride complexes with iron(III), which usually accompanies titanium. Iron(III) may be masked with phosphoric acid, but in consequence of its complexing effect on titanium, phosphoric acid diminishes the colour intensity of the Ti–H_2O_2 system. The presence of fluoride, which combines strongly with titanium, makes it impossible to use the method. Small amounts of tartaric acid do not interfere. Oxalate and citrate interfere.

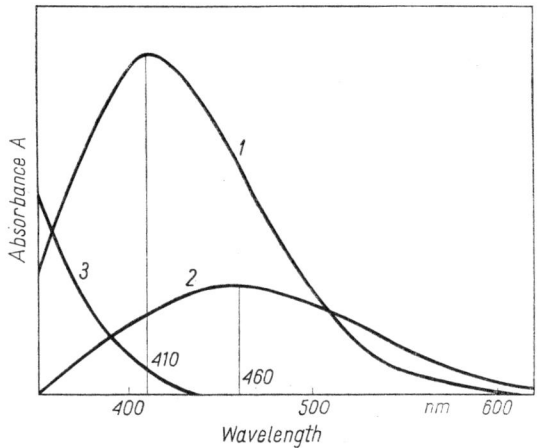

Fig. 56.1 Absorption spectra of hydrogen peroxide complexes of titanium (1), vanadium (2) and molybdenum (3)

Coloured metal ions and those which give complexes with hydrogen peroxide (V, Mo, U, Nb, Cr) also interfere, but only vanadium exerts an appreciable effect, its molar absorptivity at 410 nm being about a quarter of that of the titanium complex. The colours of the peroxide complexes of other metals are weaker. Titanium can be determined in presence of vanadium by measuring the decrease in absorbance when fluoride is added.

The sensitivity of the peroxide method is enhanced by addition of EDTA [36] or DCTA [37], which yield mixed-ligand titanium complexes.

The titanium peroxide complex can be extracted with a solution of alkylphosphonic or alkylphosphoric acids in non-polar solvents [18,38,39]. Kletenik [40] applied the peroxide reaction in benzene–ethanol medium after extraction of Ti with alkylphosphoric acid.

The hydrogen peroxide method has been used to determine titanium in steels [6,7,20], cast iron and iron ores [41], high-temperature alloys [1], uranium alloys [42], plutonium alloys [42,43], tantalum and its oxide [44], and polyethylene [8].

Differential spectrophotometry has been used for determination of titanium in uranium alloys [45], and ilmenite concentrate and ore [46,47].

Reagents

1. Hydrogen peroxide: 3% H_2O_2 aqueous solution.

2. Standard titanium solution: 1 mg/ml.
   (a) Weigh 0·834 g of suitably pure TiO_2 ignited at $\sim 900°C$ and fuse it with 8 g of $K_2S_2O_7$ in a quartz or platinum crucible. Dissolve the melt in 150 ml of hot H_2SO_4 (1+2). Dilute the solution in a volumetric flask to 500 ml with sulphuric acid (1+5).
   (b) Weigh 3 g of $K_2TiF_6 \cdot H_2O$ in a platinum evaporating dish, add 100 ml of H_2SO_4 (1+1) and evaporate the solution to white fumes. After cooling, rinse the walls of the vessel with water and evaporate to fumes again. Repeat the operation once more to completely remove HF. Dilute the cooled solution in a volumetric flask to 500 ml with sulphuric acid (1+5) and mix well. Standardize gravimetrically via TiO_2. Add enough H_2SO_4 (1+5) to the titanium solution to make the titanium concentration exactly 1 mg/ml.

Prepare working solutions by suitable dilution with $0.5M$ H_2SO_4.

Zátka has reviewed Ti standards [48], and Zátka and Hoffman [48a] recommend ammonium dioxalato-oxotitanate(IV) for the preparation of standard titanium solutions.

Procedure

Acidify the sample solution, containing not more than 1·5 mg of Ti, in a 50-ml volumetric flask with sulphuric acid so as to obtain $\sim 1M$ H_2SO_4 concentration in the final solution. Add [in order to mask iron(III)] 1 ml of conc. H_3PO_4, then 5 ml of the hydrogen peroxide solution, make up to the mark with water, and stir well. Measure the absorbance at 410 nm, using water as reference.

56.2.2 CHROMOTROPIC ACID METHOD

Chromotropic acid, i.e. 1,8-dihydroxynaphthalene-3,6-disulphonic acid, gives water-soluble, brown-red titanium complexes differing in composition and colour according to the acidity of the medium [49–53]. The pH at which the colour is developed is therefore critical.

The solutions are usually buffered with formate (pH 3–3·5) or acetate (pH 4–5). Titanium has also been determined with chromotropic acid in conc. H_2SO_4 medium [54].

The molar absorptivity of the titanium chromotropic acid complex solution at pH 3·5 and $\lambda_{max} = 460$ nm is 1.7×10^4 ($a = 0.36$).

The chief interference is from iron(III) which yields a green complex with chromotropic acid. Before the determination, therefore, larger quantities of iron should be separated or smaller ones reduced with ascorbic acid, sulphite, or dithionite. Vanadium in quantities not exceeding those of titanium is without effect [12,55].

Zirconium does not form a coloured chromotropic acid complex, but slightly lowers the absorptivity of the titanium complex. Molybdenum at concentrations below 50 μg/ml does not interfere. Fluoride interferes by masking titanium, but can be removed by fuming with sulphuric acid. Oxidants must be absent because chromotropic acid is fairly easily oxidized.

Titanium has been determined by the chromotropic acid method in cast iron and steel [55–57], copper alloys [58], magnetic alloys [59], boron [4], alkalis [10], and polyethylene [60].

Differential spectrophotometry has been used for determining titanium in products of the titanium industry [61].

A disadvantage of the reagent is that its solution grows dark on storage, because of oxidation.

Kuznetsov and Basargin [62] recommend 2,7-dichlorochromotropic acid as a reagent resistant to oxidation, its aqueous solution being stable on storage. It reacts with titanium at lower pH (~ 2) than chromotropic acid does. The absorption maximum is at 490 nm ($\varepsilon = 1\cdot12 \times 10^4$).

Dichlorochromotropic acid has been used for the determination of titanium in the presence of a considerable excess of uranium [63] and beryllium [64], and also in steels [65], aluminium alloys [66], and minerals and rocks [67].

Reagents

1. Chromotropic acid: 1% solution. Dissolve 1 g of chromotropic acid disodium salt in water, add 20 ml of 2% ascorbic acid solution, and dilute with water to 100 ml. Store in an amber-glass bottle.
2. Standard titanium solution: 1 mg/ml (p. 558).
3. Formate buffer: pH 3·5. Dissolve in water 60 ml of formic acid and 28 g of sodium hydroxide and dilute the solution to 1 litre with water.

Procedure

To the acid sample solution containing not more than 100 μg of Ti, add 2 ml of 2% ascorbic acid solution and heat the solution [to accelerate the reduction of Fe(III)]. Add to the cooled solution 2 ml of chromotropic acid solution, adjust the pH to ~ 2 with ammonia, and add 10 ml of formate buffer. Dilute to volume with water in a 50-ml volumetric flask and stir well. After 10 minutes, measure the absorbance at 460 nm, using a reagent blank as reference.

56.2.3 THIOCYANATE METHOD

In a solution containing high concentrations of thiocyanate and hydrochloric acid, titanium forms a yellow thiocyanate complex which furnishes a sensitive method for its determination [68–70]. The absorbance can be measured either on an acetone–water solution or on an MIBK extract. According to Tribalat and Calderó [70], the compound extracted into MIBK has the formula $Ti(OH)(SCN)_3$. From a solution of $5\cdot85M$ NaSCN and 0·85M HCl, MIBK extracts 95% of the titanium present, along with most of the reagent (as HSCN). The degree of extraction diminishes if the thiocyanate and the hydrochloric acid concentrations are decreased in the aqueous phase. The molar absorptivity of the complex in MIBK is $8·5 \times 10^4$ (in acetone–water solution it is $7·8 \times 10^4$). The sharp absorption maximum is at 417 nm. Interfering species include iron(III), niobium, and molybdenum if present in 10-, 25-, and 100-fold amounts, respectively, relative to Ti [70]. Thiocyanates of those metals may be extracted preliminarily from 4–5M HCl, the extraction of titanium being only negligible.

There are modifications of the thiocyanate method, in which ion-pairs of the titanium thiocyanate complex and an organic base are extracted. Young and White [71] and other authors [2,9] extract the TOPO complex into cyclohexane ($\varepsilon = 4·1 \times 10^4$ at $\lambda_{max} = 432$ nm). Triethylamine has also been used ($\varepsilon = 8 \times 10^4$) [72].

In the Tananaiko and Nebylitskaya method [73] the diantipyrylmethane (DAM) complex is extracted into chloroform. According to Babko et al. [74] the composition of the extracted complex corresponds to the formula $[HDAM^+]_2 [Ti(SCN)_6^{2-}]$. The molar absorptivity of the complex in chloroform is 8×10^4 ($6·0 \times 10^4$ according to [73]) ($a \sim 1·5$) at $\lambda_{max} = 420$ nm.

The distribution coefficient of the complex is so high that a single extraction is adequate, but the absorbance of the extract is strongly affected by the purity of the chloroform used. Ethanol reduces the absorbance.

In the DAM method there is no need for so high a thiocyanate concentration as in Tribalat and Calderó's method. The optimum acidity of the aqueous phase is 2–4M HCl. Coloured extractable ternary thiocyanate DAM complexes are also formed by Fe(III), Cu(II), Co(II), W(VI), Mo(VI) and Nb(V). Nickel and vanadium give only weakly coloured complexes and do not interfere in amounts less than 3–5 mg. Colourless ternary complexes are formed by Zn, Cd, Sn, Sb, and Zr. Fluoride, phosphate, and EDTA (if less than 10 mg) do not affect the extraction. Iron(III) and copper(II) can be reduced with thiosulphate or stannous chloride [73,75].

Reagents

1. Potassium thiocyanate: 30% solution.
2. Diantipyrylmethane (DAM): 5% solution in 2M HCl.
3. Standard titanium solution: 1 mg/ml (p. 558).
4. Stannous chloride: 10% solution in 2M HCl.

Procedure

To the acid sample solution (~ 20 ml in volume) containing not more than 25 μg of Ti, add 10 ml of the thiocyanate solution, 5 ml of the DAM solution, 5 ml of conc. HCl, and 2 ml of the $SnCl_2$ solution. Extract the titanium complex with two portions of $CHCl_3$. Dilute the extracts to volume with chloroform in a 50-ml (or smaller, depending on the colour intensity) volumetric flask, mix well, and measure the absorbance at 420 nm, using water or a reagent blank as reference.

56.2.4 OTHER METHODS

Diantipyrylmethane (DAM) gives in dilute hydrochloric acid a yellow $[Ti(DAM)_3]^{4+}$ complex, which can be used for determining titanium [76–80]. The molar absorptivity at 390 nm is 1.45×10^4 [79]. Titanium has been determined by this method in molybdenum and tungsten [15], nickel, iron, aluminium, molybdenum, niobium alloys [76], iron and steel [79], zirconium, hafnium, niobium [80], and ores, rocks, and minerals [78]. Higher titanium concentrations have been determined by differential spectrophotometry [81].

In the presence of stannous chloride a coloured complex is formed with DAM, extractable into chloroform [82,83]. DAM also gives a stable 1:1:1 complex with titanium(III) ions [after reduction of Ti(IV) with zinc amalgam] and chromotropic acid, which has been used for determining Ti(III) [84,85].

The most sensitive spectrophotometric methods for determining titanium are based on the use of the following fluorones [86]: methylfluorone [87,88], phenylfluorone [89], dihydroxyphenylfluorone [90,91], salicylfluorone [92,93], and disulphophenylfluorone ($\varepsilon = 1.08 \times 10^5$ at 570 nm) [94]. Shustova and Nazarenko [95] have found that titanium complexes with trihydroxyfluorones and phenazone are extractable into chloroform as ion-association complexes with the anions of strong acids (ClO_4^-, SCN^-, Cl^-, Br^-, I^-, $CCl_3CO_2^-$). The salicylfluorone complexes can be similarly extracted [95a].

Tiron(I) (the name indicates its use in determining titanium and iron) [96,96a] yields a yellow complex with titanium in acid medium ($\varepsilon = 1.5 \times 10^4$ at 390 nm). The system has been thoroughly studied by Sommer [97]. The method has found wide application, e.g. for determining titanium in nickel [98], silicate rocks [99], meteoritic material [100], cast iron, steel, bauxite, feldspar, glass sand, etc. [101].

For determining titanium ascorbic acid(II) is also recommended. Its complex absorbs strongly in the near ultraviolet [22,102]. The method has been used for determining titanium in steel [103], marine sediments and rocks [22], and mineral salt [104].

Sulphosalicylic acid (p. 313) [105–107] or salicylic acid [108,109] give yellow titanium complexes in acid medium, which have been used for spectrophotometric determination.

These titanium complexes may be extracted into chloroform in the presence of tributylamine [110] or pyridine [111]. Pyrocatechol in presence of oxalate [97,112] and its ternary complexes with titanium and organic bases (extractable into chloroform) [113–115] or with picolinic acid [116] are also used. Tetrabromopyrocatechol, pyrogallolsulphonic acid, and tribromopyrogallol have been used similarly [117].

Cheng [118] has determined titanium spectrophotometrically with cupferron. Related reagents are also recommended, viz., BPHA [119,120], N-furoylphenylhydroxylamine [121], N-acetylsalicyloyl-N-phenylhydroxylamine [122], salicylhydroxamic acid [123], and benzohydroxamic acid [124]. The titanium chelates are extracted into organic solvents.

A large group of reagents recommended for determining titanium consists of the azo dyes PAN [125,126], pyridylazophenols [126], PAR [126–128], Calcichrome [129], Arsenazo I [130], Solochrome Black AS [131], Chlorophosphonazo III [132], Stilbazo [133], and the triphenylmethane dyes Xylenol Orange (with and without H_2O_2) [134,135], Methylthymol Blue [135], Pyrocatechol Violet [136], and Chrome Azurol S [137]. Titanium complexes with Xylenol Orange and Methylthymol Blue can be extracted into butanol in the presence of diphenylguanidine [135].

Other organic reagents used for spectrophotometry of titanium are Bromopyrogallol Red [138,139], Eriochrome Red B [140], 8-hydroxyquinoline [141], 8-hydroxyquinaldine [142], acetylacetone [143], TTA [144], salicylamidoxime [145], 1-(o-carboxyphenyl)-3-hydroxy-3-methyltriazene [146], benzoylacetanilide [147], Khimdu [148], Tipyrogin [149], Tichromin [150,151], Dibromotichromin [152,153], tetrahydroxyphenazine ($\varepsilon = 7\cdot 95 \times 10^4$ at 580 nm) [154], Hydrazo T [155], Hydrazo II [156], and 2-methylnicotinic acid salicylhydrazide [157].

Titanium heteropoly acids are also utilized for its determination, those used being molybdotitanophosphoric acid [158–160], and molybdotitanic acid and its reduced form titanomolybdenum blue [161–163].

References

1. Silverstone, N. M. and Bach, B. B., *Metallurgia* **62**, 81 (1960).
2. Hibbits, J. O., Kallmann, S., Giustetti, W. and Oberthin, H. K., *Talanta* **11**, 1464 (1964).
3. Zinchenko, V. A. and Rudina, S. I., *Zavodsk. Lab.* **27**, 956 (1961).
4. Marczenko, Z., *Chem. Anal. (Warsaw)* **9**, 1093 (1964).
5. Přibil, R. and Veselý, V., *Talanta* **10**, 233 (1963).
6. Přibil, R. and Scheider, P., *Chem. Listy* **45**, 7 (1951).
6a. Pickering, W. F., *Anal. Chim. Acta* **9**, 324 (1953); **12**, 572 (1955).
7. Simmler, J. R., Roberts, K. H. and Tuthill, S. M., *Anal. Chem.* **26**, 1902 (1954).
8. Anduze, R. A., *ibid*. **29**, 90 (1957).
9. Hibbits, J. O., Rosenberg, A. F., Williams, R. T. and Kallmann, S., *Talanta* **11**, 1509 (1964).
10. Marczenko, Z., *Mikrochim. Acta* **1965**, 281.
11. Das, J. and Banerjee, S., *Z. Anal. Chem.* **183**, 42 (1961).
12. Ovenston, T. C., Parker, C. A. and Hatchard, C. G., *Anal. Chim. Acta* **6**, 7 (1952).
13. Alimarin, I. P. and Gibalo, I. M., *Dokl. Akad. Nauk SSSR* **109**, 1137 (1956).
14. Corbett, J. A., *Anal. Chim. Acta* **30**, 126 (1964).
15. Donaldson, E. M., *Talanta* **16**, 1505 (1969).
16. Rooney, R. C., *Anal. Chim. Acta* **19**, 428 (1958).
17. Zharovskii, F. G. and Vyazovskaya, L. M., *Ukr. Khim. Zh.* **32**, 747 (1966).
18. Kletenik, Yu. B., *Zh. Analit. Khim.* **17**, 1063 (1962); **18**, 66 (1963).

References

19. Kletenik, Yu. B. and Bykhovskaya, I. A., *ibid.* **20**, 567 (1965).
20. Tserkovnitskaya, I. A. and Novikova, E. I., *ibid.* **24**, 1160 (1969).
21. Athavale, V. T., Nadkarni, M. N. and Venkateswarlu, Ch., *Anal. Chim. Acta* **23**, 438 (1960).
22. Korkisch, J., Arrhenius, G. and Kharkar, D. P., *ibid.* **28**, 270 (1963).
23. Šimek, M., *Hutn Listy* **20**, 424 (1965)
24. Korkisch, J. and Farag, A., *Mikrochim. Acta* **1958**, 659.
25. Korkisch, J., *Z. Anal. Chem.* **178**, 39 (1960).
26. Bandi, W. R., Buyok, E. G., Lewis, L. L. and Melnick, L. M., *Anal. Chem.* **33**, 1275 (1961).
27. Dixon, E. J. and Headridge, J. B., *Analyst* **89**, 185 (1964).
28. Shakashiro, M. and Freund, H., *Anal. Chim. Acta* **33**, 597 (1965).
29. Strelow, F. W., *Anal. Chem.* **35**, 1279 (1963).
30. Kenna, B. T. and Conrad, F. J., *ibid.* **35**, 1255 (1963).
31. Dosch, R. G. and Conrad, F. J., *ibid.* **36**, 2306 (1964).
32. Babko, A. K. and Volkova, A. I., *Zh. Obshch. Khim.* **21**, 1949 (1951).
33. Babko, A. K., Volkova, A. I. and Lisichenok, S. L., *Zh. Neorgan. Khim.* **11**, 478 (1966).
34. Mori, M., Shibata, M., Kyuno, E. and Ito, S., *Bull. Chem. Soc. Japan* **29**, 904 (1956).
35. Vasil'ev, V. P. and Vorob'ev, P. N., *Zh. Analit. Khim.* **22**, 718 (1967).
36. Stolyarov, K. P. and Agrest, F. B., *ibid.* **19**, 457 (1964).
37. Lassner, E., Püschel, R. and Scharf, R., *Z. Anal. Chem.* **179**, 345 (1961).
38. Gorican, H. and Grdenic, D., *Anal. Chem.* **36**, 330 (1964).
39. Pyatnitskii, I. V., Glushchenko, L. M. and Gerasina, V. K., *Ukr. Khim. Zh.* **36**, 830 (1970).
40. Kletenik, Yu. B., *Zh. Analit. Khim.* **19**, 208 (1964).
41. Schöffmann, E. and Malissa, H., *Arch. Eisenhüttenw.* **28**, 623 (1957).
42. Evans, H. B. and Hallcock, R. R., *Anal. Chem.* **39**, 842 (1967).
43. Bergstresser, K. S., *ibid.* **29**, 532 (1957).
44. Hastings, J., McClarity, T. A. and Broderick, E. J., *ibid.* **26**, 379 (1954).
45. Neal, W. T. L., *Analyst* **79**, 403 (1954).
46. Malyutina, T. M. and Dobkina, B. M., *Zavodsk. Lab.* **27**, 650 (1961).
47. Lunina, G. E. and Romanenko, E. G., *ibid.* **34**, 538 (1968).
48. Zátka, V., *Analyst* **95**, 47 (1970).
48a. Zátka, V. and Hoffmann, O., *ibid.* **95**, 200 (1970).
49. Brandt, W. W. and Preiser, A. E., *Anal. Chem.* **25**, 567 (1953).
50. Babko, A. K. and Popova, O. I., *Zh. Neorgan. Khim.* **2**, 138 (1957).
51. Okáč, A. and Sommer, L., *Collection Czech. Chem. Commun.* **19**, 477 (1954); **20**, 1251 (1955); *Anal. Chim. Acta* **15**, 345 (1956).
52. Sommer, L., *Z. Anal. Chem.* **164**, 299 (1958); *Talanta* **9**, 439 (1962).
53. Biryuk, E. A. and Nazarenko, V. A., *Zh. Analit. Khim.* **23**, 1018 (1968).
54. Popova, O. I., *Zavodsk. Lab.* **25**, 148 (1959).
55. Rosotte, W. W. and Jaudon, E., *Anal. Chim. Acta* **6**, 149 (1952).
56. Koch, W. and Ploum, H., *Arch. Eisenhüttenw.* **24**, 393 (1953).
57. Sommer, L., *Collection Czech. Chem. Commun* **22**, 1793 (1957); **27**, 2212 (1962).
58. Wiedmann, H., *Z. Metallkunde* **48**, 410 (1957).
59. Bagdasarov, K. N. and Osmanov, Kh. A., *Zavodsk. Lab.* **34**, 1044 (1968).
60. Bolleter, W. T., *Anal. Chem.* **31**, 201 (1959).
61. Tikhonov, V. N., *Zh. Analit. Khim.* **22**, 525 (1967).
62. Kuznetsov, V. I. and Basargin, N. N., *ibid.* **16**, 573 (1961).
63. Kuznetsov, V. I., Basargin, N. N. and Kukisheva, T. N., *ibid.* **17**, 457 (1962).
64. Basargin, N. N., Kukisheva, T. N. and Solov'eva, N. V., *ibid.* **19**, 553 (1964).
65. Basargin, N. N., Tkachenko, A. N., Stupa, L. P. and Borodevskaya, L. N., *Zavodsk. Lab.* **28**, 1311 (1962).

66. Budanova, L. M. and Pinaeva, S. N., *ibid.* **29**, 149 (1963).
67. Klassova, N. S. and Leonova, L. L., *Zh. Analit. Khim.* **19**, 131 (1964).
68. Crouthamel, C. E., Hjelte, B. E. and Johnson, C, E., *Anal. Chem.* **27**, 507 (1955).
69. Mari, E. A., *Anal. Chim. Acta* **29**, 303, 312 (1963).
70. Tribalat, S. and Calderó, J.-M., *Bull. Soc. Chim. France* **1964**, 3187.
71. Young, J. P. and White, J. C., *Anal. Chem.* **31**, 393 (1959).
72. Tananaiko, M. M. and Lozovik, A. S., *Zh. Analit. Khim.* **24**, 844 (1969).
73. Tananaiko, M. M. and Nebylitskaya, S. L., *Zavodsk. Lab.* **28**, 263 (1962).
74. Babko, A. K., Tananaiko, M. M. and Lozovik, A. S., *Zh. Neorgan. Khim.* **14**, 1618 (1969).
75. Tananaiko, M. M. and Tsarenko, G. F., *Ukr. Khim. Zh.* **30**, 1213 (1964); **31**, 530 (1965).
76. Polyak, L. Ya., *Zh. Analit. Khim.* **17**, 206 (1962); **18**, 956 (1963); **19**, 1468 (1964); *Zavodsk. Lab.* **32**, 1317 (1966).
77. Lazareva, V. I. and Lazarev, A. I., *Zh. Analit. Khim.* **21**, 172 (1966).
78. Jeffery, P. G. and Gregory, G. R. E., *Analyst* **90**, 177 (1965).
79. Corbett, J. A., *ibid.* **93**, 383 (1968).
80. Wood, D. F. and Jones, J. T., *Anal. Chim. Acta* **47**, 215 (1969).
81. Malyutina, T. M., Tramm, R. S. and Pevzner, K. S., *Zavodsk. Lab.* **31**, 1054 (1965).
82. Podchainova, V. N. and Dolgorev, A. V., *Zh. Analit. Khim.* **20**, 1059 (1965).
83. Hofer, A. and Heidinger, R., *Z. Anal. Chem.* **249**, 177 (1970).
84. Busev, A. I. and Solov'eva, N. G., *Zh. Analit. Khim.* **25**, 1324 (1970).
85. Busev, A. I. and Solov'eva, N. G., *ibid.* **26**, 751 (1971).
86. Nazarenko, V. A. and Biryuk, E. A., *ibid.* **22**, 57 (1967).
87. Majumdar, A. K. and Savariar, C. P., *Anal. Chim. Acta* **21**, 584 (1959).
88. Mareček, J., Dvořak, J. and Ditz, J., *Z. Anal. Chem.* **217**, 248 (1966).
89. Pietrzak, I., *Chem. Anal. (Warsaw)* **5**, 923 (1960).
90. Asmus, E., Kurzmann, P. and Wollsdorf, F., *Z. Anal. Chem.* **197**, 413 (1963).
91. Asmus, E., Richly, W. and Wunderlich, H., *ibid.* **199**, 249 (1964).
92. Tataev, O. A., Yarysheva, E. A. and Khristosevich, O. V., *Zavodsk. Lab.* **35**, 1161 (1969).
93. Nazarenko, V. A. and Shustova, M. B., *ibid.* **37**, 146 (1971).
94. Nazarenko, V. A. and Biryuk, E. A., *Zh. Analit. Khim.* **15**, 306 (1960).
95. Shustova, M. B. and Nazarenko, V. A., *Ukr. Khim. Zh.* **33**, 623 (1967).
95a. Antonovich, V. P., Chelikhina, E. I. and Lozovaya, S. A., *Zh. Analit. Khim.* **28**, 1506 (1973).
96. Yoe, J. H. and Armstrong, A. R., *Anal. Chem.* **19**, 100 (1947).
96a. Wakamatsu, Y. and Otomo, M., *Bull. Chem. Soc. Japan* **45**, 2764 (1962).
97. Sommer, L., *Collection Czech. Chem. Commun.* **28**, 2102 (1963).
98. Andrew, T. R. and Gentry, C. H. R., *Metallurgia* **60**, 173 (1959).
99. Rigg, T. and Wagenbauer, H. A., *Anal. Chem.* **33**, 1347 (1961).
100. Easton, A. J. and Greenland, L., *Anal. Chim. Acta* **29**, 52 (1963).
101. Clark, L. J., *Anal. Chem.* **42**, 694 (1970).
102. Sommer, L., *Collection Czech. Chem. Commun.* **28**, 449 (1963).
103. Korkisch, J., *Mikrochim. Acta* **1961**, 262.
104. Bahr, H. and Jagielski, J., *Chem. Anal. (Warsaw)* **12**, 363 (1967).
105. Ziegler, M. and Glemser, O., *Z. Anal. Chem.* **139**, 92 (1953).
106. Sommer, L., *Collection Czech. Chem. Commun.* **22**, 453 (1957); **28**, 2716 (1963).
107. Qureshi, M., Rawat, J. P. and Khan, F., *Anal. Chim. Acta* **41**, 164 (1968).
108. Hultquist, A. E., *Anal. Chem.* **36**, 149 (1964).
109. Ramakrishna, R. S. and Gunawardena, H. A., *Talanta* **20**, 21 (1973).
110. Ziegler, M., Glemser, O. and Baeckmann, A., *Z. Anal. Chem.* **160**, 324 (1958); *Angew. Chem.* **70**, 500 (1958).
111. Babko, A. K. and Volkova, A. I., *Zh. Analit. Khim.* **15**, 587 (1960).

112. Pilipenko, A. T. and Lukachina, V. V., *ibid.* **25**, 2125 (1970).
113. Talipov, Sh. T. and Nigai, K. G., *ibid.* **18**, 178 (1963).
114. Tananaiko, M. M. and Vinokurova, G. N., *ibid.* **19**, 316 (1964).
115. Busev, A. I., Ali-Zade, T. D. and Solov'eva, N. G., *ibid.* **27**, 692 (1972).
116. Shnaiderman, S. Ya. and Knyazeva, E. N., *ibid.* **21**, 419 (1966).
117. Busev, A. I. and Solov'eva, N. G., *ibid.* **27**, 1100, 1283, 1529 (1972).
118. Cheng, K. L., *Anal. Chem.* **30**, 1941 (1958).
119. Zharovskii, F. G., Shpak, E. A. and Piskunova, E. V., *Ukr. Khim. Zh.* **28**, 1104 (1962).
120. Afghan, B. K., Marryatt, R. G. and Ryan, D. E., *Anal. Chim. Acta* **41**, 131 (1968).
121. Pilipenko, A. T., Shpak, E. A. and Boiko, Yu. P., *Zavodsk. Lab.* **31**, 151 (1965).
122. Savariar, C. P. and Joseph, J., *Anal. Chim. Acta* **47**, 347 (1969).
123. Alimarin, I. P., Borzenkova, N. P. and Zakarina, N. A., *Zavodsk. Lab.* **27**, 958 (1961).
124. Alimarin, I. P., Borzenkova, N. P., and Shmatko, R. I., *Zh. Analit. Khim.* **18**, 342 (1963).
125. Püschel, R. and Lassner, E., *Mikrochim. Acta* **1967**, 977.
126. Betteridge, D., John, D. and Snape, F., *Analyst* **98**, 520 (1973).
127. Ozawa, T., *Japan Analyst* **16**, 435 (1967).
128. Kho V'et Kui, Gibalo, I. M. and Lobanov, F. I., *Zh. Analit. Khim.* **29**, 269 (1974).
129. Ishii, H. and Einaga, H., *Japan Analyst* **15**, 821 (1966).
130. Nikitina, E. I., *Zh. Analit. Khim.* **14**, 431 (1959).
131. Korkisch, J., *Talanta* **8**, 583 (1961).
132. Fadeeva, V. I. and Alimarin, I. P., *Zh. Analit. Khim.* **17**, 1020 (1962).
133. Ozawa, T., *J. Chem. Soc. Japan, Pure Chem. Sect.* **92**, 522 (1971).
134. Otomo, M., *Bull. Chem. Soc. Japan* **36**, 1341, 1577 (1963).
135. Tolmachev, V. N., Gol'tsberg, I. M. and Konkin, V. D., *Zh. Analit. Khim.* **22**, 950 (1967).
136. Malát, M., *Z. Anal. Chem.* **201**, 262 (1964).
137. Horiuchi, Y. and Nishida, H., *Japan Analyst* **15**, 913 (1966).
138. Suk, V., Němcová, I. and Malat, M., *Collection Czech. Chem. Commun.* **30**, 2538 (1965).
139. Ganago, L. I. and Zharnovskaya, L. A., *Zh. Analit. Khim.* **28**, 933 (1973).
140. Tserkovnitskaya, I. A. and Perevoshchikova, V.V., *ibid.* **27**, 1111 (1972).
141. Chakrabarti, C. L., Magee, R. J. and Wilson, C. L., *Talanta* **10**, 1201 (1963).
142. Motojima, K., *Bull. Chem. Soc. Japan* **29**, 455 (1956).
143. Fujiwara, S., Nagashima, K. and Codell, M., *ibid.* **37**, 783 (1964).
144. De, A. K. and Rahaman, M. S., *Anal. Chim. Acta* **31**, 81 (1964).
145. Banerjea, D., *Z. Anal. Chem.* **159**, 123 (1957).
146. Majumdar, A. K., Bhattacharyya, B. C. and Roy, B. C., *Anal. Chim. Acta* **67**, 307 (1973).
147. Sarkar, A. K. and Das, J., *Anal. Chem.* **39**, 1608 (1967).
148. Basargin, N. N., Akhmedli, M. K. and Shirinov, M. M., *Zh. Analit. Khim.* **23**, 1813 (1968).
149. Akhmedli, M. K., Basargin, N. N. and Shirinov, M. M., *ibid.* **24**, 550 (1969).
150. Basargin, N. N., Akhmedli, M. K. and Shirinov, M. M., *ibid.* **24**, 384 (1969).
151. Nikitina, L. P. and Basargin, N. N., *ibid.* **25**, 1521 (1970).
152. Basargin, N. N., Yakovlev, P. Ya. and Deinekina, R. S., *Zavodsk. Lab.* **39**, 1043 (1973).
153. Basargin, N. N., Yakovlev, P. Ya. and Deinekina, R. S., *ibid.* **39**, 1305 (1973).
154. Asmus, E. and Peters, J., *Z. Anal. Chem.* **249**, 106 (1970).
155. Dolgorev, A. V., *Zavodsk. Lab.* **39**, 772 (1973).
156. Dolgorev, A. V., *ibid.* **38**, 1309 (1972).
157. Dolgorev, A. V., *Zh. Analit. Khim.* **28**, 1093 (1973).
158. Veitsman, R. M., *Zavodsk. Lab.* **25**, 408 (1959).

159. Shkaravskii, Yu. F., *Zh. Analit. Khim.* **18**, 196 (1963); **19**, 320, 514 (1964).
160. Murata, K., Yokoyama, Y. and Ikeda, S., *Anal. Chim. Acta* **48**, 349 (1969).
161. Guyon, J. C. and Mellon, M. G., *Anal. Chem.* **34**, 856 (1962).
162. Shakhova, Z. F. and Semenovskaya. E. N., *Vestn. Mosk. Univ. Khim.* **1968**, No. 2, 122.
163. Reznik, B. E. and Vorotyagina, V. D., *Zh. Analit. Khim.* **23**, 1230 (1968).

Chapter 57
TUNGSTEN

Tungsten (W, at.wt. 183·85) is an element similar to molybdenum and chromium. The basic properties of tungsten(VI) are not so accentuated as those of molybdenum(VI). Tungstic acid is less soluble than molybdic acid in mineral acids. Tungsten gives citrate, tartrate, oxalate, fluoride, and chloride complexes, as well as heteropoly acids with P, Si, V, etc. Besides the stable oxidation state W(VI) (in tungstate), tungsten may occur in the +V, +IV, and +III oxidation states.

57.1 Methods of Separation

57.1.1 Precipitation

To separate tungsten from metals having hydroxides insoluble in strongly basic medium (e.g. Fe, Ti, Mn, and Mg) the sample is fused with NaOH, Na_2CO_3, or Na_2O_2 and the melt is leached with water. Along with tungsten, Al, Mo, Cr, V, and As also pass into solution. Double precipitation of hydroxides with NaOH may be used to separate tungsten from cations insoluble in excess of sodium hydroxide.

Traces of tungsten can be separated by precipitation with ammonia (excess to be avoided) together with 2–5 mg of iron(III) or aluminium as collector. When zirconium is used as the collector, tungsten can be quantitatively separated from rhenium [1]. Tungsten has also been separated with hydrous MnO_2 as scavenger [2] and with ammonium molybdophosphate [3].

Precipitation of tungsten with 8-hydroxyquinoline in the presence of EDTA separates it from Pb, Bi, Hg, Cd, Cu, Fe, Al, Cr, U, Co, Mn, Ni, Zn, and V [4,5].

Kuznetsov et al. [6] coprecipitated traces of tungsten with tannin and Methyl Violet.

57.1.2 Extraction

The extraction of tungsten as the α-benzoinoximate into chloroform separates it from chromium and vanadium. Molybdenum is extracted together with tungsten [7–10].

Tungsten can be extracted as the thiocyanate complex [11] or with TBP from 8–10M HCl [11,12]. It has also been extracted with a mixture of mesityl oxide and MIBK (3+1) from 1M HCl+12M LiCl medium [13].

Kallman *et al.* [14] used counter-current extraction with MIBK to separate tungsten from molybdenum ($6M$ HCl + $2\cdot4M$ HF medium).

It is also possible to extract tungsten with cupferron [15], and 8-hydro, xyquinoline [16,17] and its derivatives, such as 8-mercaptoquinoline- 5,7-dibromo-8-hydroxyquinoline, and 2-methyl-8-hydroxyquinoline [17].

57.1.3 Ion-Exchange

Headridge and Dixon [18,19] found that tungsten is retained together with Mo, Ti, Zr, Nb, and Ta from HF–HCl solution on a strongly basic anion-exchanger, whereas Al, V(IV), Cr(III), Mn, Fe, Co, Ni, and Cu are not. The metals retained on the anion-exchanger are separated from one another by successive elution with suitably selected eluents containing HF, HCl, NH_4F, and NH_4Cl.

Traces of tungsten are separated from other elements by anion-exchange in acidic sulphate media containing hydrogen peroxide. The tungsten can easily be stripped by elution with NaOH–NaCl solution [20].

Attempts have also been made to use cation-exchangers for separating tungsten from other metals [2,18].

57.2 Methods of Determination

Tungsten is generally determined spectrophotometrically by two methods, one based on dithiol, the other on thiocyanate. These reagents are also used for determining molybdenum. Proper choice of conditions allows either to be determined in the presence of the other.

57.2.1 Dithiol Method

Tungsten, when reduced with tin(II) in acid medium, forms with dithiol (p. 360) a blue-green complex sparingly soluble in water but extractable with isoamyl, amyl, or butyl acetate, petroleum ether, or chloroform, which is suitable for spectrophotometric determination [15,21,22]. The dithiol is used either in sodium hydroxide solution or dissolved in amyl acetate.

The absorption maximum for amyl acetate solutions is at 640 nm (Fig. 32.1, p. 361), $\varepsilon = 1\cdot92 \times 10^4$ ($a = 0\cdot10$).

Molybdenum reacts like tungsten with dithiol and must either be separated, e.g. as the sulphide [23], before the determination of the latter, or extracted as the dithiolate under conditions chosen so that tungsten does not react with dithiol [2].

The dithiol method has been used to determine tungsten in molybdenum [14], steel [22], titanium, zirconium and its alloys [21–23], tantalum [21,26], niobium [25,26], titanium dioxide pigments [24], copper [27], silicate rocks [2,20], ores [28], soils and stream sediments [27a,29,30], and water [2,31].

Reagents

1. Dithiol: 0·5% solution in amyl acetate.
2. Standard tungsten solution: 1 mg/ml. Dissolve 1·79 g of sodium tungstate $Na_2WO_4 \cdot 2H_2O$ in water, and dilute the solution with water to 1 litre in a volumetric flask.
3. Stannous chloride: 10% solution in HCl (1+1).

Procedure

Separation of Mo. Evaporate to dryness the sample solution, containing not more than 250 μg of W. Dissolve the neutral residue in 1·9M HCl. Shake the solution for 30 sec with one volume of the dithiol solution. Wash the extract by shaking it for a while with 1·9M HCl. Combine the washing solution with the original aqueous solution. The extract containing the molybdenum dithiolate can be used for determining molybdenum.

Extraction and determination of W. To the aqueous phase from the separation of molybdenum (~ 15 ml in volume) add 10 ml of conc. HCl, 5 ml of stannous chloride solution, and 20 ml of dithiol solution. Heat the separating funnel in a water-bath at 80°C with occasional shaking.

After cooling, wash the extract with conc. HCl, dilute it to volume with amyl acetate in a 50-ml volumetric flask, and measure the absorbance at 640 nm against the solvent.

Note. For accurate separation of molybdenum from tungsten, strict observance of working conditions is required, especially the 1·9M HCl concentration in the solution from which molybdenum dithiolate is extracted.

57.2.2 THIOCYANATE METHOD

Tungsten(V) forms a series of yellow complexes with thiocyanate in acid medium (HCl, H_2SO_4). The mechanism of the reaction has been investigated by a number of workers [32–36]. The reaction is rapid and reproducible, if thiocyanate is added to the neutral or slightly alkaline solution followed by reducing agent and acidification. If the original solution is acidic, the tungsten exists in the form of tungstic acid polymers which are less reactive and not so rapidly reduced and complexed with SCN^- ions as the non-polymerized molecules existing immediately after acidification of tungstate. In slightly acidic medium in the presence of tartrate, tungsten forms tartrate complexes which react easily with thiocyanate.

Stannous chloride [33], titanous chloride [34], and tin amalgam [37] have been employed as reducing agents. The photochemical reduction of tungsten(VI) with ethanol has also been applied in H_2SO_4 and H_3PO_4 medium [38].

For measurement, the complex is extracted into polar organic solvents, usually isoamyl alcohol, di-isopropyl ether and MIBK; a 1:1 mixture of isoamyl alcohol and chloroform has also been used [37]. The molar absorptivity of the complex in isoamyl alcohol is $1·56 \times 10^4$ at $\lambda_{max} = 403$ nm

(a = 0·09). The absorption spectrum of the complex is shown in Fig. 57.1. Crouthamel and Johnson [33] have determined tungsten with thiocyanate in water–acetone medium.

In strongly acidic medium (8–9M HCl) containing $SnCl_2$, thiocyanate does not give a colour reaction with molybdenum [34], especially when the absorbance is measured after a lapse of a certain time (e.g. 20 minutes).

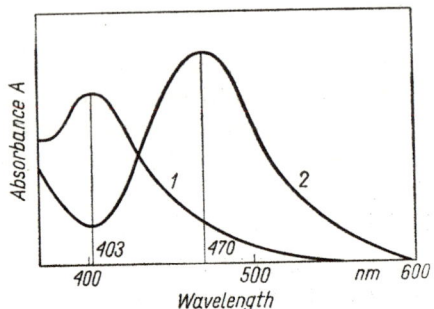

Fig. 57.1 Absorption spectra of thiocyanate complexes of tungsten (1) and molybdenum (2) in isoamyl alcohol

The colour of the tungsten complex is stable for 1–2 hours. Molybdenum can also be dealt with by prior separation as the sulphide [39].

To determine tungsten in molybdenum, Neef and Döge [40] extracted the molybdenum thiocyanate complex in the presence of thioglycollic acid, then extracted the tungsten thiocyanate complex after addition of $TiCl_3$ as reducing agent.

Iron(III) does not interfere since it is converted into Fe(II) by the reducing agent. Niobium is well masked by oxalate [41]. Fluoride and nitrite both interfere.

Before its determination by the thiocyanate method, tungsten is usually separated from the majority of metals by precipitation or extraction, often with α-benzoinoxime.

The tungsten thiocyanate complex forms ion-pairs with the tetraphenylarsonium cation, which can be extracted into chloroform [42].

A detailed discussion of existing procedures for the determination of tungsten with thiocyanate can be found in a paper by Fogg et al. [43].

The thiocyanate method has been used to determine tungsten in molybdenum and its compounds [16,39,40], cast iron and steel [12,13,41–45], high-temperature alloys [42,46], nickel alloys [37,47] titanium and its alloys [48,49], zirconium alloys [49], niobium alloys [50], tantalum [51], and silicate rocks and minerals [52].

Reagents

1. Potassium thiocyanate: 20% solution.
2. Standard tungsten solution: 1 mg/ml (p. 569).
3. Stannous chloride: 40% solution of $SnCl_2 . 2H_2O$ in conc. HCl.

Procedure

The sample solution containing not more than 300 μg of W should be slightly alkaline or neutral. If it contains tartrate, it may also be acidic. Add 5 ml of thiocyanate solution, 10 ml of $SnCl_2$ solution, and enough conc. HCl to bring its concentration in the solution to 8–9M. The total volume of the solution should be ∼ 50 ml. After 15 minutes extract the tungsten complex with two portions of isoamyl alcohol. Dilute the extract with the solvent in a 50-ml (or smaller) volumetric flask, and measure the absorbance at 403 nm, using the solvent as reference.

57.2.3 OTHER METHODS

A widely used but less sensitive method for tungsten determination [11,53–56] is based on the colour reaction of tungsten(VI) with hydroquinone in concentrated H_2SO_4 medium. Tungsten has been determined by this method in steel [54,56] and in high-temperature alloys [55].

Pyrocatechol has been used in the presence of EDTA [57]. Poluektova [58] recommends 6,7-dihydroxy-2,4-diphenylbenzpyranol as a sensitive reagent for tungsten. The molar absorptivity of the chloroform extract of the tungsten complex is $9 \cdot 3 \times 10^4$ at 535 nm. Dinitropyrocatechol [59] has been used with Brilliant Green ($\varepsilon = 1 \cdot 3 \times 10^5$ at 646 nm) and diantipyrylmethane ($\varepsilon = 2 \times 10^4$).

A study has been made of 19 derivatives of 2,3,5-trihydroxy-6-fluorone as reagents for the spectrophotometric determination of tungsten [60]; 2-hydroxyphenylfluorone ($\varepsilon = 2 \cdot 4 \times 10^4$ at 560 nm) proved to be the best.

Other organic reagents suggested for the spectrophotometric determination of tungsten include 8-hydroxyquinoline [61], 5,7-dibromo-8-hydroxyquinoline [62], 8-mercaptoquinoline ($\varepsilon = 3 \cdot 67 \times 10^3$ at 530 nm) [63], Alizarin Red S [64], pyrogallolsulphonic acid [65], Stilbazo ($\varepsilon = 6 \cdot 4 \times 10^4$ at 533 nm) [66], Stilbazogall I [67], Magneson KhS [68], Sulphonitrophenol M [69], Rhodamine B [70], and Pyrocatechol Violet [71,72].

Finally, tungsten has been determined colorimetrically as the yellow tungstovanadophosphoric acid [73], tungstomolybdic acid [74], tungstomolybdenum blue [75] or phosphotungstomolybdenum blue [76].

References

1. Plotnikov, V. I. and Kochetkov, V. L., *Zh. Analit. Khim.* **21**, 1260 (1966).
2. Chan, K. M. and Riley, J. P., *Anal. Chim. Acta* **39**, 103 (1967).
3. Khlystova, A. D., *Zh. Analit. Khim.* **23**, 211 (1968).
4. Přibil, R. and Sedlář, V., *Collection Czech. Chem. Commun.* **16**, 69 (1951).
5. Řehák, B. and Malinek, M., *Z. Anal. Chem.* **153**, 166 (1956).
6. Kuznetsov, V. I., Obozhin, V. N. and Pal'shin, E. S., *Zh. Analit. Khim.* **10**, 32 (1955).
7. Jeffery, P. G., *Analyst* **81**, 104 (1956).
8. Pfeifer, V. and Hecht, F., *Z. Anal. Chem.* **177**, 175 (1960).
9. Peng, P. Y. and Sandell, E. B., *Anal. Chim. Acta* **29**, 325 (1963).

10. Sverak, J., *Z. Anal. Chem.* **201**, 12 (1964).
11. Pfeifer, V., *Mikrochim. Acta* **1960**, 518.
12. De, A. K. and Rahaman, M. S., *Talanta* **11**, 601 (1964).
13. Shinde, V. M. and Khopkar, S. M., *ibid.* **16**, 525 (1969).
14. Kallmann, S., Hobart, E. W. and Oberthin, H. K., *Anal. Chim. Acta* **41**, 29 (1968).
15. Allen, S. H. and Hamilton, M. B., *ibid.* **7**, 483 (1952).
16. Vinogradov, A. V. and Dronova, M. I., *Zh. Analit. Khim.* **20**, 343 (1965).
17. Awad, K., Rudenko, N. P., Kuznetsov, V. I. and Gudym, L. S., *Talanta* **18**, 279 (1971).
18. Headridge, J. B. and Dixon, E. J., *Analyst* **87**, 32 (1962).
19. Dixon, E. J. and Headridge, J. B., *ibid.* **89**, 185 (1964).
20. Kawabuchi, K. and Kuroda, R., *Talanta* **17**, 67 (1970).
21. Greenberg, P., *Anal. Chem.* **29**, 896 (1957).
22. Machlan, L. A. and Hague, J. L., *J. Res. Natl. Bur. Stand.* **59**, 415 (1957).
23. Wood, D. F. and Clark, R. T., *Analyst* **83**, 326 (1958).
24. Stonhill, L. G., *Chemist-Analyst* **47**, 68 (1958).
25. Hobart, E. W. and Hurley, E. P., *Anal. Chim. Acta* **27**, 144 (1962).
26. Kallmann, S., Hobart, E. W. and Oberthin, H. K., *Talanta* **15**, 982 (1968).
27. Zopatti, L. P. and Pollock, E. N., *Anal. Chim. Acta* **32**, 178 (1965).
27a. Quin, B. F. and Brooks, R. R., *ibid.* **58**, 301 (1972).
28. Stepanova, N. A. and Yakunina, G. A., *Zh. Analit. Khim.* **17**, 858 (1962).
29. North, A. A., *Analyst* **81**, 660 (1956).
30. Bowden, P., *ibid.* **89**, 771 (1964).
31. Kawabuchi, K. and Kuroda, R., *Anal. Chim. Acta* **46**, 23 (1969).
32. Freund, H., Wright, M. L. and Brookshier, R. K., *Anal. Chem.* **23**, 781 (1951).
33. Crouthamel, C. E. and Johnson, C. A., *ibid.* **26**, 1284 (1954).
34. Finkel'shtein, D. N., *Zavodsk. Lab.* **22**, 911 (1956).
35. Babko, A. K. and Drako, O. F., *Zh. Analit. Khim.* **12**, 342 (1957).
36. Gottschalk, G., *Z. Anal. Chem.* **187**, 164 (1962).
37. Gentry, C. H. R. and Sherrington, L. G., *Analyst* **73**, 57 (1948).
38. Nemodruk, A. A. and Bezrogova, E. V., *Zh. Analit. Khim.* **24**, 404 (1969).
39. Bush, G. H. and Higgs, D. G., *Analyst* **80**, 536 (1955).
40. Neef, B. and Döge, G. H., *Talanta* **14**, 967 (1967).
41. McDuffie, B., Bandi, W. R. and Melnick, L. M., *Anal. Chem.* **31**, 1311 (1959).
42. Affsprung, H. E. and Murphy, J. W., *Anal. Chim. Acta* **30**, 501 (1964); **32**, 381 (1965).
43. Fogg, A. G., Marriott, D. R. and Burns, D. T., *Analyst* **95**, 848, 854 (1970).
44. Fogg, A. G., Jarvis, T. J., Marriott, D. R. and Burns, D. T., *ibid.* **96**, 475 (1971).
45. Luke, C. L., *Anal. Chem.* **36**, 1327 (1964).
46. Geld, I. and Carroll, J., *ibid.* **21**, 1098 (1949).
47. Eckert, G. and Bauersachs, E., *Z. Anal. Chem.* **163**, 161 (1958).
48. Norwitz, G. and Codell, M., *Anal. Chim. Acta* **11**, 359 (1954).
49. Wood, D. F. and Clark, R. T., *Analyst* **83**, 326 (1958).
50. Kharlamov, I. P., Yakovlev, P. Ya. and Lykova, M. I., *Zavodsk. Lab.* **26**, 786 (1960).
51. Akiyama, K. and Kobayashi, Y., *Japan Analyst* **14**, 292 (1965).
52. Sandell, E. B., *Ind. Eng. Chem., Anal. Ed.* **18**, 163 (1946).
53. Bricker, C. E. and Waterbury, G. R., *Anal. Chem.* **29**, 1093 (1957).
54. Pfeifer, V. and Hecht, F., *Z. Anal. Chem.* **177**, 175 (1960).
55. McKaveney, J. P., *Anal. Chem.* **33**, 744 (1961).
56. Norwitz, G. and Gordon, H., *Anal. Chim. Acta* **69**, 59 (1974).
57. Busev, A. I. and Sokolova, T. A., *Zh. Analit. Khim.* **23**, 1348 (1968).
58. Poluektova, E. N., *ibid.* **21**, 187 (1966).
59. Nazarenko, V. A., Poluektova, E. N. and Shitareva, G. G., *ibid.* **28**, 101, 1966 (1973).

60. Poluektova, E. N. and Nazarenko, V. A., *ibid.* **19**, 856 (1964).
61. Eberle, A. R., *Anal. Chem.* **35**, 669 (1963).
62. Zharovskii, F. G. and Gorina, D. O., *Zh. Analit. Khim.* **26**, 766 (1971).
63. Nazarenko, V. A. and Poluektova, E. N., *ibid.* **26**, 1331 (1971).
64. Sinha, S. N. and Dey, A. K., *Z. Anal. Chem.* **183**, 182 (1961).
65. Horák, J. and Okáč, A., *Collection Czech. Chem. Commun.* **29**, 188 (1964).
66. Kleiner, K. E. and Klibus, A. Kh., *Zh. Obshch. Khim.* **28**, 2013 (1958).
67. Ishii, H. and Einaga, H., *Bull. Chem. Soc. Japan* **39**, 193 (1966).
68. Savvin, S. B., Nambrina, E. G. and Tramm, R. S., *Zh. Analit. Khim.* **27**, 108 (1972).
69. Savvin, S. B., Nambrina, E. G. and Okhanova, L. A., *ibid.* **28**, 1119 (1973).
70. Pollock, J. B., *Analyst* **83**, 516 (1958).
71. Pashchenko, E. N. and Mal'tsev, V. F., *Zavodsk. Lab.* **34**, 12 (1968).
72. Lebedeva, L. I., Golubtsova, Z. G. and Yanklovich, N. G., *Zh. Analit. Khim.* **26**, 1962 (1971).
73. Gullstrom, D. K. and Mellon, M. G., *Anal. Chem.* **25**, 1809 (1953).
74. Reznik, B. E., Ganzburg, G. M. and Mal'tseva, G. V., *Zh. Analit. Khim.* **23**, 1848 (1968)
75. Guyon, J. C. and Marks, J. Y., *Anal. Chem.* **40**, 837 (1968).
76. Mal'tseva, G. V. and Reznik, B. E., *Zh. Analit. Khim.* **28**, 1751 (1973).

Chapter 58

URANIUM

Uranium (U, at.wt. 238·03) occurs in the +III, +IV, +V, and +VI oxidation states, compounds of uranium(VI) and (IV) being of major importance. The yellow uranyl cation precipitates as the hydroxide $UO_2(OH)_2$ at pH ~ 4. Uranium(VI) is amphoteric, sparingly soluble uranate and diuranate (Na_2UO_4 and $Na_2U_2O_7$) being formed in NaOH medium. Uranium(VI) gives peroxide, fluoride, tartrate, carbonate (pH 7–12), and nitrate complexes. Uranium(IV) and thorium have similar properties. The hydroxide $U(OH)_4$ has no amphoteric properties. In acid medium sparingly soluble UF_4 and $U(C_2O_4)_2$ occur. Uranium(IV) is oxidized slowly to U(VI) by atmospheric oxygen and rapidly by iodine, iron(III), and stronger oxidizing agents.

58.1 Methods of Separation

58.1.1 Extraction

Uranium(VI) can be selectively extracted from nitrate medium with polar solvents. The extracted species is solvated undissociated uranyl nitrate. The concentration of free nitric acid must not be too high (0·1–1M), because at higher acidities the degree of extraction of other metals [Ce(IV), Th, Zr, Au(III), As(V), Bi] is increased. To prevent the dissociation of uranyl nitrate and to enhance the distribution coefficient of uranium, considerable quantities of aluminium, calcium, or ammonium nitrate (salting-out agents) are added to the aqueous solution. Sulphate, and small amounts of phosphate and fluoride do not interfere. In the presence of chloride, iron(III) and other metals giving chloride complexes can be extracted.

The solvents used include methyl ethyl ketone in CCl_4 [1], mesityl oxide [2], ethyl acetate [3], TBP [4], TBP in iso-octane [5], TBP in xylene [6], and TOPO in cyclohexane [7]. Uranium(VI) has also been extracted from nitrate medium with triphenylarsine oxide in chloroform [8], and as the tetrapropylammonium uranyl nitrate complex, into MIBK [9].

In the presence of trioctylamine and other amines, uranium can be extracted with non-polar solvents from hydrochloric [10–12a], sulphuric [10,13–15a], or acetic acid [16] media. Nemodruk [17] extracted the anilinium salt. Ichinose [18] has extracted uranium(VI) with MIBK from chloride medium (HCl+LiCl).

Other uranium complexes used for its extraction include those with cupferron [19], acetylacetone [20], DDTC [21], 1-nitroso-2-naphthol [22], di-n-butylarsinic acid [23], HDEHP [24], phenylacetic acid [24a], and N-phenylbenzohydroxamic acid [24b].

58.1.2 Ion-Exchange

Uranium(VI) may occur in solution as cations or complex anions and so can be separated from other elements by use of cation- and anion-exchangers [25].

Unlike most metals, uranium(VI) does not form stable complexes with EDTA. On passage of EDTA solution through a column of strongly acidic cation-exchanger, uranium (and also Be and Ti) is retained in the column, the other metals passing to the eluate as anionic complexes [26,27]. Uranium has also been separated from other elements on cation-exchangers by making use of fluoride [28], chloride + dimethylsulphoxide [29], and nitrate + tetrahydrofuran alone [30] or in presence of TOPO [31].

Anionic sulphate [32], chloride [33,34], nitrate [35], carbonate [36,37], and oxalate [38] complexes have all been used for the separation of uranium by means of anion-exchangers. Uranium is eluted from the column with hydrochloric acid (0·2–1M).

Korkisch et al. [39–46b] have made numerous studies of separation of uranium from other metals (Th, Bi, Fe, Al, rare earths) on anion-exchangers, using solutions of nitric, hydrochloric, or acetic acid in methanol, ethanol, acetone, diethyl ether, dioxan, and other solvents miscible with water.

58.1.3 Precipitation

When sodium carbonate is present in excess, the uranyl ion forms a soluble carbonate complex, whereas most metals separate as carbonates, basic carbonates, or hydroxides [47,48] with the exception of vanadium, beryllium, and thorium, which remain partly in solution together with uranium. If the precipitate contains several metals, double precipitation is necessary. The amount of uranium retained by the precipitate does not exceed a few per cent. When a carbonate fusion is used for decomposition of the sample and the melt is leached with water, uranium passes into solution.

Ammonia precipitates uranium (VI) as a sparingly soluble diuranate, $(NH_4)_2U_2O_7$. As collector, aluminium or iron(III) may be used [49,50]. In the presence of EDTA, ammonia precipitates only U, Ti, Be, Sn(IV), Sb, and Nb; titanium is a suitable collector. However, if only trace amounts of uranium are processed, part remains in solution when EDTA is present.

Traces of uranium have also been coprecipitated as the phosphate with aluminium as collector [51].

The precipitation of uranium(VI) with 1-nitroso-2-naphthol at pH 6·5–8 in the presence of EDTA enables it to be separated from most other ions (including phosphate) [52].

To separate traces of uranium quantitatively from solution, coprecipitation with organic collectors [53–55] has been used. From natural waters uranium has been separated with Methyl Violet and Na–DDTC or thiocyanate [53], Arsenazo I and diphenylguanidine [54], and thiocyanate and DAM or Crystal Violet [55].

58.2 Methods of Determination

Most spectrophotometric methods for determination of uranium are based on the colour reactions of uranium(VI), but a few (such as that with Arsenazo III) are based on colour reactions of uranium(IV).

Below detailed descriptions are given of a very sensitive and selective method using Arsenazo III, moderately sensitive methods using Arsenazo I and dibenzoylmethane, and a relatively insensitive method based on thiocyanate. The influence of other metals is usually eliminated with EDTA. The spectrophotometric determination of uranium is generally preceded by extraction from nitrate medium.

58.2.1 THIOCYANATE METHOD

Thiocyanate and uranyl ions in acid medium form a stepwise series of yellow complexes such as $[UO_2SCN]^+$, $[UO_2(SCN)_2]$, and $[UO_2(SCN)_3]^-$. Higher concentrations of thiocyanate displace the equilibrium towards the last-mentioned and more intensely coloured complex. The absorption maximum of this complex lies in the near ultraviolet, at 350 nm. At wavelengths shorter than 360 nm thiocyanate ions begin to absorb.

Uranium may be determined in aqueous or aqueous acetone medium (50–60% acetone), or after extraction of the complex into an organic solvent [56–59]. The sensitivity of the method is higher if non-aqueous media are used. The uranium complex may be extracted with diethyl ether, amyl alcohol [56], or TBP in CCl_4 [60,61] or kerosene [62].

The molar absorptivity of the uranium(VI) thiocyanate complex in a mixture of TBP and CCl_4 is $2·9 \times 10^3$ ($a = 0·012$) at 380 nm.

Extraction of the complex from weakly acidic medium in the presence of EDTA prevents interference by iron(III) and many other metals. The interference of iron(III) can also be eliminated by addition of a reducing agent (stannous chloride or ascorbic acid).

Nietzel and De Sesa [63] separated uranium(VI) with MIBK from HNO_3 and $Al(NO_3)_3$ medium, and developed the colour directly in the extract by adding NH_4SCN dissolved in "butylcellosolve". Sinyakova and Klassova [64] separated uranium (from HNO_3 and NH_4NO_3 medium) with methyl ethyl ketone and added NH_4SCN solution in acetone to the extract.

The uranium thiocyanate ion-association complex with DAM or Crystal Violet may be extracted into $CHCl_3$ [55].

Because of its low sensitivity, the thiocyanate method has been used for determining uranium in ores and concentrates. The thiocyanate–acetone method is said to be excellent for uranium in presence of fluoride [64a].

Reagents

1. Potassium thiocyanate: 40% solution.
2. Standard uranium(VI) solution: 1 mg/ml. Dissolve 2·109 g of uranyl nitrate $UO_2(NO_3)_2 \cdot 6H_2O$ in water containing 1 ml of conc. HNO_3. Dilute the solution with water to 1 litre in a volumetric flask. Standardize by precipitating ammonium diuranate and igniting to U_3O_8. Prepare working solutions by dilution with $\sim 0·01 M$ HNO_3.

Procedure

To the sample solution (in HNO_3, H_2SO_4, or $HClO_4$) containing not more than 2 mg of U(VI) in 10–15 ml, add 5 ml of 10% EDTA solution and ammonia to pH 3·5–4·0. Transfer the solution to a separating funnel, add 10 ml of thiocyanate solution, and shake with two portions of 1:4 TBP–CCl_4 mixture. Dilute the combined extracts to volume with the solvent in a 50-ml or smaller (depending on the amount of uranium) volumetric flask, mix well, and measure the absorbance at 380 nm, using the solvent as a reference.

58.2.2 DIBENZOYLMETHANE METHOD

At pH 6·5–8·5 the enol form of dibenzoylmethane forms a yellow uranyl

chelate complex UO_2R_2, which can be used for spectrophotometric determination of uranium [65–69]. It dissolves in ethanol–water medium containing pyridine [67] or may be extracted with ethyl acetate [69], butyl acetate, chloroform [70], or TBP in n-hexane [71].

The absorption maximum of the complex in ethyl acetate is at 395 nm, molar absorptivity $2·0 \times 10^4$ ($a = 0·088$). Dibenzoylmethane itself absorbs at wavelengths shorter than 400 nm [65].

Phosphate, citrate, and most metals forming dibenzoylmethane complexes interfere, but EDTA is an excellent masking agent for most of the metals [66].

The dibenzoylmethane method is often used after uranium has been extractively separated from other metals as the nitrate complex [65,66,72] or by the carbonate method (see procedure below).

Francois [5] separated uranium by extracting it from nitrate medium with TBP in iso-octane and carried out an exchange reaction in the organic phase by adding a solution of dibenzoylmethane and pyridine.

Maeck et al. [68] added dibenzoylmethane in ethanol–pyridine mixture to the extract of uranyl nitrate in MIBK. A similar procedure has been adopted in other methods [7,71] except that hexamethylenetetramine is used in place of pyridine.

The dibenzoylmethane method has been used for determining uranium in ores and minerals [32,65,67] as well as in zirconium and hafnium [73].

Reagents

1. Dibenzoylmethane: 0·2% solution in ethanol.
2. Standard uranium(VI) solution: 1 mg/ml (p. 577).

Procedure

Separation of U by the carbonate method. To 20–40 ml of acidic solution (HNO_3, H_2SO_4, HCl) of uranium(VI), add 1–2 ml of ethanol and 10% Na_2CO_3 solution in excess (~5 ml more than the amount necessary for neutralization of the sample solution). Heat the solution for 15 min at 80–90°C. Filter off the precipitate on a filter paper and wash it with hot 1% Na_2CO_3 solution. Dissolve the precipitate in hot $2M$ HCl and reprecipitate with Na_2CO_3. Acidify the combined filtrates slightly with hydrochloric acid and evaporate to the desired volume.

Determination of U. To the slightly acidic solution containing not more than 300 μg of U, add 2 ml of 2% EDTA solution, 2 ml of dibenzoylmethane solution, and ammonia to pH 7–8. Extract the uranium complex with two portions of ethyl acetate. Dilute the extracts to volume with solvent in a 50-ml or smaller (depending on the amount of uranium) volumetric flask and measure the absorbance of the solution at 395 nm, using as reference the solvent or a reagent blank solution (for smaller contents of uranium).

Note. If titanium is present, add ammonium oxalate to the sample solution before applying the colour reaction.

58.2.3 Arsenazo I Method

Arsenazo I (Uranon, Neothoron, p. 48) and uranyl ions in slightly alkaline medium (pH 7·5–9·0) form a 1:1 blue complex that is utilized for the spectrophotometric determination of uranium(VI) [74–77]. The reagent itself is pink-red in slightly alkaline medium. Arsenazo I and its uranyl complex are soluble in water, owing to the sulphonate groups.

The absorption spectra for Arsenazo I and its complex with UO_2^{2+} at pH 8·5 are shown in Fig. 58.1. The molar absorptivity at $\lambda_{max} = 596$ nm is $2·3 \times 10^4$ ($a = 0·10$).

The colour of the complex is independent of temperature over the range 15–35°C. The buffer solution for pH adjustment should be added last to the reaction system, which itself should be preadjusted to pH 2·5–5·5.

In slightly alkaline medium, Arsenazo I gives colour reactions with a number of metals. EDTA (used in slight excess) and tartaric acid are the chief masking agents used in the determination of uranium [76]. Al, Bi, Co, Cu, Fe, Ni, Pb, Zn, and rare earths are masked by EDTA. Titanium is masked by tartrate. Too great an excess of EDTA somewhat lowers the colour intensity of the uranium(VI) Arsenazo I complex, and this should be taken into consideration when preparing the calibration curve under such conditions.

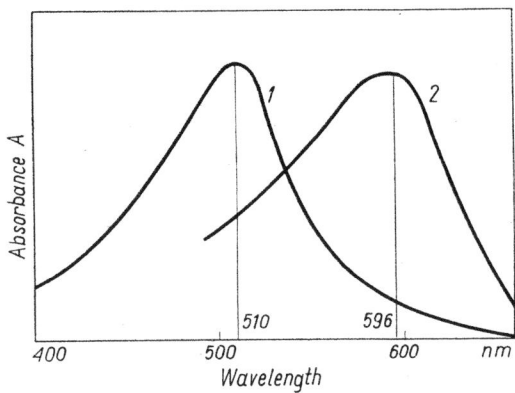

Fig. 58.1 Absorption spectra of Arsenazo I (1) and its uranium(VI) complex (2) (reagent solution as reference) at pH 8·5

Thorium interferes in the determination, but at pH 1·5 Arsenazo I does not react with U(VI) but gives a coloured complex with thorium. It is therefore possible to determine only the thorium at pH 1·5 and the sum of thorium and uranium in slightly alkaline medium [78]. Fritz and Johnson-Richard [76] separated uranium(VI) from thorium by extracting the former as a complex with DDTC.

Uranium(VI) is often determined with Arsenazo I, after separation by extraction [1,52,74,79] or ion-exchange [77,80]. Arsenazo I has been employed for the determination of uranium(VI) in natural water [54] and ores and minerals [75,78–80].

Kuznetsov and Nikol'skaya [81] have used Arsenazo I for the photometric determination of uranium(IV), with which the reagent yields in acidic medium (pH 1·5–1·8) a complex having its absorption maximum at 555 nm. The sensitivity is similar to that for U(VI). Uranium(IV) hydrolyses more readily than uranium(VI) which is why its reaction with Arsenazo I must be carried out in more acidic medium, as a result of which it is more selective. At pH 1·5 Fe(III), Al, and rare earths give no reaction with Arsenazo I. Uranium(VI) is reduced to uranium(IV) with potassium iodide.

Reagents

1. Arsenazo I: 0·05% aqueous solution.
2. Standard uranium(VI) solution: 1 mg/ml (p. 577).
3. Buffer solution: pH 8·6. Dissolve in water 2 g of boric acid and 2 g of potassium chloride, add 6·5 ml of $1M$ NaOH, and dilute to 500 ml.

Procedure

Bring the sample solution, containing not more than 300 µg of U(VI), (after separation of interfering metals) to pH 3–5 and transfer it to a 50-ml volumetric flask. Add 4 ml of Arsenazo I solution and mix. Add 10 ml of buffer solution, make up the solution with water to the mark, and measure the absorbance at 596 nm against a reagent blank.

58.2.4 ARSENAZO III METHOD

Arsenazo III (p. 48) reacts with uranium(IV) in strongly acidic medium, giving (as with thorium and zirconium ions) a green-blue complex. The spectrophotometric method based on this reaction [83–85], as reported by Savvin [82], is much superior in sensitivity and selectivity to the other spectrophotometric methods for determining uranium.

The molar absorptivity of the complex in 6–8M HCl with at least threefold molar ratio of Arsenazo III to U(IV) is $\sim 1 \cdot 27 \times 10^5$ ($a = 0 \cdot 50$) at $\lambda_{max} = 665$ nm [85]. The nature of the complex depends on the conditions [85].

Granular zinc [84] or bismuth [49,86] is used in hydrochloric acid medium to reduce U(VI) to U(IV). In 4M HCl medium, if oxalic acid is added to mask zirconium (and hafnium), only thorium interferes with the determination of uranium(IV). Of the common anions, only fluoride interferes. Uranium has been determined in this way in ores, rocks [84,86], and water [49].

Arsenazo III gives a 1:1 coloured complex with uranium(VI), which is also used for spectrophotometric determination of uranium [87–89]. Either slightly acidic (pH 2–3) or strongly acidic medium (5–6M HNO$_3$, HCl or HClO$_4$) can be used. In the latter case a considerable excess of Arsenazo III is needed. Methods based on this reaction are convenient in that they eliminate the reduction step, but their sensitivity is much lower. The molar absorptivity is $6 \cdot 0 \times 10^4$ (specific absorptivity 0·25).

The reaction can also be carried out in an organic solvent. Uranium(VI) is extracted from aqueous solution at pH 1–2 (NH$_4$NO$_3$ and EDTA) with a solution of TBP in benzyl alcohol [92], CCl$_4$ [90,91], or MIBK [93], or TOPO in benzene [93]. In this manner uranium is separated from Th, Zr, Fe(III), rare earths, and other elements. The extract is shaken with the Arsenazo III solution.

Arsenazo III has been used to determine uranium(VI) in thorium dioxide [94] and natural water [27,95]. The reagent has also been used for automatic determination of uranium in aqueous solution [91] and for determination of small amounts of uranium(VI) in the presence of large quantities of uranium(IV) [96]. Uranium(VI) is separated by extraction with TBP in the presence of EDTA, which keeps the uranium(IV) in the aqueous phase at pH 2.

Kuznetsov and Savvin [97] extracted the uranium(VI)–Arsenazo III complex with butanol from aqueous solution (pH < 3) containing EDTA and diphenylguanidine. This high molecular-weight organic base neutralizes the sulphonate groups of Arsenazo III, thus making it possible to extract the complex and the excess of reagent into the organic solvent phase.

Reagents

1. Arsenazo III: 0·25% solution. Dissolve 0·25 g of Arsenazo III and 0·5 g of sodium acetate in 100 ml of water. Store the solution in an amber-glass bottle.
2. Standard uranium(VI) solution: 1 mg/ml (p. 577).
3. Concentrated nitric acid, saturated with urea. To 500 ml of conc. HNO_3 add 5 g of urea, stir, and allow to stand overnight. Filter off undissolved urea on a sintered-glass crucible.
4. Tributyl phosphate (TBP): 20% v/v solution in toluene. Mix 20 ml of TBP with 80 ml of toluene and wash the solution twice with 40-ml portions of 5% Na_2CO_3 solution, then twice with 40-ml portions of water.
5. Wash solution. Dissolve 65 g of ammonium nitrate and 0·5 g of EDTA in 40 ml of water, adjust the pH to ~2, dilute with water to 100 ml, mix well, and filter.

Procedure

Extraction of U(VI) by TBP in toluene. To the sample solution (~25 ml) containing not more than 150 μg of U, add 25 g of NH_4NO_3 and 0·25 g of EDTA. Heat the solution to ~70°C, cool it to room temperature, and adjust the pH of the solution to 1·0 with nitric acid or ammonia. Extract uranium by shaking with 25 ml of TBP solution for 3 minutes. After separation of the phases, discard the aqueous solution and wash the extract by shaking with 2 portions (10 ml each) of wash solution for 4–5 seconds.

Determination of U. To the washed extract add 7·5 ml of Arsenazo III solution, 3 ml of water, and 1 ml of conc. ammonia solution and shake for 3 minutes. After separation of the phases transfer the aqueous phase to a 50-ml volumetric flask. Wash the organic phase with two 5-ml portions of water; shaking time 2 minutes. Add the two washes to the volumetric flask. Add 20 ml of conc. HNO_3, mix, allow to cool, make up with water to the mark, and after 5–10 minutes measure the absorbance at 655 nm against a reagent blank.

58.2.5 OTHER METHODS

Besides Arsenazo I and Arsenazo III, other azo dyes with arsonic acid groups have been suggested for spectrophotometric determination of uranium, namely Arsenazo II [98] and Thoron I [99,100] (p. 48).

Azo reagents with a phosphonic acid group, such as Chlorophosphonazo I, Phosphonazo III, and Chlorophosphonazo III (p. 50) have been suggested [17,101–102a].

Chlorophosphonazo I

Chlorophosphonazo III reacts like Arsenazo III with uranium(VI), giving a complex with molar absorptivity 7.96×10^4 at $\lambda_{max} = 670$ nm.

1-(2-Pyridylazo)-2-naphthol (PAN) (p. 46) reacts with uranium(VI) in ammoniacal medium to yield a red-violet complex [103–107] which is sparingly soluble in water and extractable into chloroform or o-dichlorobenzene. In the complex UO_2:PAN = 1:2. The molar absorptivity is 2.3×10^4 at $\lambda_{max} = 560$ nm. To mask interfering metals, EDTA and KCN have been used. Hayes and Wright [108] described an indirect determination of uranium with PAN. The U(VI)–PAN precipitate is filtered off and washed free from excess of reagent with a mixture of water and acetone, then is decomposed with 20% HCl, and the absorbance of liberated PAN is measured at 440 nm. This gives a monocolour system and enhances the sensitivity of the method ($\varepsilon = 3.3 \times 10^4$). 5-Methyl-PAN has also been used [108a].

In the method with the related reagent, 4-(2-pyridylazo)resorcinol (PAR) (p. 47), the molar absorptivity is 3.87×10^4 at $\lambda_{max} = 530$ nm [15,109–111].

Florence, and others, [112–114b] recommend 2-(2-pyridylazo)-5-diethylaminophenol (PADAP) and bromo-PADAP for determination of uranium, both reagents being twice as sensitive as PAR. The reagent can be used directly on organic extracts if dimethylformamide is added [114a].

2-(2-Thiazolylazo)-5-dimethylaminophenol (TAM) is an example of the application of an azo derivative of thiazole to the determination of

uranium [115,116]. Kasiura and Minczewski [116] extracted uranium(VI) with a mixture of TBP (20%) and chloroform and then shook the extract with 0.02% TAM solution in pyridine–water mixture. At $\lambda_{max} = 575$ nm, $\varepsilon = 4.0 \times 10^4$. The related reagents TAR [117] and 2-(2-thiazolylazo)methoxyphenols [118] have also been applied to the determination of uranium.

Korkisch and Janauer [119–121] used Solochrome Fast Grey RAS, Solochrome Black 6 BN, Solochrome Fast Red, etc. for uranium(IV and VI).

Many authors have determined larger amounts of uranium by the extractive method using 8-hydroxyquinoline and chloroform [4,8,122–125]. The optimum pH is 6–8. Addition of EDTA retains most metals (including Fe, Bi, and Al) in the aqueous phase. The absorbance is measured at ~400 nm. Uranium is also determined with chloro-oxine and bromo-oxine in similar manner [123].

The following triphenylmethane dyes have been applied to the determination of uranium: Aluminon [126], Pontachrome Azure Blue B [127], Chrome Azurol S [128], Xylenol Orange [129], and Methylthymol Blue [130].

Some flavone compounds have been recommended as reagents for uranium, in particular quercetin [131], morin [132], and 5-hydroxyflavone and 5-hydroxy-7-methoxyflavone [133].

Sensitive extractive spectrophotometric methods for uranium are based on ion-association complexes consisting of anionic uranium(VI) complexes and basis dyes such as Rhodamine B [134–138], Crystal Violet [139], Brilliant Green [140], and Methylene Blue [141]. The anionic uranium complexes are formed with benzoate [134,139–141], nitrate [135], TTA [136], and 3-pyridinecarboxylate [137].

From a number of other organic reagents used for photometric determination of uranium mention may be made of Bromopyrogallol Red [142], glyoxal bis(2-hydroxyanil) (GBHA) [143], TTA [144,145], thiothenoyltrifluoroacetone [146], thianaphthenoyltrifluoroacetone [147], diethanoldithiocarbamate [148], pyrrolidinedithiocarbamate [149], benzohydroxamic acid [150], N-phenylbenzohydroxamic acid [24b], cinnamoylphenylhydroxylamine [151], Alizarin Complexone [152], Alizarin Red S [153], sulphosalicylic acid [154], pyrogallolsulphonic acid [155], ascorbic acid [156], and pyridine-2,6-dicarboxylic acid [157]. The Chrome Azurol S complex can be extracted if cetylpyridinium bromide is added [158].

A not very sensitive peroxide method has long been known [3,47,159–161]. In alkaline H_2O_2 solution, uranium(VI) forms a yellow peruranate. Hydrogen peroxide is added to a solution of the uranium carbonate complex, or else an acidic uranium solution which has been treated with hydrogen peroxide is made alkaline with sodium hydroxide. The peroxide method is quite selective and easy to operate. Below pH 12 the colour intensity is lowered so bicarbonate and ammonium salts should be avoided in the sample solution. The absorbance is measured at 370–400 nm. Chromium(VI), molybdenum(VI), and vanadium(V) interfere.

The azide method [162,163] resembles that with thiocyanate. In a solution at pH 5–5·5 a yellow complex is obtained, which has a molar absorptivity of $4·5 \times 10^3$ at 375 nm.

Milligram quantities of uranium(VI) may be determined by means of the colour of the tetrapropylammonium nitrate complex in MIBK (at 452 nm) [68], uranyl ions in perchloric acid (415–420 nm) [164], or the uranyl sulphate complex with tri-n-octylamine in benzene solution (472 or 490 nm) [165].

In an indirect method using a redox reaction, uranium(IV) is oxidized by iron(III) in the presence of 1,10-phenanthroline. The amount of ferroin produced is proportional to the amount of uranium present [166].

Nemodruk et al. [167] have determined uranium(III) by means of its decolorization of Acid Yellow 2G (C.I.13075). It has also been determined by direct measurement at 348 nm ($\varepsilon = 1\cdot6 \times 10^3$) or 319 nm ($\varepsilon = 1\cdot1 \times 10^3$) [168].

In the Gayer and Lifshitz method [169], uranium(IV) reduces iron(III) to iron(II), and than Prussian blue is formed with ferricyanide. This provides a basis for determining uranium(IV) in presence of larger amounts of uranium(VI).

References

1. Kuznetsov, V. I. and Blekhta, V., *Collection Czech. Chem. Commun.* **26**, 1092 (1961).
2. Dhara, S. C. and Khopkar, S. M., *Mikrochim. Acta* **1965**, 931.
3. Guest, R. J. and Zimmerman, J. B., *Anal. Chem.* **27**, 931 (1955).
4. Eberle, A. R. and Lerner, M. W., *ibid.* **29**, 1134 (1957).
5. Francois, C. A., *ibid.* **30**, 50 (1958).
6. Nemodruk, A. A. and Glukhova, L. P., *Zh. Analit. Khim.* **21**, 688 (1966).
7. Horton, C. A. and White, J. C., *Anal. Chem.* **30**, 1779 (1958).
8. Keil, R., *Z. Anal. Chem.* **244**, 165 (1969).
9. Maeck, W. J., Booman, G. L., Elliott, M. C. and Rein, J. E., *Anal. Chem.* **30**, 1902 (1958).
10. Tikhomirov, V. I., Kuznetsova, A. A. and Batorovskaya, E. D., *Radiokhimiya* **6**, 173, 182, 187 (1964).
11. Hodara, I. and Balouka, I., *Anal. Chem.* **43**, 1213 (1971).
12. Vieux, A. S., *Bull. Soc. Chim. France* **1968**, 4281.
12a. Onishi, H. and Sekine, K., *Talanta* **19**, 473 (1972).
13. Boirie, C., *Bull. Soc. Chim. France* **1958**, 1088.
14. Deptuła, C. and Minc, S., *J. Inorg. Nucl. Chem.* **29**, 221 (1967).
15. Gagliardi, E. and Ilmaier, B., *Mikrochim. Acta* **1968**, 1259.
15a. Južnič, K. and Fedina, Š., *ibid.*, **1974**, 39.
16. Moore, F. L., *Anal. Chem.* **30**, 908 (1958); **32**, 1075 (1960).
17. Nemodruk, A. A., *Tr. Komis. po Analit. Khim. Akad. Nauk SSSR* **14**, 141 (1963).
18. Ichinose, N., *Talanta* **18**, 21 (1971).
19. Rulfs, C. L., De, A. K. and Elving, P. J., *Anal. Chem.* **28**, 1139 (1956).
20. Krishen, A. and Freiser, H., *ibid.* **29**, 288 (1957).
21. Clayton, R. F., Hardwick, W. H., Moreton-Smith, M. and Todd, R. *Analyst* **83**, 13 (1958).
22. Alimarin, I. P. and Zolotov, Yu. A., *Zh. Analit. Khim.* **12**, 176 (1957).
23. Pietsch, R. and Pichler, R., *Z. Anal. Chem.* **190**, 319 (1962).
24. Schmid, E. R. and Pfannhauser, W., *Mikrochim. Acta* **1971**, 434.
24a. Adam, A. and Přibil, R., *Talanta* **20**, 1344 (1973).
24b. Shukla, J. P., Agrawal, Y. K. and Bhatt, K., *Sepn. Sci.* **8**, 387 (1973).
25. Korkisch, J., *Mikrochim. Acta* **1967**, 401.
26. Krawczyk, I., *Nukleonika* **5**, 649 (1960).
27. Šulcek, Z. and Povondra, P., *Collection Czech. Chem. Commun.* **32**, 3140 (1967).
28. Krawczyk-Obojska, I., *Chem. Anal. (Warsaw)* **13**, 551 (1968).
29. Janauer, G. E., Korkisch, J. and Hubbard, S. A., *Talanta* **18**, 767 (1971).
30. Korkisch, J. and Ahluwalia, S. S., *Anal. Chem.* **38**, 497 (1966).
31. Khater, M. M. and Korkisch, J., *Talanta* **18**, 1001 (1971).
32. Seim, H. J., Morris, R. J. and Pastorino, R. G., *Anal. Chem.* **31**, 957 (1959).
33. Korkisch, J., Farag, A. and Hecht, F., *Z. Anal. Chem.* **161**, 92 (1958).
34. Nelson, F., Michelson, D. C. and Holloway, J. H., *J. Chromatog.* **14**, 258 (1964)

35. Vita, O. A., Walker, C. R., Trivisonno, C. F. and Sparks, R. W., *Anal. Chem.* **42**, 465 (1970).
36. Murthy, T. K. S., *Anal. Chim. Acta* **16**, 25 (1957).
37. Taketatsu, T., *Talanta* **10**, 1077 (1963).
38. Zaki, M. R. and Shakir, K., *Z. Anal. Chem.* **185**, 423 (1962).
39. Korkisch, J. and Janauer, G. E., *Talanta* **9**, 957 (1962).
40. Korkisch, J. and Urubay, S., *ibid.* **11**, 721 (1964).
41. Korkisch, J. and Hazan, I., *ibid.* **11**, 523, 1157 (1964); *Anal. Chem.* **36**, 2464 (1964).
42. Urubay, S., Korkisch, J. and Janauer, G. E., *Talanta* **10**, 673 (1963); *Z. Anal. Chem.* **193**, 165 (1963).
43. Korkisch, J. and Tera, F., *Z. Anal. Chem.* **186**, 290 (1962); *J. Chromatog.* **8**, 516 (1962).
44. Hazan, I., Ahluwalia, S. S. and Korkisch, J., *Z. Anal. Chem.* **206**, 324 (1964).
45. Hazan, I., Korkisch, J. and Arrhenius, G., *ibid.* **213**, 182 (1965).
46. Korkisch, J. and Arrhenius, G., *Anal. Chem.* **36**, 850 (1964).
46a. Korkisch, J. and Koch, W., *Mikrochim. Acta* **1973**, 157, 865.
46b. Koch, W. and Korkisch, J., *ibid.*, 1973, 225.
47. Wódkiewicz, L., *Chem. Anal. (Warsaw)* **3**, 789 (1958).
48. Upor, E. and Nagy, G., *Acta Chim. Acad. Sci. Hung.* **50**, 5 (1966).
49. Singer, E. and Mareček, J., *Z. Anal. Chem.* **196**, 321 (1963).
50. Upor, E., *Acta Chim. Acad. Sci. Hung.* **51**, 119, 139 (1967).
51. Hashimoto, T., *Anal. Chim. Acta* **56**, 347 (1971).
52. Titov, V. I. and Osiko, E. P., *Zh. Analit. Khim.* **17**, 129 (1962).
53. Kuznetsov, V. I. and Akimova, T. G., *ibid.* **13**, 79 (1958); *Radiokhimiya* **2**, 426 (1960).
54. Kuznetsov, V. I., Gorshkov, V. V., Akimova, T. G. and Nikol'skaya, I. V. *Tr. Komis. po Analit. Khim. Akad. Nauk SSSR* **15**, 296 (1965).
55. Babko, A. K. and Danilova, V. N., *Zh. Analit. Khim.* **18**, 1036 (1963).
56. Gerhold, M. and Hecht, F., *Mikrochemie* **36/37**, 1100 (1951).
57. Crouthamel, C. E. and Johnson, C. E., *Anal. Chem.* **24**, 1780 (1952).
58. Tucker, H. T., *Analyst* **82**, 529 (1957).
59. Moiseeva, L. M. and Tumanov, Yu. N., *Zh. Analit. Khim.* **17**, 595 (1962).
60. Clinch, J. and Guy, M. J., *Analyst* **82**, 800 (1957).
61. Koppikar, K. S., Korgaonkar, V. G. and Murthy, T. K. S., *Anal. Chim. Acta* **20**, 366 (1959).
62. Deberdeeva, R. Yu., Nemodruk, A. A. and Palei, P. N., *Radiokhimiya* **7**, 271 (1965).
63. Nietzel, O. A. and De Sesa, M. A., *Anal. Chem.* **29**, 756 (1957).
64. Sinyakova, S. I. and Klassova, N. S., *Zh. Analit. Khim.* **14**, 451 (1959).
64a. Okumura, I., Shimada, S. and Higashi, K., *Anal. Chem.* **45**, 1945 (1973).
65. Yoe, J. H., Will, F., III and Black, R. A., *ibid.* **25**, 1200 (1953).
66. Přibil, R. and Jelínek, M., *Chem. Listy* **47**, 1326 (1953).
67. Blanquet, P., *Anal. Chim. Acta* **16**, 44 (1957); *Chim. Anal. (Paris)* **41**, 247 (1959).
68. Maeck, W. J., Booman, G. L. and Rein, J. E., *Anal. Chem.* **31**, 1130 (1959).
69. Moučka, V. and Starý, J., *Collection Czech. Chem. Commun.* **26**, 763 (1961).
70. Schweitzer, G. K. and Mottern, J. L., *Anal. Chim. Acta* **26**, 120 (1962).
71. Singer, E., *Acta Chim. Acad. Sci. Hung.* **28**, 279 (1961).
72. Umezaki, Y., *Bull. Chem. Soc. Japan* **36**, 769 (1963).
73. Wood, D. F. and McKenna, R. H., *Anal. Chim. Acta* **27**, 446 (1962).
74. Holcomb, H. P. and Yoe, J. H., *Anal. Chem.* **32**, 612 (1960).
75. Shibata, S. and Matsumae, T., *Bull. Chem. Soc. Japan* **31**, 377 (1958); **32**, 279 (1959).
76. Fritz, J. S. and Johnson-Richard, M., *Anal. Chim. Acta* **20**, 164 (1959).
77. Moiseeva, L. M., Kuznetsova, N. M., Luk'yanov, V. F. and Sel'manova, G. L., *Zh. Analit. Khim.* **16**, 585 (1961).

78. Holcomb, H. P. and Yoe, J. H., *Microchem. J.* **4**, 463 (1960).
79. Kuznetsov, V. I. and Kukisheva, T. N., *Zavodsk. Lab.* **26**, 1344 (1960).
80. Luk'yanov, V. F., Moiseeva, L, M. and Kuznetsova, N. M., *Zh. Analit. Khim.* **16**, 448 (1961).
81. Kuznetsov, V. I. and Nikol'skaya, I. V., *Zavodsk. Lab.* **26**, 266 (1960).
82. Savvin, S. B., *Dokl. Akad. Nauk SSSR* **127**, 1231 (1959).
83. Savvin, S. B., *Talanta* **8**, 673 (1961); **11**, 1, 7 (1964); *Zavodsk. Lab.* **29**, 131 (1963).
84. Luk'yanov, V. F., Savvin, S. B. and Nikol'skaya, I. V., *Zh. Analit. Khim.* **15**, 311 (1960).
85. Nemodruk, A. A. and Palei, P. N., *ibid.* **18**, 480 (1963).
86. Singer, E. and Matucha, M., *Z. Anal. Chem.* **191**, 248 (1962).
87. Nemodruk, A. A. and Glukhova, L. P., *Zh. Analit. Khim.* **18**, 93 (1963).
88. Nemodruk, A. A., Palei, P. N. and Glukhova, L. P., *ibid.* **23**, 214 (1968).
89. Borák, J., Slovák, Z. and Fischer, J., *Talanta* **17**, 215 (1970).
90. Palei, P. N., Nemodruk, A. A. and Davydov, A. V., *Radiokhimiya* **3**, 181 (1961).
91. Palei, P. N., Nemodruk, A. A. and Davydov, A. V., *Tr. Komis. po Analit. Khim. Akad. Nauk SSSR* **14**, 281 (1963).
92. Nemodruk, A. A. and Glukhova, L. P., *Zh. Analit. Khim.* **23**, 552 (1968).
93. Pérez-Bustamante, J. A. and Delgado, F. P., *Analyst* **96**, 407 (1971).
94. Ichinose, N., *Z. Anal. Chem.* **255**, 109 (1971).
95. Nemodruk, A. A. and Deberdeeva, R. Yu., *Radiokhimiya* **8**, 248 (1966).
96. Nemodruk, A. A., Palei, P. N. and Glukhova, L. P., *ibid.* **7**, 372 (1965).
97. Kuznetsov, V. I. and Savvin, S. B., *ibid.* **2**, 682 (1960).
98. Kuznetsov, V. I. and Savvin, S. B., *ibid.* **1**, 589 (1959).
99. Foreman, J. K., Riley, C. J. and Smith, T. D., *Analyst* **82**, 89 (1957).
100. Mikhailov, V. A., *Zh. Analit. Khim.* **16**, 141 (1961).
101. Nemodruk, A. A., Novikov, Yu. P., Lukin, A. M. and Kalinina, I. D., *ibid.* **16**, 180, 292 (1961).
102. Luk'yanov, V. F., Lukin, A. M., Duderova, E. P. and Barabanova, T. E., *ibid.* **26**, 772 (1971).
102a. Yamamoto, T., *Anal. Chim. Acta* **65**, 329 (1973).
103. Cheng, K. L., *Anal. Chem.* **30**, 1027 (1958); *Talanta* **9**, 739 (1962).
104. Gill, H. H., Rolf, R. F., and Armstrong, G. W., *Anal. Chem.* **30**, 1788 (1958).
105. Shibata, S., *Anal. Chim. Acta* **22**, 479 (1960).
106. Zolotov, Yu. A., Alimarin, I. P. and Bagreev, V. V., *Tr. Komis. po Analit. Khim. Akad. Nauk SSSR* **15**, 59 (1965).
107. Baltisberger, R. J., *Anal. Chem.* **36**, 2369 (1964).
108. Hayes, M. R. and Wright, J. S., *Talanta* **11**, 607 (1964).
108a. Shibata, S., Furukawa, M. and Ishiguro, Y., *Mikrochim. Acta* **1974**, 129.
109. Pollard, F. H., Hanson, P. and Geary, W. J., *Anal. Chim. Acta* **20**, 26 (1959).
110. Florence, T. M. and Farrar, Y. J., *Anal. Chem.* **35**, 1613 (1963).
111. Sommer, L., Ivanov, V. M. and Novotná, H., *Talanta* **14**, 329 (1967).
112. Florence, T. M. and Farrar, Y. J., *Anal. Chem.* **42**, 271 (1970).
113. Florence, T. M., Johnson, D. A. and Farrar, Y. J., *ibid.* **41**, 1652 (1969).
114. Johnson, D. A. and Florence, T. M., *Anal. Chim. Acta* **53**, 73 (1971).
114a. Pakalns, P. and McAllister, B. R., *ibid.* **62**, 207 (1972).
114b. Pakalns, P., *ibid.* **69**, 211 (1974).
115. Sørensen, E., *Acta Chem. Scand.* **14**, 965 (1960).
116. Kasiura, K. and Minczewski, J., *Nukleonika* **11**, 399 (1966).
117. Sommer, L. and Ivanov, V. M., *Talanta* **14**, 171 (1967).
118. Sommer, L., Šepel, T. and Ivanov, V. M., *ibid.* **15**, 949 (1968).
119. Korkisch, J. and Janauer, G. E., *Z. Anal. Chem.* **182**, 26; **183**, 85 (1961).
120. Korkisch, J. and Janauer, G. E., *Mikrochim. Acta* **1961**, 537; *Anal. Chim. Acta* **25**, 463 (1961).
121. Janauer, G. E. and Korkisch, J., *Talanta* **9**, 427 (1962).

122. Silverman, L., Moudy, L. and Hawley, D. W., *Anal. Chem.* **25**, 1369 (1953).
123. Rulfs, C. L., De, A. K., Lakritz, J. and Elving, P. J., *ibid.* **27**, 1802 (1955).
124. Motojima, K., Yoshida, H. and Izawa, K., *ibid.* **32**, 1083 (1960).
125. Oki, S., *Talanta* **16**, 1153 (1969); *Anal. Chim. Acta* **44**, 315 (1969).
126. Mukherji, A. K. and Dey, A. K., *Mikrochim. Acta* **1958**, 736.
127. Katsube, Y., Uesugi, K. and Yoe, J. H., *Bull. Chem. Soc. Japan* **34**, 826 (1961).
128. Chiacchierini, E., Šepel, T. and Sommer, L., *Collection Czech. Chem. Commun.* **35**, 794 (1970).
129. Buděšínský, B., *ibid.* **27**, 226 (1962).
130. Srivastava, K. C. and Banerji, S. K., *Microchem. J*, **13**, 699 (1968).
131. Komenda, J., *Chem. Listy* **47**, 531 (1953).
132. Almássy, G., Nagy, Z. and Straub, J., *Acta Chim. Acad. Sci. Hung.* **7**, 317 (1955).
133. Dev, B. and Jain, B. D., *Z. Anal. Chem.* **196**, 178 (1963).
134. Moeken, H. H. and Van Neste, W. A., *Anal. Chim. Acta* **37**, 480 (1967).
135. Burtnenko, L. M. and Poluektov, N. S., *Zh. Analit. Khim.* **23**, 700 (1968).
136. Burtnenko, L. M., Poluektov, N. S. and Kononenko, L. I., *ibid.* **23**, 1647 (1968).
137. Poluektov, N. S. and Bel'tyukova, S. V., *ibid.* **26**, 541 (1971).
138. Poluektov, N. S., Bel'tyukova, S. V. and Otnichenko, S. F., *Ukr. Khim. Zh.* **38**, 271 (1972).
139. Kovalenko, P. N., Shchemeleva, G. G. and Sokolova, L. S., *Zh. Analit. Khim.* **22**, 1845 (1967); *Zavodsk. Lab.* **33**, 287 (1967).
140. Kovalenko, P. N., Shchemeleva, G. G. and Stepanenko, Yu. V., *Zh. Analit. Khim.* **26**, 1979 (1971).
141. Tarayan, V. M., Ovsepyan, E. N. and Petrosyan, A. A., *ibid.* **26**, 322 (1971).
142. Luk'yanov, V. F. and Duderova, E. P., *ibid.* **16**, 60 (1961).
143. Wilson, A. D., *Analyst* **87**, 703 (1962).
144. Khopkar, S. M. and De, A. K., *ibid.* **85**, 376 (1960).
145. Obrenović-Paligorić, J., Gal, I. J. and Vajgand, V., *Anal. Chim. Acta* **40**, 534 (1968).
146. Solanke, K. R. and Khopkar, S. M., *Chem. Anal. (Warsaw)* **17**, 1175 (1972).
147. Johnston, J. R., Holland, W. J. and Gerard, J., *Mikrochim. Acta* **1971**, 886.
148. Haas, W. and Schwarz, T., *ibid.* **1963**, 253.
149. Traub, A. and Boltz, D. F., *ibid.* **1969**, 749.
150. Meloan, C. E., Holkeboer, P. and Brandt, W. W., *Anal. Chem.* **32**, 791 (1960).
151. Zharovskii, F. G. and Sukhomlin, R. I., *Zh. Analit. Khim.* **21**, 59 (1966).
152. Leonard, M. A. and Nagi, F. I., *Anal. Lett.* **2**, 15 (1969).
153. Mukherji, A. K. and Dey, A. K., *Z. Anal. Chem.* **160**, 98 (1958).
154. Havel, J. and Sommer, L., *Collection Czech. Chem. Commun.* **33**, 529 (1968).
155. Horák, J. and Okáč, A., *ibid.* **33**, 304 (1968).
156. Ripan, R., Eger, I. and Bojan, N., *Rev. Roum. Chim.* **9**, 829 (1964).
157. Marangoni, G., Degetto, S. and Croatto, U., *Talanta* **20**, 1217 (1973).
158. Leong, C. L., *Anal. Chem.* **45**, 201 (1973).
159. Seim, H. J., Morris, R. J. and Frew, D. W., *ibid.* **29**, 443 (1957).
160. Smith, W. B. and Drewry, J., *Analyst* **86**, 178 (1961).
161. Boirie, C. and Platzer, R., *Acta Chim. Acad. Sci. Hung.* **33**, 275 (1962).
162. Feinstein, H. I., *Anal. Chim. Acta* **15**, 288 (1956).
163. Sherif, F. G. and Awad, A. M., *ibid.* **26**, 235 (1962).
164. Silverman, L. and Moudy, L., *Anal. Chem.* **28**, 45 (1956).
165. Deptuła, C., *Chem. Anal. (Warsaw)* **11**, 589 (1966).
166. Vydra, F. and Přibil, R., *Talanta* **9**, 1009 (1962).
167. Nemodruk, A. A., Poloznikova, E. I. and Myasoedov, B. F., *Zh. Analit. Khim.* **26**, 2388 (1971).
168. Nemodruk, A. A., Myasoedov, B. F. and Koiro, O. E., *ibid.* **28**, 1970 (1973).
169. Gayer, K. H. and Lifshitz, H. T., *Anal. Chem.* **44**, 2104 (1972).

Chapter 59

VANADIUM

Vanadium (V, at. wt. 50·94) occurs in the +V, +IV, +III, and +II oxidation states, vanadium(V) compounds being the most stable. In alkaline medium colourless vanadate VO_3^- ions exist, whereas in strongly acidic medium yellow VO_2^+ cations are present. Within the intermediate pH range polymerized orange-yellow anionic forms occur. Vanadium(V) forms heteropoly acids with phosphate, Mo(VI), and W(VI) and also peroxide complexes. Vanadium(IV) occurs as the blue vanadyl ion VO^{2+}, stable in acid solutions and readily oxidized to vanadium(V) in alkaline solution. The VO_2^{2+} cation is amphoteric. At pH ~4, $VO(OH)_2$ precipitates and at pH ~9 it dissolves. Vanadium(IV) forms fluoride, oxalate, and EDTA complexes. The strongly reducing V^{3+} (green) and V^{2+} (violet) ions are of no importance in the spectrophotometric determination of vanadium.

59.1 Methods of Separation

59.1.1 EXTRACTION

Vanadium(V) 8-hydroxyquinolinate [1] is extractable into chloroform from slightly acidic media (pH 2·5–4·5). The coloured extract provides a spectrophotometric method for determining vanadium. For details sees p. 590.

Extraction of the diethyldithiocarbamate (pH 2–3) [1–3] or of the pyrrolidinedithiocarbamate (pH 0–2) [1,4] into chloroform is applied to the separation of vanadium from aluminium and titanium. Vanadium is stripped with nitric acid (1+1) containing hydrogen peroxide.

Extraction of vanadium(V) α-benzoinoximate into chloroform from a solution at pH 2–4 enables vanadium to be separated from iron and a score of other metals [5–7]. For extractive separation of vanadium, cupferron [1,8,9], acetylacetone [10], and TTA [11] are also employed. Acetylacetone extracts vanadium(III) better than V(IV) or V(V).

Vanadium(V) is extractable with TBP [12] or mesityl oxide [13] from 5-6M HCl medium.

59.1.2 PRECIPITATION

Traces of vanadium(V) are retained quantitatively by $Fe(OH)_3$ as collector when Fe(III) (2–5 mg) is precipitated with ammonia. The final pH of the solution should be 6–7 [14].

Vanadium is very often isolated from a large number of elements by fusing the sample either in a platinum crucible with Na_2CO_3 and a small amount of KNO_3, or in a nickel crucible with Na_2O_2, the melt being leached in either case with cold water (see procedure below). The precipitate contains Fe, Cu, Ti, Ni, Co, Mn, and some Al. Besides the vanadium the filtrate contains As, Cr, P, Mo, W, the rest of the Al, and also Mn. In leaching of the melt, addition of a small amount of ethanol or H_2O_2 reduces Mn(VI and VII) and leaves all the manganese in the precipitate [14–16].

Good separation of vanadium from chromium and copper can be attained by precipitating vanadium(V) with cupferron [along with iron(III) as collector] from an acid solution (pH < 1) [17,18]. Besides the iron, titanium and zirconium are also precipitated with the vanadium.

Vanadium can be separated from metals precipitated as hydroxides by ammonia, by conversion into a peroxide complex stable in ammoniacal medium. If precipitation and separation (best of all by centrifuging) are performed twice, the losses of vanadium by retention on the precipitate do not exceed 10%.

59.1.3 Ion-Exchange

Ziegler and Rittner [19] have separated vanadium and molybdenum by means of a strongly basic anion-exchanger. From thioglycollic acid medium (pH 1–1·5), molybdenum is retained in the column while vanadium passes into the eluate.

For determination of vanadium in steel, Hall and Bryson [20] used a strongly basic anion-exchange column and an acetate–mannitol sample medium of pH 2·5–3. Vanadium(V), chromium(VI), and molybdenum(VI) were then washed out from the column with $0.6M$ NaOH, $8M$ HCl, and $1M$ HCl, respectively.

On the basis of the different stabilities of the vanadium, tungsten, and molybdenum ascorbate complexes, these metals have been separated on an Amberlite IRA 400 anion-exchange column [21].

From a hydrochloric acid–alcohol medium, the anion-exchanger Dowex 1 retains copper and iron(III), while vanadium(IV) and nickel are eluted [22].

Vanadium is eluted selectively with $0.01M$ H_2SO_4 or $HClO_4$ containing 1% H_2O_2 from a cation-exchange column, which retains many other elements [23,24]. Vanadium(V) is eluted from a Dowex 50 cation-exchange column by $0.5M$ HCl, whereas numerous other metals [e.g. Fe, Ni, Zn, Mn, Th, U(VI), and Al] are not [25].

Fritz and Topping [26] have separated V, Mo, and W with the liquid anion-exchanger Aliquat 336.

59.2 Methods of Determination

Three methods are presented in detail, two being extractive spectrophotometric methods with 8-hydroxyquinoline and benzoylphenylhydroxylamine

(BPHA) and the third involving tungstophosphovanadic acid in aqueous medium. This last is not very sensitive. The methods using azo dyes as reagents exhibit high sensitivity. Another group of methods turns to account the redox reaction of vanadium(V) with organic reagents.

59.2.1 8-Hydroxyquinoline Method

8-Hydroxyquinoline (p. 56) gives with vanadium(V) at pH 2·0–5·5 a chelate soluble in chloroform and in isoamyl alcohol, which forms the basis of an extractive-spectrophotometric method [1,27–30].

Figure 59.1 presents the absorption spectrum of vanadium(V) oxinate, which absorbs over the whole visible range. The absorption maximum at 550 nm is generally used. Absorbance measurements at the higher maximum at 380 nm are more sensitive but rather less accurate. The molar absorptivity of a chloroform solution of vanadium oxinate at 550 nm is 3.0×10^3 ($a = 0.06$). At 380 nm the corresponding values are 5.4×10^3 and 0.11. Chloroform solutions of the vanadium complex are stable in colour if the chloroform used is free from ethanol.

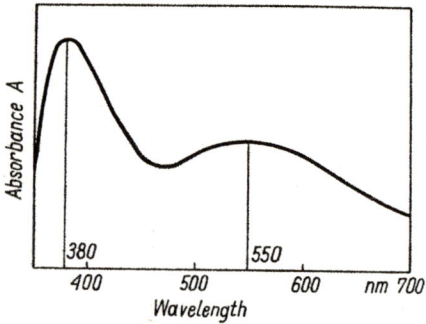

Fig. 59.1 Absorption spectrum of vanadium–8-hydroxyquinoline complex in chloroform

To eliminate the effect of other metals which form oxinates in slightly acidic medium, Talvitie [27] has recommended the following procedure: the chloroform extract which is obtained at pH 4 and contains vanadium, iron, and partly also Al, Co, Zn, Ni, Mo, W, U, Cu, Ti, and Bi oxinates, is shaken with an aqueous alkaline solution (pH 9·4), thereby stripping vanadium into the aqueous phase and leaving behind iron together with the other metals in the chloroform solution. From the acidified aqueous phase, vanadium is re-extracted with a chloroform solution of oxine and the absorbance is measured. In this manner the oxine method becomes specific for vanadium.

When separated extractively as the oxinate and stripped into the aqueous phase at pH 9·4, vanadium may also be determined by other methods which are described later in the text.

It is convenient to increase the acidity of the solution from pH 4·0 to pH 2·6–3·0 before the oxinate is extracted. The extraction rate and distribution coefficient remain the same as at pH 4, and iron is also extracted completely, but the extraction of aluminium and of other partially extractable metals is reduced by 70–80%.

Nadalin and Brozda [31] extracted vanadium(V) with oxine in chloroform solution, at pH 5·5, in the presence of Ca–EDTA. Under these conditions there is interference only from tin, titanium, and tungsten, which are not masked by EDTA. Ashbrook and Conn [32] have masked iron with EDTA at pH 4·0 and bound the excess of this reagent by adding thorium nitrate. A mixed 8-hydroxyquinoline–azide complex has been used [32a]. The 8-hydroxyquinoline method has been used to determine vanadium in natural water [14,16], biological materials [27], petrochemicals [31], uranium and its ores [32], and steel [33].

Use has also been made of related reagents such as 5,7-dibromo-8-hydroxyquinoline [34] and 5,7-di-iodo-8-hydroxyquinoline [35].

Reagents

1. 8-Hydroxyquinoline (oxine): 0·5% and 0·1% solutions in chloroform.
2. Standard vanadium solution: 1 mg/ml.
    (a) Dissolve in dilute NaOH solution 1·786 g of V_2O_5 ignited previously at ~500°C. Acidify the solution with sulphuric acid and dilute with water to 1 litre.
    (b) Dissolve 2·298 g of NH_4VO_3 in water containing 5 ml of conc. ammonia solution, acidify the solution with 10 ml of conc. HNO_3 and dilute in a volumetric flask to 1 litre.
3. Buffer solution (pH 9·4). Add 40 ml of conc. ammonia solution and 20 ml of conc. HNO_3 to 800 ml of water. Adjust the solution with ammonia or acid to pH 9·4 and dilute with water to 1 litre.
4. Chloroform, free from ethanol. Wash the commercial product 5 or 6 times with water, dry over anhydrous $CaCl_2$, and distil.

Procedure

Extractive separation of V. Adjust the pH of the sample solution containing not more than 250 µg of V(V) to 2·6–3·0, transfer the solution to a separating funnel and extract with two portions of 0·5% oxine solution (shaking time 2 minutes). Wash the extracts with water acidified with HCl to pH ~3. Strip the vanadium with two portions of the (pH 9·4) buffer solution (shaking time 5 minutes, use of a mechanical shaker is advisable).

Determination of V. Adjust the alkaline solution containing vanadium with $4M$ HCl and dilute ammonia solution to pH 2·6–3·0 and extract the vanadium with two portions of 0·1% oxine solution (shaking time 2 minutes). Dilute the extracts with chloroform to the mark in a 50-ml volumetric flask (or smaller, depending on the amount of vanadium) and measure the absorbance at 380 nm against a reagent blank.

59.2.2 Tungstophosphovanadic Acid Method

A selective but not very sensitive method for determining vanadium is based on the greenish-yellow colour of the heteropoly tungstophosphovanadic acid formed by adding phosphoric acid and sodium tungstate to the sample solution containing vanadium(V) [1–3,36–38]. According to Tikhonova [39], during the colour reaction $V_2O_6^{2-}$ groups are substituted for some $W_2O_7^{2-}$ groups in tungstophosphoric acid $H_7[P(W_2O_7)_6]$, but this formulation is open to question.

Heating of the solution to the boiling point promotes development of the colour, which is stable for at least a day. The acidity of the solution (HCl, H_2SO_4, HNO_3) has no effect on the final colour. The concentrations of phosphoric and tungstic acids may vary over a fairly wide range.

The absorption maximum of the heteropoly acid is in the ultraviolet. The absorbance of the solution is usually measured at 400 nm. At that wavelength the value due to the blank solution (colour of tungstophosphoric acid) is rather low, but it increases rapidly at shorter wavelengths. The molar absorptivity at 400 nm is 1.4×10^3 ($a = 0.03$).

The heteropoly acid may be extracted with isobutyl alcohol [37], n-hexanol [40], or MIBK [41].

Ammonium and potassium ions form sparingly soluble compounds with heteropoly acids, and interfere in this method. Larger amounts of Ti, Zr, Bi, Sb and Sn are isolated as the sparingly soluble phosphates. Molybdenum(VI) at higher concentrations interferes also by forming a coloured compound with the reagent. Reducing substances also interfere. Iron(III) in a final concentration up to 0.4 mg/ml does not interfere. Citrate, tartrate, and oxalate do not interfere.

Vanadium has been determined by the tungstophosphovanadic acid method in petroleum oils [37,42], fuel-oil ash [43], aluminium [3,40], steel [36,44], chromium [8], heat-resistant alloys [39], carbon-electrode materials [45], geochemical samples [41], and uranium minerals [41a].

Wallace and Mellon have determined vanadium in the form of tungstovanadic [46] or molybdovanadic [47] acids. Erdey *et al.* [3] utilized molybdophosphovanadic acid.

Reagents

1. Sodium tungstate: 10% solution.
2. Phosphoric acid: dilute (1+2).
3. Standard vanadium solution: 1 mg/ml (p. 591).

Procedure

Precipitation of V. Heat the acidic sample solution containing small amounts of vanadium to ~70°C and add ammonia dropwise to make the pH 6–7. If the sample solution contains no Fe, Al, or Ti, add beforehand ~2–4 mg of Fe(III) as collector. Heat the solution till the precipitate coagulates, then filter off the precipitate and wash it with hot dilute NH_4NO_3

solution. Dry the filter paper with the precipitate and ignite it in a platinum crucible.

Fuse the residue obtained (or the solid sample) in a nickel crucible with Na_2O_2. Keep the melt at dark red heat for a few minutes. Cool the melt and leach with cold water. Separate the solution (containing vanadium) from the precipitate by filtration through paper. Refilter the first portion of filtrate. Wash the precipitate on the filter paper with 1% Na_2CO_3 solution. Acidify the alkaline filtrate with sulphuric acid (to make it $0.25M$ in H_2SO_4) then heat it to boiling and cool.

Determination of V. To the sample solution in $\sim 0.25M$ H_2SO_4, containing not more than 1·0 mg of V(V), add 5 ml of phosphoric acid, 5 ml of sodium tungstate solution, and water to ~ 45 ml, then heat the solution to boiling and cool. Make up the solution to volume in a 50-ml volumetric flask and measure the absorbance at 400 nm, using a reagent blank as reference.

59.2.3 BENZOYLPHENYLHYDROXYLAMINE (BPHA) METHOD

N-Benzoyl-*N*-phenylhydroxylamine (BPHA) (p. 62), an analogue of

cupferron, gives a sparingly water-soluble violet chelate with vanadium(V) in a strongly acidic medium (2–10M HCl).

Extraction of the complex into chloroform serves as the basis of a selective method for determining vanadium [48–50]. According to Zharovskii and Pilipenko [50], the composition of the complex is $V_2O_3(C_{13}H_{10}O_2N)_4$.

The most suitable acidity is 3–4M HCl. At least a tenfold molar excess of BPHA is necessary [49]. Chloroform is best for the extraction but carbon tetrachloride, benzene, ethyl acetate, and diethyl ether can also be used. The complex forms rapidly and is stable. The order of addition of reagents is immaterial.

The molar absorptivity at $\lambda_{max} = 525$ nm is 5.1×10^3 ($a = 0.10$). The chloroform solution of BPHA does not absorb at 525 nm, so chloroform can be used as reference. The absorption spectrum of the vanadium complex with BPHA is shown in Fig. 59.2.

In strongly acidic medium BPHA forms complexes with only a few metals. Considerable quantities (20–40 mg) of Al, Co, Cr, Fe(III), Mn, Ni, Th, U, and Zn do not interfere in the determination of vanadium, but Mo(VI), Ti, and Zr form chloroform-soluble coloured (yellow, red) complexes with BPHA in strongly acidic medium and do interfere. The concentration of nitric acid in the sample solution must not be higher than 1M.

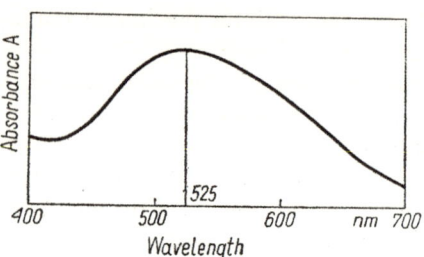

Fig. 59.2 Absorption spectrum of vanadium-benzoylphenylhydroxylamine (BPHA) complex in chloroform

Interference also arises from oxidizing agents which attack BPHA and from substances capable of reducing vanadium(V).

Vanadium has been determined by the extractive-spectrophotometric method with BPHA in steels and cast iron [48,51], iron ores [52], chromium ores [48], refractory metals [51], ilmenite [53], titanium tetrachloride (titanium is masked with fluoride) [50], silicate minerals [51,54], and petroleum coke [54a].

Extractive-spectrophotometric methods for determining vanadium may also be based upon other hydroxylamine derivatives related to BPHA, namely N-benzoyl-o-tolylhydroxylamine [55–57], N-furoylphenylhydroxylamine [58,59], N-cinnamoyl-N-phenylhydroxylamine [60,61], N-2-thiophenecarbonyl-N-phenylhydroxylamine, and N-2-thiophenecarbonyl-N-p-tolylhydroxylamine [62]. The sensitivity of the methods based on these reagents is a little higher than that with BPHA.

Reagents

1. Benzoylphenylhydroxylamine (BPHA): 0.1% solution in chloroform free from ethanol. If kept in an amber-glass bottle, the solution is stable for several days.
2. Standard vanadium solution: 1 mg/ml (p. 591).

Procedure

Acidify the sample solution containing not more than 300 μg of V(V) with hydrochloric acid so that the HCl concentration in the solution is ~4M. Transfer the solution to a separating funnel and shake with two portions of BPHA solution (shaking time 1 minute). Place the extracts in a 50-ml volumetric flask (or smaller according to the amount of vanadium), make up with chloroform to the mark, mix, and measure the absorbance at 525 nm against chloroform or a reagent blank.

59.2.4 OTHER METHODS

Benzohydroxamic acid, a hydroxylamine derivative (see p. 61), reacts with vanadium(V) in acidic medium to form a coloured complex (λ_{max} = 450 nm,

$\varepsilon = \sim 4.0 \times 10^3$) extractable with 1-hexanol, heptanol, octanol, octanol–CCl_4, or benzene [63–66]. The chief interfering metal is iron(III), which

$$\underset{\text{HN—OH}}{\underset{|}{\text{C}_6\text{H}_5\text{—C=O}}}$$

should be separated beforehand (electrolysis with an Hg cathode, extraction). Other interfering substances include Bi, Sb, Sn, Al, Ti, W, Mo, Zr, and powerful oxidants and reductants. Vanadium has been determined by this method in steel [63], plant materials [4], biological materials [65], and uranium [66].

Some other hydroxamic acids have also been used, viz. salicylhydroxamic acid [67,68], nicotinohydroxamic acid [69,70] isonicotinohydroxamic acid and quinaldinohydroxamic acid [71], 2-naphthohydroxamic acid [72], saturated [73] and unsaturated N-arylhydroxamic acids [74], and thiophene-2-hydroxamic acid [75]. p-Methoxybenzothiohydroxamic acid, suggested by Skorko-Trybuła [76], forms with vanadium(V) a green complex with molar absorptivity 2.0×10^4 at $\lambda_{max} = 372$ nm and 1.06×10^4 at $\lambda = 400$ nm. It is extractable into chloroform from $6M$ HCl medium.

Methods for vanadium determination with azo dyes are among the most sensitive. The extractive method based on PAN has $\varepsilon = 1.69 \times 10^4$ at 615 nm [77]. With PAR in aqueous medium the molar absorptivity is 3.6×10^4 at 550 nm [78]. The method has been applied to determining vanadium in petroleum fractions [79], sea-water [80,81], copper ores [7], and various alloys [82]. PAR has also been used in the presence of H_2O_2 [83], and to form the tetraphenylphosphonium complex in $CHCl_3$–acetone medium [84]. A mixed complex with PAR and hydroxylamine has been used [85]. Bromine derivatives of pyridylazo dyes have also been suggested [86]. Vanadium(IV) also reacts with PAN and PAR [87].

Other azo dyes used are: Arsenazo I [88], Nevazol NS [89], Solochrome Fast Grey [22], Solochrome Black RN [90], Fast Grey RA [91], Calcichrome [see. p. 187) [92], Sulphonazo (see p. 471) [93], Sulphonitrazo [94], Picraminazo H [95], Acid Chrome Blue K [96], Eriochrome Green B [97], and Eriochrome Red B [98].

Savvin et al. [99] compared several o,o'-dihydroxyazo compounds as reagents for vanadium. Sulphonitrophenol K proved to be the best. In the

presence of hydroxylamine it gives a ternary complex with molar absorptivity 5.5×10^4 at 645 nm; the complex is stable over the pH range 1–8.

Various authors have recommended spectrophotometric determination of vanadium with triphenylmethane dyes: Xylenol Orange ($\varepsilon = 1.3 \times$

10^4 at 530 nm) [24,100,101], Methylthymol Blue [102,103], Pyrocatechol Violet ($\varepsilon = 3.68 \times 10^4$, pH 6·2) [104], Chrome Azurol S [104,105], and Eriochrome Cyanine R [104,106].

Formaldoxime (see p. 59) gives an orange complex with vanadium in ammoniacal medium (for absorption spectrum see Fig. 2.6, p. 59) [107–109]. At $\lambda_{max} = 403$ nm the molar absorptivity is 6.6×10^3. Oxalate, fluoride, tartrate, and phosphate do not interfere.

Various organic reagents have found application in the spectrophotometric determination of vanadium, namely: pyrocatechol [15,110], 3,4-dinitropyrocatechol [111], 2-methyl-3-hydroxy-γ-pyrone (maltol) [112], tribromopyrogallol [113], haematoxylin [106,114], morin and antipyrine [115], phenylfluorone [116], TTA [11,117], thiothenoyltrifluoroacetone [118], ferron [119], 2,6-pyridinedicarboxylic acid ($+H_2O_2$) [120], 6-hydroxy-1,7-phenanthroline [121], nicotinic acid hydrazide [122], 1-(o-carboxyphenyl)-3-hydroxy-3-methyltriazene [123], β-isopropyltropolone [124], tropolone [125], phthalocyanine tetrasulphonic acid [126], 1,10-phenanthroline [127], thioglycollic acid [19,128], and disodium maleonitrile dithiolate [129].

Many photometric methods for determining vanadium are based on redox reactions, and the colour resulting from the oxidation of various organic compounds by vanadium(V). Mention may be made of 3,3'-dimethylnaphthidine [130–133], 3,3'-diaminobenzidine (DAB) (see p. 476), $\varepsilon = 3.3 \times 10^3$ at 470 nm [134–136], diphenylbenzidine [137–139], α-naphthylamine [140,141], o-dianisidine [131,142], veratrole [143], Variamine Blue [144], 2,2'-dicarboxydiphenylamine (Vanadox A), $\varepsilon = 2.3 \times 10^4$ [145], and diantipyrylmethane derivatives [146]. 1,10-Phenanthroline (+tervalent iron) reacts with vanadium(IV) [17,147]. Redox reactions have been used to determine vanadium in steels [132,147], ferrous and non-ferrous alloys [130], chromium [17], aluminium [144], silicate sediments [135], titanium tetrachloride [139], oils [138], and natural water [135,136].

In 0·5–3M H_2SO_4, hydrogen peroxide gives with vanadium(V) the orange peroxide complex which provides a rather insensitive method for determining vanadium [10,148,149]. The molar absorptivity is $\sim 2.8 \times 10^2$ at $\lambda_{max} = 460$ nm. The absorption spectrum of the complex is shown in Figure 56.1, p. 557. Titanium interferes but can be masked with fluoride.

Thiocyanate gives coloured complexes with V(IV) and V(III) [150–153]. The molar absorptivity of the vanadium(III) complex is 7.2×10^3 at 400 nm. Vanadium(III) may be determined in presence of vanadium(IV) [152]. The vanadium(IV) complex can be extracted with chloroform in the presence of pyridine [151].

Major quantities of vanadium may be determined by means of the blue vanadyl ion [154,155] or the yellow-orange acid form in H_2SO_4 medium [156].

References

1. Bock, R. and Gorbach, S., *Mikrochim. Acta* **1958**, 593.
2. Chernikhov, Yu. A. and Dobkina, B. M., *Zavodsk. Lab.* **16**, 402 (1950).
3. Erdey, L., Vigh, K. M. and Măzor, L., *Acta Chim. Acad. Sci. Hung.* **4**, 259 (1954).
4. Jones, G. B. and Watkinson, J. H., *Anal. Chem.* **31**, 1344 (1959).
5. Hoenes, H. J. and Stone, K. G., *Talanta* **4**, 250 (1960).
6. Bock, R. and Jost, B., *Z. Anal. Chem.* **250**, 358 (1970).
7. Koźlicka, M. and Wójtowicz, M., *ibid.* **257**, 191 (1971); *Chem. Anal. (Warsaw)* **16**, 739 (1971).
8. McAloren, J. T. and Reynolds, G. F., *Metallurgia* **57**, 52 (1958).
9. Crump-Wiesner, H. J. and Purdy, W. C., *Talanta* **16**, 124 (1969).
10. McKaveney, J. P. and Freiser, H., *Anal. Chem.* **30**, 526 (1958).
11. De, A. K. and Rahaman, M. S., *ibid.* **35**, 1095 (1963).
12. Majumdar, S. K. and De, A. K., *ibid.* **33**, 297 (1961).
13. Shinde, V. M. and Khopkar, S. M., *Chem. Anal. (Warsaw)* **14**, 749 (1969).
14. Sugawara, K., Tanaka, M. and Naitô, H., *Bull. Chem. Soc. Japan* **26**, 417 (1953).
15. Patrovský, V., *Z. Anal. Chem.* **144**, 140 (1955).
16. Naitô, H. and Sugawara, K., *Bull. Chem. Soc. Japan.* **30**, 799 (1957).
17. Yakovlev, P. Ya. and Razumova, G. P., *Zavodsk. Lab.* **24**, 1430 (1958).
18. Willard, H. H., Martin, E. L. and Feltham, R., *Anal. Chem.* **25**, 1863 (1953).
19. Ziegler, M. and Rittner, W., *Z. Anal. Chem.* **164**, 310 (1958).
20. Hall, F. M. and Bryson, A., *Anal. Chim. Acta* **24**, 138 (1961).
21. Korkisch, J. and Farag, A., *Mikrochim. Acta* **1958**, 646.
22. Janauer, G. E. and Korkisch, J., *Z. Anal. Chem.* **179**, 241 (1961); *Talanta* **8**, 569 (1961).
23. Fritz, J. S. and Abbink, J. E., *Anal. Chem.* **34**, 1080 (1962).
24. Janoušek, I., *Collection Czech. Chem. Commun.* **27**, 2972 (1962).
25. De, A. K. and Majumdar, S. K., *Z. Anal. Chem.* **191**, 40 (1962).
26. Fritz, J. S. and Topping, J. J., *Talanta* **18**, 865 (1971).
27. Talvitie, N. A., *Anal. Chem.* **25**, 604 (1953).
28. Nakamura, H., Shimura, Y. and Thuchida, R., *Bull. Chem. Soc. Japan* **34**, 1143 (1961).
29. Tanaka, M. and Kojima, I., *Anal. Chim. Acta* **36**, 522 (1966).
30. Kurmaiah, N., Satyanarayana, D. and Rao, V. P. R., *Talanta* **14**, 495 (1967).
31. Nadalin, R. J., and Brozda, W. B., *Anal. Chem.* **32**, 1141 (1960).
32. Ashbrook, A. W. and Conn, K., *Chemist-Analyst* **50**, 47 (1961).
32a. Rao, V. P. R. and Anjaneyulu, Y., *Mikrochim. Acta* **1973**, 481.
33. Luke, C. L., *Anal. Chim. Acta* **37**, 267 (1967).
34. Nguyen Shi Zun, Ryzhenko, V. L. and Zharovskii, F. G., *Ukr. Khim. Zh.* **35**, 206 (1969).
35. Heitner-Wirguin, C. and Gancz, M., *Talanta* **14**, 671 (1967).
36. Cooper, M. D. and Winter, P. K., *Anal. Chem.* **21**, 605 (1949).
37. Sherwood, R. M. and Chapman, F. W., Jr., *ibid.* **27**, 88 (1955).
38. Gregorowicz, Z., *Z. Anal. Chem.* **175**, 161 (1960).
39. Tikhonova, A. A., *Zavodsk. Lab.* **16**, 1168 (1950).
40. Biechler, D. G., Jordan, D. E. and Leslie, W. D., *Anal. Chem.* **35**, 1685 (1963).
41. Roberts, J. L., *Talanta* **18**, 1070 (1971).
41a. Korkisch, J. and Steffan, I., *Mikrochim. Acta* **1973**, 651.
42. Milner, O. I., Glass, J. R., Kirchner, J. P. and Yurick, A. N., *Anal. Chem.* **24**, 1728 (1952).
43. Hopps, G. L. and Berk, A. A., *ibid.* **24**, 1050 (1952).
44. Scholes, P. H., *Analyst* **82**, 525 (1957).
45. Sugawara, K., Tanaka, M. and Kozawa, A., *Bull. Chem. Soc. Japan* **28**, 492 (1955).
46. Wallace, G. W. and Mellon, M. G., *Anal. Chem.* **32**, 204 (1960).

47. Wallace, G. W. and Mellon, M. G., *Anal. Chim. Acta* **23**, 355 (1960).
48. Ryan, D. E., *Analyst* **85**, 569 (1960).
49. Priyadarshini, U. and Tandon, S. G., *Anal. Chem.* **33**, 435 (1961).
50. Zharovskii, F. G. and Pilipenko, A. T., *Ukr. Khim. Zh.* **25**, 230 (1959).
51. Donaldson, E. M., *Talanta* **17**, 583 (1970).
52. Hofer, A. and Heidinger, R., *Z. Anal. Chem.* **246**, 125 (1969).
53. Pilkington, E. S. and Wilson, W., *Anal. Chim. Acta* **47**, 461 (1969).
54. Patrovský, V., *Chem. Listy* **60**, 1545 (1966).
54a. Hulanicki, A. and Karwowska, R., *Chem. Anal. (Warsaw)* **18**, 709 (1973).
55. Majumdar, A. K. and Das, G., *Anal. Chim. Acta* **31**, 147 (1964); **36**, 454 (1966).
56. Jeffery, P. G. and Kerr, G. O., *Analyst* **92**, 763 (1967).
57. Majumdar, A. K. and Bhowal, S. K., *ibid.* **96**, 127 (1971).
58. Pilipenko, A. T., Sereda, I. P. and Shpak, E. A., *Zavodsk. Lab.* **32**, 660 (1966).
59. Pilipenko, A. T., Shpak, E. A. and Kurbatova, G. T., *Zh. Analit. Khim.* **22**, 1014 (1967).
60. Zharovskii, F. G. and Sukhomlin, R. I., *ibid.* **21**, 59 (1966).
61. Priyadarshini, U. and Tandon, S. G., *Analyst* **86**, 544 (1961).
62. Tandon, S. G. and Bhattacharyya, S. C., *Anal. Chem.* **33**, 1267 (1961).
63. Wise, W. M. and Brandt, W. W., *ibid.* **27**, 1392 (1955).
64. Gorczyńska, K., Walędziak, H. and Ciecierska-Stokłosa, D., *Chem. Anal. (Warsaw)* **4**, 809, 883 (1959).
65. Hulcher, F. H., *Anal. Chem.* **32**, 1183 (1960).
66. Kuehn, P. R., Howard, O. H. and Weber, C. W., *ibid.* **33**, 740 (1961).
67. Bhaduri, A. S. and Rây, P., *Z. Anal. Chem.* **154**, 103 (1957).
68. Pilz, W., *Mikrochim. Acta* **1958**, 789.
69. Dutta, R. L., *J. Indian Chem. Soc.* **35**, 243 (1958).
70. Minczewski, J. and Skorko-Trybuła, Z., *Chem. Anal. (Warsaw)* **6**, 377 (1961).
71. Dutta, R. L., *J. Indian Chem. Soc.* **36**, 285, 339 (1959).
72. Bass, V. C. and Yoe, J. H. *Anal. Chim. Acta* **35**, 337 (1966); *Talanta* **13**, 735 (1966).
73. Gupta, Y. K. and Tandon, S. G., *Anal. Chim. Acta* **66**, 39 (1973).
74. Bhura, D. C. and Tandon, S. G., *ibid.* **53**, 379 (1971).
75. Minczewski, J. and Skorko-Trybuła, Z., *Talanta* **10**, 1063 (1963).
76. Skorko-Trybuła, Z., *Nukleonika* **10**, 559 (1965); *Chem. Anal. (Warsaw)* **10**, 831 (1965); **12**, 815 (1967).
77. Staten, F. W. and Huffman, E. W. D., *Anal. Chem.* **31**, 2003 (1959).
78. Budevsky, O. and Johnova, L., *Talanta* **12**, 291 (1965).
79. Steinke, I., *Z. Anal. Chem.* **233**, 265 (1968).
80. Kiriyama, T. and Kuroda, R., *Anal. Chim. Acta* **62**, 464 (1972).
81. Nishimura, M., Matsunaga, K., Kudo, T. and Obara, F., *ibid.* **65**, 466 (1973).
82. Gagliardi, E. and Ilmaier, B., *Mikrochim. Acta* **1967**, 180.
83. Bagdasarov, K. N., Akhmedova, Kh. A. and Tataev, O. A., *Zavodsk. Lab.* **35**, 12 (1969).
84. Široki, M. and Djordjevic, C., *Anal. Chim. Acta* **57**, 301 (1971).
85. Lukachina, V. V., Pilipenko, A. T. and Karpova, O. I., *Zh. Analit. Khim.* **28**, 86 (1973).
86. Gusev, S. I. and Shalamova, G. G., *ibid.* **22**, 1357 (1967); **23**, 686 (1968).
87. Mushran, S. P., Prakash, O. and Verma, T. R., *Bull. Chem. Soc. Japan* **45**, 1709 (1972).
88. Lozanovskaya, I. N. and Petrashen', V. I., *Zh. Analit. Khim.* **22**, 1196 (1967).
89. Basargin, N. N., Yakovlev, P. Ya., Busev, A. I. and Zanina, I. A., *Zavodsk. Lab.* **35**, 411 (1969).
90. Janauer, G. E., Tera, F. and Korkisch, J., *Mikrochim. Acta* **1961**, 599.
91. Khalifa, H. and Farag, A., *Z. Anal. Chem.* **158**, 109 (1958).
92. Ishii, H. and Einaga, H., *Japan Analyst* **19**, 371 (1970).

93. Golubtsova, R. B., Savvateeva, S. N. and Yaroshenko, A. D., *Zavodsk. Lab.* **34**, 1184 (1968).
94. Barenbaum, M. E., Dedkov, Yu. M. and Orlova, E. S., *Zh. Analit. Khim.* **27**, 1967 (1972).
95. Zadumina, E. A. and Cherkesov, A. I., *Izv. Vyssh. Ucheb. Zaved., Khim. Khim. Tekhnol.* **12**, 1483 (1969).
96. Morachevskii, Yu. V. and Tserkovnitskaya, I. A., *Zh. Analit. Khim.* **16**, 106 (1961).
97. Tikhonov, V. N., *ibid.* **26**, 2142 (1961).
98. Tserkovnitskaya, I. A. and Perevoshchikova, V. V., *ibid.* **27**, 1111 (1972).
99. Savvin, S. B., Mineeva, V. A., Okhanova, L. A. and Pachadzhanov, D. N., *ibid.* **26**, 2364 (1971).
100. Otomo, M., *Bull. Chem. Soc. Japan* **36**, 137 (1963).
101. Budevsky, O. and Přibil, R., *Talanta* **11**, 1313 (1964).
102. Tikhonov, V. N., Grankina, M. Ya. and Vernigora, V. P., *Zh. Analit. Khim.* **22**, 359 (1967).
103. Srivastava, K. C. and Banerji, S. K., *Chim. Anal. (Paris)* **52**, 973 (1970).
104. Janssen, A. and Umland, F., *Z. Anal. Chem.* **254**, 286 (1971).
105. Sanyal, P., Sangal, S. P. and Mushran, S. P., *Bull. Chem. Soc. Japan* **40**, 217 (1967).
106. Prakash, O., Awasthi, J. N. and Mushran, S. P., *Chim. Anal. (Paris)* **51**, 125 (1969).
107. Tanaka, M., *Mikrochim. Acta* **1954**, 701.
108. Marczenko, Z. and Stępień, A., *Chem. Anal. (Warsaw)* **8**, 705 (1963).
109. Marczenko, Z., *Roczniki Chem.* **38**, 187 (1964); *Anal. Chim. Acta* **31**, 224 (1964).
110. Nestler, C. G. and Nobis, M., *Z. Anal. Chem.* **167**, 81 (1959).
111. Busev, A. I. and Karyakina, Z. P., *Zh. Analit. Khim.* **22**, 1506 (1967).
112. Jungnickel, H. E. and Klinger, W., *Z. Anal. Chem.* **203**, 257; **206**, 275 (1964).
113. Busev, A. I. and Karyakina, Z. P., *Zh. Analit. Khim.* **22**, 1350 (1967).
114. MacMillan, E. and Samuel, B. W., *Anal. Chem.* **38**, 250 (1966).
115. Shnaiderman, S. Ya. and Prokof'eva, G. N., *Zh. Analit. Khim.* **25**, 2368 (1970).
116. Verma, J. R., Prakash, O. and Mushran, S. P., *Anal. Chim. Acta* **52**, 357 (1970).
117. Ikehata, A. and Shimizu, T., *Bull. Chem. Soc. Japan* **38**, 1385 (1965).
118. Solanke, K. R. and Khopkar, S. M., *Talanta* **21**, 245 (1974).
119. Kurmaiah, N., Satyanarayana, D. and Rao, V. P. R., *Anal. Chim. Acta* **35**, 484 (1966).
120. Pearse, G. A., Jr., *Anal. Chem.* **34**, 536 (1962).
121. Dougherty, J. A. and Mellon, M. G., *ibid.* **37**, 1096 (1965).
122. Krych, Z. and Lipiec, T., *Chem. Anal. (Warsaw)* **12**, 535 (1967).
123. Majumdar, A. K., Bhattacharyya, B. C. and Roy, B. C., *Anal. Chim. Acta* **67**, 307 (1973).
124. Menis, O. and Iyer, C. S. P., *ibid.* **55**, 89 (1971).
125. Rizvi, G. H. and Singh, R. P., *Talanta* **19**, 1198 (1972).
126. Tserkovnitskaya, I. A. and Perevoshchikova, V. V., *Zh. Analit. Khim.* **26**, 1527 (1971).
127. Bhadra, A. K., *Talanta* **20**, 13 (1973).
128. Jacobsen, E. and Strøm, P., *Anal. Chim. Acta* **53**, 309 (1971).
129. Chatterjee, A. B., Basu, A. and Bag, S. P., *Mikrochim. Acta* **1974**, 275.
130. Milner, G. W. and Nall, W. R., *Anal. Chim. Acta* **6**, 420 (1952).
131. Klug, O. N. and Metlenko, A., *Chem. Anal. (Warsaw)* **10**, 819 (1965).
132. Rosotte, R., *Chim. Anal. (Paris)* **49**, 512 (1967).
133. Bannard, L. G. and Burton, J. D., *Analyst* **93**, 142 (1968).
134. Cheng, K. L., *Talanta* **8**, 658 (1961).
135. Chan, K. M. and Riley, J. P., *Anal. Chim. Acta* **34**, 337 (1966).
136. Riley, J. P. and Taylor, D., *ibid.* **41**, 175 (1968).
137. Eeckhout, J. and Weynants, A., *ibid.* **15**, 145 (1956).
138. Agazzi, E. J., Burtner, D. C., Crittenden, D. J. and Patterson, D. R., *Anal. Chem.* **35**, 332 (1963).

139. Radcliffe, N. C. and Parker, J. R., *Anal. Chim. Acta* **52**, 9 (1970).
140. Senetskaya, L. P. and Teplyakov, M. M.. *Zh. Analit. Khim.* **16**, 731 (1961).
141. Albert, F. M. and Stoia, M., *Z. Anal. Chem.* **202**, 420 (1964).
142. Ariel, M. and Manka, J., *Anal. Chim. Acta* **25**, 248 (1961).
143. Waechter, M. J., Hamon, M. and Guernet, M., *Analusis* **1**, 439 (1972).
144. Erdey, L. and Szabadváry, F., *Z. Anal. Chem.* **155**, 90; **159**, 429 (1957).
145. Frumina, N. S., Mustafin, I. S., Nikurashina, M. L. and Vechera, M. K., *Talanta* **16**, 138 (1969).
146. Podchainova, V. N., Dolgorev, A. V. and Dergachev, V. Ya., *Zavodsk. Lab.* **31**, 790 (1965); *Zh. Analit. Khim.* **21**, 53 (1966).
147. Rosotte, R. and Jaudon, E., *Chim. Anal. (Paris)* **36**, 160 (1954).
148. Hartkamp, H., *Z. Anal. Chem.* **169**, 339 (1959).
149. Schwarz, H., *ibid.* **176**, 241 (1960).
150. Crouthamel, C. E., Hjelte, B. E. and Johnson, C. E., *Anal. Chem.* **27**, 507 (1955).
151. Ayres, G. H. and Scroggie, L. E., *Anal. Chim. Acta* **26**, 470 (1962).
152. Zolotavin, V. L., Levashova, L. B. and Dolgarev, A. V., *Zh. Analit. Khim.* **17**, 336 (1962).
153. Karyakin, Yu. V. and Zaval'skaya, A. V., *ibid.* **23**, 1742 (1968).
154. Santini, R., Jr., Hazel, J. F. and McNabb, W. M., *Anal. Chim. Acta* **6**, 368 (1952),
155. Kranz, M. and Krzyżaniak, J., *Chem. Anal. (Warsaw)* **5**, 243 (1960).
156. Sarma, P. L., *Anal. Chem.* **36**, 1076 (1964).

Chapter 60
ZINC

Zinc (Zn, at.wt. 65·37) occurs exclusively in the +II oxidation state and is amphoteric. The hydroxide $Zn(OH)_2$ is precipitated at pH 6·8; in excess of alkali it dissolves to form tetrahydroxyzincate anions. It is readily soluble in ammonia, yielding the ammine complex. White zinc sulphide begins to precipitate at pH 1·2. Zinc forms stable complexes with cyanide and EDTA; its chloride and thiocyanate complexes are rather weak.

60.1 Methods of Separation

60.1.1 Precipitation

Small amounts of zinc are separated from solution as the sulphide in slightly acidic tartrate medium. As the collector, cadmium, copper, and mercury [1] are most often applied. Mercury is easily removed by igniting the precipitate. To separate traces of zinc from cadmium, the latter is complexed with iodide while the former is precipitated at pH 8·3 as the hydroxide together with aluminium hydroxide as the collector. For isolating zinc as hydroxide from a not too alkaline medium, $Mg(OH)_2$ has been used as collector [2]. In this way zinc is separated from larger quantities of antimony and tin. Zinc has also been coprecipitated with $Fe(OH)_2$ as collector [3].

The anionic zinc thiocyanate complex forms sparingly soluble compounds with basic dyes such as Methylene Blue, Methyl Violet, Rhodamine B and Malachite Green [4], which are useful for the separation of small amounts of zinc from other metals, in particular from cadmium [5].

60.1.2 Extraction

Among numerous extractive methods for separation of zinc, the best seems to be that with dithizone [6,7] (see below for details). Diethyldithiocarbamate is also useful for the preliminary separation of zinc [8,9].

The thiocyanate complex of zinc can be extracted by isoamyl alcohol or MIBK [10–11a]. Nazarenko *et al.* [12] have extracted the zinc thiocyanate–pyridine complex into chloroform at pH 5. By extracting the zinc thiocyanate complex with a 2% solution of tribenzylamine in $CHCl_3$, it is possible to separate zinc from cadmium [13].

Zinc chloride complexes in the presence of amines such as trioctylamine [14,15], dioctylmethylamine [16,17], and tribenzylamine [16] have found application in the separation of zinc. Chloroform, xylene, MIBK, and trichloroethylene are used as solvents. Zinc is separated from cadmium by extraction with Aliquat 336-S-1 in xylene from aqueous iodide solutions [17a].

Pietsch and Pichler [18] have separated zinc from many metals by extracting with chloroform its complex with di-n-butylarsinic acid. Schweitzer and Clifford have used aliphatic monocarboxylic acids for the extraction of zinc [19].

60.1.3 Ion-Exchange and Other Methods

In ion-exchange methods [20–24], use is made of differences in the stability of the chloride complexes of zinc and of other metals. According to Nishimura and Sandell [24], the strongly basic anion-exchanger Dowex 1 sorbs zinc from $2M$ HCl medium. Zinc is eluted from the column with $0.001M$ HCl.

Hunter and Miller [20] retained zinc and cadmium on the anion-exchanger Amberlite IRA-400 by passing through the column a solution in $2M$ HCl. They kept cadmium on the column by means of hydriodic acid and eluted zinc with water and $0.25M$ HNO$_3$. In acidic medium containing iodide and sulphate, this anion-exchanger retains only cadmium, whereas zinc passes to the eluate [25].

Geilmann and Neeb [26,27] have isolated small amounts of zinc from pyrite ores and bauxite by sublimation in a stream of hydrogen at 1100°C.

60.2 Methods of Determination

None of the other spectrophotometric methods for zinc determination matches the dithizone method. Zincon and PAN are also used, but are considerably inferior to dithizone in sensitivity, and have no compensating advantages.

Comparative investigations have been made of the more important spectrophotometric methods of zinc determination [28,29].

60.2.1 Dithizone Method

This is one of the most sensitive spectrophotometric methods. The molar absorptivity of zinc dithizonate in CCl$_4$ solution (at $\lambda_{max} = 538$ nm) is 9.26×10^4 ($a = 1.42$). If the correct pH and masking agents are used, the dithizone method is specific for zinc [30–32].

On shaking an aqueous zinc solution (pH 4–11) with a CCl$_4$ solution of dithizone (cf. p. 42), zinc dithizonate, Zn(HDz)$_2$, is formed and the organic layer changes colour from green to pink. The extraction is relatively slow.

Thiosulphate is most commonly used as a masking agent. At pH 4·0–5·5 (acetate buffer), thiosulphate forms stable complexes with Cu, Ag, Hg, Bi, Pb and Cd, thus preventing the reactions of those metal ions with dithizone. Thiosulphate also masks small quantities of nickel and cobalt. At higher concentrations of those metals, it is advisable to add small amounts of cyanide as masking agent. In the dithizone method for zinc, iodide, thiourea, and dithiocarbamates can also be used for masking interfering metals [33–36].

When there is more cadmium present than zinc in the solution analysed, the former is slightly extracted in spite of the masking effect of thiosulphate, but can be scrubbed from the extract by shaking it with a dilute Na_2S solution, cadmium sulphide being more stable than cadmium dithizonate.

In the presence of Fe, Al, Ti, and other easily hydrolysed metals, the extraction of zinc from acetate medium should be preceded by the addition of tartrate or citrate.

The zinc can be stripped from the CCl_4 layer with 0·01–0·02M mineral acid. The accurate determination of zinc requires a double extraction to be applied. The first extract is stripped with dilute acid, the resulting aqueous solution is treated with acetate buffer and thiosulphate, and the extraction with dithizone is repeated.

The excess of dithizone is removed from the extract by shaking it with very dilute ammonia solution. However, when determining trace amounts of zinc, it is important to remember that traces of this metal can often be found in ammonia solutions. Before the elimination of free dithizone, the zinc dithizonate extract is washed with thiosulphate solution to decompose the dithizonates of other metals present in the extract.

From the carbon tetrachloride extract containing Zn, Cd, Ni, and Co (all of which form soluble ammine complexes), Zn and Cd can be separated quantitatively by utilizing the differences in resistance of the four metal dithizonates to acid treatment. Ni and Co dithizonates begin to form at pH ~4, but once formed are rather hard to decompose even with quite concentrated mineral acids. This is not the case with Zn and Cd dithizonates, which are easily decomposed by dilute hydrochloric acid. Shaking the CCl_4 extract with two portions of dilute HCl (pH 1·5) for 30 sec transfers Zn and Cd quantitatively to the aqueous phase, leaving Ni and Co dithizonates in the CCl_4 phase. After the reaction with the first portion of dilute HCl, the dithizone from the decomposition of Zn and Cd dithizonates should be removed from the CCl_4 phase with dilute ammonia solution.

McClellan and Sabel [37] have separated zinc from nickel and cobalt by taking advantage of the different rate of extraction of these metals with dithizone. Manita [38] has determined zinc with dithizone in the one-phase alcohol–water system at pH 11, using an alcoholic solution of dithizone.

Instead of dithizone, its analogue di-2-naphthylthiocarbazone can be used [39].

The dithizone method has been used to determine zinc in cadmium and its compounds [5,25,33,40], indium and thallium [12], gallium [41], nickel and its alloys [36,42], cobalt [15], copper salts [3], tin [43], silver [44], gold [45], aluminium and its alloys [46–48], iron oxide [11], meteorites [24] silicate rocks [49], soil [50], boron [51], tellurium [52,53], alkalis [6], lubricating oils [54], sewage and effluents [55–57], foods [58,59], plant material [60], organic material [61], and urine [62].

Reagents

1. Dithizone: 0·002% solution in CCl_4. For preparation see p. 492.
2. Standard zinc solution: 1 mg/ml. Dissolve 1·000 g of zinc metal in 15 ml of HCl (1+1) and dilute the solution with water to 1 litre.
3. Acetate buffer: pH 5. Dissolve in water 50 g of anhydrous sodium acetate and 30 g of glacial acid and dilute with water to 250 ml. Purify the solution from traces of metals by shaking with portions of dithizone solution in CCl_4. Keep the solution in a polyethylene bottle.
4. Sodium thiosulphate: 10% solution. Purify the solution from traces of metals by shaking with dithizone solution in CCl_4. Keep the solution in a polyethylene bottle.
5. Wash solution: 10 ml of solution (3) and 10 ml of solution (4) diluted with water to 100 ml.
6. Ammonia solution. Saturate water, distilled twice in quartz apparatus, with ammonia gas from a cylinder. Use a polyethylene bottle and cool the water in an ice-water bath. Make sure that water cannot suck back up the delivery tube.

Procedure

Place the slightly acid (pH 2–3) solution containing not more than 20 μg of Zn and not larger than 25 ml in volume, in a separating funnel, add 5 ml of acetate buffer and 5 ml of thiosulphate solution, and shake with portions of the dithizone solution in CCl_4 (1 ml of 0·002% H_2Dz solution corresponds to 2·6 μg of Zn) until the green carbon tetrachloride layer no longer changes colour. Each shaking should last not less than 2 minutes (a mechanical shaker is advisable). Shake the combined extracts with two 5-ml portions of wash solution. Wash out free dithizone from the carbon tetrachloride layer with dilute ammonia solution (1 drop of conc. NH_3 solution in 25 ml of water). Dilute the pink solution of zinc dithizonate with CCl_4 in a 50-ml or smaller volumetric flask (according to colour intensity) and mix well. If the solution is turbid (owing to formation of an emulsion) filter it through a paper filter previously washed with the dilute dithizone solution and carbon tetrachloride. Measure the absorbance of the clear solution at 538 nm, using the solvent as reference.

Note. In the determination of traces of zinc it is essential to take into account a blank test of the content of zinc in the reagents, water, and vessels.

60.2.2 OTHER METHODS

2-Carboxy-2'-hydroxy-5'-sulphoformazylbenzene, a common reagent for the spectrophotometric determination of zinc, also called Zincon, was introduced by Yoe and Rush [63,64].

[Structure of Zincon showing OH and SO₃H substituted benzene ring connected via N=N–C(Ph)=N–NH to a COOH-substituted benzene ring]

In slightly alkaline solution (pH 9) zinc forms with Zincon a blue complex ($\lambda_{max} = 625$ nm, $\varepsilon = 2\cdot 0 \times 10^4$). A large number of metals interfere, making it necessary to separate the zinc first [14,65]. Zincon has been used for zinc determination in biological materials [9,66], plant materials [22], water [67–69], and lubricating oils [70].

1-(2-Pyridylazo)-2-naphthol (PAN) (p. 45) gives a red chelate with zinc, extractable into chloroform ($\varepsilon = 5\cdot 2 \times 10^4$ at 550 nm). In the presence of tartrate, iodide, thiosulphate, cyanide, and ascorbic acid, zinc has been determined in the presence of large quantities of Pb, Hg, Cu, and Ag [71,72]. PAN has been adopted for the determination of zinc in nickel and its alloys [17,73] and in copper [73].

Among other azo reagents the following are suggested for zinc determination: PAR [74], 1-(2-thiazolylazo)-2-naphthol (TAN) [75], 2-(2-quinolylazo)-1-naphthol (QAN) [76], 4-(2-thiazolylazo)-resorcinol (TAR) [77,78], Magneson IREA [29,79], 5-nitrophenol-(2-azo-1')-2'-(β-acetylhydrazino)-naphthalene (NAAN) ($\varepsilon = 3\cdot 8 \times 10^4$ at 646 nm) [29,80], Azoazoxy BN [81], 5-(2-pyridylazo)-p-cresol, 5-(2-pyridylazo)-2-monoethylamino-p-cresol and 2-(5-bromo-2-pyridylazo)-5-diethylamino-m-phenol [82], 1-(5-chloro-2-pyridylazo)-2-naphthol (5-Cl-PAN) ($\varepsilon = 8\cdot 4 \times 10^4$ at 564 nm) [83], Sulpharsazen [29], and Arsenazo III [84].

The anionic zinc thiocyanate complex forms ion-pairs with basic dyes, extractable by e.g. diethyl ether, benzene or chloroform. Use has been made of Rhodamine B [85,86], Methyl Violet [87], Victoria Blue [88], and 6-methoxy-3-methyl-2-[4-(N-methylanilino)phenylazo]benzothiazolium chloride [89]. It is also possible to measure the absorbance of the gelatine-stabilized coloured suspensions given by the zinc thiocyanate complex with derivatives of antipyrine (chrome-pyrazoles) [90,91]. The complex with Malachite Green has $\varepsilon = 5–12 \times 10^4$ depending on the solvent used [92]. Oxine is also used for extractive determination of zinc. The presence of alkylamines or 1,10–phenanthroline facilitates the extraction [93–95].

Other organic reagents used include Xylenol Orange [29,96,97], Semixylenol Orange [98], Methylthymol Blue [99], Semimethylthymol Blue

[100], Na–DDTC [101], Murexide [102], diphenylcarbazone [103], tetraphenylporphine [104], and thiothenoyltrifluoroacetone [105].

Small amounts of metallic zinc in zinc oxide can be determined by indirect methods. Zinc is oxidized with dichromate (in H_2SO_4 and H_3PO_4 media) whereupon the non-reacted portion of Cr(VI) is determined spectrophotometrically with diphenylcarbazide [106]. It is also possible to dissolve the sample in dilute H_2SO_4 in the presence of iron(III) ions and 1,10-phenanthroline. The red Fe(II)–1,10-phenanthroline complex is formed in amount corresponding to the content of zinc in the zinc oxide analysed [107].

References

1. Chuiko, V. T. and D'yachenko, N. P., *Zh. Neorgan. Khim.* **7**, 903 (1962).
2. D'yachenko, N. P., *Tr. Komis. po Analit. Khim; Akad. Nauk SSSR* **15**, 271 (1965).
3. D'yachenko, N. P., Negrebetskaya, I. V. and Chuiko, V. T., *Zh. Analit. Khim.* **22**, 1425 (1967).
4. Babko, A. K. and Marchenko, P. V., *Tr. Komis. po Analit. Khim. Akad. Nauk SSSR* **9**, 65 (1958).
5. Marchenko, P. V., *Zavodsk. Lab.* **26**, 532 (1960).
6. Marczenko, Z., *Mikrochim. Acta* **1965**, 281.
7. Marczenko, Z., Mojski, M. and Kasiura, K., *Zh. Analit. Khim.* **22**, 1805 (1967).
8. Jones, G. B., *Anal. Chim. Acta* **11**, 88 (1954).
9. Stewart, J. A. and Bartlet, J. C., *Anal. Chem.* **30**, 404 (1958).
10. Kataev, G. A. and Shpaer, I. S., *Izv. Vyssh. Ucheb. Zaved., Khim. Khim. Tekhnol.* **7**, 891 (1964).
11. Minczewski, J. and Różycki, C., *Z. Anal. Chem.* **239**, 158 (1968).
11a. Różycki, C., Lachowicz, E. and Jodełka, J., *Chem. Anal. (Warsaw)* **19**, 639 (1974).
12. Nazarenko, V. A., Fuga, N. A., Flyantikova, G. V. and Esterlis, K. A., *Zavodsk. Lab.* **26**, 131 (1960).
13. Marchenko, P. V. and Voronina, A. I., *Ukr. Khim. Zh.* **35**, 652 (1969).
14. Scroggie, L. E. and Dean, J. A., *Anal. Chim. Acta* **21**, 282 (1959).
15. Uny, G., Mathien, C., Tardif, J. P. and Tran Van Danh, *ibid.* **53**, 109 (1971).
16. Mahlman, H. A., Leddicotte, G. W. and Moore, F. L., *Anal. Chem.* **26**, 1939 (1954).
17. Andrew, T. R. and Nichols, P. N. R., *Analyst* **90**, 161 (1965).
17a. McDonald, C. W. and Rhodes, T., *Anal. Chem.* **46**, 300 (1974).
18. Pietsch, R. and Pichler, E., *Mikrochim. Acta* **1961**, 914.
19. Schweitzer, G. K. and Clifford, F. C., *Anal. Chim. Acta* **45**, 57 (1969).
20. Hunter, J. A. and Miller, C. C., *Analyst* **81**, 79 (1956).
21. Kallmann, S., Steele, C. G. and Chu, N.Y., *Anal. Chem.* **28**, 230 (1956).
22. Maier, R. H. and Bullock, J. S., *Anal. Chim. Acta* **19**, 354 (1958).
23. Zvereva, M. N., *Zavodsk. Lab.* **24**, 387 (1958).
24. Nishimura, M. and Sandell, E. B., *Anal. Chim. Acta* **26**, 242 (1962).
25. Baggott, E. R. and Willcocks, R. G. W., *Analyst* **80**, 53 (1955).
26. Geilmann, W. and Neeb, R., *Angew. Chem.* **67**, 26 (1955).
27. Geilmann, W., Neeb, R. and Eschnauer, H., *Z. Anal. Chem.* **154**, 418 (1957).
28. Margerum, D. W. and Santacana, F., *Anal. Chem.* **32**, 1157 (1960).
29. Kamaeva, L. V., Podchainova, V. N. and Fedorova, N. D., *Zavodsk. Lab.* **37**, 258 (1971).
30. Schweitzer, G. K. and Honaker, C. B., *Anal. Chim. Acta* **19**, 224 (1958).
31. Starý, J. and Růžička, J., *Talanta* **8**, 296 (1961).
32. Subbaraman, P. R., Cordes, S. M. and Freiser, H., *Anal. Chem.* **41**, 1878 (1969).

33. D'yachenko, N. P. and Chuiko, V. T., *Tr. Komis. po Analit. Khim. Akad.Nauk SSSR* **14**, 303 (1963).
34. Hulanicki, A., *Chem. Anal. (Warsaw)* **11**, 1081 (1966).
35. Hulanicki, A. and Minczewska, M., *Talanta* **14**, 677 (1967).
36. Galik, A., *ibid.* **14**, 731 (1967).
37. McClellan, B. E. and Sabel, P., *Anal. Chem.* **41**, 1077 (1969).
38. Manita, M. D., *Tr. Komis. po Analit. Khim. Akad. Nauk. SSSR* **7**, 194 (1956).
39. Martin, A. E., *Anal. Chem.* **25**, 1853 (1953).
40. D'yachenko, N. P. and Chuiko, V. T., *Zavodsk. Lab.* **29**, 522 (1963).
41. Monnier, D. and Prod'hom, G., *Anal. Chim. Acta* **31**, 101 (1964).
42. Ott, W. L., MacMillan, H. R. and Hatch, W. R., *Anal. Chem.* **36**, 363 (1964).
43. Marczenko, Z. and Kasiura, K., *Chem. Anal. (Warsaw)* **10**, 449 (1965).
44. Marczenko, Z. and Kasiura, K., *ibid.* **9**, 87 (1964).
45. Marczenko, Z., Kasiura, K. and Krasiejko, M., *ibid.* **14**, 1277 (1969).
46. Jean, M., *Anal. Chim. Acta* **7**, 338 (1952).
47. Monnier, D. and Prod'hom, G., *ibid.* **30**, 358 (1964).
48. Różycki, C., *Chem. Anal (Warsaw)* **14**, 459 (1969).
49. Stanton, R. E., McDonald, A. J. and Carmichael, I., *Analyst* **87**, 134 (1962).
50. Koter, M., Krauze, A. and Bardzicka, B., *Chem. Anal. (Warsaw)* **10**, 1247 (1965).
51. Marczenko, Z., *ibid.* **9**, 1093 (1964).
52. Wilczewski, T., Kozera, F. and Dobrowolski, J., *ibid.* **18**, 137 (1973).
53. Wilczewski, T., *ibid.* **18**, 897 (1973).
54. Marple, T. L., Matsuyama, G. and Burdett, L. W., *Anal. Chem.* **30**, 937 (1958).
55. Committee on Methods for Analysis of Trade Effluents, *Analyst* **82**, 443 (1957).
56. Christie, A. A., Kerr, J. R. W., Knowles, G. and Lowden, G. F., *ibid.* **82**, 336 (1957).
57. Mills, E. V. and Brown, B. L., *ibid.* **89**, 551 (1964).
58. Francis, A. C. and Pilgrim, A. J., *ibid.* **82**, 289 (1957).
59. Duffield, W. D., *ibid.* **83**, 503 (1958).
60. Page, E. R., *ibid.* **90**, 435 (1965).
61. Westöö, G., *ibid.* **88**, 287 (1963).
62. Kägi, J. H. R. and Vallee, B. L., *Anal. Chem.* **30**, 1951 (1958).
63. Yoe, J. H. and Rush, R. M., *Anal. Chim. Acta* **6**, 526 (1952).
64. Rush, R. M. and Yoe, J. H., *Anal. Chem.* **26**, 1345 (1954).
65. Frierson, W. J., Rearick, D. A. and Yoe, J. H., *ibid.* **30**, 468 (1958).
66. McCall, J. T., Davis, G. K. and Stearns, T. W., *ibid.* **30**, 1345 (1958).
67. Platte, J. A. and Marcy, V. M., *ibid.* **31**, 1226 (1959).
68. Sadílková, M., *Mikrochim. Acta* **1968**, 934.
69. Matsui, H., *Anal. Chim. Acta* **66**, 143 (1973).
70. Anand, K. S., Dayal, P. and Anand, O. N., *Z. Anal. Chem.* **239**, 33 (1968).
71. Flaschka, H. and Weiss, R., *Microchem. J.* **14**, 318 (1969); **15**, 653 (1970).
72. Bykhovtsova, T. T., *Zavodsk. Lab.* **40**, 512 (1974).
73. Berger, W. and Elvers, H., *Z. Anal. Chem.* **171**, 255 (1959); **199**, 166 (1964).
74. Goldstein, G., Maddox, W. L. and Kelley, M. T., *Anal. Chem.* **46**, 485 (1974).
75. Kawase, A., *Talanta* **12**, 195 (1965).
76. Kawase, A., *Anal. Chim. Acta* **58**, 311 (1972).
77. Marshall, B. S., Telford, I. and Wood, R., *Analyst* **96**, 569 (1971).
78. Evans, W. H. and Sayers, G. S., *ibid.* **97**, 453 (1972).
79. Vysokova, N. N., Lukin, A. M. and Smirnova, K. A., *Zavodsk. Lab.* **34**, 930 (1968).
80. Kamaeva, L. V., Podchainova, V. N., Fedorova, N. D., Dunaevskaya, K. A. Zelichenok, S. L. and Dziomko, V. M., *Zh. Analit. Khim.* **25**, 1718 (1970).
81. Gorbenko, F. P. and Degtyarenko, L. I., *Tr. Vses. Nauchn.-Issled. Inst. Khim. Reakt.* No. 29, 69 (1966); *Anal. Abstr.* **15**, 3196 (1968).
82. Gusev, S. I., Nikolaeva, E. M. and Pirozhkova, E. A., *Zh. Analit. Khim.* **26**, 1740 (1971).

83. Shibata, S., Furukawa, M. and Sasaki, S., *Anal. Chim. Acta* **51**, 271 (1970).
84. Michaylova, V. and Yuronkova, L., *ibid.* **68**, 73 (1974).
85. Tvaroha, B. and Malá, O., *Mikrochim. Acta* **1962**, 634.
86. Slovák. Z. and Přibyl, M., *Collection Czech. Chem. Commun.* **31**, 1742 (1966).
87. Klyachko, I. R. and Vinogradova, A. D., *Zavodsk. Lab.* **24**, 540 (1958).
88. Pilipenko, A. T., Kish, P. P. and Zimomrya, I. I., *Ukr. Khim. Zh.* **37**, 186 (1971).
89. Kish, P. P. and Zimomrya, I. I., *Zavodsk. Lab.* **35**, 541 (1969); **36**, 526 (1970).
90. Zhivopistsev, V. P., Selezneva, E. A., Lipchina, A. P. and Bragina, Z. I., *Zh. Analit. Khim.* **21**, 28 (1966).
91. Zhivopistsev, V. P., Selezneva, E. A. and Bragina, Z. I., *Zavodsk. Lab.* **35**, 1156 (1969).
92. Kish, P. P., Zimomrya, I. I. and Zolotov, Yu. A., *Zh. Analit. Khim.* **28**, 252 (1973).
93. Umland, F. and Hoffmann, W., *Z. Anal. Chem.* **168**, 268 (1959).
94. Schweitzer, G. K. and van Willis, W., *Anal. Chim. Acta* **30**, 114 (1964).
95. Woodward, C. and Freiser, H., *Anal. Chem.* **40**, 345 (1968).
96. Študlar, K. and Janoušek, I., *Talanta* **8**, 203 (1961).
97. Ishihara, Y., Naniwa, T., Yokokura, S. and Uchida, S., *Japan Analyst* **17**, 991 (1968).
98. Yoshino, T., Murakami, S., Kagawa, M. and Araragi, T., *Talanta* **21**, 79 (1974).
99. Yoshino, T., Imada, H., Murakami, S. and Kagawa, M., *ibid.* **21**, 211 (1974).
100. Yoshino, T., Murakami, S. and Kagawa, M., *ibid.* **21**, 199 (1974).
101. Kress, K. E., *Anal. Chem.* **30**, 432 (1958).
102. Tolmachev, V. N. and Kirzhner, O. M., *Zh. Analit. Khim.* **13**, 430 (1958).
103. Einaga, H. and Ishii, H., *Analyst* **98**, 802 (1973).
104. Banks, V. V. and Bisque, R. E., *Anal. Chem.* **29**, 522 (1957).
105. Solanke, K. R. and Khopkar, S. M., *Bull. Chem. Soc. Japan* **46**, 3082 (1973).
106. Norman, V. J., *Analyst* **89**, 261 (1964).
107. Kruse, J. M., *Anal. Chem.* **43**, 771 (1971).

Chapter 61
ZIRCONIUM AND HAFNIUM

Zirconium (Zr, at.wt. 91·22) occurs exclusively in the +IV oxidation state; it is similar to titanium in its properties. In HNO_3 and $HClO_4$ solutions zirconium Zr^{4+} and zirconyl ZrO^{2+} ions occur. They tend to polymerize as the concentration increases. Hydrolysis of zirconium ions begins at pH 1–1·5. Zirconium hydroxide has practically no amphoteric properties. Zirconium forms stable complexes with fluoride, EDTA, and hydroxy acids.

Hafnium (Hf, at.wt. 178·49) is much the same as zirconium in chemical properties. It usually accompanies zirconium to the extent of 1·5–2%. The methods for separation and determination of zirconium outlined below, also apply for hafnium.

61.1 Methods of Separation

61.1.1 PRECIPITATION

For precipitation of zirconium as the hydroxide, iron(III) is used as scavenger [1,2]. Depending on the elements from which the zirconium is to be separated, $Zr(OH)_4$ is precipitated with either ammonia or an alkali. Zirconium is separated from titanium by precipitation in the presence of H_2O_2.

The phosphate is sparingly soluble (e.g. in $2M$ HCl). If zirconium phosphate is dissolved in oxalic acid and the solution made alkaline with sodium hydroxide, $Zr(OH)_4$ is precipitated (more sparingly soluble than the phosphate) [3]. However, Stanton [4] experienced difficulty with the dissolution step.

After fusion of the sample or of the ignited hydroxide or phosphate precipitate with Na_2CO_3 or $NaOH+Na_2O_2$, followed by the leaching of the melt with water, zirconium is in the precipitate, whereas phosphate, fluoride, and sulphate as well as As, V, Cr, Mo, W, and Al are in solution.

In acid solution zirconium gives sparingly soluble compounds with arsonic acids; this has been used for its separation from other elements [5–7]. Cupferron has been used to separate zirconium traces, with iron(III) as collector [8,9].

Zirconium (hafnium) traces can be coprecipitated with organic collectors [10,11].

61.1.2 Extraction

Extraction of zirconium with thenoyltrifluoroacetone (TTA) in xylene or benzene from 4–6M HCl or 3–4M HClO$_4$ medium enables it to be separated from many metal ions, including titanium, rare earth elements, thorium, aluminium, uranium, and iron(III) [12–16a]. TTA has also been used for fractional separation of zirconium from hafnium [17,18].

Zirconium can be selectively extracted with BPHA solutions in inert solvents [19–21].

Some authors propose the separation of Zr and Hf from other metals by extraction with TBP or solutions of dialkylphosphoric acids from nitric acid solutions [22–25]. Huré et al. [26] have separated Zr from Hf with nitric acid as medium and TBP as extractant.

Fischer and Pohlmann [27] have separated Zr from Hf by means of the thiocyanate complexes. Mention should also be made of the extraction of zirconium from HCl solutions by high molecular-weight amines [28].

61.1.3 Ion-Exchange

Anion-exchangers are used to separate zirconium as anionic complexes with fluoride [29,30], oxalate [31,32], ascorbate [29], sulphate [29], and chloride [15]. Most metals are not retained by the column; those that are can be separated from zirconium by use of appropriate eluents [30]. By way of illustration, the retained zirconium and titanium fluoride complexes are separated by eluting first titanium with 0·05M H$_2$SO$_4$ containing H$_2$O$_2$ and then zirconium with 4M HCl [29].

In separating zirconium and titanium from molybdenum, advantage is taken of the higher stability of the molybdenum oxalate complex, which remains on the column while zirconium and titanium are eluted with dilute H$_2$SO$_4$ [32]. The existence of the fairly stable zirconium sulphate complex, retained by Amberlite IRA-400, enables zirconium to be separated from Th, Ti, Fe, and Al [29].

Forsling [33] has separated zirconium from hafnium by sorbing the fluoride complexes on a strongly basic anion-exchanger and eluting with 0·22M HCl and 0·005M HF (zirconium is eluted first).

Zirconium (and hafnium) have been separated from other metals by means of strongly acidic cation-exchangers, use being made of hydrochloric [34], perchloric [35], nitric (+TOPO and methanol) [36], and formic acid [37] media. In formic acid medium metal ions form positive ions except for zirconium, which seems to produce anionic complexes [37].

Zirconium has been separated from hafnium on cation-exchangers by virtue of the differences in stability of their citrate [38,39] and formate complexes [40].

Blasius and Kynast [41] have separated hafnium from zirconium by means of a chelating resin.

61.2 Methods of Determination

Until quite lately Alizarin S was the only well-known reagent for spectrophotometric determination of zirconium. Nowadays many methods are known. The Alizarin S method is still of frequent use although it is inferior to many other methods in sensitivity. Exceptionally high sensitivity and selectivity are displayed by the Arsenazo III method. The method based on Pyrocatechol Violet differs from other methods in the relatively high pH of the reaction medium (pH 5·2).

Babko and Vasilenko [42] and other authors [43,44] have made comparative studies of the more important spectrophotometric methods for determining zirconium.

61.2.1 ALIZARIN S METHOD

Alizarin S (Alizarin Red S) reacts with zirconium (hafnium) ions in acid medium (pH 0·5–1·0) to form a purple-red compound which is sparingly soluble in water. In the presence of a protective colloid, it remains in the form of a perfectly dispersed suspension. The reagent is soluble, to give a yellow colour. In more acid media a 1:1 compound is mainly formed. As the pH increases, 1:2 and 1:3 complexes are produced. In the colorimetric method [45–48], 0·1–0·2M HCl (HNO$_3$ or HClO$_4$) medium is used. In such a medium the effect of other metals is fairly insignificant and the method exhibits considerable selectivity.

The sensitivity of the method is not high. The molar absorptivity at λ_{max} = 520 nm is 7·0 × 10^3 (a = 0·08). The colour develops within 15 minutes and persists for 45 minutes.

Moderate quantities (1–10 mg) of Ti, Al, Th, and Fe do not interfere at pH ~ 1. Larger amounts of iron(III) should be reduced to Fe(II), best of all with ascorbic acid. Fluoride, phosphate, and large quantitites of sulphate prevent the colour reaction.

Drăgulescu *et al.* [49] have extracted the coloured complex with butanol in the presence of trichloroacetic acid.

Zirconium has been determined by the Alizarin S method in nickel alloys [50], copper alloys [51], magnesium alloys [46,52], uranium alloys [53,54], thorium [55], titanium alloys [56], phosphate ores [3], rutile [4], and rocks, minerals and water [1].

The coloured complex of zirconium with Alizarin S has also been made the basis of differential spectrophotometric methods for determining zirconium [57,58].

Reagents

1. Alizarin S: 0·05% aqueous solution.
2. Standard zirconium solution: 1 mg/ml.
    (a) Dissolve 3·9 g of zirconyl chloride (hafnium-free), $ZrOCl_2.8H_2O$ in $2M$ HCl and dilute the solution with the acid to 1 litre. Standardize gravimetrically by precipitation with ammonia and ignition to ZrO_2. Dilute the zirconium solution with $2M$ HCl to a Zr concentration of exactly 1 mg/ml.
    (b) Dissolve 0·1000 g of metallic zirconium by heating with 10 ml of conc. H_2SO_4 and 1 ml of conc. HF to fumes in a platinum vessel. Allow to cool, rinse the walls of the vessel with water and evaporate again to white fumes. Cool the residue, add 25 ml of water, heat until clear, and dilute the solution with water in a volumetric flask to 100 ml. Working solutions are obtained by suitable dilutions of the stock solution with $1M$ HCl or HNO_3.
3. Gum arabic: 1% solution.

Procedure

To the sample solution ($0·1M$ with respect to HCl) in a 50-ml volumetric flask, containing not more than 500 μg of Zr, add 1 ml of gum arabic solution, 5 ml of Alizarin S solution, and $0·1M$ HCl to the mark. Mix and let the solution stand for 15 minutes. Measure the absorbance at 520 nm, using a reagent blank as reference.

61.2.2 PYROCATECHOL VIOLET METHOD

Pyrocatechol Violet (PV) (p. 51) yields stable coloured complexes with zirconium (hafnium) ions in acetate medium and in the presence of EDTA. This was utilized by Flaschka and Farah [59] for the spectrophotometric determination of zirconium. The Zr–PV system is rather complicated. Depending on the concentration ratio of zirconium and reagent (pH 5·0–5·4, EDTA), complexes with various ratios of Zr to PV occur in the solution, and the absorption maximum varies from ~ 520 to 620 nm. A blue complex (with absorption maximum at 620 nm) arises with a deficiency of PV. Pyrocatechol Violet itself at pH 5·0–5·4 is orange, with absorption maximum at 445 nm.

Under the conditions given below, the molar absorptivity is ~ $2·5 \times 10^4$ ($a = 0·27$).

In the absence of EDTA (at pH 5·2) the absorption maximum of the blue complex shifts to 650 nm [60,61]. This is proof that EDTA takes part in the formation of the zirconium complexes with PV. The determination of zirconium with Pyrocatechol Violet without EDTA is more sensitive but not so selective. The EDTA also masks interfering metals.

In the presence of EDTA no interference was noted with 5 mg of Fe, 5 mg of Al, 1 mg of Th, or 0·5 mg of Ti or rare earths. The concentration of EDTA (1–5 ml of 0·1M solution in 50 ml of sample solution) has no effect on the colour obtained, which increases in intensity for a few minutes then remains constant for a long time.

The Pyrocatechol Violet method is noteworthy in that the presence of sulphate even in considerable quantities does not interfere. Fluoride interferes, however.

Young and White [62] and others [8,63] combine the Pyrocatechol Violet method with the preliminary extraction of zirconium with TOPO in cyclohexane. Pyrocatechol Violet in ethanol is added to the extract. The complex thus formed has its absorption maximum at 655 nm ($\varepsilon = 4 \cdot 0 \times 10^4$).

Zirconium has been determined by the Pyrocatechol Violet method in steel [64,65], niobium [63], phosphate ores [66], titanium and its alloys [67], and cerium and lanthanum oxides [60].

Reagents

1. Pyrocatechol Violet: 0·04% aqueous solution.
2. Standard zirconium solution: 1 mg/ml (p. 612).
3. Acetate buffer: pH 5.2. Dissolve 27 g of anhydrous sodium acetate in 900 ml of water, add glacial acetic acid ($\sim$ 3 ml) to obtain pH 5·2, and make up the solution with water to 1 litre.

Procedure

To the acid sample solution containing not more than 80 μg of Zr, add 5 ml of 5% EDTA solution, water to $\sim$ 30 ml, and 5 ml of acetate buffer. Adjust the pH with ammonia (1+1) to 5·2 ± 0·2. Add 2 ml of Pyrocatechol Violet solution and then buffer solution (pH 5·2) to the mark in a 50-ml volumetric flask. Mix and allow the solution to stand for 10 minutes. Measure the absorbance at 530–550 nm, using a reagent blank as reference.

Note. A portion of sample solution which has been treated with $\sim$ 10 mg of ammonium fluoride after the colour has been developed (to form the colourless zirconium fluoride complex which is more stable than the zirconium Pyrocatechol Violet complex) may also be used as reference.

61.2.3 XYLENOL ORANGE METHOD

Xylenol Orange (XO) (p. 51) reacts in acid medium with zirconium (hafnium) ions to form a purple-red water-soluble 1:1 complex. The complex has been recommended as a basis for spectrophotometric determination of zirconium and hafnium [68–72].

The intensity of the colour obtained depends on the type and concentration of the acid in the solution. The most intense colour is obtained in 0·5–1M HClO$_4$ and 0·5–0·8M HCl media.

The molar absorptivity of the zirconium Xylenol Orange complex in 0·8M HClO$_4$ solution is $3 \cdot 5 \times 10^4$ ($a = 0 \cdot 38$) at 535 nm. The absorption maximum of Xylenol Orange in this medium is at 440 nm. The absorption spectra of both the reagent and the zirconium complex are shown in Fig.61.1.

In the determination of zirconium with Xylenol Orange, fluoride, phosphate, and oxalate interfere. Larger amounts of sulphate, depending on their concentration, decrease the colour intensity of the complex.

Under the conditions given below, the following amounts of the species shown were found not to interfere: 5 mg of Fe(II), 5 mg of Al, 1 mg of Ti, 0·5 mg of Th. Iron(III) should be reduced before the colour development.

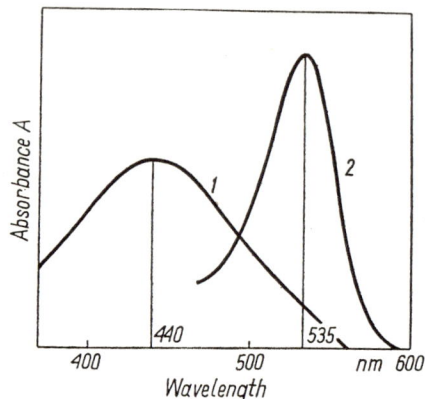

Fig. 61.1 Absorption spectra of Xylenol Orange (1) and its zirconium complex (2) in 0·8M HClO$_4$

Larger amounts of Bi, Sn(II), Mo, and Nb interfere. Xylenol Orange does not react with zirconium in the presence of hydrogen peroxide. Hafnium is incompletely masked with H_2O_2 which enables Zr to be distinguished from Hf in the Xylenol Orange method [68,73].

Cerrai and Testa [74] have extracted zirconium from 6–12M HCl medium with trioctylamine in benzene, and carried out the Xylenol Orange colour reaction in the organic phase. In the presence of diphenylguanidine the zirconium Xylenol Orange complex can be extracted with butanol [75].

Challis [76] has described the simultaneous determination of Zr and Hf with Xylenol Orange. By measurement of the absorbances at three different acidities, chosen such that both Hf and Zr complexes were dissociated to various degrees, the amounts of both metals could be determined.

Zirconium has been determined by the Xylenol Orange method in steels [6,77], niobium and tantalum compounds [78,79], high-temperature alloys [68], nickel alloys [76,80], ores [81], and table salt [82].

Reagents

1. Xylenol Orange: 0·05% aqueous solution.
2. Standard zirconium solution: 1 mg/ml (p. 612).
3. Ascorbic acid: 1% aqueous solution.

Procedure

Dissolve in hot ~ 2M HClO$_4$ or HCl the precipitate containing Zr(OH)$_4$ (either the precipitate which remains undissolved on leaching the sodium carbonate melt with water, or the precipitate obtained from the sample solution with sodium hydroxide). Dilute the cooled solution so as to make the concentration of $HClO_4$ therein $0.8\ (\pm 0.1)M$, or of HCl $0.6\ (\pm 0.1)M$.

Place the solution or an aliquot containing not more than 70 μg of Zr in a 50-ml volumetric flask, add 1 ml of ascorbic acid solution to reduce iron(III), and 2 ml of Xylenol Orange solution, dilute the solution with $0.8M$ $HClO_4$ or $0.6M$ HCl to the mark, and mix well. After 10 minutes measure the absorbance at 535 nm, using a reagent blank solution as reference.

61.2.4 ARSENAZO III METHOD

Zirconium (like hafnium) reacts with Arsenazo III (p. 48) in strongly acidic medium (2–10M HCl) to form an emerald green water-soluble complex. At this acidity hydrolysis and polymerization of zirconium ions no longer occur, which secures good reproducibility.

The sensitivity of the method depends very much on the acidity of the medium [43,83–87]. The maximum colour intensity can be obtained in 8–10M HCl (with an excess of Arsenazo III). As the HCl concentration is reduced, the colour intensity diminishes, in 2–4M HCl being only 33–50% of that obtained in 8–10M HCl. The complexes have a metal:reagent ratio of 1:2 if reagent is present in excess, and 1:1 if the metal is in excess. For accuracy it is essential to have the same concentration of acid in the sample and the standard solutions used for the calibration curve.

The molar absorptivity of the zirconium Arsenazo III complex in 9M HCl medium at $\lambda_{max} = 665$ nm is $\sim 1.2 \times 10^5$ (specific absorptivity 1.30). At this wavelength the excess of Arsenazo III does not absorb (see Fig. 61.2).

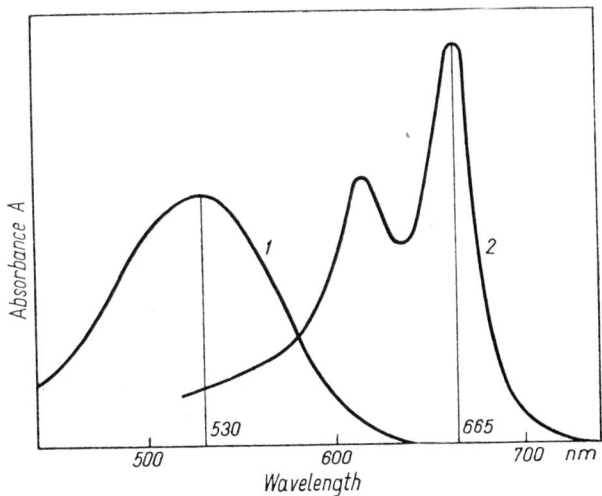

Fig. 61.2 Absorption spectra of Arsenazo III (1) and its zirconium complex (2) in 9M HCl

Over the acidity range 2-10M HCl only thorium and uranium(IV) interfere; other metals do not react with Arsenazo III in strongly acid medium. In most materials, therefore, zirconium can be determined in the solution immediately after the sample has been dissolved. If there is a need to concentrate the zirconium, titanium can be used as the collector in precipitation, since it does not interfere in the colour reaction.

Sulphate and phosphate have little effect on the colour reaction, but fluoride and oxalate must be absent.

Arsenazo III can be used to determine zirconium in hafnium [88]. The method is based on the fact that the molar absorptivities of the Zr and Hf complexes respond differently to change in acidity. The method is not very precise.

Zirconium has been determined by the Arsenazo III method in cast iron and steels [9,89–93], hafnium [88], various metal alloys [43], nickel alloys [94], uranium [20], molybdenum alloys [95,96], aluminium alloys [97], minerals [83], silicates [98], and zinc and cadmium salts [99].

Hafnium has been determined in steel [100].

Reagents

1. Arsenazo III: 0·05% aqueous solution.
2. Standard zirconium solution: 1 mg/ml (p. 612).

Procedure

To the acid sample solution in a 50-ml volumetric flask, containing not more than 40 μg of Zr, add 4 ml of Arsenazo III solution and 37 ml of conc. HCl, dilute with water to the mark, and mix well. Measure the absorbance at 665 nm, using a reagent blank as reference.

Notes. (1) On account of the high concentration of HCl in the sample solution the cuvettes should be suitably covered before being placed in the photometric apparatus.
(2) If the highest sensitivity is not needed, 6M HCl medium can be used.

61.2.5 OTHER METHODS

The most numerous group of reagents for spectrophotometric determination of zirconium is that of the azo dyes, but all of them are inferior to Arsenazo III in sensitivity and selectivity. Here mention may be made of Arsenazo I [101–103], which can be used to determine both Zr and Hf by variation of acidity [102], Thoron I [15,104], Chlorophosphonazo III [105], Stilbazogall I and Stilbazogall II [106], Stilbazo [107], PAN [23,25,108], Chlorosulphophenol S [109], Nitrosulphophenol S ($\varepsilon = 3.9 \times 10^4$, can be used in the presence of Th^{4+} and UO_2^{2+}) [110], Picramine R [110], Picramine CA [111], Picramine-epsilon (molar absorptivity 3.73×10^4 at 540 nm) [112,113], Picraminazo H [114,115], Picraminazochrome [115,116], SPADNS [117], Solochrome Violet R [118], Solochrome Black RN and Solochrome Dark Blue B [119], 2,2′,4′-trihydroxyazobenzene-5-sulphonic acid [120], and 2′-bromo-4′,5′-dihydroxyazobenzene-4-sulphonic acid [121]. It is also possible to precipitate zirconium with *p*-(*p*-dimethylaminophenylazo) benzenearsonic acid, then to decompose the precipitate with fluoride, and to measure the absorbance of free reagent [122].

Some very sensitive spectrophotometric methods are based on fluorones, namely phenylfluorone [2,123,124], m-nitrophenylfluorone [125,126], methylfluorone [127], and 2′-quinolylfluorone [128]. They are, however, not so selective as the Arsenazo III method.

Flavone dyes can also be used for spectrophotometric determination of zirconium, viz. morin (I) [129–131], quercetin (II) [5,132], robinetin

[133], quercetagetin [134], flavonol and myricetin [135]. The zirconium compounds formed with flavone reagents or the fluorones are water-insoluble.

Besides Xylenol Orange and Pyrocatechol Violet (discussed above) other triphenylmethane dyes have been proposed for zirconium determination, viz. Eriochrome Cyanine R [7], Methylthymol Blue [73], Semixylenol Orange [136], Chrome Azurol S [137], and Solochrome Azurine BS [138].

Several other organic reagents have been used for spectrophotometric zirconium determination. Some of them are worth notice, in particular Bromopyrogallol Red [139], Zirconin (Gallocyanine MS) ($\varepsilon = 4\cdot 0 \times 10^4$ at 625 nm) [140,141], purpurogallin [142], 8-hydroxyquinoline ($\varepsilon = 1\cdot 4 \times 10^4$ at 390 nm, xylene) [21,32,143], indoferron [144], 3-thianaphthenoyltrifluoroacetone [145], Phthalexone S [146], chloranilic acid [147,148], haematoxylin [149], and quinalizarin sulphonic acid [150]. An extractive photometric method using benzoic acid and Rhodamine B has been suggested for determining zirconium in the presence of hafnium, the latter being masked with fluoride [151].

Several absorptiometric methods have been proposed on the basis of heteropoly acids. They use molybdozirconic acid and its blue reduced form [152,153], molybdophosphozirconic acid [154], and the corresponding blue form [155], and reduced molybdosulphatozirconic acid [156,157]. Methods have also been developed for determining hafnium, use being made of the Hf-Mo heteropoly acid reduced with stannous oxalate ($\varepsilon = 6\cdot 7 \times 10^3$ in aqueous solution and $7\cdot 7 \times 10^3$ at 750 nm in n-butanol) [158,159]. Hafnium has also been determined by means of its reaction with molybdophosphoric acid [160].

References

1. Degenhardt, H., *Z. Anal. Chem.* **153**, 327 (1956).
2. Nazarenko, V. A. and Poluektova, E. N., *Zavodsk. Lab.* **28**, 656 (1962).

3. Kononenko, L. I. and Poluektov, N. S., *ibid.* **25**, 1050 (1959).
4. Stanton, N. B., *Analyst* **93**, 802 (1968).
5. Grimaldi, F. S. and White, C. E., *Anal. Chem.* **25**, 1886 (1953).
6. Čechová, D., *Chemist-Analyst* **56**, 94 (1967).
7. Mather, D. M., Millar, F. and Pollock, A. F., *Analyst* **96**, 393 (1971).
8. Hibbits, J. O., Rosenberg, A. F., Williams, R. T. and Kallmann, S., *Talanta* **11**, 1509 (1964).
9. Pakalns, P., *Anal. Chim. Acta* **57**, 51 (1971).
10. Zaikovskii, F. V., Furtova, E. V., Ivanova, V. N. and Sadova, G. F., *Zh. Analit. Khim.* **23**, 206 (1968).
11. Moroshkina, T. M. and Savinova, G. G., *ibid.* **24**, 1165 (1969).
12. Huffman, E. H., Iddings, G. M., Osborne, R. N. and Shalimoff, G. V., *J. Am. Chem. Soc.* **77**, 881 (1955).
13. Moore, F. L., *Anal. Chem.* **28**, 997 (1956).
14. Marsh, S. F., Maeck, W. J., Booman, G. L. and Rein, I. E., *ibid.* **33**, 870 (1961).
15. Sugawara, K. F., *ibid.* **36**, 1373 (1964).
16. Hála, J., *J. Inorg. Nucl. Chem.* **29**, 187, 1317, 1777 (1967).
16a. Onishi, H. and Sekine, K., *Talanta* **19**, 473 (1972).
16b. Erten, H. N., *J. Radioanal. Chem.* **14**, 343 (1973).
17. Huffman, E. H. and Beaufait, L. J., *J. Am. Chem. Soc.* **71**, 3179 (1949).
18. Schultz, B. G. and Larsen, E. M., *ibid.* **72**, 3610 (1950).
19. Fouché, K. F., *Talanta* **15**, 1295 (1968).
20. Vita, O. A., Levier, W. A. and Litteral, E., *Anal. Chim. Acta* **42**, 87 (1968).
21. Villarreal, R., Young, J. O. and Krsul, J. R., *Anal. Chem.* **42**, 1419 (1970).
22. Scadden, E. M. and Ballou, N. E., *ibid.* **25**, 1602 (1953).
23. Rolf, R. F., *ibid.* **33**, 125, 149 (1961).
24. Šraier, V. and Čakrt, E., *Collection Czech. Chem. Commun.* **29**, 2738 (1964).
25. Ross, L. E., Drabek, V. M. and Larsen, R. P., *Talanta* **16**, 748 (1969).
26. Huré, J., Rastoix, M. and Saint-James, R., *Anal. Chim. Acta* **25**, 1, 118 (1961).
27. Fischer, W. and Pohlmann, H. P., *Z. Anorg. Chem.* **328**, 252 (1964).
28. Sato, T. and Watanabe, H., *Anal. Chim. Acta* **49**, 463 (1970); **54**, 439 (1971).
29. Korkisch, J. and Farag, A., *Z. Anal. Chem.* **166**, 81, 170, 181 (1959).
30. Dixon, E. J., and Headridge, J. B., *Analyst* **89**, 185 (1964).
31. Bandi, W. R., Buyok, E. G., Lewis, L. L. and Melnick, L. M., *Anal. Chem.* **33**, 1275 (1961).
32. Shakashiro, M. and Freund, H., *Anal. Chim. Acta* **33**, 597 (1965).
33. Forsling, W., *Arkiv Kemi* **5**, 489, 503 (1953).
34. Strelow, F. W. E., *Anal. Chem.* **31**, 1974 (1959).
35. Dosch, R. G. and Conrad, F. J., *ibid.* **36**, 2306 (1964).
36. Korkisch, J. and Orlandini, K. A., *Talanta* **16**, 45 (1969).
37. Qureshi, M., Husain, W. and Israili, A. H., *ibid.* **15**, 789 (1968).
38. Benedict, J. T., Schumb, W. C. and Coryell, C. D., *J. Am. Chem. Soc.* **76**, 2036 (1954).
39. Bukhtiarov, V. E., *Zavodsk. Lab.* **34**, 1297 (1968).
40. Qureshi, M. and Husain, K., *Anal. Chem.* **43**, 447 (1971).
41. Blasius, E. and Kynast, G., *J. Radioanal. Chem.* **2**, 55 (1969).
42. Babko, A. K. and Vasilenko, V. T., *Zavodsk. Lab.* **27**, 640 (1961); *Zh. Analit. Khim.* **18**, 71 (1963).
43. Goryushina, V. G., Romanova, E. V. and Archakova, T. A., *Zavodsk. Lab.* **27**, 795 (1961).
44. Dedkov, Yu. M., Ryabchikov, D. I. and Savvin, S. B., *Zh. Analit. Khim.* **20**, 574 (1965).
45. Green, D. E., *Anal. Chem.* **20**, 370 (1948).
46. Wengert, G. B., *ibid.* **24**, 1449 (1952).
47. Gübeli, O. and Jacob, A., *Helv. Chim. Acta* **38**, 1026 (1955).

48. Zittel, H. E. and Florence, T. M., *Anal. Chem.* **39**, 320 (1967).
49. Drăgulescu, C., Simonescu, T. and Policec, S., *Talanta* **11**, 747 (1964).
50. Bach, B. B. and Francis, J. T., *Metallurgia* **62**, 281 (1960).
51. Stern, D. G., *ibid.* **71**, 51 (1965).
52. Mayer, A. and Bradshaw, G., *Analyst* **77**, 476 (1952).
53. Larsen, R. P., Ross, L. E. and Kesser, G., *Talanta* **4**, 108 (1960).
54. Buchanan, R. F., Hughes, J. P. and Bloomquist, C. A., *ibid.* **6**, 100 (1960).
55. Silverman, L. and Hawley, D. W., *Anal. Chem.* **28**, 806 (1956).
56. Wood, D. F. and McKenna, R. H., *Analyst* **87**, 880 (1962).
57. Manning, D. L. and White, J. C., *Anal. Chem.* **27**, 1389 (1955).
58. Freund, H. and Holbrook, W. F., *ibid.* **30**, 462 (1958).
59. Flaschka, H. and Farah, M. Y., *Z. Anal. Chem.* **152**, 401 (1956).
60. Young, J. P., French, J. R. and White, J. C., *Anal. Chem.* **30**, 422 (1958).
61. Vladimirova, L. M., Yagodin, G. A. and Chekmarev, A. M., *Zh. Analit. Khim.* **22**, 1345 (1967).
62. Young, J. P. and White, J. C., *Talanta* **1**, 263 (1958).
63. Wood, D. F. and Jones, J. T., *Analyst* **90**, 125 (1965).
64. Staats, G. and Brück, H., *Z. Anal. Chem.* **230**, 271 (1967).
65. Ratcliffe, D. B. and Byford, C. S., *Anal. Chim. Acta* **58**, 223 (1972).
66. Chernikhov, Yu. A., Luk'yanov, V. F. and Knyazeva, E. M., *Zh. Analit. Khim.* **14**, 207 (1959).
67. Chernikhov, Yu. A., Dobkina, B. M. and Petrova, E. I., *Zavodsk. Lab.* **26**, 529 (1960).
68. Cheng, K. L., *Talanta* **2**, 61, 186, 266; **3**, 81 (1959).
69. Babko, A. K. and Shtokalo, M. I., *Ukr. Khim. Zh.* **27**, 566 (1961).
70. Champion, P. M., Crowther, P. and Kemp, D. M., *Anal. Chim. Acta* **36**, 413 (1966).
71. Koźlicka, M., *Chem. Anal. (Warsaw)* **13**, 1117 (1968).
72. Karlysheva, K. F., Koshel', A. V. and Vashul', L. F., *Ukr. Khim. Zh.* **38**, 493 (1972).
73. Cheng, K. L., *Anal. Chim. Acta* **28**, 41 (1963).
74. Cerrai, E. and Testa, C., *ibid.* **26**, 204 (1962).
75. Tolmachev, V. N., Gol'tsberg, I. M. and Konkin, V. D., *Zh. Analit. Khim.* **22**, 950 (1967).
76. Challis, H. J. G., *Analyst* **94**, 94 (1969).
77. Keller, H., and Hennesen, K., *Arch. Eisenhüttenw.* **39**, 921 (1968).
78. Elinson, S. V. and Nezhnova, T. I., *Zavodsk. Lab.* **30**, 396 (1964).
79. Nevzorov, A. N. and Ganenko, Z. G., *ibid.* **33**, 285 (1967).
80. Ilina, L. I., *ibid.* **39**, 410 (1973).
81. Luk'yanov, V. F. and Knyazeva, E. M., *Zh. Analit. Khim.* **16**, 248 (1961).
82. Kröller, E., *Z. Anal. Chem.* **226**, 199 (1967).
83. Goryushina, V. G. and Romanova, E. V., *Zavodsk. Lab.* **26**, 415 (1960).
84. Savvin, S. B., *Talanta* **8**, 673 (1961); **11**, 1, 7 (1964).
85. Mel'chakova, N. V., Stanislavskaya, M. N. and Peshkova, V. M., *Zh. Analit. Khim.* **19**, 701 (1964).
86. Pakalns, P., *Anal. Chim. Acta* **44**, 73 (1969).
87. Muk, A. and Savvin, S. B., *Zh. Analit. Khim.* **26**, 98 (1971).
88. Elinson, S. V. and Mirzoyan, N. A., *Zavodsk. Lab.* **27**, 798 (1961).
89. Savvin, S. B., Kadaner, D. S. and Ryabova, A. S., *Zh. Analit. Khim.* **19**, 561 (1964).
90. Kammori, O., Takuchi, I. and Komiya, R., *Japan Analyst* **14**, 106 (1965).
91. Savvin, S. B., Dedkov, Yu. M. and Romanov, P. N., *Zh. Analit. Khim.* **22**, 65 (1967).
92. Busev, A. I., Kharlamov, I. P. and Byzova, E. P., *Zavodsk. Lab.* **38**, 1433 (1972).
93. Ashton, A., Fogg, A. G., and Burns, D. T., *Analyst* **99**, 108 (1974).
94. Onishi, H. and Sekine, K., *Anal. Chim. Acta* **62**, 204 (1972).
95. Polyak, L. Ya., *Zavodsk. Lab.* **32**, 1317 (1966).

96. Dupraw, W. A., *Talanta* **19**, 807 (1972).
97. Mustafin, I. S., Shchukina, V. S. and Malinina, I. V., *Zh. Analit. Khim.* **21**, 1136 (1966).
98. Šulcek, Z., Kremer, J. and Doležal, J., *Collection Czech. Chem. Commun.* **34**, 1720 (1969).
99. Chmilenko, F. A., Smirnaya, V. S. and Chuiko, V. T., *Zh. Analit. Khim.* **28**, 2176 (1973).
100. Kammori, O., Taguchi, I. and Komiya, R., *Japan Analyst* **14**, 249 (1965).
101. Kuznetsov, V. I., Budanova, L. M. and Matrosova, T. V., *Zavodsk. Lab.* **22**, 406 (1956).
102. Kononenko, L. I., Lauer, R. S. and Poluektov, N. S., *Ukr. Khim. Zh.* **25**, 633 (1959).
103. Shibata, S., Ishiguro, Y. and Matsumae, T., *Anal. Chim. Acta* **23**, 384 (1960).
104. Horton, A. D., *Anal. Chem.* **25**, 1331 (1953).
105. Fadeeva, V. I. and Alimarin, I. P., *Zh. Analit. Khim.* **17**, 1020 (1962).
106. Cherkesov, A. I. and Pushinov, Yu. V., *ibid.* **20**, 665 (1965).
107. Toshio, O., *Japan Analyst* **20**, 1132 (1971).
108. Crawley, R. H. A., *Anal. Chim. Acta* **26**, 281 (1962).
109. Dedkov, Yu. M., Kadaner, D. S., Pisarenko, N. D., Ryabova, A. S. and Savvin, S. B., *Zavodsk. Lab.* **30**, 654 (1964).
110. Savvin, S. B. and Dedkov, Yu. M., *ibid.* **30**, 645 (1964).
111. Goyal, S. S. and Tandon, J. P., *Talanta* **15**, 895 (1968).
112. Dedkov, Yu. M., Ermakov, A. N. and Korsakova, N. V., *Zh. Analit. Khim.* **25**, 1912 (1970); *Zavodsk. Lab.* **37**, 1411 (1971).
113. Dymova, M. S., Rybina, T. F., Yakovlev, R. Ya., Fridlyandskaya, E. I. and Nelina, E. A., *ibid.* **39**, 1307 (1973).
114. Cherkesov, A. I. and Zadumina, E. A., *Zh. Analit. Khim.* **24**, 941 (1969).
115. Busev, A. I., Cherkesov, A. I. and Zadumina, E. A., *Izv. Vyssh. Ucheb. Zaved., Khim. Khim. Tekhnol.* **12**, 1649 (1969).
116. Zadumina, E. A. and Cherkesov, A. I., *Zavodsk. Lab.* **35**, 10 (1969).
117. Banerjee, G., *Anal. Chim. Acta* **16**, 62 (1957).
118. Korkisch, J. and Osman, M., *Z. Anal. Chem.* **171**, 107 (1959).
119. Korkisch, J., *ibid.* **182**, 253 (1961).
120. Fletcher, M. H., *Anal. Chem.* **32**, 1827 (1960).
121. Basargin, N. N. and Davydova, R. T., *Zh. Analit. Khim.* **29**, 275 (1974).
122. Eberle, A. R., Pinto, L. and Lerner, M. W., *Anal. Chem.* **34**, 1176 (1962).
123. Kimura, K. and Sano, H., *Bull. Chem. Soc. Japan* **30**, 80 (1957).
124. Zharovskii, F. G. and Pilipenko, A. T., *Zavodsk. Lab.* **23**, 1407 (1957).
125. Sano, H., *Talanta* **2**, 187 (1959).
126. Sano, H., *Bull. Chem. Soc. Japan* **32**, 299 (1959).
127. Majumdar, A. K. and Savariar, C. P., *Z. Anal. Chem.* **178**, 352 (1961).
128. Asmus, E. and Klank, W., *ibid.* **265**, 260, 267 (1973).
129. Tůma, H. and Tietz, N., *Collection Czech. Chem. Commun.* **23**, 142 (1958).
130. Tůma, H. and Kabický, V., *Talanta* **8**, 749 (1961).
131. Blank, A. B., Mirenskaya, I. I. and Eksperiandova, L. P., *Zh. Analit. Khim.* **28**, 1331 (1973).
132. Polyak, L. Ya. and Bashkirova, I. S., *ibid.* **19**, 842 (1964).
133. Katyal, M., Trikha, K. C. and Singh, R. P., *Z. Anal. Chem.* **230**, 107 (1967).
134. Katyal, M., Gupta, B. P., Bhardwaj, D. K. and Singh, R. P., *Anal. Chim. Acta* **42**, 173 (1968).
135. Hörhammer, L., Hänsel, R. and Hieber, W., *Z. Anal. Chem.* **148**, 251 (1955).
136. Olson, D. C. and Margerum, D. W., *Anal. Chem.* **34**, 1299 (1962).
137. Mustafin, I. S. and Shchukina, V. S., *Zh. Analit. Khim.* **21**, 309 (1966).
138. Tandon, S. N., Sharma, C. L., Gill, J. S. and Sabharwal, A. K., *ibid.* **28**, 382 (1973).

139. Sakai, T. and Funaki, Y., *Bull. Chem. Soc. Japan* **42**, 2272 (1969).
140. Mustafin, I. S. and Shchukina, V. S., *Zavodsk. Lab.* **33**, 12, 294 (1967); *Zh. Analit. Khim.* **22**, 1338 (1967).
141. Mustafin, I. S. and Shchukina, V. S., *Tr. Komis. po Analit. Khim. Akad. Nauk SSSR* **17**, 345 (1969).
142. Alimarin, I. P., Puzdrenkova, I. V. and Dol'nikova, S. Ya., *Zh. Analit. Khim.* **17**, 700 (1962).
143. Van Santen, R. T., Schlewitz, J. H. and Toy, C. H., *Anal. Chim. Acta* **33**, 593 (1965).
144. Sakai, T., *Bull. Chem. Soc. Japan* **43**, 3171 (1970).
145. Johnston, J. R., Holland, W. J. and Gerard, J., *Mikrochim. Acta* **1972**, 608.
146. Cherkesov, A. I., Zadumina, E. A. and Arzamastseva, S. F., *Zh. Analit. Khim.* **28**, 1513 (1973).
147. Frost-Jones, R. E. U. and Yardley, J. T., *Analyst* **77**, 468 (1952).
148. Hahn, R. B. and Johnson, J. L., *Anal. Chem.* **29**, 902 (1957).
149. Teketatsu, T., *J. Chem. Soc. Japan, Pure Chem. Sect.* **74**, 1011 (1953).
150. Culkin, F. and Riley, J. P., *Anal. Chim. Acta* **32**, 197 (1965).
151. Bel'tyukova, S. V., Poluektov, N. S. and Meshkova, S. B., *Zh. Analit. Khim.* **27**, 191 (1972).
152. Guyon, J. C. and Clowers, C. C., *Anal. Chim. Acta* **37**, 401 (1967).
153. Shakhova, Z. F. and Semenovskaya, E. N., *Vestn. Mosk. Univ. Khim.* **1968**, No. 2, 122.
154. Murata, K., Yokoyama, Y. and Ikeda, S., *Anal. Chim. Acta* **48**, 349 (1969).
155. Veitsman, R. M., *Zavodsk. Lab.* **26**, 927 (1960).
156. Dehne, G. C. and Mellon, M. G., *Anal. Chem.* **35**, 1382 (1963).
157. Clowers, C. C., Jr. and Guyon, J. C., *ibid.* **41**, 1140 (1969).
158. Shakhova, Z. F., Semenovskaya, E. N., Sokovikova, N. K. and Koval'chuk, V. A., *Zh. Analit. Khim.* **25**, 485, 490 (1970).
159. Alimarin, I. P., Semenovskaya, E. N. and Sokovikova, N. K., *ibid.* **26**, 126 (1971).
160. Clowers, C. C., Jr. and Guyon, J. C., *Mikrochim. Acta* **1969**, 989.

INDEX

NOTE. Individual reagents are collected in alphabetical order under the general heading *Reagents* in the index. Compounds are given the names used in the text, which in general are those used by the original authors, and the same compound may appear under various names (these are generally given in parentheses to facilitate cross-reference).

absorbance 7
absorptiometry 3
absorption coefficients 7
   laws 6
   spectrum 5, 32
absorptivity 7, 103
   and band-width 20
   and slit-width 21
   molar 7, 103
   specific 19, 103
accuracy 24
adsorption compounds 29
aluminium 110
   determination 111, 116
     Chrome Azurol S 115
     Eriochrome Cyanine R 113
     8-hydroxyquinoline 111
   separation 110
ammonia 391
   determination 391, 395
     indophenol method 392
     Nessler method 394
   separation 391, 393
amplification methods 30, 126
antimonomolybdophosphate 128
antimony 121
   chloride 122
   determination 123, 127
     iodide 125
     Rhodamine B 123
     starch-iodine 126
   hydride 122
   separation 121, 124, 125
arsenic 131
   determination 132, 138
     arsenomolybdenum blue 133
     Gutzeit method 136
     silver diethyldithiocarbamate 135
   separation 131, 134

arsenomolybdenum blue 133, 138
arsenovanadomolybdic acid 138
arsine 131, 136
auxochromic groups 5
azide determination 400

band-width and Beer's law 8
   and sensitivity 20
barium 499
   determination 501
   separation 182, 499
bathochromic shift 5, 6
Beer, 4, 6
Beer's law 6
   and band-width 8
   deviations 8, 9, 33
beryllium 145
   determination 143, 146
     Chrome Azurol S 143
     Eriochrome Cyanine R 145
   separation 145
bismuth 149
   determination 150, 155
     diethyldithiocarbamate 153
     dithizone 150
     iodide 152
     Xylenol Orange 154
   separation 149
blank tests 23, 34
boron 159
   determination 160, 165
     carmine 163
     curcumin 161
   separation 159, 163
borosalicylate 166
Bouguer 4, 6
bromate determination 174

Index

bromine 171
   determination 171, 173
      Phenol Red 172
   separation 171

cadmium 176
   determination 177, 179
      dithizone 177
   separation 176
caesium 105
   determination 108
   separation 105, 106
calcium 182
   determination 183, 187
      glyoxal bis(2-hydroxyanil) 183
      murexide 186
   separation 182, 187
calibration curves 32
carbon 191
   determination 195
   dioxide 192, 195
   disulphide 195
   monoxide 195
cementation 70, 74, 121, 150, 268, 281, 351, 432, 490
cerium 198, 438
   determination 199, 201, 440, 443
      Arsenazo I 440
      Arsenazo III 442
      formaldoxime 200
      8-hydroxyquinoline 199
   separation 198, 200, 438, 442
chelates 29
chlorate 210
chloride 204
   determination 204, 208
      Methyl Red 206
      turbidimetry 205
   separation 204, 205, 208
chlorine 204
   determination 204, 208
      Methyl Red 206, 209
      turbidimetry 205
   dioxide 210
   separation 204, 208
chlorite 210
chromate 217
chromium 213
   determination 215, 220
      chromate 217
      dichromate 217
      diphenylcarbazide 215
      EDTA 219
      pyrophosphate 220

chromium (*contd*)
   separation 213
chromophoric groups 5
chromyl chloride 215
cobalt 224
   determination 225, 231
      1-nitroso-2-naphthol 225
      2-nitroso-1-naphthol 225
      nitroso-R salt 227
      thiocyanate 229
   separation 224
colorimetric titration 10
colorimetry 3
   visual 10
colour 4
   and molecular structure 4
   of visible radiation 4
complexes, composition 31
copper 238
   determination 239, 247
      cuprizone 246
      cuproine 243
      dithiocarbamate 241
      dithizone 239
   separation 238
coprecipitation 67, 69
coprecipitants
   active carbon 329
   aluminium hydroxide 68, 70, 83, 142, 150, 214, 225, 267, 289, 300, 305, 422, 475, 567, 575, 601
   aluminium oxinate 555
   aluminium phosphate 575
   antimony 546
   antimony sulphide 358, 361
   arsenic 474, 521
   arsenic sulphide 275, 350, 358, 449
   barium chromate 71, 322, 500, 504
   barium fluoride 198
   barium oxalate 198
   barium sulphate 70, 214, 322, 538
   beryllium 546
   beryllium phosphate 198
   bismuth sulphide 530
   cadmium 369
   cadmium diethyldithiocarbamate 70
   cadmium sulphide 238, 306, 350, 547, 601
   calcium carbonate 322
   calcium fluoride 198, 439, 468, 537
   calcium hydroxide 421
   calcium oxalate 198, 438, 500
   cerium fluoride 537
   cerium iodate 537
   cinchonine 71

coprecipitants (*contd*)
  cobalt phosphate 70, 150
  copper sulphide 70, 72, 121, 149, 322, 350, 358, 432, 490, 547, 601
  p-dimethylaminoazobenzene 530
  iron(III) cupferronate 358, 555, 589, 609
  iron(II) hydroxide 490, 601
  iron(III) hydroxide 68, 69, 70, 110, 132, 142, 150, 198, 214, 275, 281, 289, 300, 329, 338, 380, 413, 422, 432, 442, 458, 475, 530, 537, 546, 555, 567, 575, 588, 609
  iron(III) oxinate 338
  lanthanum fluoride 198, 537
  lanthanum hydroxide 68, 70, 83, 110, 150, 214, 225, 238, 281, 289, 306, 309, 322, 338, 370, 530, 555
  lanthanum oxalate 198, 468
  lanthanum phosphate 72, 182, 329
  lead arsenate 358
  lead iodide 530
  lead phosphate 358
  lead sulphate 214, 475
  lead sulphide 238, 306, 413, 432, 490
  magnesium ammonium phosphate 132
  magnesium hydroxide 70, 225, 256, 338, 370, 380, 555, 601
  manganese dioxide (hydrous) 70, 121, 124, 132, 150, 267, 306, 358, 380, 500, 546, 549, 567
  manganese hydroxide 530, 555
  mercuric iodate 537
  mercuric sulphide 70, 238, 275, 322, 490, 601
  Methyl Orange 530
  Methyl Violet 71, 358, 567, 576
  Methyl Violet + iodide 150
  Methylene Blue 142, 289, 358
  molybdenum sulphide 121, 132, 547
  molybdophosphate 70, 275, 567
  nickel dimethylglyoximate 70, 71, 412, 416
  Nile Blue A 289
  niobic acid 482, 486
  palladium dimethylglyoximate 70, 71
  PAN 71
  rhenium sulphide 432
  selenium 432, 521
  silver chloride 296
  silver iodide 351
  silver sulphide 268, 322
  stannic sulphide 289
  strontium sulphate 322
  tannin 289, 358, 555, 567
  tellurium 281, 283, 413, 432, 474, 490

coprecipitants (*contd*)
  tellurous acid 121
  thallous chloride 490
  thionalide 132
  thorium hydroxide 358
  titanium hydroxide 70, 110, 142, 198, 575
  zinc hydroxide 214
  zinc phosphate 238
  zinc sulphide 70, 176, 238
  zirconium arsenate 555
  zirconium hydroxide 83, 110, 198, 380, 567
cupellation 282, 413, 431, 463
cyanate determination 194
cyanide 191
  determination 192, 193
  separation 191

dichromate 217
distillation
  ammonia 391, 393
  antimony 122
  arsenic 131, 421
  boron 159, 421
  fluorine 254, 258
  germanium 274, 421
  iodine 296
  lead 323
  manganese 339
  mercury 351
  osmium 403, 432
  rhenium 449
  ruthenium 432, 462, 464
  selenium 474, 476
  silicon 421, 481
  sulphur 504, 508
  thallium 530
  tin 546, 551
distribution coefficient 75
Duboscq colorimeter 11

electrolysis 73, 239, 323, 351
  mercury cathode 74, 110, 142, 182, 329, 421
  extinction 7
extraction 31, 75, 80
  efficiency 75
  reagents 77, 80
  solvents 76
  titration 44
filters 13
fire assay 282, 413, 431, 458, 463

fluorine 254
  determination 256, 262
    Alizarin Complexone 260
    Eriochrome Cyanine R 256
    sulphosalicylic acid 259
  separation 254, 258

gallium 267
  determination 268, 269
    Rhodamine B 268
  separation 267, 269
germanium 274
  determination 275, 277
    phenylfluorone 275
  separation 274, 277
germanomolybdenum blue 277
gold 281
  determination 282, 284
    bromide 283
    Rhodamine B 282
  separation 281, 283
Griess method 107, 108, 395
Gutzeit method 136

hafnium 609
  determination 611, 616
    Alizarin S 611
    Arsenazo III 615
    Pyrocatechol Violet 612
    Xylenol Orange 613
  separation 609
Hehner tubes 10
heteropoly acids (*see also* the individual compounds) 30, 62, 79, 80, 132, 133, 138, 147, 202, 277, 278, 421, 422, 423, 424, 481, 482, 484, 486, 487, 519, 526, 534, 542, 562, 571, 592, 617
hexachloro-osmate(IV) 406
hexafluoroarsenate 139
hydrazine determination 399
hydrogen peroxide determination 409
hydrogen sulphide determination 506, 512
hydroxylamine determination 399
hyponitrite determination 400
hypophosphite determination 426, 427
hypsochromic shift 5, 6

indium 288
  determination 289, 292
    bromo-oxine 290
    PAR 291

indium (*contd*)
  separation 288
indirect methods 30
interference filters 13
iodate determination 299
iodide determination 296, 298
iodine 296
  determination 296, 298
    extraction method 298
    starch-iodine 296
ion-association complexes 29, 78
ion-exchange (*see* separation of individual elements) 85
  capacity 91
  distribution coefficient 91
iridium 302
  determination 302, 303
    tin(II) bromide 302
  separation 302, 431, 457
iron 305
  determination 306, 315
    bathophenanthroline 311
    bipyridyl 309
    phenanthroline 309
    sulphosalicylate 313
    thiocyanate 307
  separation 305, 308
Irving reversion technique 22, 44
isopiestic distillation 34

Kjeldahl method 391

lakes 29
Lambert 4, 6
lanthanum, lanthanides, *see* rare earths
lead 322
  determination 323, 326
    dithizone 326
  separation 322
light filters 13
limits of determination 22
lithium 105
  determination 106
  separation 105, 106

magnesium 329
  determination 330, 334
    Eriochrome Black T 332
    Titan Yellow 330
  separation 329

manganese 338
  determination 339, 346
    formaldoxime 342
    PAN 344
    permanganate 339
  separation 338
masking agents 27, 28, 71
mercury 350
  cathode electrolysis 74, 110, 142, 182, 329, 421
  determination 351, 354
    dithizone 351
  separation 350
  mixed colour method 44
molar absorptivity 7, 103
mole, definition 22
molybdenum 358
  determination 360, 364
    dithiol 360
    thiocyanate 362
  separation 358, 363
molybdoarsenic acid 132, 133, 138, 421
molybdoceric acid 202
molybdocerophosphoric acid 202
molybdohafnic acid 617
molybdogermanic acid 277, 278
molybdophosphoric acid 147, 421, 422, 542, 617
molybdophosphovanadic acid 592
molybdophosphozirconic acid 617
molybdophosphozirconium blue 617
molybdosilicic acid 421, 481, 482, 486
molybdostannosilicic acid 487
molybdosulphatozirconic acid 617
molybdotitanic acid 562
molybdotitanophosphoric acid 562
molybdovanadogermanic acid 277
molybdovanadophosphoric acid 423, 424
molybdovanadosilicic acid 487
monochromators 14

Nessler cylinders 10
  method 394
  reagent 394, 399
nickel 369
  determination 370, 374
    dimethylglyoxime 370
    α-furildioxime 373
  separation 369, 373
niobium 389
  determination 381, 385
    Bromopyrogallol Red 383
    PAR 384
    thiocyanate 382

niobium (*contd*)
  separation 380
nitrate determination 397, 399
nitric oxide determination 399
nitrite determination 395, 396
nitrogen 391
  dioxide determination 399

optical density 7
osmium 403
  determination 404, 405
    diphenylcarbazide 404
  separation 403, 405, 431, 462
  tetroxide 403
oxygen 408
ozone determination 409

palladium 412
  determination 413, 416
    dithizone 415
    iodide 413
  separation 412, 414, 416, 431
perbromate determination 174
perchlorate determination 210
perchromic acid 220
periodate determination 299
persulphate determination 513
phosphates
  condensed 426
  determination 423
phosphine determination 426
phosphite determination 427
phosphoantimonomolybdenum blue 422
phosphomolybdenum blue 422, 534
phosphorus 421
  determination 422, 426
    molybdovanadophosphoric acid 424
    phosphomolybdenum blue 422
  separation 421, 424
phosphotungstomolybdenum blue 571
photoelectric apparatus 11, 14
photoemissive cells 12
photometric titration 16
photomultipliers 12
photovoltaic cells 11
platinum 431
  determination 433, 434
    stannous chloride 433
  separation 431
polychromatic light 8
polymetaphosphate determination 426
polynuclear complexes 29
polythionates, determination 512, 513

628   Index

potassium 105
  determination 107
  separation 105
precision 24
precipitation 67, 73
preconcentration 67
Pulfrich photometer 11
pyrophosphate determination 426

rare earths 438
  determination 440, 443
    Arsenazo I 440
    Arsenazo III 442
  separation 438, 442
reagent purity 34
reagents
  acenaphthenequinone monoxime 406, 435, 465
  acetophenone 475
  acetylacetone 77, 85, 86, 88, 111, 143, 176, 198, 214, 224, 248, 262, 267, 281, 289, 305, 316, 329, 359, 449 465, 562, 575, 588
  acetylacetone dioxime 247
  N-acetylsalicyloyl-N-phenylhydroxyl-amine 382, 562
  Acid Alizarin Black SN 247, 542
  Acid Chrome Blue K 188, 595
  Acid Chrome Dark-green C 270
  Acid Chrome Violet K 210, 386
  Acid Monochrome Green S 232
  Acid Yellow 2G 584
  Acridine Orange 519
  Adogen-364 267, 288
  ajatin 194
  Alberon, see Chrome Azurol S
  Alcian Blue 513
  Aliquat-336 79, 198, 214
  Aliquat-336-S-1 602
  alizarin 166, 262, 511
  Alizarin Complexone 117, 260, 583
  Alizarin (Red) S 62, 117, 166, 220, 233, 262, 293, 417, 444, 471, 542, 571, 583, 611
  alkylamines 605
  alkyl ketoximes 417
  alkylphosphonic acids, 557
  alkylphosphoric acids 267, 530, 557
  Allthiox 459
  4-allyl-2-methoxyphenol 409
  Aluminon 116, 147, 248, 316, 417, 444, 583
  Amaranth 262, 511
  Amberlite LA-1 79

reagents (contd)
  Amberlite LA-2 547
  o-aminoazo compounds 417
  4-aminoazobenzene 396
  5-amino-2-benzimidazolethiol 417, 459
  4-aminobenzenesulphonic acid 395
  m-aminobenzoic acid 406
  o-aminobenzoic acid 174
  2-amino-4-chlorobenzenethiol 364, 426, 487
  4-amino-4'-chlorobiphenyl 512
  2-amino-1-cyclopentene-1-dithio-carboxylic acid 375
  p-aminodimethylaniline 505
  aminodiphenylamine 534
  6-amino-4-hydroxy-2-mercapto-5-nitrosopyrimidine 231
  Aminomethylazo III 247
  2-amino-6-methylthio-4-pyrimidine carboxylic acid 496
  2-amino-8-naphthol-3,6-disulphonic acid 406
  2-amino-8-naphthol-6-sulphonic acid 465
  2-aminoperimidine 512
  p-aminophenol 299, 534
  o-aminophenoldithiocarbamic acid 220
  o-aminophenol-p-sulphonic acid 406
  1-aminopyrene 399
  8-aminoquinoline 396, 417
  ammonia 72, 83, 248
  n-amyl-2-pyridylketoxime 247
  aniline 574
  p-anisidine 534
  anthranilic acid 176, 285, 406, 435, 465, 538
  anthraquinone derivatives 62
  anthraquinone-2-sulphonate 408
  anthrarufin-2,6-disulphonic acid 471
  antipyrine 79, 174, 596
    derivatives 79, 180, 202, 605
    dyes 127, 166, 354, 454, 552
  2-antipyrylazo-5-diethylaminophenol 354, 534
  APANS, see Thoron I
  o-arsanilic acid 174
  Arsazen 326, 417
  Arsenazo I 45, 48, 116, 146, 156, 166, 188, 262, 292, 335, 386, 440, 471, 519, 542, 562, 576, 578, 595, 616
  Arsenazo II 542, 581
  Arsenazo III 45, 48, 106, 146, 179, 188, 417, 442, 471, 500, 501, 540, 581, 605, 615
  Arsenazo IV 542

Index

reagents (contd)
Arsenazo DAL 542
Arsenazo M 442
Arsenazokhimdu 49
Arsenazo-p-NO$_2$ 444
N-arylhydroxamic acids 595
ascorbic acid 561, 583
Astrazon Blue B 270
azide 248, 316, 418, 460, 583, 591
Azo-azoxy BN 83, 183, 185, 188, 247, 499, 605
Azo-azoxy PMP 500
Azo dyes 45
Azomethine H 166
Azonol A 232, 471
Azovan Blue 334
Azoxin-C 354
Azoxin-H 354
Azure C 166
barbituric acid 192, 208
barium chloranilate 511
  chloride 505
  iodate 511
  rhodizonate 511
Basic Blue K 299
bathocuproine 243
bathocuproinedisulphonic acid 245
bathophenanthroline 312
  iron(II) complex 409
bathophenanthrolinedisulphonic acid 313
5-benzamidoanthraquinone-2-sulphonic acid 107
benzhydrazide 300
benzidine 202, 210
benzidine-pyridine 173, 192, 194, 208
α-benzildioxime 375, 417
benzil mono-(2-pyridyl)hydrazone 233
benzohydroxamic acid 61, 346, 386, 562, 583, 594, 595
benzoic acid 583, 617
α-benzoinoxime (cupron) 247, 359, 363, 570, 588
benzophenylhydroxamic acid, see BPHA
benzoselenadiazole 417
benzotriazole 176, 369
benzoylacetanilide 562
benzoylhydroxamic acid, see benzohydroxamic acid
N-benzoyl-N-phenylhydroxylamine, see BPHA
o-(2-benzoylthiourido)benzoic acid 405
N-benzoyl-o-tolylhydroxylamine 594
2-benzylaminopyridine 382
N-benzylaniline 267, 288, 529, 538

reagents (contd)
Beryllon I–IV 146
Beryllon II 166
biacetyl monoxime 453
bianthronyl 399
Bindschedler's Green 354
2,2′-bipyridyl 62, 309, 355, 383, 444, 496
  iron(II) complex 180, 210, 453, 454, 534
2,2′-biquinolyl (cuproine) 238, 243
bis-acetaldehyde-oxalyldihydrazone 246
N,N'-bis(o-aminoacetophenone)-ethylenedi-imine 376
N,N'-bis(o-aminobenzylidene)ethylenediamine 233
2,7-bisazochromotropic acid derivatives 247, 417
1,1-bis(6-chloroanthraquinolyl)amine 166
bis-cyclohexamine-oxalylhydrazine (cuprizone) 246
N,N'-bis(3-dimethylaminopropyl)dithiooxamide 232, 435, 459, 465
bis(ethylacetoacetate)oxalyldihydrazone 246
N,N'-bis-(β-hydroxypropyl)-o-phenylenediamine 202, 299
  4-bromo derivative 299
bis-(2-methyl-2-pyridyl)glyoxal dihydrazone 248
1,3-bis[8′-mercaptotheophyllinyl-(7′)-propane 156
bismuth 426, 512
Bismuthiol I 416
Bismuthiol II 406, 416, 478, 524, 525
bis(neocuproine) copper(I) complex 298
bis-pyrazolone 395
2,3-bis(salicylideneamino)benzofuran 248
bithionol 316
biuret 233
BPHA 62, 77, 111, 122, 149, 202, 274, 316, 359, 380, 381, 382, 469, 547, 593, 610
Brilliant Green 52, 127, 166, 210, 270, 278, 285, 293, 299, 354, 426, 454, 519, 532, 552, 571, 583
Brilliant Yellow 334
bromide 63, 72, 83, 84, 121, 128, 131, 132, 149, 156, 177, 224, 248, 284, 288, 302, 316, 350, 354, 406, 418, 432, 435, 458, 463, 465, 522, 529, 546, 552
Bromobenzothiazo 179

reagents (*contd*)
  bromobenzothiazolylazocresol 179
  2'-bromo-4',5'-dihydroxyazobenzene-4-sulphonic acid 616
  2-bromo-4,5-dihydroxybenzene-(1-azo-1')-benzene-4'-sulphonic acid 270
  9-(5-bromo-2-hydroxyphenyl)-2,3,7-trihydroxy-6-fluorone 117
  bromo-oxine (*see also* dibromo-oxine) 58, 183, 290, 365, 517, 552, 583
  5-bromo-PAAP (5-Br-PAAP) 232, 326, 605
  bromo-PADAP 582
  Bromophenol Blue 311
  2-(5-bromo-2-pyridylazo)-5-diethylamino-*m*-phenol (5-bromo-PAAP, 5-Br-PAAP) 232, 326, 605
  Bromopyrogallol Red 52, 128, 156, 278 365, 383, 444, 471, 496, 542, 552, 562, 583, 617
  brucine 210, 396, 397, 399
  n-butylamine 490
  4-butylnioxime 375
  Butylrhodamine B (C or S) 53, 127 139, 269, 285, 454, 519, 526, 552
  butyric acid 538
  cacotheline 316
  Cadion 179, 354
  Cadion 2B 179
  Calcein 52, 188
  Calcichrome 116, 187, 232, 247, 335, 346, 375, 562, 595
  Calcion 187
  Calcon 188, 335
  Calmagite 116, 188, 335, 542
  Capri Blue 166, 270
  7-[α-(*o*-carbomethoxyanilino)benzyl]-8-hydroxyquinoline (CMAB-oxine) 156, 326, 335
  carbonate 72
  Carboxyarsenazo 501
  2-carboxy-2'-hydroxy-5'-sulphoformazylbenzene (Zincon) 247, 354, 605
  carboxylic acids 85
  4-carboxynioxime 375
  Carboxynitrazo 444
  1-(*o*-carboxyphenyl)-3-hydroxy-3-methyltriazene 562, 596
  2-(2-carboxypyridyl-3-azo)-chromotropic acid 116
  1-(4-carboxy-2-thiazolylazo)-2-naphthol 292
  carmine 160, 163, 444
  catechol 263, 364, 381
  cetylpyridinium bromide 444, 583

reagents (*contd*)
  cetyltrimethylammonium bromide (CTAB) 116, 365, 550
  chloramine-T 395
  chloranilic acid 166, 188, 262, 263, 326, 364, 427, 501, 617
  chloride 72, 83, 84, 87, 88, 121, 125, 131, 132, 149, 177, 224, 233, 248, 263, 267, 316, 350, 354, 418, 432, 457, 458, 463, 479, 526, 529
  2-chloro-5-cyano-3,6-dihydroxybenzoquinone 188
  chlorocyanoformazan 116
  chloro-oxine (*see also* dichloro-oxine) 58, 583
  4-(5-chloro-2-phenylazo)-1,3-diaminobenzene (5-Cl-PADAB) 232
  Chlorophosphonazo I 581
  Chlorophosphonazo III 49, 50, 188, 335, 444, 471, 500, 501, 542, 562, 582, 616
  Chlorophosphonazo DAL 444
  Chlorophosphonazo R 49
  2-chloropyridine 412
  1-(5-chloro-2-pyridylazo)-2-naphthol 179, 605
  Chlorosulphophenol S 616
  Chlorpromazine 156, 285
  Chromal Blue 146, 471
  chromate 534
  Chrome Azurol S (Alberon) 51, 115, 143, 144, 220, 233, 248, 262, 263, 270, 292, 316, 365, 376, 417, 444, 471, 542, 562, 583, 596, 617
  chrome-pyrazoles 605
  Chromotrope 2B 50, 166, 172, 444, 542
  Chromotrope 2C 116, 146, 542
  Chromotrope 2R 335, 444
  chromotropic acid 263, 316, 364, 399, 444, 558
    bisazo derivatives 247, 417
  Chromoxane Pure Blue B 116
  Chromoxane Violet R 116, 316
  Chrompyrazole 487
  Chrompyrazole I 285
  chrysazin 166
  N-cinnamoylphenylhydroxylamine 594
  citrate 72
  Clayton Yellow 330
  5-Cl-PADAB 232
  5-Cl-PAN 179, 605
  CMAB-oxine (7-[α-(*o*-carbomethoxyanilino)benzyl]-8-hydroxyquinoline) 156, 326, 335
  Cobaltone (*see* 1-nitroso-2-naphthol)

Index 631

reagents (contd)
 copper 195
 acetate 195
 diethyldithiocarbamate (Cu–DDTC) 144, 326, 354, 495
 neocuproine complex 80, 243
 pyridine 194
 o-Cresolphthalein Complexone (Metalphthalein) 500, 501
Crystal Violet 53, 127, 139, 166, 174, 180, 210, 270, 285, 298, 316, 354, 363, 426, 435, 454, 465, 496, 519, 534, 552, 576, 577, 583
cupferron 69, 77, 80, 81, 85, 122, 149, 267, 305, 306, 329, 358, 359, 380, 381, 439, 449, 547, 555, 562, 568, 575, 588, 609
cuprizone (bis-cyclohexanone-oxalylhydrazone) 246
cuproine 238, 243
Cupron (α-benzoinoxime) 247, 359, 363
Cuprotest 245
curcumin 139, 160, 161, 376
cyanate, 233
cyanide 72, 496, 512
C-cyano-N,N'-di(2-hydroxyphenyl)-formazan 271
cyclohexanone 479
1,2-cycloheptanedionedioxime 375
1,2-cyclohexanedionedioxime (nioxime) 375, 417
1,2,3-cyclohexanetrionetrioxime 232, 375
DAB (3,3'-diaminobenzidine) 220, 476, 596
Dalzin 496
DAM (diantipyrylmethane) 79, 80, 177, 202, 230, 267, 303, 308, 316, 375, 406, 413, 418, 435, 459, 522, 534, 560, 561, 576, 577
DAMM (diantipyrylmethylmethane) 156
DAPM (diantipyrylpropylmethane) 156, 284, 285, 302, 303, 406, 435, 450, 454, 457, 522, 526, 534, 571
 derivatives 596
DAXIM 248
DCTA 214, 219, 233, 248, 408
DDTC 54, 77, 122, 149, 153, 176, 182, 224, 232, 238, 241, 329, 350, 359, 364, 375, 412, 432, 490, 529, 555, 575, 576, 588, 601, 606
Devarda's alloy 391, 399
diacetyl 399
diacetyldioxime, see dimethylglyoxime

reagents (contd)
 diacetylmonoxime p-nitrophenylhydrazone 232
 dialkylphosphoric acids 556, 610
 diallyldithiocarbamate (Dalzin) 496
3,3'-diaminobenzidine (DAB) 220, 476, 596
3,4-diaminobenzoic acid 435, 465
diaminochrysazin 166
p-diaminodiphenylsulphone 399
2,3-diaminopyridine 465
o-dianisidine 195, 202, 220, 300, 304, 316, 410, 596
dianthrimide 62, 160, 165
1,1'-dianthrimide 278, 479, 512, 526
2,2'-dianthrimide 479
1,5-diantipyrinyl-3-cyanoformazan 471
diantipyrylmethane, see DAM
diantipyrylmethylmethane (DAMM) 156
diantipyrylpropylmethane, see DAPM
Diantipyrylazo 444
diazosulphanilic acid 426
2,3,8,9-dibenzo-4,7-dimethyl-5,6-dihydro-1,10-phenanthroline (Cuprotest) 245
dibenzoylmethane 316, 444, 577
dibenzyldithiocarbamate 56, 154, 242, 322
dibenzyldithio-oxamide 435
Dibromoarsenazo II 542
dibromogallic acid 386, 519
5,7-dibromo-8-hydroxyquinoline, see dibromo-oxine and bromo-oxine
dibromo-oxine 81, 270, 444, 471, 542, 568, 571, 591
dibromo-PAAP 534
dibromopyrogallolsulphophthalein 117
Dibromotichromin 562
di-n-butylarsinic acid 575
dibutyldithiocarbamate 56
di-n-butyldithiophosphoric acid 78
di-n-butylphosphoric acid 359
Dicarboxyarsenazo III 444
2,2'-dicarboxydiphenylamine (Vanadox A) 596
3,4-dichlorobenzyltriphenylphosphonium 406
2,7-dichlorochromotropic acid 316, 559
5,7-dichloro-8-hydroxyquinoline, see chloro-oxine and dichloro-oxine
dichloro-oxine 346, 471
2,6-dichlorophenolindophenol 512
dichromate 410, 512, 526
dicinchoninic acid 248

reagents (contd)
dicupral 247, 495
diethanoldithiocarbamate 242, 583
2-diethylaminoethanethiol 459
5-diethylamino-2-nitrosophenol 231
diethylammonium diethyldithiocarbamate 55, 132, 241, 339
diethyldithiocarbamate, see DDTC
diethyldithiophosphoric acid 454
diethylenetriamine 376
diethylenetriaminepenta-acetic acid 248
di-(2-ethylhexyl)dithiophosphoric acid 547
di-(2-ethylhexyl)phosphoric acid (HDEHP) 78, 556
dihexylamine 308
dihydroquercetin 364
2,4-dihydroxyacetophenone 263
3,4-dihydroxyanthraquinone-2-sulphonic acid 271
o,o'-dihydroxyazobenzene 335
3,4-dihydroxyazobenzene-2'-carboxylic acid 270
4,4'-dihydroxy-2,2'-biquinoline 245
o-dihydroxychromenols 278, 444
6,7-dihydroxy-2,4-diphenylbenzpyranol 571
6,7-dihydroxy-2,4-diphenylbenzopyri- lium chloride 364
2-(1,8-dihydroxy-3,6-disulpho-2-naphthylazo)-phenoxyacetic acid 262
1,8-dihydroxynaphthalene-3,6-disul- phonic acid, see chromotropic acid
4,7-dihydroxy-1,10-phenanthroline 315
1-(2,4-dihydroxyphenylazo)-2-naphthol- 4-sulphonic acid 270
α(2,4-dihydroxyphenylazo)pyridine 471
dihydroxyphenylfluorone 561
di(o-hydroxyphenylimino)ethane 188
o-dihydroxythiobenzhydrazide 435
1,5-di-(2'-hydroxy-3',5',6'-trichloro-phenyl)-3-acetylformazan 471
5,7-di-iodo-8-hydroxyquinoline, see di-iodo-oxine and iodo-oxine
di-iodo-oxine 471, 591
di-isoamyl methylphosphonate 439
β-diketones 69, 86
dilituric acid 107, 501
dimedone 193
   dioxime 232
2,2'-dimercaptodiethylsulphide 375
2,6-dimercapto-3,5-diphenylthiopyran-4-one 156
2,3-dimercaptopropanol 248

reagents (contd)
dimercaptothiopyrones 248
   derivatives 552
o-dimethoxybenzene 202
4,4'-dimethoxystilbene 409
p-dimethylaminoazobenzenearsonic acid 262
p-dimethylaminobenzaldehyde 399
p-dimethylaminobenzylidenerhodanine, see rhodanine
5-dimethylamino-2-nitrosophenol 231
p-(p-dimethylaminophenylazo)benzene- arsonic acid 616, 617
dimethylaminophenylfluorone 276, 519
2,4-dimethylbenzamidoxime 232
2,9-dimethyl-4,7-dihydroxy-1,10-phenanthroline 245
3,3'-dimethylene-4,4'-diphenylbi- quinolyl 245
dimethylglyoxime 62, 81, 107, 231, 247, 316, 369, 370, 376, 412, 414, 417, 453, 552
1,3-dimethyl-4-imino-5-oximinoalloxan (DAXIM) 248
3,3'-dimethylnaphthidine 208, 408, 596
3,5-dimethylpyrazole 248
Dimethylsulphonazo III 500
Dimethylsulphonazo DAL 500, 501
dinaphthizone 43, 353, 492, 603
N,N'-di(2-naphthyl)-p-phenylenediamine 304
di-2-naphthylthiocarbazone (dinaphth- zone) 43, 353, 492, 603
dinapthylthiourea 526
3,5-dinitrocatechol 552
3,4-dinitropyrocatechol 596
3,5-dinitropyrocatechol 571
dioctylamine 423
dioctylmethylamine 384, 601
dioxan 106
dioximes 77
diphenylamine 399
   sulphonate 409
diphenylbenzidine 596
diphenylboric acid 376
diphenylcarbazide 215, 247, 285, 404, 454, 501, 511, 526, 534
diphenylcarbazone 247, 271, 293, 304, 326, 354, 460, 465, 512, 606
diphenyldithiophosphoric acid 417
diphenylglyoxime 375
diphenylguanidine 270, 293, 303, 364, 372, 459, 562, 576, 614
diphenyliodonium 230
diphenylselenium oxide 460

Index

reagents (contd)
1,4-diphenylsemicarbazide 463
2,4-diphenylsemicarbazide 464
N,N'-diphenylthiocarbamohydroxamic acid 364
diphenylthiocarbazone, see dithizone
1,4-diphenylthiosemicarbazide 435, 454, 478, 526
2,4-diphenylthiosemicarbazide 453
diphenylthiourea 465, 526
diphenylthiovioluric acid 465
dipicolinate 346
dipicrylamine 107, 108, 534
1,2-di(4-pyridyl)ethylene 409
di(2-pyridyl)glyoxime 417
2,2'-dipyridyl-α-glyoxime 285, 316
2,2'-dipyridyl-β-glyoxime 316
2,2'-dipyridylketoxime (di-2-pyridyl-ketoxime) 232, 285, 316, 417
8,8'-diquinolyl disulphide 248
2,2'-diquinolylketoxime 232, 417
disodium maleonitrile dithiolate 596
disulphophenylfluorone 277, 561
3,6-disulpho-TAN 270, 292, 534
5,5'-dithiobis(2-nitrobenzoic acid) 512
dithiocarbamates 80, 81, 386, 534, 547
dithiofluorescein 194
dithiol (4-methyl-1,2-dimercapto-benzene) 360, 426, 454, 475, 496, 568
dithiolthione derivatives 417
β-dithionaphthoic acid 375
dithio-oxamide (see also rubeanic acid) 232, 375
   derivatives 375
dithiopyrilmethane 156
dithizone (diphenylthiocarbazone) 32, 40, 41, 43, 44, 45, 80, 81, 82, 106, 107, 149, 150, 176, 177, 224, 238, 239, 281, 285, 289, 293, 322, 323, 346, 350, 351, 369, 375, 415, 432, 434, 478, 491, 492, 511, 529, 531, 534, 552, 602
   assay 492
   preparation of solutions 492
   purification 44, 491
   standardization of solutions 45
ditolylthiourea 405
EDTA 72, 108, 219, 233, 248, 263, 315, 346, 365, 376, 417, 444, 465, 500, 512
EGTA 188
Eriochrome Azurol G 471
Eriochrome Black T 83, 188, 232, 332, 444, 542

reagents (contd)
Eriochrome Brilliant Violet B 147, 471
Eriochrome Cyanine R (Solochrome Cyanine R) 51, 83, 113, 145, 220, 256, 262, 270, 292, 316, 444, 460, 534, 596, 617
Eriochrome Green B 595
Eriochrome Red B 562, 595
Erio SE 188
ethyl-4,6-dihydroxy-5-nitrosonicotinate 316
N,N'-ethylene-bis(o-mercaptobenzamide) 232
ethylenediamine 248
ethylenediaminedipropionic acid 248
Ethyl Rhodamine B (Rhodamine 3B) 53, 519
ethylsulphonaphtholazoaminocresol 270
ethyl xanthate 132
eugenol 409
Fast Green FCF 208
Fast Green Sulphon Black S 146
ferricyanide 195, 410, 584
ferroin 80, 139, 166, 193, 194, 202, 209, 210, 232, 284, 285, 298, 409, 427
ferron (7-iodo-8-hydroxyquinoline-5-sulphonic acid) 117, 262, 315, 417, 596
ferrous sulphate 399
ferrozine 315
flavone 62
flavonols 62, 617
fluorescein 172, 299, 496
Fluorexone 52
fluoride 72, 84, 86, 87, 88
formaldehyde 509
formaldoxime 59, 316, 342, 375, 596
formamidinesulphinic acid 458
formic acid hydrazide 285
fuchsin(e) 172, 174, 454
α-furildioxime 62, 83, 210, 232, 247, 373, 417, 451
β-furildioxime 374
γ-furildioxime 374
α-furilmonoxime 232
N-furoylphenylhydroxylamine 61, 562, 594
2-furoyltrifluoroacetone 248
Gallein 293, 365, 551
gallic acid 364
Gallion 270, 292
Gallocyanine MS (Zirconin) 617
GBHA [glyoxal bis(2-hydroxyanil)] 83, 180, 183, 293, 471, 583

reagents (contd)
glycerol 233
Glycinecresol Red 220, 270
glycinedithiocarbamate 56, 154
Glycinethymol Blue 220
glyoxal bis(2-hydroxyanil), see GBHA
glyoxime 417
haematein 271
haematoxylin 117, 166, 262, 278, 386, 519, 552, 596, 617
haemoglobin 194
halides 72, 80, 88
HDEHP 78, 149, 308, 538, 575
4-heptanone oxime 417
heptoxime 375
hexafluoroacetylacetone 143
hexafluorotitanic acid 487
H-resorcinol 166
hyamine 230
Hydrazo T 562
Hydrazo II 562
hydrogen peroxide 63, 213, 263, 365, 386, 465, 497, 556, 583, 596
hydroquinone 571
hydroxamic acids 61, 346, 380
hydroxide 72
o-hydroxyacetophenone oxime 417
3-hydroxy-1,3-diphenyltriazine 534
p-hydroxydithiobenzoic acid 395
o-(2-hydroxy-5-dodecylphenylazo)-arsonic acid 143
o-(2-hydroxy-5-dodecylphenylazo)-benzoic acid 143
5-hydroxyflavone 583
hydroxylamine 595
derivatives 61
5-hydroxy-7-methoxyflavone 583
o-(2-hydroxy-5-methylphenylazo)-benzoic acid 143
2-hydroxy-3-naphthoic acid 316
6-hydroxy-1,7-phenanthroline 315, 596
2-(2-hydroxyphenyl)benzothiazoline 552
2-(o-hydroxyphenyl)benzothiazoline 326
2-(o-hydroxyphenyl)benzoxazole 176
2-hydroxyphenylfluorone 571
3-hydroxypyridine-2-thiol 316
8-hydroxyquinaldine 111, 147, 220, 346, 562
8-hydroxyquinoline (oxine) 56, 58, 69, 70, 77, 80, 81, 107, 108, 110, 111, 176, 180, 182, 188, 199, 220, 224, 233, 267, 270, 281, 289, 290, 302, 303, 305, 306, 322, 326, 329, 335, 338, 339, 346, 358, 365, 376, 381, 386, 409, 426, 444, 465, 471, 479, 522,

reagents (contd)
526, 529, 542, 547, 552, 555, 562, 568, 571, 582, 588, 590, 605, 617
8-hydroxyquinoline-N-oxide 304, 465
8-hydroxyquinoline-5-sulphonic acid 365
3-hydroxy-1-p-sulphonatophenyl-3-phenyltriazine 365, 417
2-(2-hydroxy-5′-sulphophenylazo)-chromotropic acid 335
o-hydroxythiobenzhydrazide 465
1-hydroxy-4-p-toluidinoanthraquinone 166
hypobromite 460
hypochlorite 395, 460
indanetrione 395
Indigo Carmine 408
indoferron 316, 471, 617
indo-oxine 542
iodide 63, 72, 84, 85, 88, 122, 125, 131, 132, 149, 152, 176, 180, 194, 209, 224, 278, 288, 322, 326, 350, 354, 409, 432, 435, 458, 479, 522, 526, 529, 533, 546, 547, 550, 552, 602
iodine, see starch-iodine
iodine pentoxide 195
7-iodo-8-hydroxyquinoline-5-sulphonic acid (ferron) 117, 262, 315, 417, 596
iodo-oxine (see 5,7-di-iodo-8-hydroxyquinoline) 591
iron(II) 409, 552
iron(III) 194, 209, 400, 426, 512, 526, 583, 584, 596, 605
isonicotinic acid hydrazide 299
isonicotinohydroxamic acid 595
isonitrosoacetophenone 376, 417
isonitrosoacetylacetone 417
isonitrosodimedone 231
isonitrosomalonylguanidine 231
4-isopropylnioxime 375
β-isopropyltropolone 316, 596
kaempferol (tetrahydroxyflavone) 128, 271, 293
Khimdu 562
lanthanum chloranilate 426
lead 426, 512
citrate 512
dibenzyldithiocarbamate 242
diethyldithiocarbamate 241
leuco-Berbelin Blue 408
leuco-Crystal Violet 304
leuco-Malachite Green 285, 409
leuco-phenolphthalein 409
Lumogallion 270, 292, 386, 471, 552
magnesium 426

Index 635

reagents (*contd*)
  Magneson I 334
  Magneson II 334
  Magneson IREA 334, 364, 386, 471, 605
  Magneson KhS 247, 571
  Magon 334, 335
    sulphonate 334, 335
  Malachite Green 52, 176, 230, 263, 270, 293, 426, 454, 519, 552, 601, 605
  maltol 596
  manganese(II) 408, 409
  manganate 410
  mannitol 87, 160
  meconic acid 316
  MEDTA 500
  Meldola's Blue 270, 519, 534
  mercaptoacetic acid, *see* thioglycollic acid
  N-mercaptoacetyl-*p*-anisidine 478
  2-mercaptobenzimidazole 281, 405, 465, 478
  o-mercaptobenzoic acid 417
  mercaptobenzothiazole 281, 299, 375, 406, 417, 459, 478, 526
  5-mercapto-3-(4-bromophenyl)-1,3,4-thiadiazole-2-thione 156
  β-mercaptocinnamic acid 375
    derivatives 375
  2-mercapto-4,5-dimethylthiazole 459
  β-mercaptohydrocinnamic acid 417
  5-mercapto-3-(2-naphthyl)-1,3,4-thiadiazole-2-thione 525
  5-mercapto-3-phenyl-1,3,4-thiadiazole-2-thione (Bismuthiol II) 524
  mercaptopropionic acid 364, 417
    derivatives 364
  2-mercaptopyridine-N-oxide 316
  mercaptoquinoline (thio-oxine) 58, 128, 139, 248, 281, 290, 302, 346, 364, 416, 454, 530, 568, 571
  5-mercapto-1,3,4-thiadiazolidine-2-thione 435
  Mercupral 354, 495
  mercury chloranilate 194, 209, 512
    p-dimethylbenzylidenerhodanine complex 193
    diphenylcarbazide complex 299
    diphenylcarbazone complex 209
    diphenylthiocarbazone (dithizone) complex 173, 193, 299
    Metalphthalein complex 194
    Methylthymol Blue complex 173, 194
    quinoline complex 194
    thiocyanate 209
  mesityl oxide 267, 588

reagents (*contd*)
  Metalphthalein (*o*-Cresolphthalein Complexone) 188, 354, 500, 501
  metamizol 316
  p-methoxybenzothiohydroxamic acid 149, 595
  6-methoxy-3-methyl-2-[4-(N-methylanilino)phenylazo]benzothiazolium chloride 605
  3-methyl-2-benzothiazolinone hydrazone (MBTH) 409
  4-methyl-1,2-dimercaptobenzene, *see* dithiol
  Methylene Blue 127, 166, 176, 194, 202, 210, 270, 278, 285, 299, 354, 408, 418, 512, 519, 534, 583, 601
  4,4'-methylenediaminediantipyrine 522
  methylfluorone 127, 364, 561, 617
  Methyl Green 270, 354, 454, 519
  2-methyl-3-hydroxy-γ-pyrone (maltol) 364, 596
  2-methyl-8-hydroxyquinoline, *see* 8-hydroxyquinaldine
  2-methylnicotinic acid salicylhydrazide 562
  4-methylnioxime 375, 453
  Methyl Orange 173, 208, 311
  5-methyl-PAN 582
  1'-(3-methyl-1-phenyl-5-pyrazolone-4-azo)-4'-nitrophenyl-2'-sulphonic acid 107
  6-methylpicolinic acid thioamide 248
  3-methyl-5-propylpyrrole-(2-azo-2')-phenol 232
  6-methylpyridyl-2-aldoxime 247
  methyl-2-pyridylketoxime 247, 316, 453
  Methyl Red 202, 206, 209, 409
  Methylthymol Blue 51, 116, 147, 156, 220, 262, 263, 270, 292, 316, 326, 335, 354, 386, 444, 471, 542, 562, 583, 596, 617
  Methyl Viologen 408
  Methyl Violet 53, 80, 127, 150, 166, 176, 218, 285, 298, 426, 454, 517, 534, 538, 601, 605
  molybdate 188, 479, 482, 512
  molybdenum blue 166
  molybdophosphoric acid 233, 248, 376, 386, 444
  molybdophosphotungstic acid 365
  molybdosilicic acid 552
  mono-2-ethylhexylphosphoric acid 359
  monomethylthionine 166
  morin 62, 166, 262, 271, 293, 316, 364, 471, 519, 542, 583, 596

reagents (contd)
  morpholine-N-dithiocarbamate 454
  morpholinium morpholine-N-dithiocarbamate 242
  morpholinium 3-oxapentamethylenedithiocarbamate 346
  p-(morpholino)-N-(4'-hydroxy-3'-methoxy)benzylideneaniline 406
  murexide 186, 187, 376, 471, 500, 606
  myricetin 617
  Naphthachrome Green G (phenoxydinaphthofuchsonedicarboxylic acid) 117, 147
  naphthalhydroxamic acid 188
  Naphthazarin 542
  2-naphthohydroxamic acid 595
  α-naphthol 534
  naphthoselenadiazole 417
  α-naphthylamine(1-naphthylamine) 395, 396, 596
  1-naphthylamine-7-sulphonic acid 396
  1-naphthylamine-3,5,7-trisulphonic acid 406, 465
  1-naphthylamine-4,6,8-trisulphonic acid 406
  neocuprizone [bis(ethylacetoacetate)-oxalyldihydrazone] 246
  neocuproine 80, 210, 243
    copper complex 399
  Neonickelone, see α-furildioxime
  Neothoron, see Arsenazo I
  Neutral Red 210, 298
  Nevazol NS 595
  nickelone, see α-benzildioxime
  nicotinamidoxime 232, 375
  nicotinic acid hydrazide 596
  nicotinohydroxamic acid 595
  Nile Blue 173
  Nile Blue A 166, 278, 454, 519
  nioxime (1,2-cyclohexanedionedioxime) 375, 417
  nitrate 85, 538, 574, 583
  nitrite 432
  Nitroanthranylazo 106, 107
  Nitrobromoarsenazo 471
  nitroethane 198
  5-nitro-2-furfurylidenesemicarbazone 354
  nitron 448
  Nitro-orthanilic S 500, 501
  5-nitrophenol-(2-azo-1')-2'-(β-acetylhydrazino)naphthalene (NAAN) 605

reagents (contd)
  p-nitrophenylazochromotropic acid 471
  p-nitrophenylazo-orcinol 146
  5-(3-nitrophenylazo)salicylic acid 335
  nitrophenylfluorone 277
  m-nitrophenylfluorone 617
  o-nitrophenylfluorone 364, 386
  p-nitrophenylfluorone 548
  2-(5-nitro-2-pyridylazo)-1-naphthol 247
  nitropyrocatechol 166
  nitroresorcinol monoethyl ether 231
  3-nitrosalicylic acid 231
  p-nitrosodiethylaniline 434
  p-nitrosodimethylaniline 303, 417, 434, 460, 465
  p-nitrosodiphenylamine 460
  3-nitroso-4-hydroxy-5,6-benzocoumarin 465
  3-nitroso-4-hydroxycoumarin (oximidobenzotetronic acid) 304, 406, 412, 435, 460, 465
  1-nitroso-2-naphthol (Cobaltone) 81, 224, 225, 316, 460, 465, 575
  2-nitroso-1-naphthol 83, 225, 417, 465
  2-nitroso-1-naphthol-4-sulphonic acid 229, 248
  nitrosophenols 316
  3-nitroso-2,6-pyridinediol 231, 406, 417, 465
  Nitroso-R salt 107, 108, 224, 227, 316, 406, 417, 460, 465, 534
  Nitrosulphophenol S 616
  NTA 417
  Orthanilic S 501
  oxalate 72, 182, 188
  oxamidoxime 232
  oximidobenzotetronic acid (3-nitroso-4-hydroxycoumarin) 303, 304, 406, 412, 435, 460, 465
  PAAC [5-(2-pyridylazo)-2-monoethylamino-p-cresol] 292
    bromo-derivatives 292
  PAAP [5-(2-pyridylazo)-5-diethyl-m-aminophenol] 534
  5-Br-PAAP (2-[(5-bromo-2-pyridyl)azo]-5-diethylaminophenol) 232
  5-Cl-PADAB (4-[(5-chloro-2-pyridyl)-azo]-1,3-diaminobenzene) 232
  PADAP [2-(2-pyridylazo)-5-diethylaminophenol] 582
  Palladiazo 417
  palladium α-furildioxime complex 194
    molybdenum mixture 195
    phenanthroline complex 195

reagents (*contd*)
  PAMB [5-(2-pyridylazo)-4-ethoxy-2-monoethylamino-1-methylbenzene] 292
  PAN [1-(2-pyridylazo)-2-naphthol] 45, 46, 83, 128, 179, 232, 247, 270, 292, 303, 305, 316, 344, 375, 386, 417, 435, 444, 460, 534, 552, 562, 582, 595, 605, 616
  o-α-PAN [2-(2-pyridylazo)-1-naphthol] 247
  PAQH (pyridine-2-aldehyde-2-quinolylhydrazone) 376
  PAR [4-(2-pyridylazo)resorcinol] 45, 46, 47, 128, 156, 179, 220, 232, 247, 270, 291, 316, 326, 346, 354, 364, 375, 384, 417, 421, 444, 460, 519, 534, 542, 552, 562, 582, 595, 605
  pararosaniline 409, 509
  peroxide (*see also* hydrogen peroxide) 72
  phenanthraquinone monoxime 232
  1,10-phenanthroline 62, 156, 194, 195, 309, 355, 365, 375, 382, 444, 465, 496, 512, 556, 583, 596, 605, 606
  1,10-phenanthroline-iron(II) complex (*see also* ferroin) 435
  phenazine 561
  Phenazo 334
  1-phenol-2,4-disulphonic acid 397
  phenolphthalin 194
  Phenol Red 172, 192
  phenoxydinaphthofuchsonedicarboxylic acid (Naphthachrome Green G) 117, 147
  2-phenoxyquinizarin-3,4'-disulphonic acid 117, 147
  phenylacetic acid 238, 575
  phenylanthranilic acid 202
  phenylarsonic acid 380
  N-phenylbenzohydroxamic acid 62, 575, 583
  o-phenylenediamine 285, 435, 478
  p-phenylenediamine 192
  phenylene thiourea (2-mercaptobenzimidazole) 281, 405, 465, 478, 526
  phenylfluorone 262, 275, 276, 293, 316, 364, 426, 487, 519, 547, 561, 596, 617
  phenylhydrazine 365
  phenylhydrazinedithiocarbamate 56
  phenylhydrazine-*p*-sulphonic acid 478
  1-phenyl-3-methyl-4-benzoyl-5-pyrazolone 183

reagents (*contd*)
  phenyl-2-(6-methylpyridyl)ketoxime 247
  phenyl-α-pyridylketoxime 285, 417
  phenyl-2-pyridylketoxime 315, 316, 375, 453
  5-phenylsalicylic acid 263
  1-phenyltetrazoline-5-thione 417, 435
  1-phenylthiosemicarbazide 232
  1-phenyl-2-thiourea 453
  phosphate 439
  phosphine oxides 399
  phosphomolybdenum blue 106, 108, 316, 359, 386
  Phosphonazo III 582
  Phosphonazo R 106
  Phthalein Violet 166
  Phthalexone S 617
  phthalocyanine tetrasulphonic acid 596
  picolinealdehyde-2-quinolylhydrazone 417
  Picraminazo 334
  Picraminazo H 595, 616
  Picraminazochrome 616
  Picramine CA 616
  Picramine R 247, 386, 616
  Picramine RG 270
  Picramine-epsilon 50, 247, 386, 616
  picric acid 108
  picrolonic acid 188
  picryl chloride 399
  piperazine-bis-dithiocarbamate 242
  piperidinedithiocarbamate 457
  polyethylene glycol 284
  polymethine dyes 334
  Pontachrome Azure Blue B (Solochrome Azurine BS) 316, 417, 583
  Pontacyl Violet 4BSN 335
  potassium ethyl xanthate 364
  propylene carbonate 423, 485
  propylfluorone 471
  Prussian blue 193
  Purpurin 278
  Purpurogallin 278, 617
  pyrazine-2,3-dicarboxylic acid 364
  pyrazolinedithiocarbamate 56, 154, 417, 454
  pyrazolone 192, 193, 395
  α-pyridildioxime 453
  pyridine 308, 417, 444, 448
    derivatives 364, 448
  pyridine-2-aldehyde-2-pyridylhydrazone 417
  pyridine-2-aldehyde-2-quinolylhydrazone 233, 376, 417
  2-pyridinealdoxime 232, 316, 417

reagents (*contd*)
2,6-pyridinediacetoxime 247, 375
2,6-pyridinediamidoxime 316
pyridine-2,6-dicarboxylic acid 220, 316, 496, 583, 596
pyridine-*N*-oxide 284
pyridyl-2-aldoxime (*see also* 2-pyridinealdoxime) 285
2-pyridylazo derivatives, 128, 417
 bromo-derivatives 386
pyridyl-2-azochromotropic acid 375
5-(2-pyridylazo)-*p*-cresol 605
2-(2-pyridylazo)-5-diethylaminophenol (PADAP) 582
5-(2-pyridylazo)-5-diethylaminophenol (PAAP) 534
5-(2-pyridylazo)-4-ethoxy-2-monoethylamino-1-methylbenzene (PAMB) 292
5-(2-pyridylazo)-2-monoethylamino-*p*-cresol (PAAC) 292, 605
 bromo-derivatives 292
pyridylazonaphthols, *see* PAN
pyridylazophenols 562
 bromo-derivatives 595
4-(2-pyridylazo)resorcinol, *see* PAR
3-(2-pyridyl)-5,6-diphenyl-1,2,4-triazine (PDT) 248
3′-pyridylfluorone 548
pyridylpyridinium dichloride 399
2-pyridylquinoxalines 245
2-pyridyl-2-thienyl-β-ketoxime 232
pyrocatechol 386, 406, 561, 571, 596
Pyrocatechol Violet 51, 116, 128, 156, 166, 220, 263, 270, 278, 292, 316, 365, 386, 444, 471, 549, 562, 571, 596, 612
pyrogallol 406, 408
Pyrogallol Red 52, 116, 496
pyrogallolsulphonic acid 364, 517, 562, 571, 583
Pyronine G 127
pyrophosphate 248, 346
2-pyrrolaldehyde-ethylenedi-imine 376
pyrrolidinedithiocarbamate 56, 128, 154, 242, 381, 583, 588
QAN [2-(2-quinolylazo)-1-naphthol] 605
QAQH (quinoline-2-aldehyde-2-quinolylhydrazone) 246
quercetagetin 617
quercetin 62, 139, 166, 271, 293, 316, 364, 471, 519, 534, 542, 552, 583
quinaldinohydroxamic acid 595

reagents (*contd*)
quinalizarin 62, 147, 160, 165, 278, 444, 512
quinalizarinsulphonic acid 617
quinisatinoxime 316, 406
2-quinizarinsulphonic acid 117
Quinolinazo 106
Quinolinazo R 232
quinoline-2-aldehyde-2-quinolylhydrazone (QAQH) 246
8-quinoline carboxylic acid 316
6-(2-quinolylazo)-3,4-dimethylphenol 128, 534
2-(2-quinolylazo)-1-naphthol (QAN) 605
2′-quinolylfluorone 617
N-8-quinolyl-4-toluenesulphonamide (QTS) 248
quinoxaline-2,3-dithiol 156, 232, 248, 375, 416, 435
Reinecke's salt 176
Rezarson 278, 364, 386
rhenium 194
 α-furildioxime complex 399
Rhodamine B (C, S) 53, 83, 123, 127, 166, 180, 188, 194, 268, 278, 282, 293, 316, 354, 444, 454, 486, 530, 542, 552, 571, 583, 601, 605, 617
Rhodamine 3B (Ethyl Rhodamine B) 53, 83, 519
Rhodamine G 127
Rhodamine 6G 53, 127, 166, 269, 278, 293, 346, 426, 519
rhodanine (*see also* *p*-dimethylaminobenzylidenerhodanine) 284, 354, 416, 435, 490, 493
 azo-derivatives 435
rhodizonate 500
rhodizonic acid 188
robinetin 617
rosaniline 173
Rose Bengal 243
Rose Bengal Extra 417
rubeanic acid (*see also* dithio-oxamide) 248, 332, 375, 406, 417, 465
 derivatives 375
rufigallol 147
ruthenium(III) chloride 395
saccharin 490
Safranine O 408
Safranine T 127, 408, 426, 454, 534
salicylaldoxime 247
salicylamidoxime 562
salicylate 238
salicylfluorone 293, 364, 444, 561

reagents (*contd*)
  salicylhydroxamic acid (salicylo-
    hydroxamic acid) 61, 386, 562, 595
  salicylic acid 259, 399, 444, 490, 561
  salicylidene-*o*-aminophenol (2-salicyl-
    ideneaminophenol) 117, 376
  salicylideneamino-2-thiophenol 552
  sebacic acid 538
  selenium(IV) 399
  selenourea 405, 526
  Semimethylthymol Blue 605
  Semixylenol Orange 52, 605, 617
  silicomolybdenum blue 263
  silicotungsten blue 108
  silver diethyldithiocarbamate 128, 135, 299
    dithizonate 139, 173, 194, 512
    phenanthroline-Bromopyrogallol complex 194
    p-sulphamidobenzoate (*p*-sulpha-
      moylbenzoate) 128, 139, 195
    thiocyanate 173
    thiofluorescein complex 194, 512
  SNADNS 542
  sodium tetrachloromercurate 509
  Solochrome Azurine BS (Pontachrome
    Azure Blue B) 147, 583, 617
  Solochrome Black AS 562
  Solochrome Black 6 BN 582
  Solochrome Black PV 335
  Solochrome Black RN 595, 616
  Solochrome Black WDFA 542
  Solochrome Cyanine R, *see* Eriochrome
    Cyanine R
  Solochrome Dark Blue B 616
  Solochrome Fast Grey 542, 595
  Solochrome Fast Grey RA 595
  Solochrome Fast Grey RAS 582
  Solochrome Fast Red 542, 582
  Solochrome Fast Red ERS 375
  Solochrome Violet R 364, 616
  SPADNS 116, 166, 262, 444, 538, 542, 616
  stannous halides 458
  starch 296, 299, 511, 533
  starch-iodine 126, 209
  Stilbazo 116, 166, 292, 444, 562, 571, 616
  Stilbazochrome 116, 471
  Stilbazogall I 116, 364, 571, 616
  Stilbazogall II 616
  sulphanilic acid 202, 395
  Sulpharsazen 49, 232, 354, 496, 605
  sulphide 72, 326
  sulphite 174
  Sulphoallthiox 459

reagents (*contd*)
  Sulphochlorophenol R 270, 292, 421
  Sulphochlorophenol S 50, 364, 385, 421
  Sulphochlorophenolazorhodanine 285, 435, 460
  Sulphochrome 116
  5-sulpho-4′-diethylamino-2,2′-
    dihydroxyazobenzene 116
  Sulphonazo 270, 292, 471, 595
  Sulphonazo III 500, 501
  Sulphonitrazo 364, 595
  Sulphonitrophenol K 364, 395
  Sulphonitrophenol M 50, 385, 417
  Sulphonitrophenol R 270
  Sulphonitrophenol S 386, 426
  Sulphonitrophenolazurin-1,4 270
  p-sulphophenylazosalicylic acid 146
  sulphosalicylate 427
  sulphosalicylic acid 313, 561, 583
  sulphuric acid 526
  TAAC [5-(2-thiazolylazo)-2-monoethyl-
    amino-*p*-cresol] 386
  TAM [2-(2-thiazolylazo)-5-dimethyl-
    aminophenol] 47, 247, 375, 582
  TAMP [*o*-(2-thiazolylazo)-4-methoxy-
    phenol] 354, 582
  TAN (1-thiazolylazo-2-naphthol) 47, 232, 247, 375, 417, 460, 605
  tannin 70, 274
  TAR [4-(2-thiazolylazo)resorcinol] 47, 232, 292, 364, 386, 406, 421, 460, 465, 519, 534, 542, 582, 605
    bromo-derivatives 232
  tartaric acid 316
  tartrate 72
  TBA, *see* tribenzylamine
  TBP (tributyl phosphate) 79, 85, 183, 198, 214, 267, 288, 305, 322, 350, 439, 448, 457, 468, 475, 490, 500, 521, 522, 526, 529, 556, 574, 580, 581, 588, 610
  TEHPO 547, 550
  terpyridyl (2,2′,2″-terpyridine) 233, 315, 496
  tetrabromochrysazin 166
  tetrabromofluorescein 444
    ethyl ester 435
  tetrabromopyrocatechol 562
  tetrabutylammonium 213, 285
  tetraethyleneglycol dimethyl ether 413
  tetraethylthiuram disulphide (dicupral) 247, 495
  3,5,7,4′-tetrahydroxyflavone (kaemp-
    ferol) 128, 271, 291
  tetrahydroxyphenazine 562

reagents (contd)
  tetraiodobismuthite 106
  tetramethyl-*p*-phenylenediamine 409
  N,N'-tetramethyl-*o*-tolidine (Tetron) 202, 285
  tetramethylthiuram sulphide 526
  tetraphenylarsonium 80, 159, 230, 267, 339, 340, 381, 382, 406, 418, 448, 450, 570
  tetraphenylborate 106, 107
  tetraphenylphosphonium 80, 303, 522
  tetraphenylporphines 606
  tetraphenylstibonium 256
  tetrapropylammonium 210, 574, 583, 595
  Tetron (*N,N'*-tetramethyl-*o*-tolidine) 202, 285
  thenoyltrifluoroacetone, *see* TTA
  3-thianaphthenoyltrifluoroacetone 202, 220, 248, 316, 583, 617
  Thiazole Yellow 330
  thiazolidine-2-thione derivatives 248
  2-(2-thiazolylazo)-5-dimethylamino-phenol (TAM) 47, 247, 375, 582
  thiazolylazo compounds 47, 128, 417
  o-(2-thiazolylazo)-4-methoxyphenol [2-(2-thiazolylazo)methoxyphenol, TAMP] 354, 582
  5-(2-thiazolylazo)-2-monoethylamino-*p*-cresol (TAAK) 386
  1-thiazolylazo-2-naphthol, *see* TAN
  4-(2-thiazolylazo)resorcinol, *see* TAR
  thioacetanilide 458
  thioacetic acid, *see* thioglycollic acid
  p-thiocresol 454
  thiocyanate 60, 83, 84, 156, 176, 229, 248, 262, 267, 307, 339, 362, 382, 406, 412, 426, 427, 432, 439, 450, 460, 465, 468, 559, 567, 569, 576, 596, 601, 605, 610
  thiodibenzoylmethane 180, 232, 248, 354, 534
  thioglycollic acid (thioacetic acid, mercaptoacetic acid) 220, 232, 316 364, 396, 416, 478, 526, 596
  thiolactams 156
  thiomalic acid 364, 416, 459
  thio-Michler's ketone 285, 354, 417, 496
  thionalide 69, 70, 132, 416
  thionaphthenic acid 526
  2-thione-5-mercapto-1,3,4-thiadi-azolidine 406
  thionine dyes 534
  thio-oxine, *see* mercaptoquinoline allyl ether (Allthiox) 459

reagents (contd)
  N-2-thiophenecarbonyl-*N*-phenyl-hydroxylamine 594
  N-2-thiophenecarbonyl-*N*-*p*-tolyl-hydroxylamine 594
  thiophene-2-hydroxamic acid 595
  thiorhodanine azo-derivatives 435
  thiosalicylamide 285, 406, 435, 459, 465, 534
  thiosulphate 359
  thiothenoyltrifluoroacetone 180, 232, 248, 293, 326, 354, 375, 583, 596, 606
  thiotropolone 232
  thiourea 155, 396, 405, 416, 453, 464, 526
    derivatives 454
    N-phenyl derivatives 156
  thiovioluric acid 316, 465
  Thoron I 48, 106, 156, 262, 292, 417, 443, 460, 465, 539, 581, 616
  Thoron II 443, 542
  Thoronol 539
  THP 439
  Thymol Blue 192
  Thymolphthalexone 188, 335, 444
  Tichromin 386, 562
  tin(II) 285
  Tipyrogin 386, 562
  Tiron 202, 248, 316, 364, 386, 406, 444, 561
  Titan Yellow (Clayton Yellow, Thiazole Yellow) 330
  titanium(IV) 409
    peroxide complex 166
  o-tolidine 173, 202, 208, 210, 233, 299, 346, 396, 409
  p-tolueneamidoxime 232
  2-(4-toluenesulphonamido)aniline (TSA) 248
  p-toluidine 534
    derivatives 232
  TOPO (trioctylphosphine oxide) 79, 213, 284, 305, 439, 448, 538, 547, 560, 574, 580, 613
  TPTZ [2,4,6-tri(2-pyridyl)-1,3,5-triazine] 233, 315, 465, 512
  triazine derivatives 406
  1,2,4-triazoline-3-thione 435
  tribenzylamine (TBA) 79, 177, 213, 303, 459, 601
  tribenzylammonium 448
  tribromoanthrarufin 166
  tribromopyrogallol 386, 562, 596

reagents (*contd*)
  tri-n-butylamine 308, 364, 382, 490
  tributylammonium 230, 448, 450
  tributylphosphine 288
  tricaprylmethylammonium 230
  triethanolamine 346
  triethylamine 560
  triethylenetetramine (TRIEN) 248
  trifluoroacetylacetone 111, 143, 238, 305
  2,2′,4′-trihydroxyazobenzene-5-sulphonic acid 616
  2,3,4-trihydroxybenzoic acid 278
  3,4,5-trihydroxybenzoic acid 278
  trihydroxyfluorones 53, 117, 277, 364, 386, 561, 571
  trinitrobenzoic acid 512
  trioctylamines (TIOA and TOA) 79, 156, 213, 224, 230, 308, 364, 418, 423, 433, 457, 481, 485, 574, 583, 601
  trioctylammonium 381
  trioctylphosphine oxide, *see* TOPO
  triphenylarsine 305, 413
    oxide 574
  triphenylbenzylphosphonium 174, 299
  triphenylmethylarsonium 230, 308, 375
  triphenylphosphine 395
  triphenyl phosphite 238
  triphenylselenonium 218
  2,3,5-triphenyltetrazolium 299
  2,4,6-tri(2′-pyridyl)-*s*-triazine (TPTZ) 233, 315, 465, 512
  tris(2-diethylaminoethoxy)benzene 174
  Tropaeolins 417
  tropolone 460, 465, 596
  TSA [2-(4-toluenesulphonamido)-aniline] 248
  TTA (thenoyltrifluoroacetone) 77, 106, 176, 183, 188, 202, 220, 233, 248, 267, 285, 303, 305, 316, 346, 359, 376, 381, 386, 406, 417, 435, 439, 444, 449, 460, 465, 469, 500, 530, 538, 562, 588, 610
  tungstomolybdic acid 365
  tyrosine 210
  unithiol 316, 364
  uranium benzohydroxamate 409
  Uranon, *see* Arsenazo I
  uranyl 426
  vanadium benzohydroxamate 409
  Vanadox A (2,2′-dicarboxydiphenyl-amine) 596
  Variamine Blue 194, 248, 285, 299, 596
  Variamine Blue B 354

reagents (*contd*)
  veratrole 596
  Victoria Blue 605
  Victoria Blue B 354, 519
  Victoria Blue 4R 270, 285, 454, 526, 534
  Victoria Violet 166
  xanthate 259, 264
  Xylenol Orange 51, 116, 147, 154, 180, 220, 270, 292, 316, 326, 365, 376, 386, 417, 444, 460, 465, 469, 511, 534, 542, 552, 562, 583, 595, 605, 613
  xylenols 397, 399
  Xylidyl Blue I 334, 335
  Xylidyl Blue II 334, 335
  zinc dibenzyldithiocarbamate 242
  Zincon (2-carboxy-2′-hydroxy-5′-sulphoformazylbenzene) 247, 354, 605
  Zirconin (Gallocyanine MS) 617
rhenium 448
  determination 450, 453
  α-furildioxime 451
  thiocyanate 450
  separation 448, 451
rhodium 457
  determination 458, 459
  stannous halide 458
  separation 431, 457
rubidium 105
  determination 108
  separation 105
ruthenium 462
  determination 463, 464
  diphenylthiosemicarbazide 463
  separation 431, 462, 464

Sandell sensitivity index 20
scandium 468
  determination 469, 471
  Xylenol Orange 469
  separation 468
selectivity 26
  masking 27
  temperature effects 28
  two-wavelength method 28
selenium 474
  determination 475, 478
  3,3′-diaminobenzidine 476
  sol method 475
  separation 474, 476
sensitivity 19
  and blanks 23

642 Index

sensitivity (contd)
  and reagent purity 22
  index (Sandell) 20
silicomolybdenum blue 481, 484
silicon 481
  determination 482, 486
    molybdosilicic acid 482
    silicomolybdenum blue 484
  separation 481, 486
silver 490
  determination 491, 495
    dithizone 491
    rhodanine 493
  separation 490
sodium 105
  determination 106, 107
  separation 105
solvents 76
specific absorptivity 19, 103
spectrophotometers 14
spectophotometric titration 16
spectrophotometry 3
  automated 18
  differential 16
  sensitivity 19
spectrum stripping 21
standard curves 15, 33
standard series method 10
standard solutions 31
starch-iodine method 296, 408, 413, 414, 533
stray light 9
strontium 499
  determination 500
  separation 182, 499
sub-boiling technique 34
sulphate determination 505, 511, 512
sulphide determination 506
sulphite determination 509, 512
sulphur 504, 512
  determination 504, 511, 512
    barium sulphate 505
    Methylene Blue 506
    pararosaniline 509
  dioxide determination 509, 512
  separation 504, 508

tantalum 516
  determination 516, 519
    Methyl Violet 517
    pyrogallol 516
  separation 380
tantalomolybdenum blue 519

tellurium 521
  determination 522, 526
    Bismuthiol II 524
    sol method 522
  separation 521, 523
tetrafluoroborate 159, 160, 166
tetrathionate determination 512
thallium 529
  determination 530, 534
    Brilliant Green 532
    Rhodamine B 530
    starch-iodine 533
  separation 529, 531, 532
thiocyanate 191
  determination 194
  separation 194
thiosulphate determination 512
thorium 537
  determination 538, 542
    Arsenazo III 546
    Thoron I 539
  separation 537, 541
tin 546
  determination 547, 552
    dithiol 550
    phenylfluorone 547
    Pyrocatechol Violet 549
  separation 546, 549, 550, 551
titanium 555
  determination 556, 561
    chromotropic acid 558
    hydrogen peroxide 556
    thiocyanate 559
  separation 555
transmittance 7
trimethylborate 159, 163
trithionate determination 512
tungsten 567
  determination 568, 571
    dithiol 568
    thiocyanate 569
  separation 567, 569
tungstomolybdenum blue 571
tungstomolybdic acid 571
tungstophosphovanadic acid 592
turbidimetry 18

uranium 574
  determination 576, 581
    Arsenazo I 578
    Arsenazo III 580
    dibenzoylmethane 577
    thiocyanate 576
  separation 574, 578, 581

Index

vanadium 588
  determination 589, 594
    benzoylphenylhydroxylamine 593
    8-hydroxyquinoline 590
    tungstophosphovanadic acid 592
  separation 588, 591,
  volatilization (*see also* distillation) 85

Winkler method 408

yttrium 438
  determination 440, 443
    Arsenazo I 440
    Arsenazo III 442

yttrium (*contd*)
  separation 438

zinc 601
  determination 602, 605
    dithizone 602
  separation 601
zirconium 609
  determination 611, 616
    Alizarin S 611
    Arsenazo III 615
    Pyrocatechol Violet 612
    Xylenol Orange 613
  separation 609